S0-AKJ-163

The Dynamics of Small Bodies in the Solar System

NATO ASI Series

Advanced Science Institute Series

A Series presenting the results of activities sponsored by the NATO Science Committee, which aims at the dissemination of advanced scientific and technological knowledge, with a view to strengthening links between scientific communities.

The Series is published by an international board of publishers in conjunction with the NATO Scientific Affairs Division

A Life Sciences **B Physics**	Plenum Publishing Corporation London and New York
C Mathematical and Physical Sciences **D Behavioural and Social Sciences** **E Applied Sciences**	Kluwer Academic Publishers Dordrecht, Boston and London
F Computer and Systems Sciences **G Ecological Sciences** **H Cell Biology** **I Global Environment Change**	Springer-Verlag Berlin, Heidelberg, New York, London, Paris and Tokyo

PARTNERSHIP SUB-SERIES

1. Disarmament Technologies	Kluwer Academic Publishers
2. Environment	Springer-Verlag / Kluwer Academic Publishers
3. High Technology	Kluwer Academic Publishers
4. Science and Technology Policy	Kluwer Academic Publishers
5. Computer Networking	Kluwer Academic Publishers

The Partnership Sub-Series incorporates activities undertaken in collaboration with NATO's Cooperation Partners, the countries of the CIS and Central and Eastern Europe, in Priority Areas of concern to those countries.

NATO-PCO-DATA BASE

The electronic index to the NATO ASI Series provides full bibliographical references (with keywords and/or abstracts) to about 50,000 contributions from international scientists published in all sections of the NATO ASI Series. Access to the NATO-PCO-DATA BASE is possible via a CD-ROM "NATO Science and Technology Disk" with user-friendly retrieval software in English, French, and German (©WTV GmbH and DATAWARE Technologies, Inc. 1989). The CD-ROM contains the AGARD Aerospace Database.

The CD-ROM can be ordered through any member of the Board of Publishers or through NATO-PCO, Overijse, Belgium.

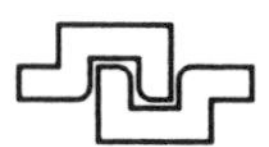

Series C: Mathematical and Physical Sciences – Vol. 522

The Dynamics of Small Bodies in the Solar System

A Major Key to Solar System Studies

edited by

Bonnie A. Steves

Department of Mathematics,
Glasgow Caledonian University,
Glasgow, United Kingdom

and

Archie E. Roy

Department of Physics and Astronomy,
University of Glasgow,
Glasgow, United Kingdom

Kluwer Academic Publishers

Dordrecht / Boston / London

Published in cooperation with NATO Scientific Affairs Division

Proceedings of the NATO Advanced Study Institute on
The Dynamics of Small Bodies in the Solar System: A Major Key to Solar System Studies
Maratea, Italy
29 June – 12 July 1997

Library of Congress Cataloging-in-Publication Data

The dynamics of small bodies in the solar system : a major key to solar system studies / edited by Bonnie A. Steves and Archie E. Roy.
p. cm. -- (NATO ASI series Series C: Mathematical and physical sciences ; vol. 522)
"Proceedings of the NATO Advanced Study Institute, held in Maratea-Acquafredda, Italy, between June 29 and July 12, 1997, entitled The Dynamics of Small Bodies in the Solar System: a Major Key to Solar System Studies--Preface.
Includes index.
ISBN 0-7923-5466-4 (alk. paper)
1. Solar system--Congresses. 2. Asteroids--Congresses. 3. Few-body problem--Congresses. 4. Astrophysics--Congresses.
I. Steves, Bonnie A. II. Roy, A. E. (Archie E.), 1924- .
III. The Dynamics of Small Bodies in the Solar System: a Major Key to Solar System Studies (1997 : Acquafredda di Maratea, Italy)
IV. Series: NATO ASI series. Series C, Mathematical and physical sciences ; no. 522.
QB500.5.D92 1999
523.5--dc21 98-31687

ISBN 0-7923-5466-4

Published by Kluwer Academic Publishers,
P.O. Box 17, 3300 AA Dordrecht, The Netherlands.

Sold and distributed in North, Central and South America
by Kluwer Academic Publishers,
101 Philip Drive, Norwell, MA 02061, U.S.A.

In all other countries, sold and distributed
by Kluwer Academic Publishers,
P.O. Box 322, 3300 AH Dordrecht, The Netherlands.

Printed on acid-free paper

Printed in the Netherlands

These Proceedings of the 1997 NATO Advanced Study
Institute held in Maratea, Italy
between June 29 and July 12
are dedicated to the
memory of

Professor Victor Szebehely
(1921 - 1997)

renowned celestial mechanician, cherished teacher and dear friend.

CONTENTS

SECTION FIVE: Studies of Dynamical Systems

EPILOGUE: The Shape of Things to Come

PREFACE

The reader will find in this volume the Proceedings of the NATO Advanced Study Institute held in Maratea-Acquafredda, Italy, between June 29 and July 12, 1997, entitled THE DYNAMICS OF SMALL BODIES IN THE SOLAR SYSTEM: A MAJOR KEY TO SOLAR SYSTEM STUDIES.

This Advanced Study Institute was the latest in the 'Cortina' series of NATO ASI's begun in the early 1970's firstly under the directorship of Professor Victor Szebehely and subsequently under Professor Archie Roy. All, except the latest, were held at the Antonelli Institute, Cortina d'Ampezzo, Italy. Many of those now active in the field made their first international contacts at these Institutes. The Institutes bring together many of the brightest of our young people working in dynamical astronomy, celestial mechanics and space science, enabling them to obtain an up-to-date synoptic view of their subjects delivered by lecturers of high international reputation. The proceedings from these institutes have been well-received in the international community of research workers in the disciplines studied.

The present institute included 15 series of lectures given by invited speakers and some 45 presentations made by the other participants. The majority of these contributions are included in these proceedings.

The theme of the latest ASI in the series, viz. *The Dynamics of Small Bodies in the Solar System: A major key to solar system studies* was chosen because of the widespread recognition that the small natural bodies of the solar system bear an importance in solar system studies far outweighing their size and mass relative to the planets. The study of the nature and dynamics of asteroids, comets, meteor streams, natural satellites and ring systems currently provides a wealth of information concerning the history and dynamical evolution of the solar system as a whole. Such objects also provide excellent test-beds for the study of resonance and chaos.

It has become clear that the evidence of the bombardment of planet Earth by comets and asteroids with the subsequent extinction of a proportion of living species is not only a topic of historical interest for the astronomer, biologist, botanist and geologist, but is also an ongoing process that must be addressed by any scientist responsible for the survival of terrestrial civilisation. The spectacular bombardment of Jupiter in 1995 by the fragments of Comet Shoemaker-Levy underlines the necessity of international support for operations such as Spaceguard for finding, listing and assessing all Earth orbit crossing asteroids and comets within the Solar System.

The dynamics of the numerous large asteroidal bodies recently found in the outer parts of the solar system are a current subject of interest, raising questions of whether they have originated there or can have been sent there from the classical asteroidal belt. Their relation to the hypothetical Kuiper Belt is also of considerable interest.

It is also clear that the careful design of spacecraft trajectories for artificial satellites not only of planets, but also of asteroids and natural satellites, is of major significance in enlarging our knowledge of the solar system. Artificial statellites of asteroids and irregularly shaped natural satellites can be used to obtain their mass distributions and shapes. In addition, recent theoretical results obtained from an alternative representaion of the gravitational field of non-spherical bodies are of relevance to such studies.

All these topics were addressed not only in the courses of lectures given by the invited speakers, but also in the seminar presentations made by participants. It is evident that the majority of those attending the Institute are collaboratively engaged in many of these topics. Many new research bonds were formed among the participants which will be developed during the years to come.

The Organising Committee (Professor A. E. Roy (Director), Professor V. Szebehely (Associate Director), Dr P. J. Message, Professor A. Milani and Dr B. A. Steves) is grateful to the NATO Scientific Affairs Division, not only for its guidance, counsel and support, but also for providing the opportunity to invite colleagues from a number of NATO's designated Cooperative Countries. The warm international friendship demonstrated by the participants during the duration of the NATO ASI was a most valuable feature of the meeting. The Organising Committee would like to take this opportunity to thank all those whose contributions made the ASI a success and such a very happy occasion. In particular the Committee would like to thank the staff of the Hotel Villa del Mare, Acquafredda, Maratea where the ASI was held. They were unstintingly helpful and friendly. The editors would also like to thank Mr. A. D. Barnett of the Department of Mathematics of Glasgow Caledonian University for his valuable help in the preparation of this volume.

Each of the book's sections has an introduction specially written by renowned experts in their fields. It is hoped that the book will be regarded as a worthy text and helpful reference volume for all students and researchers involved in the study of the nature and dynamics of asteroids, comets, meteor streams, natural satellites and ring systems.

It is appropriate that the book is dedicated to one of the foremost researchers in Celestial Mechanics, Professor Victor Szebehely, who although present at Maratea among his many friends, died shortly after his return to the United States. His books, his many papers and above all his enthusiasm and help given to so many in the Celestial Mechanics community over many years will be forever remembered.

Bonnie A. Steves
Dept. of Mathematics
Glasgow Caledonian University
Glasgow G4 0BA
UNITED KINGDOM

Archie E. Roy
Dept. of Physics and Astronomy
Glasgow University,
Glasgow G12 8QQ
UNITED KINGDOM

IN MEMORY OF VICTORY SZEBEHELY

K. ZARE
Austin, Texas

The Escape Of A Dominant Star From the Cluster of Celestial Mechanics

On September 13, 1997 Victor Szebehely passed away in Austin, Texas. We had planned to be co-authors of a paper in this proceeding. Instead, under the circumstances, I decided to write a short 'memoir' in honor of my distinguished co-author, mentor and friend.

On August 10, 1921, a *star* was born in Budapest, Hungary. He received his Ph.D in 1946 at the University of Budapest, writing a dissertation on the *problem of three bodies.* He was *ejected* to the United States in 1947 where he was *temporarily captured* by many academic institutions including Yale University (1963-1968) and the University of Texas (1968-1997). His research activities had an *interplay* between ship dynamics (1947-1957) for which he was knighted by Queen Juliana of the Netherlands (1956) and space dynamics (1957-1997) for which he received numerous national and international awards (1978-1997). During his *lifetime* he had many *encounters* with his international colleagues and friends, and in particular nine encounters (1972-1997) with the 'Cortina group'. His characteristic was not only his unusual *brightness* as a leading researcher in his field, but also his kindness, gentleness, sense of humour, and recognizable accent. His enthusiasm for celestial mechanics and in particular for the three body problem was *observed* by many any time he talked on the subject. His encouragement was always a *source of energy* for his students and colleagues to do research in celestial mechanics.

His *last encounter* with the 'Cortina group' occurrred at Maratea where he walked with his friends through the streets of Pompeii. My last conversation with him was on the phone when he told me about his cancer and his inevitable *escape.* That was the saddest day in my life. A month later the *final escape* occurred on Saturday, September 13 as it was *observed* by his daughter Julie. " ... As I sit here writing to you, there is another *star* in the *sky*... My father died quietly at home today... On the table beside my Dad was a scientific journal, folded open, with his glasses on it... "

The dominant star *escaped*, but he has *left behind* many enthusiastic researchers inspired by him to keep his fascination with the three body problem alive.

SECTION ONE: ASTEROIDS AND TRANS-NEPTUNIAN OBJECTS

INTRODUCTION

E. BELBRUNO
Innovative Orbital Design, Inc.
410 Saw Mill River Road,
Ardsley, New York 10502, USA

AND

I.P. WILLIAMS
Astronomy Unit,
Queen Mary and Westfield College,
University of London, London, E1 4NS, UK

The study of asteroids started on the first day of the 19th century when Piazzi, observing from Palermo, discovered a new object in the Solar System. Though the specific discovery by Piazzi was serendipidous, a campaign had been initiated by von Zach, whereby the zodiac was divided into 24 zones and a different astronomer was assigned to search each zone for a suspected planet. Piazzi was not one of these 24 astronomers. The expectation of finding a planet was based on the belief that the Titius-Bode law that predicted planetary distances was correct (see Nieto 1973 for a discussion of this law). We must remember that its correctness had only recently in 1781 been demonstrated through the discovery of the planet Uranus by Herschel. The only unexpected element in the discovery by Piazzi was that the new planet, named Ceres, was rather faint, much fainter than expected, indicating that the body was somewhat smaller than the "predicted" planet. Within the next four years, three further similar objects were discovered, Pallas, Juno and Vesta, two by Olbers and Juno by Harding. No further objects of this class were discovered for 40 years and it was during this period that the group were called minor planets - for they clearly were not proper planets. It was also during this time interval that the hypothesis was put forward first that these minor planets were remnants of a proper planet that had been broken up by some mechanism. At about the same time William

B.A. Steves and A.E. Roy (eds.), The Dynamics of Small Bodies in the Solar System, 1–5.

Herschel felt that the family deserved a more up-beat name than "minor planets" and suggested the name "asteroids" a name that caught on with the public, though according to the International Astronomical Union, they are still officially "Minor Planets".

The story repeats itself nearly two centuries later as the twentieth century draws to a close. In the 1940's and 50's, theories for the formation of the planets by Edgeworth (1943, 1949) and Kuiper (1951) had pointed out that there was no reason to suppose that the Solar system finished with the known planets. Indeed they went further and suggested that a swarm of smaller objects should exist beyond Pluto. Observational technology at the time was not capable of detecting this family of objects, but as soon as it was, searches began. In 1992, the searchers were rewarded when Jewitt and Luu (1992) discovered an object given the temporary designation $1992QB_1$. By the latter half of the twentieth century, scientific progress was more rapid and five further objects had been discovered within twelve months, and discoveries have regularly been made ever since. The story of the discovery of this family of objects and the current state of knowledge is described in the Chapter by Williams.

Returning to the story of the asteroids, once the forty years was over and Astraea was discovered, the floodgates opened and by 1900 there were 450 known, 1000 by 1923, 5000 by 1991 and it will be a close run thing as to whether or not 10000 is reached by the new millenium. As the number of known asteroids increased, it became possible to investigate global properties and distributions. It was very evident that the vast majority of the population was distributed with semi-major axis between 2 and 3.2 AU, but in 1867, Kirkwood pointed out that there were gaps within this distribution (now commonly called the Kirkwood gaps) and that these gaps corresponded to the positions of resonances with Jupiter.

The structure of these resonance gaps is still not fully known, and remains a formidable problem. An analytical approach to this problem in a simplified idealized setting can be put within the framework of the so called Komogorov-Arnold-Moser (KAM) theory. This can be done if the asteroids are assume to move in the same plane as Jupiter, and that Jupiter's motion about the Sun is assumed to be uniformly circular. Also, Jupiter's mass is assumed to be 'sufficiently small'. In this case it can be proven that the motion within the resonance gaps is chaotic in nature, and infinitely complex. In this simplified case, the gaps are generally seperated from one another, and moving from one gap to another - that is, from one resonance to another, is difficult and generally not possible. However, in the real world, KAM theory is not applicable. Jupiter's mass cannot be assumed to be 'sufficiently small', and further, it moves in a noncircular orbit about the Sun. Also, the motion of the asteroids are not coplanar with Jupiter.

In this case it is possible to move from one resonance to another. The gaps are no longer seperated from one another, and the routes between them are extemely complicated. A major problem at this time is to understand the interconnections between the different resonance gaps and how an asteroid moves between them. It is generally a slow diffusive motion taking millions of years. An example of a paper which discusses this is by Nesvorny and Ferraz-Mello (1997). This diffusive motion has been studied by many other authors. Another outstanding problem is to understand the stability of the regions surrounding the different resonances. In some cases the motion near a resonance can be quite unstable, and an asteroid moving in such a region can rapidly move away from the resonance. Close to other resonances on the other hand, an asteroid tends to linger in the vicinity and move among other higher order resonances. In this sense the resonance is more stable. An example of this situation is discussed by Milani *et al.* (1997).

Further complicating the motion between the resonance gaps are other perturbations whose effect is not fully understood. These include higher order resonances due to the gravitational influence of other bodies such as Mars or Saturn. There are also nongravitational perturbations such as thermal effects and the Yarkovsky effect (according to Opik 1951, this effect was described by Yarkovski around 1900 in a "lost" paper) .

At the current time there are no analytic or mathematical theories which can answer these questions, and long duration numerical integrations are necessary. This makes it possible to readily study these problems by interested students. Aspects are discussed in this volume in papers by Simula Ferraz-Mello and Giordani, Roig and Ferraz-Mello, Nesvorny and Ferraz-Mello, Mitchenko and Ferraz-Mello.

In 1906, asteroid Achilles was discovered and the existence of this asteroid indicated that it was not just internal resonances with Jupiter that were important. It was proven by Lagrange in the 1700's that the three-body problem has five equilibrium points, where a body at rest relative to the other two remains at rest. In the case of a planet moving about the Sun in an approximate circle, say Jupiter, three of these points will lie on the Sun-Jupiter line, and the other two form the vertex of an eqilateral triangle with the Sun and Jupiter. This is true for the Earth, Mars, and the other planets. Lagrange hypothesized that the equilibrium points could be candidate locations for other mass points to lie. Achilles was the first such body to be found and it was to be the first of a large group of asteroids that now go under the collective name of Trojan asteroids, the numbers librating about the two points (leading and trailing) being roughly equal. In 1990, the first Martian Trojan was discovered, indicating that Jovian resonances are not the only important resonances within the asteroidal population.

The developments in the study of the Edgeworth-Kuiper Objects followed a remarkably similar pattern to that of the asteroids, though on a much more compressed time-scale. Prior to the initial discovery, a search campaign had been running for a few years, based on theoretical considerations, that these objects existed. Within eighteen months, objects had been discovered that were moving on orbits that were in mean motion resonance with Neptune, in this case the 3:2 external resonance. Since then objects have been discovered on other resonances, but the 2:1 appears to be very under-populated compared to the 3:2. The problem of populating these resonances is of great dynamical importance and is discussed in the chapter by Belbruno.

There it is shown that a basic energy relation is minimized for objects moving about the Sun in resonance with Neptune at the 2:3 resonance. The minimization of this relation gives a possible reason why the 2:3 resonance is preferred over all other resonances. These results predict eccentricities in good agreement with the eccentricities of the orbits thus far observed. In this analysis it was assumed that in the distant past the objects moving about the Sun in resonance with Neptune were temporarily captured by Neptune. This assumption was motivated by the work of Morbidelli (1997) who numerically demonstrated the existence of a very slow diffusive process responsible for Neptune encountering bodies in the 2:3 resonance. Duncan*et al* (1995) estimate that the number of comet sized bodies in the 2:3 resonance which encounter Neptune to be 200 million per billion years. The general structure of the resonance gaps for Kuiper-Edgeworth Objects is not understood and represents an open problem which has been studied by many people including Duncan *et al* (1995), Morbidelli *et al* (1996), Malhotra (1996). It is also discussed by Hadjidemetriou and Hadjifotinou in this volume.

Also of great significance is the inter-relationships between Kuiper-Edgeworth Objects and, comets, Near Earth Asteroids and Centaurs. This is a very exciting and fast developing field where exciting new developments are continually occurring and we look forward to more exciting chapters in future NATO ASI's.

References

1. Duncan, M.J., Levison, H.F. & Budd S.M., 1995, The Dynamical Structure of the Kuiper Belt, *Astron. J.*, **110**, 3073-3081
2. Edgeworth K.E., 1943, The evolution of our Planetary System, *J. of the British Astron. Assoc.*, **53**, 181-188
3. Edgeworth K.E. 1949, The origin and evolution of the solar system, *Mon. Not. R. Astr. Soc.*, **109**, 600-609
4. Jewitt D.C. & Luu J.X., 1992, *I.A.U. Circular no 5611*
5. Kuiper G.P., 1951, On the origin of the Solar System. In *Astrophysics, A Topical*

Symposium, (J.A. Hynek ed.), 357-424, McGraw-Hill, New-York
6. Malhotra R., 1996, The Phase Space structure near Neptune Resonances in the Kuiper Belt, *Astron. J.*, **111**, 504-516
7. Milani A., Nobili A. et al , 1997, Stable Chaos in the Asteroid Belt, *Icarus*, **125**, 13-31
8. Morbidelli A., Thomas F. & Moons M., 1996, The Resonant Structure of the Kuiper Belt and the Dynamics of the first 5 Trans-Neptunian Objects, *Icarus*, **118**, 322-440
9. Nesvorny D. & Ferraz-Mello S., 1977, Chaotic Diffusion in the 2:1 Asteroidal Resonance,*Astron. Astrophys.*, **320**, 672-680
10. Nieto M., 1973, *Titius-Bode Law of Planetary Distances*, Pergamon Press, Oxford
11. Opik E.J., 1951, Collisional probabilities with the planets and the distribution of interplanetary matter, *Proc. Roy. Irish Acad*, **54**, 165-199

ON HIGH-ECCENTRICITY SMALL-AMPLITUDE ASTEROIDAL LIBRATIONS

A. SIMULA, S.FERRAZ-MELLO
Instituto Astronômico e Geofísico. Universidade de São Paulo
Av. Miguel Stéfano 4200, São Paulo, SP, Brazil

AND

C. GIORDANO
Universidad Nacional de La Plata - La Plata, Argentina

Abstract. We study a modelling of the averaged librations of asteroids in a first order resonance with Jupiter based on an asymmetric expansion of the non planar disturbing potential about a libration center with non zero eccentricity. A derivation of an abridged dynamical system that represents the averaged motion of an asteroid in a first order resonance with Jupiter is presented. In this system the perturber lies on an elliptic orbit and the model is more suited for the analysis of the gravitational phenomena in the asteroidal belt than the classical integrable models imposing a circular motion on Jupiter. The libration laws with the proper frequencies that characterize the small amplitude solutions are obtained without any restriction to asteroid eccentricity.

1. Introduction

An early modelling of the planar motion of the Hildas (Ferraz-Mello, 1988; Gallardo and Ferraz-Mello, 1995) showed that the averaged high-eccentricity librations of asteroids in a first-order resonance with Jupiter follow more or less closely three main laws:

(a) *The law of structure.*– A closed formula relating the averaged semi-major axis (or mean motion) and eccentricity of the stationary solutions of the averaged circular problem, known as *libration centers.* The actual averaged motions of a librating asteroid are, in this case, free oscillations – the so-called *librations* – about these stationary solutions.

B.A. Steves and A.E. Roy (eds.), The Dynamics of Small Bodies in the Solar System, 7–12.

(b) *The law of periods.*– A formula giving the period of the librations of small amplitude about the libration center.

(c) *The laws of the second forced mode.*– These laws complete the law of structure giving the amplitude of the main forced oscillations due to the eccentricity of the orbit of Jupiter (The forced eccentricity e_c given by the law of structure is independent on Jupiter's eccentricity). While these laws are not as analytically rigorous as the first ones, they have an upmost importance in the shaping of the distribution of the eccentricities of the actual asteroids and Kuiper belt objects (see Gallardo & Ferraz-Mello, 1995).

The key element in the reaching of these early results was the use of the averaged expansion of the planar potential of the disturbing forces in Taylor series about centers with non-zero eccentricity (see Ferraz-Mello & Sato, 1989). This model has been extended (Roig *et al.*, 1997) for small to moderate inclinations and has been used for the construction of a mapping which has proved to be useful to improve the knowledge of the resonant structure of the asteroidal belt (see Ferraz-Mello *et al.* 1998).

Now, we detail here the construction of the explicit spatial model in regular canonical variables. We show how a rotation in phase space is necessary to show the formal integrability of an approximate intermediate dynamical system. This abridged model has a topology that represents in good approximation the complete model: it is actually what is called in literature an Andoyer Hamiltonian (see also Ferraz-Mello S., 1985 for an extended report), whose solution requires the use of elliptic integrals. We show that this model is close to the complete dynamical system and able to reproduce the main topological characteristics of the asteroid librations.

2. Expansion of the Hamiltonian

The motion of a resonant asteroid is usually described using functions of the usual Delaunay variables $P = -J - J_z - \frac{L}{p}$, $J = L - G$, $J_z = G - H$ and the associate angles $\psi = (p+1)\lambda_1 - p\lambda$, $\sigma = \psi - \varpi$, $\sigma_z = \psi - \Omega$. The Hamiltonian is given by

$$\mathcal{H} = -\frac{\mu^2}{2L^2} - n_1\frac{(p+1)}{p}L + R; \tag{1}$$

μ is the gravitational constant, n_1 the mean-motion of Jupiter and R is the potential of the disturbing force acting on the asteroid. The term $n_1\frac{p+1}{p}L$ is introduced by the non conservative transformation $(L, G - L, \lambda, \varpi) \rightarrow (P, J, \psi, \sigma)$. The averaged system is an autonomous three-degrees-of-freedom one. In order to simplify the system, we follow the same steps as in the high-eccentricity theory and introduce Sessin's transformation, with the substitution of $L - G, \sigma$ by the variables

$$K = \sqrt{2\,J}\cos\sigma + \beta e_1 \cos(\psi - \varpi_1)$$
$$H = \sqrt{2\,J}\sin\sigma + \beta e_1 \sin(\psi - \varpi_1), \tag{2}$$

(where β is a constant to be chosen later), and $(G - H,\, \sigma_z)$ by the couple $V = \sqrt{2\,J_z}\cos\sigma_z$ and $U = \sqrt{2\,J_z}\sin\sigma_z$. The rotation in phase space introduced by Sessin's transformation will permit the elimination of the terms linear in Jupiter's eccentricity which appear in the lower order terms in the perturbing potential. In order to formally integrate the system we have to introduce Sessin's momentum $\mathcal{G}$

$$\mathcal{G} = \frac{J_2}{p} + P + \frac{1}{2}\beta^2 e_1^2 - \beta e_1 \left[K\cos(\psi - \varpi_1) + H\sin(\psi - \varpi_1) \right] \tag{3}$$

(Sessin, 1981; see Ferraz-Mello 1987), instead of P, and consider the new set of variables $K, H, V, U, \mathcal{G}, \psi$. The constant parameter J_2 is the same one introduced in the preliminar averaging of the potential as an arbitrary constant. Here, it will be used exclusively in order to simplify the expression of $\mathcal{G}$. Some easy bracket calculations are enough to show that this new set is canonical. We may use the immediate relation

$$L = J_2 - p\left[\mathcal{G} + \frac{1}{2}\left(K^2 + H^2\right) + \frac{1}{2}\left(V^2 + U^2\right)\right] \tag{4}$$

and obtain the expression of J from (2), which will be needed for the expansion of the perturbing function. Indeed, to have the potential R expressed explicitly in terms of the new variables one needs to calculate the relationship between k, h and K, H; it is easy to invert Sessin's transformation and obtain

$$k = B[K - \beta e_1 \cos(\psi - \varpi_1)]$$
$$h = B[H - \beta e_1 \sin(\psi - \varpi_1)], \tag{5}$$

where $B = \sqrt{\frac{1}{L}\left(1 - \frac{J}{2\,L}\right)}$ is the ratio of the eccentricity to $\sqrt{2J}$. One can find similar relations for the variables connected with the inclination. The expansion of the Hamiltonian may be achieved by expanding all functions of the semi-major axis about the constant value:

$$L_c = \sqrt{\mu\, a_c} = J_2 - \frac{p}{2}K_c^2 \tag{6}$$

(K_c is obtained by inversion of $k_c = B\,\sqrt{2\,J}$ which has to be computed in the expansion's center). In order to compute the frequencies we need just the quadratic terms in the variables K, H, U, V and the linear terms in $\mathcal{G}$. We assume e_1, η_1 (respectively eccentricity and sine of half inclination of the perturber) $= \mathcal{O}(\varepsilon^{1/3})$, where ε is the ratio of the mass of Jupiter to that of the Sun (1/1047.355). We choose to expand the Keplerian term up to the order of $(L - L_c)^2$, which is more than enough to get all the quadratic terms in the expansion, that is:

$$\frac{\mu^2}{2L^2} = \frac{\mu^2}{2\,L_c^2} - n_c\,(L - L_c) + \frac{m_c}{p^2}\,\frac{(L - L_c)^2}{2}; \quad m_c = \frac{3\mu^2 p^2}{L_c^4} \tag{7}$$

and n_c is the mean motion frequency calculated in the center of expansion. In order to expand the potential one needs also to expand B (which appears in the inverse of Sessin's transformation) about $L = L_c$. As B depends also on K and H through J we should also expand it in the powers of $(K - K_c)$ and H. For sake of simplicity, this step will be substituted by an expansion in the powers of $J - J_c$ where $J_c = \frac{1}{2}K_c^2$ is our center of expansion.

3. The abridged system

The lower order terms in the perturbing potential are[1] $A_{26}\,(K - K_c) + A_{25}\,\mathcal{G}$. The terms $A_{19}\,(K - K_c)^2 + A_{17}\,H^2 + A_{23}\,U^2 + A_{24}\,V^2$ are smaller but contribute to the frequencies and have to be added to eqn. (7) in order to respect the frequencies of the global system. The reduction to the plane of the system obtained by the substitution of these terms for the potential and of the expansion (7) for the Keplerian term in the Hamiltonian (7) is in this way integrable. Terms smaller than these ones, but depending on the angle $\psi - \varpi_1$, break the integrability of the intermediate system, if added. The third dimension carries a coupling among the variables in the higher order terms of the Hamiltonian, which also breaks the integrability. If we consider the equations of the equilibrium, in the plane (K, H), we find the particular solutions (with $e_1 = 0$)

$$K = K_c,\; H = 0,\; \mathcal{G} = \text{const},\; \psi = \dot{\psi}_c t + \text{const}. \tag{8}$$

The frequency $\dot{\psi}_c = (p + 1)\,n_1 - n_c\,p + A_{25}$ may be obtained from the derivative of the Hamiltonian with respect to $\mathcal{G}$ evaluated in the center ($K = K_c$, $H = 0$, $\mathcal{G} = 0$). The implicit relation a_c *vs.* $e_c = k_c$, which caracterizes the stationary solutions, defines the *law of structure*, which is obtained from the equation $\dot{H} = 0$. It is formally given by $A_{26} + K_c\,[n_1\,(p + 1) - n_c\,p] = 0$.

[1]The coefficients A_{-} are function of the center. For $a_c = .7628\;a_J$ we have $A_{25} = -1.804\;10^{-2}$, $A_{26} = 7.569\;10^{-4}$, $A_{17} = -8.946\;10^{-3}$, $A_{19} = -7.421\;10^{-2}$, $A_{23} = 4.091\;10^{-3}$, $A_{24} = 6.92\;10^{-3}$: they have to be multiplied by the mass factor divided by $a_J = 5.2$ A.U. With this center, we have also $\beta = -.336$.

The *law of periods* gives the frequency of σ and may be calculated by the Hessian of the Hamiltonian, in the variables (H, K). The eigenvalues of the associated matrix are $f_k = n_1\,(p+1) - n_c\,p + 2\,A_{19} - m_c\,K_c^2$ and $f_h = n_1\,(p+1) - n_c\,p + 2\,A_{17}$. The frequency of libration will be given by $f_\sigma = \sqrt{f_h\,f_k}$; the frequency of the longitude of perihelion on average is given by $f_\varpi = \dot{\psi}_c$. In the (U, V) space, we obtain $f_u = n_1\,(p+1) - n_c\,p + 2\,A_{23}$ and $f_v = n_1\,(p+1) - n_c\,p + 2\,A_{24}$. The frequency of the node is $f_\Omega = \sqrt{f_u\,f_v}$.

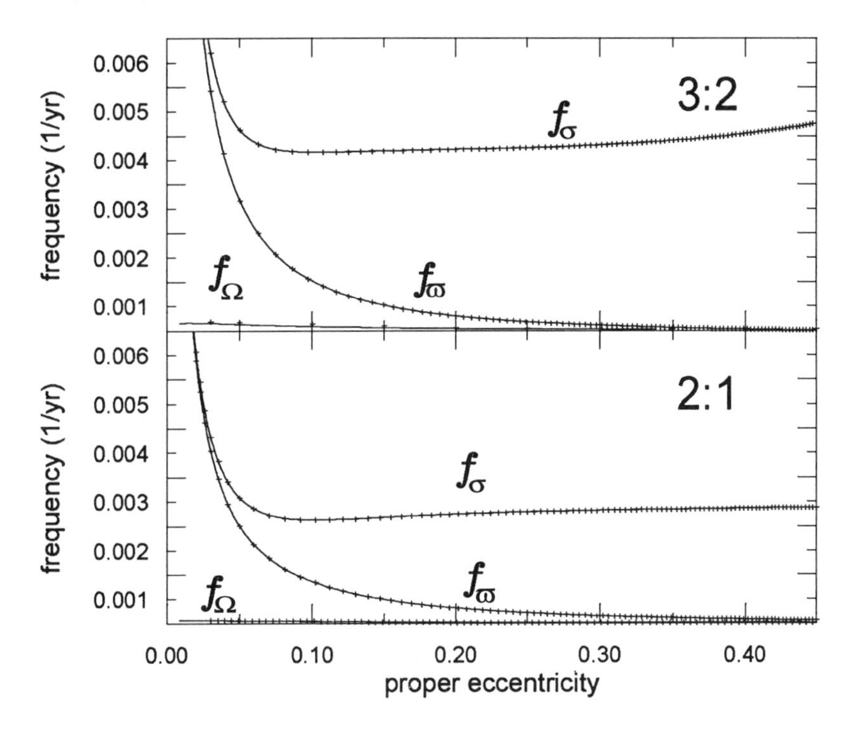

Figure 1. Comparison of frequencies calculated on the pericentric branches for 2:1 and 3:2 resonance. The crosses represent the values of f_σ, f_ϖ and f_Ω computed using numerical methods.

The law of structure and the law of periods are obtained in the reduction to the case $e_1 = 0$. We note that this is not a true reduction to the circular case since the main effects due to the eccentricity of Jupiter are already embedded in the definition of the variables K and H. This pseudo-reduction to a circular case is restricted to terms in which e_1 appears multiplied by one of the small quantities $(K - K_c)$ and H. From now on, we will call this reduced problem 'pseudo-circular' approximation.

We show in *Figure 1* a simple test of the frequencies f_σ and f_ϖ, compared with the ones calculated by Nesvorný using the method of Henrard (1990). In the same figure, the frequency of the node, f_Ω, is compared with the one computed by Michtchenko (1997) with a complete numerical

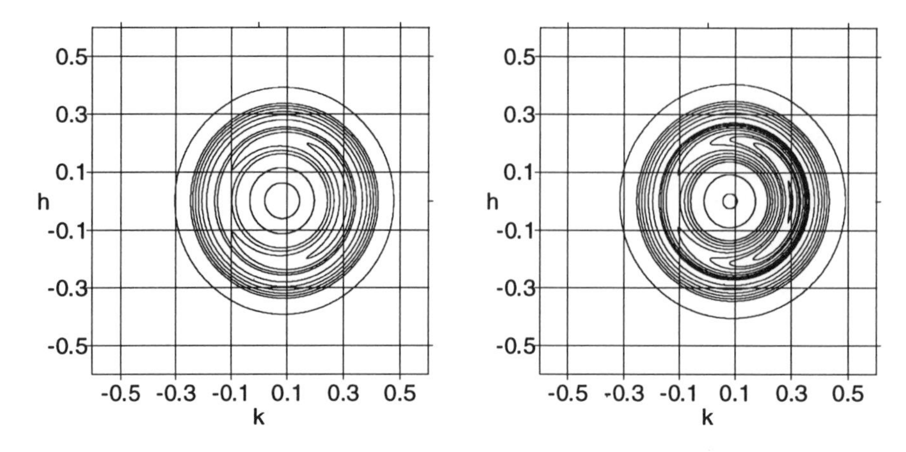

Figure 2. A section of our intermediate 'pseudo circular' system (right) and the one obtained using the full asymmetric expansion (left) as a model for the potential. The libration center is ($e_c = .2993$, $a_c = .7628\ a_J$).

method. In *Figure 2* we show the topology of the solutions of our intermediate system (right). On the left it is shown a section obtained using the full asymmetric expansion. As the frequencies near the pericentric branch and the topology of the solutions are preserved, we can use our simple abridged Hamiltonian to model the small amplitude librations.

References

1. Ferraz-Mello S.: 1985, 'Resonance in regular variables I. Morphogenetic analysis of the orbit in the case of first-order resonance', *Cel. Mech.* **35**, 209-234.
2. Ferraz-Mello S.: 1987, 'Averaging the Elliptic Asteroidal Problem near a First-Order Resonance', *Astron. J.* **94**, 208-212.
3. Ferraz-Mello S.: 1988, 'The High-Eccentricity Libration of the Hildas', *Astron. J.* **96**, 400-408.
4. Ferraz-Mello S., Michtchenko T. A., Nesvorný D., Roig F. and Simula A.: 1998, 'The depletion of the Hecuba gap *vs.* the long-lasting Hilda group', submitted to *Planetary and Space Sciences.*
5. Ferraz-Mello S. and Sato M.: 1989, 'The very-high-eccentricity expansion of the disturbing function near resonances of any order', *Astron. Astrophys.* **225**, 541-547.
6. Gallardo T. and Ferraz-Mello S.: 1995, 'The high-eccentricity libration theory revisited', *Cel. Mech. Dyn. Astr.* **62**, 145-165.
7. Henrard J.: 1990, 'A semi-numerical perturbation method for separable Hamiltonian systems', *Cel. Mech. Dyn. Ast.* **49**, 43-67.
8. Michtchenko T.A.: 1997, personal communication.
9. Nesvorný D.: 1997, personal communication.
10. Roig F., Simula A., Ferraz-Mello, S. and Tsuchida M.: 1997, 'The high-eccentricity asymmetric expansion of the disturbing function for non-planar resonant problems', *Astron. Astrophys.*, (in press).
11. Sessin W.: 1981, '*Estudo de um Sistema de dois Planetas com Periodos commensuraveis*', Ph.D. Thesis, IAG-USP.

A SYMPLECTIC MAPPING APPROACH FOR THE STUDY OF STOCHASTICITY IN THREE DIMENSIONAL ASTEROIDAL RESONANCES

F. ROIG AND S. FERRAZ-MELLO
Instituto Astronômico e Geofísico, Universidade de São Paulo
Av. Miguel Estéfano 4200, São Paulo, SP, Brazil

1. Introduction

In recent years many efforts have been done in order to understand the dynamics of main-belt asteroids at first-order mean-motion resonances with Jupiter. The problem of interest is the almost absence of asteroids at the 2/1 resonance (Hecuba gap), and the existence of a well defined group at the 3/2 resonance (Hilda group). Although the basic dynamics of these two resonances is similar, we know now that their behavior is strongly associated to the secondary and secular resonances web inside the mean-motion resonance. The overlap of these secular and secondary resonances can explain the chaos observed at very high and very low eccentricities ($e > 0.4$ and $e < 0.2$) in both resonances (Morbidelli & Moons, 1993; Michtchenko & Ferraz-Mello, 1995). However, for intermediate eccentricities there exist a region of regularity, which is almost empty of asteroids in the 2/1 resonances, but it is populated in the 3/2 resonance.

A few years ago, acurate numerical simulations for times of the order of the age of the Solar System were not practicable, and even simplified models (averaged problems) were not enough to describe the very-long-term evolution of the resonant dynamics. It was Wisdom (1982) who first introduced the idea of using a mapping approach of the Hamiltonian flux to explain the formation of the gap at the 3/1 resonance, but Wisdom's map could not be extended to study the 2/1 gap due to the convergence problems of the Laplacian expansion of the disturbing function in which it was based. To avoid these convergence problems, Ferraz-Mello (1997) introduced the idea to combine an implicit first-order symplectic integrator (Hadjidemetriou, 1991) with the high-eccentricity asymmetric expansion of the disturbing function, averaged near a resonance. The original map-

B.A. Steves and A.E. Roy (eds.), The Dynamics of Small Bodies in the Solar System, 13–18.

ping of Ferraz-Mello was planar, and had a set of swiches allowing him to turn off and on several perturbations of Jupiter's orbit. With such mapping, Ferraz-Mello dicovered the important role played by the short-periodic variations of Jupiter's orbit in the acceleration of diffusion mechanisms. These short-period perturbations are mainly due to the mean motion 5/2 commensurability between Jupiter and Saturn, usually known as the "great inequality" (GI).

In this paper, we extend the results obtained with the planar mapping to the spatial case. To do this, we construct a mapping model based on the spatial asymmetric expansion of the disturbing function recently develloped by Roig *et al.* (1997). We also make some important improvements in the formulas which modelate GI perturbations obtaining a more realistic mapping model. We stress the fact that our results are only of exploratory character, needing confirmation through precise numerical integrations.

2. The mapping model

For a resonance of the form $(p+q)/p$ (p, q integers), we introduce the classical action-angle variables

$$\begin{array}{llll} P = -\sqrt{\mu a}\left[\dfrac{p+q}{p} - \sqrt{1-e^2}\cos I\right] & , & \psi = \dfrac{p+q}{q}\lambda' - \dfrac{p}{q}\lambda & \\ J = \sqrt{\mu a}\left[1 - \sqrt{1-e^2}\right] & , & \sigma = \psi - \varpi & (1) \\ J_z = \sqrt{\mu a}\sqrt{1-e^2}\left[1-\cos I\right] & , & \sigma_z = \psi - \Omega & \end{array}$$

where μ is the gravitational constant, a, e, I are the semi-major axis, eccentricity and inclination, and λ, ϖ, Ω are the mean longitude, the longitude of perihelion and the longitude of node, respectively. Primed variables correspond to the disturbing body.

Then, we consider the Hamiltonian $\mathcal{H}$ of the restricted three-body problem, and we average it over the fast variable $Q = \frac{\lambda-\lambda'}{q}$. This averaged system is decribed by the Hamiltonian

$$\mathcal{H}^* = -\frac{\mu^2}{2L^2} - \frac{p+q}{p}n'L - \mu m' R^*, \tag{2}$$

where n' and m' are the mean motion and the mass of the disturbing body, L is the classical Delaunay momentum: $L = \sqrt{\mu a} = -\frac{p}{q}(P + J + J_z)$, and R^* is the averaged disturbing function:

$$R^* = \frac{1}{2\pi}\int_0^{2\pi} R\, dQ. \tag{3}$$

The disturbing function is replaced by the high-eccentricity asymmetric expansion for non-planar resonant problems (Roig *et al.*, 1997). This is a

Taylor expansion in the regular variables $k = e\cos\sigma, h = e\sin\sigma$ around the point (k_0, h_0, a_0), truncated at the suitable order and averaged over Q.

The orbital elements of the disturbing body (Jupiter in this case) appear in R^* and can be considered as known functions of the time. Thus, the main secular variations in Jupiter's orbit are taken from the syntetic secular theory LONGSTOP1B of Nobili *et al.* (1989). We considered also the most important short periodic variations in e_{Jup}, ϖ_{Jup} and λ_{Jup}, arising from the "great inequality" and associated perturbations in Jupiter's motion. The secular and short periodic variations of the perturber's orbit can be added to the model in different ways, allowing us to study their particular influence in the dynamics of the problem.

Once we have constructed our averaged Hamiltonian in action-angle variables $(\boldsymbol{\theta}, \mathbf{J})$, the symplectic mapping is simply defined as the canonical transformation generated implicitly by the function

$$S(\boldsymbol{\theta}^n, \mathbf{J}^{n+1}) = \sum_{i=1}^{3} \theta_i J_i + \tau \mathcal{H}^*(\boldsymbol{\theta}^n, \mathbf{J}^{n+1}). \tag{4}$$

The equations of the transformation are

$$\begin{aligned} J_i^{n+1} &= J_i^n - \tau \frac{\partial \mathcal{H}^*(\boldsymbol{\theta}^n, \mathbf{J}^{n+1})}{\partial \theta_i^n} \\ \theta_i^{n+1} &= \theta_i^n + \tau \frac{\partial \mathcal{H}^*(\boldsymbol{\theta}^n, \mathbf{J}^{n+1})}{\partial J_i^{n+1}} \end{aligned} \tag{5}$$

and the map-step is selected as the period of the synodic variable Q, i.e.

$$\tau = \frac{2\pi q}{n - n'} \tag{6}$$

(see Hadjidemetriou, 1991).

3. Results

Figures 1(a)-(e) show some results of the application of our mapping model to the 2/1 asteroidal resonance. In those figures the pericentric branch (C) and the separatrixes of the resonance (S_1, S_2) are indicated, together with some secondary resonances (f_σ/f_ϖ) and secular resonances (ν_5, ν_{16}). The curve L represents the lowest limit of the strong-chaotic high-eccentricity region described by Morbidelli & Moons (1993). The shadowed area is the grid of initial conditions of the 165 solutions simulated by 10^8 yr. with the mapping. In all cases, $I_0 = 0°$, $\sigma_0 = 0°$, $\varpi_0 - \varpi_0' = 0°$ and $\omega_0 = 90°$. Full circles represents initial conditions of the solutions reaching an eccentricity

$e > 0.65$ during the simulation, while open circles are those conditions for which $\sigma > 90°$ with $e > 0.5$. In both cases we considered that the object "escaped" from the resonance.

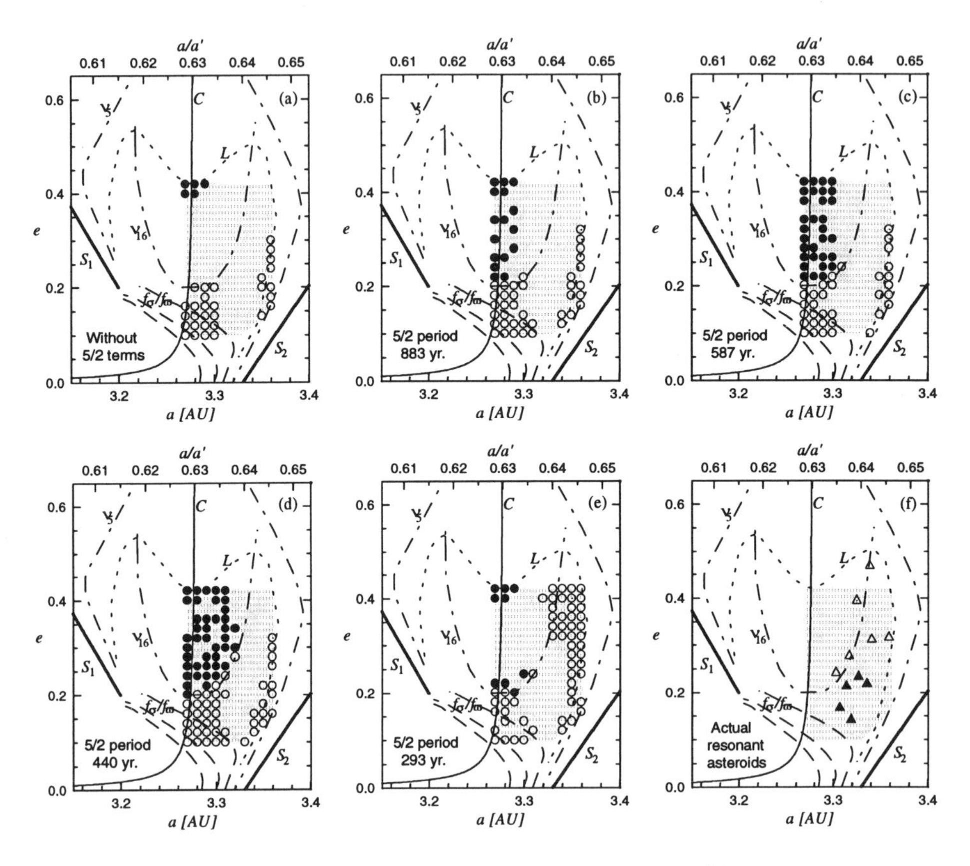

Figure 1. (a)-(e): Results of the mapping for the 2/1 resonance. (f): Actual distribution of real objects in the 2/1 resonance (see text).

Figure 1(a) corresponds to an integration in which we only take into account the main secular perturbations in Jupiter's orbit, without the short-periodic variations due to the GI. In Figs. 1(b)-(e) the short-period terms of the GI were added, considering different values of the critical frequency $5n_{Sat} - 2n_{Jup}$ in each case. We see that, in the central region of the resonance, the value of the GI-period has a notorious influenece in the number of "escapes", which is greater when the GI-frequency approximates from the libration frequency (f_σ) of the resonance (recall that the libration period is about 440 yr.). This is evident from Fig. 1(d), but the case of Fig. 1(e) is also interesting, since in that case "escapes" occur in the region where the libration frequency is about 300 yr. (i.e., similar to the GI-period).

The above result can be explained on the basis of the beating between the GI-frequency and f_σ, which contributes to accelerate the diffusion mechanisms in the central part of the resonance, i.e.: between the pericentric branch and the ν_{16} resonance (Ferraz-Mello *et al.*, 1998). Such a beating can deplete the region in a few tens of million years, and a very slight change of the mean semi-major axis of Jupiter and Saturn can provide the necesary change in the GI-frequency to provoke a stronger beating. A simple migration model of the major planets, changes the GI-period over a wide range of values. In fact, only a change of 0.3% in a_{Jup} and a_{Sat} is enough to decrease the GI-period to half its actual value of 880 yr. These changes modify just the GI-frequency, since it is formed by the critical combination $5n_{Sat} - 2n_{Jup}$. Other long-period and secular frequencies do not have significant changes.

Figure 1(f) shows the actual distribution of real asteroids with well-known orbits inside the 2/1 resonance. Full triangles correspond to Zhongguo group while open triangles are Griqua type objects. It is worth noting that the group of Zhongguo is located in a region of the resonance where the value of the GI-frequency seems to have no effect in accelerating the diffusion, allowing these object to survive for a longer time.

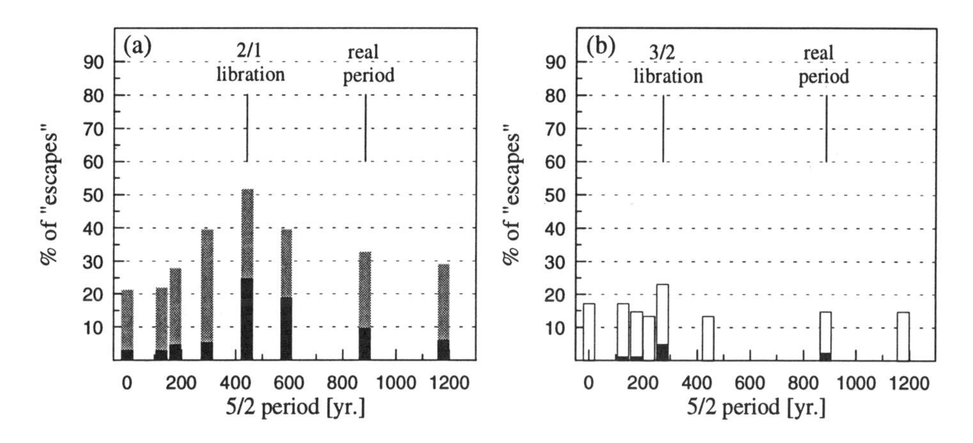

Figure 2. (a): Percent of "escapes" (over 165 simulations) as a function of the GI-period for the 2/1 resonance. Black bars correspond to the full circles of Figs. 1(a)-(e), and gray bars to the open circles. (b): Percent of "escapes" (over 82 simulations) for the 3/2 resonance. In this case, black bars correspond to objects reaching $e > 0.45$ during the simulation, and white bars are solutions that the mapping cannot model properly (see text).

The results for the 3/2 resonance do not show a similar clear dependence of the number of escapes with the value of the GI-period. We tested 82 real asteroids of the Hilda group and almost all of them remain in the resonance for the total time-span (10^8 yr.). A few percent of objects present a very-high amplitude of libration ($> 120°$), and there are also some low-eccentricity circulators. Our mapping cannot model these cases properly, so the simulations were interrupted. In the most pesimist hypothesis these objects could be considered as "escaped" ones, but in fact we cannot say anything about their behavior. Figure 2(a)-(b) compares the number of escapes in 2/1 and 3/2 resonances as a function of the GI-period. In the 3/2 resonance a maximum of escapes (black bars) seems to happen when the GI-period is close to 270 yr. (the libration period), but this result is not statistically significant as in the case of the 2/1 resonance. This supports the statement that the mechanism able to deplete the 2/1 resonance may not be efficient enough to deplete the 3/2 one.

4. Conclusions

Our results confirm the determinant role played by the GI in the depletion of the central region of the 2/1 asteroidal resonance. Recently, this was also confirmed with precise numerical integrations (Michtchenko & Ferraz-Mello, 1997) showing that, when the GI-frequency $\nu_{5:2}$ beats with the libration frequency f_σ the diffusion is accelerated. This is not observed in the 3/2 resonance since in such case the beating is more difficult to occur (see discussion in Ferraz-Mello *et al.*, 1998).

References

1. Ferraz-Mello S. (1997) A symplectic mapping approach to the study of the stochasticity of asteroidal resonances, *Cel. Mech, Dyn. Astr.* **65**, 421-437.
2. Ferraz-Mello S., Michtchenko T.A., Nesvorný D., Roig F. and Simula A. (1998) The depletion of the Hecuba gap *vs.* the long-lasting Hilda group, submitted to *Planetary and Space Science.*
3. Hadjidemetriou J.D. (1991) Mapping models for Hamiltonian systems with application to ressonant asteroidal motion, in *Predictability, Stability and Chaos in N-body Dynamical Systems*, A.E. Roy (ed.), Plenum Press, pp. 157-175.
4. Michtchenko T.A. and Ferraz-Mello S. (1995) Comparative study of the asteroidal motion in the 3/2 and 2/1 resonances with Jupiter I. Planar model, *Astron. Astrophys.* **303**, 945-963.
5. Michtchenko T.A. and Ferraz-Mello S. (1997) Escape of asteroids from the Hecuba gap, *Planetary and Space Science*, in press.
6. Morbidelli A. and Moons M. (1993) Secular resonances in mean motion commensurabilities: the 2/1 and 3/2 cases, *Icarus* **102**, 316-332.
7. Nobili A.N., Milani A. and Carpino M. (1989) Fundamental frequencies and small divisors in the orbits of the outer planets, *Astron. Astrophys.* **210**, 313-336.
8. Roig F., Simula A., Ferraz-Mello S. and Tsuchida M. (1997) The high-eccentricity asymmetric expansion of the disturbing function for non-planar resonant problems, *Astron. Astrophys.*, in press.
9. Wisdom J. (1982) The origin of the Kirkwood gaps: a mapping for asteroidal motion near the 3/1 commensurability, *Astron. J.* **87**, 577-593.

CHAOTIC DIFFUSION IN THE 2/1, 3/2 AND 4/3 JOVIAN RESONANCES

NESVORNÝ, D. AND FERRAZ-MELLO, S.
Universidade de São Paulo
Instituto Astronômico e Geofísico
Av. Miguel Stefano 4200
04301 - São Paulo
Brazil
E-mail: david@orion.iagusp.usp.br

1. Introduction

The determination of the maximum Lyapunov exponents in the 2/1 and 3/2 Jovian resonances (Ferraz-Mello 1994) suggested that the chaotic diffusion in the 2/1 resonance may be much faster than in the 3/2. As this fact could explain why the primordial population of the 2/1 resonance was removed (forming the Hecuba gap) and the 3/2-population was kept (Hilda group), we have decided to evaluate the diffusion speed for a large number of initial conditions using the frequency map analysis of Laskar (1988) (see also: Laskar 1996), adapted for the particular problem of Jovian resonances (Nesvorný and Ferraz-Mello 1997a). The 'diffusion maps' were also computed for the 4/3 Jovian resonance as it is, due to its similarity to the 3/2 resonance, a potential candidate for stable motion.

2. The 2/1 resonance

Figure 1 shows the diffusion maps in two planes of initial conditions: $i = 0$ and $a = 0.64$. Initial angles were $\sigma = \Delta\varpi = \Delta\Omega = 0$, where Δ is the difference between asteroidal and Jupiter quantities. The integration was performed with Jupiter and Saturn using the symmetric multi-step integrator (Quinlan and Tremaine 1990). The data were submitted to the low-pass filter (Quinn *et al.* 1991) and were Fourier analyzed (Šidlichovský and Nesvorný 1997). This allowed us to determine the relative change of the

B.A. Steves and A.E. Roy (eds.), The Dynamics of Small Bodies in the Solar System, 19–24.

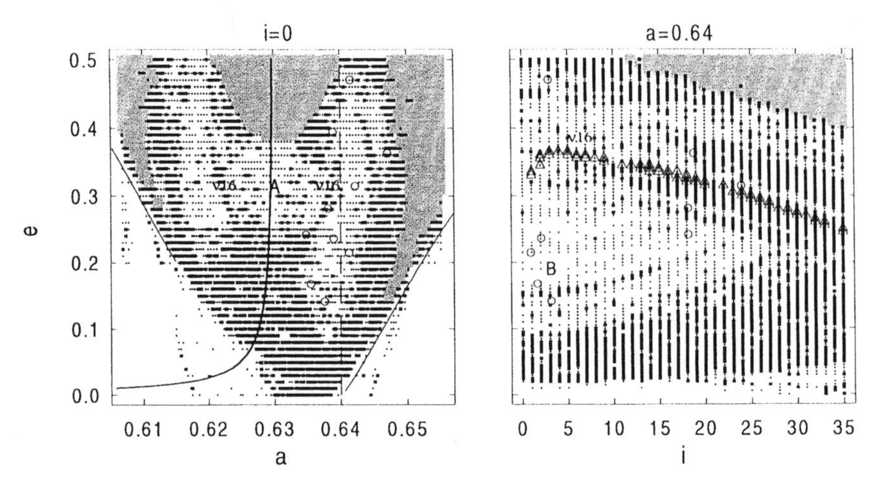

Figure 1. The diffusion maps of the 2/1 resonance: $i = 0$ on the left and $a = 0.64$ on the right.

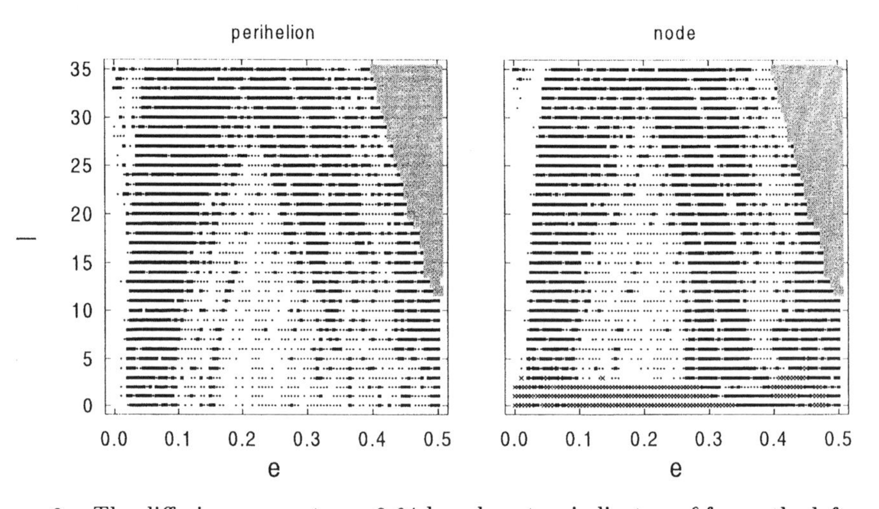

Figure 2. The diffusion maps at $a = 0.64$ based on two indicators: δf_{ϖ} on the left and δf_{Ω} on the right.

perihelion frequency δf_{ϖ} in the interval 2×10^5 years, which has been used as a measure of the local diffusion speed (see Nesvorný and Ferraz-Mello 1997a, 1997b for more details). The black squares in Fig. 1 correspond to the trajectories with $\delta f_{\varpi} > 10^{-2}$ and the small crosses to $10^{-3} < \delta f_{\varpi} < 10^{-2}$ (the merged gray squares were the orbits which escaped from the resonance in less then 6×10^5 years). The rest, which were left blank, were the most stable orbits with $\delta f_{\varpi} < 10^{-3}$. The separatrices and the libration centers of the 2/1 resonance were plotted as full lines. The dashed vertical line in the left figure is at $a = 0.64$, where the diffusion map on the right side reveals the structure in inclinations. Moreover, ten resonant asteroids (Bowell *et al.* 1994) were integrated into the plane $\sigma = \Delta\varpi = 0$ and $\dot{\sigma} > 0$ and were projected in the diffusion maps (circles).

A clear correspondence can be found between the most unstable regions (i.e. the fast escaping and $\delta f_{\varpi} > 10^{-2}$ orbits) and the overlapping inner resonances: 1) the secondary resonances in low-eccentricities (Lemaître and Henrard 1990), 2) the secular resonances ν_5 and ν_6 in high eccentricities and near the separatrices (Morbidelli and Moons 1993), and 3) the secular resonance ν_{16} forming an arc in the plane $i = 0$ with the lower limit slightly above $e = 0.2$ at the libration centers. The plane $a = 0.64$ shows how both the low and high-eccentricity unstable regions enlarge in higher inclinations and wipe out (at about $i = 25°$) the relatively stable area centered at $e = 0.2$ (B). This is the bridge between the secondary and secular resonant complexes described by Henrard *et al.* (1995). The secular resonance ν_{16} was computed numerically in this figure (triangles).

Four of the resonant asteroids: 3789 Zhongguo, 1975 SX, 1990 TH7 and 1993 SK3, are found in the area with the slowest diffusion speed at about $a = 0.64$ (for $\dot{\sigma} > 0$), $e = 0.2$, $i < 5°$. These asteroids may be the product of the collision event which formed the Themis family as suggested by Morbidelli (1996). However, the exact scenario of the formation of the contemporary resonant population is not clear since the long-term integration seems to suggest the stability of the central area (A) on the time scales of 10^9 years, which contradicts the fact that no primordial asteroids are observed there at the present. The effect of the short-period perturbing terms of the Jupiter-Saturn pair (Michtchenko and Ferraz-Mello 1997) may be important here. These terms considerably accelerate the diffusion speed (Ferraz-Mello and Michtchenko 1997, Nesvorný and Ferraz-Mello 1997b) and might have significantly contributed to the removal of the primordial population from the 2/1 resonance.

In Fig. 2 we compare the diffusion maps based on two different indicators: δf_{ϖ} and the relative change of the node frequency δf_{Ω}. In addition to the coding used in Fig. 1, the big crosses in low inclinations in the figure on the right denote the node libration, where the indicator was not evaluated due to the degraded precision. The similar structure of high and low-speed diffusion regions in both figures documents the fact that if the diffusion in f_{ϖ} happens, there might be expected the diffusion of the same order in f_{Ω}. A partial difference can be observed at $e = 0.3$ and $i < 10°$, where the diffusion in f_{Ω} is faster.

3. The 3/2 and 4/3 resonances

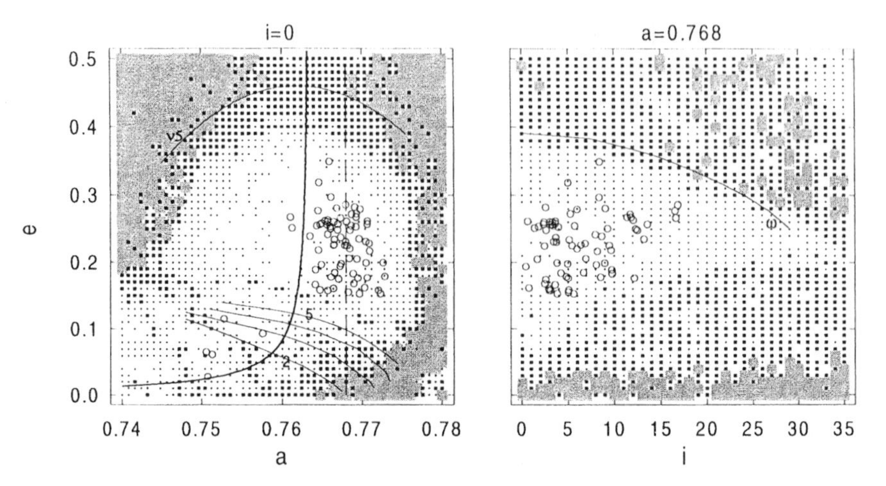

Figure 3. The diffusion maps of the 3/2 resonance: $i = 0$ on the left and $a = 0.768$ on the right.

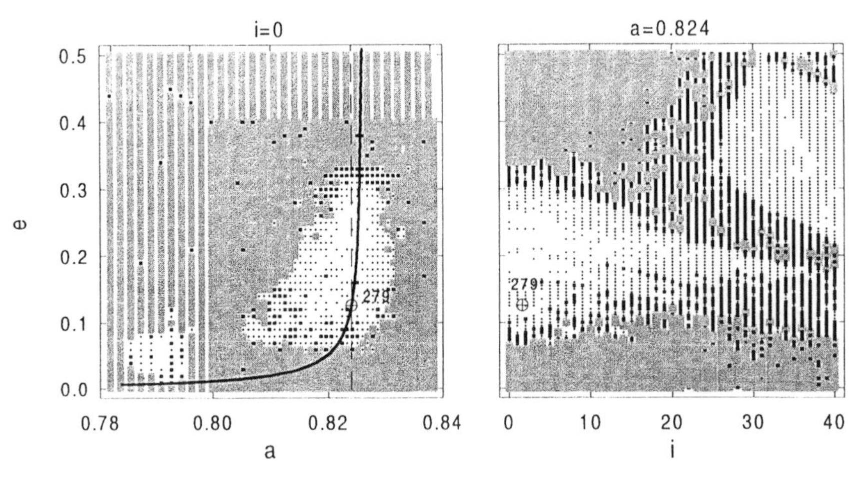

Figure 4. The diffusion maps of the 4/3 resonance: $i = 0$ on the left and $a = 0.824$ on the right.

The same procedure as in the case of the 2/1 resonance was performed for the 3/2 and 4/3 resonances (Figs. 3 and 4, respectively). The only difference in the coding of the diffusion maps is that the small crosses correspond to $10^{-4} < \delta f_{\varpi} < 10^{-2}$. In spite of this, the central regions of both the 3/2 and 4/3 resonances are almost blank in the figures and witness to the very slow diffusion there. Nevertheless, there is a significant difference in the asteroidal population of both resonances. While there is the Hilda group of asteroids in the central region of the 3/2 resonance, the 4/3 resonance hosts only 279 Thule ($\oplus$).

In Fig. 3, several inner resonances are shown for the 3/2 resonance. ν_5 and secondary resonances ($f_\sigma/f_\varpi = 2,3,4,5$) were computed for $i = 0$ using the Henrard semi-numeric method (Henrard 1990, Morbidelli and Moons 1993). They are plotted as thin lines in the figure on the left. The location of the Kozai resonance (denoted by ω) at the σ-libration centers was computed by the same method. Its stable stationary points are the thick line in the figure on the left, separatrices are the thin lines. A clear correspondence between the location of the inner resonances in the 3/2 resonance and the fast-diffusion regions is notable.

However, it is not always true that the inner resonances destabilize the motion, as can be seen in the case of the 4/3 resonance (Fig. 4 on the right), where the stable region at about $a = 0.824$, $e = 0.38$ and $i = 35°$ is associated with the Kozai resonance. This region and the central region of the 4/3 resonance are, with the exception of 279 Thule, completely void of asteroids, which is an interesting fact since there the diffusion speed is of comparable value to that in the 3/2 resonance, where 79 members of the Hilda group are found.

Our long-term integrations (several 10^8 years and four outer planets) with the initial conditions in both the low-diffusion regions of the 4/3 resonance did not show any irregularities of orbits and strengthened the idea of the resonance stability. Presumably, some other mechanism was at work in removing the primordial population of the 4/3 resonance.

4. Conclusions

Figure 5 shows the mean frequency change in the most stable regions of 1) the 2/1 resonance: $a = 0.64$ and $i = 5°$, 2) the 3/2 resonance: $a = 0.768$ and $i = 5°$, and 3) the 4/3 resonance: $a = 0.824$ and $i = 3°$. In this figure, δf_ϖ was estimated on the basis of integrations with four outer perturbing planets. The figure clearly demonstrates the fact that the diffusion in the 3/2 resonance (and also 4/3) is much slower than in the 2/1 resonance. The asteroids of the Hilda group are located at the place where the frequency change is under the level 10^{-4}. As there is no such stable region in the 2/1 resonance, this difference in the diffusion rates may explain the puzzling difference between the populations of two most prominent first-order resonances in the asteroidal belt.

References

Bowell, E., Muinonen, K. and Wasserman, L. H. (1994) A public-domain asteroid orbit database, *Asteroids, Comets, Meteors, editors: A. Milani et al.*, Kluwer, Dordrecht, pp. 477–481

Ferraz-Mello, S. (1994) Dynamics of the asteroidal 2/1 resonance, *Astron. J.* **108**, pp. 2330–2337

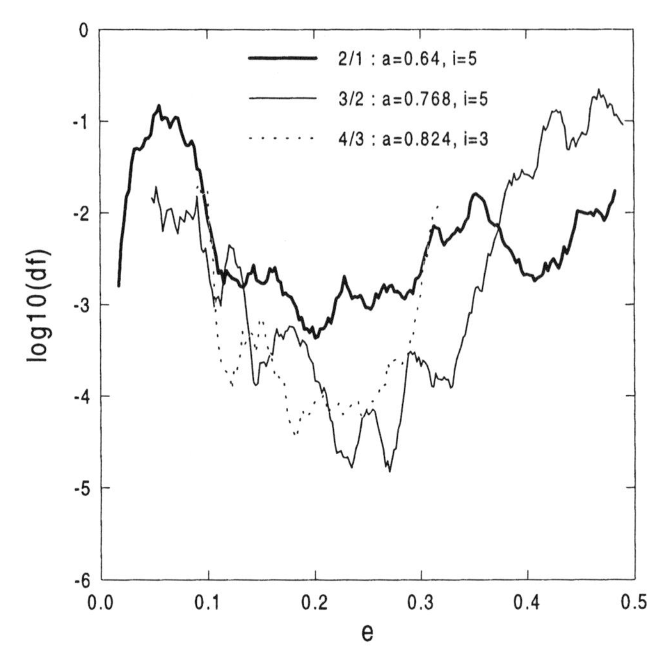

Figure 5. Comparison of the diffusion speeds in the most stable regions of the 2/1, 3/2 and 4/3 resonances.

Ferraz-Mello, S. and Michtchenko, T. A. (1997) Orbital evolution of asteroids in the Hecuba gap, The Dynamical Behavior of Our Planetary System, editors: R. Dvorak et al., Kluwer, Dordrecht

Henrard, J. (1990) A semi-numerical perturbation method for separable hamiltonian systems, *Celest. Mech.* **49**, pp. 43-68

Henrard, J., Watanabe, N. and Moons, M. (1995) A bridge between secondary and secular resonances inside the Hecuba gap, *Icarus* **115**, pp. 336-346

Laskar, J. (1988) Secular evolution of the solar system over 10 million years, *Astron. Astrophys.* **198**, pp. 341–362

Laskar, J. (1996) Introduction to frequency map analysis, *Proceedings of the NATO Advanced Study Institute 3DHAM95*, Kluwer, Dordrecht, in press

Michtchenko, T. A. and Feraz-Mello, S. (1997) Escape of asteroids from the Hecuba gap, *Planetary and Space Science*, submitted

Morbidelli, A. (1996) The Kirkwood gap at the 2/1 commensurability with Jupiter: new numerical results, *Astron. J.* **111**, pp. 2453–2461

Morbidelli, A. and Moons, M. (1993) Secular resonances in mean motion commensurabilities: the 2/1 and 3/2 cases, *Icarus* **102**, pp. 316–332

Nesvorný, D. and Ferraz-Mello, S. (1997a) Chaotic diffusion in the 2/1 asteroidal resonance, *Astron. Astrophys.* **320**, pp. 672–680

Nesvorný, D. and Ferraz-Mello, S. (1997b) On the asteroidal population of the first-order jovian resonances, *Icarus*, in press

Quinlan, G. D. and Tremaine, S. (1990) Symmetric multistep methods for the numerical integration of planetary orbits, *Astron. J.* **100**, pp. 1694–1700

Quinn, T. R., Tremaine, S. and Duncan, M. (1991) A three million year integration of the Earth's orbit, *Astron. J.* **101**, pp. 2287–2305

Šidlichovský, M. and Nesvorný, D. (1997) Frequency modified Fourier transform and its application to asteroids, *Cel. Mech. and Dyn. Astron.* **65**, pp. 137–148

ON THE SIMILARITIES AND DIFFERENCES BETWEEN 3/2 AND 2/1 ASTEROIDAL RESONANCES

T. A. MICHTCHENKO AND S. FERRAZ-MELLO
Instituto Astronômico e Geofísico,
Universidade de São Paulo,
Caixa Postal 3386, CEP 01060–970
São Paulo, SP, Brazil.
E-mail address: tatiana@orion.iagusp.usp.br

Abstract. The phase spaces of both 3/2 and 2/1 mean motion asteroidal resonances were studied by means of numerical integration and Fourier and wavelet analyses. The measurement of the proper asteroidal frequencies allowed to construct the webs of the inner resonances (secular, secondary and Kozai resonances) on the (a,e)–planes representing the phase space of each resonance. A comparison between the 3/2 and 2/1 resonances reveals qualitative similarities, which are due to their common first order resonance character. However, differences were also detected and seem to originate the opposite distribution characteristics of both mean motion resonances: the existence of the Hilda group inside the 3/2 resonance and the existence of the Hecuba gap associated with the 2/1 resonance.

1. Introduction

In this paper, the problem concerning the origin of the Hecuba gap is investigated by comparing the phase space structures of the 2/1 and 3/2 resonances. It is well-known that these resonances are of first order and that the leading terms in the expansion of the disturbing function are almost identical differing only in the numerical coefficients. The purpose of this paper is to understand how qualitatively similar descriptions of both resonances can give rise to opposite distribution characteristics: the presence of the real bodies in the 3/2 resonance (Hilda group) and the presence of the Hecuba gap associated with 2/1 resonance.

B.A. Steves and A.E. Roy (eds.), The Dynamics of Small Bodies in the Solar System, 25–30.

We draw phase space portraits of both resonances by calculating the webs formed by secular, secondary and Kozai resonances. The positions of the actual bodies inside the boundaries of both 3/2 and 2/1 resonances are superimposed over these frames of resonant lines. Analysis of the distribution of the asteroids inside the mean motion resonances points out the importance of the action of the inner resonances on the depletion processes. Special attention is paid to the middle-eccentricity depleted region of the 2/1 resonance and the role played by the frequency of the great inequality of the Jupiter–Saturn system on accelerating the diffusion processes in this region.

2. Comparison of phase spaces of the 2/1 and 3/2 resonances

All results presented here were obtained by numerical integrations of the exact equations of motion of the four-body problem (Sun–Jupiter–Saturn–asteroid). In order to cut out the short-period terms (less than 80–100 years), the solutions of the numerical integration were submitted to a low-pass filtering. The initial conditions were chosen in such way that they allowed to represent the results on the (a,e)–plane with fixed initial values of the angular elements, $\sigma_0 = \Delta\varpi_0 = \Delta\Omega_0 = 0$. Our interest was restricted to orbits of low inclination; thus, I_0 was fixed at 3°.

In order to evaluate the proper asteroidal frequencies, the output data of the numerical integrations were Fourier and wavelet analyzed. The three proper frequencies of the asteroidal motion chosen for subsequent analysis were: (1) the frequency of libration, f_σ, (2) the frequency of the longitude of perihelion, $f_{\Delta\varpi}$, and (3) the frequency of the longitude of the ascending node, $f_{\Delta\Omega}$. For fixed initial values of the inclination, these frequencies are smooth functions of two parameters: the initial eccentricity and semi-major axis. Figure 1 shows the proper frequencies calculated near stable pericentric solutions for both resonances. It is noted that in both resonances the corresponding frequencies show qualitatively similar behaviour, whereas their numerical values are different.

Based on the evaluation of the proper frequencies, the webs of the main secular and secondary resonances were constructed in both 3/2 and 2/1 mean-motion resonances. Figure 2 (a-b) shows the location of the secular and secondary resonances and also the boundaries of some main inner resonances on the (a,e)–planes of initial conditions. The thick lines indicate the borders of the mean-motion resonances. The curves C represent the pericentric branches of stationary solutions of the restricted problem for each resonance.

The position of minor bodies obtained from Bowell's catalogue (Bowell *at al.* 1994) which remain inside the resonance boundaries over, at least,

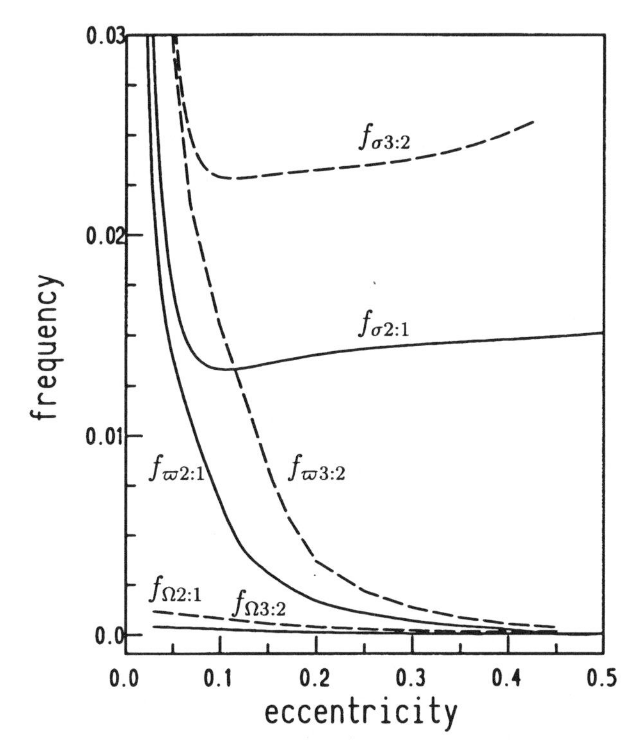

Figure 1. The proper asteroidal frequencies calculated on the pericentric branch as a function of the initial eccentricity. Solid curves correspond to 2/1 resonance and dashed curves correspond to 3/2 resonance.

some thousand years, were recalculated on the corresponding (a,e)–planes. There are about 86 numbered and multi–opposition members of the Hilda group and 13 numbered and multi–opposition asteroids inside the 2/1 resonance. In Fig. 2 (a-b), their positions are marked either by crosses if the asteroidal motion seems to be regular over 10 Myrs years, or by full circles when the asteroidal motion is chaotic. Each asteroid is indicated by two points corresponding to $\dot{\sigma} < 0$ (left-hand side of the pericentric branch) and $\dot{\sigma} > 0$ (right-hand side of the pericentric branch).

A comparison between the 3/2 and 2/1 resonances reveals their similarities. First, it is clearly noted in Fig. 2 (a-b) that the high-eccentricity regions of both resonances are dominated by secular resonances. We show the locations of the ν_5 secular resonances (*cf.* Morbidelli and Moons 1993). The ν_6 secular resonances and Kozai resonances are not plotted in this figure due to the fact that their numerical evaluation is complicated by strongly chaotic motion in these regions. Instead, we draw the lower boundaries of the zone of overlapping of the ν_5, ν_6 and Kozai resonances by L in Fig. 2 (a-b). All solutions with initial conditions chosen inside this complex show strongly chaotic motion characterized by great enhancement of eccentricities and inclinations; then the asteroid undergoes close approaches with Jupiter and is expelled from the resonance. According to Fig. 2 (a-b), the actual asteroids avoid the high-eccentricity regions in both resonances.

The other similarity between 3/2 and 2/1 resonances is their depleted low-eccentricity regions. Both low-eccentricity regions are characterized by presence of the well-known secondary resonances of the kind $f_\sigma / f_{\Delta\varpi}$. Their lower boundaries are plotted by dashed lines. The rapid transition across the overlapping zones produces the slow chaotic diffusion of the orbits along a band of overlapping resonances (see Ferraz-Mello *et al.* 1998). Starting near the pericentric branch of the corresponding resonance and at small inclinations, the orbits suffer a random walk in amplitude of libration and inclination. The escape of the asteroids occurs when the orbits approach the boundaries of the overlapping complex indicated by the L curves. Due to the random character of this process, escape times vary between a few ten million and a few hundred millions years.

There are no actual 2/1 resonant asteroids inside the secondary resonance region and only two actual 3/2 resonant asteroids inside this region which show weakly chaotic behaviour. There are also 6 members of the Hilda group situated below the secondary resonance complex; two asteroids of this group show chaotic behaviour.

The most important distinction between the 3/2 and 2/1 resonances may be clearly noted when the middle-eccentricity regions of the phase spaces in Fig. 2 (a-b) are compared. In the case of the 2/1 resonance, there is a very important ν_{16} secular resonance, which is a resonance between the oscillation mode of the longitude of the ascending node of the asteroid and the s_6 mode of the planetary theory. In the case of the 3/2 resonance, the ν_{16} secular resonance lies inside the high-eccentricity chaotic zone. The evolution of orbits near the ν_{16} is characterized by a large increase of inclination (up to 20°) and chaotic diffusion along the ν_{16} resonant line; during this diffusion, the solution is driven to the domain of the ν_5 and Kozai resonances, where a huge increase in eccentricity occurs and the asteroid is expelled from the resonance.

The majority of the Hilda group is concentrated in the middle-eccentricity region, free from the action of the secular and secondary resonances. A few actual asteroids inside 2/1 resonance stay close the ν_{16} secular resonance, have large inclinations and show chaotic behaviour. In a zone below the secular resonance ν_{16}, there are 5 actual asteroids on regular orbits (depicted by crosses in Fig. 2 a), known as Zhongguo-type asteroids.

The depleted central region in the 2/1 resonance, corresponding to solutions with small amplitude of libration, cannot be explained solely by actions of the secondary and secular resonances. Our recent works (Michtchen-

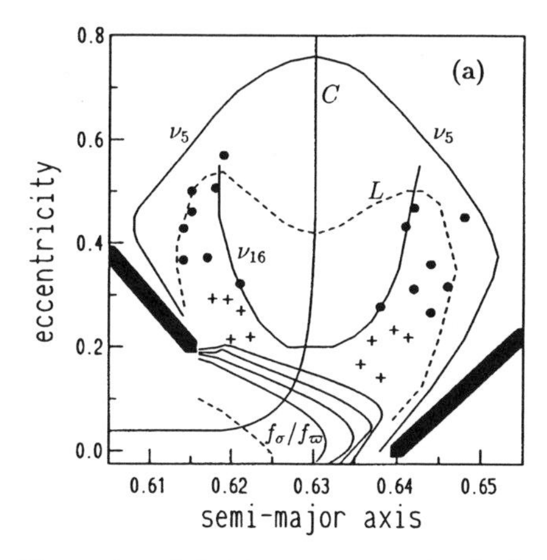

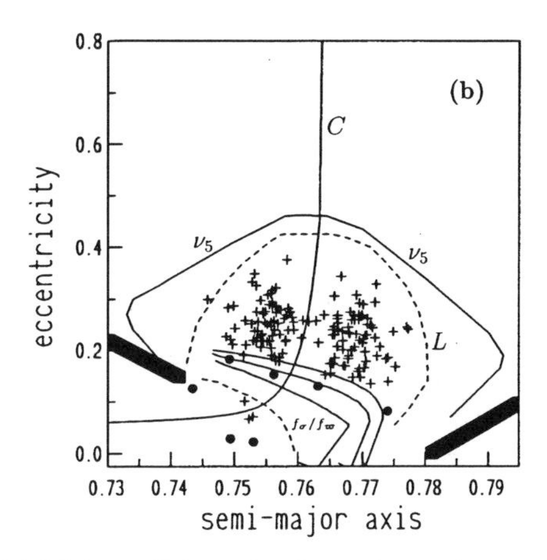

Figure 2. **(a)** 2/1 resonance: (a,e)–plane of initial conditions for fixed $I_0 = 3°$. The initial values of the angular elements are fixed at $\sigma_0 = \Delta\varpi_0 = \Delta\Omega_0 = 0$. The thick lines are borders of the resonant zone. The curve C is the pericentric branch of libration. The positions of the inner resonances (ν_5, ν_{16} and $f_\sigma/f_{\Delta\varpi} = \frac{2}{1}, \frac{3}{1}, \frac{4}{1}, \frac{5}{1}$) are indicated. Crosses and full circles depict the positions of the actual asteroids from Bowell's catalogue with regular and chaotic motion, respectively. **(b)** 3/2 resonance: same as **(a)**, except $f_\sigma/f_{\Delta\varpi} = \frac{2}{1}, \frac{3}{1}, \frac{4}{1}$.

ko and Ferraz-Mello 1997, Ferraz-Mello *et al.* 1998) show the determinant role played by the perturbations associated with Jupiter's Great Inequality on the diffusion enhancement in this region of the 2/1 resonance. The dynamical mechanism responsible for the diffusion is "diffusive streaming" (Tennyson 1982) which results in the migration of the libration center along a pericentric branch of the 2/1 resonance. When one asteroid starts its motion very close to the pericentric branch, the effect of the beat between frequency of libration and frequency of Great Inequality is to increase the libration amplitude of asteroid. This libration amplitude increase drives the asteroid towards the ν_{16} resonance, where the asteroidal inclination rises largely. Then the asteroid diffuses along ν_{16} resonance line towards the strongly chaotic complex of the secular resonances.

On the other hand, our simulations have shown that the same diffusion mechanism is inefficient for the depletion of asteroids in the case of the 3/2 resonance. This fact may explain the opposite distribution characteristics of the 2/1 and 3/2 mean motion resonances: the existence of the Hilda group inside the 3/2 resonance and the existence of the Hecuba gap associated with the 2/1 resonance.

Acknowledgments

The authors acknowledge the decisive support of the Research Foundation of São Paulo State (FAPESP) to this investigation.

References

1. Bowell, E., Muinonen, K., Wasserman, L.H: 1994, "A public-domain asteroid orbit database" , In *Asteroids, Comets, Meteors*, edited by A.Milani *et al.* 477-481.
2. Ferraz-Mello, S., Michtchenko, T.A., Nesvorný, D., Roig, F. & Simula, A.: 1997, "The depletion of the Hecuba gap *vs.* the long lasting Hilda group", (submitted).
3. Ferraz-Mello, S., Michtchenko, T.A. & Roig, F.: 1998, "The determinant role of Jupiter's Great Inequality in the depletion of the Hecuba gap", (submitted).
4. Michtchenko, T.A. & Ferraz-Mello, S.: 1997, "Escape of asteroids from Hecuba gap", *Planet. Space Sci.* (in press).
5. Morbidelli, A. and Moons, M.: 1993, "Secular resonances in mean-motion commensurabilities. The 2/1 and 3/2 cases". *Icarus* **102**, 316-332.
6. Tennyson, J.: 1982, "Resonant transport in near-integrable systems with many degrees of freedom", *Physica***5D**, 123-135.

A STUDY OF CHAOS IN THE ASTEROID BELT

M. ŠIDLICHOVSKÝ
Astronomical Institute of the Academy of Sciences of the Czech Republic, Boční II 1401, 141 31 Praha 4, Czech Republic

AND

D. NESVORNÝ
Universidade de São Paulo, Instituto Astrônomico e Geofísico, Av. Miguel Stefano 4200, 04301 São Paulo, Brazil

1. Introduction

Laskar's results [1] on the chaotic motion of the inner planets, the existence of chaotic regions inside orbital resonances of asteroids, the movement of the rotational axis of Mars [2] and ultimately Earth's unstable obliquity [3] brought the interest of astronomers back to the problem of chaos in the solar system. The possibilities of studies of chaos were recently widened from the original Poincaré mapping and Lyapunov characteristic coefficient (LCE) to Laskar frequency analysis [1] and [4], sup-map analysis [5], fast Lyapunov indicators [6] and local Lyapunov numbers distribution [7], [8] and [9].

In this paper we try to apply three of the aforementioned methods to the first one hundred of numbered asteroids. We try to compare the LCE, our modification of Laskar method and local Lyapunov numbers distribution.

2. Numerical Procedures

We used the MSI integrator [10] based on the symmetric multistep method of the twelfth order [11]. Seven planets except Mercury and Pluto were taken into accout (the mass of Mercury was added to the Sun). A step of 4 days was used and digital filtering of short periodic frequencies (with periods shorter than 1 200 yr)[12] was applied. The output of filtered $k+ih = \exp i\tilde{\omega}$ and $p + iq = sinI/2 \exp i\Omega$ was registered in intervals of 120 000 days. Simultaneously, the equations of variations were solved and local Lyapunov numbers for each asteroids were counted in each of the predefined bins to get

B.A. Steves and A.E. Roy (eds.), The Dynamics of Small Bodies in the Solar System, 31–36.

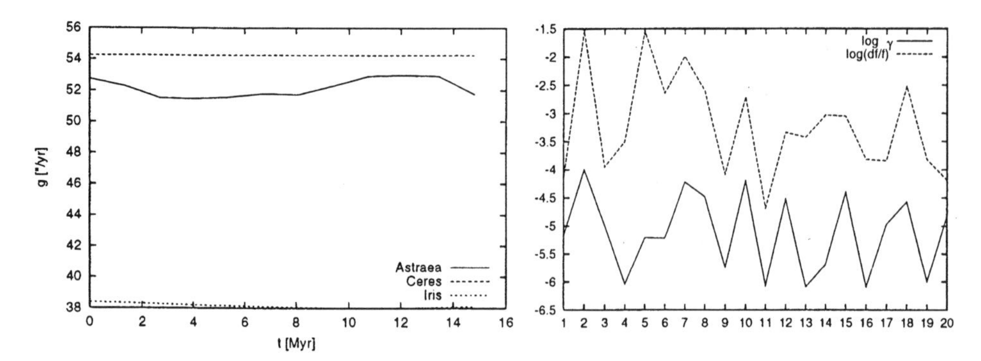

Figure 1. a) An example of the frequncy variations for three asteroids. b) $\log df/f$ and $\log \gamma$ for the first twenty numbered asteroids.

their distribution. The mean value was registered for obtaining LCE. These calculations were performed without filtering. Data $k + ih$ and $q + ip$ were than analyzed with our Frequency Modified Fourier Transform (FMFT) [12] which is essentially Laskar's method with additional corrections of frequencies (and recalculation of amplitudes and phases for new frequencies).

TABLE 1. Basic frequencies g, s for planets compared with Longstop 1A calculations where only outer planets were taken into account.

planet	g["/y]	g["/y] Longstop	s["/y]	s["/y] Longstop
Venus	7.1767		-6.1745	
Earth	17.2776		-18.7757	
Mars	17.8039		-17.7266	
Jupiter	4.2564	4.2447	0.	0.
Saturn	28.2398	28.2386	-26.3476	26.3392
Uranus	3.0865	3.0870	-2.9921	-2.9926
Neptun	0.6718	0.6727	-0.6917	0.6914

3. Variations of Frequencies

Frequencies were determined from 32 768 records corresponding to about 11 Myr. For obtaining Lyapunov coefficients we integrated for a somewhat longer interval of 15 Myr. For the first twenty asteroids we made computations for 25 Myr to be able to move an 11 Myr window (in which the frequency is determined). The variation of frequencies is then registered.

We determined ten frequencies with the most prominent amplitudes for

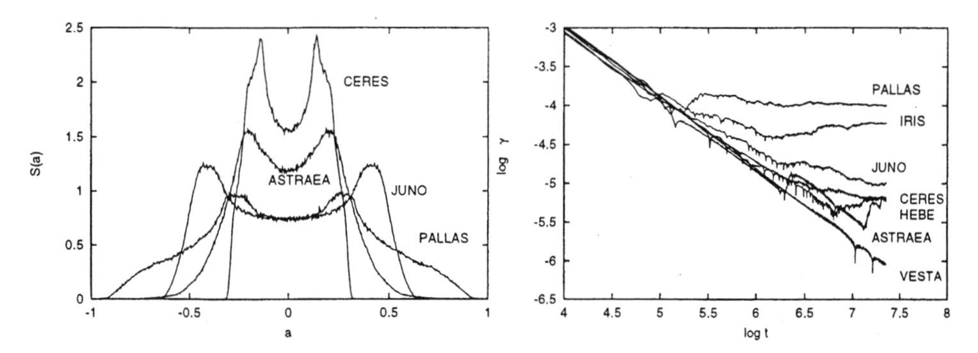

Figure 2. a) Local Lyapunov numbers distribution for Ceres, Juno, Pallas and Astrea. b) $\log \gamma$ versus $\log t$ for asteroids No.1–7. This plot makes it possible to see if the saturation of the curve has been achieved.

each planet and asteroid. Usually the first asteroidal frequency obtained from $k + ih$ analyses denoted g is the frequency of perihelion motion, the first frequency from $p + iq$ denoted s is the frequency of the motion of the node s. There are exceptions when the most prominent frequency is equivalent to some of the planetary frequencies. When we detect that, we list the second frequency (the new frequency which is not present in the planetary system without asteroids) in Tab. 2. Tab. 1 shows the g and s for planets from our calculations and for Longstop 1A [13]. The difference is caused by the inner planets which are not taken into account (in Longstop). It is impossible to give all frequencies here (Tab. 2 shows only g and s), but these results can be used as rough representations of the long period variations of elements of the planets and asteroids [12]. For the first 20 asteroids we calculated frequencies using the 11 Myr window. This window was shifted (eleven times roughly 1.35 Myr) and frequencies g and s were determined for each of the twelve positions of the window. Fig. 1a shows an example of the frequency g variations for three asteroids. We introduce

$$\frac{df}{f} = \left| \frac{f_{\max} - f_{\min}}{f_{\max}} \right| \tag{1}$$

where $f_{\max}$ and $f_{\min}$ are the maximum and minimum values of the twelve frequencies we found in the described process. Fig. 1b shows the value of $\log df/f$ for first 20 asteroids together with $\log \gamma$ (see the next section) which is the approximmation of log of the largest LCE. The rough agreement of both methods is clear, although some differences will become smaller if we

TABLE 2. Basic frequencies g, s and Lyapunov times L_T for first one hundred numbered asteroids.

No.	$g["/y]$	$s["/y]$	$L_T[10^3y]$	No.	$g["/y]$	$s["/y]$	$L_T[10^3y]$
1	54.251	-59.236	151	51	34.323	-39.274	662
2	-1.315	-46.503	10	52	102.557	-87.884	635
3	43.840	-61.466	99	53	53.194	-57.610	19
4	36.866	-39.564	>1098	54	47.424	-60.067	40
5	51.507	-50.587	160	55	57.854	-59.902	588
6	31.483	-41.798	168	56	48.569	-57.968	89
7	38.266	-46.576	17	57	90.567	-74.667	91
8	31.976	-35.673	30	58	56.612	-56.017	>1268
9	38.755	-42.014	549	59	53.201	-54.881	451
10	128.763	-97.124	16	60	39.804	-46.531	27
11	40.758	-43.167	>1175	61	44.358	-65.505	127
12	34.146	-40.901	33	62	124.030	-100.884	>990
13	35.932	-45.445	>1211	63	38.258	-41.782	315
14	48.839	-56.281	495	64	57.380	-60.185	41
15	42.706	-52.037	25	65	205.226	-174.111	52
16	76.931	-73.303	>1222	66	54.907	-57.014	>1007
17	43.758	-46.313	94	67	39.197	-44.435	245
18	32.593	-39.409	37	68	59.237	-64.706	>1103
19	41.872	-45.192	>1000	69	76.483	-86.115	256
20	40.871	45.107	65	70	45.123	-52.449	24
21	41.499	-44.621	690	71	16.248	-45.618	168
22	56.877	-63.494	42	72	33.911	-35.538	469
23	48.688	-64.641	25	73	55.795	-52.937	877
24	132.105	-103.435	463	74	63.716	-72.573	357
25	17.667	-38.616	97	75	56.591	-73.808	16
26	55.377	-55.885	>758	76	251.021	-213.657	115
27	38.562	-26.348	60	77	56.178	-56.364	>971
28	57.537	-66.819	>980	78	49.323	-62.440	13
29	49.710	-47.278	>855	79	41.794	-47.243	32
30	38.585	-40.531	>1126	80	32.847	-38.314	52
31	31.539	-65.005	134	81	66.736	-73.749	614
32	50.234	-50.691	>662	82	64.588	-80.187	63
33	83.732	-108.338	10	83	40.345	-43.572	>950
34	55.363	-58.885	>847	84	35.045	-43.096	43
35	84.042	-108.823	20	85	43.905	-53.130	200
36	34.172	-66.714	4	86	117.887	-107.553	6
37	54.783	-58.099	429	87	134.871	-130.858	>883
38	56.055	-63.544	367	88	59.656	-64.058	869
39	53.884	-56.667	>870	89	35.324	-43.908	>316
40	28.240	-35.204	>1037	90	140.782	-105.338	81
41	44.376	-76.604	14	91	52.699	-49.701	>949
42	40.058	-47.098	62	92	175.762	-84.671	42
43	32.632	-35.605	65	93	55.044	-61.206	175
44	41.297	-46.406	70	94	126.277	-84.044	>753
45	56.634	-58.404	675	95	70.434	-76.001	>793
46	58.881	-48.973	30	96	57.429	-77.506	239
47	70.375	-69.796	304	97	46.062	-63.260	74
48	105.419	-82.091	>931	98	40.624	-61.865	234
49	109.721	-107.831	284	99	44.598	-61.027	97
50	56.489	-68.554	10	100	106.387	-94.847	176

also take into account changes in s (e.g. the value $\log df/f$ for asteroid 20 will increase).

4. Maximum LCE and Local Lyapunov Numbers Distribution

We used the MSI package for solving equations of variations together with original equations. The solution is performed in 6-dimensional phase space of cartesian coordinates and velocities. In these calculations we made no filtering and the local Lyapunov numbers were recorded after 4 000 days. The spectra $S(a)$ [8] are shown in Fig. 2a. We were not able to find anything interesting from them apart from confirming their independence of the choice of initial vector for solution of equations of variation and their independence on time (after some saturation time the spectrum does not change any more if we integrate for a longer time interval). The byproduct of the local Lyapunov numbers calculation is the possibility to get maximum LCE. Their mean value of calculated at time t (and divided by time step) is the value γ;

$$\gamma(t) = \frac{1}{t} \ln \left| \frac{\mathbf{v}(t)}{\mathbf{v}(0)} \right|, \tag{2}$$

where $\mathbf{v}(t)$ is the solution of variational equations. Plotting $\log \gamma$ versus $\log t$ in Fig. 2b allows an estimate of maximum LCE, which will be denoted λ. For most asteroids the curve in Fig. 2b is saturated and maximum LCE may be approximated by $\gamma(t_{\max})$. In our calculations for the first twenty asteroids $t_{\max} \sim 25Myr$ but usually much shorter integration would be sufficient. For some asteroids (e.g. Vesta in Fig. 2b) saturation is not achieved and we can expect $\lambda < \gamma(t_{\max})$. In Tab. 2 we give the Lyapunov time $L_T = 1/\lambda$ (character $>$ means there that saturation was not achieved).

5. Conclusions

We found that 42 of the first hundred asteroids have Lyapunov time L_T shorter than 100 000 yrs. For at least 25Myr, which is 250 L_T, there is no strong 'event'. Macroscopic instability time [14] is very long. The frequencies and, therefore, proper elements undergo very small variations and we see that stable chaos [15] is very common even in the large asteroid set.

We could compare L_T with other results [15] for three asteroids where our sets intersect. We obtained practically the same values for 10 Hygiea and 522 Helga ($L_T \sim$ 6 450 yr) but for 65 Cybele we have L_T =52 000 yr against 6 662 yr of [15]. We believe this discrepancy is caused by different perihelion motion of outer planets as in [15] inner planets were not taken into account. In high order resonances responsible for stable chaos the order of resonance multiplied by the Jupiter's longitude of perihelion is typically part of the resonant variable. As the order is high the rate of change of resonant variable may be different and the same result in L_T cannot be guaranteed.

Acknowledgements

The support by grant 205/95/0184 from the Grant Agency of the Czech Republic is acknowledged (M.Š.). The second author (D.N.) thanks the Research Foundation of the State of São Paulo for sponsorship of his work.

References

1. Laskar, J. (1990) The Chaotic Motion in the Solar System. A Numerical Estimate of the Size of the Chaotic Zones, *Icarus* **88**, pp. 266–291
2. Laskar, J. and Robutel, P. (1993) The Chaotic Obliquity of the Planets, *Nature* **361**, pp. 608–612
3. Néron de Surgy, O. and Laskar, J. (1997) On the Long Term Evolution of the Spin of the Earth, *Astron. Astrophys* **318**, pp. 975–989
4. Laskar, J., Froeschlé, Cl. and Celetti, A. (1992)The Measure of Chaos by Numerical Analysis of the Fundamental Frequencies. Application to Standard Mapping, *Physica D* **56**, pp. 253–269
5. Froeschlé, Cl. and Lega, E. (1997) On the Measure of the Structure around the Last KAM Torus before and after its Break-up, *Celest. Mech. Dynam. Astron.* **64**, pp. 21–31
6. Lega, E. and Froeschlé Cl. (1997) Fast Lyapunov Indicators Comparison with Other Chaos Indicators Application to Two and Four Dimensional Maps, *The Dynamical Behaviour of our Planetary System, Proceedings of the Fourth Alexander von Humboldt Colloquium on Celestial Mechanics, editors: R. Dvorak, J. Henrard*, Kluwer Academic Publishers, Dordrecht, pp. 257–275
7. Froeschlé, Cl., Froeschlé, Ch. and Lohinger, E. (1993) *Celest. Mech. Dynam. Astron.* **56**, pp. 307–314
8. Voglis, N. and Contopoulos, G. (1994) Invariant Spectra of Orbits in Dynamical Systems, *J. Phys A: Math. Gen.* **27**, pp. 4899–4909
9. Contopoulos, G. and Voglis, N. (1966) Spectra of Stretching Numbers and Helicity Angles in Dynamical Systems, it Celest. Mech. Dynam. Astron. **64**, pp. 1–20
10. Šidlichovský, M., Nesvorný, D. (1994) Temporary Capture of Grains in Exterior Resonances with the Earth,*Astron. Astrophys.* **289**, pp. 972–982
11. Quinlan, G.D. and Tremaine, S. (1990) Symmetric Multistep Methods for the Numerical Integration of Planetary Orbits, *A.J.* **100**, pp. 1694–1700
12. Šidlichovský, M. and Nesvorný, D. (1997) Frequency Modified Fourier Transform and its Application to Asteroids, *The Dynamical Behaviour of our Planetary System, Proceedings of the Fourth Alexander von Humboldt Colloquium on Celestial Mechanics, editors: R. Dvorak, J. Henrard*, Kluwer Academic Publishers, Dordrecht, pp. 137–148
13. Carpino, M., Milani, A., Nobili A. M. (1987) Long-term Numerical Integrations and Synthetic Theories for the Motion of Outer Planets, *Astron. Astrophys.* **181**, pp. 182–194
14. Morbidelli, A. and Froeschlé, Cl. (1996) On the Relationship between Lyapunov Times and Macroscopic Instability Times, *Celest. Mech. Dynam. Astron.* **63**, pp. 227–239
15. Milani, A., Nobili, A.M. and Knežević, Z. (1997) Stable Chaos in Asteroid Belt, *Icarus*, **125**,pp. 13–31

HOPPING IN THE KUIPER BELT AND SIGNIFICANCE OF THE 2:3 RESONANCE

EDWARD BELBRUNO
Innovative Orbital Design, Inc.
410 Saw Mill River Road,
Ardsley, New York 10502

Abstract. A proposed mechanism responsible for causing certain resonant motions in the Kuiper belt is described. It is the so called 'hop' with Neptune where an object dramatically moves from one resonance to another. This implies, among other things, that in the distant past, the currently resonant Kuiper belt objects, including Pluto, may have been weakly captured by Neptune. In another result, it is indicated why the 2:3 resonance is prevalent over all other being a minimum of a special energy relation. This paper can be viewed in part as an extension of a recent paper by Belbruno and Marsden(1997).

1. Introduction

In the past four years, a few dozen objects have been observed moving at distances just beyond Neptune, in the Kuiper belt(Jewitt and Luu, 1995). Approximately one half of these objects are moving in resonance with Neptune. Of these resonances, the 2:3 is predominant(Marsden, 1997,1996a; Morbidelli 1997). (An m:n resonance is one where the Kuiper Belt object does m revolutions about the sun in the time it takes Neptune to do n revolutions, m,n=1,2,3,...; $m \neq n$.) There are two questions that can be asked about this situation. One is why the 2:3 is prevelant. The other is what type of motion took place so that these objects are in resonance with Neptune? These two questions are addressed in this paper, and, with an analytic model suggested in a recent paper by Belbruno and Marsden(1997), they can both be reasonably addressed. The analysis done here does not require knowledge of the resonant dynamical structure of the transneptunian region (Duncan *et al*, 1995; Malhotra, 1996). We make use of some recent

B.A. Steves and A.E. Roy (eds.), The Dynamics of Small Bodies in the Solar System, 37–49.

results of Morbedelli(1997) on the existence of a slow diffusive process responsible for Neptune ecountering bodies in the 2:3 resonance. The number of such bodies is estimated to be approximately 5 X 10^7 per billion years. An estimate of the number comet sized bodies in the 2:3 resonance which encounter Neptune per billion years yields a number of 2 X 10^8, (Duncan *et al*, 1995).

This paper for the most part can be viewed as an extension of a result in Belbruno and Marsden(1997). Because that paper is referred to often, it is labeled as BM97 for convenience. In BM97, the abrupt changes in resonances from one value to another with respect to Jupiter for nine short period comets are analyzed. This abrupt resonance change is caste in the framework of so called 'hop' dynamics independently discovered by Belbruno(1990). Using this framework, the hopping of a comet from one resonance situation to another can be better theoretically understood. This understanding is related to the mathematical work of Mather(1982). The 'fuzzy boundary' concept is useful in studying the cometary hops. This boundary about Jupiter is a location, in the the position-velocity space, where the motion of a comet is sensitive, and as it moves it is feeling the gravitational forces of Jupiter and the sun in a nearly equaly fashion. An object moving in this region is said to be 'weakly captured' by Jupiter. When in weak capture, the two-body Kepler energy of the comet with respect to Jupiter is negative and nearly zero in value. In BM97 a sketch of a mathematical dynamical proof is given to explain why the hop occurs in a more physical way. The statement being described in BM97 is labeled BM1: *If a comet, or any other small body, is in weak capture at Jupiter, then it will escape from Jupiter in a relatively short period of time, in forwards or backwards time, onto a resonant orbit about the sun, in resonance with Jupiter.* This observation plays a key role in this paper. A hop at Jupiter is formally defined as a quick change from one resonance(m:n) to another (i,j), i,j = 1,2,3,..., $i \neq j$, with respect to Jupiter, through weak capture.

The main idea of this paper is to assume that BM1 is valid also at Neptune. That is, if a small body is weakly captured at Neptune, then it is quickly ejected onto a resonance orbit about the sun in resonance with Neptune. A main result of this paper is to be able to translate this into a useful set of equations using a set of assumptions listed in Section 2. These equations result into an explicit relation that must be satisfied. This is a relation between the periapsis distance from Neptune that the small body was initially in weak capture, and the resonance type of the orbit about the sun. From this, the key result is obtained, that is, the eccentricity of the resonance orbit itself. These eccentricities provide a way to check the validity of the assumptions. The applicability of the assumptions are discussed in Sections 2,3, and in the case of the Kuiper belt objects, appear to be very good.

It turns out that the predicted eccentricities are quite accurate. When applied to all the relevant resonant Kuiper belt objects, there is close agreement with the observed eccentricities. This implies that the original assumption of weak capture at Neptune is probably valid and yields the first main result of this paper referred to as Result 1: *At some time in the distant past, the Kuiper belt objects currently moving in resonance with Neptune may have been weakly captured by Neptune, and may have performed a hop at that time from one resonance to the observed one.* In this sense it is more likely to occur. This is discussed in Section 3.

Most of the resonant Kuiper belt objects turn out to be in the 2:3 resonance. The algebraic relationship between the resonance values and the periapsis distance from Neptune at weak capture yields a curve. The significance of the 2:3 resonance is seen when viewing this curve since it has a unique minimum at this resonance. This yields the second main result, Result 2: *The 2:3 resonance among all possible resonances is the one which minimizes the two-body Kepler energy of a small body with respect to Neptune, under the assumption that it was at its periapsis with respect to the sun at the time of weak capture.* See Section 3.

In the last section, a hop is forced to occur from the Kuiper belt object, 1995 HM_5 by modifying its elements. This is done to illustrate that hops at Neptune are not too unlikely. The hop itself takes only two years to perform. Thus, this shows that resonance changes at Neptune can occur very quickly, and need not occur as a gradual process over millions of years. The speed and behavior of this hop is consistant with those that occur at Jupiter for the short period comets studied in BM97. This indicates that as a process, it may play a significant role in solar system dynamics. Throughout this paper, by resonant it is meant *near* resonant. This is elaborated upon in BM97.

2. An Estimate of Eccentricities for Orbits of Resonant Kuiper Belt Objects

We restrict ourselves to the set $\mathcal{R}$ of resonant orbits about the sun, of type m:n, in resonance with Neptune. Under the following assumptions, we will show that the eccentricities of these orbits are uniquely determined. The resulting eccentricities show a remarkable agreement with the observed values of the observed Kuiper belt objects thus validating our assumptions.

Let $\mathcal{K}$ represent a candidate *resonant* Kuiper belt object. Following the proof sketch in Section 4.1 of BM97, the following assumptions are made:

A1. The orbit for $\mathcal{K}$ is approximated by a Keplerian ellipse.

A2. The orbit for $\mathcal{K}$ lies either completely outside of Neptune's orbit, or completely inside Neptune's orbit. These are referred to as *outer*, or *inner* orbits, respectively.(See Figure 8 in BM97)(The Kuiper belt objects are outer orbits by definition. We allow for inner orbits for complete generality.)

A3. Weak capture can be approximated with zero Kepler energy for $\mathcal{K}$ with respect to Neptune.

A4. There existed a time when $\mathcal{K}$ was weakly captured at Neptune at a periapsis with respect to the sun of an outer orbit, or at an apoapsis with respect to the sun for an inner orbit. (This assumption is reasonable based on the results of Duncan *et al*(1995) and Morbidelli (1997) mentioned in the Introduction.

A5. The orbit of Neptune is approximated as a circle of mean radius 30.11 AU, and mean orbital velocity of 5.43 km/s.

Using A1, the Kepler energy of $\mathcal{K}$ with respect to the sun, H_S, is computed by making use of Kepler's equation for the period P of the orbit for $\mathcal{K}$.

Set $\gamma = n/m$, and let a be the semi-major axis of the orbit. Kepler's equation is given by,

$$P = 2\pi\mu^{-1/2}a^{3/2}, \tag{1}$$

where $\mu = GM_S$, where G is the gravitational constant, and M_S is the mass of the sun. P_N is the period of Neptune of approximately 164.8 years. A1 implies the condition,

$$P = \gamma P_N. \tag{2}$$

Setting (1) equal to (2) yields,

$$a = a(\gamma) = c\gamma^{2/3}, \tag{3}$$

where $c = [(\mu^{1/2}/2\pi)P_N]^{2/3}$. Thus,

$$H_S = F(\gamma) = \frac{-\mu}{2a(\gamma)}. \tag{4}$$

Equation 4 is one measure of the energy. Another energy relation can be found by making use of A3, A4. This is found as follows:

Using A4, now assume $\mathcal{K}$ is in weak capture at Neptune at it's periapsis or apoapsis.

According to A3, at weak capture, $\mathcal{K}$ has the same Kepler energy as Neptune, with respect to the sun. This means, according to A4, that $\mathcal{K}$ is at rest with respect to Neptune and moving with it on the sun-Neptune line. It's velocity on this line is given by,

$$\nu = (\frac{r_N \pm r_e}{r_N})v_N,$$

where + sign means that $\mathcal{K}$ is on the sun-Neptune line at a distance r_e beyond the center of Neptune, at the periapsis of it's outer orbit with respect to the sun, r_p. and the - sign means that $\mathcal{K}$ is on the sun-Neptune line at a distance r_e in front of the center of Neptune, at the apoapsis of it's inner orbit with respect to the sun, r_a. r_N, v_N are the distance and velocity, respectively, of Neptune with respect to the sun. For the inner orbit, $r_e < r_N$.

When $\mathcal{K}$ escapes from Neptune at either of these two positions, it's velocity with respect to Neptune, according to A3, is obtained by setting it's Kepler energy with respect to Neptune equal to zero and solving for the velocity v_e, which is given by,

$$v_e = (2\mu_N/r_e)^{1/2},$$

where $\mu_N = GM_N$, and M_N is the mass of Neptune.

Since the direction of v_e is normal to the sun-Neptune line and therefore parallel to the direction of ν, then the magnitude of the velocity of $\mathcal{K}$ with respect to the sun at r_a or r_p is given by

$$v = \nu + v_e. \tag{5}$$

It is remarked that

$$r_a = r_N - r_e, \quad r_p = r_N + r_e. \tag{6}$$

With the value of v given by (5), the energy of $\mathcal{K}$ with respect to the sun can also be expressed as,

$$H_S = G(r_e) = v^2 - \frac{\mu}{r_N \pm r_e}, \tag{7}$$

where r_e is restricted to the domain $\mathcal{D}$ where $G < 0$.

By A1, H_S of $\mathcal{K}$ remains constant. Thus, by Equations (4),(7),

$$G(r_e) = F(\gamma).$$

The last equation yields,

$$\gamma = \gamma(r_e) = [-\frac{\mu}{2cG(r_e)}]^{3/2}. \tag{8}$$

This is the basic equation relating the resonance type to the escape distance from Neptune. It turns out to be accurate.

Equation 8 represents the orbits of the set $\mathcal{R}$ whose apoapsis or periapsis are in weak capture at Neptune. It is valid for $r_e \in \mathcal{D}$, and, for the case of inner orbits, $r_e < r_N$. Also, it is required that $r_e > 0$.

Although γ is a discrete variable defined over the rational numbers, it is extended into the set of all positive real numbers. Then, the function $\gamma(r_e)$ can be viewed as a smooth function. As r_e is allowed to vary over a given domain of values, γ then spans a set of values. The function $\gamma(r_e)$ can then be graphed, and the behavior of the graph can be used to determine the admissible values of r_e. For the case of the inner orbits, it is necessary that $\gamma \leq 1$, and for the outer orbits, $\gamma \geq 1$. Only those values of r_e yielding γ in these respective ranges are acceptable. Moreover, r_e are further restricted to correspond to rational or near rational values of γ. If the assumptions are realistic, then a majority of the observed Kuiper belt resonances should be recovered.

This is indeed the case as is seen in in the graph of $\gamma = \gamma(r_e)$ shown in Figure 1, corresponding to the outer orbits, which is the relevant situation. The set of Kuiper belt objects with reliable orbital elements are listed by Marsden(1997). This list includes 10 objects in resonance. These resonances are 2:3, 3:5, 3:4, where $\gamma = 1.5, 1.67, 1.33$, respectively. The 2:3 includes 8 of these objects. Referring to Figure 1, the 2:3, and 3:5 are included in our modeling which comprises nearly all the cases. These are listed in Table 1, with their eccentricities, Because their are several 2:3 cases, the mean eccentricity of .234 is also listed.

Summary

Object	Observed e	Mean Observed e	Predicted e	Error (Actual)	Error (From Mean)
1994 TB	.304	0.234	.222	.082	.012
1993 SB	.316			.094	
1995 HM5	.243			.021	
1993 RO	.202			.020	
1993 SC	.191			.031	
Pluto	.254			.032	
1994 JR1	.127			.095	
1994 JS	.238		.233	.005	
Post Hop Simulated	.174		.090	.084	

Table 1. The first group of objects are 2:3. The second case is 3:5. The last case is 5:4 produced from a simulated hop in Section 3 (inner orbit).

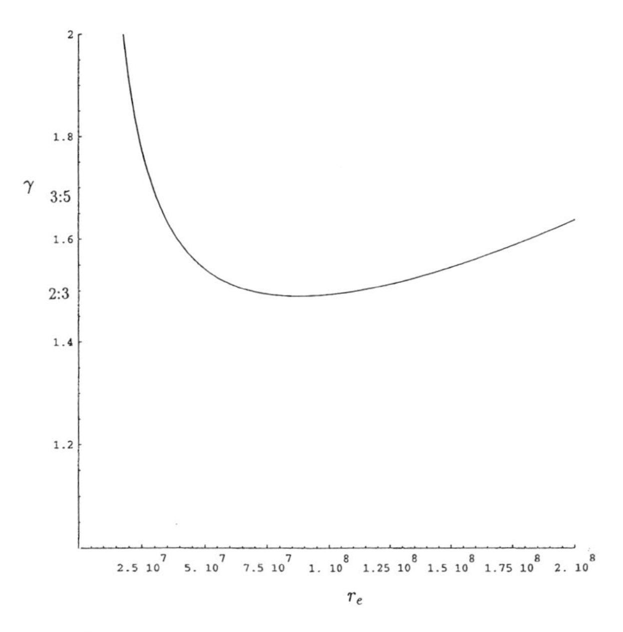

Figure 1. Resonance curve for outer orbits

The values of r_e corresponding to the different resonances of 2:3, 3:5 are easily read off of Figure 1. The graph is seen to be minimum at approximately $\gamma = 1.5$. The minimum precisely occurs at $\gamma = 1.491$ which is sufficiently close. The value of $r_e = .167$ AU is obtained for the 3:5 resonance. For the 2:3 case, the r_e value corresponding the minumum of the graph is .578 AU. Instead of using the exact minimum, if one uses the value of $\gamma = 1.5$ which is slightly higher by .09 unit, then there is a small spread for r_e from .455 AU to .749 AU.

The values of the corresponding eccentricities, e, are obtained from the r_e values by first using Equation 6 to determine r_p, and Equation 3 to determine a from γ. Then, e is determined from

$$e = 1 - \frac{r_p}{a}. \tag{9}$$

For 2:3, $.218 \leq e \leq .225$. At the minimum of the graph for $\gamma(r_e)$, $e = .222$ which is also the mean between .218 and .225. For 3:5, $e = .233$. These are listed in Table 1 in the right hand column.

There is good agreement with the observed values of e. For the observed 3:5 case, $e = .238$ which differs from our value by only .005. Using the mean value of .234 for the observed cases, and our median of .222, the error is .012. Because there are eight observed 2:3 objects with a small variation, it is reasonable to consider the mean value of .238. If the mean observed value is not used, the errors are still quite small, and are within .09. These

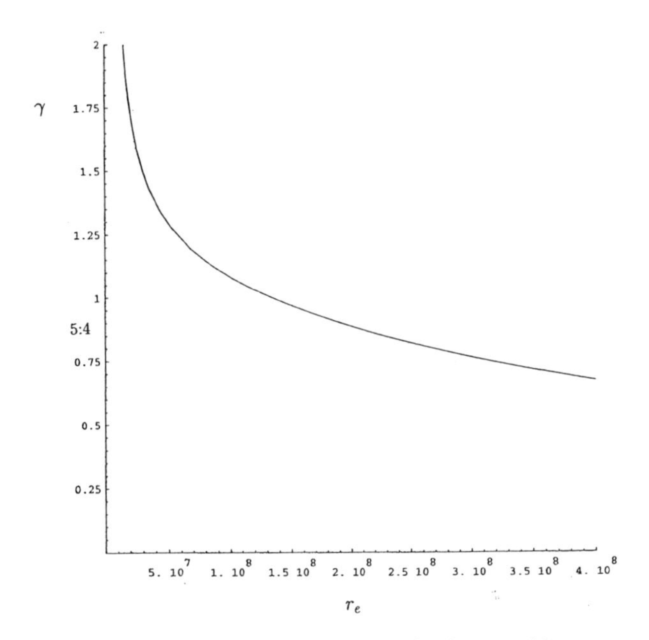

Figure 2. Resonance curve for inner orbits

small errors indicate that our modeling and assumptions are reasonable. The consequences of this are discussed in the next section.

It is remarked that the values of the eccentricities for the nonresonance objects listed by Marsden(1996) range from .02 to .13 and are therefore substantially different from the resonance values. The resonance of 3:4 which is not included by our graph, also has a substantially different eccentricity of .10.

The inner orbits were not analyzed here since the Kuiper belt objects are beyond Neptune's orbit. If, however, the graph of $\gamma(r_e)$ is constructed in this case, one obtains $.75 \leq \gamma \leq 1$, for .67 AU $\leq r_e \leq$3.0 AU. This graph has no relative minimum. See Figure 2. The results in this case are also in close agreement with with the simulation in the next section, where a hop was performed at Neptune from a 2:3 resonance to a 5:4. This 5:4 has an eccentricity of .174. The value of e computed using Figure 2 for $\gamma = .8$ is .090, differing only by .084 units. The value of e=.090 is determined by first noting that $r_e = 1.87$ AU corresponds to $\gamma = .8$. r_a is obtained from (6), and a from (3). Together they yield $e = (r_a/a) - 1)$.

3. Conclusions: Kuiper Belt Hopping and the 2:3 Resonance Preference

The results of the pervious section imply that nearly all the observed resonance Kuiper belt objects may have been weakly captured by Neptune in the distant past. This is deduced because from the last section, our modeling yielded an accurate prediction of the eccentricities of most of the observed resonance objects. Because of this agreement, this implies that our assumptions were valid. A key assumption was A4, thus justifying our conclusion. This justifies the first part of Result 1 in the Introduction. The second part follows from BM1 of BM97 described in the Introduction. That is, when an object is in weak capture at Jupiter, it moves onto a resonance orbit with respect to the sun in both forward and backwards time, which was justified by all of the observed cases under numerical simulation. Thus, in forwards time starting from the first resonance, there is a hop from one resonance to another. This is likely to happen here for Neptune, and the simulation done in Section 4 is an example of this. In that case an object at weak capture goes to a 5:4 in forwards time and a 2:3 in backwards time.

Therefore, dynamically, one would have the following situation. In the distant past, an object is moving in an elliptic orbit outside Neptune's orbit. As sufficient time goes by, it gradually moves in resonance with Neptune where it is pulled to weak capture. It then moves quickly from weak capture onto another resonance observed today from the hop.

Another conclusion from Section 2 results from the fact that the graph shown in Figure 1 of $\gamma = \gamma(r_e)$ for the outer orbits has a minimum at approximately the resonance 2:3. This graph results from Equation 8. This equation, in turn, results from considering all resonant orbits which are in weak capture at some time at a periapsis. Since the 2:3 resonance is a minimum of $\gamma = \gamma(r_e)$, then

$$\gamma \geq 1.5.$$

Equation 4 then implies that

$$\min_{1.5 \leq \gamma} H_S = H_S(1.5).$$

Thus, the energy, H_S, is minimized at approximately the 2:3 resonance. This yields Result 2 in the Introduction. Under the assumption, in general, that minimal energy is preferred, then the 2:3 resonance is the preferred value.

Hence, dynamically, one would expect to see a majority of resonance objects in the Kuiper belt moving with this resonance. This is clearly the case which comprises approximately 80% of all resonances.

There are several reasons why our model is a reasonable one. When a body is weakly captured at Neptune, for example, it is approximately parabolic with respect to it. Thus, it feels the gravitational perturbation of Neptune to a small degree. Once it quickly escapes Neptune, it is essentially a two-body problem between the body and the sun. Because Neptune has a small mass ratio with respect to the sun of .00005 and because the time in weak capture is very small, approximating the Kuiper belt objects as Keplerian ellipses is reasonable. Also, weak capture as defined in BM97 is not actually zero two-body energy with respect to Neptune. It is a slightly below zero, which for the short time spans for weak capture, is a reasonable to assume it is zero. Assumption 2 is closely related to the fact that a body leaves weak capture at a periapsis or apoapsis. It also implies that the body will not be significantly perturbed by Neptune unless it is near an apoapsis or periapsis of the body's orbit.

Because of these assumptions, there should be a little discrepancy between our predicted value and what is observed. This discrepancy turns out to be small.

4. Simulation of a Hop for 1995 HM$_5$

The hop has been proposed in this paper as a mechanism for causing resonances for a large set of observed Kuiper belt objects. Thus, it would be instructive to cause one to occur using precision integration and modeling. This is done in this section. It is shown to have similiar characteristics to those of the short period comets in BM97. The hop is from a 2:3 $\rightarrow 5:4$ which is from the outer to the inner obits. As was discussed at the end of the last section, the eccentricity of the inner orbit has good agreement with the predicted value. The precision integration modeling follows BM97, and is only briefly discussed.

The differential equations used to integrate the motion of $\mathcal{K}$ are given by

$$\ddot{\mathbf{x}} = \sum_{i=1}^{10} Gm_i(\mathbf{x} - \mathbf{x}_i)|\mathbf{x} - \mathbf{x}_i|^{-3}, \tag{10}$$

where $\dot{} \equiv d/dt$, $\mathbf{x} = (x_1, x_2, x_3) \in \Re^3$ is the position of $\mathcal{K}$, $\mathbf{x} = \mathbf{x}(t)$ and m_i and $\mathbf{x}_i \in \Re^3 (i = 1, 2, 3, ..., 10)$ are the masses and positions of the planets and sun given by the ephemeris DE403 (Standish *et al.* 1995). An eleventh-order numerical integrator is used.

An orbit of a Kuiper belt object is suitably modified to facilitate hop dynamics. The object chosen from the list(Marsden 1997) is 1995 HM$_5$ because it's periapsis distance of 29.80 AU is approximately that of Neptune's and it has a value of e of .243 which is .021 units from our predicted value of .222. The inclination, i, is also small at 4.81^o(J2000). This object was recovered by Gladman(1996). For notation, $\mathcal{K}$ represents 1995 HM$_5$.

The hop is forced to occur near the last periapsis of Neptune with respect to the sun on February 22, 1881. The elements of Neptune are $a = 30.056AU, e = .085, i = 1.770^o, \Omega = 131.786^o, \omega = -86.545^o, M =$

0.000^o(J2000). The values of $a = 39.345AU, e = .243$ for $\mathcal{K}$ are not altered. It's other orbital elements are chosen to match Neptunes. Thus, on Feb. 22, 1881, $\mathcal{K}$ is at it's periapsis, and therefore close to Neptune's position. It turns out to be approximately 678,486 km distant. To force a hop, the starting position of $\mathcal{K}$ is taken 6 years prior to it's periapsis on it's orbit. It is then numerically integrated forward using (10). This facilitates a nonlinear gravitational interaction with Neptune, which, in general, should capture the object for a short period of time, and then be ejected. This is observed, and when it is ejected it is moving with the 5:4 resonance. Also, when it is captured it is in the weak capture state, and thus a hop has occured. As in BM97, the start of the hop is taken when the object first enters Neptune's classical sphere of influence at 80,000,000 km distance from Neptune, and the hop end is when it leaves it. The respective dates for this are October 23, 1876, November 26, 1882. Thus, the hop duration is 6 years which is typical for those in BM97 at Jupiter. The elements of $\mathcal{K}$ in the post hop state on October 5, 1945 at solar periapsis are $a = 25.66AU, e = .174, i = 1.77^o, \Omega = 132.66^o, \omega = 91.59^o, M = 0.00^o$. The period is 47,480 days. In Figure 3, a plot of the hop near Neptune is shown. The other parameters studied in BM97 also show typical behavior.

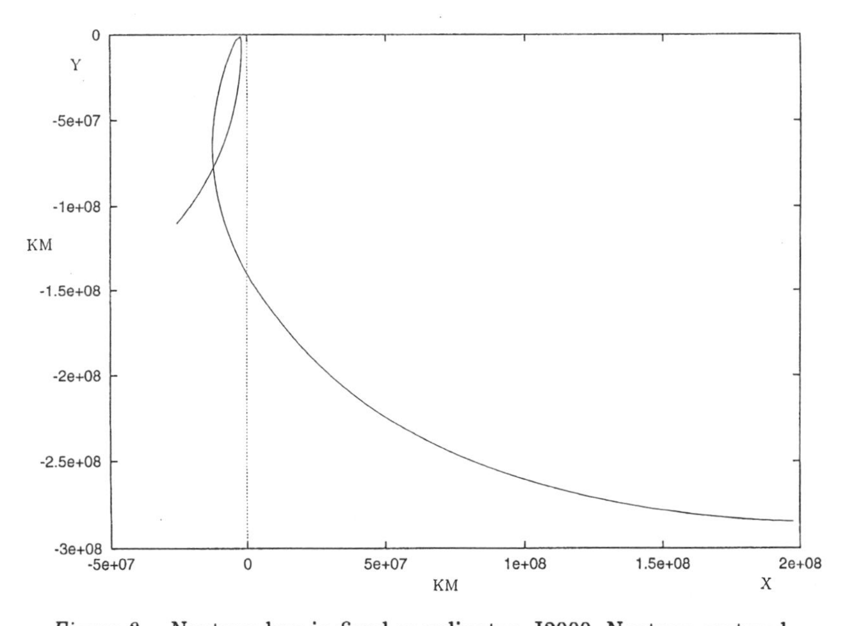

Figure 3. Neptune hop in fixed coordinates, J2000. Neptune centered.

The Kepler energy of $\mathcal{K}$ with respect to the sun is given by

$$H_S = \frac{1}{2}|\mathbf{v}|^2 - \frac{\mu}{|\mathbf{x}|},$$

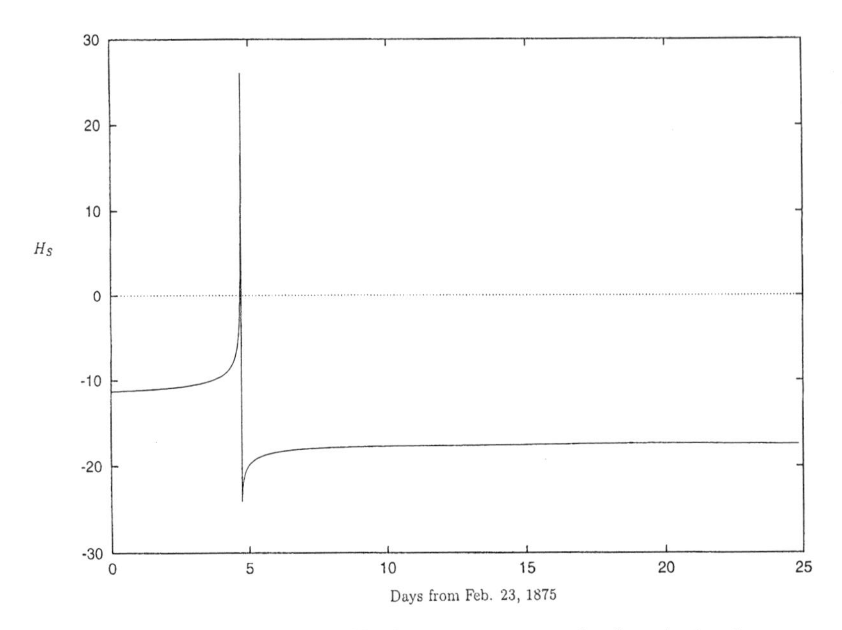

Figure 4. Behaviour of Kepler energy w.r.t. the Sun during hop.

where **v** is the three-dimensional velocity vectors of K with respect to the sun. Typically, during weak capture, H_S varies in a sensitive way over values that are negative but very near zero. For the short period comets studied in BM97, H_S has typically large nonlinear fluctuations while the comets are weakly captured and nearly constant otherwise. This is also the case here as is seen in Figure 4. Similiar to the comets at Jupiter in BM97, the Kepler energy with respect to Neptune H_N for $\mathcal{K}$ is negative and near zero during weak capture. For example, at the periapsis with respect to Neptune on November 10, 1879, $H_N = -.126km^2/s^2$. The distance to Neptune on this date is 86,007 km.

5. Conclusions

Using a model deduced from Belbruno and Marsden(1997), it is possible to predict the eccentricities with good precision of most of the Kuiper belt objects. The model implies that in the distant past, these Kuiper belt objects were likely to have been weakly captured by Neptune. This implies that they hopped into their current resonances. Also, the model yields a way to understand why the 2:3 resonance is preferred to over all the others. Using these ideas, a hop is simulated at Neptune using precision numerical integration by modifying an observed Kuiper belt object. These results support the notion that the hop is an important dynamic consideration in solar system evolution.

Acknowledgments

I would like to thank Brian Marsden for many interesting discussions, and the support of The Geometry Center.

References

1. Belbruno, E., & Marsden B. G. 1997, AJ, 113, 1433.
2. Belbruno, E. 1997, Proc. Inter. NEO Conf. (Ann. NY Acad. Sciences, New York)
3. Duncan, M. J., Levison, H. F., & Budd, S. M. 1995, AJ, 110, 3073.
4. Gladman, B. 1996, Minor Planet Electronic Circ., No. M07.
5. Jewitt, D. C., & Luu, J. X. 1995, AJ, 109, 1867.
6. Malhotra, R. 1996, AJ, 111, 504.
7. Malhotra, R. 1995, AJ, 110, 420.
8. Marsden, B. G. 1997, Asteroids, Comets, Meteors 1996 (Elsevier Science Ltd., Oxford)
9. Marsden, B. G., 1996a, Completing the Inventory of the Solar System, A.S.P. Conference Series.
10. Marsden, B. G., & Williams, G. V. 1996b, Catalogue of Cometary Orbits (Smithsonian Astrophysical Observatory, Cambridge, Mass.)
11. Mather, J. N. 1982, Topology, 21, 457.
12. Morbidelli, A. 1997, Icarus, 127, 1.
13. Siegel, C. L., & Moser, J. 1971, Lectures on Celestial Mechanics, Grundlehren Series (Springer, Berlin).
14. Standish, E. M., Newhall, X. X., Williams, J. G., & Folkner, W. M. 1995, JPL IOM 314.10-127.
15. Weissman, P. R. 1995, Ann. Rev. Astron. Astrophys., 33, 527.
16. Wisdom, J. 1980, AJ, 85, 1122.

THE SOLAR SYSTEM BEYOND NEPTUNE

I.P. WILLIAMS
Astronomy Unit
Queen Mary
University of London
London E1 4NS
UK

The possible existence of objects beyond the known Planets in the Solar system has fascinated humankind for a long time. In 1992, the first such object, given the temporary designation 1992 QB_1, was found orbiting the Sun on a near circular orbit at a distance of nearly $45AU$. Since that date, the number of discovered objects has been steadily growing and now stands at around 60. This paper summarizes the known facts regarding these.

1. Introduction

Ever since the beginning of the Copernican era, which place the Sun rather than the Earth at the centre of the Solar System, so that the planets moved on elliptical orbits about the mass centre, humankind has been intrigued by the possibility that unknown objects may exist beyond the known planets. The discovery of such new objects has generated a great deal of interest amongst both the general public and the astronomical community. In 1781 Herschel discovered Uranus at a distance of $19.2AU$, thus with one discovery, the size of the Solar System doubled. The story of the discovery of Neptune is well known and, with the $150th$ anniversary of the discovery only recently past, both the story and the controversy over who actually should be given credit for the discovery, has been given many an airing. For the purpose of this paper, it is sufficient to point out that astrometric observations of the recently discovered Uranus indicated that its motion did not strictly conform to that predicted by orbital theory, even when the gravitational perturbations from all the the other known planets were included. Independent calculations by Adams (1847) and LeVerrier (1849) both suggested that a solution to the anomalous motion of Uranus was provided by the existence of an other planet beyond Uranus. Neptune was

B.A. Steves and A.E. Roy (eds.), The Dynamics of Small Bodies in the Solar System, 51–64.

discovered very close to the predicted position in September 1846, again independently by Galle and Challis. Babinet had pointed out that the orbital elements of the newly discovered Neptune were not in total agreement with the predictions of LeVerrier and Adams and postulated that an other planet, which he called Hyperion, must exist further out. The belief in the existence of a planet beyond Neptune persisted and, after a long and thorough search, Tombaugh discovered a planet in 1930, which was subsequently named Pluto rather than Hyperion, orbiting the Sun at a mean distance of 39.4AU, much further out than Neptune at a mean distance of 30.1AU. It soon became apparent that Pluto was far to small a body to account for the alleged anomalies in the orbit of Uranus on which the belief of its existence had been founded. It was argued that a tenth planet must be awaiting discovery somewhere beyond Pluto. At the Lowell observatory, Tombaugh spent many years systematically searching for the tenth planet. Finally he concluded (Tombaugh 1961) that no unknown Solar system object brighter than a blue (B) magnitude of 16 existed within the entire sky north of a declination of -40^0. In addition he ruled out bodies brighter that a visual(V) magnitude of 17.5 in the ecliptic. These surveys more or less marked the end of search campaigns for planets of a size comparable to Uranus and Neptune beyond the limits of the existing solar system. It should also be noted for completeness that, according to Standish (1993), there are no discrepancies between theory and observations of the motion of Uranus that requires the existence of a trans Plutonian planet in order to explain them.

2. Cosmogony and the Outer Solar System

After the second world war, the old cosmogonic theories based on the tidal action of a passing star removing material from the Sun, associated mainly with the names of Jeans (1928) and Jeffreys (1929), were falling into disrepute and other theories following the original ideas of Kant (1755) and Laplace (1796) were being proposed. According to these ideas, the raw material out of which the planets formed was initially in the form of a disk or nebula about the young Sun. The planets formed by accumulation of what became known as *planetesimals*, in reality small solid bodies composed of rock and metals in the inner solar system, but also of ices further out than roughly the asteroidal belt, since the temperature was low enough at such distances from the Sun for ices to exist in solid form. Two such theories were proposed by Edgeworth (1943, 1949) and Kuiper (1951). They both concluded that nebular material could exist beyond Neptune, though accumulation into large planets would be difficult as the accumulation time is very long. However, a swarm of smaller bodies could orbit beyond Neptune and Edgeworth suggested that these could be observed as comets if they ever were to be perturbed into the inner solar system.

3. Comets and their Origin

At the time, the standard model for a comet was the *sand bank* model. According to this model, the comet nucleus was a loose agglomeration of individual dust grains, all moving about the Sun on nearly identical , but independent, orbits (see Lyttleton 1953). Support for this model appeared to come from the observed behaviour of meteor streams and by the appearance of a very strong concentration of dust present following the disintegration of comet $3D/Biela$ in the mid nineteenth century. However, when Whipple (1950) proposed the icy conglomerate model for a cometary nucleus, where the nucleus was a single identity with dust grains embedded in an icy matrix, it was realized that these features of meteor showers as well as other features could be accommodated in the new model. In this new model, comets lost material at each apparition and so could not survive as active comets for a time interval anywhere near the age of the Solar system. Hence dormant comets had to be stored somewhere and new ones would from time to time enter the inner solar system. Oort (1950) hypothesised that storage place for cometary nuclei was at a great distance from the Sun in a near spherical cloud and that stellar perturbations were responsible for the random injection of new comets. This hypothesis had the great merit that new comets could appear from any direction, just as was observed in reality. These ideas of Oort were published soon after the works of Edgeworth and Kuiper had appeared and became almost universally accepted. In contrast, for nearly four decades, the works of Kuiper and Edgeworth were ignored by the majority of astronomers, though there were a few exceptions most notable being being Whipple (1964) and Fernandez (1980). Oort had pointed out that the cometary cloud had to be further than 10^4AU from the Sun in order that the randomizing effect of stellar perturbations could produce a near spherical distribution. (at distances closer than this the gravitational field of the Solar System is very dominant). In contrast to the new, or long period comets, the inclination of the orbits of short period comets is heavily concentrated around zero, that is almost all short period comets move in the ecliptic plane. It was generally argued that this low inclination was a consequence of planetary perturbations, mainly from Jupiter, since the short period comets experience repeated relatively close encounters with these planets. However, by the late 1980's numerical simulations of these encounters orbits had demonstrated that such perturbations could not produce low inclination comets in sufficient numbers to match the observations (see for example Duncan *et al.* 1988, Stagg and

Bailey 1989, Quinn *et al.* 1990). These results suggest that a distribution of comets moving on low inclination orbits must also exist to feed the short period comet population. Hence, there was a growing belief that some form of a Kuiper-Edgeworth belt of comets must exist, culminating by the late eighties in searches starting to be conducted to look for Kuiper-Edgeworth objects.

4. The first searches

One early search was by Kowal (1989) who surveyed $6400deg^2$ of sky down to a V magnitude of 20. He failed to find any Trans-Neptunian objects, but the survey discovered the first member, namely 2060 *Chiron*, of a new class of solar system objects which has been given the collective name of *Centaurs*. The orbit of 2060 *Chiron* has a perihelion distance of 8.46AU and an aphelion distance of 18.96AU. It is thus a Saturn crosser and is also nearly an Uranian crosser. Since that date six other similar bodies have been discovered, 5145 *Pholus*, 7066 *Nessus*, 1994 TA, 1995 DW_2, 1995 GO and 1997 CU_{26}. Three of these, 5145 *Pholus*, 7066 *Nessus* and 1995 GO have their aphelia beyond Neptune. This class of objects are of considerable interest as Solar System bodies, possibly being a population of dormant or near dormant cometary nuclei, but they are not the Trans-Neptunian objects that were the subject of the search. Other unsuccessful searches from this period were Luu and Jewitt (1988), Levison and Duncan (1990), Cochran *et al.* (1991) and Tyson *et al.* (1992). In order to understand why these searches from the late 1980's to the early 1990's were not successful, it is useful to discuss the difficulties involved in a successful detection of objects beyond Neptune.

5. The Difficulties in detecting Trans-Neptunian Objects

The aim is to discover postulated members of a Kuiper-Edgeworth belt of dormant cometary nuclei orbiting beyond Neptune or even beyond the mean distance of Pluto. Of course, many periodic comets spend significant parts of their lives in the region beyond Neptune. A comet like $109P/Swift{-}Tuttle$, having a period of 135 years will spend about 50 percent of its life beyond Neptune, while comet Hale-Bopp will spend more than 99 percent of its life there. Such comets, moving on their present orbits, are not the objects of interest either. To be members of the Kuiper-Edgeworth belt, they must be Trans-Neptunian, having their perihelion essentially beyond Neptune as well as their aphelion. Locating such objects present special difficulties and it is instructive to discuss these and also the ways in which these difficulties were overcome. Comet $1P/Halley$ is probably the most well known comet. It was first recovered on October 16th 1982 by Jewitt and Danielson (1984) as it approached the Sun prior to perihelion in 1986, with the 200 inch telescope on Palomar Mountain using a CCD detector. At the time the comet was $11AU$ from the Sun and had a visual magnitude, V of 24.2, implying that its absolute magnitude at that time was about 14. Placed at a distance of $40AU$, that is just beyond Pluto, with the same absolute magnitude of 14 as at recovery, it would appear to Earth based observers as a magnitude 30 object, beyond the capabilities of present day ground based telescopes. $Comet 1P/Halley$ has a nucleus of dimensions $15 \times 8km$ (Keller *et al.* 1987). The largest measured cometary nuclei are $29P/Schwassmann - Wachmann\,1$ at $30km$ diameter (Meech *et al.* 1993) and $109P/Swift - Tuttle$ at $24km$ diameter (O'Ceallaigh *et al.* 1995). A potential target of this size at $40AU$ but otherwise similar to the nucleus of comet Halley would be 1.5 magnitudes brighter, still beyond the capabilities of current ground based telescopes. Thus, in order that an object in this region has the same apparent brightness as comet Halley at discovery, namely 24, its cross-sectional area would need to be about 300 times that of comet $1P/Halley$ or 40 times that of comet $29P/Schwassmann - Wachmann\,1$.

The apparent brightness of an object depends also on the amount of light it reflects rather than absorbs (its albedo). Estimating the albedo of an undetected object is rather problematic. Measured albedos are only available for a few cometary nuclei but the values obtained are consistently small and range from about 0.02 to 0.05 (see figure 6 of Fitzsimmons and Williams 1994). Since potential Kuiper-Edgeworth belt objects would have spent their entire life away from the Sun and hence experienced no coma activity, then any estimate of their albedo can only be an educated guess. If we assume that the albedo is within the cometary range, then the magnitudes given above are valid. However, if the albedo were to be the same as 2060 $Chiron$, then the required diameter could be reduced slightly and the object would still have the same brightness as comet $1P/Halley$ at detection. Thus the diameter of any body that might be detected would have to be considerably larger than the nucleus of any known comet. Since within any size distribution there is a bias in favour of smaller bodies, the expected number of $200km$ objects is significantly smaller than the number of $30km$ objects and a much larger area of sky will have to be searched before any success might be expected. The first technical requirement is thus to have a CCD detector with a sufficiently large field of view that a reasonable area of sky can be searched in a reasonable time. With hindsight, we know that the frequency is of order one per square degree so that, realistically, the CCD and telescope must be capable of covering this area in about a week of observing. Such CCD's started to become generally available in the beginning of the 1990's.

The main problem associated with the attempted detection of a faint object is one of obtaining a high enough signal to noise ratio. The standard way of improving this ratio is to increase the integration time. By integrating for several hours, images of astronomical objects with brightness as low as 28th magnitude have been obtained. Unfortunately, in the search for Trans-Neptunian objects, integrating for a long time is not an option for the simple reason that the target is moving. At a heliocentric distance of about $40AU$, an object on a circular orbit would display a retrograde motion of about 3 $arcsec$ per hour when observed at opposition, the retrograde motion arising because of the motion of the Earth. Of course, it is fortunate that this motion exists, for if it did not, there would be no way of distinguishing between a Trans-Neptunian object from a distant star. It is not an option either to move the telescope at this rate when attempting to discover an object, if the rate is not correct the situation is made worse. However, this can be done if one is attempting to recover a known object, for the rate is then known. It can also be used if the expected number of objects to be discovered is significant, on average then, some will be enhanced through doing this. A better way is to track at the siderial rate, but to co-add frames shifted by a few arcseconds. Typical seeing has the effect of spreading a point source of light into a disk with a diameter of the order of one arc second. that with a point source moving at 3 $arcsec$ per hour, light will significantly falling on new pixels after about 20 minutes and there is no gain in the signal to noise ratio to be had through integrating for periods longer than this. Using a 20 minute exposure on a typical $2m$ class telescope with dark skies and 1 $arcsecond$ seeing, an object at V magnitude of 23 would produce a signal to noise ration of about 30. However, this increases to about 40 if the R band is used and the object has solar colours. This increased efficiency is largely due to the greater sensitivity of the CCD, which peaks between 600 and $700nm$, and to the wider bandwidth at R compared to V, even though the dark sky is generally one magnitude brighter at R. A further gain through using the R band rather than the V band may come from the optical characteristics of Trans-Neptunian objects. Distant solar system bodies exhibit a range of colors, from essentially neutral (2060 *Chiron*) to extremely red (5145 *Pholus*). The general expectation was that any detected objects would be redder than solar colors so that the gain through using the R band is increased. Consequently the R-band has been used for most searches. It also follows that the first Trans-Neptunian objects to be detected would be the brighter ones and so might have a magnitude in the region of 23. They would thus either be significantly larger than normal cometary nuclei or have a significantly higher albedo or a combination of both.

In a single frame exposed for 20 minutes, a Trans-Neptunian object will be indistinguishable from a background star, that is, it will not exhibit any noticeable trailing. To identify a slow moving object, a second image must be obtained a few hours later. A problem that can arise is that observing conditions may change on this time scale so that a faint image is missed in the worst of the two sets of conditions. In consequence no identification of a moving object is made. Detection by itself is not however sufficient, it is also necessary to obtain a rough orbit for two important reasons. First, unless an orbit is obtained there is no way of knowing whether this object is a normal comet close to aphelion or a genuine Trans-Neptunian object. Second, unless a rough orbit is obtained it will be impossible to recover the object again or to recognize it as the same object in future observations. It is thus essential to obtain other images within days of the first pair and preferably others at regular intervals over a period of months. An orbit obtained from observations spread over a few month should be adequate to predict a position that is accurate enough to allow recovery of the object at the next opposition (in roughly twelve months time). However, failure to recovery the object at the next opposition will almost certainly mean that the object is lost. For this reason, there is a need for follow up astrometric observations for several years after discovery.

Finally, it should be noted that the new mosaics of CCD detectors with a large field of view significantly increase the efficiency of the search procedure. Such mosaics cover an area more than ten times that covered in a single frame by Jewitt and Luu (1995) in their early surveys.

6. The First Discoveries

The first detection of a potential member of the Kuiper-Edgeworth belt was made at the end of August 1992 by Jewitt and Luu (1992). The object had an R magnitude close to 23 and, assuming a circular orbit, a heliocentric distance $43.8AU$, clearly making it a Trans-Neptunian object. It has been given the temporary designation 1992 QB_1. The observations were taken at the f/10 Cassegrain focus of the University of Hawaii 2.2m telescope located on Mauna Kea, Hawaii, using a coated Tektronic CCD detector with 2048×2048 pixel and a broad R band filter centered at $650nm$ with a full width at half maximum of $120nm$. In the same way that one swallow does not make a summer, one object does not make a Kuiper-Edgeworth belt, but as Jewitt pointed out soon after the discovery, speculating that a belt of objects existed was somewhat more profitable after one object had been discovered than speculation based on no known objects. Luu and Jewitt (1993a) discovered a second object early in 1993, also with an R magnitude close to 23. If a circular orbit was assumed, this object, called 1993 FW, was at a heliocentric distance of $43.9AU$. If the albedo of these objects is about 0.04, that is similar to comets, then their diameters are in the 200km range, considerably larger than cometary nuclei and comparable to some of the larger asteroids. It was likely that these two object were at the upper end of the size range with a distribution of smaller objects down to standard cometary nuclei sizes being far more numerous but so far undetected.

Following the impetus given by these discoveries, it was natural that further searches would be conducted. In consequence, four further new objects were discovered in 1993. The four new objects were given the temporary designations 1993 RO (Jewitt and Luu 1993), 1993 RP (Luu and Jewitt 1993b) 1993 SB and 1993 SC (Williams *et al.* 1993). Assuming circular orbits, the heliocentric distances for these four new objects were respectively 32.3, 35.4, 33.1 and 34.5AU, much smaller than for the initial two objects and indeed, placing all four well within the orbit of Pluto. The search area had been chosen so that objects on a heliocentric orbit would be close to opposition and was by coincidence about 60^0 from the position of Neptune on the sky. This, together with the lower values of the heliocentric distances lead to speculation that the four new objects may be Neptune *Trojans*, that is objects librating about one of the Lagrangian triangle equilibrium points. This speculation proved to be incorrect. However, the true nature of the orbits of these objects proved to be perhaps equally remarkable. The circular orbit radii for these four objects were only slightly larger than the orbit of Neptune so that close encounters with that planet would occur at frequent intervals, leading inevitably to major perturbations of the orbits. Circular orbits had to be incorrect. 1993 SC was considerably brighter in apparent magnitude, at an R magnitude of 21.5, than the five other objects discovered at that point in time, and indeed, is still one of the brightest known Trans-Neptunian object. This brightness allowed it to be observed by other smaller telescopes so that a number of astrometric positions became available for it in the months following discovery. While a circular orbit with a radius of around 34AU was still consistent with the observations, orbits with significant eccentricities were also possible and probably more realistic. In particular, a solutions with a semi-major axis of about 40AU also fitted the observations. Numerical integrations showed that most such orbits would also be unstable, because of the close encounters with Neptune. If 1993 SC is on a stable orbit, then it must avoid close encounters with Neptune, and it can do this by being in some resonance. One possibility consistent with the observational constraints was that 1993 SC moved on a Pluto-type orbit very close to the $2:3$ mean motion resonance with Neptune. Such an orbit would have a period of order 248 years, and a semi-major axis of around 39.5AU, the eccentricity being around 0.13, a perfectly sensible value but significantly different from zero. Computations showed that the minimum distance between 1993 SC

moving on such an orbit and Neptune was greater than $14AU$ so that close encounters were avoided (Marsden 1994a, Williams *et al.* 1995, Williams and Collander-Brown 1997). No additional observations for the other three objects discovered in September 1993 were available, but Marsden (1994a) postulated that these three bodies also had stable orbits of this resonant type.

The recovery when near opposition in 1994 of 1993 SB, 1993 SC and 1993 RO near the positions predicted assuming Pluto-type resonant orbits lent further credence to this hypothesis. The three objects, 1993 RO, 1993 SB and 1993 SC were also recovered in 1995, very close to the positions predicted by the same near resonance model and it is now generally assumed that such solutions are correct. These three objects have also been recovered in 1996 and 1997. However it will take many more years of observations before the orbital elements will be known with certainty. Sadly, to date 1993 RP has not been recovered and it is probable that it has been lost. This emphasizes the need for observations in the weeks and months following the initial discovery.

The initial discoveries indicated that two distinct dynamical classes of Trans-Neptunian objects exist, a set moving on near circular orbits well beyond $40AU$, and a set moving on a resonant orbit with Neptune which could have a perihelion distance as small as about $30AU$ and significant eccentricities giving rise to aphelion distances in the region of $45AU$. Following the initial discovery in 1992 of 1992 QB_1, the rate of discovery has increased dramatically with twelve discoveries in 1994, and fourteen in each of the years 1995, 1996 and 1997. These numbers show that there is a significant population of bodies in the outer Solar System.

7. The current known members of the Kuiper-Edgeworth belt: orbit and location.

At the time of writing, there are about 60 known members of the Kuiper-Edgeworth belt. Almost all of them fall into one or other of the groups mentioned above, that is, they are either *Plutinos* moving on resonant orbits with Neptune or they are on near circular orbits of greater semi-major axis. The few exceptions will be discused later. The names of the discoverers, the orbital data, the absolute magnitude and the number of length of time that the object has been observed can all be found on the Worl Wide Web page of the Central Bureau for Astronomical Telegrams (http://cfa-www.harvard.edu/iau/lists/TNOs.html) For this reason we do not reproduce this data here, but instead produce a broad-brush picture of the general characteristics of the two groups of Trans-Neptunian objects.

7.1. THE MAIN KUIPER-EDGEWORTH BELT OBJECTS

Based on the object population known to date, this is the most populous of the groups with about 60% of the total belonging to it, at present amounting to 35 objects. Within this set of 35 objects discovered to date, the semi-major axes range from around $41AU$ to $47AU$ with a mean of $43.4AU$. The eccentricities, having excluded all objects where a circular orbit has been assumed, because of a lack of observational data, range from 0.006 to 0.222, with a mean of 0.083. It is therefor reasonable to describe this set as consisting of objects moving on nearly circular orbits at around $43AU$. The inclinations range from 0.6^0 to 31.6^0, with a mean inclination of 8.35^0. It is thus not true to describe this as a system lying in the plane of the ecliptic. The absolute magnitude ranges from 8.0 to 4.5 with a mean of 6.5. These are thus substantial sized objects with diameters of the order of several hundred kilometers.

7.2. THE PLUTINOS

Since our definition of this group is the set of objects that move near the $3:2$ mean motion resonance with Neptune, it is not surprising that the semi-major axes of the group is tightly bound around $39.44AU$, the semi-major axis of Pluto's orbit. The mean value for the set of 21 objects discovered to date in this group is $39.58AU$ with a range from $39.31AU$ to $39.94AU$. What is surprising is that the two groups are well defined with only one object with a semi-major axis between $39.94AU$ and $41AU$.The eccentricities of this group range from 0.079 to 0.335, with a mean value of 0.185, about twice the mean value for the first group. At such values of the eccentricity, there is a noticeable difference between perihelion distance and aphelion distance of nearly $15AU$, hence these objects can not be regarded as moving on nearly circular orbits. The inclination of the orbits range from 0.4^0 to 30.9^0 with a mean of 8.86^0, essentially the same as for the main group. The absolute magnitude ranges from 9.0 to 6.5 with a mean value of 7.7. On average these Plutinos are thus 1.2 magnitudes fainter in absolute terms than the main belt. The distance element would account for 0.4 magnitudes, hence the Plutinos appear to be on average intrinsically less bright than the main belt. This could be either due to a different albedo or a different size distribution.

7.3. THE OTHERS

At present there are four objects known that do not fit into one or other of the above two categories. They are, 1995 DA_2, 1994 JV, 1995 QY_9 and 1996 TL_{66}. Two of them,1995 DA_2 and 1994 JV, respectively have semi-major axes of $36.181AU$ and $35.251AU$. These are very close to the 4 : 3 and 5 : 4 mean motion resonances with Neptune respectively, and perhaps are other examples of the same phenomenon as described above for the Plutinos, namely a capture into mean motion resonance with Neptune. They have inclinations respectively of 6.6^0 and 18.1^0. 1995 QY_9 seems to be intermediate between the two main groups with a semi-major axis of $40.115AU$ and an eccentricity of 0.271.The inclination of the orbit of 1994 QY_9 is 4.8^0.

The fourth Trans-Neptunian Object, 1996 TL_{66} appears to be on a very different type of orbit. It has a semi-major axis of $84.5AU$ and an eccentricity of 0.585. At aphelion, it is thus more that $130AU$ from the Sun while the perihelion distance is only $35AU$, similar to the Plutinos. It also has a high inclination of 24^0.

7.4. GENERAL COMMENTS ON ORBITS

The above gives a general summary of the known orbital facts. One problem with analyzing the orbits of Trans-Neptunian Objects is that the orbital period is several hundred years. Hence, in the six years since the first discovery, the fraction of true anomaly covered is miniscule. All objects have almost been seen at one point on their orbit and deducing orbital parameters from data covering such a short arc is difficult. In addition, many of the objects were not re-observed at the first available opportunity and most have not subsequently been re-observed. It is thus unfortunately true that we should regard these as lost. The objects number about 14 and includes 1994 JV, mentioned above. Many of the high inclination objects are also in this list, though it is important to note the the object with the highest inclination to date, 1996 RQ_{20} has been seen at two oppositions and so is fairly secure. This may be significant since a number of numerical simulations of the Kuiper-Edgeworth belt have been run (*eg* Duncan *et al.* 1995, Malhotra 1995 1996, Morbidelli *et al.* 1995 , Weissman and Levison 1996). In general, these models predicted distributions with a lower inclination than possessed by 1996 RQ_{20} At present, a topic of major dynamical interest is the migration of bodies within this general region, are Plutinos formed in resonant orbits or captured there from the general distribution, is 1996 TL_{66} an example of migration from the Oort cloud to the Kuiper-Edgeworth belt or will it eventually end up as a Plutino?

8. Optical and Spectral Properties

Over the last few years a number of non astrometric observations of the brighter Trans-Neptunian objects, particularly 1993 SC have been carried out with a view to determine composition and physical properties. Based on a few observations soon after discovery, Williams *et al.* (1995) suggested that there might be a periodic variation in the light-curve. Later observations, (Davies *et al.* 1997, Tegler and Romanishin 1997) failed to detect such variations and there is thus currently no measurement of lightcurve variations that can give any information on the rotation period or shape of any Trans-Neptunian Object. A number of observers have also attempted to obtain broad-band colours of individual Kuiper-Edgeworth objects. Luu and Jewitt (1996) give some of the colours for two other Trans Neptunian Objects, 1992 QB_1 with $B-V$ of 0.65, $V-R$ of 0.6 and 1993 FW with $V-R$ of 0.4. Magnusson *et al.* (1997) obtained respectively a $V-R$ of 0.7, 0.3 and 0.45 for 1994 JQ_1, 1995 DC_2 and 1995 SW_2. Green *et al.* (1997) give $V-R$ colours of 0.77 for 1995 DC_2 0.54 for 1993 SC, 0.75 for 1994 JR_1, 0.55 for 1995 DA_2 Jewitt and Luu (1998) found $V-R$ values of 0.68 for 1993 SC, 0.32 for 1996 TO_{66}, 0.65 for 1996 TP_{66}, 0.43 for 1996 TS_{66} and 0.13 for 1996 TL_{66}. The main conclusion seems to be that considerable diversity exists in the measured colours of Trans-Neptunian objects, with $V-R$ lying in a range from around 0.13 to 0.75. There is some consistency between observers regarding individual object, 1993 SC, having values reported for $V-R$ of 0.68, 0.75, and 0.54. This spread is much smaller than the object to object spread so that there is some basis for believing that diversity exists. The lower values suggest fairly neutral colours, examples being 1996 TL_{66} and 1996 TO_{66}. These are very similar to the Centaur 2060 *Chiron* . Others, such as 1993 SC, 1995 DA_2 and 1996 TP_{66} are much redder, more similar to the Centaur 5145 *Pholus*. Such observations are at their early stages and it perhaps rather soon to make definitive statements of a more precise nature than the general ones we have made above.

Luu and Jewitt (1996) and Brown *et al.* (1997) obtained a spectrum of 1993 SC. The Spectrum of Luu and Jewitt suggested a $V-R$ color of 0.57, consisted with the values given above. The spectrum of Brown *et al.* also indicated that 1993 SC is very red.

9. Conclusions

The whole study of objects beyond Neptune is still in its infancy and it is instructive to note that none of the known objects have orbits that could be classified as **secure**, judged by the standards of other orbits in the solar system. Thus further astrometric observations of all the known objects at regular intervals, at least at every opposition, for the next few years is required.

Optical and infra-red spectral information on the brighter objects is also called for if we are to understand the inter-relations between these and other classes of objects in the outer solar system.

Finally, the search for further objects should continue, the number of semi-secure high inclination objects is still very small, too small to make any sensible statistical statements about their distribution. Similarly the number moving on resonant orbits other than the 3 : 2 is tiny and many more discoveries are called for before we can conclude that some of these resonances may be empty.

References

1. Adams J.C., 1847, On the perturbations of Uranus, *Mem. R. astr. Soc.*, **16**, 427-460
2. Cochran A.L. Cochran W.B. & Torbett M.V., 1991, A deep image search for the Kuiper disk of comets, *Bull. Amer. Astron. Soc.*, **23**, 1314
3. Davies J.K. McBride N. & Green S.F., 1997, Optical and infrared photometry of Kuiper belt object 1993 *SC*, *Icarus*, **125**, 61-66
4. Duncan, M.J., Levison, H.F. & Budd S.M., 1995, The Dynamical Structure of the K uiper Belt, *Astron. J.*, **110**, 3073-3081
5. Duncan M. Quinn T. & Tremaine S., 1988, The origin of short period comets *Astrophys. J.*, **328**, L69-73.
6. Edgeworth K.E., 1943, The evolution of our Planetary System, *J. of the British Astron. Assoc.*, **53**, 181-188
7. Edgeworth K.E. 1949, The origin and evolution of the solar system, *Mon. Not. R. a str. Soc.*, **109**, 600-609
8. Fernandez J.A., 1980, On the existence of a comet belt beyond Neptune, *Mon. Not. R. astr. Soc.*, **192**, 481-489
9. Green S.F. McBride N. O'Ceallaigh D.P. Fitzsimmons A. Williams I.P. & Irwin M., 1997, *Mon. Not. R. astr. Soc.*, **290**, 186-192
10. Fitzsimmons A.F. & Williams I.P., 1994, The nucleus of Comet P/Levy 1991 XI *Astron. Astrophys.*, **289**, 304-310
11. Jeans J.H., 1928, Astronomy and Cosmogony, *Cambridge Univ Press*, London
12. Jeffreys H., 1929, *Mon. Not. R. astr. Soc.*, **89**, 636
13. Jewitt D.C. & Danielson G.E., 1984, *Icarus*, **60**, 435-444
14. Jewitt D.C. & Luu J.X., 1992, *I.A.U. Circular no 5611*
15. Jewitt D.C. & Luu J.X., 1993, *I.A.U. Circular no 5865*
16. Jewitt D.C. & Luu J.X., 1995, The Solar System Beyond Neptune, *Astron. J.*, **109**, 1869-1876
17. Jewitt D.C. & Luu J.X., 1998, Optical-Infrared spectral diversity in the Kuiper Belt, *Astron. J.*, (in press)
18. Kant I., 1755, Allgemeine Naturgeschichte und Theorie des Himmels,
19. Keller H.U. & 21 others, 1987, Comet Halley's nucleus and activity, *Astron. Astrophys.*, **187**, 807-823
20. Kowal C., 1989, A Solar System Survey, *Icarus*, **77**, 118-123
21. Kuiper G.P., 1951, On the origin of the Solar System. In *Astrophysics, A Topical Symposium*, (J.A. Hynek ed.), 357-424, McGraw-Hill, New-York
22. Laplace P.S., 1796, Exposition di systeme de Monde, Paris
23. LeVerrier U.J.J.,1849, Nouvelles recherches sur les mouvements des planetes, *Comp. Rend. Acad. Sci. Paris*, **29**, 1-3.
24. Levison H.F. & Duncan M.J., 1990, A search for Protocomets in the Outer Regions of the Solar System, *Astron. J.*, **100**, 1669-1675.
25. Luu J.X. & Jewitt D.C., 1988, *Astron. J.*, **95**, 1256
26. Luu J.X. & Jewitt D.C., 1993a, *I.A.U. Circular no 5730*
27. Luu J.X. & Jewitt D.C., 1993b, *I.A.U. Circular no 5867*

28. Lyttleton R.A., 1953, The Comets and their Origin, Cambridge Univ. Press, London
29. Malhotra R., 1995, The Origin of Plutos Orbit-Implications for the Solar System beyond Neptune, *Astron. J.*, **110**, 420-429
30. Malhotra R., 1996, The Phase Space structure near Neptune Resonances in the Kuip er Belt, *Astron. J.*, **111**, 504-516
31. Marsden B.G., 1994, *I. A. U. Circular no 5983*
32. Meech K.J. Belton M.J.S. Mueller B.E.A. Dickinson M.W & Li H.R., 1993, Nucleus Properties of P/Schwassman-Wachmann 1, *Astron. J.*, **106** , 122-
33. Morbidelli A. Thomas F. & Moons M., 1995, The Resonant Structure of the Ku iper Belt and the Dynamics of the first 5 Trans-Neptunian Objects, *Icarus* , **118**, 322-440
34. O'Ceallaigh D.P. Fitzsimmons A. & Williams I.P., 1995, The CCD Photometry of Co met-109P Swift-Tuttle *Astron. Astrophys.*, **297**, 17-20
35. Oort P., 1950, The structure of the cloud of comets surrounding the Solar System and a hypothesis concerning it's origin *Bull. Astron. Inst. Netherlands*, **11**, 91-110
36. Quinn T. Tremaine S. & Duncan M., 1990, Planetary Perturbations and the ori gin of Short Period Comets, *Astrophys. J.*, **355**, 667-679
37. Stagg C.R. & Bailey M.E., 1989, Stochastic Capture of Short Period Comets, *Mon . Not. Roy. astr. Soc.*, **421**, 507-541
38. Standish, E.M., 1993, Planet X: No dynamical evidence in optical observations, *Astron. J.*, **105**, 2000-2005
39. Tegler S.C. & Romanishin W., 1997, The extraordinary colours of Trans-Neptunian objects 1994 *TB* and 1993 *SC*, *Icarus*, **126**, 212-217
40. Tombaugh C., 1961, The Trans-Neptunian planet search, In *Planets and Satellites*, eds G. Kuiper & B. Middleton, Chicago Univ Press, Chicago 12-30
41. Tyson J.A. Guhathakurta P. Bernstein G. & Hut P., 1992, Limits of the surf ace density of faint Kuiper Belt objects. *Bull. Amer. Astron. Soc.*, **24**, 1127
42. Weissman P. R.& Levison H.F., 1996, The Population of the Trans-Neptunian Region, in *Pluto-Charon*, University of Arizona Press, Tucson
43. Whipple F.L., 1950, A comet model 1, The acceleration of comet Encke, *Astrophys. J.*, **111**, 375-394
44. Whipple, F.L., 1964, Evidence for a comet belt beyond Neptune. *Proc. Nat. Acad. Sci*, **51**, 711-718
45. Williams I.P. Fitzsimmons A. & O'Ceallaigh D., 1993, *I.A.U. Circular no 5869*
46. Williams I.P. Fitzsimmons A. O'Ceallaigh D. & Marsden B.G., 1995, The slow-moving objects 1993SB and 1993SC, *Icarus*, **116**, 180-185
47. Williams I.P. & Collander-Brown S.J., 1997, Trans-Neptunian Objects, *Surveys in Geophysics*, **18**, 341-361

DYNAMICAL STUDY OF TRANS-NEPTUNIAN OBJECTS AT THE 2:3 RESONANCE

JOHN HADJIDEMETRIOU AND K.G. HADJIFOTINOU
Department of Physics, University of Thessaloniki,
540 06 Thessaloniki, Greece

1. Introduction

The orbital evolution of Kuiper belt objects has lately been a subject of intense research. The interest in this field is continuously growing due to recent observations of small objects in low inclination orbits beyond Neptune (see Jewitt *et al.* 1996). One of the main scopes of contemporary research is to determine the mechanisms that lead trans-Neptunian objects to either remain in place, orbiting regularly for time-scales of the age of the solar system, or, through chaotic evolution, diffuse to the inner solar system and eventually become short period comets.

Apart from several studies on the global structure of the Kuiper belt (for a review see e.g. Duncan *et al.* 1995), the recent discovery of the 2:3 resonant Kuiper belt members has risen the interest on studying the resonant structure of the Kuiper belt and the particular dynamics of the 2:3 as well as other exterior resonances to Neptune. The work of Duncan *et al.* (1995) is based on numerical integrations. The whole region between 30 and 50 AU is investigated, with particular attention to the 3:4 and 2:3 mean motion resonances with Neptune. Furthermore, Morbidelli *et al.* (1995) study exclusively the resonant structure of the Kuiper belt by both analytic and numerical means. Finally, Malhotra (1996) studies the dynamics of the major mean motion resonances with Neptune from the 5:6 (at 34 AU) up to the 1:3 resonance (at 62.6 AU). Her study is based on computing Poincaré surfaces of section of the planar circular restricted three-body problem near these resonances.

In this work we aim to investigate the stability of Kuiper belt objects near the 2:3 resonance with Neptune. We use a semi-analytic approach based on the construction of a symplectic mapping model that is valid near the 2:3 resonance. Adapting the same methodology as Hadjidemetriou

B.A. Steves and A.E. Roy (eds.), The Dynamics of Small Bodies in the Solar System, 65–70.

(1991, 1993) for the 3:1 resonance, we construct a four-dimensional mapping in order to study the dynamics of the planar elliptic restricted three body problem near the 2:3 resonance, with the Sun and Neptune as primaries. Our mapping is based on the averaged Hamiltonian of the elliptic restricted problem, expressed in 2:3 resonant action-angle variables.

In order for the mapping model to be realistic, its fixed points should coincide, qualitatively and quantitatively, with the fixed points of the Poincaré map of the real model on a suitably defined surface of section. The fixed points of the Poincaré map correspond to the periodic orbits of the real model (the restricted three body problem). Thus, by computing the main families of resonant periodic orbits of the restricted problem, at the 2:3 resonance, we can easily check the validity of our mapping model. In the region of phase space where the above conditions are fulfilled, our mapping model is similar to the real system and consequently its study will give much insight on the dynamical evolution of a Kuiper belt object.

2. The Restricted Three Body Problem at the 2:3 Resonance

The resonant structure of the planar restricted problem at the 2:3 resonance is described by the families of periodic orbits, and their stability character. As we mentioned in the previous section, this is important in order to check the validity of our model.

The coordinate system we use is the rotating reference frame xOy, with origin at the center of mass of Sun and Neptune and the x-axis defined by the line Sun-Neptune. The units are normalized such that the sum of masses of the Sun and Neptune, the semimajor axis of Neptune's orbit and the gravitational constant are all equal to unity. In these units, the mass of Neptune is $\mu = 5.178 \times 10^{-5}$ and the orbital period of Neptune around the Sun (i.e. the period of rotation of the rotating frame) is equal to 2π.

We present first the families of the *circular* case at the 2:3 resonance, computed for the value of $\mu = 5.178 \times 10^{-5}$. There are two families of resonant periodic orbits (periodic orbits of the second kind), family I and family II, along which the eccentricity varies, but the semimajor axis is almost constant, corresponding to the 2:3 resonance (Figure 1a). For the orbits of family I, at $t = 0$, the value of the resonant angle σ (defined in Eq. 1 of Sect. 3) is $\sigma = 0$ and the test particle lies at the perihelion of its orbit, while for the orbits of family II it is $\sigma = \pi$ and the particle lies at aphelion. In both cases, Neptune is aligned with Sun and the particle.

Family I is unstable for small values of the particle's eccentricity e. However, for orbits with eccentricity from 0.22 to 0.26, the particle, when at perihelion, is within the Hill radius of Neptune and therefore is subjected to a close approach - an actual collision occurs around $e = 0.24$. This reflects

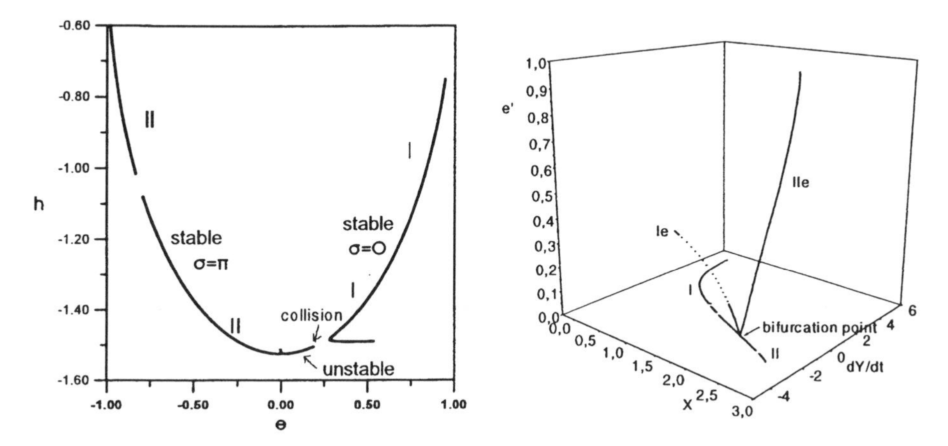

Figure 1. (a) The families of 2:3 resonant periodic orbits of the circular problem, in an eccentricity-energy diagram. Positive eccentricity means position at pericenter (at $t = 0$) and negative eccentricity position at apocenter. (b) The two families of the elliptic restricted three body problem at the 2:3 resonance. The curves on the $e' = 0$ plane represent the families of the circular problem, in the space $x\ \dot{y}$.

to a discontinuity of periodic orbits around that area. Beyond the collision, the stability index of the periodic orbits changes drastically, and finally for eccentricities greater than 0.3 the family becomes stable. We also observe that beyond the "collision area" the multiplicity of the periodic orbits has changed from 1 to 3.

Family II is stable throughout the variation of the particle's eccentricity. However, around e=0.8, there is a discontinuity in the family and we observe a small unstable region within which occurs a jump in the multiplicity of the periodic orbits from 1 to 5.

The families of the elliptic problem bifurcate from the above mentioned families of the circular problem at those points where the period is exactly 6π. It turns out that only two families exist, along which the parameter that varies is Neptune's eccentricity e'. These two families bifurcate from an orbit of the stable family II of the circular problem with eccentricity $e = 0.47$. We call the resonant families of the elliptic problem that bifurcate from this point families I_e and II_e (Figure 1b).

Family I_e has values of the resonant angles $\sigma = \pi$, $\nu = 0$ (ν is defined in Eq. 2 of Sect. 3), while family II_e has $\sigma = \pi$, $\nu = \pi$. Family I_e is unstable (single instability) throughout the variation of e', while family II_e is stable.

3. The averaged Hamiltonian

We describe here briefly the averaged Hamiltonian at the 2:3 resonance, because our mapping model is based on it. It is given in the 2:3 resonant action-angle variables:

$$S = \sqrt{a}(1 - \sqrt{1 - e^2}), \qquad \sigma = -2\lambda' + 3\lambda - \tilde{\omega}, \tag{1}$$

$$N = \sqrt{a}(2/3 - \sqrt{1 - e^2}), \quad \nu = 2\lambda' - 3\lambda + \tilde{\omega}', \tag{2}$$

where $a, e, \lambda, \tilde{\omega}$ are the semimajor axis, the eccentricity, the mean longitude and the longitude of perihelion of the test particle, and the corresponding primed quantities refer to Neptune. In these variables, the averaged Hamiltonian of the elliptic restricted three body problem is (Šidlichovský, p.c.)

$$H = H_0(S, N) + \mu H_1(\sigma, S, N) + \mu e' H_2(\sigma, S, \nu, N), \tag{3}$$

where

$$H_0 = 2(N - S) - \frac{1}{18(N - S)^2},$$

$$H_1 = -\{eA_{23} \cos\sigma + e^2(A_3 + A_{35} \cos 2\sigma) + e^3(A_{24} \cos\sigma + A_{45} \cos 3\sigma)\}$$

and

$$H_2 = -\{A_{13} \cos\nu + e[A_1 \cos(\sigma + \nu) + A_{34} \cos(\sigma - \nu)]\}.$$

The eccentricity e is a function of the actions S, N and

$$A_{23} = 1.895649, \qquad A_3 = 0.879749, \qquad A_{35} = 6.305147,$$
$$A_{24} = 0.999088, \qquad A_{45} = 25.802899, \qquad A_{13} = -1.545532,$$
$$A_1 = -1.526682, \quad A_{34} = -10.151440.$$

In order for the above averaged Hamiltonian to be realistic, its fixed points, both for the circular problem ($e' = 0$) and the elliptic problem, should correspond to the resonant periodic orbits described in the previous section. It turns out that the bifurcation point of the families of the elliptic problem, as obtained from the averaged Hamiltonian, is not in the correct position. In order to restore it in the right place, we added a small *correction term* $H_c(N)$. Its exact form will be discussed in the next section.

4. The symplectic mapping model

For the construction of the mapping to model the planar elliptic problem we use the averaged Hamiltonian described in the previous section, with two modifications:

$$H' = H_0(S, N) + \mu H_{1c}(\sigma, S, N) + \mu e' H_2(\sigma, S, \nu, N) + H_c, \tag{4}$$

where instead of H_1 we have H_{1c},

$$H_{1c} = -\{eA_{23} \cos\sigma + e^2(A_3 + A_{35} \cos 2\sigma) + \epsilon e^3(A_{24} \cos\sigma + A_{45} \cos 3\sigma)\}$$

and the term H_c is added:

$$H_c = \beta(\epsilon)N^2 + \alpha(\epsilon)N.$$

These corrections are necessary in order to have all the fixed points in their correct position and with the correct stability. The values of the constants β ,α and ϵ are empirically found to be $\beta = -0.003287$, $\alpha = -0.001966$, $\epsilon = 0.5$.

The mapping is obtained from the generating function

$$W = \sigma_n S_{n+1} + \nu_n N_{n+1} + TH'(\sigma_n, S_{n+1}, \nu_n, N_{n+1})$$

through the equations:

$$\begin{aligned} \sigma_{n+1} &= \partial W/\partial S_{n+1}, \quad S_n = \partial W/\partial \sigma_n, \\ \nu_{n+1} &= \partial W/\partial N_{n+1}, \quad N_n = \partial W/\partial \nu_n, \end{aligned}$$

where $T = 6\pi$ is the synodic period of a 2:3 resonant periodic orbit. By its construction this mapping is symplectic and can be proved (Hadjidemetriou, 1993) that it has the same fixed points as the Hamiltonian H', given by Equation (4).

5. The dynamics of Kuiper belt objects at the 2:3 resonance

Using our four-dimensional mapping model we integrated for about 10^8 years particle orbits with semimajor axes ranging from 37.5 to 41.5 AU and initial eccentricities ranging from 0.05 to 0.47. Two different cases were considered: orbits starting near perihelion ($\sigma_0 = 0$) and near aphelion ($\sigma_0 = \pi$). As we have mentioned in Sect. 2, the latter are phase-protected from close encounters with Neptune. The results within the time-span of 100 million years are shown in Figs. 2a, b for σ_0 near 0 and near π respectively. In both figures, the vertical dashed line shows the exact position of the 2:3 resonance, while the solid line denotes the lower border of the (a_0, e_0) area where orbits are initially Neptune-crossing.

From Fig. 2a it is obvious that, most orbits from 37.5 to 41.5 AU with e_0 greater than 0.2 that are not phase-protected, are chaotic and become Neptune-crossing within 10^8 years. Therefore, regular motion is restricted to smaller eccentricities. Finally, a slight displacement from the exact resonance at $e = 0.47$ gives rise to chaotic orbits.

All the above results are supportive of the idea that, while objects at the 2:3 resonance far from phase-protection mechanisms have strongly chaotic orbits and quickly develop cometary behaviour, phase-protected 2:3 resonant objects may orbit serenely for billions of years. This is consistent with the recent observations of Kuiper belt members at the 2:3 resonance with

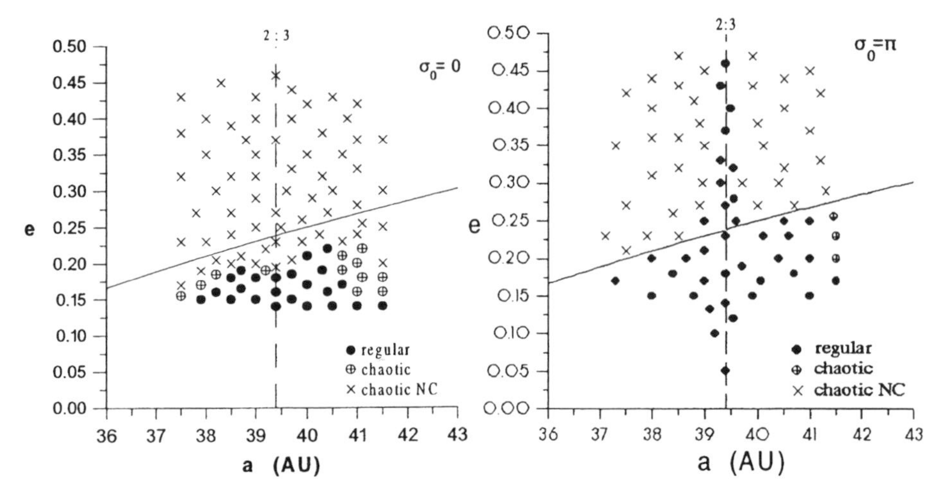

Figure 2. Distribution of orbits in the $a - e$ plane produced from integration with our four-dimensional mapping. The vertical dashed line shows the exact position of the 2:3 resonance, while the solid line denotes the lower border of the Neptune-crossing area. (a) Orbits with $\sigma_0 \simeq 0$. (b) Orbits with $\sigma_0 = \pi$ (phase-protected).

eccentricities up to 0.32. According to our findings, it is possible that even higher eccentricity objects may be discovered existing at this resonance, unless it is proved that the effect of other planets removes them from that region.

Acknowledgements

The authors would like to thank Dr. M. Šidlichovský for his provision of the averaged Hamiltonian at the 2:3 resonance. K.G. Hadjifotinou is financially supported by the Hellenic Scholarship Foundation (I.K.Y.).

References

Duncan, M.J., Levison, H.F. and Budd, S.M. (1995): *The dynamical structure of the Kuiper Belt*, Astron. J. **110 (6)**, December 1995, 3073-3081.

Hadjidemetriou, J.D. (1991): *Mapping Models for Hamiltonian Systems with application to Resonant Asteroid Motion*, in *Predictability, Stability and Chaos in N-Body Dynamical Systems*, A.E. Roy (ed.), 157-175, Kluwer Academic Publishers, Dordrecht.

Hadjidemetriou, J.D. (1993): *Asteroid Motion near the 3:1 Resonance*, Celest. Mech. **56**, 563-599.

Jewitt, D., Luu, J. and Chen, J. (1996): *The Mauna Kea-Cerro-Tololo (MKCT) Kuiper Belt and Centaur survey*, Astron. J. **112 (3)**, September 1996, 1225-1238.

Malhotra, R. (1996): *The Phase Space structure near Neptune resonances in the Kuiper belt*, Astron. J. **111 (1)**, January 1996, 540-516.

Morbidelli, A., Thomas, F. and Moons, M. (1995): *The Resonant Structure of the Kuiper Belt and the Dynamics of the First Five Trans-Neptunian Objects*, Icarus **118**, 322-340.

Šidlichovský, M. (1996): private communication.

SUPEROSCULATING INTERMEDIATE ORBITS AND THEIR APPLICATION IN THE PROBLEM OF INVESTIGATION OF THE MOTION OF ASTEROIDS AND COMETS

V.A. SHEFER
Research Institute of Applied Mathematics and Mechanics
Tomsk State University, Tomsk 634050, Russia

The approach applied in the present paper for thr construction of intermediate orbits is based on the idea of introducing of a fictitious attracting centre, following the ideas of Shaikh [1] that have received further development in a number of works [2, 3, 4]. Common to them is the fact that the intermediate orbits constructed by their authors give the motion of the unperturbed Keplerian orbit with respect to the fictitious attracting centre the mass of which is constant on the initial part of trajectory. In this work we propose a generalized method which allows one to construct the new classes of intermediate orbits with closer approximation to the real motion in comparison with approaches of the above-mentioned authors. The fundamental concept of the method implies that the mass of the fictitious centre for the initial part of motion is chosen not as a constant but is sought in the form of a function of time minimizing the perturbing factor.

Let us consider a body of negligible mass (small body) moving under the action of the Newtonian attraction of a system of point masses and other forces of an arbitrary nature. The differential equations of relative motion will be written in the commonly used form

$$\ddot{\mathbf{x}} = -\frac{k^2 M}{r^3}\mathbf{x} + \mathcal{F} \equiv \mathbf{G}, \tag{1}$$

where $\mathbf{x}$ is the position vector of the small body, $r = |\mathbf{x}|$, k^2 is the gravitational constant, M is the mass of the central body, $\mathcal{F}$ is the vector of the perturbing acceleration. Let the position and velocity vectors of the small body be known at the initial instant of time $t = t_0$: $\mathbf{x}(t_0) = \mathbf{x}_0$, $\dot{\mathbf{x}}(t_0) = \dot{\mathbf{x}}_0$. We refer to the motion described by the equations (1) as the real motion.

B.A. Steves and A.E. Roy (eds.), The Dynamics of Small Bodies in the Solar System, 71–76.

We represent the equations of motion (1) with respect to the new coordinate system connected with a fictitious attracting centre in the form

$$\ddot{\mathbf{q}} = -\frac{\nu}{R^3}\mathbf{q} + \mathbf{F}, \tag{2}$$

where $\mathbf{q} = \mathbf{x} - \mathbf{Z}$, $\mathbf{Z}$ is the position vector of the fictitious centre, $R = |\mathbf{q}|$, ν is the gravitational parameter of the fictitious centre, $\mathbf{F}$ is the new vector of the perturbing acceleration. The initial conditions are of the form: $\mathbf{q}(t_0) = \mathbf{q}_0 = \mathbf{x}_0 - \mathbf{Z}_0$, $\dot{\mathbf{q}}(t_0) = \dot{\mathbf{q}}_0 = \dot{\mathbf{x}}_0 - \dot{\mathbf{Z}}_0$.

We assume that generally the gravitational parameter ν is variable and represents some continuous function of time. The parameter ν is defined from the condition of minimum of the quantity $\Phi = \mathbf{F}^2$. Let us assume next that the fictitious centre located at $\mathbf{Z}$ moves with constant velocity i.e.,

$$\mathbf{Z} = \mathbf{Z}_0 + \dot{\mathbf{Z}}_0(t - t_0). \tag{3}$$

Then the equations (2) and the parameter ν are written correspondingly as

$$\ddot{\mathbf{q}} = \mathbf{G}, \qquad \nu = -(\mathbf{G} \cdot \mathbf{q})R. \tag{4}$$

In the coordinate system in which the real motion (2) is considered we consider an intermediate motion defined by the equations

$$\ddot{\mathbf{q}}^* = -\frac{\mu}{{R^*}^3}\mathbf{q}^* \tag{5}$$

and the initial conditions: $\mathbf{q}^*(t_0) = \mathbf{q}_0^*$, $\dot{\mathbf{q}}^*(t_0) = \dot{\mathbf{q}}_0^*$. Here μ is the gravitational parameter of the fictitious centre defining this intermediate motion, $R^* = |\mathbf{q}^*|$. Let $\mu = \mu(t)$ be a function continuously changing in time. Then the equations (5) which are called the equations of the Gylden-Mestschersky problem [5, 6] give the motion which is not Keplerian in the general case.

We shall choose the parameter $\mu = \mu(t)$ as some approximation to the function ν such that the equations (5) of the intermediate motion become strictly integrated ones.

Following Mestschersky [6] we consider the two approximations

$$\mu = \frac{\mu_0^2}{\mu_0 - \dot{\mu}_0(t - t_0)}, \tag{6}$$

and

$$\mu = \frac{\mu_0^2}{\sqrt{\mu_0^2 - 2\mu_0\dot{\mu}_0(t - t_0) + (3\dot{\mu}_0^2 - \mu_0\ddot{\mu}_0)(t - t_0)^2}}, \tag{7}$$

where $\mu_0 = \nu_0$, $\dot{\mu}_0 = \dot{\nu}_0$, $\ddot{\mu}_0 = \ddot{\nu}_0$. They correspond to the Mestschersky's first law of change of mass and Mestschersky's universal law, repectively.

Thus the position of the fictitious centre (3) and the intermediate motion (5) are completely determined by the vectors $\mathbf{Z}_0$, $\dot{\mathbf{Z}}_0$, $\mathbf{q}_0^*$, $\dot{\mathbf{q}}_0^*$ and scalars μ_0, $\dot{\mu}_0$ and $\ddot{\mu}_0$.

We take in (3) $\dot{\mathbf{Z}}_0 = \mathbf{0}$ and require

$$\mathbf{q}_0^* = \mathbf{x}_0 - \mathbf{Z}_0, \qquad \dot{\mathbf{q}}_0^* = \dot{\mathbf{x}}_0, \qquad \ddot{\mathbf{q}}_0^* = \mathbf{G}_0. \tag{8}$$

These conditions ensure second-order osculation to the trajectory of the real motion at initial time.

By (5) the last condition (8) yields

$$\mu_0 = |\mathbf{G}_0| R_0^{*^2}, \qquad \mathbf{q}_0^* = -\frac{R_0^*}{|\mathbf{G}_0|}\mathbf{G}_0. \tag{9}$$

Hence the fictitious centre is positioned on the vector $\mathbf{G}_0$ or on its extension. We choose the arbitrary parameter R_0^* such that the main part of deviation of the intermediate motion from the real one is minimized. For this we consider the condition for a minimum of the quantity $\chi_3 = (\mathbf{q}_0^{*(3)} - \mathbf{q}_0^{(3)})^2$ with respect to the variable R_0^*. Here and below the indices in parentheses denote the order of a derivative with respect to time t.

We consider the following cases.

Case 1a. The parameter μ is chosen in the form of Mestschersky's first law. The deviation $\mathbf{q}_0^{(3)} - \mathbf{q}_0^{*(3)}$ in this case represents a component of the vector $\dot{\mathbf{G}}_0$ which is normal to the plane formed by the vectors $\mathbf{q}_0^*$ and $\dot{\mathbf{q}}_0^*$. Thus, if the vector $\dot{\mathbf{G}}_0$ is located on this plane the magnitude of χ_3 vanishes and we achieve an osculation of the third order.

Case 1b. The parameter μ is of the form of Mestschersky's universal law. The values of the quantity χ_3 for this orbit and for orbit *1a* coincide.

Let us take now $\mathbf{Z}$ in the general form (3) and impose the conditions

$$\mathbf{q}_0^* = \mathbf{x}_0 - \mathbf{Z}_0, \qquad \dot{\mathbf{q}}_0^* = \dot{\mathbf{x}}_0 - \dot{\mathbf{Z}}_0, \qquad \ddot{\mathbf{q}}_0^* = \mathbf{G}_0, \qquad \mathbf{q}_0^{*(3)} = \dot{\mathbf{G}}_0. \tag{10}$$

These conditions ensure third-order osculation between the real trajectory and the intermediate orbit at the instant $t = t_0$. As in the case of second-order osculation we have the relations (9). The main part of the deviation of the intermediate motion from the real one is characterized by the quantity $\chi_4 = (\mathbf{q}_0^{*(4)} - \mathbf{q}_0^{(4)})^2$.

In a similar manner we study the two particular cases:

Case 2a. The parameter μ varies according to (6). From the last condition of (10) and in view of (5) and (9) we obtain

$$\dot{\mathbf{q}}_0^* = \frac{R_0^*}{|\mathbf{G}_0|}\left(\frac{1}{2}\alpha\mathbf{G}_0 - \dot{\mathbf{G}}_0\right), \tag{11}$$

where R_0^* and $\alpha$are arbitrary parameters. The minimum of χ_4 is to be taken with respect to both of these variables. Then we obtain the result that the vector $\mathbf{q}_0^{(4)} - \mathbf{q}_0^{*(4)}$ represents the component of $\ddot{\mathbf{G}}_0$ which is normal to the plane formed by the vectors $\mathbf{q}_0^*$ and $\dot{\mathbf{q}}_0^*$. Thus, whenever the vector $\ddot{\mathbf{G}}_0$ is located on this plane the magnitude of χ_4 vanishes. That is, the order of osculation increases to 4.

Case 2b. The parameter μ is taken in the form (7). The quantity χ_4 in this case depends only on the parameter α. The condition for a minimum of χ_4 with respect to α yields the same result as in *Case 2a.*

As soon as the constants R_0^* and α have been determined all the parameters of the fictitious centre and the intermediate orbits are uniquely found.

The integration of equations (5) can be carried out by the special space-time transformations:

$$\mathbf{u} = \frac{\mu}{\mu_0}\mathbf{q}^*, \qquad d\theta = \left(\frac{\mu}{\mu_0}\right)^2 dt. \tag{12}$$

The equations (5) with μ from (6) are transformed into the following system of equations describing the intermediate motion in the parametric space of $\mathbf{u}$-variables:

$$\mathbf{u}'' = -\frac{\mu_0}{|\mathbf{u}|^3}\mathbf{u}, \qquad t' = [1 - \frac{\dot{\mu}_0}{\mu_0}(t - t_0)]^2, \tag{13}$$

where accents denote differentiation with respect to the new variable θ.

If μ is of the form (7) the equations (5) by means of the transformations (12) are reduced to the following system of equations:

$$\mathbf{u}'' = -\frac{\mu_0}{|\mathbf{u}|^3}\mathbf{u} - \frac{2\dot{\mu}_0^2 - \mu_0\ddot{\mu}_0}{\mu_0^2}\mathbf{u}, \tag{14}$$

$$t' = 1 - 2\frac{\dot{\mu}_0}{\mu_0}(t - t_0) + \frac{3\dot{\mu}_0^2 - \mu_0\ddot{\mu}_0}{\mu_0^2}(t - t_0)^2. \tag{15}$$

The solutions of the equations (13) and (14)–(15) can be found in terms of elementary and elliptic functions [6].

Using the inverse transformations we find the position and the velocity vector of the body on the intermediate orbit in the physical space at any given instant of the time t.

The conditions presented in this work which ensure a high order touch between the intermediate orbit and the real motion, are referred to as the conditions of superosculation, and the corresponding orbits are called the superosculating orbits.

To test the efficiency of constructed orbits on an example of the motion of small bodies in the Solar system computer programs have been developed for the numerical integration of the differential equations for the deviation of the real motion from the intermediate one, $\boldsymbol{\rho} = \mathbf{q} - \mathbf{q}^* = \mathbf{x} - \mathbf{x}^*$:

$$\ddot{\boldsymbol{\rho}} = \mathbf{G} + \frac{\mu}{R^{*3}}\mathbf{q}^*. \tag{16}$$

The parameter μ is taken according to Mestschersky's first law. The programs make use of Everhart's algorithm of 15th order [7]. The unusual minor planet Icarus and the short-period comet Honda-Mrkos-Pajdusakova were chosen as the objects to be investigated. The initial epochs were chosen near the dates of aphelion passages. In the case of the comet the initial epoch corresponds to the position of the body on the sphere of influence of Jupiter. Perturbations due to all major planets and the Moon are taken into account. In the numerical experiments, absolute values of the vectors of deviation, $|\boldsymbol{\rho}|$, were compared. The results of the calculations are presented in Tables 1–2. The tables give a clear notion of the approximation accuracy and of the time interval in which the orbit with second-order osculation (ORBIT II) and the orbit with third-order osculation (ORBIT III) approximate the real motion better than the osculating Keplerian orbit (ORBIT I).

TABLE 1. Deviations in AU of the intermediate orbits from the real trajectory of Icarus ($t_0 =$ 2440190.5 JED)

Time in days from t_0	ORBIT I	ORBIT II	ORBIT III
1	1.2×10^{-9}	7.7×10^{-14}	1.0×10^{-14}
2	4.7×10^{-9}	8.4×10^{-13}	1.7×10^{-13}
5	2.9×10^{-8}	2.8×10^{-11}	6.7×10^{-12}
10	1.2×10^{-7}	4.3×10^{-10}	1.2×10^{-10}
20	4.5×10^{-7}	6.5×10^{-9}	2.4×10^{-9}
40	1.8×10^{-6}	9.4×10^{-8}	5.3×10^{-8}
80	8.0×10^{-6}	1.8×10^{-6}	1.2×10^{-6}
120	2.2×10^{-5}	9.8×10^{-6}	1.3×10^{-5}

The constructed superosculating orbits can be effectively used as reference orbits in Encke's method [8], integrating the equations for deviations (16) by means of some method of numerical integration. Because the constructed orbits are closer to the real orbits in comparison with the osculating orbits at least on an initial part of the trajectory, the accuracy and speed of calculation on a computer is increased.

TABLE 2. Deviations in AU of the intermediate orbits from the real trajectory of the comet Honda-Mrkos-Pajdusakova (t_0 = 2428080.5 JED)

Time in days from t_0	ORBIT I	ORBIT II	ORBIT III
1	2.3×10^{-6}	2.8×10^{-9}	3.3×10^{-11}
2	9.1×10^{-6}	2.2×10^{-8}	5.2×10^{-10}
5	5.5×10^{-5}	3.6×10^{-7}	1.9×10^{-8}
10	2.1×10^{-4}	3.3×10^{-6}	3.2×10^{-7}
20	7.6×10^{-4}	3.3×10^{-5}	5.9×10^{-6}
40	2.6×10^{-3}	3.6×10^{-4}	1.3×10^{-4}
80	8.0×10^{-3}	4.0×10^{-3}	3.5×10^{-3}

The constructed orbits can also be effectively applied in the problem of the determination of the preliminary orbits because of the highly accurate approximation of the obtained observations.

The method of the constructed superosculating orbits is especially effective in studying the motion of the small body (asteroid, comet, spacecraft) near large perturbing masses (planets, their satellites).

References

1. Shaikh, N.A. (1966) A new perturbation method for computing Earth-Moon trajectories, *Astronautica Acta* **12**, 207–211.
2. Skripnichenko, V.I. (1970) On Shaikh method for computing trajectories allowing close approach to the perturbing body, in *Materials of the Symposium "Dynamics of Small Bodies of the Solar System"*, ELM, Baku, pp. 9–10 (in Russian).
3. Batrakov, Yu.V. (1981) Intermediate orbits approximating the initial part of perturbed motion, *Bul. ITA AS USSR* **15**, 1–5 (in Russian).
4. Sokolov, V.G. (1982) Intermediate orbits with the fourth order touch to trajectories of perturbed motion, *Bul. ITA AS USSR* **15**, 176–181 (in Russian).
5. Gylden, H. (1884) Die Bahnbewegungen in einem Systeme von zwei Koerpern in dem Falle, dass die Massen Veraenderungen unterworfen sind, *Astronomische Nachrichten* **109**, 1–6.
6. Mestschersky, I.W. (1952) *Works in Mechanics of Bodies with Variable Masses*, GITTL, Moscow (in Russian).
7. Everhart, E. (1974) Implicit single-sequence methods for integrating orbits, *Celestial Mechanics* **10**, 35–55.
8. Roy, A.E. (1978) *Orbital Motion*, Adam Hilger LTD, Bristol.

TESTS OF GENERAL RELATIVITY USING SMALL BODIES OF THE SOLAR SYSTEM

C. MARCHAL
O.N.E.R.A. D.E.S. BP 72
92322 CHATILLON CEDEX
FRANCE

Abstract. The tests of general relativity have two main purposes:

A) To verify, or not, some consequences of the general relativity. The STEP experiments (space tests of the equivalence principle) belong to this first class of tests.

B) To allow us to choose among the numerous relativity theories that have been proposed since 1916 and that are competing with that of Einstein.

The different theories are characterized by their "post-Newtonian parameters" and especially by their coefficients "beta" and "gamma".

Several moderately accurate measures of the coefficient gamma have already been obtained - and they still agree with the Einsteinian theory - but the measure of beta is much more difficult.

The discussion of the accuracies of the possible experiments shows the great interest of a Shapiro experiment using an asteroid such as Icarus.

1. On the Future of Celestial Mechanics

The future of "celestial mechanics" depends essentially, of course, on the definition of these two words.

If celestial mechanics only refers to the classical three and n-body problems in the Newtonian absolute space with its absolute time and the Newtonian laws of gravity, its future is rather limited even if it remains very interesting.

In this direction we can hope for progress in the following domains.

B.A. Steves and A.E. Roy (eds.), The Dynamics of Small Bodies in the Solar System, 77–92.

A) Mathematical understanding of the n-body problem.
The Poincaré conjecture on periodic orbits.
The two Chazy conjectures.
The everywhere dense character of escape orbits.
The fractal or Cantor structure of the set of bounded orbits.
The analysis of temporary chaotic motions and temporary captures.
The oscillatory motions of the first and second kind.
The Arnold diffusion conjecture.
The total dispersion conjecture.
Etc.... [1, pages 519-526].

B) Progress in physical understanding.
B.1) Limits on the stability of a planetary system (such as the solar system).

We must therefore give a practical and physical meaning to the word "stability", for instance "no variations larger than 10% of the semi-major axes and no variations larger than 0.1 of the eccentricities during several billions of years".

If the variations become larger, the system certainly becomes rapidly unstable, except for the eccentricity of the innermost planet that can have large variations without much disturbances of the other planets.

In such a study we will meet classical limits such as the total mass of planets (relative to their Sun), the mutual inclinations, the importance of eccentricities and of successive separations between planets - all this drawing a picture not very different from our own planetary system.

But we will also meet unexpected effects coming from chaotic motions, horizon of prediction, time of divergence, importance of temporary chaotic motions, of resonances etc...

We will then perhaps meet strange propositions such as the following:"There is a 92% probability that our solar system survives for two billion years within the above definition of stability". (Because of its future sensitivity to present conditions).

These stability studies will give us a better idea of the possible number of planetary systems in our galaxy.

B.2) We can also study the stability of a planetary system about a strong binary (there are many such binaries).
B.3) Stability of a planetary system about a weak binary (such as Alpha Centauri with two binaries of similar masses, an 80 year period and 0.55 eccentricity).
B.4) Stability inside a triple or multiple system.
B.5) Disturbances given in a planetary system by the passage of a star.

For instance will our system remain in the domain of long-term stability after the passage of a star of Sun-mass at two or three times the Neptune distance with a velocity of 50 or 100 km/s ?

B.6) Finally, if we are able to deal with systems with a very large number of bodies, we can look for the long term evolution of a galactic system.

What proportion of stars does it loose in outer space per billion years? What is the influence of a large central "black hole", a Newtonian mass of about 200 million Sun-masses ? What is the gravitational influence of "dark matter" ? (A more or less simple supplementary potential).

If celestial mechanics goes beyond these pure Newtonian n-body studies there seems to be two major directions of research.

C) Influence of dissipative or conservative small forces: radiation pressure, tidal forces, electromagnetic forces, Poynting - Robertson effect, collisions inside a cloud of dust, etc...

These forces certainly have an essential effect in the long tem evolution of many systems:

C.1) Systems with small distances (Earth - Moon system, strong binaries...)
C.2) Motions about the center of mass.
C.3) Systems with many small bodies (contraction of clouds of dust and birth of new stars).

The study of all these effects will certainly reveal several yet unknown small forces.

D) The second major direction of development of celestial mechanics is the theory of relativity.

We can consider two levels.

D.1) At the small-level, the relativity is considered as a small perturbation of the Newtonian theory. Its "relativistic effects" lead to various conservative small forces and few extremely small dissipative forces.

We must notice that these effects depend upon the theory of relativity under consideration; there are indeed more than thirty different theories in competition with that of Einstein[1]. Hence these first level studies have three types of interest.

1. **I)** They give a better knowledge of the small forces considered in section C.

   II) They allow a better prediction of future evolutions considered in sections B and C.

[1]See next section.

III) They allow the best possible choice of space experiments allowing to choose between the different possible theories of relativity.

D.2) The second level of study of relativity is a general one.

What happens, what are the motions of celestial bodies in the regions of large space curvature, in the vicinity of black holes or neutron stars ?

Let us consider only one example. There are many famous "jets" of matter at very fast velocity (up to 80 000 km/s). These jets appear to be related to the existence of black holes and it is considered that they correspond to the polar direction of a fastly rotating black hole: a small percentage of the falling matter escapes in the polar directions.

This doesn't correspond to the usual gravitational possibilities of a rotating black hole with its Kerr matrix of curvature, and so electromagnetic forces are called for the explanation of the phenomenon.

Another possibility would be the existence of binary black holes or of strong binary neutron stars. When such a binary enters into a cloud of dust a small part of the falling matter escapes at very large velocities because this corresponds to normal three-body relativistic trajectories.

Thus celestial mechanics has a very bright future and the main reason for this situation is because it analyses on the largest scales of space and time many extremely sensitive phenomena with ever more accurate and sensitive instruments.

Furthermore, the progress of astronautics gives now to celestial mechanics the status of an experimental science, a considerable improvement with respect to its former status of observational and theoretical science.

2. On the Different Theories of Relativity

We must make a clear distinction between the unique "special theory of relativity" and the numerous general theories of relativity.

The special theory of relativity is esssentially the work of Hendrik Antoon Lorentz, Henri Poincaré, Albert Einstein and Hermann Minkowski [2] - [7]. Its main pillar is the "principle of relativity" given for the first time at its true level by Henri Poincaré at the world scientific conference of Saint-Louis (Missouri) in September 1904: "The laws of physical phenomena must be the same for a fixed observer and for an observer in rectilinear and uniform motion so that we have no possibility to perceive if we are, or not, dragged in such a motion" [4, page 306].

This principle, and the Maxwell equations of electromagnetism, requires us to forsake the Newtonian notions of absolute time, absolute space and ether [3, pages 111, 245, 246]. It leads very naturally to the Lorentz transfomation of reference frames and to the physical character of the transformed parameters of space and time. This principle of relativity was based on the negative results of the numerous ether experiments of the years 1880-1900, including the famous Michelson-Morley experiment of 1887 and we can write that the special theory of relativity has a very strong experimental background and is very accurately verified especially in the cyclotron experiments in which particles can easily have velocities larger than 99 % of that of light.

The general relativity had not at all this strong experimental background. It comes from the incomplete character of the special relativity that doesn't include the gravitational effects. Einstein was forced to choose without proof some simple hypotheses such as the equivalence principle, the isotropy of space-time, the invariance of fundamental physical constants, etc... but the analysis of newly discovered astronomical phenomena (pulsars, quasars, solar neutrinos...) the technological progress and their applications (satellites, space probes, lasers, atomic clocks and masers, etc..) have led some theoreticians to doubt Einstein general relativity and to propose alternative theories.

The accurate verification of the equivalence principle by Eötvös, Dicke and Braginsky [8] - [10] has ruled out some alternative theories, for instance the early Poincaré general relativity theory, but there remain more than twenty theories facing that of Einstein and the situation must be clarified [27].

Several moderately accurate tests-experiments have already been realised and we propose here an accurate measure of the parameter "beta".

3. The Parameters "beta" and "gamma".

The comparison of the different theories of general relativity (also called theories of gravitation) is based on some parameters called "post-Newtonian parameters". These parameters, $\beta, \gamma, \alpha_1, \alpha_2, \alpha_3, \zeta_1, \zeta_2, \zeta_3, \zeta_4$ etc...[2] model the different relativistic effects; for instance the relativistic advance of the perihelion of Mercury is proportional to the sum $2 + 2\gamma - \beta$.

In these conditions, the measure of the post-Newtonian parameters through some suitable space test-experiment allows us to choose among the different theories of gravitation. We, of course, hope that only one theory will survive the current test experiments and we fear that no one will survive and that the truth still escapes.

The simplest way for the understanding of the parameter β and γ is the following.

About a celestial body of spherical symmetry the Einstein equations lead to the following expression of the proper time s and its differential

[2] In the Einsteinian case $\beta = \gamma = 1$, the other parameters are equal to zero.

element ds given by the "Schwarzschild ds^2";

$$ds^2 = F(r)dt^2 - \frac{1}{c^2}\left\{G(r)dr^2 + r^2\left(d\phi^2 + \cos^2\phi dL^2\right)\right\} \tag{1}$$

with:

$$\left.\begin{array}{rcl} t & = & \text{"cosmic time" (imagine an atomic clock very far} \\ & & \text{away and without velocity with respect to the} \\ & & \text{body of interest).} \\ r & = & \text{"radial distance"} \\ \phi & = & \text{latitude} \\ L & = & \text{longitude} \\ c & = & \text{velocity of light} = 299\ 792\ 458 \text{ m/s} \\ F(r) & = & \dfrac{1}{G(r)} = 1 - \dfrac{2m}{r} \text{ in the Einsteinian case.} \\ m & = & \text{relativistic radius of the body of interest} = \dfrac{GM}{c^2} \\ & & (m = 1.477\text{km for the Sun.}) \\ G & = & \text{constant of the law of universal attraction} \\ & & (= 6.672 \times 10^{-11}\text{m}^3/\text{s}^2\text{kg.}) \\ M & = & \text{mass of the body of interest} \\ & & (= 1.989 \times 10^{30}\text{kg for the Sun.}) \end{array}\right\} \tag{2}$$

With the spherical symmetry, the angles ϕ and L are ordinary geometrical angles that can be obtained by suitable astronomical observations. On the contrary, the Schwarzschild radial distance r is not directly measurable, it is deduced from orbital computations.

The physical meaning of the Schwarzschild ds^2 is the following: if we imagine a test-body at the space-time point (t, r, ϕ, L) and moving with the velocity $(dr/dt, d\phi/dt, dL/dt)$, its proper time s, given by its atomic clock, is also given by (1).

An equivalent expression is the "Robertson ds^2". Robertson uses the cosmic time t of Schwarzschild and the three coordinates x, y, z similar to the Cartesian coordinates. The radial distance of Robertson is then:

$$\rho = (x^2 + y^2 + z^2)^{\frac{1}{2}} \tag{3}$$

ρ, ϕ and L are related to x, y, z by the usual Euclidian expressions:

$$\left.\begin{array}{rcl} x & = & \rho\cos\phi\cos L \\ y & = & \rho\cos\phi\sin L \\ z & = & \rho\sin\phi \end{array}\right\} \tag{4}$$

The radial distance ρ is chosen by Robertson so that the expression (1) becomes:

$$ds^2 = f(\rho)dt^2 - \frac{1}{c^2}g(\rho)\left\{dx^2 + dy^2 + dz^2\right\} \tag{5}$$

This requires of course

$$f(\rho) = F(r) \;\; ; \;\; g(\rho)d\rho^2 = G(r)dr^2 \;\; ; \;\; \rho^2 g(\rho) = r^2 \tag{6}$$

In the Einsteinian case; with $F(r) = 1/G(r) = 1-(2m/r)$ we obtain easily:

$$\left.\begin{aligned} r &= \rho + m + \left(\frac{m^2}{4\rho}\right) \\ f(\rho) &= \frac{(2\rho - m)^2}{(2\rho + m)^2} = 1 - \frac{2m}{\rho} + \frac{2m^2}{\rho^2} - \frac{3m^3}{2\rho^3} + \ldots \\ g(\rho) &= \left\{1 + \frac{m}{2\rho}\right\}^4 = 1 + \frac{2m}{\rho} + \frac{3m^2}{2\rho^2} + \ldots \end{aligned}\right\} \tag{7}$$

The other theories of gravitation (Ni, Jordan, Thiry, Brans-Dicke, etc ... [27]) are based on very different notions of matter and energy. They lead to various functions $f(\rho)$ and $g(\rho)$ that were originally approximated by:

$$\left.\begin{aligned} f(\rho) &= 1 - \frac{2m\alpha}{\rho} + \frac{2m^2\beta}{\rho^2} + O\left(\frac{m^3}{\rho^3}\right) \\ g(\rho) &= 1 + \frac{2m\gamma}{\rho} + O\left(\frac{m^2}{\rho^2}\right) \end{aligned}\right\} \tag{8}$$

Notice that:

A) In the Einsteinian case $\alpha = \beta = \gamma = 1$, but for instance $\gamma \sim 0.85$ in the Brans-Dicke theory.

B) The coefficient α is useless, the case $\alpha = 2; \beta = 4; \gamma = 2$ is obviously equivalent to the case $\alpha = \beta = \gamma = 1$ (with a twice smaller m and the same product αm that remains equal to GM/c^2: the first order motion is the Newtonian one and the mass M is the gravitational mass known only by its gravitational effects).

Hence it has been decided that the coefficient α will systematically be chosen equal to unity and for this reason the two first post-Newtonian parameters are called β and γ.

For some theories (Yilmaz, Papapetrou, etc...), where the spherical symmetry is no longer respected, there are some small effects related to the velocity of the Sun with respect to the Galaxy and the analysis becomes more complex but remains similar to that developed in the next section.

4. Motion in a Central Field. The Effects of the Parameters β and γ

The relativistic effects related to β and γ allow the measure of these two parameters.

The free motions of test-bodies and light beams lead to the best examples of relativistic effects. These free motions have a very simple definition: let us consider in space-time an initial and a final point: t_0, r_0, ϕ_0, L_0 and t_f, r_f, ϕ_f, L_f (or t_0, x_0, y_0, z_0 and t_f, x_f, y_f, z_f). A space probe that goes from the first point to the second with the largest possible proper time $s_f - s_0$ follows a trajectory of free motion. If, furthermore, that proper time $s_f - s_0$ is zero, the trajectory is that of a light beam.

The free-motion problem is thus an ordinary problem of optimization. Let us consider it with the Robertson coordinates t, x, y, z.

$$\left.\begin{aligned} ds^2 = f(\rho)dt^2 - \frac{1}{c^2}g(\rho)\left\{dx^2 + dy^2 + dz^2\right\} \\ \vec{\rho} = \overrightarrow{(x, y, z)} \quad ; \quad \rho = \|\vec{\rho}\| = \{x^2 + y^2 + z^2\}^{\frac{1}{2}} \end{aligned}\right\} \tag{9}$$

The "cosmic time" t is the description parameter while s and $\vec{\rho}$ are the state parameters. We will call p_s and $\vec{p_\rho}$ their conjugate parameters.

The Hamiltomian H of the control is

$$H = p_s \frac{ds}{dt} + \vec{p_\rho}\frac{d\vec{\rho}}{dt} \tag{10}$$

Let us put $d\vec{\rho}/dt = \vec{V}$. The optimal Hamiltonian H^* is then the maximum of H with respect to the "control" $\vec{V}$ when we take account of (9):

$$H^* = \sup_{\vec{V}}\left\{p_s \underbrace{\sqrt{\left(f(\rho) - \frac{V^2}{c^2}g(\rho)\right)}}_{ds/dt} + \vec{p_\rho}\frac{d\vec{\rho}}{dt}\right\} \tag{11}$$

This leads to :

$$\left.\begin{aligned} H^* &= H^*(p_s, \vec{p_\rho}, \rho) = \{fp_s^2 + (fc^2p_\rho^2/g)\}^{\frac{1}{2}} \\ \vec{V} &= \vec{p_\rho}c^2\sqrt{f}\{g^2p_s^2 + gc^2p_\rho^2\}^{-\frac{1}{2}} = \frac{\vec{p_\rho}fc^2}{gH^*} \end{aligned}\right\} \tag{12}$$

The usual Pontryagin optimality equations then give:

$$\left.\begin{aligned} \frac{ds}{dt} &= \frac{\partial H^*}{\partial p_s} = p_s\sqrt{f}\{p_s^2 + (c^2p_\rho^2/g)\}^{-\frac{1}{2}} = \frac{p_sf(\rho)}{H^*} \\ \frac{dp_s}{dt} &= -\frac{\partial H^*}{\partial s} = 0 \\ \frac{d\vec{\rho}}{dt} &= \vec{V} = \frac{\partial H^*}{\partial \vec{p_\rho}} = \vec{p_\rho}c^2\sqrt{f}\{g^2p_s^2 + gc^2p_\rho^2\}^{-\frac{1}{2}} = \frac{\vec{p_\rho}fc^2}{gH^*} \end{aligned}\right\} \tag{13a}$$

This equation has already been obtained in (12).

$$\left.\begin{array}{l} \dfrac{d\vec{p_\rho}}{dt} = -\dfrac{\partial H^*}{\partial \vec{p_\rho}}, \text{ hence } \dfrac{d\vec{p_\rho}}{dt} \text{ is parallel to } \vec{\rho} \text{ and the} \\ \text{vector product } \ \vec{\rho} \times \vec{p_\rho} \ \text{ is constant.} \\ \dfrac{dH^*}{dt} = \dfrac{\partial H^*}{\partial t} = 0 \end{array}\right\} \qquad (13b)$$

Thus $p_s, \vec{\rho} \times \vec{p_\rho}$ and H^* are constant and the corresponding integrals of motions are the following:

$$\left.\begin{array}{l} f(\rho)\left(\dfrac{dt}{ds}\right) = \dfrac{H^*}{p_s} = k_1 = \text{constant} \\ \dfrac{\vec{\rho} \times \vec{V} g(\rho)}{f(\rho)} = \vec{k_2} = \text{constant vector} \\ \left(\dfrac{f(\rho)}{k_1^2}\right) + \left(\dfrac{g(\rho)V^2}{f(\rho)c^2}\right) = 1 \end{array}\right\} \qquad (14)$$

With the Schwarzschild parameters, these integrals become:

$$\left.\begin{array}{l} F(r)\left(\dfrac{dt}{ds}\right) = k_1 = \text{constant} \\ \vec{r} \times \left(\dfrac{d\vec{r}}{ds}\right) = k_1\vec{k_2} = \text{constant vector} \\ \left(\dfrac{dr}{ds}\right)^2 = \dfrac{1}{G(r)}\left\{\dfrac{k_1^2c^2}{F(r)} - c^2 - \dfrac{k_1^2k_2^2}{r^2}\right\} \end{array}\right\} \qquad (15)$$

The problem of the free motions in a Robertson or a Schwarzschild ds^2 is thus integrable for any functions $f(\rho)$ and $g(\rho)$ (or $F(r)$ and $G(r)$). For a slow body its free motion is almost Keplerian and we will express it with the help of a neighbouring suitable auxiliary Keplerian motion and its usual parameters the constants n, a, e, p and the anomalies v, E, M:

$$\left.\begin{array}{rl} n & = \text{mean angular motion} \\ a & = \text{semi-major axis} \\ n^2a^3 & = GM = mc^2 = \text{gravitational constant} \\ & (= 1.3271 \times 10^{20}\text{m}^3/\text{s}^2 \text{ for the Sun}) \\ e & = \text{eccentricity} \\ p & = a(1-e^2) = \text{semi-latus rectum} \\ v & \text{and } E : \text{true and eccentric anomalies ;} \\ \tan(E/2) & = [(1-e)/(1+e)]^{\frac{1}{2}} \tan(v/2) \\ M & = E - e\sin E = \text{mean anomaly} \end{array}\right\} \qquad (16)$$

This auxiliary Keplerian motion is related to the integrals of motion (14) by the following complex expressions that allow a simple description of the neighbouring free motion.

$$\left.\begin{aligned} k_1^2 &= 1-\left(\frac{m}{a}\right)+\left\{\frac{(1+\gamma)m^2}{a^2}\right\} \\ k_2^2 &= mc^2\left\{p+m(4+3\gamma+\gamma e^2-2\beta)\right\} \end{aligned}\right\} \tag{17}$$

The free motion is planar, the orbital plane contains the centre of attraction and we will assume that it is the equatorial plane $\phi = 0$ (or $z = 0$). The parameters ρ, L (longitude), s and t are then given by the following:

$$\left.\begin{aligned} \rho &= a(1-e\cos E)=\frac{p}{(1+e\cos v)} \\ L &= L_0+v\left\{1+(2+2\gamma-\beta)\frac{m}{p}\right\} \\ s &= s_0+\frac{M}{n}\left\{1+(1+5\gamma)\frac{m}{2a}\right\}+(2\gamma me\sin E/na) \\ t-s &= t_0-s_0+\frac{3Mm}{2na}+\frac{2me\sin E}{na} \end{aligned}\right\} \tag{18}$$

These first order expressions (in m) are valid for any eccentricity and, with the spherical symmetry, for any inclination.

L_0, s_0, t_0 are three constants of integration and M/n is the proper time of the auxiliary Keplerian motion.

Let us note that:

A) The neglected second and upper order terms give errors of the order m^2/p which are always less than 1cm in the solar system.

B) During a revolution, the relativistic perturbations are of the order of the relativistic radius m, we must then expect effects of a few kilometers.

C) The difference $t-s$ is independent of β and γ, hence a clock experiment cannot give these parameters.

D) The only visible long period effect is related to the expression of L. It is the secular advance of perihelion the angular velocity of which is:

$$\frac{(2+2\gamma-\beta)mn}{p} \tag{19}$$

E) This advance of perihelion is the only relativistic effect that contains the parameter β.

We also need the motion of a light beam, for instance a light beam starting at $(0, y_0, 0)$ at the time $t = 0$ in the direction of O_x.

The trajectory of the photon is then given by:

$$\left.\begin{aligned} x &= ct - m(1+\gamma)\mathrm{Ln}\left\{\frac{\left(ct + \sqrt{(c^2t^2 + y_0^2)}\right)}{y_0}\right\} + O\left(\frac{ctm^2}{y_0^2}\right) \\ y &= y_0 - m(1+\gamma)\left\{\sqrt{\left(1 + \frac{c^2t^2}{y_0^2}\right)} - 1\right\} + O\left(\frac{ctm^2}{y_0^2}\right) \\ z &= 0 \end{aligned}\right\} \quad (20)$$

Hence the total deflection of a light beam is $2m(1+\gamma)/y_0$ and thus, if $\gamma = 1$, a light beam arriving at 90° from the Sun into the astrometric satellite Hipparcos has already a deflection of $0.004''$ that must be taken into account ![3]

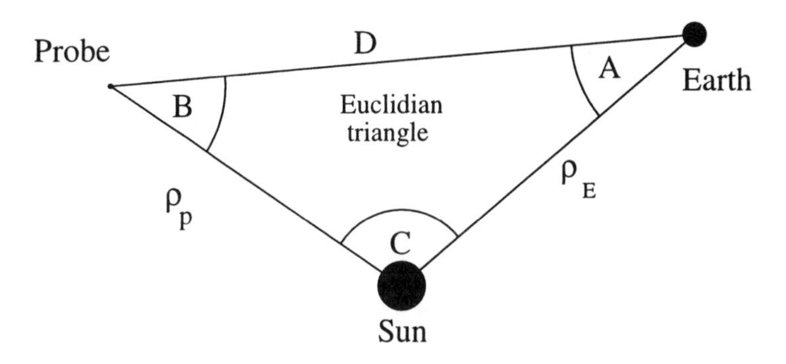

Figure 1. Analysis of a light beam.

On the other hand the transit time between the Earth and a space probe (Figure 1) is modified by the space curvature. This transit time $(t_f - t_0)$ is

[3]The accuracy of the Hipparcos astrometric experiment is about $0.001''$ and γ has been considered to be equal to one in the analysis of the results.

given by the following expression that can easily be deduced from (20):

$$\left.\begin{array}{l} c(t_f - t_0) = D - m(1+\gamma)\mathrm{Ln}\left(\tan\dfrac{A}{2}\tan\dfrac{B}{2}\right) + O\left(\dfrac{m^2 D}{\rho_E^2 \sin^2 A}\right) \\ \text{with} \\ D = \text{Euclidian Earth-probe distance} = \left\{\rho_E^2 + \rho_p^2 - 2\rho_E\rho_p\cos C\right\}^{\frac{1}{2}} \\ A = \text{angle (Sun-Earth-probe)} \\ B = \text{angle (Earth-probe-Sun)} \\ C = \text{angle (Earth-Sun-probe)} \\ \text{Euclidian triangle Sun-Earth-probe} : A + B + C = 180^\circ \ ; \\ \dfrac{\sin A}{\rho_p} = \dfrac{\sin B}{\rho_E} = \dfrac{\sin C}{D} \end{array}\right\} \quad (21)$$

5. Relativistic Experiments. The Measure of β and γ

We have already met the Eötvös, Dicke and Braginsky experiments on the equivalence principle [8] - [10], and they will be improved, by a factor of 10^5 or 10^6, by the STEP experiment (Space Test of Equivalence Principle [11] - [13]). These accurate experiments allow us to exclude the non-metric theories but don't allow to choose among the metric theories.

The clock experiments [14] - [19], are exactly in the same situation. Up to now they are always favourable to the metric theories.

The motion of the Moon has a small periodic component, of a few meters, depending on the theory of gravitation (Nordvedt effect [20]. The laser ranging measure of the Earth-Moon distance with an accuracy that will soon reach 1 cm is an excellent element of choice even if its accuracy remains small. Today this element remains favourable to the Einstein theory.

The deflection of light beams was one of the first three tests of General Relativity but it remains small and difficult to measure, its accuracy is generally weak. However, notice that the Hipparcos astrometric experiment has taken account of the deflection of light beams and has, for the best coherence of the experiment, led to (γ - 1) = a few thousandths.

The analysis of strong gravitational effects: vicinity of black holes, binary pulsars, etc..., is obviously an excellent source of information about the theories of gravitation even if its interpretation is generally very difficult. It must be considered more as a verification and an observation than as a measure and an experiment.

The detection of gravitational waves will give a lot of information about the strong gravitational phenomena and the theories of gravitation. This detection is actively prepared both on Earth [21] - [22], and in space [23] - [24], but it is extremely difficult; it corresponds to very small effects with many sources of perturbations. All past attempts of detection have been unsuccessful.

A very beautiful relativistic experiment has been proposed by F. Everitt [28]: the gyroscope experiment. Four very sensitive and extremely well protected gyroscopes will be put into a satellite on a low Earth orbit. The space curvature will imply deviations of the gyroscope axes of the order of $7''$ per year... and it is useless to emphasize the difficulty of this mission!

We must not forget the indirect relativistic experiments such as the measure of the eventual variations of the velocity of light, of the coefficient G of the law of universal attraction and of the other physical "constants". Up to now no variation has been detected, but the accuracy of these measures (relative variation of 10^{-8} or 10^{-9} per year) remains below the level that would be of interest for the theoreticians of gravitation.

Finally, we must consider the direct measure of β and γ on the motion of space probes and celestial bodies.

5.1. THE DIRECT MEASURE OF THE PARAMETERS β AND γ ON THE MOTION OF SPACE PROBES AND CELESTIAL BODIES.

Let us consider the figure 1 and the equation (21): the transit time $(t_f - t_0)$ between the Earth and a space probe has large variations when the angles A and B vary in the vicinity of zero (passage of the probe beyond the Sun). This "Shapiro effect" has been used several times for the measure of the parameter γ and we now know that this parameter is very close to one with an accuracy better than one per cent.

However, notice that these Shapiro experiments face many sources of perturbations.

The relativistic term $m(1+\gamma)\mathrm{Ln}\left(\tan\frac{A}{2}\tan\frac{B}{2}\right)$ of (21) can exceed 30 km when A and B are small but the corresponding light beam then crosses the Solar corona where it undergoes unknown phenomena. Lasers and double radar frequencies have been tried to overcome these difficulties.

The motion of the test space probe itself undergoes many perturbations. The planetary perturbations can be modelled accurately, the effect of the Solar oblateness is less known and the surface perturbations (radiation pressure, solar wind, etc.) give a parasite acceleration of the order of a few $10^{-8}\mathrm{m/s}^2$ that forbids all accurate measures.

In order to fight this last source of perturbation, the use of drag-free probes has been proposed. The parasite acceleration then falls to a few

10^{-10}m/s^2 which is still insufficient for an accurate measure of the parameter β. Nevertheless, for the parameter γ, this method associated with a laser beam can lead to an accuracy of 10^{-6} or even 10^{-7} [25] - [26].

Another possibility is the use of natural celestial bodies as test bodies. The radiation pressure parasite acceleration falls then to a few 10^{-12}m/s^2 for asteroids and even less for planets which is sufficiently small. However, in the absence of radar-transponders, the natural bodies are bad and weak reflectors with many sources of inaccuracies.

Hence, the ideal solution seems to be to send a radar-transponder on some suitable celestial body and to use this body as a test-body.

We have already noticed that the measure of the parameter β is much more difficult than that of the parameter γ (and thus β is much less known than γ). Since β only appears in the relativistic advance of perihelion it requires a long-term analysis and preferably a test-body with large eccentricity.

With the following orbit and characteristics, the asteroid Icarus seems excellent for this mission:

Semi-major axis   a $= 1.0777\text{UA} = 161,200,000\text{km}$
eccentricity   e $= 0.8266 \Rightarrow p = a(l - e^2) = 0.3413\text{UA}$
inclination $i = 22.97°$
longitude of node $\Omega = 87.70°$
argument of perihelion $\omega = 30.95°$
diameter: about 1.5 km
mass : between 5 and 10 billion tons

With this large eccentricity and inclination the asteroid Icarus cannot be reached directly by the usual chemical rockets and several planetary gravitational assistances must be used. It is also possible to use a "solar sail": this kind of mission is in their domain of excellence and a sail of 2000 m^2 and 100 kg can reach Icarus in about 4 years .

When we compare a radar-transponder on Icarus and on Mercury, we find for Icarus at least the four following major advantages.

A) The relativistic advance of perihelion gives measurable effects of 130 km/year for Icarus and 49.6 km/year only for Mercury.

B) The arrival on Icarus, with a negligible gravity field, is much easier than on Mercury; especially for a solar sail.

C) With a radius less than 1 km, the rotation of Icarus can be modelled easily and accurately. This is not the case for the slow rotation of Mercury.

D) On the solar equatorial plane, the inclination of Mercury's orbit is only 3.4°, while that of Icarus' orbit is 16° and allows a decoupling between the effects of the parameter β and those of the coefficient J_{20} of the dynamical oblateness of the Sun.

In these conditions, with a rough model of the effect of the solar radiation pressure on Icarus, it is possible to measure in a few years the parameter β of theories of gravitation with an accuracy of 10^{-3} and the coefficient J_{20} of the dynamical oblateness of the Sun with an accuracy of 10^{-7}.

Notice the importance of this coefficient J_{20} : if it is of the order of -10^{-6}, the Sun has an almost uniform rotation (slightly faster at equator, period 25 days, than near the poles, period 34 days); but if J_{20} is of the order of -10^{-5}, as assumed by some theories, the Sun has a large differential rotation and its average rate of rotation is twice that of the surface.

6. Conclusion

The numerous small bodies of the solar system are excellent test-bodies for many applications. They offer a wide variety of major-axes, eccentricities and inclinations and have much less disturbing perturbations than the space probes.

Equipped with radar or laser transponders they will give excellent opportunities to improve our knowledge in many domains.

The space-time curvature about a spherical body is essentially characterized by two parameters, β and γ, that are both equal to unity in the Einstein general relativity but have other values in the alternative theories of gravitation.

The experimental measure of γ is much easier than that of β and the experimenters have more or less neglected this latter parameter while we now know that γ is within 1% of its Einsteinian value.

The analysis of the various sources of perturbation shows the possibility of an accurate measure of β (to 0.1 %) with a radar-transponder on the asteroid Icarus.

That measure will also give an accurate value of the dynamical oblateness of the Sun, a major parameter governing its differential rotation.

References

1. Marchal C. (1990), *The three-body problem*, Elsevier Science Publishers BV, Amsterdam.
2. Lorentz H.A. (1904), Electromagnetic phenomena in a system moving with any velocity less than that of light. *Proc. Royal Acad. Amsterdam* **6**, p.809.
3. Poincaré H. (1902), La science et l'hypothèse. Ed. Flammarion, Paris.
4. Poincaré H. (1904), L'état actuel et l'avenir de la physique mathématique. *Bulletin des Sciences Mathématiques* **28- 2**^{e} série (reorganized 39-1).

5. Poincaré H. (Jun. 1905), Sur la dynamique de l'électron. *Comptes rendus de l'Académie des Sciences de Paris* **140**, pp. 1504-1508.
6. Poincaré H. (Jan. 1906), Sur la dynamique de l'électron. *Rendiconti del Circolo Matematico di Palermo* **21**, pp. 129-175; received July 23, 1905.
7. Einstein A. (Sep. 1905), Zur Elektrodynamik der bewegten Körper. *Annalen der Physik* **17**, pp. 891-921; received June 30, 1905.
8. Eötvös R.V., Pekar D., Fekete E. (1922), *Ann Physik* **68**, p. 11.
9. Roll P.G., Krotov R., Dicke R.H. (1964), *Ann Physik* **28**, p. 442.
10. Braginsky V.B., Panov V.I. (1971), *Zh. Eksp. Teor. Fiz* **61**, p. 875.
11. Barlier F., Blaser J.P., Cavallo G., Damour T., Decher R.; Everitt C.W.F., Fuligni F., Lee M., Nobili A., Nordtvedt K., Pace 0., Reinhard R., Worden P., (Jan. 1991), Satellite test of the equivalence principle - Assessment study report. ESA-NASA-SCI (91).
12. Nobili A.M. et al. (May 1993), Galileo Galilei - Test of equivalence principle at room temperature with masses mechanically suspended inside a spinning non drag free spacecraft. Proposal for the M3 medium size mission of ESA.
13. Touboul P., Rodrigues M., Wiliermenot E.. Bernard A. (Nov. 1996), Electrostatic accelerometers for the equivalence principle test in space. *Classical and quantum gravity*, Vol.**13**, no. 11A. Institute of Physics Publishing, Bristol, UK.
14. Pound R.V., Rebka G. (1960), *Phys. Rev. Letters* **4**, p. 337
15. Pound R.V., Snider J.L. (1965), *Phys.Rev.* **140**-B, p. 788.
16. Hafele J.C., Keating R.E. (1972), Around the world atomic clocks -Predicted and observed time gains. *Science*, pp. 177-242.
17. Briatore L., Leschiutta S. (1977), Evidence for Earth gravitational shift by direct atomic scale comparison. *Il nuovo cimento*, Vol. **37**, BN2, p. 219.
18. Iijima S., Fujiwara K. (1978), An experiment for the potential blue shift at the Norikura Corona Station. *Annals of the Tokyo Astr. Obs.* **2, 17, 1**, p. 68.
19. Vessot R.F.C., Levine M.W. (1977), Experimental gravitation. *Proceedings of the Academia Nazionale dei Lincei*, Vol. **34**, pp. 371-391.
20. Nordtvedt K. (1997), La Lune au secours d'Einstein. *La Recherche*, pp. 70-76, Février.
21. Vogt R.E. et al. (1992), *The US LIGO project.* Proceedings of the Sixth Marcel Grossmann meeting on General Relativity (Kyoto 1991), Sato and Nakamura (ed.), pp. 244-266.
22. Vinet J.Y. (Jun. 1995), Quand vibre l'espace-temps (les projets LIGO et VIRGO). *La Recherche* No. 277, Vol. **26**, pp. 634-639.
23. Bender P., Ciufolini I., Danzmann K., Folkner W., Hough J., Robertson D., Rudiger A., Sandford M., Schilling R., Schutz B., Stebbins R., Summer T., Touboul P., Vitale S., Ward H., Winkler W., (Feb. 1996), LISA Laser Interferometer Space Antenna for the detection and observation of gravitational waves "A Cornerstone project in ESA's long term space science programme "Horizon 2000 Plus. *MPQ* **208**.
24. Hellings R.W., Ciufolini I., Giampieri G., Martin Lo, Gonzalez J., Marchal C., Touboul P., Robertson D., Vaillon L. (1993), Mission concept study for SAGITTARIUS, a space-borne astronomical gravity-wave interferometer. JPL Engineering Memorandum 314-569.
25. Veillet C. (1994), "Solar Orbit Relativity Test"; A proposal in the discipline area of fundamental physics in response to ESA's call for mission concepts for the follow- up to horizon 2000.
26. Melliti T., Fridelance P., Samain E. (To appear 1998), Study of gravitational theories and of the solar quadrupole moment with the SORT experiment: Solar Orbit Relativity Test. *Astronomy and Astrophysics.*
27. Will C.M. (1981), *Theory and experiment in gravitational physics*, Cambridge University Press, Cambridge, England.
28. Everitt C.W.F. (1971), *The Stanford gyroscope experiment.* Proceedings of the conference on experimental tests of gravitation theories. R.W. Davies ed., pp. 68- 81.

SECTION TWO: NEAR EARTH OBJECTS

INTRODUCTION

K. MUINONEN
Observatory, P.O. Box 14,
FIN-00014 University of Helsinki, Finland

AND

A. MILANI
Department of Mathematics, University of Pisa,
via Buonarroti 2, I-56127 Pisa, Italia

Near–Earth objects (NEOs) are our closest celestial companions in the Solar System—yet there are major open questions around NEOs, their orbital dynamics and physical and chemical properties as celestial bodies. The difficulties mostly arise because of the close approaches, which introduce mathematical singularities in the equations of motion, require very accurate orbits to be modelled, and result in strongly chaotic behaviour; yet it is precisely because of the possibility of close approaches, and even of collisions, that these objects are of special interest. The following seven articles offer views to the celestial mechanics of asteroids, comets, and meteoroids approaching our planet.

A. Milani reviews the numerical integrations and semianalytical theories to explain the dynamical behaviours of planet-crossing asteroids, including near-Earth asteroids. These are divided into seven dynamical classes depending on various dynamical criteria that include the capability of Earth crossing and Jupiter crossing, perihelion distance, protection mechanisms, and the dominance of close approaches to the planets. A. Milani further reviews statistical theories to describe the expected frequency of close encounters, over time scales of the order of several thousand years or more.

P. Michel focuses in on secular resonances as contributing to the dynamical transport of main-belt asteroids to Earth-crossing orbits. In particular,

B.A. Steves and A.E. Roy (eds.), The Dynamics of Small Bodies in the Solar System, 93–94.

suggestions are put forward to explain the population of near-Earth asteroids on high-inclination orbits.

K. Muinonen reviews the populations of near-Earth asteroids and comets, and the impact hazard imposed by them. Theoretical and computational tools (including the first non–Gaussian methods) are provided for the analysis of the collision probability between small and large solar system bodies based on the Bayesian a posteriori probability density of the orbital elements. The methods cover observational time arcs from few hours upward. A detailed application to the near-Earth object 1997 XF_{11} shows how the collision probability evolves while more and more observations are included in the analysis.

B. Conway describes optimum orbital interception strategies for near–Earth asteroids concentrating on minimum-time very-low-thrust trajectories for interception using electric propulsion. A detailed application is provided to the near-Earth asteroid 1991 RB as a function of time prior to the date of interception.

I. P. Williams brings up the open questions in studies of meteoroids and meteor streams. In particular, the meteoroid ejection and evolution processes, and their realization as major meteor outbursts, are summarized. Outbursts are then described for the Perseid, Lyrid, and Leonid meteor showers. In addition, insight is offered to the interesting history of falling stars.

G. B. Valsecchi provides a summary of cometary orbital evolution, in particular, the effects from planetary close encounters. Special emphasis is given for the encounters with the giant planets. Various cometary classes of object are reviewed. Test comets similar to comet 39P/Oterma are integrated over a close approach to Jupiter, and the resulting changes in orbital elements are presented in detail. A very long satellite capture bears intriguing similarities to that for comet D/Shoemaker-Levy 9 that impacted on Jupiter in July 1994.

H. Prętka studies numerically the effect of galactic perturbations on the motion of Oort-cloud comets. Three different models for the galactic potential are used and their implications are compared. Long-term evolutions of both individual comets and a population of comets are shown. The most detailed galactic potential model allows an intriguing transport of comets from prograde to retrograde orbits.

The discussion of the dynamics of NEOs presented in these papers is by no means complete; however, it can well be used as an indication of the open problems and the range of topics currently being investigated.

DYNAMICS OF PLANET-CROSSING ASTEROIDS

A. MILANI
Space Mechanics Group,
Department of Mathematics, University of Pisa,
Via Buonarroti 2, I-56127 PISA, Italy
E-mail: milani@dm.unipi.it

Abstract. A small, but by no means negligible, fraction of the small bodies of the Solar System is on planet-crossing orbits, including Earth-crossing ones. The dynamics of planet-crossing asteroids/comets is strongly controlled by the occurrence of close approaches. The node crossing cycle, resulting from the secular evolution of the orbital elements, especially the argument of perihelion, is apparent in the evolution of all the elements, including the semimajor axis. The most common type of orbits defines the Geographos class, in which close approaches occur at random whenever they are made possible by the distance of the orbits. Other orbits are protected from close approaches either by mean motion resonances (Toro class) or by secular perturbations (Kozai class). The Alinda class is defined by the presence of a mean motion resonance with Jupiter, which can change over a comparatively short time the eccentricity and therefore the crossing behavior. In all cases the orbits are chaotic, and in the long run transitions between the different orbit types can occur. This paper summarizes the experimental evidence, resulting from numerical integrations, and the semianalytical theories (based upon the adiabatic invariant and upon the Kozai approximation) which can explain in a satisfactory way most of the dynamical behaviors found in the experiments.

1. THE ASTEROID COMPLEX, NEAR EARTH OBJECTS

The asteroids are a population of small bodies, whose distribution in space is concentrated in a *main belt* between the orbits of Mars and Jupiter (Figure 1). A significant number of objects is also found in the two Trojan clouds, roughly at the same distance from the Sun as Jupiter. A compara-

B.A. Steves and A.E. Roy (eds.), The Dynamics of Small Bodies in the Solar System, 95–126.

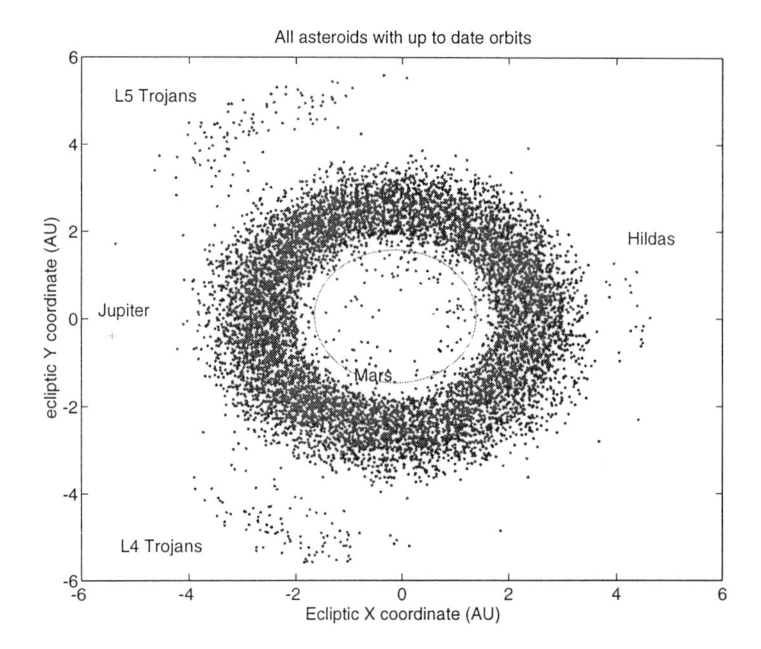

Figure 1. The asteroid complex projected on the ecliptic plane; the x, y coordinates are in astronomical units (AU). Most asteroids are found between the orbits of Jupiter (whose current position is the cross at the left) and the orbit of Mars (the line near the center), but a small fraction of the asteroids are planet-crossers.

tively small fraction of this population has orbits which can cross the orbit of some major planet; this is apparent from a plot, such as Figure 2, showing only asteroids down to given size. However the number of asteroids grows in a steep way as the minimum size considered decreases; thus at smaller sizes a comparatively large number of asteroid orbits with perihelia closer to the Sun than Mars, and even the Earth and Venus, have been discovered.

Figure 3 shows the orbital elements (a, e) of these smaller asteroids, and also of the short periodic comets, which are bodies of sizes comparable to the asteroids but more easily visible because of the release of gas and dust; these comets have often orbits crossing the orbit of Jupiter. A figure such as this one, showing the orbital elements of many small asteroids, gives a good intuitive understanding of the transport routes between the main belt and the Earth neighbourhood: these routes appear as trails in Figure 3; as an example, one route exits the main belt through the $3:1$ mean motion resonance at $a \simeq 2.5\, AU$, another one exploits the secular resonance around $a \simeq 2.2\, AU$.

There is a very large literature on the dynamics of main belt asteroids (also of Trojans; see [17] and references therein); the dynamical behavior of planet-crossing orbits has been studied only in comparatively recent times,

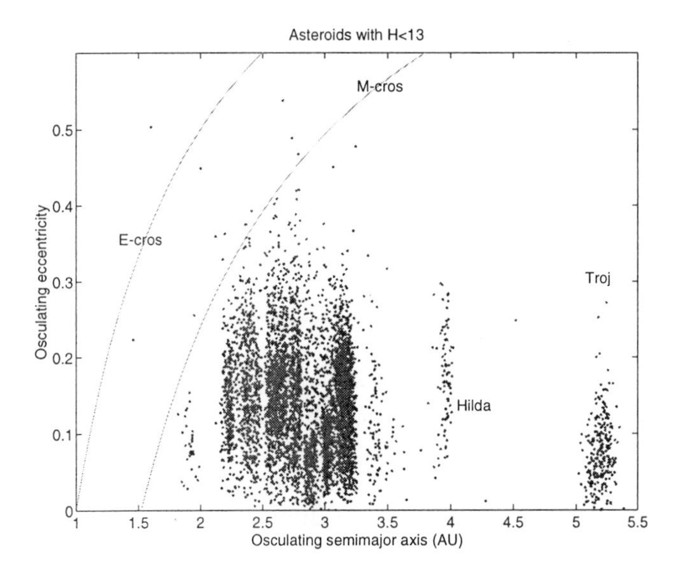

Figure 2. The asteroid complex represented in the semimajor axis a and eccentricity e plane; most asteroids are *main belt*: their orbits do not cross the orbits of Jupiter and Mars. The Trojans cross the orbit of Jupiter, very few asteroids brighter than absolute magnitude 13 ($\simeq 10\,km$ in diameter) are beyond the Mars-crossing and the Earth crossing lines.

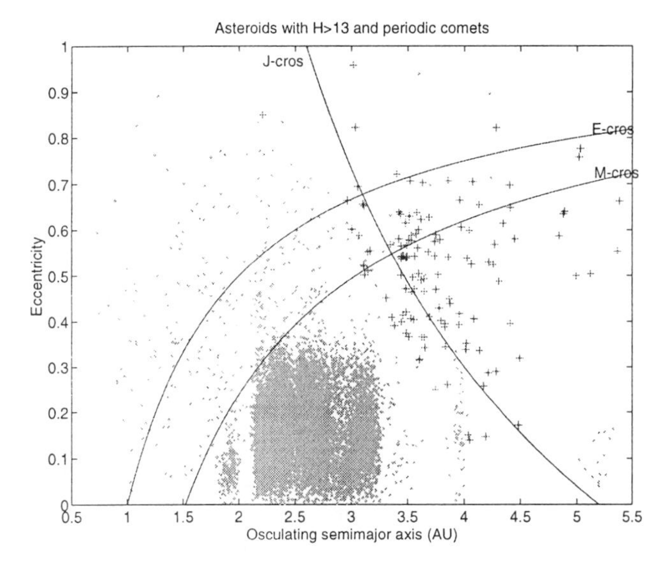

Figure 3. The asteroids smaller than absolute magnitude 13 (dots) show a distribution, in the same a, e plane of Figure 2, extending well beyond the Mars-crossing line and even the Earth-crossing line. The short periodic comets (crosses), that is the minor planets which have, at least at some apparition, detectable emissions of gas and dust, are often found in Jupiter-crossing orbits.

and most of the studies are only descriptions of the output of numerical experiments. Clearly the numerical integration of a large sample of orbits is a necessary but not sufficient condition for understanding the complex problem of planet-crossing dynamics: the familiarity with a large set of examples needs to be supported by a number of theoretical tools. The purpose of this paper is to introduce at least some of the examples of dynamical behavior and some of the conceptual tools to understand them.

2. NORMAL PLANET-CROSSING: GEOGRAPHOS CLASS

Let us consider a planet, e.g. the Earth, on a roughly circular orbit with radius a', and an asteroid with an orbit having the perihelion distance $q = a(1-e) < a'$ and aphelion distance $Q = a(1+e) > a'$. Under these conditions it is clear that the two orbits could have a common point, but we need to consider the geometry in 3 dimensions.

Each of the two osculating orbits lies in a plane; these two planes being, in general, distinct, have a common line, the *line of nodes*. The two points of the asteroid orbit on the orbit plane of the Earth are the *ascending* and the *descending node*, with distances from the sun:

$$r_+ = \frac{a(1-e^2)}{1+e\cos\omega} \quad ; \quad r_- = \frac{a(1-e^2)}{1-e\cos\omega}$$

with a minimum of q and a maximum of Q, depending upon the value of the *argument of perihelion* ω. For some values of ω a *node crossing* occurs, when one of the two *nodal distances* $r_+ - a'$ and $r_- - a'$ is zero, and one of the two nodes of the asteroid orbit belongs to the osculating orbit of the Earth.

In practice, the orbital elements of the asteroid orbit are by no means constant, they slowly change, mostly as a result of *secular perturbations*. Over time scales of $10,000 \simeq 100,000$ years, in many cases, the eccentricity e and inclination I undergo small relative changes, while the argument of perihelion ω performs several revolutions. In this case the asteroid is a *quadruple crosser*, that is there are four times during each period of circulation of ω when node crossings occur, one for each quadrant (Figure 4). In other cases, the changes in e, I are not negligible, and more complicated sequences of node crossings can occur, with up to 8 node crossings per period of ω.

The existence of node crossing has three main implications. First, the classical analytical theories to compute short periodic perturbations, secular perturbations, resonances, and solutions of any kind, are either not available or do not work properly. This results from the presence of the singularity of collision, which can occur if the Earth and the asteroid are

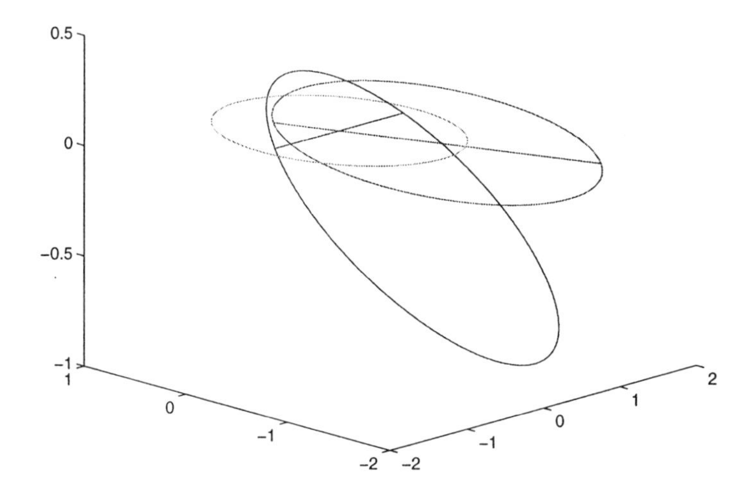

Figure 4. Node crossing: the two elliptic orbits in this plot have the same semimajor axis a, eccentricity e and inclination I, but different values of the longitude of the node Ω and of the argument of perihelion ω; the circular orbit has a semimajor axis a' such that $a(1-e) < a' < a(1+e)$. For $\omega = 0$ the elliptic orbit is linked to the circular orbit (like two consecutive links in a chain); for $\omega = \pi/2$ the ellipse and the circle are unlinked, thus for some intermediate value of ω there has been a crossing of the two orbits along the line of mutual nodes.

passing through the common node at the same time. The classical perturbative theories use either series expansion, which are in this case divergent, or averaging, which results in improper (often divergent) integrals.

Second, the possibility of close approaches of the asteroid to the Earth (also to other planets) results in chaotic motion. The correlation between the Lyapounov time, over which two nearby orbits increase their distance on average by a factor exp(1), and the average time between close approaches, is well understood and has been tested in a quantitative way by [30] and by [27]. To explain in a simple way this phenomenon we can resort to a piecewise 2-body model: during the time span when the nodal distance is small enough for close approaches, if the phases of the Earth and of the asteroid along their respective orbits are independent, there is a finite chance that an encounter occurs. When this happens, we use a 2-body asteroid–Earth model, in which the asteroid follows an hyperbolic orbit around the Earth (Figure 5). A bundle of nearby orbits, with changing impact parameter with respect to the Earth, gets scattered with distances increased by a factor of 2 to 3, if the closest approach is inside 0.1 AU. This increase of the mutual distances of nearby orbits is repeated at each occurrence of a close approach, and each time the factors of increase multiply the previous

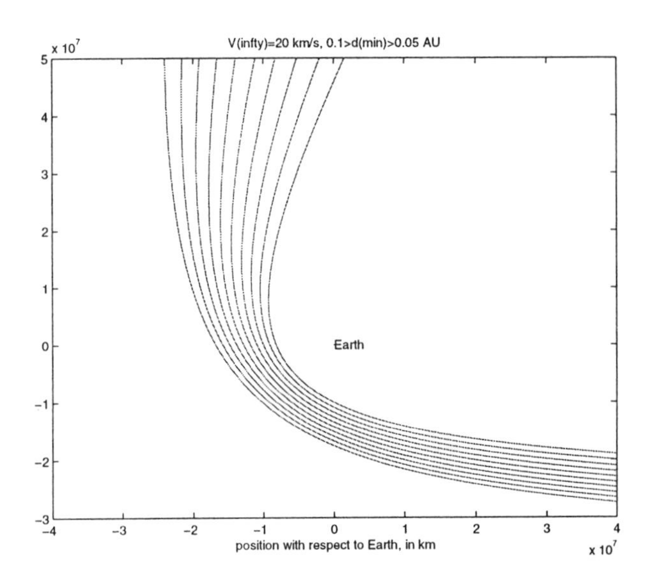

Figure 5. The increase in distance between nearby orbits during a close approach is illustrated by these hyperbolic orbits around the Earth. In this example, the asymptotic velocity at infinity is $20\,km/s$ and the minimum distance between 0.05 and 0.1 AU.

increase; this generates an exponential divergence, with Lyapounov time roughly equal to the average time span between two close approaches.

As it is well known from both theory and numerical experiments, chaotic orbits exhibit random behavior, alternating between different states, and also diffusion in the phase space which could result in large relative changes of the orbital elements; all the planet-crossing region of the phase space is indeed a single large chaotic sea, within which every transition if possible, although not necessarily probable. Typically cometary orbits and typical near Earth asteroid orbits can be connected by an evolutionary path including deep close approaches; a good example is the numerically computed orbit of *(1862) Apollo* shown in [20], Figures 5 and 6.

Third, the occurrence of node crossings only for some values of the argument of perihelion, which has "secular" periods of many thousands of years, results in intermittent behavior. When the nodal distance is small, frequent close approaches result in a random walk of all the orbital elements, even the semimajor axis, with occasional large jumps. When on the contrary the minimum distance between the two orbits is high (say $> 0.2\,AU$), the asteroid orbit shows all the features of a regular orbit, which could be represented by a Fourier series, over which averaging is meaningful, for which proper elements could be computed. However, this regular behavior only lasts until the next node crossing, when the orbital elements get random walked to some other value, from which a new regular orbit segment can begin.

Thus it is very difficult to make use of any statistical representation of the chaotic motion of an Earth-approaching body. If some kind of ergodic principle does hold for this class of orbits, that is if it is possible to exchange averages over regions of the phase space with time averages, this can occur only over times many orders of magnitude longer than the time span between two node crossings; these time spans are in practice too long to be explored numerically, even with the fastest computers of today, unless approximations are introduced which can not represent in a reliable way the real behavior of the exact orbits.

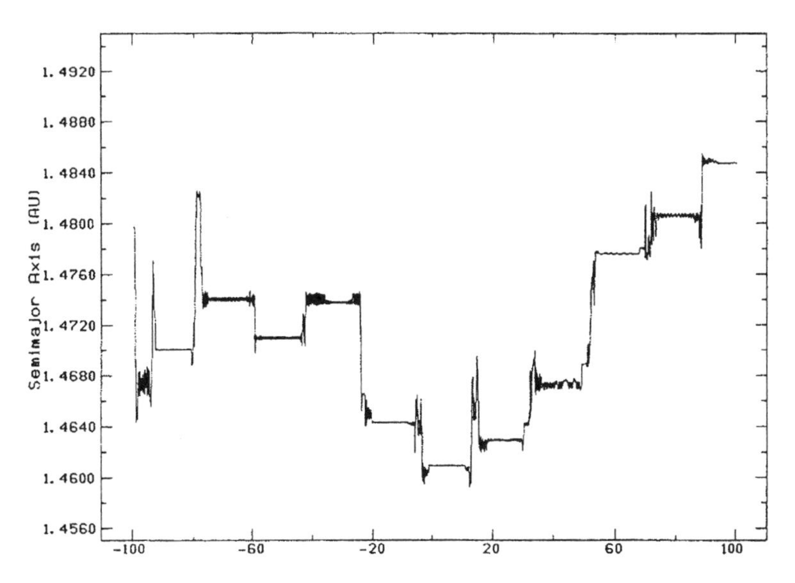

Figure 6. The semimajor axis of *(1864) Daedalus* as a function of time, from the output of the numerical integrations of *PROJECT SPACEGUARD*, over a time span of 200, 000 years. The time is is marked on the horizontal axis in thousands of years, in this as well as in the following figures. Jumps occur at node crossings with the Earth (see Figure 7) and are the result of close approaches (see Figure 8); when the nodal distance is comparatively large, the semimajor axis undergoes only small oscillations, and has negligible secular variations.

The orbits which show this intermittent behavior, alternating between random occurrence of close approaches near node crossings and regular evolution of the orbital elements (with a almost constant) when nodes are far apart, belong to the *Geographos* class, according to the classification proposed by [20]. This classification is based on the analysis of a set of 89 Earth-crossing orbits numerically computed for 200, 000 years; the Geographos class was the most populated one in that sample, which included orbits corresponding -although only approximately- to all the Near-Earth Asteroids known at the time. Thus we can say that the Geographos are

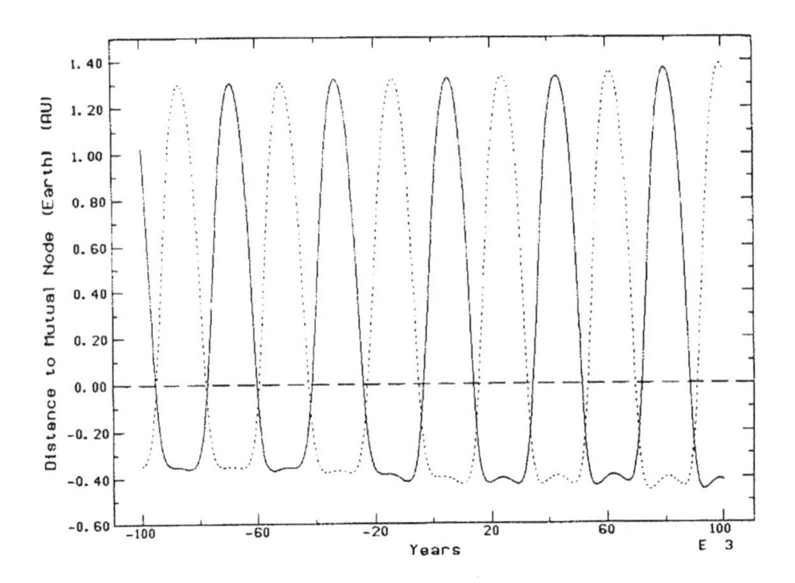

Figure 7. Nodal distance of the asteroid *(1864) Daedalus* with the Earth, as a function of time; the continuous line refers to the ascending node, the dotted one to the descending node. Four node crossings occur for every revolution of the argument of perihelion ω.

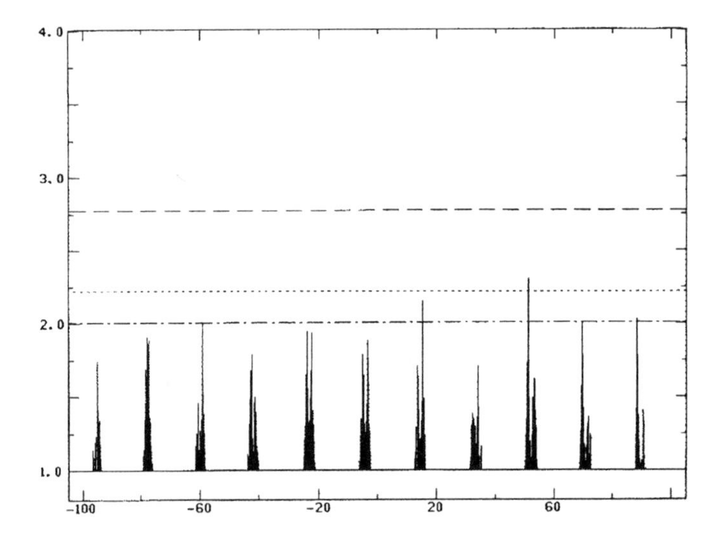

Figure 8. Close approaches of *(1864) Daedalus* to the Earth; in this *comb plot* each black line represents a close approach, and the height is the base 10 logarithm of the inverse of the minimum distance (in AU). By comparing with Figure 7, it is clear that close approaches occur at random whenever they are made possible by the distance of the orbits.

the "normal" Earth-crossing orbits, although their dynamical behavior is abnormal enough when compared to the one of the main belt asteroids.

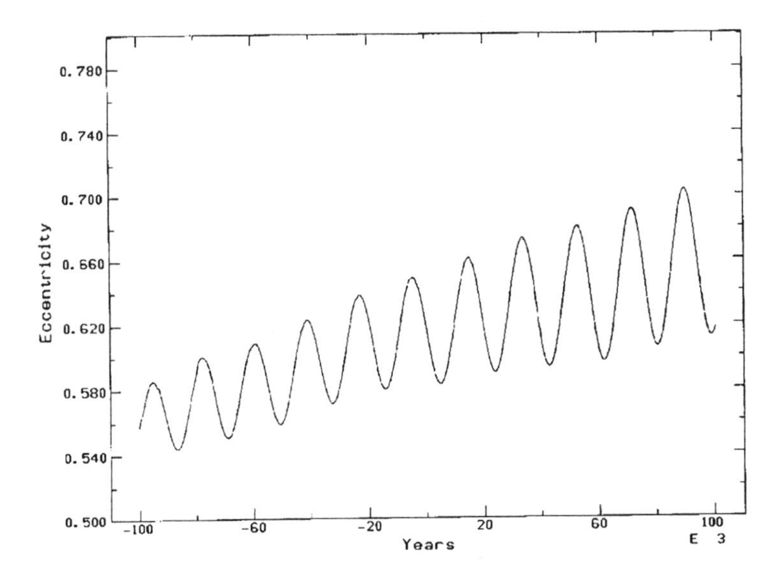

Figure 9. The eccentricity of *(1864) Daedalus* undergoes smooth long term perturbations, which do not appear to be strongly affected by the close approaches occurring near the node crossings. In a sense to be defined, the secular perturbations exist and control the evolution of the elements different from the semimajor axis; in this case, a secular increase in eccentricity is the most important dynamical evolution taking place.

As an example, Figures 6-9 show the behavior of *(1864) Daedalus*, a typical Geographos class asteroid. The semimajor axis (Figure 6) is almost constant over time spans of several thousand years, then jumps to significantly different values, in such a way that on the long term it undergoes a kind of random walk. The nodal distance (Figure 7) crosses the zero lines near the times of the jumps in a; the actual occurrence of close approaches follows an intermittent pattern (Figure 8), which can be described as random occurrence of encounters whenever the distance between the two orbits is small enough. The eccentricity (Figure 9) changes in an almost smooth way, as if secular perturbations exist in some sense and control the evolution of the orbital elements e, ω; the same is true for I, Ω.

Two main questions arise from this empirical description of the dynamical behavior of most Earth-crossing orbits. The first question is whether it is possible to exploit this alternating, but in some sense qualitatively uniform, dynamical behavior of the Geographos class orbits to give some statistical description, as an example to compute the probability of collision with the Earth. The second question is whether it is possible to develop an analytical (or at least semianalytical) theory to account for the ostensibly regular behavior of the orbital elements e, I, ω, Ω even in the presence of node crossings and close approaches.

3. STATISTICAL THEORIES FOR ENCOUNTERS

The attempts to model the occurrence of both close approaches and collisions for planet–crossing orbits have been based upon a statistical approach, starting from the pioneering effort by Öpik, see [25]. A statistical theory can be constructed by adopting three simplifying assumptions:

[1] The orbits between two consecutive close approaches are assumed to be regular and modeled in some simplified way, such that an explicit computation is possible. In the simplest possible approach, the orbits are modeled as keplerian ellipses with constant semimajor axis a, eccentricity e and inclination I, uniformly precessing with time; more elaborate models exploit secular perturbation theories, obtained by either analytical or semianalytical methods, in which the constants of the motion are "proper" a, e and I as opposed to the corresponding instantaneous, or "osculating", elements.

[2] Close approaches and collisions are modeled as purely random events. More specifically, whenever the keplerian ellipses representing the osculating orbits of one asteroid and one planet are close enough to allow for a close approach, the position of the planet and of the asteroid on the orbit (i.e. the anomalies) are assumed to be randomly distributed and uncorrelated; mean motion resonances, which constrain the two anomalies, are not allowed in the model. Each close approach is seen as a random event uncorrelated with any previous one.

[3] The orbital elements a , e, I of the asteroid are assumed to change only as a result of close approaches; in the more refined theories, this applies to the proper elements. Since at deep close approaches these changes can be computed e.g. by a hyperbolic two–body encounter approximation, the orbital evolution can be described as a random walk. For each present state, the probability (hence the expected waiting time) for a given change in the elements can be explicitly computed; this makes possible the use of Monte Carlo simulations. In some more sophisticated theories, other events which can result in changes of the elements, such as the capture in some resonance, are taken into account.

The simplest theories in this class, such as the one by Kessler, estimate the frequency of the occurrence of close approaches by modeling the asteroid orbit as if it were a probability cloud. By averaging over the angular variables ℓ (mean anomaly), ω and Ω, the probability density of the asteroid position is a function of a, e, I; because of the averaging over Ω, the probability is independent from longitude. The resulting density $S(r, \beta)$ at a radius r from the Sun and at an ecliptic latitude β is given by [10], [11]:

$$S(r, \beta) = \frac{1}{2\pi^3 r a \sqrt{(\sin^2 I - \sin^2 \beta)(r-q)(Q-r)}}$$

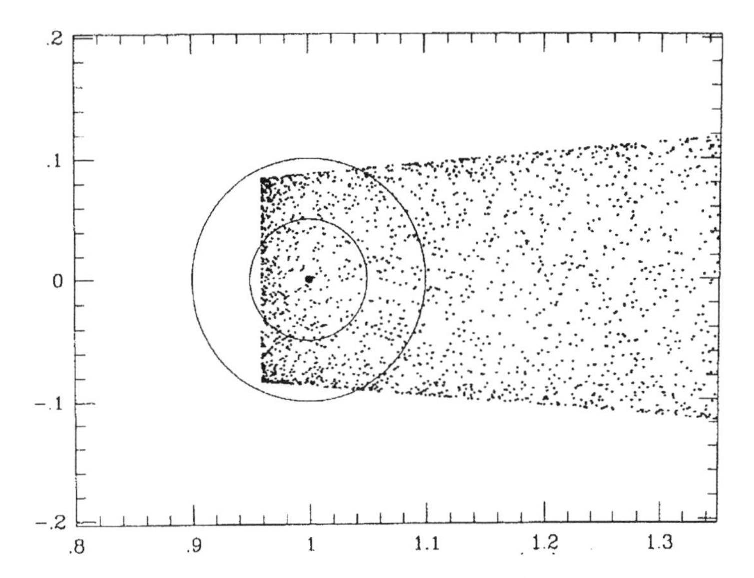

Figure 10. The probability density of the asteroid position has been simulated by a random sample of points on orbits with random values of ℓ, ω, Ω; here they are plotted on the r, β plane (radius from the sun in AU, on the horizontal axis; ecliptic latitude in radians, on the vertical one). In the simplest Kessler type formula, the number of encounters within an impact parameter D is estimated by multiplying the probability density at the position of the Earth $\beta = 0, r = 1$ by the cross section πD^2. A more sophisticated formula takes into account the changes of the probability density with the position across the cross section disk, which is especially important when the cross section disk contains part of the boundary of the region supporting the probability cloud, as in the two cases shown in this plot.

provided r is larger than $q = a(1-e)$ and smaller than $Q = a(1+e)$, and $|\beta|$ is smaller than I; S is zero if any of these constraints is not satisfied. Then the most likely number of close approaches per unit time through a cross section A can be computed from S and from the relative velocity V of the two orbits:

$$P(r, \beta) = S(r, \beta) V A$$

The above formula needs to be averaged over the target orbit; in the approximation of a circular orbit for the target planet, with $e' = I' = 0$and with $r = a'$ the semimajor axis of the planet:

$$P = \frac{VA}{2\pi^3 a a' \sin I \sqrt{(a'-q)(Q-a')}}$$

where V can be computed from the known value of the velocity of the asteroid orbit when node crossing occurs.

The cross section A for a close approach within an impact parameter smaller than D is just πD^2, and the relationship between impact parameter and minimum distance d_{MIN} is defined by the two–body hyperbolic encounter formula:

$$D^2 = \left(1 + \frac{2GM'}{d_{MIN}V^2}\right) d_{MIN}^2$$

with GM' the gravitational constant of the planet.

More complicated formulas have been derived to take into account the eccentricity e' of the orbit of the planet [26]; but the main problem is not in the approximation $e' = 0$. For either $r = q$ or $r = Q$ or $\beta = \pm I$ the probability density is singular (infinite); nevertheless the probability of collision is finite, when computed as an improper integral over the cross section. Thus it is possible to regularise the apparent singularity of the probability density, e.g. with the semianalytic method of [21].

In practice, the orbits with close approaches near either perihelion ($q \simeq a'$) or aphelion ($Q \simeq a'$), and/or with very small I can have a very large number of close encounters, and the asteroids whose orbits evolve through one such state have probabilities of collision comparatively high and difficult to compute exactly (because of the instability of the computation). If these difficult cases are handled with care, it is possible to achieve a good accuracy in the prediction of the deep close approach frequency for a large population of orbits.

As an example, in [21] the average probability of collision with the Earth for a Near Earth orbit (with $q < 1\,AU$ at least for some time) is estimated by extrapolating from the sample of close approaches found in a $200,000$ years integration at 3.9×10^{-9} per year per object. The same computation done on the basis of a modified Kessler theory (with regularisation) gives a probability of 3.6×10^{-9} per year per object. Other authors give not very different results [?]. This means that the probabilities of collision, when averaged over a long time span and over a large number of orbits, can be computed in sufficiently accurate way. The main source of uncertainty on the probability of collision is in our very incomplete knowledge of the population of planet crossing objects, and does not depend upon lack of mathematical knowledge.

4. THE PROTECTED ONES: THE TORO CLASS

An Earth-crossing orbit of the "normal" type, that is of the Geographos class, undergoes close approaches at random, during comparatively short time spans around the epochs of the node crossings. However, there are orbits which are Near Earth in that they satisfy $q < 1\,AU < Q$, and nevertheless close approaches either do not take place at all, or take place more seldom and with larger distances from the Earth than the statistical theories would predict.

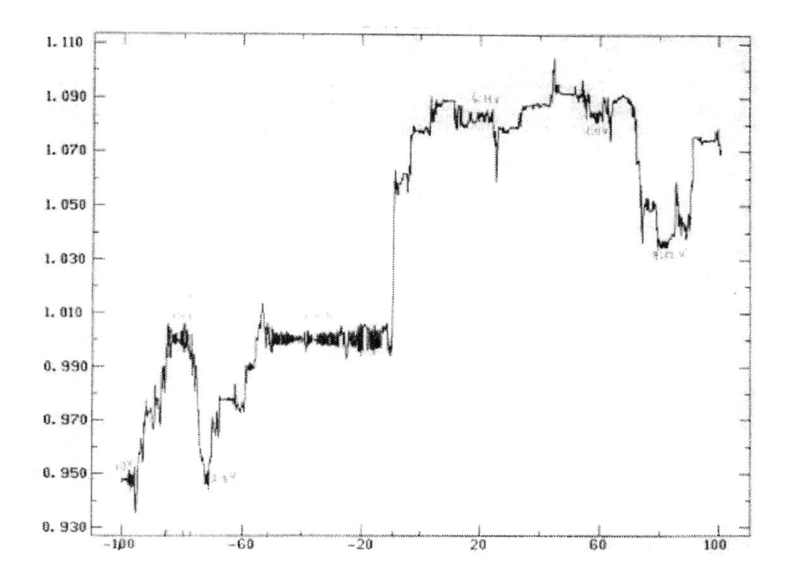

Figure 11. Semimajor axis of the asteroid *(2063) Bacchus*, as a function of time, from the output of the numerical integrations of *PROJECT SPACEGUARD*, over a time span of 200,000 years. A transition between an *Aten* type orbit (with $a < 1\,AU$) and an *Apollo* type (with $a > 1\,AU$) occurs at $\simeq -10,000$ years from present time, as a result of a very close approach ($\simeq 2$ radii of the planet) with Venus.

The *protection mechanisms* which can be responsible for this "Earth-avoidance" behavior are essentially two: either node crossings occur, but close approaches do not take place even when the distances between the two orbits would allow, or node crossings do not occur. In this Section we discuss the first case, that is the *Toro class*, according to the classification of [20].

The empirical evidence for the existence of a Toro class of dynamical behavior is in a number of examples found by several authors, starting from [9], and in a systematic way in the large database of planet-crossing orbits generated by the *PROJECT SPACEGUARD*, which contain thousands of figures like Figures 11-13. The Figures we have selected refer to the asteroid *(2063) Bacchus*, and show the 'normal' Geographos class behavior, with semimajor axis jumping at random as a result of close approaches to both the Earth and Venus (Figure 11) around the node crossing epochs (e.g. Figure 12 shows the node crossings with the Earth). However, there are some node crossings with the Earth which do not correspond to close approaches: in the Figure 13 these "safe intervals" are apparent as gaps in the comb. In this example the protection mechanism is active when the semimajor axis of *(2063) Bacchus* is close to $1\,AU$, and therefore must be related to the $1:1$ resonance, that is to a state of "Earth's Trojan".

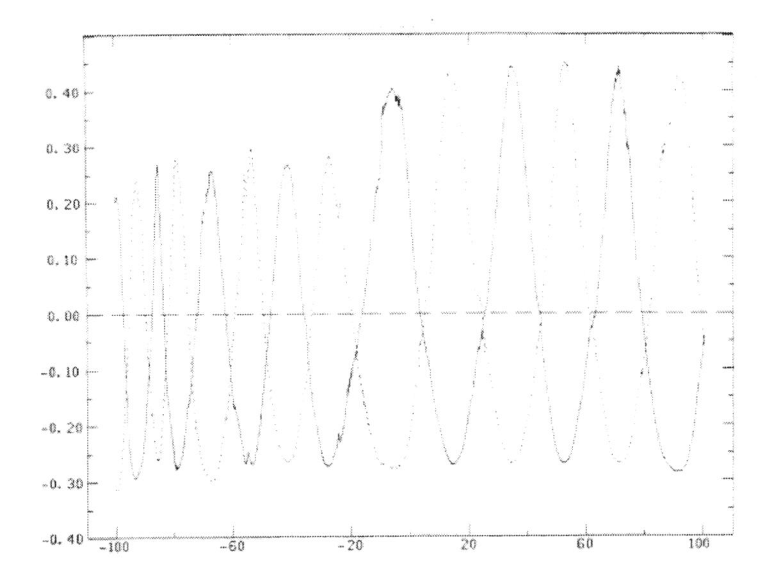

Figure 12. Nodal distance of the asteroid *(2063) Bacchus* with the Earth; the continuous line refers to the ascending node, the dotted one to the descending node. The node crossing are comparatively frequent, and when the nodal distance with respect to the Earth is close to $-0.3\,AU$, almost tangent encounters with Venus can take place.

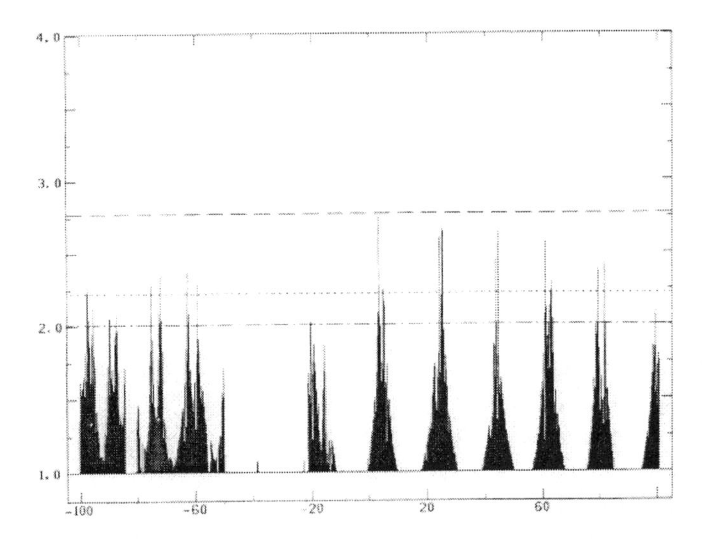

Figure 13. Close approaches of *(2063) Bacchus* to the Earth; in this *comb plot* each black line represents a close approach. By comparing with Figure 11, it is clear that close approaches do not occur near eight node crossings: two around $-80,000$ years, and a sequence of six between $-50,000$ and $-20,000$ years.

To model in the simplest way this behavior, let us assume the planet –e.g. the Earth– is in a circular orbit with radius a', and the asteroid has or-

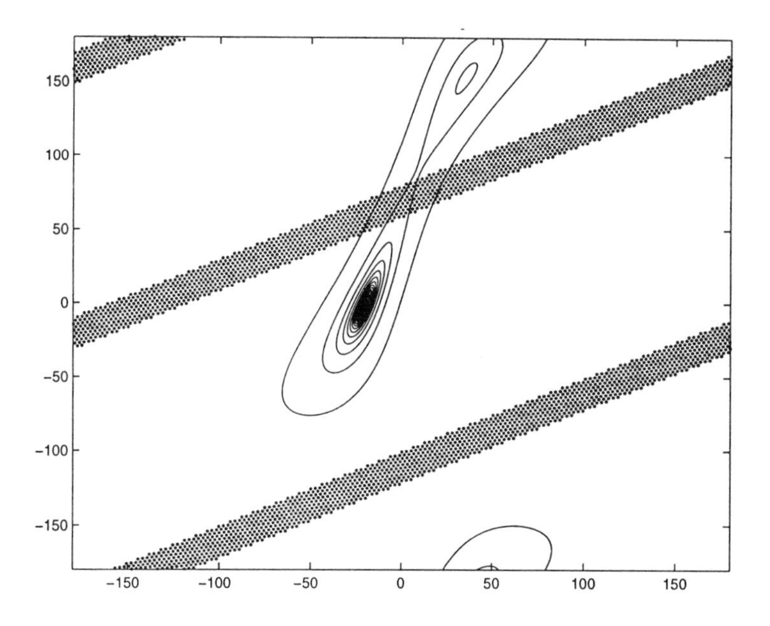

Figure 14. Level curves of the inverse distance between the asteroid and the Earth –that is, apart from a constant factor, the direct perturbing function– as a function of the two mean anomalies. The perturbing function has a very sharp maximum near the ascending node, with both ℓ and ℓ' near $0°$, because the orbital elements (a, e, I, ω) are such that the ascending node is near $1\,AU$. On the same plot, the mean anomalies evolving linearly with time, with a ratio of mean motions close to $1:2$, and a phase such that the region of close approaches is avoided.

bital elements (a, e, I, ω) such that –for some time around the node crossing epoch– the distance from the Sun at one node is close to a': let us say

$$r_+ = \frac{a(1-e^2)}{1+e\cos\omega} \simeq a' \;;$$

note that the value of the longitude of the node Ω does not matter. For a time span short with respect to the frequencies of the secular perturbations, but longer than the orbital period, the only elements which change in a significant way are the anomalies ℓ, ℓ' of both the asteroid and the planet. Since these are angle variables, the short term dynamics can be described on a phase space which is a *torus*, the Cartesian product of two circles; a torus can be represented, as in Figure 14, as a square with the opposite sides identified.

Figure 14 shows the level lines of the gravitational potential of the Earth, as felt by an asteroid in an Earth-crossing orbit very close to a node crossing (the distance at the ascending node is less than $0.1\,AU$). The gravitational potential has a very sharp maximum near the node, while the perturbation due to the Earth is very small elsewhere. As time goes by, the changes in the anomalies ℓ, ℓ' of both orbits can be described approximately by the linear functions of time $\ell = nt + \ell_0$; $\ell' = n't + \ell'_0$, with n, n' the mean motions. For values of a such that the ratio n/n' is close to a fraction q/p with q, p small integers, that is near a *mean motion resonance*, the orbit does not spread uniformly on the torus, but can avoid significant portions of it, possibly including the region around the maximum, that is avoiding close approaches.

That such a temporary protection mechanism can occur at some node crossings is clear, but the question is: does this happen by chance? By means of a purely cinematical description, as given above, we would expect that if a is close enough to a resonant value, the orbit will have the right phase ℓ_0 for protection at some node crossing, and then have a wrong phase in some later node crossing, essentially at random. This is not the case because the mean anomaly of the asteroid, and therefore the phase of the resonance (some *critical argument* of the form $p\ell - q\ell'$+some combination of $\omega, \Omega, \omega', \Omega'$), does not change linearly but is subject to a kind of restoring force. The restoring force is of course stronger the closer are the two orbits, and has very high values near the node crossings, and near the maximum point on the torus.

The behavior of the critical argument of a Toro class orbit is shown as a function of time in Figure 15, from [19]. The plot shows "avoidance", in the sense that the dangerous position with both the Earth and the asteroid near the ascending node is avoided, but only when the nodal distance is small (less than $0.13\,AU$ in this example). This Figure is typical of the Toro class orbits, including the case of the namesake asteroid *(1685) Toro*, Figure 6 in [20].

Thus the protective effect of mean motion resonance occurs much more often than it would take place if it was controlled only by chance. Figure 15 is enough to understand that the restoring force due to the perturbations from the Earth acting on the asteroid mean anomaly ℓ is pushing the orbit away from the collision.

This "negative attraction" effect is somewhat counter our earth-bound intuition, but in fact is another form of the first paradox of astrodynamics, well known to astronauts. If an asteroid (or spacecraft) is pushed forward along its orbit by some perturbing acceleration acting along track (it does not matter if this acceleration is due either to the attraction of a third body, or to the action of rocket engines), the orbital energy increases, the semimajor axis increases, and the mean motion decreases, thus the asteroid (spacecraft) is pushed backward; the displacement produced by an along track acceleration Δf is $\Delta x = -3/2\, t^2\, \Delta f$.

To transform this intuitive explanation in a rigorous mathematical ar-

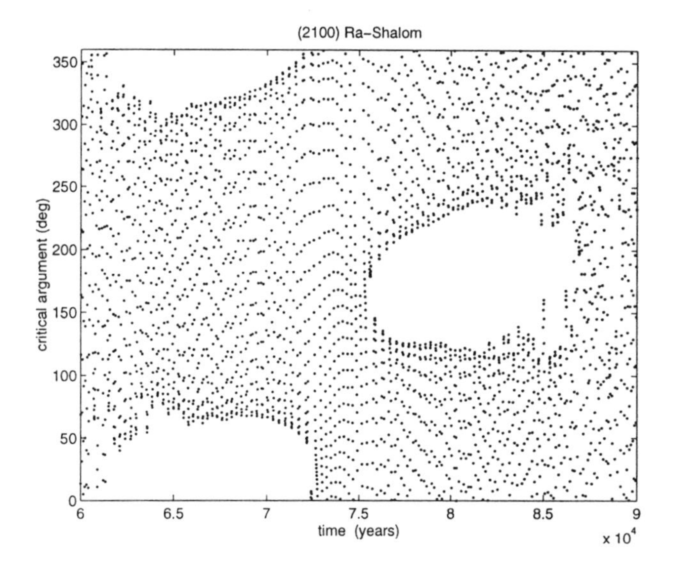

Figure 15. Critical argument $-5\lambda + 7\lambda' - 2\varpi$ (in degrees, between $0°$ and $360°$), over the $100,000$ years time span of numerical integration, for the asteroid *(2100) Ra-Shalom*. The two gaps occurring around $0°$ between $\simeq 62,000$ and $72,000$ years, and around $180°$ between $\simeq 76,000$ and $86,500$ years, indicate the avoidance of the closest approach state by means of some protection mechanism. On the contrary, between $\simeq 72,000$ and $76,000$ years the critical argument circulates, and the semimajor axis does not undergo resonant perturbations. Node crossings occur at $\simeq 66,000$ and at $\simeq 82,000$ years.

gument we need to resort to a semianalytical theory, obtained by averaging [19]. If $\sigma = p\ell - q\ell'$ is a slow variable, because of $n/n' \simeq q/p$, then there is another variable $\tau = c\ell - d\ell'$ obtained together with σ by a unimodular transformation, that is c, d are integers and $pd - qc = 1$. Such τ is a fast angle, and averaging with respect to τ is a good approximation by the *averaging principle* [1]; if the averaging is performed over a single angle variable, the averaging principle is a rigorous theorem, with estimate of the error done in neglecting the short periodic terms. Then we are left with the semi-averaged Hamiltonian, still depending upon one angle variable:

$$\overline{H}(\sigma, \Sigma, T) = H_0(\Sigma, T) - \frac{1}{2\pi}\int_0^{2\pi} R(\sigma, \tau, \Sigma, T)d\tau$$

where Σ, T are the action variables conjugate to the angle variables σ, τ, H_0 is the Hamiltonian of the 2-body unperturbed problem transformed to the new variables, and R is the usual perturbing function of the 3-body problem, also depending upon the other orbital elements, especially ω which controls the nodal distance.

For the semi-averaged Hamiltonian $\overline{H}$, the fast angle τ is a cyclic variable, hence T is an integral of motion; once the value of T is fixed by the initial conditions, H_0 is a function of Σ only. If we avoid being confused by too many changes of variables, and write the result as a function of some more usual variable, such as the Delaunay $L = k\sqrt{a}$, with $k = \sqrt{GM_\odot}$ the gravitational constant in Gauss' form,

$$\overline{H} = H_0 - \overline{R} = -\frac{k^4 M_\odot^2}{2L^2} - \frac{q}{p} n' L - \overline{R}(\sigma, L)$$

with $\overline{R}$ the averaged perturbing function. We can now understand the "repulsive" restoring force: the first derivative of H_0 with respect to L is $n - (q/p)n'$, zero at the exact resonance; the second derivative is always negative.

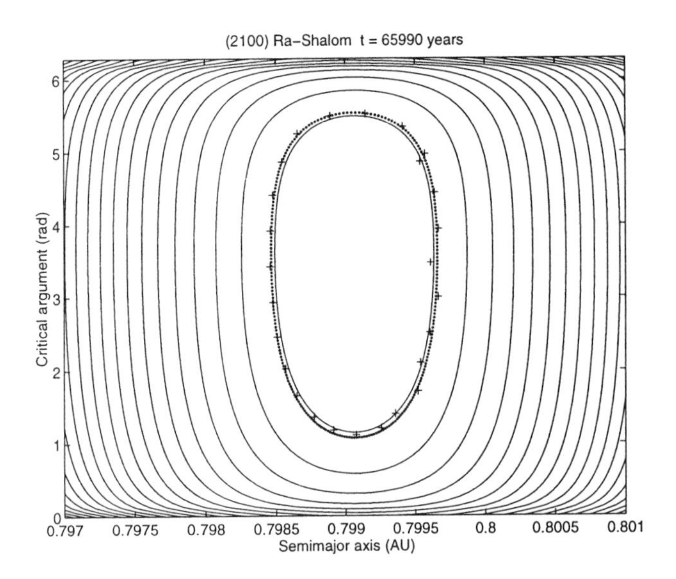

Figure 16. Level lines, in the plane with coordinates a and the critical argument (in radians), of the semi-averaged Hamilton function at the time of the node crossing; the crosses are the orbit of *(2100) Ra-Shalom*, as computed numerically by the non-averaged equations of motion. Libration of the critical argument, in this case of the $5:7$ resonance, is forced by the "infinite" potential energy maximum resulting from the averaging of the collision, in this case at the value $0°$ of the critical argument.

Thus the qualitative behavior of the semi-averaged Hamiltonian system can be understood by comparing with the simple Hamiltonian

$$K(x, y) = -y^2 + V(x) \ ;$$

when the nodal distance is small, $V(x)$ has a sharp minimum near the node. But the Hamiltonian is concave with respect to y, and therefore the minima of the potential energy are avoided, exactly as the maxima in

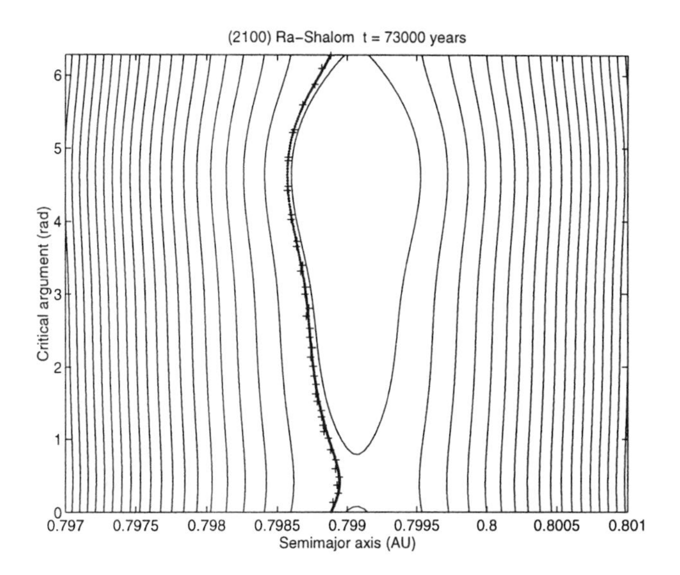

Figure 17. Level lines of the semi-averaged Hamilton function at a time when the nodal distance has increased to about 0.13 AU. For *(2100) Ra-Shalom* (line with crosses) the libration has changed into circulation, because the (suitably normalized) area enclosed by the curve in the previous figure is larger than the area enclosed by the separatrix in this figure.

the more familiar case with convexity with respect to y. When the nodal distance $\to 0$, the value of the minimum $\to -\infty$, and whatever the initial value of K there is an avoidance region near the singularity.

This behavior is shown in Figure 16, based on an explicit computation of $\overline{H}$ by numerical quadrature, for values of the other elements e, ω, I such that the nodal distance is very small. The changes in the elements different from a, ℓ are driven by the secular perturbations, rather than by the resonant interaction with the Earth. Thus we can consider the problem to be defined by the semi-averaged Hamiltonian $\overline{H}$, depending upon time through the slow variables e, ω, I, Ω. This is the nominal situation to apply the adiabatic invariant theory [8], by which the solutions follow closely the *guiding trajectories* provided by the solution of the system $\overline{H}$ in the (σ, Σ) plane for fixed values of the other elements. That is, there is an "almost integral", the adiabatic invariant, which can be computed by means of the area enclosed by the level lines of $\overline{H}$ in the (σ, Σ) plane.

When the area enclosed by the separatrix curve (which is the curve through the saddle point, corresponding to the minimum distance) becomes smaller than the area required by the adiabatic invariant, then the guiding trajectory changes topology, and the solution switches from libration to circulation, as shown in Figure 17. This explains the apparently "astute" behavior of the Toro class asteroids: whenever the perturbation due to close approaches is strong, the orbit switches to a libration state, avoiding encounters at the node; when the nodal distance increases again, the potential well is not deep enough and confinement in the libration region does not occur.

The Toro state in most cases does not last for a very long time, because resonances can protect from close approaches to one planet but can not protect for a significant span of time from close approaches to two planets, for the simple reason that the planets are not resonant among them. Thus all the Toro class asteroids change their dynamical state as a result of a close approach to a planet different from the one they are protected from, typically after a time span of a few $10,000$ years. The only known exceptions are orbits which cross only the orbit of Mars, such as the Eros clones studied by [16]; if the perihelion is well above $1\,AU$, a Toro-like state with Mars can protect from close approaches, even for millions of years.

It is important to remember again that all the planet-crossing orbits are strongly chaotic, thus the long term behavior of all the orbits can not be predicted; in particular, the time span of residence of a specific orbit in a Toro state with a specific resonance is unpredictable. As an example, the time in which *(2063) Bacchus* either has been, or will be, an Earth Trojan can not be predicted, although it has to be expected that it will get there eventually.

5. THE PROTECTED ONES: THE KOZAI CLASS

The other case of ostensibly protected behavior occurs for the near-Earth orbits of the *Kozai class*, according to the classification of [20]. The name comes from a namesake asteroid, *(3040) Kozai*, but also from the author of the seminal work on this subject [12].

The evidence from numerical integrations is that of a dynamical behavior as in the example of Figures 18-19. There are many asteroids with perihelion distance q which can, as a result of the secular perturbations to the eccentricity, decrease below $1\,AU$, and nevertheless there are no node crossings, that is, the nodal distance is always positive. This happens only for orbits with significant inclination, because in this case the evolution of e and ω are strongly coupled. In the example of the Figures 18-19, that is the Mars-crossing asteroid *(1866) Sisyphus*, q can be as low as $0.8\,AU$, and at the same time the nodes are above $1.1\,AU$, because the argument of perihelion has a value either around $\pi/2$ or around $3\pi/2$ when the eccentricity is at a maximum.

To understand this mechanism of avoidance of node crossings we need to model the secular perturbations on (e, ω, I, Ω); for this we resort again

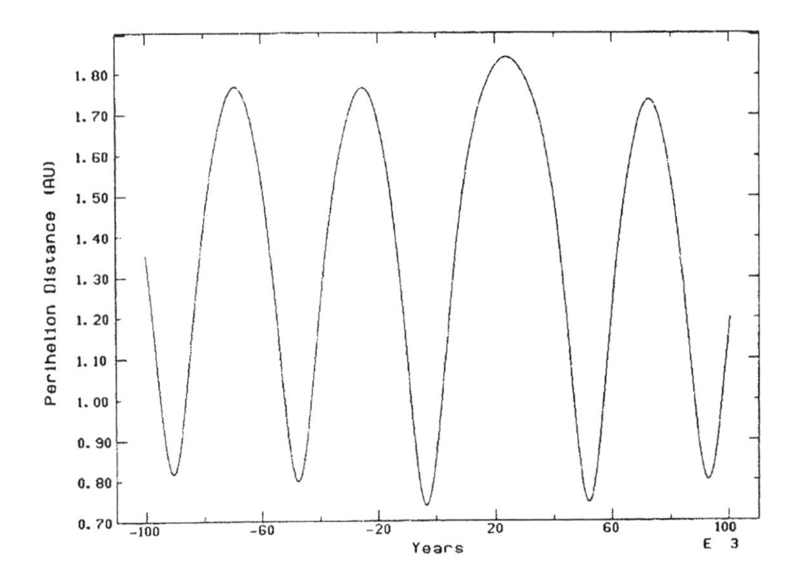

Figure 18. The perihelion distance of *(1866) Sisyphus*, when the eccentricity is near its maximum over the secular perturbation cycle, decreases well below 1 AU.

to the *averaging principle*, but in this case the average is performed on both the fast angles ℓ, ℓ', assuming that there is no mean motion resonance, that is the torus of the mean anomalies is in this case covered uniformly (more exactly, in an ergodic way, see [1]). The averaged Hamiltonian is

$$\overline{H} = \frac{1}{(2\pi)^2} \int_{-\pi}^{\pi} \int_{-\pi}^{\pi} [H_0 - R_{dir} - R_{ind}] \, d\ell d\ell'$$

where H_0 is the unperturbed 2-body Hamiltonian, $R_{dir} = k^2 m/D$ is the direct perturbing function (with D the distance between the asteroid and the planet, m the mass of the perturbing planet) and R_{ind} is the indirect part of the perturbing function. By a classical result, the average over the anomalies of the indirect perturbing function is zero, thus

$$\overline{H} = H_0 - \frac{k^2 m}{(2\pi)^2} \int_{-\pi}^{\pi} \int_{-\pi}^{\pi} \frac{1}{D} \, d\ell d\ell' = H_0 - \overline{R} \, .$$

There are two main methods to compute the averaged perturbing function $\overline{R}$. It is possible to expand the perturbing potential in series, with the eccentricities and inclinations as small parameters. That is, R is expanded as a Taylor series in e, I, e', I', and for each order as a Fourier series in the angles ℓ, ω, Ω. The classical D'Alembert rules, which result from the invariance of the problem with respect to orthogonal transformations, strongly constrain the terms which can appear with non zero coefficient in such a

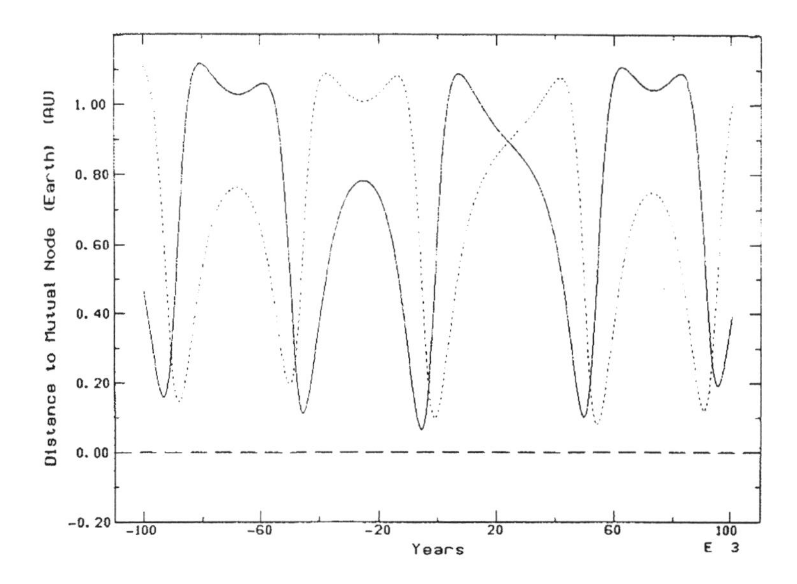

Figure 19. The nodal distance between *(1866) Sisyphus* and the Earth is always positive, that is whenever the perihelion of Sisyphus is closer than 1 *AU* to the Sun, this occurs either well above or well below the ecliptic plane.

series, and it turns out that only the even order terms are allowed; see e.g. [18]. Thus the lowest order terms containing the eccentricities and the inclinations are quadratic, and if the theory is truncated to degree two the secular perturbations can be described, in suitable variables, as the solutions of a system of linear differential equations.

The method outlined above is the classical one, introduced by Laplace and others more than 200 years ago, and it gives a reasonable first approximation of the secular perturbations of the orbits of moderately perturbed planets with low eccentricities and inclinations, such as the major planets Venus to Neptune. This approximation fails when eccentricity and inclination are large; in this case the largest term in the perturbing function neglected by the Laplace linear theory is the one with $e^2 \sin^2 I \cos(2\omega)$.

An integrable first approximation, different from a linear system, was introduced by [12]. Let us assume that the perturbing planet is on a circular orbit, and on the reference plane: $e' = I' = 0$. The argument applies also to the case of many perturbing planets, provided the orbits are all circular and all on the same plane. Then the problem averaged over ℓ' can be described in the way already introduced by Gauss, as the gravitational problem defined by a mass distributed in rings along the planetary orbits; the perturbing potential is axisymmetric with respect to the axis through the Sun and orthogonal to the plane of the planets, and the component of the angular momentum of the asteroid orbit along the same axis is preserved.

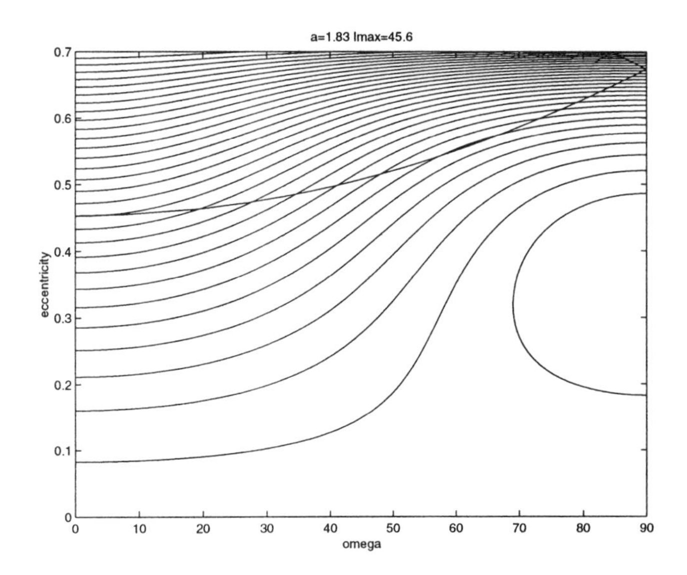

Figure 20. Level lines of the averaged perturbing function for a Kozai class orbit, in the (ω, e) plane; ω on the horizontal axis is in degrees. The node crossing lines with the Earth, where one of the two nodal distances is zero, are also drawn: the libration keeps the secular evolution of the orbits well clear of the node crossings.

In terms of $\overline{H}$ and of the keplerian orbital elements $(a, e, I, \omega, \Omega, \ell)$, the averaged Hamiltonian is independent of ℓ and of Ω, and the two conjugate variables are integrals: they are proportional to $L = k\sqrt{a}$ and to

$$Z = L\,\sqrt{(1-e^2)}\,\cos I$$

respectively. In conclusion

$$\overline{H} = \overline{H}(\omega, G; L, Z)$$

is a one degree of freedom Hamiltonian, with two parameters L and Z. This averaged problem is integrable; once a and $\sqrt{1-e^2}\,\cos I$ are fixed, the secular evolution can be described as a curve in the (e, ω) plane; I is a function of e, which can be deduced from $Z = const$, and Ω does not matter. This curve can be drawn as level curve of the function $\overline{H}(\omega, e)$.

For low inclination the $\overline{H} = const$ curves in the (e, ω) plane are not very different from $e = const$ lines; but for large I, the coupling terms such as the one with $e^2 \sin^2 I\, \cos(2\omega)$ become dominant, and the eccentricity undergoes large relative changes as ω circulates. Below a critical value of Z/L, the topology of the level curves of $\overline{H}$ changes and a separatrix appears, bounding a region where ω librates, typically around either $\pi/2$ or $3\pi/2$. This *Kozai resonance* is especially effective in keeping the perihelion of the

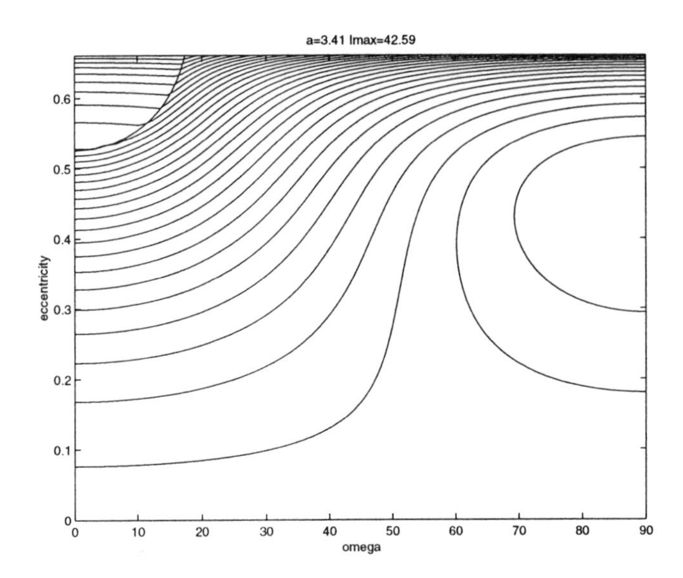

Figure 21. Level lines of the averaged perturbing function of an orbit which could cross the orbit of Jupiter. The Jupiter-crossing line is also drawn, and the level lines are non smooth along it, because of the divergence phenomenon discussed in the text.

asteroid out of the plane of the perturbing planets. Figure 20 shows the phase space of the averaged problem, for values of the initial conditions consistent with those of the asteroid *(3040) Kozai*. A large libration region is "safe" from node crossings; there are also many solution curves where a strong $e - \omega$ coupling achieves the same result, because ω can be near 0, but only for e near its minimum value.

The same mechanism protects the high inclination asteroids at the outer edge of the main belt from close approaches to Jupiter, as in Figure 21, because $\omega = \pi/2$ implies that not only the perihelion, but also the aphelion is out of the plane of the planetary orbits. When on the contrary the planet with which collision could occur has a semimajor axis a' very close to a, then the node crossings are avoided when the nodal points are at perihelion and at aphelion, that is for $\omega \simeq 0$ and $\omega \simeq \pi$, as in Figure 22 and in [13]. For very high inclination the same averaged dynamics can result in increases of the eccentricity up to values very close to 1, and then the fate of the orbit can be to encounter the surface of the Sun; however this requires a very high initial inclination, and occurs almost only to comets [2].

The behavior of the Kozai class asteroids implies that the Öpik-Kessler methods to compute probability of collision might fail, in that they give a finite probability to collisions which can not occur at all. A more flexible method to compute frequency of close approaches and probability of collision, which could take into account the actual occurrence of node crossings, was developed over many years, beginning with the basic formulas devised by [25], later extended and developed by [29], [6], and many others. The basic geometric idea is as follows. When two osculating orbits have a small nodal distance, it is possible to approximate a short span of both ellipses with their tangent lines at the respective nodal points, parametrized in such a way that the velocity at the nodal point is the same on the ellipses and on the straight lines. Then the distance, as a function of the parameters on these straight lines, is an easily computed quadratic form. From this approximate distance function it is possible to compute analytically the time span spent by the two orbits in the region where the distance is less than a given impact parameter.

The large scale tests performed in [21] show that the Öpik-Wetherill method is very effective in predicting the number of close approaches which is possible at each actual node crossing, thus it solves the problem of computing the probability of collision for the Kozai class, provided the secular evolution of the orbital elements e, I, ω is available, e.g. from numerical integrations. It is also a very effective method to detect Toro class orbits, which by definition violate the formula, by avoiding close approaches at actual node crossings. This method can not be used for low inclination and almost tangent encounters, the same difficulties found with the Kessler method.

The Figures 20, 21 and especially 22 show that the level curves of the averaged Hamiltonian $\overline{H}$ are not smooth on the node crossing lines. Indeed the integral over the torus of the variables ℓ, ℓ' is an improper one, and it is convergent because the singularity of collision is a pole of order one. On the contrary, if the averaging is applied to the right hand side of the equation of motion, the singularity of collision is a pole of order two, and the improper integral over the torus is not convergent. Thus the very meaning of the secular perturbation equations is uncertain.

Recently it has been shown [7] that the averaged Hamiltonian $\overline{H}$, although not smooth, has enough regularity –in a neighbourhood of the node crossing line– to allow a generalized definition of secular perturbation equations. This has been obtained by using the Öpik-Wetherill approximation of a short stretch of the orbits near the mutual node with straight lines. The approximate distance function d between the points on the straight lines is a quadratic form, and then the integral of $1/d$ over the torus can be computed either analytically or semi-analytically; the singularity of $1/d$ at exact node crossing is used to remove the singularity from the improper integral (Kantorovic method to compute improper integrals).

In this way it is possible to compute secular orbits even in the presence of node crossings, and thus to compute the secular frequency of circulation/libration of ω and of circulation of Ω. As a matter of principle, this

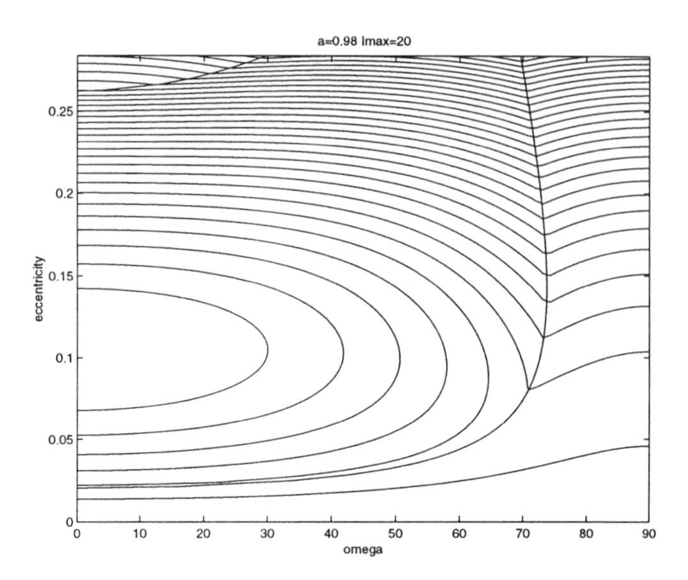

Figure 22. Averaged perturbing function of an orbit with $a = 0.98\,AU$, which could cross both the orbit of the Earth and the orbit of Venus. The safe Kozai class state is obtained for libration of ω around 0°. Along the node crossing lines with the Earth (on the right) and with Venus (on the left) the solution curves of the averaged problem are not smooth; to define a secular frequency of libration/circulation of ω requires a suitably generalized definition of secular perturbations.

allows one to use the Kozai theory as a first integrable step in a perturbation theory, to study secular resonances, and to compute proper elements even for planet-crossing orbits; the theory developed in [7] contains explicit algorithms for this, but most of the work remains to be done.

6. THE SPACEGUARD DYNAMICAL CLASSIFICATION

The current empirical knowledge, based mostly on numerical integrations, on the dynamical behavior of Earth-crossing orbits –over time scales of a few 10,000 to a few 100,000 years– can be summarized in a classification of the orbit types following [20]. For most, but by no means all, of the classes some theory is available, at least as a tool to understand the qualitative features, and as a first approximation. For longer time spans (millions of years) other dynamical phenomena, such as the secular resonances, are relevant, as well as the complex effects of chaotic diffusion.

The classification uses as criteria the main features of the planet-crossing dynamics: node crossings and close approaches to the Earth, mean motion resonances with the Earth, the value of q below/above $1\,AU$, mean motion resonances and approaches with Jupiter. The logic is summarized in the block diagram of Figure 23.

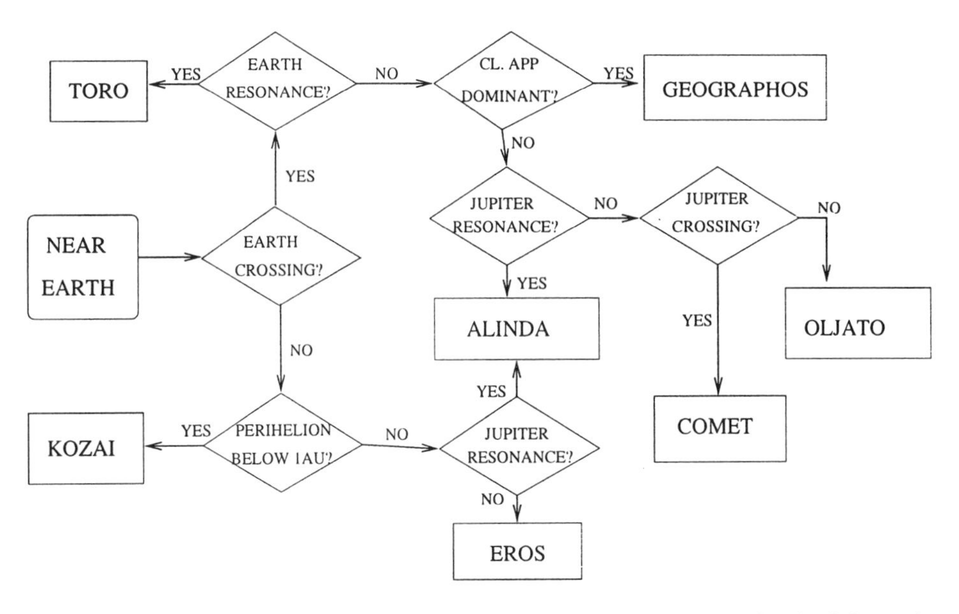

Figure 23. The SPACEGUARD dynamical classification of Near Earth Orbits: the procedure by which the orbit of a Near Earth Asteroid is assigned to a class is described as a block diagram; the questions asked in the decision boxes are explained in the text.

For each asteroid which comes "Near Earth" in some sense (say $q < 1.1\, AU$) the nodal distance is computed as a function of time; the first question is whether it has node crossings with the Earth. If there are no node crossings, but the perihelion distance is –at least for part of the time, during a period of ω– below $1\, AU$, the orbit is protected by secular perturbations and is classified as *Kozai class.* If the perihelion is above $1\, AU$, and the orbit is not involved in any major mean motion resonance with Jupiter, then it belongs to the *Eros class*; as it is the case with the namesake asteroid *(433) Eros*, it could become an Earth-crosser and have a significant probability of impacting the Earth, but only as a result of secular resonances, acting over time scales of millions of years [15], [16].

On the other hand, an asteroid which is not presently Earth-crossing could be strongly perturbed by a mean motion resonance with Jupiter; in this case, it belongs to the *Alinda class.* The behavior of the semimajor axis of an Alinda is easily identified, because the resonant perturbations by Jupiter generate large and rapid changes in a, as in the example in Figure 24. However, the reason for placing these orbits in a separate class is the very rapid secular evolution of the eccentricity shown in Figure 25; as a result, the value of q can decrease below $1\, AU$ in a very short time span, typically of the order of a few $10,000$ years.

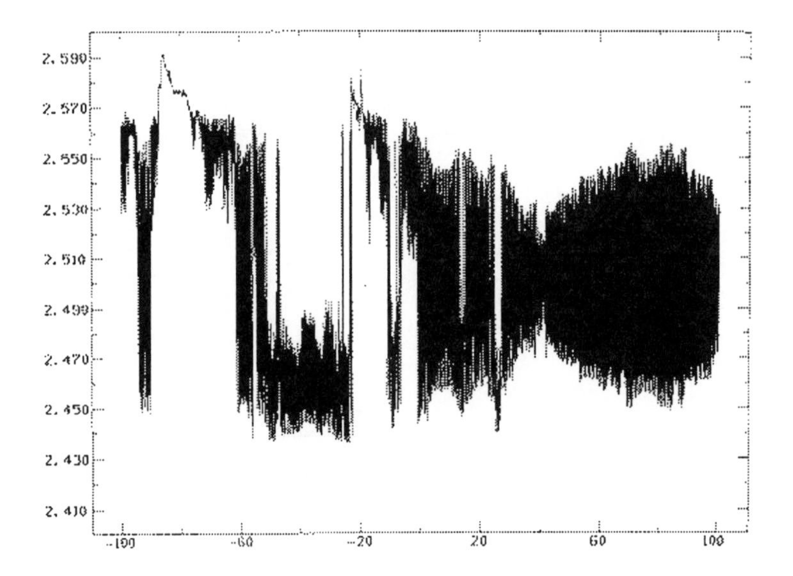

Figure 24. Semimajor axis of the asteroid *(2608) Seneca* from the output of the PROJECT SPACEGUARD numerical integration. The large changes in semimajor axis are due to a mean motion resonance with Jupiter, in this case the 3 : 1.

The superposition of the large changes in e and of the resonant changes in a results in a nodal distance which changes in a significant way both over time scales of a few hundred years –the typical period of the resonance critical arguments– and over a time scale of several thousand years –the period of the secular perturbations. This can generate the so called *super-crossing* behavior, in which node crossings take place at intervals of a few hundred years, but also a comparatively rapid change between *Apollo* orbits (with $q < 1\,AU$) and *Amor* orbits (with $q > 1\,AU$), as shown in Figure 26.

The dynamical behavior of the orbits in mean motion resonance with Jupiter has been studied with semianalytic techniques, e.g. in [22], [23], and is understood at least in qualitative terms. However, the interaction between the resonant, but long distance, perturbations by Jupiter and the short range interactions with the terrestrial planets results in a very complicated behavior of these Kirkwood gap asteroids. The evidence from numerical integrations indicates that there is a synergic effect, by which the orbital elements of the Alinda can change very rapidly, to the point that the distinction between orbits which either are or are not presently Earth-crossing is irrelevant, even over time scales of a few 10,000 years.

Going back to the first decision box in Figure 23, if node crossings with the Earth do occur, the orbit is *Toro class* if it is sometimes locked in a mean motion resonance, and this resonance is effective to avoid (at least decrease in frequency and depth) the close approaches to our planet. If this is not the

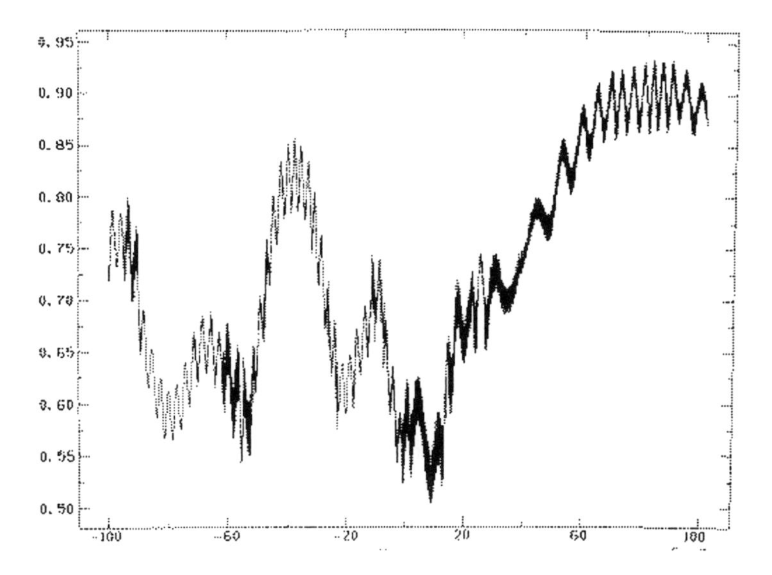

Figure 25. The eccentricity of the asteroid *(2608) Seneca* undergoes very large changes, especially when the locking in the mean motion resonance is maintained for many thousands of years, as it is the case in the positive time portion of this numerical integration. As a result, in this example the perihelion distance change from values above 1 *AU* to Mercury-crossing values.

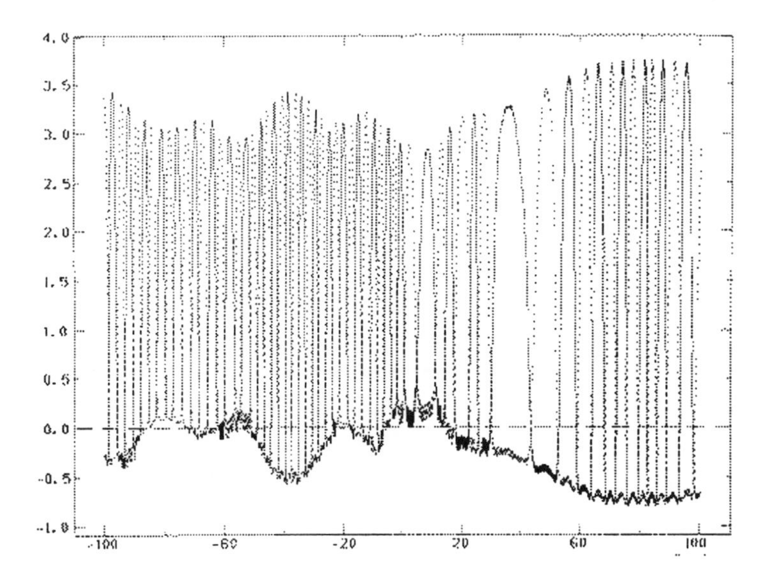

Figure 26. Nodal distance of the asteroid *(2608) Seneca* with respect to the Earth. The orbit changes between Earth-crossing and not crossing states over a few thousands of years, and also shows the peculiar super-crossing behavior, in which the nodal distance goes through zero during each cycle of the critical arguments of the 3 : 1 resonance with Jupiter, that is over a time span of a few hundred years.

case, the time series of a should be examined to decide if the contributions from short range interactions with the terrestrial planets are the dominant source of changes. A formal test based upon correlation analysis could be used, as in [21], Section 5; but in most cases visual inspection of the orbital elements plots is enough to decide. The Earth-crossing orbits for which the evolution of a is dominated by close approaches to the Earth (and in some cases to Venus) are *Geographos class*; the ones where resonant perturbations from Jupiter are dominant are again *Alinda class.*

The classification scheme described above is logical and for each class there are analytical tools to understand the dynamical behavior. Now come the cases for which the theories are still inadequate, although they can be used to describe the behavior in a qualitative sense. If the evolution of the orbital elements, in particular for the semimajor axis, is neither dominated by the close approaches nor by the mean motion resonances with Jupiter, and nevertheless node crossings take place with the terrestrial planets, the orbit is too strongly perturbed to be described by the currently available theories. If this is due to the occurrence of close approaches to Jupiter, this is a typical cometary orbit (of the so-called *Jupiter family*); the fact that objects in such orbits can be found in catalogues of asteroids should not be surprising. If it is indeed physically a comet, with a high content of volatiles, this means only that when the perihelion descends below some critical value for sublimation (depending upon the surface composition) the occurrence of an outburst is not necessary, but may depend upon the state of the surface, which in turn may depend upon the dynamical and physical history. On the other hand the same orbit could result from dynamical evolution either from an Alinda class of from a marginally unstable Trojan, and could correspond to an object with asteroidal composition.

There is still one case, corresponding to the last box on the right in Figure 23: there are orbits which are not resonant with Jupiter (at least not for a significant span of time), do not undergo orbital changes dominated by close approaches to the terrestrial planets, and are neither Jupiter crossing nor Jupiter approaching; according to the classification of the *PROJECT SPACEGUARD* they form the *Oljato class*, from the namesake asteroid *(2201) Oljato*; see [20], Figure 14. These orbits have an eccentricity very large and changing secularly, low to moderate inclination, and very frequent close approaches to the terrestrial planets; they can be described as comet-like orbits with aphelia just decoupled from Jupiter, and therefore are likely to be a very important step in the dynamical pathway of transition between Jupiter family comets and Near Earth asteroids.

The problem is, these are the orbits we are not yet able to model analytically in all their complexity, because the combined effects of secular resonances and close approaches to the terrestrial planets play an essential rôle in the evolution of the Oljato class orbits. Theories are available for secular resonances, especially the most important one, resulting from the argument of perihelion $\varpi = \omega + \Omega$ of the asteroid revolving around the Sun with the same average rate as the one of Saturn [24], but these theories are not consistently applicable when close approaches with the Earth can take place; numerical experiment confirm the existence of the secular resonances even where node crossing can occur [5]. In the near-Earth and near-Mars region other secular resonances, locking the asteroid perihelium and node to the ones of the terrestrial planets, are known to occur [14], [16], but the theory exists so far only for low eccentricities and non-crossing orbits. The Oljato class orbits are often found in the Taurid region [28]; large scale changes in the orbital elements, resulting from resonances, provide the entrance path into this region, with orbits coming both from the asteroid main belt and from the Jupiter family of comets, and the exit path: collision with the Sun [4], hyperbolic expulsion after a close approach to Jupiter, and less frequently collision with one of the terrestrial planets.

I would like to conclude by saying that the dynamical classification of [20] is an useful tool to understand the dynamics of the Earth-approaching objects, but it needs to be stressed that all these are strongly chaotic orbits, and that all the orbital elements can undergo large relative changes, as a result of either deep close approaches or the action of some resonance. Thus transitions can occur (see [20], Figure 16), and indeed often occur, between the dynamical classes. The Toro and Alinda classes are especially unstable, with typical residence times in the class $< 100,000$ years. As an example, transitions can occur between the Alinda, comet-like and Oljato classes; that is, an asteroid coming from some Kirkwood gap can be injected into a comet-like orbit as well as ending up as a meteorite.

Acknowledgements

The author wishes to thank A. E. Roy, P. Farinella and G.-F. Gronchi for their assistance in revising the manuscript.

References

1. Arnold, V.: 1976, *Les méthodes mathématiques da la mécanique classique*, Editions MIR, Moscou.
2. Bailey, M.E., Chambers, J.E., Hahn, G: 1992, 'Origin of sungrazers: a frequent cometary end state', *Astron. Astropys.* **257**, 315–322.
3. Bottke, W. F. Jr., Nolan, M.C., Greenberg, R., Kolvoord, R. A.: 1994, 'Collisional lifetimes and impact statistics of near-Earth asteroids', in *Hazards due to comets and asteroids*, T. Gehrels ed., Univ. of Arizona press, 337–358
4. Farinella, P., Froeschlé, Ch., Froeschlé, C., Gonczi, R., Hahn, G., Morbidelli, A., Valsecchi, G.B.: 1994, 'Asteroids falling into the Sun', *Nature* **371**, 314–317
5. Froeschlé, Ch., Hahn, G., Gonczi, R., Morbidelli, A., Farinella, P.: 1995, 'Secular

resonances and the dynamics of Mars–crossing and near-Earth asteroids', *Icarus* **117**, 45–61
6. Greenberg, R.: 1982, 'Orbital interactions: a new geometrical formalism', *Astr. J* **87**, 184–195
7. Gronchi, G.F., Milani, A.: 1998, 'Averaging on Earth-crossing orbits', submitted.
8. Henrard J.:1993, 'The adiabatic invariant in classical mechanics', *Dynamics Reported* **2**
9. Janiczek, P. M., Seidelmann, P. K. and Duncombe, R. L.: 1972, 'Resonances and encounters in the inner Solar System', *Astron. J.* **Vol. 77**, pp. 764–773
10. Kessler, D.J., Cour-Palais, B.G.: 1978, 'Collision frequency of artificial satellites: the creation of a debris belt' *J. Geophys. Res* **83**, 2637–2646
11. Kessler, D.J.: 1981, 'Derivation of the collision probability between orbiting objects: The lifetimes of Jupiter's outer moons', *Icarus* **48**, 39–48
12. Kozai, Y.: 1962, 'Secular perturbation of asteroids with high inclination and eccentricity.', *Astron.J.* **67**, 591-598
13. Michel, P., Thomas, F. C.: 1996, 'The Kozai resonance for near-Earth asteroids with semimajor axes smaller than 2 AU', *Astron. Astrophys.* **307**, 310–318
14. Michel, P., Froeschlé, Ch.: 1997, 'The location of linear secular resonances for semimajor axes smaller than 2 AU', *Icarus* **128**, 230–240
15. Michel, P., Froeschlé, C., Farinella, P.: 1996, 'Dynamical evolution of two near-Earth asteroids to be explored by spacecraft: (433) Eros and (4660) Nereus', *Astron.Astrophys.* **313**, 993–1007
16. Michel, P., Farinella, P., Froeschlé, C.: 1996, 'Dynamics of Eros', submitted.
17. Milani, A.: 1993, 'The Trojan asteroid belt: proper elements, stability, chaos and families', *Celestial Mechanics* **57** 59–94
18. Milani, A. : 1994, 'Proper elements and stable chaos', in *From Newton to chaos: modern techniques for understanding and coping with chaos in N–body dynamical systems*, A. E. Roy & B.A. Steves eds., Plenum, New York, pp. 47-78.
19. Milani, A., Baccili, S.: 1998, 'Dynamical classification of Earth-crossing orbits: the dance of the Toro asteroids', submitted.
20. Milani, A., Carpino, M., Hahn, G. and Nobili, A. M.: 1989, 'Dynamics of Planet-crossing Asteroids: Classes of Orbital Behavior', *Icarus* **Vol. 78**, pp. 212–269
21. Milani, A., Carpino, M. and Marzari, F.: 1990, 'Statistics of Close Approaches between Asteroids and Planets:', *Icarus* **Vol. 88**, pp. 292–335
22. Moons, M., Morbidelli, A. : 1995, 'Secular resonances in mean motion commensurabilities; the 4/1, 3/1, 5/2 and 7/3 cases', *Icarus* **114**, 33-50
23. Morbidelli, A., Moons, M. ; 1993, 'Secular resonances in mean motion commensurabilities; the 2/1 and 3/2 case', *Icarus* **102**, 316-332
24. Morbidelli, A.; 1993, 'Asteroid secular resonant proper elements', *Icarus* **105**, 48–66
25. Öpik, E. J.: 1951, 'Collision probabilities with the planets and the distribution of interplanetary matter', *Proc. R. Ir. Acad.* **54A**, 165-199
26. Steel, D.I., Baggaley, W.J.: 1985, 'Collisions in the solar system–I. Impacts of Apollo-Amor-Aten asteroids upon the terrestrial planets' *Mon. Not. R. Astr. Soc.* **212**, 817–836
27. Tancredi, G.: 1998, 'Chaotic dynamics of planet-encountering bodies', *Celestial Mechanics*, in press.
28. Valsecchi, G.B., Morbidelli, A., Gonczi, R., Farinella, P., Froeschlé, Ch., Froeschlé, C.: 1995, 'The dynamics of objects in orbits resembling that of P/Encke', *Icarus* **118**, 169–180
29. Wetherill, G.W. : 1967, 'Collisions in the asteroid belt', *J. Geophys. Res.* **72**, 2429–2444.
30. Whipple, A.: 1995, 'Lyapounov times of the Inner Asteroids', *Icarus* **115**, 347-353
31. Williams, J.G., Faulkner, J.: 1981, 'The position of secular resonance surfaces', *Icarus* **46**, 390–399.

ASTEROID AND COMET ENCOUNTERS WITH THE EARTH

Impact Hazard and Collision Probability

K. MUINONEN
Observatory, P.O. Box 14,
FIN-00014 University of Helsinki, Finland;
Astronomical Observatory, Box 515,
S-75120 Uppsala, Sweden

1. Abstract

While progressing in its orbit about the Sun, planet Earth encounters large numbers of small solar system bodies—asteroids, comets, and meteoroids. In the course of time, some of these small bodies collide with the Earth and are destroyed in the atmosphere, while some prevail until a catastrophic impact on the surface. There is vast agreement among scientists that the interplay between the Earth and the small bodies poses a significant, natural hazard for the humankind. The near–Earth asteroids and comets have affected the evolution of life on the Earth in the past. It is relevant to work toward preventing the asteroids and comets from colliding with the Earth and, at the same time, toward protecting the humankind from the cosmic impacts. Here the populations of near–Earth asteroids and comets and their impact hazard are reviewed. The review is succeeded by Gaussian and rigorous methods for the calculation of the collision probability of a large and a small body, and for the derivation of upper bounds. Finally, the close approach to the Earth in year 2028 by the small body 1997 XF_{11} is analysed in detail starting from its discovery in 1997 Dec. 6.

2. Introduction

Near–Earth objects (NEOs) are asteroids, comets, and meteoroids with perihelion distances less than 1.3 AU and aphelion distances larger than 0.983 AU. The lifetimes of NEOs are typically 10–100 Ma and thus considerably shorter than 4.6 Ga, the age of the solar system. NEOs are supplied by various regions of the solar system, such as the main belt of asteroids through mutual collisions among its members with subsequent transport

B.A. Steves and A.E. Roy (eds.), The Dynamics of Small Bodies in the Solar System, 127–158.

to orbital resonances with Jupiter, and the Oort cloud of comets through galactic tides and encounters with nearby stars and massive molecular clouds. NEOs are removed by impacts onto the Sun and the terrestrial planets, tidal disruptions, and dynamical interactions with Jupiter. Asteroids, comets, and meteoroids provide insight into the origin of the planetary system, and are important resources of near–Earth space.

The lunar landscape is saturated by impact craters with sizes varying from submicrometers to hundreds of kilometres; that is, more than 12 orders of magnitude. Up to date, 160 impact craters have been identified on the Earth (Grieve 1997), and approximately 30 % of these are buried and thereby unresolvable on the ground. Examples of craters are, in the order of increasing age,

- Kaalijärvi, Estonia; diameter 0.11 km, age 4000 ± 1000 yr,
- Meteor Crater, Arizona, U.S.; diameter 1.2 km, age 50000 yr,
- Chicxulub, Yucatan, Mexico; diameter > 180 km, age 65 Myr,
- Lappajärvi, Finland; diameter 23 km, age 77 Myr.

The mass extinction at the K–T boundary between the Cretaceous and Tertiary periods some 65 million years ago was caused by an impact of a large comet or asteroid (Alvarez et al. 1980), the Chicxulub crater being identified as one of the craters connected to the impact event.

Near–Earth asteroids, comets, and meteoroids are reviewed in a considerable number of recent books and papers, and it is evident that not all of them can be presently acknowledged. Lewis et al. (1993), Gehrels (1994), and Remo (1997) have edited volumes entitled *Resources of Near–Earth Space*, *Hazards due to Comets and Asteroids*, and *Near–Earth Objects—The United Nations International Conference*, respectively, that thoroughly collect and summarise most of the NEO research topics.

As for near–Earth asteroids, Rabinowitz et al. (1994) describe their population in terms of their orbital elements, sizes, and physical properties, whereas Milani et al. (1989) provide dynamical classification and Bottke et al. (1994) consider their collisional lifetimes and impact statistics. As for near–Earth comets, Shoemaker et al. (1994) concentrate on the flux of periodic comets near Earth, whereas Bailey et al. (1994) address the hazards due to giant comets with special attention to asteroids (2060) Chiron and (5145) Pholus. Weissman (1997) reviews the long–period comets and the Oort cloud, and Steel (1997) analyses cometary impacts on the biosphere. Ceplecha et al. (1997) correlate groundbased and spaceborne observations of bolides, and deduce the implications on the population of meteoroids.

Tables 1 and 2 include the closest approaches to the Earth by asteroids and comets, respectively, as analysed and published by the *Minor Planet Center*. The record closest approach by an asteroid was that by 1994 XM_1 at approximately 112000 km, whereas the closest approach by a comet was

TABLE 1. The closest asteroid Earth–approaches by 1998 May 24 as recorded by the *Minor Planet Center*. H is the absolute magnitude (Bowell et al. 1989).

Distance (AU)	Date (TDT)	Permanent designation	Provisional designation	H (mag)
0.0007	1994 Dec. 9.8		1994 XM_1	28.0
0.0010	1993 May 20.9		1993 KA_2	29.0
0.0011	1994 Mar. 15.7		1994 ES_1	28.5
0.0011	1991 Jan. 18.7		1991 BA	28.5
0.0029	1995 Mar. 27.2		1995 FF	26.5
0.0030	1996 May 19.7		1996 JA_1	20.5
0.0031	1991 Dec. 5.4		1991 VG	28.8
0.0046	1989 Mar. 22.9	(4581) Asclepius	1989 FC	20.5
0.0048	1994 Nov. 24.8		1994 WR_{12}	22.0
0.0049	1937 Oct. 30.7	(Hermes)	1937 UB	18
0.0050	1995 Oct. 17.2		1995 UB	27.5
0.0067	1993 Oct. 18.8		1993 UA	25.0
0.0069	1994 Apr. 12.1		1994 GV	27.5
0.0071	1993 May 17.9		1993 KA	26.0
0.0071	1997 Oct. 26.2		1997 UA_{11}	25.0
0.0074	1997 Feb. 9.8		1997 CD_{17}	27.5
0.0078	1976 Oct. 20.7	(2340) Hathor	1976 UA	20.26
0.0099	1988 Sept. 29.0		1988 TA	21.0

that by D/1770 L_1 (Lexell) at 2.3×10^6 km. Note that 1991 VG could be a piece of space debris rather than a natural object, and that the cometary list is complete for objects coming within 0.1020 AU after the year 1700.

Comet D/1770 L_1 Lexell, the first comet with an elliptic orbit, was transported to the outskirts of the solar system after gravitational interactions with Jupiter, and is an early, perhaps the earliest example of chaotic dynamics we have observed. What can be considered a remarkable achievement at that time, Anders Lexell succeeded in theoretically predicting a close encounter with Jupiter as well as its dramatic consequences on the orbital elements of the comet. More than two centuries has elapsed since comet Lexell's dance with Jupiter and, as underscored by the forecoming close approach to the Earth in October 2028 of the near–Earth object 1997 XF_{11}, we continue to be puzzled by the challenging problem of predicting and assessing future close approaches.

Theoretical research on the average, long–term planetary collision probability of small bodies has been carried out by, e.g., Öpik (1951, 1976),

TABLE 2. The closest cometary Earth–approaches by 1998 May 24 as recorded by the *Minor Planet Center*.

Distance (AU)	Date (TDT)	Permanent designation
0.0151	1770 July 1.7	D/1770 L_1 (Lexell)
0.0229	1366 Oct. 26.4	55P/1366 U_1 (Tempel-Tuttle)
0.0312	1983 May 11.5	C/1983 H_1 (IRAS-Araki-Alcock)
0.0334	837 Apr. 10.5	1P/837 F_1 (Halley)
0.0366	1805 Dec. 9.9	3D/1805 V_1 (Biela)
0.0390	1743 Feb. 8.9	C/1743 C_1
0.0394	1927 June 26.8	7P/Pons-Winnecke
0.0437	1702 Apr. 20.2	C/1702 H_1
0.0617	1930 May 31.7	73P/1930 J_1 (Schwassmann-Wachmann 3)
0.0628	1983 June 12.8	C/1983 J_1 (Sugano-Saigusa-Fujikawa)
0.0682	1760 Jan. 8.2	C/1760 A_1 (Great comet)
0.0839	1853 Apr. 29.1	C/1853 G_1 (Schweizer)
0.0879	1797 Aug. 16.5	C/1797 P_1 (Bouvard-Herschel)
0.0884	374 Apr. 1.9	1P/374 E_1 (Halley)
0.0898	607 Apr. 19.2	1P/607 H_1 (Halley)
0.0934	1763 Sept. 23.7	C/1763 S_1 (Messier)
0.0964	1864 Aug. 8.4	C/1864 N_1 (Tempel)
0.0982	1862 July 4.6	C/1862 N_1 (Schmidt)
0.1018	1996 Mar. 25.3	C/1996 B_2 (Hyakutake)
0.1019	1961 Nov. 15.2	C/1961 T_1 (Seki)

Wetherill (1967), and Kessler (1981). In the latter part of the present article, a collision probability is defined that is a generalisation of these works. The present treatment is based on the a posteriori probability density of the orbital elements described in Muinonen and Bowell (1993ab) and Muinonen (1996), and relates imminently to the short–term impact hazard. For the computation of the collision probability, the Sun and the planets are assumed to be finite spheres that move in deterministic orbits, whereas the small bodies are taken as massless probability densities in the phase space of the orbital elements. Collisions bring along time–irreversibility that changes the time–reversible philosophy of traditional orbit determination.

Chodas (1993) and Yeomans and Chodas (1994) have computed short–term collision probabilities in the linear approximation using spatial error ellipsoids. It will be seen that the linear approximation can provide valuable first insight into the collision probability, but nonlinear analyses are required to establish rigorous collision probabilities.

In Section 3, the near–Earth asteroid and comet populations are characterised by their numbers, sizes, and physical properties. Section 4 deals with the hazard of asteroid and comets impacts. In Section 5, methods are outlined for computing the short–term collision probability for asteroids and comets. Upper bounds are derived for the collision probability, and the methods are applied to the near–Earth object 1997 XF_{11}. Conclusions are presented in Section 6.

3. Near–Earth Objects

NEOs are asteroids, comets, and meteoroids with perihelion distances less than 1.3 AU and aphelion distances larger than 0.983 AU. Some near–Earth asteroids (NEAs) and comets (NECs) are capable of, currently or at some time in the next 10000–30000 years, intersecting the capture cross section of the Earth, and have been termed Earth–crossing asteroids (ECAs) or comets (ECCs), or collectively Earth–crossing objects (ECOs). It has been agreed that potentially hazardous objects (PHOs), asteroids (PHAs) or comets (PHCs), are those with minimum orbital intersection distances (MOIDs) less than 0.05 AU from the Earth's orbit. Only the acronyms NEO and MOID are used in the present context. Note that some near–Earth asteroids are protected from collision by orbital resonances (e.g., Wiegert et al. 1997).

A closer study of the definition of the NEO class of object reveals that, for objects observed at single apparitions only, it can be exceedingly difficult to decide whether a certain asteroid or comet is an NEO. The orbital elements of asteroids and comets are random variables obeying certain a posteriori probability densities (Muinonen and Bowell 1993). The probabilistic approach allows one to compute the probability that a certain small body belongs to the NEO population or is even potentially hazardous. In Section 5, similar techniques are described for the computation of upper bounds for the collision probability.

There are several astronomical, both professional and amateur NEO search programs that promise to enhance the NEO discovery rates significantly in the near future, fulfilling the goals of the Spaceguard Survey (Morrison 1992). The reader is refered to the WWW–pages of the *Minor Planet Center* (`http://cfa-www.harvard.edu/iau/NEO/TheNEOPage.html`) and *The Spaceguard Foundation* (`http://www.brera.mi.astro.it/SGF/`) for detailed information and links to the ongoing NEO surveys.

3.1. NEAR–EARTH ASTEROIDS

Near–Earth asteroids consist of Aten, Apollo, and Amor asteroids with the following definitions: Atens have semimajor axes $a < a_{\oplus}$ and aphelion

TABLE 3. Known Aten, Apollo, and Amor asteroids in 1998 May 24 based on the statistics by the *Minor Planet Center*. Two numbers are given for Amors: one for those with perihelia less than 1.13 AU, the other for the remaining Amors. Note that the lost Amor asteroid (719) Albert is a single–apparition asteroid.

	All	Numbered	Multi–apparition	Single–apparition
Atens	30	9	7	14
Apollos	241	67	35	139
Amors	120	38	15	67
	111	37	13	62
Σ	391	114	57	220
	502	151	70	282

distances $Q > q_\oplus$; Apollos have $a > a_\oplus$ and perihelion distances $q < Q_\oplus$; and Amors have perihelion distances $Q_\oplus < q < 1.3$ AU. For the present purposes, the semimajor axis, and perihelion and aphelion distances of the Earth are fixed at $a_\oplus = 1.000$ AU, $q_\oplus = 0.983$ AU, and $Q_\oplus = 1.017$ AU, respectively.

Table 3 shows the statistics of the 502 known near–Earth asteroids as of 1998 May 24. The known population consists of 30 Atens, 241 Apollos, and 231 Amors. Amors with perihelion distances less than 1.13 AU, roughly the perihelion distance of (433) Eros, are Earth–crossing and thereby particularly interesting. Note that more than half of the known near–Earth asteroids have been observed at single apparitions only. In 1998 May 24, E. Bowell and B. Koehn (`http://asteroid.lowell.edu/`) list 184 potentially hazardous asteroids, an increase of 35 members since 1997 June 1.

The known population of near–Earth asteroids allows, after debiasing procedures, the estimation of the true population. According to Rabinowitz et al. (1994), there are 1500, 5600, 140000, and 10^6 near–Earth asteroids larger than 1 km, 500 m, 100 m, and 50 m with perihelion distances less than 1.13 AU, respectively. These model numbers agree within a factor of two, down to tens of meters diameter, with the results of Spacewatch and other surveys.

The near–Earth asteroids are envisaged to be irregularly shaped, as shown by groundbased radar observations and the subsequent elaborate modeling (e.g., Hudson and Ostro 1994). According to the hypothesis by Muinonen (1998) and subsequent estimation of statistical parameters by Muinonen and Lagerros (1998), the shapes of small solar system bodies can

TABLE 4. Known near–Earth short–period and long–period comets (Marsden and Williams 1997). The first rows correspond to comets with perihelion distances $q < 1.017$ AU, whereas the second rows show the numbers for 1.017 AU $< q <$ 1.3 AU.

	All	Numbered	Single–apparition
Jupiter–family	15	11	4
	19	13	6
Halley–type	13	8	5
	6	2	4
Long–period	457	0	457
	97	0	97
Σ	485	19	466
	122	15	107

be described, in the first approximation, by lognormal statistics (Gaussian random sphere). Most near–Earth asteroids are probably dark C–type asteroids, brighter S–type asteroids, or dormant cometary nuclei (Luu and Jewitt 1989) with rotational periods varying from a few hours to several days. *The Spaceguard Foundation* (`http://www.brera.mi.astro.it/SGF/`) and the European Asteroid Research Node (`http://earn.dlr.de`) provide access to various databases of near–Earth asteroids.

3.2. NEAR–EARTH COMETS

Near–Earth comets are divided into short–period (or periodic) comets with orbital periods $T_{\text{orb}} < 200$ a and long–period comets with $T_{\text{orb}} > 200$ a. Furthermore, the short–period comets are either Halley–type comets with 20 a $< T_{\text{orb}} <$ 200 a or Jupiter–family comets with $T_{\text{orb}} < 20$ a. It is probable that many of the near–Earth asteroids described earlier are either dormant or extinct periodic comets.

Table 4 shows the statistics of known near–Earth short–period and long–period comets by Marsden and Williams (1997). There are 15 Jupiter–family, 13 Halley–type, and 457 long–period comets that can come or have come within the Earth's aphelion distance. In addition, there are 19 Jupiter–family, 6 Halley–type, and 97 long–period comets with 1.017 AU $< q <$ 1.3 AU. The number of near–Earth comets totals 607. Minimum orbital intersection distances are currently not available for cometary NEOs.

According to Shoemaker and Wolfe (1982), the flux of Earth–crossing comets can be described by the power law ($[N] = \mathrm{a}^{-1}$)

$$N(D) \quad = \quad N(1\ \mathrm{km}) \left(\frac{D}{1\ \mathrm{km}} \right)^{-1.97} . \qquad (1)$$

As for long–period comets, based on Weissman (1991) and on Shoemaker and Wolfe (1982), Bowell and Muinonen (1994) estimated that $14\ \mathrm{a}^{-1} < N(1\ \mathrm{km}) < 44\ \mathrm{a}^{-1}$ and $N(1\ \mathrm{km}) \approx 184\ \mathrm{a}^{-1}$, respectively. These numbers reflect the enormous uncertainties in relating the apparent brightness of a comet to the size of its nucleus. From Shoemaker et al. (1994) and Bowell and Muinonen (1994), about 5 active Jupiter–family comets and 2 active Halley–type comets cross the Earth's orbit per year.

Most comets are thought to be loosely bound, icy dirt–balls or dirty iceballs (Whipple 1950). Comet 1 P/Halley is so far the only near–Earth object ever examined by spacecraft. The images obtained by the cameras on board the European Space Agency's Giotto spacecraft revealed an irregularly shaped nucleus with several jets emerging from the sunlit side of the nucleus (Keller et al. 1986).

4. Asteroid and Comet Impact Hazard

The impact hazards due to asteroids and comets have been reviewed by Chapman and Morrison (1994) and Morrison et al. (1994). They evaluate and categorise the hazard by introducing classes in which the impact consequences vary from harmless atmospheric explosions to mass extinctions:

1. bolides or upper atmospheric explosions,
2. land impacts causing local or regional devastation,
3. ocean impacts capable of resulting in regional disasters,
4. global ecological catastrophes,
5. mass extinctions.

The impact hazard classes 1–5 describe the overall, qualitatively different characteristics of impact events, but large uncertainties are still involved in the determination of threshold impactor energies. Morrison et al. (1994) describe the environmental effects in more detail as a function of the impactor size and explosive energy.

To quantify the frequency of impacts in different classes, Chapman and Morrison (1994) and Morrison et al. (1994) give average time intervals between impacts. Such averages hardly ever coincide with the true intervals between two impacts. Instead, a characteristic time scale of impacts, an order of magnitude estimate of the interval between two impacts, is utilised. Furthermore, a cumulative number of impacts is given for a reference time interval $T_{\mathrm{ref}} = 1$ Ga long enough to include a statistically meaningful number of impacts in all classes above. The predicted numbers of impacts randomly timed within T_{ref} can depend on the starting epoch, much like the orbital elements of small bodies changing due to gravitational interactions.

Consider next the five impact hazard classes defined above. Class–1 bolides are harmless, unless they are misinterpreted as nuclear explosions during political crises. The cumulative number of class 1–5 impacts in 1 Ga is approximately 2×10^7 and strongly dominated by class–1 bolides from 10–20 m small bodies. The characteristic time scale is 10 a. The TNT–equivalent explosive energy as per event is typically larger than 1 MT. For comparison, the Hiroshima atomic bomb corresponded to a 12–kT TNT–explosion.

The 15–MT Tunguska explosion in 1908, though having occurred in the Earth's atmosphere, belongs to the impact hazard class 2 that would cause local or regional devastation. Such impacts, either land impacts or atmospheric explosions above land areas, would destroy of the order of 2000 km^2 area in the vicinity of the explosion. The cumulative number of class 2–5 impacts within the reference time interval, dominated by the class–2 impacts, would be 2×10^6, and the small–body sizes fall into the range of 20–100 m. The characteristic time scale is 10^2 a. The numbers of similar impact events as per the entire surface of the Earth and the urban areas only are 5×10^6 and 10^5, respectively, with characteristic time scales of 10^2 a and 10^4 a.

The impact hazard class 3 includes regionally–devastating impacts that would destroy areas equivalent to entire countries. The primary destruction mechanism would be the tsunami so coastal areas would be particularly endangered. The cumulative number of class 3–5 impacts within the reference interval is 10^4, the characteristic time scale being 10^5 a. Class–3 impact hazard would already cause a moderate risk relative to other natural disasters. The TNT–equivalent explosive energy as per event would be typically 10^4 MT, and the impactor size is 100 m–1 km.

The most significant hazard comes from the class–4 impacts that would cause global ecological catastrophes. By definition, one quarter of the population would be killed in the impact event, whose energy threshold would be close to 10^6 MT of TNT. At the current rate, 1000–2000 global ecological catastrophes would occur as per 1 Ga, and the impactor size is about 2 km. The characteristic time scale is 10^6 years. The risk due to global ecological catastrophes is of the same order as to natural hazards like earthquakes or severe storms. However, class–4 impact catastrophes are capable of destroying the entire human civilisation.

The primary global effects of class–4 impacts would be caused by dust and smoke transported into the stratosphere and, subsequently, distributed around the entire planet by atmospheric circulation. The fine dust particles would remain in the atmosphere for months, blocking sunlight and reducing

TABLE 5. Average annual risks of death in U.S.A. and Canada (Morrison 1997, Chapman and Morrison 1994).

Probability (10^{-6} a^{-1})	Hazard type
300	accidents (excluding motor vehicle)
200	homicide and suicide
160	motor vehicle accidents
10	fire
5	electrocution
1	airplane accidents
1	impacts by small bodies, classes 1–4
0.3	storms and floods
0.1	earthquakes
0.1	impacts by small bodies, classes 1–3
0.01	impacts by small bodies, classes 1–2
< 0.01	nuclear accidents, design goal

temperatures on the surface. The human mortality would mostly be due to starvation and disease, the former caused by a world–wide crop loss.

The class–5 impact events would cause mass extinctions and, in case such an event is ever allowed to occur, would kill almost the total human population. The predicted cumulative number of such $> 10^8$–MT impact events as per the reference time interval is 10–100 with the characteristic time scale of 10^7 a, and the impactor size range is 10–15 km. An example is the K–T extinction event, which destroyed more than 99 % of the biomass 65 Ma ago.

Table 5 compares the impact hazard to other human hazards using U.S. and Canadian statistics (Morrison 1997, Chapman and Morrison 1994). The impact hazard risk is currently much smaller than the risk of dying, for example, in motor vehicle accidents, but of the same order as risks caused by other natural hazards like storms, floods, and earthquakes. It is worth noting that the statistics are poor for earthquakes, and practically non–existing for impacts.

To conclude, the impact hazard is a low–probability high–consequence hazard that, unlike other natural hazards, can endanger the survival of the human civilisation. However, the impact hazard also differs from other natural hazards in that it can be mitigated by, first, discovering the near–Earth objects causing the impact hazard and, second, altering the orbit of any object established to be on a collision course with the Earth.

5. Collision Probability

Theoretical and computational methods are here reviewed and further developed for the assessment of the collision probability of a small body with one or more large bodies. Different tools are required for small bodies with short and long observational time arcs, respectively, and the tools are described in the order of increasing time arc starting from the moment of discovery.

Section 5.1 defines the collision probability theoretically. In Sect. 5.2 that concerns initial orbit determination and short observational time arcs, an approximate upper bound for the collision probability is established with the help of the minimum orbital intersection distance. Section 5.3 describes a computational method applicable to orbital element probability densities that are moderately constrained in the six–dimensional phase space. In Sects. 5.2–3, Monte Carlo techniques are required for the computation of the minimum MOID probability density and the actual collision probability.

In case no reliable assessment of the collision probability is obtained using Monte Carlo analyses, upper bounds can be derived using the so–called collision orbit method in Sect. 5.4. For well–constrained orbital element probability densities, as described in Sect. 5.5, approximate upper bounds can be obtained for the collision probability using Gaussian spatial probability densities propagated through the close–approach time interval. In Sect. 5.6, the methods are applied to the near–Earth object 1997 XF_{11}, called hereafter XF_{11}.

5.1. DEFINITION

Assume that the Sun and the planets, the $J+1$ large (or target) bodies, are finite spheres moving in deterministic orbits. For the present analysis, the small (or projectile) body can be located either in free space (state A) or on the finite spheres (state B), but not inside the spheres. Collision refers to the small body crossing the surface of any one of the finite spheres.

In orbit determination (e.g., Muinonen and Bowell 1993a), both random and systematic errors in astrometric observations introduce uncertainties into the orbital elements. The systematic errors can be due to reference star catalogue errors, timing errors, non–gravitational forces, or even unknown close encounters among small bodies. Nevertheless, it is here assumed that the observations have been corrected for systematic errors so the following analysis concerns the random errors.

Denote the osculating Keplerian orbital elements of the small body at epoch t_0 by

$$\boldsymbol{P} \;=\; (a, e, i, \Omega, \omega, M_0)^T, \tag{2}$$

where a is the semimajor axis, e the eccentricity, i the inclination, Ω the longitude of ascending node, ω the argument of perihelion, and M_0 is the mean anomaly. The three angular elements ω, Ω, and i are referred to the ecliptic at a specified equinox (here J2000.0).

The random observational errors introduce statistical uncertainties into the orbital elements. Let $p_p(\boldsymbol{P}, t_0)$ be the a posteriori probability density of the orbital elements in the six–dimensional phase space (Muinonen and Bowell 1993a), and let the forbidden interiors of the finite spheres be taken into account by an a priori probability density implicit in $p_p(\boldsymbol{P}, t_0)$.

With the help of $p_p(\boldsymbol{P}, t_0)$, the solar–planetary collision probability of the small body on a given time interval $\tau = \{t \mid t \in [t_1, t_2]\}$ equals the integral

$$P_c(\tau) = \int_{V_c(\tau, t_0)} d\boldsymbol{P}\, p_p(\boldsymbol{P}, t_0), \qquad (3)$$

where the phase space domain $V_c(\tau, t_0)$ contains the orbital elements leading to a collision on the interval τ. Collisions are irreversible in time, so the epoch t_0 must precede τ, that is, $t_0 < t_1$.

Instead of the singular Keplerian orbital elements, the position $\boldsymbol{r}$ and velocity $\boldsymbol{v}$ at the epoch t_0 are introduced,

$$p_p(\boldsymbol{r}, \boldsymbol{v}, t_0) = \int d\boldsymbol{P}\, p_p(\boldsymbol{P}, t_0)\, \delta_D(\boldsymbol{r} - \boldsymbol{r}(\boldsymbol{P}))\, \delta_D(\boldsymbol{v} - \boldsymbol{v}(\boldsymbol{P})), \qquad (4)$$

where δ_D is Dirac's function. Since the small body can be located either in free space (A) or on the surface of a large body (B), p_p is composed of two probability densities p_A and p_B with occupation probabilities P_A and P_B,

$$\begin{aligned} p_p(\boldsymbol{r}, \boldsymbol{v}, t_0) &= P_A(t_0)\, p_A(\boldsymbol{r}, \boldsymbol{v}, t_0) + P_B(t_0)\, p_B(\boldsymbol{r}, \boldsymbol{v}, t_0), \\ P_A(t_0) + P_B(t_0) &= 1. \end{aligned} \qquad (5)$$

The small body cannot be located inside the large bodies so

$$p_p(\boldsymbol{r}, \boldsymbol{v}, t_0) = 0, \quad |\boldsymbol{r} - \boldsymbol{r}_j(t_0)| < R_j, \quad j = 0, \ldots, J, \qquad (6)$$

where $\boldsymbol{r}_j$ and R_j are the positions and radii of the large bodies. The occupation probability P_B is the sum of the occupation probabilities on the surfaces of the large bodies,

$$\begin{aligned} P_B(t_0) &= \sum_{j=0}^{J} P_{B,j}(t_0), \\ p_B(\boldsymbol{r}, \boldsymbol{v}, t_0) &= \frac{1}{P_B(t_0)} \sum_{j=0}^{J} P_{B,j}(t_0)\, p_{B,j}(\boldsymbol{R}_j/R_j, t_0) \cdot \\ &\quad \frac{1}{R_j^2}\, \delta_D(|\boldsymbol{r} - \boldsymbol{r}_j(t_0)| - R_j)\, \delta_D(\boldsymbol{v} - \boldsymbol{v}_j(t_0)), \end{aligned} \qquad (7)$$

where $\boldsymbol{R}_j$ ($j = 0, \ldots, J$) give the spherical surfaces of the large bodies with respect to their centers, and $\boldsymbol{v}_j$ are the velocities (if desired, rotation of the large bodies can be accounted for).

The total solar–planetary collision probability $P_{\rm c}(\tau)$, and the collision probabilities $P_{{\rm c},j}(\tau)$ with the individual $J+1$ large bodies equal

$$\begin{aligned} P_{\rm c}(\tau) &= P_A(t_1) - P_A(t_2) = \sum_{j=0}^{J} P_{{\rm c},j}(\tau), \\ P_{{\rm c},j}(\tau) &= P_{B,j}(t_2) - P_{B,j}(t_1), \quad j = 0, \ldots, J. \end{aligned} \tag{8}$$

Although the collision probability was defined above for an arbitrary number of large bodies, practical applications usually concern only a single large body. It is hereafter assumed that that large body is the Earth noting that the methods can be readily applied to any number of large bodies. Furthermore, if $[t_{\rm i}, t_{\rm f}]$ is the time interval of the astrometric observations of the small body, it is assumed that

$$P_A(t_0) = 1, \quad t_0 \leq t_{\rm f}; \tag{9}$$

that is, the observations enforce state A in free space. It is reasonable to assume that a collision with the Earth can be observed so that, in fact, $P_{B,\oplus} = 0$ for all epochs in the past as long as no collision has been recorded.

5.2. INITIAL UPPER BOUNDS

For small bodies with short observational arcs, the a posteriori probability density $p_{\rm p}(\boldsymbol{P}, t_0)$ can be highly complicated and drastically non–Gaussian (Muinonen et al. 1997). Orbital eccentricity can obtain values across the range of $]0.0, 1.0[$, and the semimajor axis can correlate strongly with the eccentricity. Cometary–type orbital solutions are often allowed by short observational arcs, so that the semimajor axis can reach out to several dozens of AU. According to Marsden (1991), the initial orbit determination problem is characterised by extreme indeterminacy of the orbital elements.

Although the complicated probability density of the orbital elements makes it difficult to compute, for example, the uncertainties of the orbital elements, it is of great importance to develop methods for the evaluation of the collision probability starting from the very moment of the discovery of the small body. The minimum orbital intersection distance (MOID) between two elliptic orbits (hyperbolic and parabolic orbits can be treated likewise) is the key parameter when establishing initial upper bounds for the collision probability. The utilisation of MOIDs for close encounter analyses has been suggested, in particular, by Bowell and Muinonen (1994), who presented the MOIDs of multiapparition near–Earth asteroids (known at that time) with all the planets from Mercury to Jupiter. It is important to note that MOID is computed using the osculating elements, and thus depends on the epoch.

Let $S = S(\boldsymbol{P}, \tau)$ be the *minimum* MOID for the Keplerian orbital elements $\boldsymbol{P}$ within the time interval $\tau = \{t \mid t \in [t_1, t_2]\}$. The probability density $p_{\rm p}(S, \tau)$ follows formally from $p_{\rm p}(\boldsymbol{P}, t_0)$ in a straightforward way,

$$p_{\rm p}(S, \tau) = \int d\boldsymbol{P}\, p_{\rm p}(\boldsymbol{P}, t_0)\, \delta_{\rm D}(S - S(\boldsymbol{P}, \tau)). \tag{10}$$

The integral on the right–hand side is six–dimensional for small bodies experiencing gravitational perturbations whereas, in the two–body approximation, the integral becomes independent of the mean anomaly M_0.

In practice, the computation of $p_{\rm p}(S, \tau)$ must be carried out numerically via a systematic study of $p_{\rm p}(\boldsymbol{P}, t_0)$. In the present context, the task is accomplished with a Monte Carlo technique by systematically varying the distances between the observer and the small body at two observation dates, allowing for random deviations of the observed and theoretical positions on the sky–plane, and accepting only orbital solutions with sky–plane residuals smaller than a given threshold value large enough to encompass practically all realistic orbital fits.

An initial upper bound for the collision probability can be computed from $p_{\rm p}(S, \tau)$ by determining the probability within the radius $R_\oplus$ of the large body,

$$P_{\rm c}(\tau) < \int_0^{R_\oplus} dS\, p_{\rm p}(S, \tau). \tag{11}$$

To obtain the upper bound, it is thus conservatively assumed that the closest approach distances theoretically possible would always be realised within the time interval τ. Allowing for planetary or other perturbations, the upper bound in Eq. (11) provides a fairly robust estimation of the collision probability of any small body. The definition in Eq. (11) guarantees that the upper bound cannot decrease when τ is extended to start earlier or end later.

In the two–body approximation for the orbital evolution of the small body, the upper bound in Eq. (11) will be practically independent of τ: small changes are due to the changes in the orbital elements of the large body. The two–body approximation yields approximate upper bounds with acceptable computational efforts, and can be routinely applied to, for example, all single–apparition asteroids.

Additionally, it is enlightening to compute probability bounds for close–approach distances less than the lunar mean distance $\Delta = 0.0026$ AU and less than $\Delta = 0.05$ *AU*, the threshold MOID for a small body to be potentially hazardous (Morrison 1992). These bounds are obtained from Eq. (11) by simply computing the integral for the desired Δ,

$$P_{\rm p}(\Delta, \tau) \;<\; \int_0^{\Delta} dS\, p_{\rm p}(S, \tau). \tag{12}$$

Finally, as brought up in Sect. 3, similar methods yield the probability for an object to belong to the NEO class of object.

5.3. COLLISION PROBABILITY COMPUTATION

It is evident that the rigorous computation of the two probability density components $P_A p_A$ and $P_B p_B$ in Eq. (5) at an arbitrary future epoch is difficult, notwithstanding the irreversible propagation of the probability density forward in time. However, the collision probability can be treated computationally, for example, by using Monte Carlo simulations for large numbers of sample orbits (Muinonen and Bowell 1993a).

Let $\rho = \rho(\boldsymbol{P}, \tau)$ be the impact parameter, the minimum distance of the small body from the center of the large body, for the orbital elements $\boldsymbol{P}$ within the time interval $\tau = \{t \mid t \in [t_1, t_2]\}$. The probability density $p_{\rm p}(\rho, \tau)$ follows from $p_{\rm p}(\boldsymbol{P}, t_0)$,

$$p_{\rm p}(\rho, \tau) \;=\; \int d\boldsymbol{P}\, p_{\rm p}(\boldsymbol{P}, t_0)\, \delta_{\rm D}(\rho - \rho(\boldsymbol{P}, \tau)), \tag{13}$$

and can be computed with the help of the Monte Carlo method described by Muinonen and Bowell (1993a) for both Gaussian and rigorous non–Gaussian probability densities of orbital elements. The collision probability can be computed from the impact parameter distribution,

$$P_{\rm c}(\tau) \;=\; \int_0^{R_\oplus} d\rho\, p_{\rm p}(\rho, \tau). \tag{14}$$

In a similar fashion, the impact parameter distribution allows the determination of the probabilities for, first, miss distance smaller than the lunar distance and, second, miss distance smaller than the potentially hazardous distance of 0.05 AU. The computation of the impact parameter distribution is recommendable even if the actual collision probability is too small to be reliably determined with the Monte Carlo method.

5.4. UPPER BOUNDS USING COLLISION ORBITS

At an epoch t_0 close to the observations, the Gaussian probability density often sufficiently accurately describes the a posteriori probability density

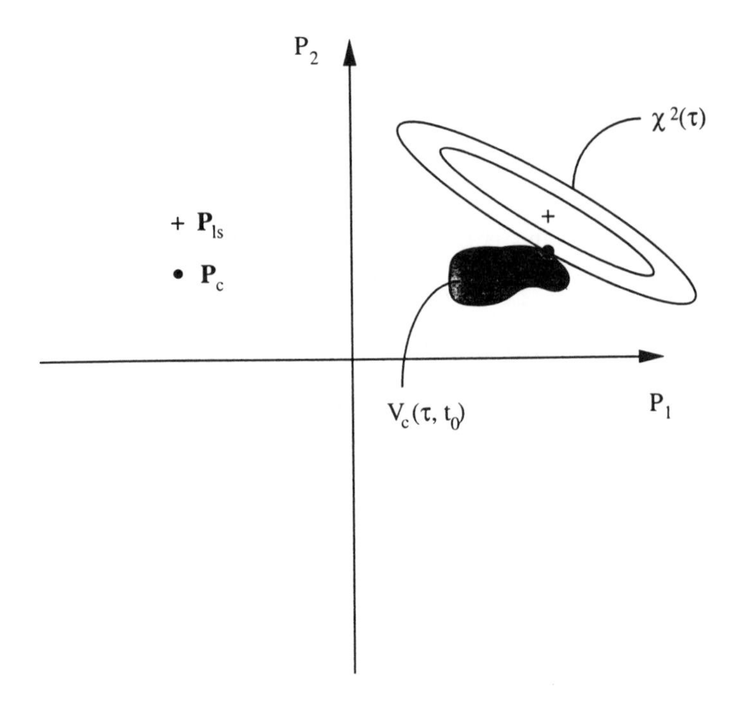

Figure 1. Upper bound of the collision probability for a Gaussian probability density of orbital elements. For simplicity, let P_1 and P_2 stand for the orbital elements at epoch t_0. Let $\tau = \{t \mid t \in [t_1, t_2]\}$ be the time interval for collision probability computation, and let $\boldsymbol{P}_c$ be the orbit that lies closest (on the $\chi^2(\tau)$ confidence ellipse) to the least—squares orbit $\boldsymbol{P}_{ls}$ and still results in collision within τ. The collision probability on τ must be smaller than the probability outside the $\chi^2(\tau)$ confidence ellipse (the survival ellipse).

of the orbital elements. For small collision probabilities that cannot be efficiently analysed using the Monte Carlo –method in Sect. 5.3, upper bounds can be established using collision orbits. Such bounds can be established for both Gaussian and rigorous non–Gaussian probability densities.

Given the time interval τ for the collision study, collision–yielding orbital elements can be searched in the neighborhood of the least–squares orbit at epoch t_0 close to the observation dates. Once such orbital elements are found, one can further search for the collision orbit that minimises the residuals of the least–squares fit or, in other words, lies probabilistically closest to the maximum likelihood orbital solution. In the linear approximation, the method entails the computation of the collision orbit that lies closest to the least–squares elements $\boldsymbol{P}_{ls}$ in terms of the covariance matrix Σ (Fig. 1).

Denote the observational residuals by $\Delta\psi$, and the covariance matrix of the observations by Λ (Muinonen and Bowell 1993a). The maximum likelihood collision orbit yields the metric

$$\begin{aligned}
\chi^2(\tau) &= \min_{\boldsymbol{P}\in V_c(\tau,t_0)} \left[\Delta\psi^T(\boldsymbol{P})\Lambda^{-1}\Delta\psi(\boldsymbol{P}) - \Delta\psi^T(\boldsymbol{P}_{\rm ls})\Lambda^{-1}\Delta\psi(\boldsymbol{P}_{\rm ls})\right] \\
&\approx \min_{\boldsymbol{P}\in V_c(\tau,t_0)} \Delta\boldsymbol{P}^T(t_0)\Sigma^{-1}(t_0)\Delta\boldsymbol{P}(t_0), \\
& \quad \Delta\boldsymbol{P}(t_0) = \boldsymbol{P} - \boldsymbol{P}_{\rm ls}(t_0),
\end{aligned} \tag{15}$$

where the observational residual term corresponding to the least–squares orbital solution has been included for convenience. Thereafter, an upper bound for the collision probability in τ follows from the computation of the probability outside *the survival ellipsoid* defined by the confidence boundary $\chi^2(\tau)$ of the closest collision orbit. What is exactly true for a Gaussian probability density, we obtain

$$\begin{aligned}
P_{\rm c}(\tau) &\leq Y_6(\chi(\tau)), \\
Y_6(\chi(\tau)) &= \frac{1}{8}\left[(\chi^2(\tau)+2)^2+4\right]\exp\left[-\frac{1}{2}\chi^2(\tau)\right],
\end{aligned} \tag{16}$$

a result that is remarkably simple. For a rigorous non–Gaussian probability density, Eq. (16) can be taken to be approximately true: extensive analyses of the probability density would be required to improve the accuracy of the upper bound. If the collision orbit has been computed by accounting for the perturbations by the large bodies, the upper bound in Eq. (16) fully accounts for the nonlinearities present in the propagation of the orbital elements from t_0 forward through τ. Close encounters can be characterised with the superexponential metric

$$X_6 = \log_{10}(-\log_{10} Y_6), \tag{17}$$

so that

$$P_{\rm c}(\tau) \leq 10^{-10^{X_6}}. \tag{18}$$

The least–squares and collision orbits allow a straightforward computation of the upper bounds. The survival ellipsoid can be computed numerically, for example, with the help of the downhill simplex minimisation method both for Gaussian and non–Gaussian probability densities. The phase space domain of collision orbits can be complicated, perhaps not even mathematically single–connected, so caution is advisable when searching for the global extremum.

Theoretically, the metric X_6 can assume formal values within $]-\infty, \infty[$. The most relevant domain of values surrounds $X_6 \approx 0$, in which case collision can neither be ascertained or ruled out. Depending on the orbital element probability density, for sufficiently large X_6–values, the Gaussian assumption of the confidence boundary breaks down. In such occasions, the collision probability will typically be extremely low, and no further collision assessment is then needed.

5.5. APPROXIMATE BOUNDS FOR GAUSSIAN DISTRIBUTIONS

Upper bounds of the collision probability can be approximated largely analytically for small bodies, whose three–dimensional spatial a posteriori probability densities within τ are well described by the Gaussian distribution (Muinonen and Bowell 1993b, Muinonen 1996). In comparison to the collision orbit method above, the spatial method is considerably more approximate: it assumes, first, a Gaussian a posteriori probability density of the orbital elements at epoch t_0 and, second, linear propagation of the probability density through the interval τ. Nevertheless, the method is fast, can yield important first insight into the collision probability computation, and yields a true upper bound for the approximate collision probability computed in the linear approximation (Chodas 1993, Yeomans and Chodas 1994).

Recall that the Earth is a finite sphere, and conceptualise the small body as a spatial probability ellipsoid (Fig. 2). Compute the probability inside the unambiguous spatial ellipsoid that touches the finite sphere. Within τ, compute the touching ellipsoid with minimum probability. In other words, look for the minimum–probability ellipsoid (again the survival ellipsoid) that touches the surface of the finite sphere,

$$\begin{aligned} \chi^2(\tau) &= \min_{\Omega_\oplus, t \in \tau} \Delta \boldsymbol{r}(\Omega_\oplus, t)^T \Sigma_{\boldsymbol{r}}^{-1}(t) \Delta \boldsymbol{r}(\Omega_\oplus, t) \\ \Delta \boldsymbol{r}(\Omega_\oplus, t) &= \boldsymbol{r}_\oplus(t) + \boldsymbol{R}_\oplus(\Omega_\oplus, t) - \boldsymbol{r}_{\mathrm{ls}}(t), \end{aligned} \tag{19}$$

where $\boldsymbol{r}_{\mathrm{ls}}$ and $\Sigma_{\boldsymbol{r}}$ denote the least–squares position and covariance matrix of the small body, and $\Omega_\oplus$ denotes a location on the spherical surface.

Conservatively, assume next that all the probability outside the survival ellipsoid corresponds to a collision with the finite sphere. The probability constrained by the three–dimensional $\chi^2(\tau)$–ellipsoid can be calculated analytically, and its complement appears on the right–hand side of the following equation that gives an approximate upper bound for the collision probability:

$$P_{\mathrm{c}}(\tau) \lesssim Y_3(\chi(\tau)),$$

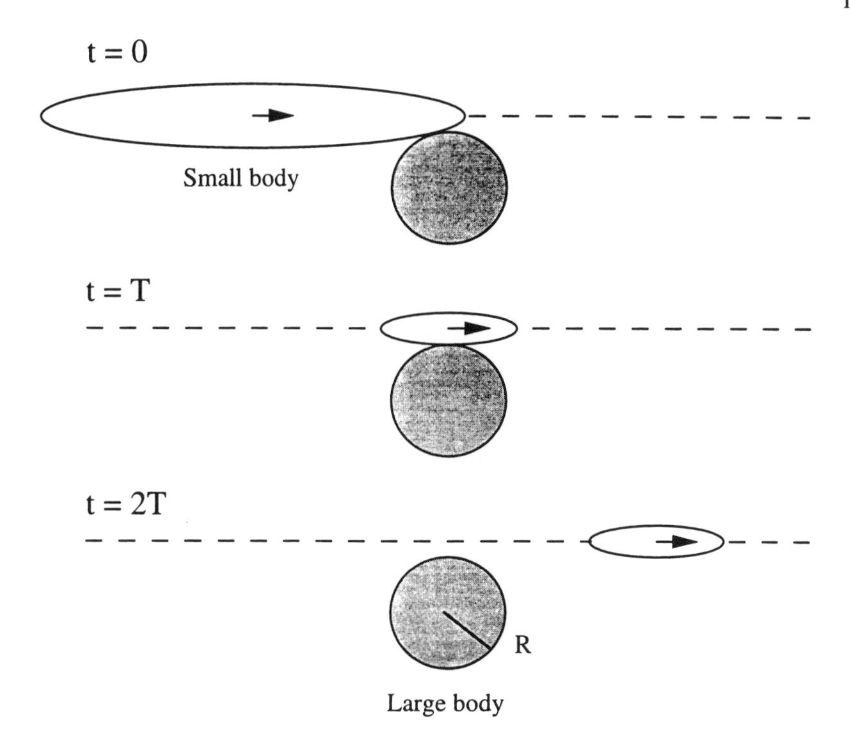

Figure 2. Close encounter between the probability ellipse of the small body and the spherical large body. In the linear approximation, an approximate upper bound for the collision probability on $\tau = \{t|\, t \in [0, 2T]\}$ can be obtained by computing the ellipse that survives the encounter after touching the large body.

$$Y_3\left(\chi(\tau)\right) = 2Q_1\left(\chi(\tau)\right) + \sqrt{\frac{2}{\pi}}\chi(\tau)\exp\left[-\frac{1}{2}\chi^2(\tau)\right], \qquad (20)$$

where Q_1 is the Gaussian cumulative distribution function

$$Q_1(x) = 1 - P_1(x) = 1 - \int_{-\infty}^{x} dy\, \frac{1}{\sqrt{2\pi}}\exp\left(-\frac{1}{2}y^2\right). \qquad (21)$$

Note the intriguing feature of multivariate Gaussian statistics that the six–dimensional upper bound in Eq. (16) is mathematically simpler than the three–dimensional one in Eq. (20). For $\chi(\tau) >> 1$, that is, for statistically distant encounters between the small body and the large body,

$$Y_3\left(\chi(\tau)\right) \approx \sqrt{\frac{2}{\pi}}\exp\left[-\frac{1}{2}\chi^2(\tau)\right]\left[\chi(\tau) + \frac{1}{\chi(\tau)}\right].$$

$$\approx \sqrt{\frac{2}{\pi}}\chi(\tau)\exp\left[-\frac{1}{2}\chi^2(\tau)\right]. \tag{22}$$

Close encounters can be characterised with the metric

$$X_3 = \log_{10}(-\log_{10} Y_3), \tag{23}$$

in which case

$$P_c(\tau) \lesssim 10^{-10^{X_3}}. \tag{24}$$

X_3 is closely related to the encounter metric presented by Muinonen and Bowell (1993b) and Muinonen (1996). Its advantage is the simple relation to the approximate bound of the collision probability.

5.6. COLLISION PROBABILITY OF 1997 XF_{11}

In the orbital analysis, DE–405 positions and velocities (Standish et al. 1995) are used for the planets with the Earth and Moon as separate bodies, while the perturbations due to the largest asteroids (1) Ceres, (2) Pallas, and (4) Vesta are excluded. Parts of the `OrbFit 1.1` free orbit determination software package are used for the orbital analyses. `OrbFit` derives from the work by Milani et al. (1995), Carpino and Knězević (1996), and Muinonen and Bowell (1993a) and, currently, `OrbFit 1.6.0` is available at `ftp://copernico.dm.unipi.it/pub/orbfit/`.

XF_{11} received public attention, when its close approach to the Earth in 2028 was published in *IAUC No. 6837* (International Astronomical Union Circular). Soon thereafter, archive observations of XF_{11} were located on photographic films in 1990, and the subsequent new orbit in *IAUC No. 6839* appeared to rule out the collision with the Earth. The case of XF_{11} was addressed in both *Science* (**279**, 1843, 1998) and *Nature* (**392**, 215, 1998; **392**, 639, 1998). More information on the early developments around XF_{11} was published in *IAUC No. 6879.*

As of 1998 May 24, there were 105 astrometric positions observed for XF_{11} at 13 different observing sites around the world. XF_{11} was discovered in 1997 Dec. 6 by Spacewatch, and the crucial follow–up observations succeeded on Dec. 8–9, allowing further observations from 1997 Dec. 18 to 1998 March 4. On 1998 March 12, March and April 1990 observations were located on films taken at Palomar Mountain Observatory.

The collision probability is analysed based on five gradually growing sets of observations described in Table 6. Set I contains 10 positions on three nights close to the discovery date, and leads us to study the difficult problem of initial orbit determination and collision assessment based on highly indeterminate orbital elements. Set II includes 20 observations with

TABLE 6. The observation sets I–V for the near–Earth asteroid XF_{11} (*Minor Planet Center*). Here t_i and t_f are the starting and ending dates of the observations, T_{arc} is the time arc of the observations, N_{tot} and N_{obs} are the total number of Right Ascension (α) and Declination (δ) observations and observations included in orbit determination, respectively, σ_α and σ_δ are the rms–values for α and δ, and β_Λ is the slippage factor due to unacceptable correlations and systematics in the residuals.

Set	t_i (UTC)	t_f (UTC)	T_{arc} (d)	N_{tot}	N_{obs}	$\sigma_\alpha('')$	$\sigma_\delta('')$	β_Λ
I	97–12–06.47	97–12–09.69	3.2	10	10	0.17	0.29	1.33
II	97–12–06.47	97–12–21.65	15.2	20	16	0.45	0.55	1.27
III	97–12–06.47	98–02–04.36	59.9	94	88	0.49	0.45	1.39
IV	97–12–06.47	98–03–04.17	87.7	98	92	0.51	0.47	1.52
V	90–03–22.44	98–03–04.17	2903.7	105	102	0.59	0.60	1.74

TABLE 7. The least–squares orbital element sets I–V with 1–σ slippage–corrected uncertainties (in units of the last digit shown, computed in the linear approximation) for the near–Earth asteroid XF_{11} based on the observations described in Table 6. The epoch is 1998–01–15.0 TDT (JD 2450828.5), and the coordinate system is J2000.0. We give the $\Delta\chi^2$ values and the complementary probability percentages Q pertaining to the six–dimensional confidence ellipsoids, on which the set–V orbital elements lie.

	Set I	Set II	Set III	Set IV	Set V
a (AU)	1.325	1.4416(14)	1.441728(17)	1.441760(15)	1.44172206(62)
e	0.536	0.48303(96)	0.483728(31)	0.483795(17)	0.4837555(43)
i (°)	4.285	4.0902(51)	4.09454(12)	4.0948(71)	4.094792(22)
Ω (°)	214.462	214.077(30)	214.1255(17)	214.1275(17)	214.13088(54)
ω (°)	87.810	102.58(24)	102.4759(21)	102.4713(16)	102.46692(56)
M_0 (°)	139.207	112.59(33)	112.6179(28)	112.6148(30)	112.6217(11)
$\Delta\chi^2$	8.4×10^8	14.49	28.63	16.94	0.00
Q (%)	$10^{-10^{8.3}}$	2.46	0.00715	0.95	100.00

an observational arc of little more than two weeks. Set III consists of 94 1997–1998 observations, excluding four observations in March 1998. Set IV contains all 98 1997–1998 observations, spanning from 1997 Dec. 6 to 1998 March 4. Based on set–IV observations, *IAUC No. 6837* was issued with a suggestion of an intriguingly close approach to the Earth in 2028 Oct. 26. Finally, set V includes all 105 1990–1998 observations.

Initial orbit determination using Gauss's method succeeded for set I, and the subsequent differential correction yielded least–squares orbital elements for all observation sets I–V. Table 7 presents the elements corresponding

TABLE 8. Set–V correlation matrix for XF_{11}. Epoch as in Table 7.

	a	e	i	Ω	ω	M_0
a	1.000000	0.958099	0.080818	-0.037035	0.187977	-0.614421
e		1.000000	0.224101	-0.161194	0.042716	-0.367209
i			1.000000	0.168311	-0.620557	0.357528
Ω				1.000000	-0.551409	-0.331031
ω					1.000000	-0.498416
M_0						1.000000

to each observation set. Also shown are the 68.3 % orbital uncertainties computed for the set II–V orbital elements in the linear approximation. In that approximation, the six–dimensional Gaussian probability density of the orbital elements is fully described by the least–squares orbital elements and their covariance matrix. For example, Table 8 includes the set–V correlation matrix: the semimajor axis and eccentricity are rather strongly correlated.

The validity of the linear approximation can be evaluated by computing a $\Delta\chi^2$–metric between the set–V least squares orbital elements (the most accurate elements) and those of the other sets, by using the covariance matrices of the latter sets. Table 7 gives the $\Delta\chi^2$–values and complementary probabilities outside the confidence ellipsoid defined by the set–V orbital elements. Immediately, the covariance matrix method fails to describe the uncertainties of the set–I orbital elements: statistically, the elements V lie far away from the elements I. The failure is due to the complicated, non–Gaussian structure of the rigorous probability density of the elements (Muinonen et al. 1997). The set–V elements lie within the 99.7 % probability ellipsoids of the set–II and set–IV elements, but not within the corresponding probability ellipsoid of the set–III elements. To conclude, the assumption of the linear approximation is unjustified, and a rigorous non–Gaussian treatment is required (Muinonen and Bowell 1993a).

As for observation set I, a systematical study of the phase space of the orbital elements was carried out by varying the topocentric distances between 0.45 AU and 3.45 AU for the first and last set–I observations, and randomly sampling orbits with Right Ascension and Declination residuals less than 4.0 arcseconds. The study revealed that the set–I observations allow practically arbitrary semimajor axes larger than 1.2 AU, eccentricities larger than 0.35 and, since also any inclination larger than 3.6° is possible, both prograde and retrograde orbits fit the observations. The lower bound on the eccentricity immediately suggests that the object be a potential

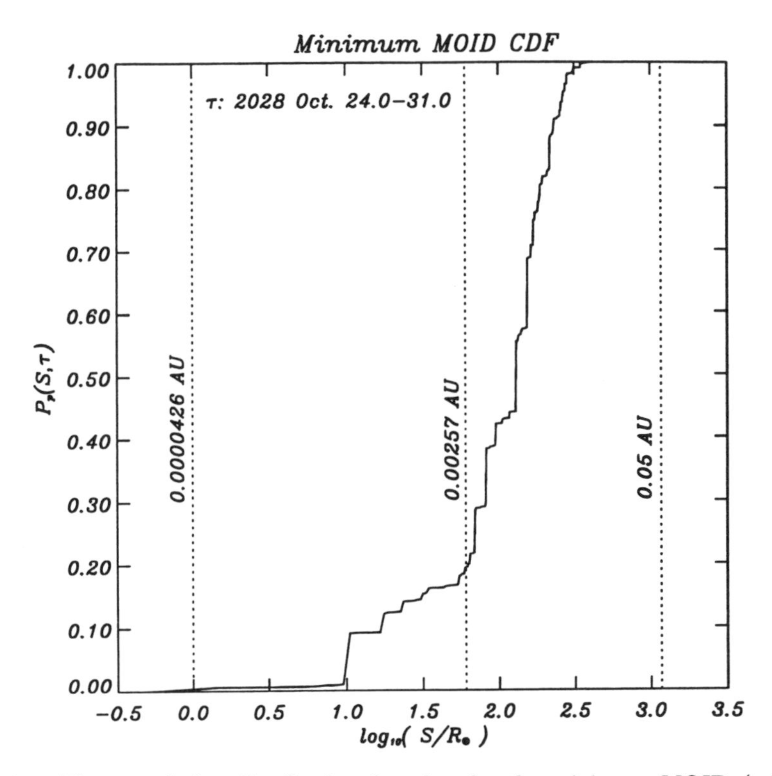

Figure 3. The cumulative distribution function for the minimum MOID (minimum orbital intersection distance, Earth radius $R_\oplus$ as unit distance) between the asteroid XF_{11} and the Earth (assumed to move in a circular orbit) based on orbit solution I consisting of 34950 two–body orbits. The Earth radius, lunar mean distance, and distance of 0.05 AU are illustrated by the dotted vertical lines.

planet–crosser.

Figure 3 shows the minimum MOID cumulative distribution function (Eq. 10) for the small body on a logarithmic geocentric distance scale with the Earth radius as the distance unit. Altogether 34950 orbits, computed as described above and in Sect. 5.2, were used in the estimation of the minimum MOID distribution. The distribution shows that the asteroid stood an almost 100 % probability for being potentially hazardous, a maximum 20 % probability for coming within the lunar distance, and a maximum 0.7 % probability for collision with the Earth some time in the future. These numbers suggest that XF_{11} be a potentially hazardous small body and an exceptionally interesting object already in early December 1997.

In Fig. 4a, a Monte Carlo simulation is shown for the geocentric distance of XF_{11} in 2025–2035 using 1000 sample orbits based on the rigorous orbit solution II. Though the geocentric distance is mostly confined

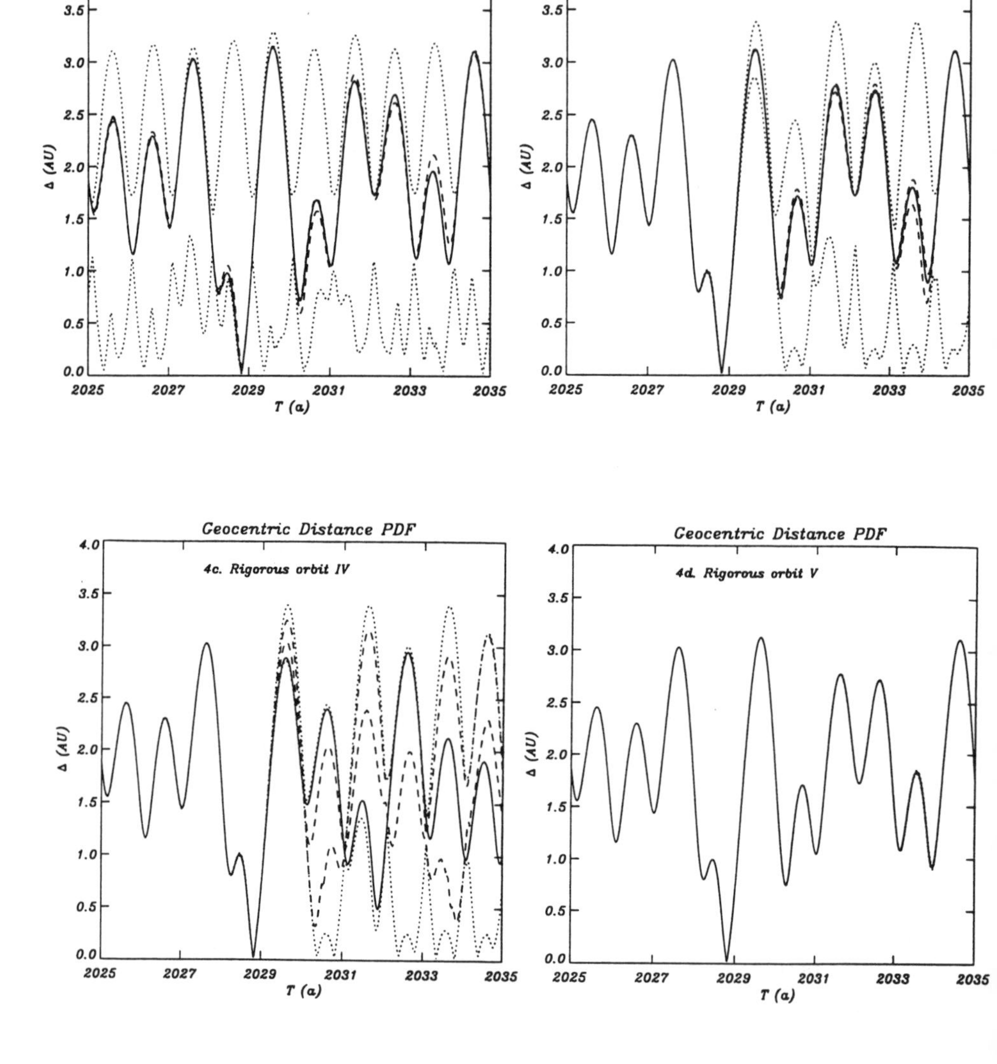

Figure 4. The probability densities for the geocentric distance of XF_{11} across the time interval 2025–2035. The solid line denotes the nominal least–squares prediction, the dashed lines bracket the central 68.3 % of the probability densities, and the dotted lines give the full ranges of geocentric distances of the 1000 Monte Carlo sample orbits. Predictions based on rigorous non–Gaussian orbit solutions (a) II, (b) III, (c) IV, and (d) V (Tables 6–7).

TABLE 9. The X_6 and X_3–metrics (Eqs. 17 and 23) for the 2028 close approach with rms–values (σ_α^c and σ_δ^c) and extremum residuals ($\Delta\alpha_e \cos\delta$ and $\Delta\delta_e$) for R.A. and Dec.

	Set I	Set II	Set III	Set IV	Set V
X_6	-0.048	1.625	2.985	3.048	3.990
X_3	–	1.442	2.956	3.012	3.957
$\sigma_\alpha^c('')$	0.71	0.88	2.14	3.81	10.3
$\sigma_\delta^c('')$	0.77	2.43	4.06	4.28	19.3
$\Delta\alpha_e \cos\delta('')$	-1.19	3.12	5.71	15.9	-42.1
$\Delta\delta_e('')$	1.36	-4.93	9.36	9.62	-109

to a narrow range around the nominal least–squares prediction, the entire ranges of geocentric distance are usually several AU, underscoring the fact that the position of the asteroid is poorly determined in 2028. Intriguingly, a similar computation using the Gaussian probability density of the orbital elements results in wider distributions for the geocentric distance. Figures 4b and 4c show the evolution of the probability density of the geocentric distance for rigorous orbit solutions III and IV. The geocentric distance remains constrained until the 2028 close encounter that results in considerable gravitational scattering and subsequent chaotic behaviour of the geocentric distance. Finally, Fig. 4d shows that, for the entire observation set V, the gravitational scattering is greatly suppressed. For orbit solutions III–V, the geocentric distance computations based on the Gaussian and rigorous orbital element probability densities are in fair agreement with each other.

For each orbit solution II–V, Monte Carlo orbits were generated and propagated through 2028 Oct. 24.0–31.0 TDT to derive the impact parameter distributions (Eq. 13) shown in Fig. 5 both for rigorous (1000 sample orbits) and Gaussian probability densities (5000 orbits). None of the orbits led to a collision with the Earth. As evident from Fig. 5a, the Gaussian approximation fails to provide the correct impact parameter distribution in the case of observation set II. Based on the rigorous solution II (Fig. 5a), XF_{11} stood the chances of 3.8 % and 60 % for close–approach distances smaller than the lunar distance and 0.05 AU, respectively. The rigorous and Gaussian solutions III result in similar impact parameter distributions (Fig. 5b) and show a high probability of close to 100 % for a close approach within 0.05 AU, and a 13 % probability for an approach within the lunar distance. For solutions IV, the probability for an approach within the lunar distance has risen to 72 % (Fig. 5c), while it continues to be

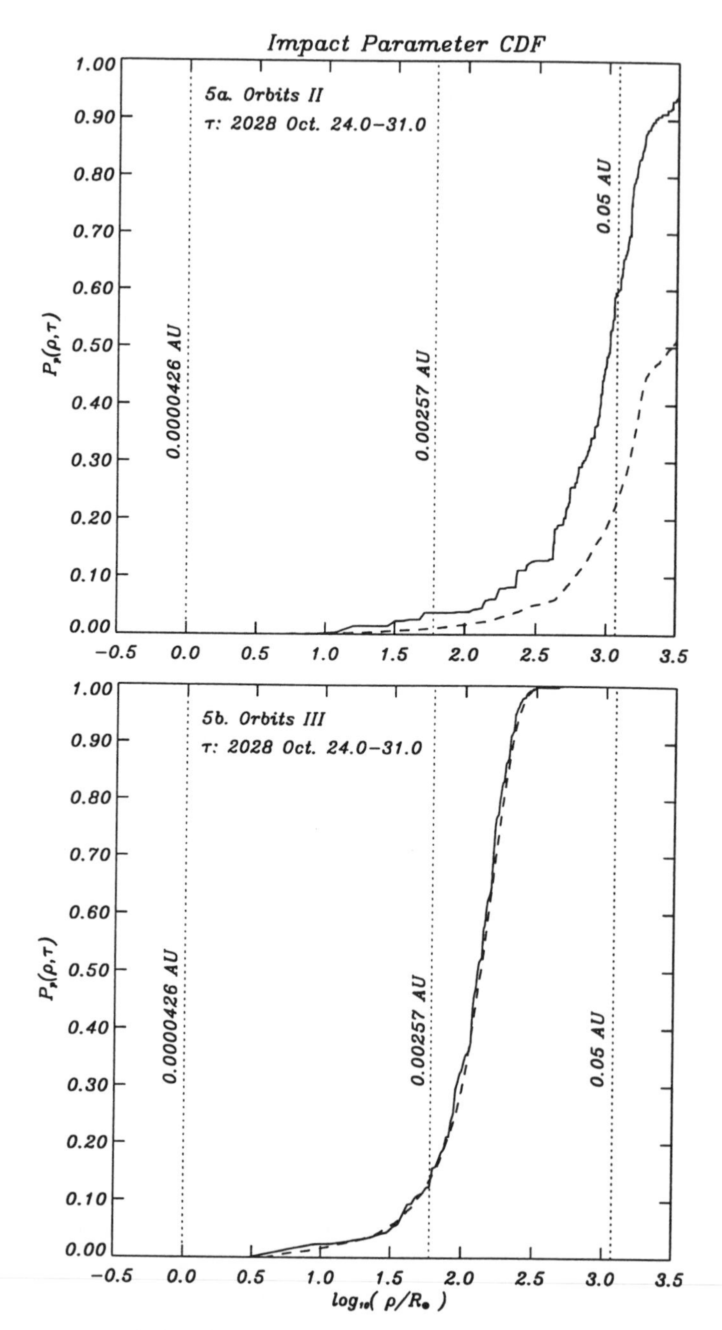

Figure 5. The cumulative distribution function for the impact parameter of XF_{11} within the time interval TDT 2028 Oct. 24.0 to 2028 Oct. 31.0. 5000 and 1000 sample orbits were generated to compute the impact parameter distributions based on the Gaussian (dashed line) and rigorous non–Gaussian orbit solutions (solid line): (a) II, (b) III, (c) IV, and (d) V (Tables 6–7).

highly probable that the miss distance is less than 0.05 AU. For solutions V, the maximum likelihood miss distance computed from the steep impact parameter distribution (Fig. 5d) is 0.00652 ± 0.00011 AU.

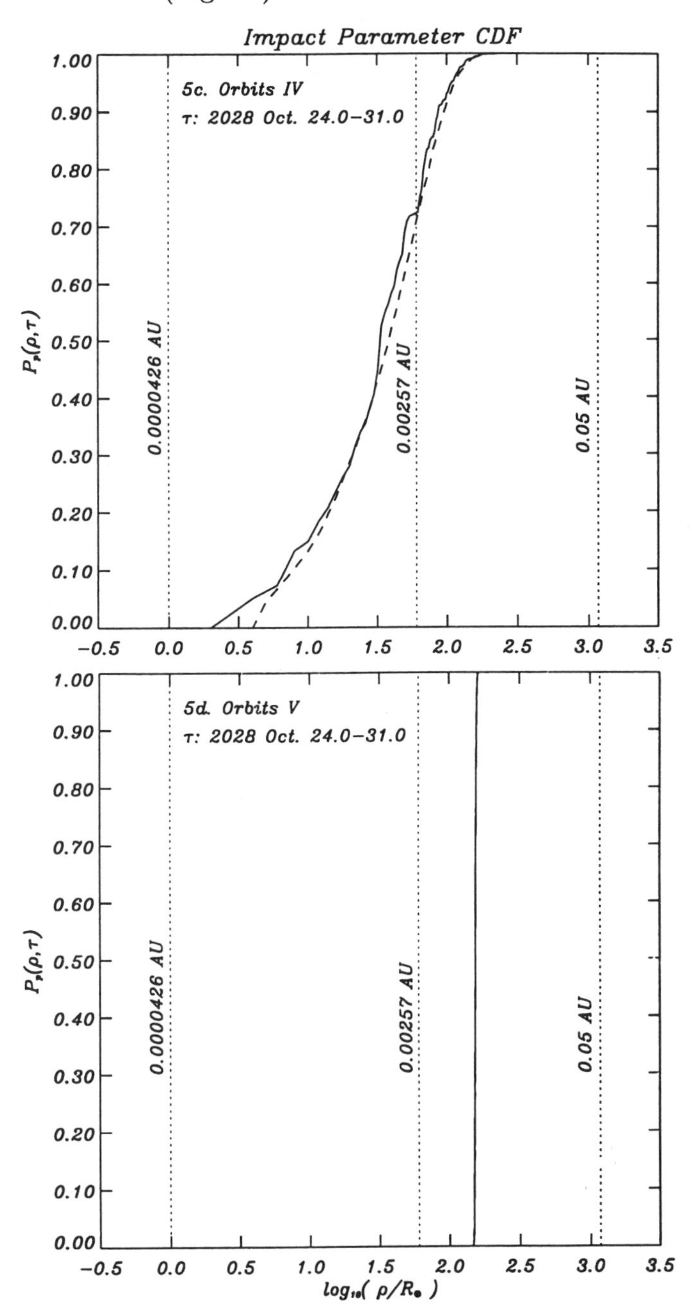

Note the inevitable consequence of nonlinearity that the nominal close–

approach distances computed using the least–squares orbital elements II–V do not coincide with the maximum likelihood close–approach distances derived from the impact parameter distributions. There is little relevant information in nominal close–approach distances.

Table 9 shows the X_6–metrics (Eq. 17) computed using the rigorous orbit solutions I–V, that is, by finding the collision orbits with minimum observational residuals. Notice that the reliable computation of these collision orbits required a one–second time resolution in 2028. The metric increases steadily from -0.048 (orbit solution I) to 3.990 (orbit solution V), and signals of increasing probability of a miss distance safely larger than the radius of the Earth. Note, however, that for orbit solution I, the upper bound of the collision probability is loose at 13 %, when the X_6–metric is mapped onto a six–dimensional Gaussian probability ellipsoid. The collision probability is already very small for orbit solution II, and becomes practically negligible for orbit solutions III–V.

It is important to note that the upper bound is only slightly affected by the inclusion of the March 1998 observations in the data set, although the nominal closest encounter distance moves within the lunar distance. Adding the 1990 observations into the data set increases X_6 more noticeably, the changes being roughly equal between sets II and III and sets IV and V. As for the X_3–metrics computed in the linear approximation, they move closer to the more rigorous X_6–metrics with increasing observational data sets and, in general, are intriguingly close to the X_6–values. While the nominal close approach distance fluctuates (IAUC No. 6837 and No. 6839), the superexponent X_6 increases monotonically signalling of decreasing upper bound for the collision probability. Based on applying the new methods to XF_{11}, the superexponent X_6 is a robust measure of the collision probability.

The first, cautious upper bound of 1:50000 for the collision probability (observation set IV) published in *Science* (**279**, 1843, 1998) was computed by extrapolating the cumulative distribution function of the impact parameter down to the Earth radius using a best–fit lognormal distribution function, and noting that the tail of the lognormal distribution systematically exceeded the tail resulting from the Monte Carlo computations.

Table 9 further includes the rms–values of the observational residuals for the maximum likelihood collision orbits, and the overall extremum residuals. The observational residuals show significant systematic trends for sets II–V, but the smallness of the residuals is nevertheless striking.

The nominal geocentric distance and touching spatial probability ellipsoid using the Gaussian orbit solution V are shown in Fig. 6 during the 21st century. The geocentric distance in Fig. 6a reveals numerous close approaches to the Earth. In Fig. 6b, the superexponent X_3 (Eq. 23) gradually decreases when the time interval is expanded from 2000 to 2100. For a

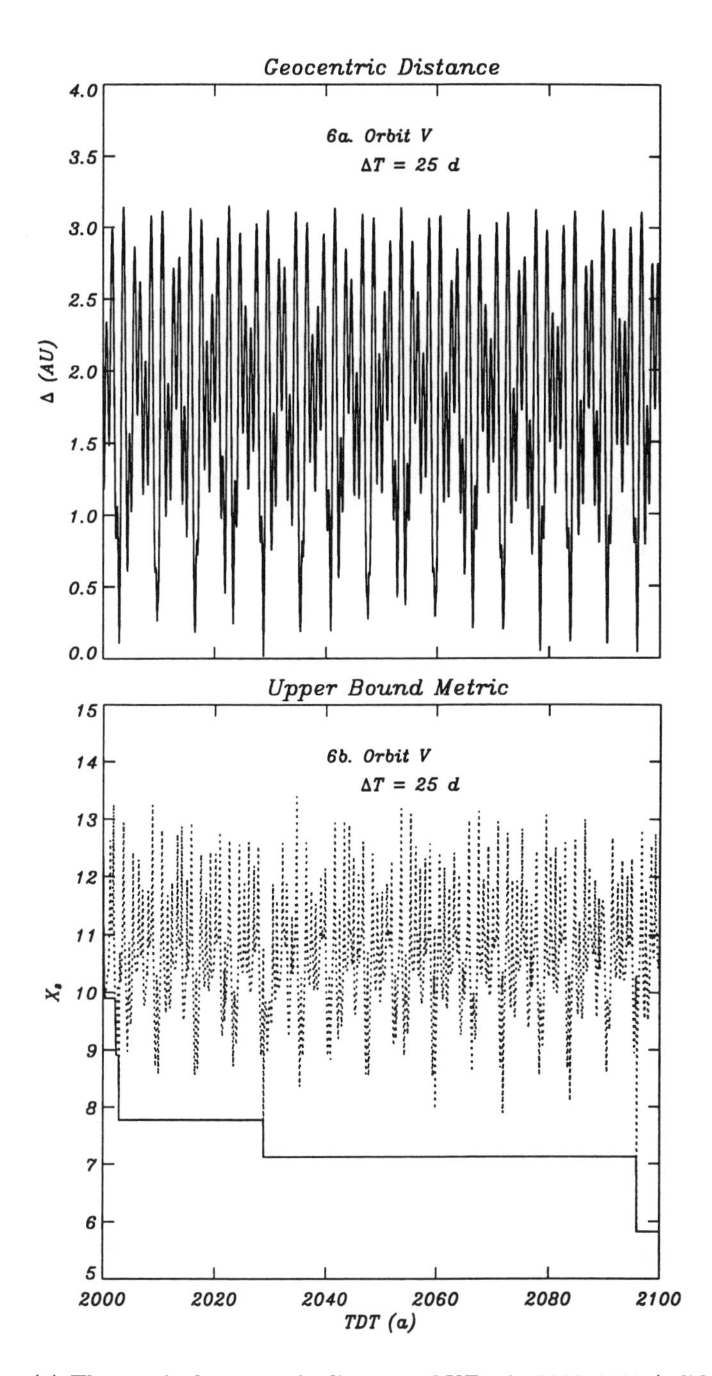

Figure 6. (a) The nominal geocentric distance of XF_{11} in 2000–2100 (solid line) based on orbit solution V. (b) The spatial probability ellipsoid touching the surface of the Earth (dotted line), and the close encounter metric X_3 (solid line, Eq. 23) with 25 d time resolution.

coarse time resolution of 25 d (compare to the 1 s tolerance for Table 9), the 2028 close encounter is seen to dictate the size of the survival ellipsoid until 2095.

To re–evaluate the case of XF_{11} in light of the new methods, the impact hazard in 2028 could not have been ruled out immediately after the discovery in 1997 Dec. 6–9. In Dec. 21, it would have become evident that XF_{11} could have a close approach to the Earth in 2028. However, the observations collected by Dec. 21 would have supported the conclusion that the collision probability was very small, less than 10^{-42}. The upper bound would have continued to tighten, when new 1997–1998 observations were incorporated and, by 1998 Feb. 4, the upper bound for the collision probability would have already been negligible. Finally, common sense is required in assessing the collision probability: although state–of–the–art computational methods have here been utilized to study XF_{11}, unknown factors could alter the current conclusions (see Section 5).

6. Conclusion

The various populations of near–Earth objects and their impact hazard have been reviewed in detail. A miscellany of new methods has been provided for the estimation of collision probabilities between small and large bodies. The methods cover near–Earth objects with both short and long observational time arcs, constituting the first complete treatment of the collision probability estimation from indeterminate orbits at the time of discovery to ones with observational data spanning several apparitions. The 2028 close approach of the near–Earth object 1997 XF_{11} has been thoroughly analysed noting that the collision probability had diminished to a small number by 1997 Dec. 21, roughly two weeks after discovery in 1997 Dec. 6.

Acknowledgements

The author is grateful to A. E. Roy and B. Steves for the opportunity to attend an innovative and pleasant *Advanced Study Institute* in Maratea, Italy, and wishes to thank M. Kaasalainen and J. Virtanen for constructive comments. `OrbFit` has been developed in collaboration with A. Milani, M. Carpino, and Z. Kněžević.

References

Alvarez, L.W., Alvarez, W., Asaro, F., and Michel, H.V. (1980) Extraterrestrial cause for the Cretaceous–Tertiary extinction, *Science* **208**, 1095–1108.

Bailey, M.E., Clube, S.V.M., Hahn, G., Napier, W.M., and Valsecchi, G.B. (1994) Hazards due to giant comets: climate and short–term catastrophism, in T. Gehrels (ed.),

Hazards due to comets and asteroids, University of Arizona Press, Tucson, pp. 479–533.

Bottke, Jr., W.F., Nolan, M.C., Greenberg, R., and Kolvoord, R.A. (1994) Collisional lifetimes and impact statistics of near-Earth asteroids, in T. Gehrels (ed.), *Hazards due to comets and asteroids*, University of Arizona Press, Tucson, pp. 337–357.

Bowell, E., and Muinonen, K. (1994) Earth-crossing asteroids and comets: groundbased search strategies, in T. Gehrels (ed.), *Hazards due to comets and asteroids*, University of Arizona Press, Tucson, pp. 149–197.

Bowell, E., Hapke, B., Domingue, D., Lumme, K., Peltoniemi, J., and Harris, A.W. (1989) Application of photometric models to asteroids, in R.P. Binzel, T. Gehrels, and M.S. Matthews (eds.), *Asteroids II*, University of Arizona Press, Tucson, pp. 298–315.

Carpino, M., and Knežević, Z. (1996) Determination of the mass of (1) Ceres from close approaches of other asteroids, in S. Ferraz-Mello, B. Morando, J.-E. Arlot (eds.), *IAU Symposium 172, Dynamics, Ephemerides, and Astrometry of Solar System Bodies* Kluwer Academic Publishers, Dordrecht, pp. 203–206.

Ceplecha, Z., Jacobs, C., and Zaffery, C. (1997). Correlation of ground- and spacebased bolides, in J.L. Remo (ed.), *Near-Earth Objects, The United Nations International Conference, Annals of the New York Academy of Sciences* **822**, pp. 145–154.

Chapman, C.R., and Morrison, D. (1994) Impacts on the Earth by asteroids and comets: assessing the hazard, *Nature* **367**, 33–39.

Chodas, P.W. (1993) Estimating the impact probability of a minor planet with the Earth, *Bull. Amer. Astron. Soc.* **25**, 1226.

Gehrels, T. (ed.) (1994) *Hazards due to Comets and Asteroids*, University of Arizona Press, Tucson.

Grieve, R.A.F. (1997) Target Earth: Evidence for Large-scale impact events, in J.L. Remo (ed.), *Near-Earth Objects, The United Nations International Conference, Annals of the New York Academy of Sciences* **822**, pp. 319–352.

Hudson, R.S., and Ostro, S.J. (1994) Shape of asteroid 4769 Castalia (1989 PB) from inversion of radar images, *Science* **263**, 940–943.

Keller, H.U., Arpigny, C., Barbieri, C., Bonnet, R.M., Cazes, S., Coradini, M., Cosmovici, C.B., Delamere, W.A., Huebner, W.F., Hughes, D.W., Jamar, C., Malaise, D., Reitsema, H.J., Schmidt, H.U., Schmidt, W.K.H., Seige, P., Whipple, F.L., and Wilhelm, K. (1986), First Halley multicolour camera imaging results from Giotto, *Nature* **321**, 320–326.

Kessler, D.J. (1981) Derivation of the collision probability between orbiting objects: the lifetimes of Jupiter's outer moons, *Icarus* **48**, 39–48.

Lewis, J.S., Matthews, M.S., and Guerrieri, M.L. (eds.) (1993) *Resources of Near-Earth Space*, University of Arizona Press, Tucson.

Luu, J., and Jewitt, D. (1989) On the relative numbers of C types and S types among near-Earth asteroids, *Astron. J.* **98**, 1905–1911.

Marsden, B.G. (1991) The computation of orbits in indeterminate and uncertain cases, *Astron. J.* **102**, 1539–1552.

Marsden, B.G., and Williams, G.V. (1997) *Catalogue of Cometary Orbits*, 12th Edition, Smithsonian Astrophysical Observatory, Cambridge, Massachusetts, 118 pp.

Milani, A., Carpino, M., Hahn, G., and Nobili, A.M. (1989) Dynamics of planet-crossing asteroids: classes of orbital behaviour (Project SPACEGUARD), *Icarus* **78**, 212–269.

Milani, A., Carpino, M., Rossi, A., Catastini, G., and Usai, S. (1995) Local geodesy by satellite laser ranging: a European solution, *Manuscr. Geod.* **20**, 123–138.

Morrison, D. (ed.) (1992) *The Spaceguard Survey*, Report of the NASA International Near-Earth-Object Detection Workshop. (Prepared at the Jet Propulsion Laboratory, Pasadena, California, for NASA's Office of Space Science and Applications, Solar System Exploration Division, Planetary Astronomy Program.)

Morrison, D. (1997) The terrestrial impact flux and the associated impact hazard, presentation at *International Astronomical Union, The XXIIIrd General Assembly, Joint Discussion 6, Interactions between Planets and Small Bodies.*

Morrison, D., Chapman, C.R., and Slovic, P. (1994) The impact hazard, in T. Gehrels (ed.), *Hazards due to comets and asteroids*, University of Arizona Press, Tucson, pp. 59–91.

Muinonen, K. (1996) Spaceguard Survey for near–Earth objects: collision probability assessment, *Mem. S. A. It.* **67**, 999–1004.

Muinonen, K. (1998) Introducing the Gaussian shape hypothesis for asteroids and comets, *Astron. Astrophys.* **332**, 1087–1098.

Muinonen, K., and Bowell, E. (1993a) Asteroid orbit determination using Bayesian probabilities, *Icarus* **104**, 255–279.

Muinonen. K., and Bowell, E. (1993b) Collision probability for Earth–crossing asteroids on stochastic orbits, *Bull. Amer. Astron. Soc.* **25**, 1116.

Muinonen, K., and Lagerros, J. (1998) Inversion of shape statistics for small solar system bodies, *Astron. Astrophys.* **333**, 753–761.

Muinonen, K., Milani, A., and Bowell, E. (1997) Determination of initial eigenorbits for asteroids, in I.M. Wytrzyszczak, J.H. Lieske, and R.A. Feldman (eds.), *Dynamics and Astrometry of Natural and Artificial Celestial Bodies*, Kluwer Academic Publishers, Dordrecht, pp. 191–198.

Öpik, E.J. (1951) Collisional probabilities with the planets and the distribution of interplanetary matter, *Proc. Roy. Irish Acad. A.* **54**, 165–199.

Öpik, E.J. (1976) *Interplanetary Encounters, Close–Range Gravitational Interactions*, Elsevier, Amsterdam.

Rabinowitz, D.L., Bowell, E., Shoemaker, E.M., and Muinonen, K. (1994) The population of Earth–crossing asteroids, in T. Gehrels (ed.), *Hazards due to comets and asteroids*, University of Arizona Press, Tucson, pp. 285–312.

Remo, J.L. (ed.) (1997) *Near–Earth Objects, The United Nations International Conference, Annals of The New York Academy of Sciences* **822**, New York.

Shoemaker, E.M., Weissman, P.R., and Shoemaker, C.S. (1994) The flux of periodic comets near Earth, in T. Gehrels (ed.), *Hazards due to comets and asteroids*, University of Arizona Press, Tucson, pp. 313–335.

Shoemaker, E.M., and Wolfe, R.F. (1982) Cratering time scales for the Galilean satellites, in D. Morrison (ed.), *Satellites of Jupiter*, University of Arizona Press, Tucson, pp. 277–339.

Standish, E.M., Newhall, X.X., Williams, J.G., and Folkner, W.F. (1995) JPL Planetary and Lunar Ephemerides, DE403/LE403. JPL IOM 314.10–127.

Steel, D. (1997) Cometary impacts on the biosphere, in P.J. Thomas, C.F. Chyba, and C.P. McKay (eds.), *Comets and the Origin and Evolution of Life*, Springer, pp. 209–241.

Weissman, P.R. (1991) Dynamical history of the Oort cloud, in R.L. Newburn, Jr., M. Neugebauer, and J. Rahe (eds.), *Comets in the Post–Halley Era*, Kluwer Academic Publishers, Dordrecht, pp. 463–486.

Weissman, P.R. (1997) Long–period comets and the Oort cloud, in J.L. Remo (ed.), *Near–Earth Objects, The United Nations International Conference, Annals of the New York Academy of Sciences* **822**, pp. 67–95.

Wetherill, G.W. (1967) Collisions in the asteroid belt, *J. Geophys. Res.* **72**, 2429–2444.

Whipple, F.L. (1950) A comet model I. The acceleration of comet Encke, *Astrophys. J.* **112**, 464–474.

Wiegert, P.A., Innanen, K.A., and Mikkola, S. (1997) An asteroidal companion to the Earth, *Nature* **387**, 685–686.

Yeomans, D.K., and Chodas, P.W. (1994) Predicting close approaches of asteroids and comets to Earth, in T. Gehrels (ed.), *Hazards due to comets and asteroids*, University of Arizona Press, Tucson, pp. 241–258.

OPTIMAL LOW-THRUST INTERCEPTION AND DEFLECTION OF EARTH-CROSSING ASTEROIDS

Bruce A. Conway

Dept. of Aeronautical & Astronautical Engineering
306 Talbot Laboratory
University of Illinois
Urbana, IL 61801, USA
bconway@uiuc.edu

ABSTRACT

The spectacular collision of the Shoemaker-Levy 9 asteroid with Jupiter in July 1994 was a dramatic reminder of the inevitability of such catastrophes in the Earth's future unless steps are taken to develop methods for Earth-approaching object detection and possible interdiction. In this work optimal (minimum-time) very-low-thrust trajectories using electric propulsion are determined for the interception of asteroids which pose a threat of collision with the Earth. At the time of interception the system state transition matrix is found and used to indicate the direction in which an impulse should be applied to the asteroid to maximize the subsequent deflection.

INTRODUCTION

The spectacular collision of the Shoemaker-Levy 9 asteroid with Jupiter in July 1994 was a dramatic reminder of the fact that the Earth has and will continue to experience such catastrophic events. While the frequency of such massive collisions is very low, smaller objects collide with the Earth regularly and do damage which would be intolerable in any populated region. A consensus is developing that while the probability for collision is low the potential for destruction is immense and thus some resources should be devoted to threat detection and possible interdiction.

The population of Earth-crossing objects is significant and continuously increasing by virtue of discovery. As of 1994, 163 Earth-crossing asteroids are known [1], the largest being 1627 Ivar with a dimension of ~8 km and mass of ~10^{15} kg. (For comparison the K/T or Cretaceous/Tertiary boundary impact of 65 million years ago which left a ~200 km crater, releasing an estimated 10^8 MT equivalent energy in the Gulf of Mexico, would have been caused by a body ~10 km in size.) The census of Earth approaching asteroids is believed complete only for objects of Ivar's size; for bodies on the order of ~1 km the completeness of

B.A. Steves and A.E. Roy (eds.), The Dynamics of Small Bodies in the Solar System, 159–169.

the census is estimated to be only ~10%, based on the rate at which such objects continue to be discovered [1].

Of course it is the prevention of a catastrophic impact with which this paper is concerned. The dangerous object must be intercepted, at the earliest possible time, and then deflected or destroyed. This paper is not concerned with how the latter is accomplished; one strategy may be to detonate a large nuclear weapon near the surface of the asteroid or comet. In this work optimal trajectories for the interception of dangerous asteroids are found. Low-thrust, high specific impulse propulsion is used because of the significant advantages it provides in propulsive mass required for a given mission. However, a low-thrust departure from Earth would require many revolutions of the Earth which would consume a lot of time. It seems reasonable then to use an impulsive velocity change for the initial departure, followed by continuous low-thrust propulsion.

The analysis is necessarily three-dimensional as many of the Earth-crossing asteroids have significant inclinations. However, it simplifies the analysis to assume that before the application of the departure impulse the intercepting vehicle is in a low-Earth circular orbit which lies in the ecliptic plane. The magnitude of this departure impulse is something which might in real life depend on the target and/or on the capability of the launch vehicle. For this work the departure impulse chosen is less than or equal in magnitude to that required for the most commonly employed interplanetary trajectory, i.e. the impulse required for escape from the Earth onto a Hohmann transfer ellipse to Mars, which yields a hyperbolic excess velocity ($v_{\infty/E}$) of 2.94 km/sec. The optimizer may choose the point in the initial low-Earth orbit from which to begin the departure. It may also choose the direction in which it is applied via in-plane and out-of-plane thrust pointing angles. The Earth is assumed to be in a circular orbit about the sun and its true longitude on the departure date is found using the ephemeris program MICA from the US Naval Observatory [2]. Specific impulse for the low-thrust motor is chosen from a range of values (2000 - 4000 sec) representative of current technology.

METHOD

A complete description of the the formulation and solution of the problem of minimum-time asteroid interception may be found in [3]. Only the outline of the method will be described here. The variational equations used are those for the equinoctial elements. This avoids singularity of the elements for circular or equatorial orbits. The equinoctial elements and their relationship to the conventional elements are [4]:

$$
\begin{aligned}
a &= a \\
P_1 &= e \sin \tilde{\omega} \\
P_2 &= e \cos \tilde{\omega} \\
Q_1 &= \tan \frac{i}{2} \sin \Omega \\
Q_2 &= \tan \frac{i}{2} \cos \Omega \\
\ell &= \tilde{\omega} + M = \Omega + \omega + M
\end{aligned} \tag{1}
$$

In these variables the variational equations become [4]:

$$\frac{da}{dt} = \frac{2a^2}{h}\left[(P_2 \sin L - P_1 \cos L)\, R + \frac{p}{r}\, T\right]$$

$$\frac{dP_1}{dt} = \frac{r}{h}\left\{-\frac{p}{r} \cos L\, R + \left[P_1 + \left(1 + \frac{p}{r}\right)\sin L\right] T - P_2\,(Q_1 \cos L - Q_2 \sin L)\, N\right\}$$

$$\frac{dP_2}{dt} = \frac{r}{h}\left\{\frac{p}{r} \sin L\, R + \left[P_2 + \left(1 + \frac{p}{r}\right)\cos L\right] T - P_1\,(-Q_1 \cos L + Q_2 \sin L)\, N\right\} \qquad (2)$$

$$\frac{dQ_1}{dt} = \frac{r}{2h}\left(1 + Q_1^2 + Q_2^2\right) \sin L\, N$$

$$\frac{dQ_2}{dt} = \frac{r}{2h}\left(1 + Q_1^2 + Q_2^2\right) \cos L\, N$$

$$\frac{d\ell}{dt} = n - \left[\frac{a}{a+b}\left(\frac{p}{h}\right)(P_1 \sin L + P_2 \cos L) + \frac{2b}{a}\right] R - \left[\frac{a}{a+b}\left(\frac{r}{h} + \frac{p}{h}\right)(P_1 \cos L - P_2 \sin L)\right] T - \left(\frac{r}{h}\right)(Q_1 \cos L - Q_2 \sin L)\, N$$

where:

T = tangential component of thrust acceleration = $F \cos\beta \cos\gamma$

R = radial component of thrust acceleration = $F \sin\beta \cos\gamma$

N = normal component of thrust acceleration = $F \sin\gamma$

F = acceleration produced by the low-thrust motor

$$b = \text{semiminor axis} = a\sqrt{1 - P_1^2 - P_2^2}$$

$$h = \text{angular momentum} = n\, a\, b$$

$$n = \text{mean motion} = \sqrt{\frac{\mu}{a^3}}$$

$$\frac{a}{a+b} = \frac{1}{1 + \sqrt{1 - P_1^2 - P_2^2}} \qquad (3)$$

$$\frac{p}{r} = 1 + P_1 \sin L + P_2 \cos L$$

$$\frac{r}{h} = \frac{h}{\mu\,(1 + P_1 \sin L + P_2 \cos L)}$$

The control variables in the problem are the thrust pointing angles β and γ. Angle β is the in-plane thrust pointing angle; it is measured from the normal to the radius vector and is a positive angle if it yields a component of thrust pointing radially-outward. Angle γ is the out-of-plane thrust pointing angle. It is positive if it yields a component of thrust in the direction of the orbital angular momentum.

The true longitude L is obtained from the mean longitude ℓ by first solving Kepler's equation in equinoctial variables, as described in Battin's book [4].

The thrust acceleration F varies as propellant is consumed; assuming a constant thrust motor,

$$\frac{dF}{dt} = \frac{F^2}{c} = \frac{F^2}{g\, I_{sp}} \,, \tag{4}$$

where g is the acceleration of gravity at the Earth's surface, c is the motor exhaust velocity, and I_{sp} is the motor specific impulse (in sec.).

The problem is then to choose the time history of the thrust pointing angles β and γ in order to minimize the performance index, which is the time of flight, subject to satisfaction of the system variational equations (2), the system initial condition constraints,

$$(a, P_1, P_2, Q_1, Q_2, \ell \,) \text{ must satisfy (at } t = 0\text{):}$$

$$\bar{r} = \bar{r}_E \tag{5}$$

$$\bar{v} = \bar{v}_E + \bar{v}_{\infty/E}$$

(where $\bar{r}_E$ and $\bar{v}_E$ represent the position and velocity of the Earth; $\bar{v}_{\infty/E}$ is the hyperbolic excess velocity of the spacecraft with respect to the Earth) and the terminal constraint (of interception):

$$(a, P_1, P_2, Q_1, Q_2, l) \text{ must satisfy (at } t = t_{final}\text{):}$$

$$\bar{r} = \bar{r}_A \tag{6}$$

where the subscripts E and A refer to the Earth and the asteroid respectively. The optimizer is free to choose two parameters which do not explicitly appear in the variational equations; two pointing angles, in-plane (β_0) and out-of-(ecliptic) plane (γ_0) pointing angles which describe the direction of $\bar{v}_{\infty/E}$ following the impulsive Δv which allows the vehicle to escape from low-Earth orbit. The initial condition constraints (5) yield 6 nonlinear scalar constraint equations and the terminal constraint (6) yields 3 scalar constraint equations.

NUMERICAL OPTIMIZATION

The problem is solved using the method of direct collocation with nonlinear programming (DCNLP) [5, 6, 7]. In this solution method the continuous problem is discretized by dividing the total time into "segments" whose boundaries are termed the system "nodes". Each state is known only at discrete points; at the nodes and, depending on how the problem is formulated, at zero, one, or more points interior to a segment. A quadrature rule is used to relate state variables at successive nodes; this rule becomes a nonlinear constraint equation. The system nonlinear constraints are these quadrature rules which enforce satisfaction of the system differential equations, the initial condition constraints (5) and the conditions for interception (6). The discretized problem thus becomes a nonlinear programming problem. The parameters are the state variables (which are the 6 spacecraft orbit equinoctial variables + the thrust acceleration magnitude) at the nodes and center points of the segments and the control variables (the two thrust pointing angles) at the nodes, center point, and collocation points of each segment. There are three additional NLP variables; the final time t_f and the in-plane (β_0) and out-of-(ecliptic) plane (γ_0) pointing angles which describe the direction of $\bar{v}_{\infty/E}$ following the impulsive Δv which allows the vehicle to escape from low-Earth orbit.

The NLP problem solver works best when all variables have nearly the same order of magnitude. Thus normalized units are used in which the distance unit is 1 astronomical unit (1 AU), 2π time units (TU) are one orbit period at $a = 1$ AU, i.e. 365.25 days, and the sun's gravitational parameter is $\mu = 1$.

The NLP problem is solved using the program NZSOL, an improved version of the program NPSOL [8]. For the example trajectory determined in the next section 20 segments were used to discretize the trajectory yielding an NLP problem with 452 variables. More segments could be used but the results for the state and control variable time histories, presented in the next section, show that all vary slowly and only over a small range, so that 20 segments capture quite well the system time history. This program requires an initial guess of the vector of parameters. The numerical optimization method used here has been quite robust when applied to a variety of orbit transfer problems [6, 7, 9, 10].

MAXIMIZATION OF THE DEFLECTION

After the optimal trajectory is found we determine the optimal direction in which an impulse should be applied to the asteroid at interception in order to maximize the close-approach distance of the asteroid (to the Earth) months later. At the time of interception t_0 the system state transition matrix $\Phi(t, t_0)$ determines the perturbation in position and velocity which will result at time t due to a perturbation in position and velocity applied at t_0, i.e.,

$$\begin{bmatrix} \delta\bar{r} \\ \delta\bar{v} \end{bmatrix} = \Phi(t, t_0) \begin{bmatrix} \delta\bar{r}_0 \\ \delta\bar{v}_0 \end{bmatrix} = \begin{bmatrix} \tilde{R} & R \\ \tilde{V} & V \end{bmatrix} \begin{bmatrix} \delta\bar{r}_0 \\ \delta\bar{v}_0 \end{bmatrix} \tag{7}$$

therefore

$$\delta\bar{r}(t) = R\, \delta\bar{v}_0(t_0) \tag{8}$$

where the time of interest, t, is the time of close approach to Earth, and

$$[R] = \frac{r_0}{\mu}(1 - F)\left[(\bar{r} - \bar{r}_0)\bar{v}_0^T - (\bar{v} - \bar{v}_0)\bar{r}_0^T\right] + \frac{C}{\mu}\,\bar{v}\,\bar{v}_0^T + G\,[I]\,,$$

$$F = 1 - \frac{r}{p}(1 - \cos\theta)\,,\quad \cos\theta = \frac{\bar{r}\cdot\bar{r}_0}{r\, r_0}\,,$$

$$G = \frac{1}{\sqrt{\mu}}\left[\frac{r\, r_0}{\sqrt{p}}\sin\theta\right]\,,\ \text{with} \tag{9}$$

$$C = \frac{1}{\sqrt{\mu}}\left[3\, U_5 - \chi\, U_4 - \sqrt{\mu}\,(t - t_0)\, U_2\right]\,,\ \text{where } \chi = \sqrt{a}\,(E - E_0) \text{ and}$$

$U_1(\chi, \alpha)$, $U_2(\chi, \alpha)$, $U_3(\chi, \alpha)$, $U_4(\chi, \alpha)$, $U_5(\chi, \alpha)$ are the "Universal Functions", cf. [4] where $\alpha = 1/a$.

We want to maximize $|\delta\bar{r}(t)| = \max\left([R]\,\delta\bar{v}_0\right)$, which is the equivalent to maximizing $\delta\bar{v}_0^T[R]^T[R]\,\delta\bar{v}_0$. This quadratic form is maximized, for given $|\delta\bar{v}_0|$, if $\delta\bar{v}_0$ is chosen parallel to the eigenvector of $[R]^T[R]$ which is conjugate to the largest eigenvalue of $[R]^T[R]$. This yields the optimal direction for the perturbing velocity impulse $\delta\bar{v}_0$, which will be expressed on the space fixed basis since this is the basis on which [R] is implicitly expressed. $\delta\bar{v}_0$ may then be expressed in an asteroid fixed radial, transverse, normal basis as,

$$\delta\bar{v}_{0\,RTN} = \begin{bmatrix} c\theta c\Omega - cis\Omega s\theta & c\theta s\Omega + cic\Omega s\theta & sis\theta \\ -s\theta c\Omega - cis\Omega c\theta & -s\theta s\Omega + cic\Omega c\theta & sic\theta \\ sis\Omega & -sic\Omega & ci \end{bmatrix} \delta\bar{v}_{0\,XYZ} \,. \qquad (10)$$

EXAMPLES

Several optimal trajectories have been found for the interception of actual Earth-crossing asteroids. As an example optimal trajectories have been found for the interception of Earth-approaching asteroid 1991RB. Its orbit elements are [1], as of 15 September 1991,

$$a = 1.4524 \text{ AU}, \; e = .4846, \; i = 19.580°$$

$$\Omega = 359.599°, \; \omega = 68.708°, \; M = 328.08°$$

This asteroid will approach the Earth to within .04 AU, or 15 lunar distances, on 19 September 1998. If it is assumed that launch from Earth takes place 6 months prior to the close approach, that is, on 19 March 1998, and if we assume that

$$F_{initial} = 0.14 \text{ AU/TU}^2 = 84.6 \cdot 10^{-6} \text{ g}, \; I_{SP} = 4000 \text{ sec},$$
$$v_{\infty/E} = .0333 \text{ AU/TU} = .9807 \text{ km/sec}$$

then the minimum time of flight is found to be 2.5096 TU = 145.9 days. The resulting optimal trajectory is shown in Figure 1. The time histories for the optimal in-(orbit) plane and out-of-plane thrust pointing angles are shown in Figures 2 and 3. In these figures the abscissa indicates node position, but this is linearly proportional to time, with the 80th node corresponding to the final time of 145.9 days. One can see from Figure 2 that the out-of-orbit plane thrust angle is large and negative. This is consistent with the spacecraft trajectory seen in Figure 1; the inclination of 1991RB's orbit is large and, to intercept the asteroid months in advance of the close approach, the spacecraft must travel a considerable distance below the ecliptic plane.

Optimal trajectories for interception of 1991 RB have been determined for many different combinations of specific impulse, initial thrust acceleration, date of departure, and hyperbolic excess velocity. The $V_{\infty/E}$ or hyperbolic excess velocity of escape is always chosen to be less than or equal to .0988 AU/TU = 2.94 km/sec, i.e. just what is required for a Hohmann transfer to Mars, as mentioned previously. It results from an impulse applied in low Earth orbit, usually by the upper stage of the launch vehicle, and was chosen so that we may assume that the asteroid interceptor may be launched with an existing rocket.

These different cases yielded optimal trajectories with different times of interception. Figure 4 shows the maximum amount of deflection which can be obtained, at what would otherwise be the time of close approach to the Earth, as a function of the interval between interception and close approach (on 19 September 1998). The impulse is assumed to be applied to the asteroid in the direction chosen, as described in the previous section, to maximize the deflection at the subsequent close approach. The figure shows that, if the asteroid is reached several months before the time of collision, each 1 m/sec of velocity change imparted to the asteroid may yield a deflection distance comparable to the width of the Earth. The arrival date vs. the launch date is shown in Figure 5, as a function of initial thrust acceleration and hyperbolic excess velocity at departure from Earth. A specific impulse of 4000 sec for the spacecraft's electric propulsion is assumed for all of the results shown in the figure.

The maximum deflection for a given impulse applied to the asteroid as a function of launch date and initial thrust acceleration is shown in Figure 6 (assuming again that $I_{SP} = 4000$ sec and also that $V_{\infty/E} = .0988$ AU/TU). It is clear from the figure that the improvement in deflection distance obtained from increasing the engine thrust, and hence decreasing the time of flight, is significant, e.g. increasing the thrust acceleration by approximately 60%, from 61 micro-g to 97 micro-g, increases the deflection obtained by almost 100%.

The advantage of using continuous low-thrust propulsion is most apparent when the impulsive Δv required to duplicate the mission is determined. For the case described in the previous paragraphs, the interception of 1991 RB, the time of flight, the date of departure, (given in the previous section) and the positions of departure and arrival are known. Determining the conic which connects these two points, in the same time of flight as for the low-thrust trajectory, but using impulsive velocity changes, is a Lambert problem. Solving Lambert's equation yields the semimajor axes of the elliptic sections connecting these two points in the given flight time [11]. It is then straightforward to find the minimum velocity change required to transfer onto one of these elliptical trajectories.

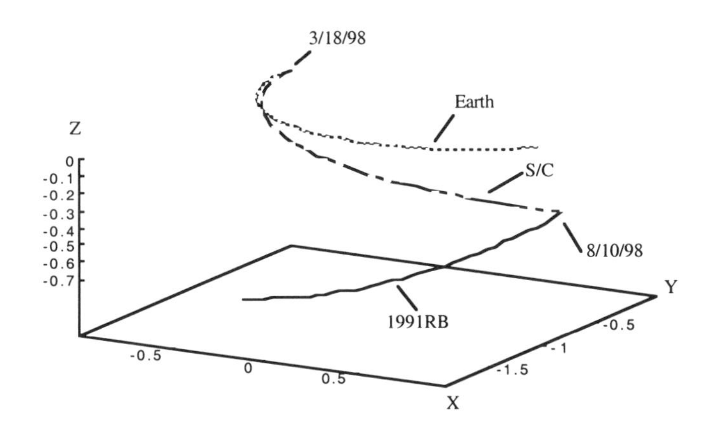

Figure 1. Optimal trajectory for the interception of asteroid 1991RB

The impulsive Δv's required to duplicate the mission of the low-thrust spacecraft are found to be quite large, on the order of 10 km/sec and larger. This is a very substantial Δv, too large to be performed by a single-stage vehicle. A two-stage rocket using present technology can achieve such a Δv only if its payload mass is on the order of 1-3% of the launch mass [3]. In comparison, near-term ion propulsion technology should allow the payload mass to be approximately 10-12% of the launch mass if the optimal low-thrust trajectory is used [3].

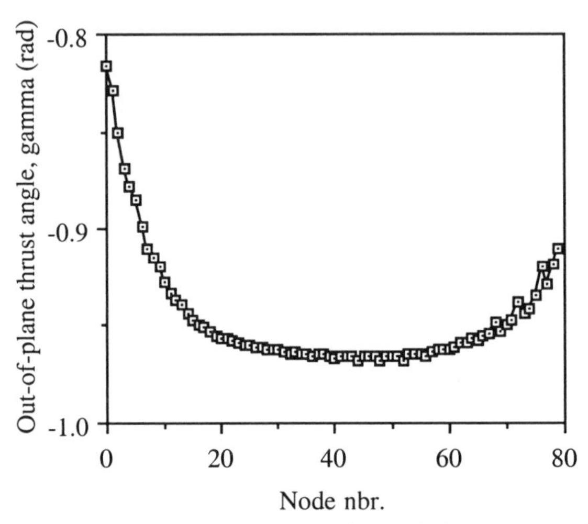

Figure 2. Time history of the optimal out-of-plane thrust pointing angle

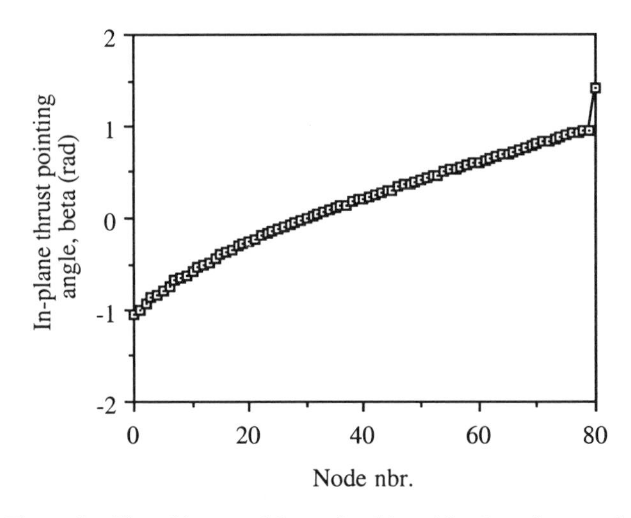

Figure 3. Time history of the optimal in-orbit-plane thrust pointing angle

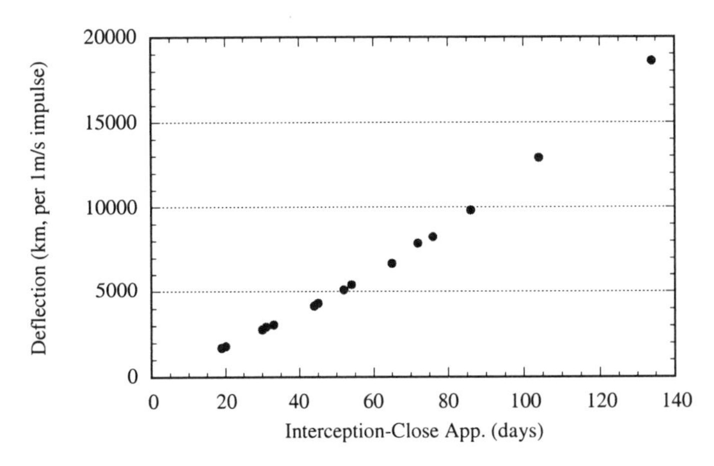

Figure 4. Specific deflection as a function of date of interception

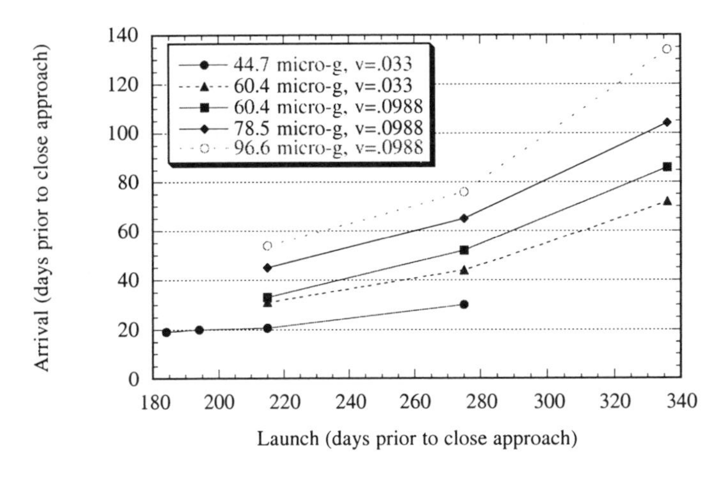

Figure 5. Arrival (interception) date vs. launch date for various engine thrust accelerations and hyperbolic excess velocity of Earth escape

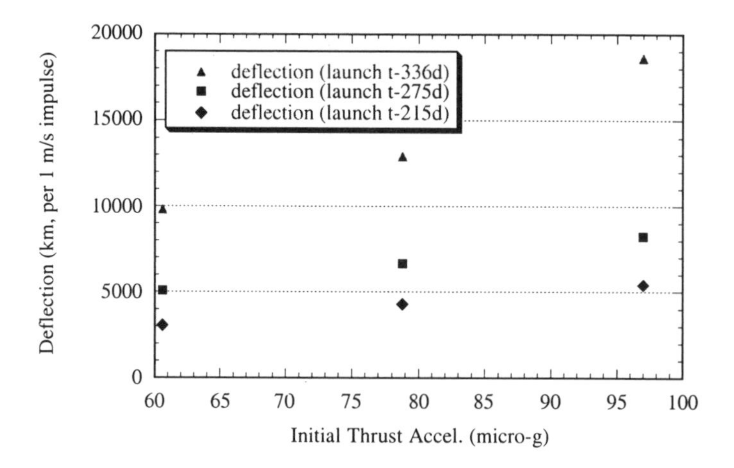

Figure 6. Maximum specific deflection as a function of launch date and engine thrust acceleration

CONCLUSIONS

The method of collocation with nonlinear programming has successfully solved the problem of finding minimum-time low-thrust trajectories for the interception of Earth-approaching asteroids. Many of the strategies for amelioration of the danger of an asteroid's collision with the Earth involve, at the time of interception, applying a small impulsive velocity change to the asteroid. In this work we show how, that is, in what direction, this impulse should be applied in order to maximize the deflection of the asteroid at the time of close approach. While the objective function of the optimizer is the flight time, we see from Figure 4 that as the time of flight is reduced (or conversely when the period between interception and potential collision with Earth is increased) the amount of deflection of the asteroid obtained from a given impulse always increases. Thus the minimum-time trajectories are also deflection-maximizing.

One important result, illustrated in Figure 5, is that, if the interceptor spacecraft is launched less than a year in advance of the asteroid's possible collision with Earth, the deflection obtained will be on the order of an Earth diameter for every 1 m/sec velocity change applied to the asteroid. Considering how many millions of kg of mass some of the Earth-crossing asteroids possess it would be very difficult to change their velocity by even 1 m/sec. Thus it may only be feasible to deflect asteroids of moderate size if they can be reached several years in advance of potential collision.

The combined impulsive/low-thrust trajectory used in this analysis, in which the impulse is used only for Earth departure to reduce the long flight time which would be required for a low-thrust escape from Earth orbit, appears to be very suitable for the cases used as examples here, in which launch takes place 4-6 months before the predicted close approach of the asteroid. If the danger of collision is known much earlier a trajectory which uses only low-thrust propulsion may be feasible and would likely have additional advantages in total propellant required.

REFERENCES

1. Rabinowitz, D. L. et al. (1994) The population of Earth-crossing asteroids, *Hazards Due to Comets & Asteroids*, T. Gehrels (ed.), Univ. of Arizona Press, Tucson, pp. 285-312.
2. MICA, an Interactive Astronomical Almanac, (1989) U.S. Naval Observatory, Washington DC.
3. Conway, B. A. (1997) Optimal low-thrust interception of Earth-crossing asteroids, *J. of Guidance, Control, and Dynamics*, **20**, No. 5, 995-1002.
4. Battin, R. H. (1987) *An Introduction to the Mathematics and Methods of Astrodynamics*, AIAA Education Series, AIAA Publ., New York.
5. Hargraves, C. R. and Paris, S. W. (1987) Direct trajectory optimization using nonlinear programming and collocation, *Journal of Guidance, Control, and Dynamics*, **10**, No. 4, 338-342.
6. Enright, P. J. and Conway, B. A. (1992) Discrete approximations to optimal trajectories using direct transcription and nonlinear programming, *Journal of Guidance, Control, and Dynamics*, **15**, No. 4, 994-1002.
7. Herman, A. L. and Conway, B. A., (1996) Direct optimization using collocation based on high-order Gauss-Lobatto quadrature rules", *J. of Guidance, Control, and Dynamics*, **19**, No. 3, 592-599.
8. P. E. Gill, et al. (1993) *User's Guide for NZOPT 1.0: A Fortran Package For Nonlinear Programming*, McDonnell Douglas Aerospace.
9. Scheel, W. A. and Conway, B. A. (1994) Optimization of very-low-thrust, many revolution spacecraft trajectories, *J. of Guidance, Control, and Dynamics,* **17**, No. 6, 1185-1192.
10. Tang, S. and Conway, B. A. (1995) Optimization of low-thrust interplanetary trajectories using collocation and nonlinear programming, *J. of Guidance, Control, and Dynamics*, **18**, No. 3, 599 - 604.
11. Prussing, J. E. and Conway, B. A. (1993) *Orbital Mechanics*, Oxford University Press, New York.

SECULAR RESONANCES: TRANSPORT MECHANISM TO EARTH–CROSSING ORBITS

P. MICHEL
Observatoire de la Côte d'Azur
CNRS/UMR 6529, B.P. 4229, 06304 Nice Cedex 4, France

1. Introduction

The population of Near–Earth Asteroids (NEAs) is composed of bodies ranging from nearly 40 km to sub–kilometer in diameter with orbits which cross those of the terrestrial planets. Though the identification of their exact birth place is still not determined, many of them have been recognized to originate in the asteroid main belt and two main dynamical mechanisms have been identified as efficient transport mechanisms to the planet–crossing regions: mean motion resonances with Jupiter and secular resonances. Mean motion resonances correspond to commensurabilities between the orbital periods of a small body and Jupiter. We refer to Moons (1997) for a detailed review of the results concerning these resonances. In this paper, we concentrate on secular resonances and we summarize the new results of recent studies (Michel et al., 1996a, b, Michel and Froeschlé, 1997, Michel, 1997) which have shown that NEAs can also be transported from their current positions to other zones of the planet–crossing regions not only as a consequence of close approaches to planets but also due to the occurence of secular resonances.

It is first important to recall that the planets give rise not only to mutual perturbations but also to secular ones on any small body orbiting around the Sun and force the precession and the deformation of its orbit. We denote by g the precession of the asteroid's longitude of perihelion and by s that of its node. When one of these frequencies is nearly equal to one of the fundamental frequencies g_j or s_j (j is the index of the considered planet) of the Solar System, a secular resonance takes place and causes the divergence of the solution of the linearized averaged equations for the elements of the small body (see e.g. Williams, 1969, Bretagnon, 1974, Morbidelli and Henrard, 1991). Following Williams' notation (1969), the resonances with

B.A. Steves and A.E. Roy (eds.), The Dynamics of Small Bodies in the Solar System, 171–177.

respectively, $g = g_5$, $g = g_6$ and $s = s_6$ are usually called ν_5, ν_6 and ν_{16} (the index 5 and 6 refer to Jupiter and Saturn, respectively). We will keep the same notations in the following.

The role of secular resonances as efficient transport mechanisms from the main belt has been studied by many authors. We refer to Froeschlé and Morbidelli (1994) and Froeschlé (1997) for detailed reviews. Recently, we have found that these resonances can also act in the terrestrial planet–crossing regions and that they can for instance affect the orbit of a small body so that it is transported from a Mars–crossing to an Earth–crossing orbit or to a highly inclined orbit, providing a possible explanation to the peculiar orbital distributions of some NEAs such as the SEAs (Small Eart–Approachers, see Sec. 3). These mechanisms have first been demonstrated by long–term numerical integrations of NEAs' orbits (Michel et al. 1996a, b) and then by a semi–analytical method which gave for the first time the location and effects of the 1-st order secular resonances in the region of semimajor axes < 2 AU (Michel and Froeschlé, 1997, Michel, 1997). It has thus been shown that the orbital evolution of NEAs is not only influenced by close approaches to planets but rather results from the action of different dynamical mechanisms.

2. Transport mechanisms to the Earth–crossing region by secular resonances

Amor asteroids are NEAs with current perihelion distances ranging from 1.017 AU to 1.3 AU and whose orbital semimajor axes are greater than 1 AU. These objects are solely Mars–crossers and can approach but do not cross the Earth's orbit. Recently, we have found that these asteroids could be transported from their Mars–crossing orbits to an Earth–crossing one due to the effect of a secular resonance with Mars, namely ν_4, corresponding to the commensurability of the mean precession frequency of their perihelion longitude and the fundamental frequency g_4. Figure 1 shows the already published result concerning the asteroid (433) Eros (Michel et al., 1996a, b). Though many other numerical integrations have shown similar behaviors, we choose to show again the case of this asteroid since, with a diameter of ≈ 23 km, it is the second largest known NEA and it will be observed by the *NEAR* space mission in 1999. The result of the numerical integration of Eros' orbit (Fig. 1) shows that, though the probability is very small, this asteroid could end its life by an impact with our planet. But more important is to understand how an object which only crosses the orbit of Mars can be led to an Earth–crossing orbit. Here, the transport mechanism is the ν_4 resonance which secularly increases the orbital eccentricity of the small body (see Fig. 1), consequently decreasing its perihelion distance

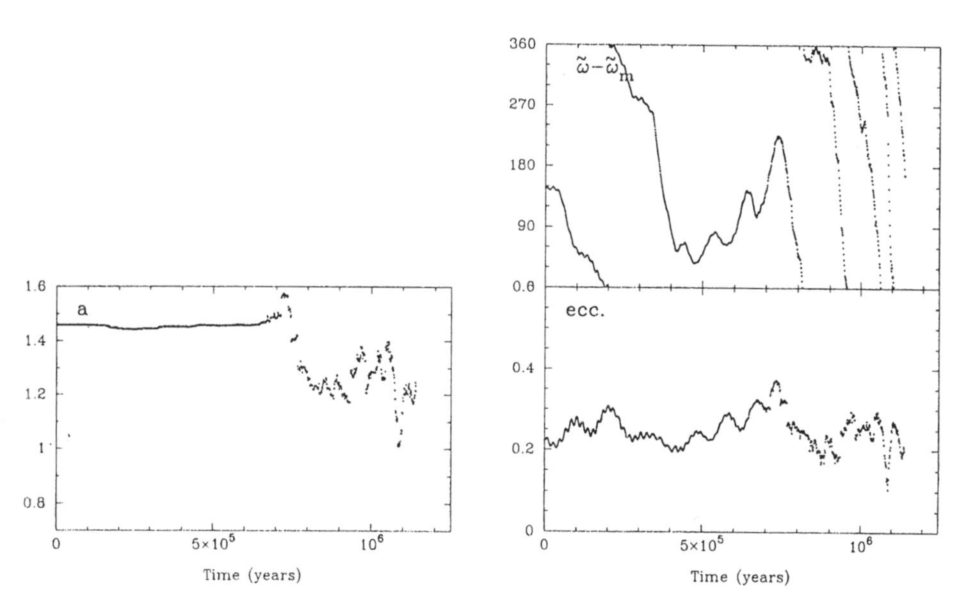

Figure 1. Evolution of the orbital elements of Eros until it collides with the Earth. Besides semimajor axis, eccentricity and inclination, the figure shows also the critical argument $\varpi - \varpi_M$ of the ν_4 secular resonances (ϖ_M being the perihelion longitude of Mars).

down to a value smaller than the orbital radius of the Earth. Then, close approaches to this planet result in a random walk of its semimajor axis and eventually the small body collides with our planet. Therefore, secular resonances with the inner planets appear to be efficient transport mechanisms in the region where these planets evolve.

A semi–analytical study has then confirmed this result by getting the location and effects of the 1st–order secular resonances in the region of semimajor axes < 2 AU (Michel and Froeschlé, 1997, Michel, 1997). Figure 2 shows that all the resonances involving the mean precession frequencies of the longitude of perihelion are present in this region. Moreover, since the semi–analytical method is also very suitable to study the regions in which the dynamics related to the argument of perihelion is strongly non–linear, it allows to obtain the location of the Kozai resonance. This last is indicated on Figures 2 and 3 by the zones marked KOZAI and corresponds to a libration of the perihelion argument of the small body around 0° or

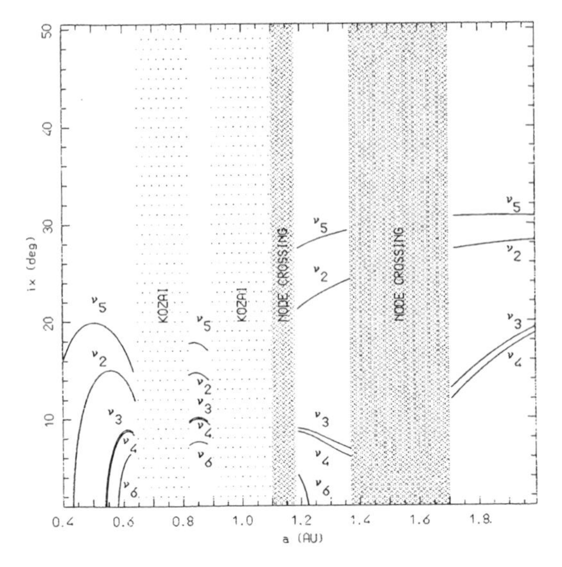

Figure 2. The location in the (a (AU), i (degrees)) plane of the secular resonances involving the precession rates of the perihelion longitudes of a small body and the planets, for $e = 0.1$. We follow the notations of Williams (1969) in naming the resonances. The regions marked KOZAI correspond to the regions of libration of the perihelion argument of the small body around 0° or 180°. In the regions marked NODE–CROSSING, the calculation of the free frequencies is not possible (see Michel and Froeschlé (1997) for explanations).

180°. Contrary to the secular resonances, the Kozai resonance provides an efficient protection mechanism from close approaches to planets (Michel and Thomas, 1997). In the zones marked NODE–CROSSING (Fig. 2 and 3), the computation of the free frequencies is not possible since there is an intrinsic problem with calculating the secular perturbations of planet–crossing orbits (See Michel and Froeschlé, 1997).

The study of the effects of secular resonances (Michel, 1997) has shown that the ones with the inner planets are the most efficient to cause drastic eccentricity changes and that the overlapping of two of them, namely ν_3 and ν_4 with, respectively, the Earth and Mars, even leads to a chaotic evolution of the orbital eccentricity (see Michel, 1997, Fig. 11). Therefore, secular resonances are not only a transport mechanism from the main belt to the planet–crossing region, they have also a role to play in the orbital evolution of NEAs and their transport from Mars–crossing to Earth– or even Venus–crossing orbits.

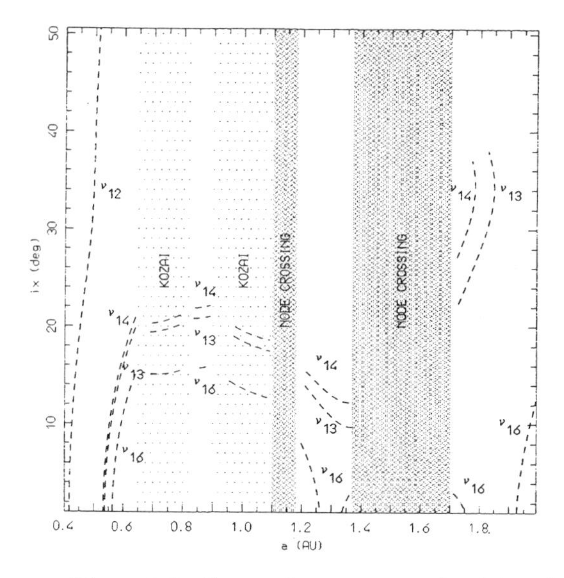

Figure 3. The same as Figure 2 for the secular resonances involving the precession rates of the nodal longitudes of a small body and the planets.

3. Transport mechanisms to highly inclined orbits

The location and effects of secular resonances involving the mean precession frequencies of the nodal longitudes of small bodies and planets have also been determined by Michel and Froeschlé (1997), and Michel (1997) with their semi–analytical method. Figure 3 shows that all these resonances are also present in the region of semimajor axes < 2 AU. Moreover, Michel (1997) has shown that the overlapping of the ν_{13} and ν_{14} resonances is a source of large scale chaos, leading to a chaotic behavior of the orbital inclination in the range from 5° to 25°. Thus, NEAs on highly inclined orbits could have been transported to these orbits by this efficient mechanism.

As an example, Figure 4 shows the orbital evolution of one of the SEAs (Small Earth–Approachers), computed by means of a numerical integration. The SEAs are objects with diameters smaller than 50 m. They have been discovered by the Spacewatch telescope (Rabonowitz et al., 1993) and have a perihelion distance close to 1 AU, orbital eccentricities $e < 0.35$ and inclinations ranging from 0° to $\approx 30°$. Rabinowitz et al. estimate that there is an excess of these small bodies with respect to the predictions based on the size distribution of larger objects in this region. Figure 4 shows that the overlapping of resonances results in a drastic decrease of the orbital inclination from $\approx 25°$ to 10° whereas the ν_{16} resonance with Saturn does not seem to have a significant effect. Up to now, no study has been able to reproduce or explain the highly inclined orbits of some of the SEAs and

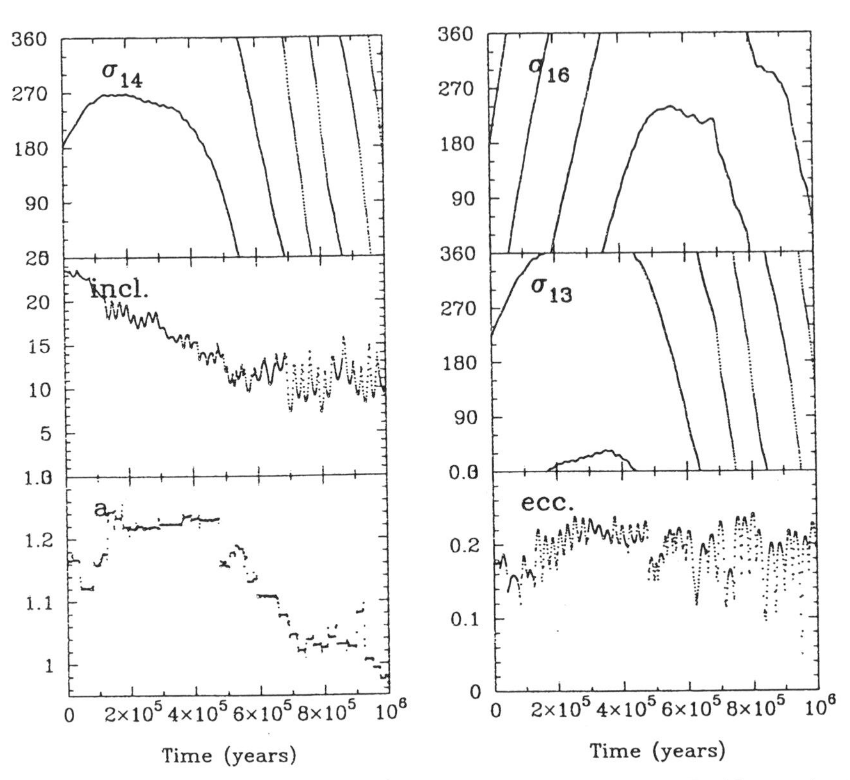

Figure 4. Evolution of the orbital elements of 1992 DU over 1 Myr. Besides semimajor axis a (UA), eccentricity e and inclination i (degrees), the figure shows the evolutions of the two critical arguments of the ν_{13}, ν_{14} and ν_{16} resonances: $\sigma_{13} = \Omega - s_3 t - \beta_3$, $\sigma_{14} = \Omega - s_4 t - \beta_4$ and $\sigma_{16} = \Omega - s_6 t - \beta_6$, where Ω is the longitude of node of the asteroid's orbit, s_3, s_4 and s_6 are the proper frequencies corresponding to each resonance, and β_3, β_4 and β_6 are the proper phases associated to the Earth, Mars and Saturn at time $t = 0$.

we can see here that this resonance overlapping may provide a transport mechanism to this kind of orbit. Thus, while an asteroid is already inside the terrestrial planet–crossing region, its orbital inclination can suffer a change of 15°, either decreasing or increasing, due to the secular resonances with the inner planets. This could be the reason why Bottke et al. (1996) could not produce highly inclined NEAs using a Monte–Carlo code accounting only for close approaches to planets.

References

Bottke, W.F., Nolan, M.C., Melosh, H.J., Vickery, A.M. and Greenberg, R. (1996) Origin of the Spacewatch small Earth-approaching asteroids, *Icarus* **122**, 406–427.

Bretagnon, P. (1974) Termes à longues périodes dans le système solaire, *Astron. Astrophys.* **30**, 141–154.

Froeschlé, Ch. and Morbidelli, A. (1994) The secular resonances in the Solar System, in Milani et al. (eds.), *Asteroids, Comets, Meteors*, pp. 189–204.
Froeschlé, Ch. (1997) Dynamical transport to planet–crossing orbits, *Celest. Mech.* **65**, 165–173.
Michel, P. and Thomas, F. (1996) The Kozai resonance for near–Earth asteroids with semimajor axes smaller than 2 AU, *Astron. Astrophys.* **307**, 310–318.
Michel, P., Froeschlé, Ch. and Farinella, P. (1996) Dynamical evolution of two near–Earth asteroids to be explored by spacecraft: (433) Eros and (4660) Nereus, *Astron. Astrophys.* **313**, 993–1007.
Michel, P. and Froeschlé, Ch. (1997) The location of linear secular resonances for semi-major axes smaller than 2 AU, *Icarus* **128**, 230–240.
Michel, P. (1997) Effects of linear secular resonances in the region of semimajor axes smaller than 2 AU, *Icarus* **129**, 348–366.
Moons, M. (1997) Review of the dynamics in the Kirkwood gaps, *Celest. Mech.* **65**, 175–204.
Morbidelli, A. and Henrard, J. (1991) Secular resonances in the asteroid belt: theoretical perturbation approach and the problem of their location, *Celest. Mech.* **51**, 131–167.
Rabinowitz, D.L., Gehrels, T., Scotti, J.V., McMillan, R.S., Perry, M.L., Wisniewski, W., Larson, S.M., Howell, E.S. and Mueller, B.E.A. (1993) Evidence for a near–Earth asteroid belt, *Nature* **363**, 704–706.
Williams, J.G. (1969) Secular perturbations in the Solar System, *Ph.D. dissertation*, University of California, Los Angeles.

METEOROID STREAMS AND METEOR SHOWERS

I.P. WILLIAMS
Astronomy Unit
Queen Mary
University of London
London E1 4NS
UK

Since time immemorial, humans have recognized streaks of fire moving fast across the sky, though the understanding that these were caused by small particles burning through friction with the upper atmosphere came somewhat later. The scientific study of meteors is perhaps no more than 200 years old, but in that time considerable progress has been made. By now, most of the important physical progesses are known, though of course we still do not understand everything about meteor showers. For example, making an accurate prediction of what the Leonid display will be like in 1998 or 1999 is still a great challenge.

1. Introduction

Humans must have been aware of the streaks of fire crossing the sky that we now call meteors ever since they became capable of being aware of anything. They were often referred to as *falling stars* and in many ancient Chinese, Japanese and Korean records, mention can be found of "stars falling like rain", or "many falling stars" (see Hasegawa, 1993). It must have been the same general thought that gave rise to the English colloquial name for them of *Shooting Stars.* The usually spectacular display from the Perseids in early August was referred to by Irish country folk as "the burning tears of St Lawrence" (see Yeomans 1991). Within the Western Christian world, there was a belief that the Universe was perfect so that these displays could not of course be actually related to falling stars and for a very long time they were regarded as Atmospheric phenomena. Indeed, the name *meteors* implies this. In the book of Revelations in the Bible, the opening of a sequence of seven seals fortells the end of the world and the description of the events following the opening of the sixt seal talks of the stars falling from

B.A. Steves and A.E. Roy (eds.), The Dynamics of Small Bodies in the Solar System, 179–185.

the sky like leaves off a tree. Possibly as a consequence of this spectacular meteor showers in the past have triggered many a speculation that the end of the world is nigh. (see Hughes 1996, for an interesting discussion of this in connection with perhaps the most famous meteor shower engraving that exists). Steel (1998) has discussed the possibility that part of *The Rime of the Ancient Mariner* by Coleridge was inspired by a display from the Leonid shower.

The observational facts concerning meteors in their simplest form are quite straight forward. Meteors can be seen at any time, though the mean sporadic rate is very low, no more than a few per hour. This is not the stuff to inspire either Coleridge or religeous fanatics. At certain times the meteor rate climbs regularly each year, for example in early August the rate of meteor influx climbs to one or two per minute. This is of course the Perseid meteor shower. During such time periods, the meteors do not appear uniformly distributed across the sky but appear to flow, or radiate out of a fixed point. Not surprisingly, this point is called the *radiant* of the shower and the radiant of the Perseid shower lies in the Constellation of Persius, hence the name. This behaviour is interpreted as implying that the meteoroids are moving on parallel courses and that the existence of a radiant point is due to parallax. In other words,there is a stream of meteoroids impinging the Earth and generating the meteor shower. The first persons on record to have noticed that shower meteors radiated from a point were Olmstead (1834) and Twining (1834). A few years later, Herrick (1837, 1838) pointed out that the annual showers were periodic on a siderial rather than a tropical year, in other words they were extra-terrestrial in origin. Some indication of this had come some thirty years earlier when Benzenberg and Brandes (1800) had simultaneously observed meteors from two different locations and through parallax determined their height to be about $90km$. These annual showers, may be spectacular enough to generate names for them in folklore, especially when they coincide with famous saint days, but again this is hardly the stuff to base predictions for the end of the world on. Fortunately, some showers appear to generate a very enhanced display at regular intervals. The most well known of these is the Leonids, where truly awsome displays are recorded as having occurred. For example, in 1966, the rate was tens of meteors per second, a truly falling of the stars from the sky. such a display lasted for under an hour, but records show that such displays may be seen at intervals of time that are multiples of about 33 years. Two such recorded displays were in 1799 and 1833 and these helped Adams (1867), LeVerrier (1867) and Schiaparelli(1867) to conclude that the orbit of the Leonid meteors were very similar to that of comet $55P/Tempel-Tuttle$ and that 33 years were very close to the orbital period of this comet. Since then comet-meteor stream pairs have been identified for virtually all recognizable significant stream, and a list of pairings was produced by Cook (1973). Many of the gaps in Cook's list have since been filled.

These simple facts allows a straightforward model of meteor showers and associated meteoroid streams to be constructed. solid particles, which we shall call *meteoroids* are lost from a comet. Since any relative speed between comet and meteoroid will be much less than the orbital speed, the meteoroids will move on orbits that are only slightly perturbed from the cometary orbit. If a large number of meteoroids are lost, this will of course form a cloud about the comet, co-moving with it. As the semi-major axes of each meteoroid will be slightly different, each will have a slightly different orbital period, resulting in a drift in the epoch of return to perihelion. After many orbits this results in meteoroids effectivly being located at all points around the orbit. There will also be some reduction in the space density of meteorods with the passage of time for two main reasons, the small changes in semi-major axes of individual meteoroids due to gravitational perturbations will continue, so that the total volume occupied by a stream increase, while also meteoroids will be lost from the stream through collisions, radiation pressure and gravitational perturbations. A normal stream is thus middle-ages, with meteoroids all around the orbit so that a shower is seen every year. Gravitational perturbations from the planets will also cause a steady evolution of the mean orbit of the meteoroid stream, this being most noticable in the longitude of the ascending node, that is the time of appearance of the associated shower. This drift is usually slightly under a day per century. Such evolution may also cause a change in the number of meteors seen as the heliocentric distance of the ascending node changes and the densest part of the stream moves away from the Earth's orbit. In a very old stream, the number density of meteoroids will be low so that the stream is never very noticeable, but again constant each year. A very young stream on the other hand will only show activity at certain years, and that at a much enhanced level whenever the Earth passes through the cloud of meteoroids which is still surrounding the cometary nucleus since insufficient time has passed for it to spread about the orbit.

This picture of meteoroid stream evolution and the associated behaviour of meteor showers was firmly established by the 1950s. Indeed so firmly was it established that most astronomers came to the view (which is still widely held) that there was nothing much further to be learnt from the study of meteor showers. Though I believe that the basic underlying physics implied in the above model is still true, I also believe that there is a considerable amount that we do not fully understand. I will discuss some of these in the following sections.

2. The Meteoroid Ejection Process

Though the general description given above regarding the formation process is almost certainly correct and that any ejection velocity of a meteoroid relative to the parent will be small, the subsequent stream evolution will depend on this ejection velocity. Clearly, if this is nearly equal to the orbital speed, meteoroids can essentially reach all parts of the Solar System so that no stream will be seen. Equally obviously, if the ejection speed is zero, the meteoroids will co-move with the parent for all time (since gravitational perturbations will also be identical). There is presumably a non-linear relationship between the ejection velocity and the spread in orbital elements of meteoroids. Initial changes occurr directly because of this velocity, but the consequence is that the meteoroid then experiences different perturbations from the parent. Early on, the ejection velocity effect dominates, while later the differential gravitational perturbations dominates. The problem is deciding what is early and what is late. A numerical simulation can of course answer this question for an individual stream provided the ejection velocity is known. Unfortunately, it is not. The escape velocity from a cometary nucles is of order a few meters per second and presumably the escape velocity is greater than this. Similarly gas outflow speed from comets is of the order of a few kilometers per second and the meteoroid ejection velocity is presumably less than this. But these general considerations give a range of a factor of a thousand. Observations of dust in comet Hale-Bopp suggested outflow speeds of the order of $300ms^{-1}$, bang in the middle of this range. This observed dust is smaller than conventional meteoroids and so may have been ejected with a higher speed. Williams (1996) reversed the argument, asking what the ejection velocity would be if all the dispersion found in meteor streams was due to the ejection velocity. This clearly gives an upper limit and it turned out to be in the range $160 - 880ms^{-1}$, in agreement with the above arguments but not of great help in narrowing the range. Theoretical considerations of the ejection process similarly do not help. Different models all produce values within the above range (*eg* Whipple 1951, Gustafson 1989, Harris *et al.* 1995) but between them essentially cover the whole available range. Solving this problem is very important, It can be done either through a study of meteor showers, in which case we learn about cometary nuclei or through the study of cometary nuclei, in which case great advances will be made in the study of meteoroid streams.

3. Outbursts in Meteor showers

As already mentioned, we should expect young meteoroid streams to produce showers of uneven strength from year to year, being very strong when the parent is close to the Earth and much weaker otherwise. Such strong meteor displays we shall call outbursts. Outbursts are a regular feature of the Leonid stream, indeed it is these outbursts that have made the Leonids famous and in the early days helped towards our understanding of meteoroid streams. However, the situation within individual streams is not quite as simple as it looks at first sight, and in fact few streams behave like the simple picture when an outburst is observed. In this section we shall look briefly at three meteor showers that each illustrate a potential problem.

3.1. THE PERSEIDS

The Perseids is one of the few major and regular meteor showers that produces a display of roughly the same strength every year. It is also a fairly old shower with many record of it existing in ancient documents. Its parent is also well known, comet109$P/Swift-Tuttle$. In the early 90's, a second peak in the activity curve of the Perseids was noticed (The optical observations behind these activity profiles are summarized by Brown and Rendtel, 1997). Since the appearance of this peak was roughly coincident in time with the return of the parent comet to the Earth's locality, this peak was interpreted as being due to new meteoroids, recently ejected from the comet. Detailed models by Williams and Wu (1994) confirmed that this peak could be associated primarily with meteoroids ejected at the last (rather than current) apparition of the comet. This new peak in the Perseid shower activity profile is thus not such a mystery but its existence does remind us that meteotoid streams are not perhaps as static as had been thought. Not all the meteoroids may be of the same age and new meteoroids are added to the stream at each apparition of the comet. If significant changes in the cometary orbit occurrs, then a meteoroid stream is not so much a single coherent stream as a number of similar filaments.

3.2. THE LYRIDS

In contrast to the Perseids where a strong display is seen each year, the Lyrids are almost non exisent in most years, but outbursts are seen which, in relation to the normal activity are quite strong, reaching a Zenithal hourly rate of several hundred, an increase over the norm of perhaps 30 or so. Such an increase in the Perseid stream for example would lead to the event being labelled a major storm. Lindblad and Porubcan (1992) have chronicled the all the recorded outbursts in the Lyrids and , not surprising perhaps in view of the actual weakness of the whole event, found that most records were recent with several being recorded this century. The parent comet of the Lyrids, comet Thatcher, has a period of order 400 years (it has actually only been observed once, at the epoch of discovery, so there is considerable uncertainty about the actual period of comet Thatcher, beyond the fact that it is long). These recorded outbursts can not therfore be associated with the return of comet Thatcher to the inner solar system. Arter and Williams (1997) have produced a computer model of the stream, suggesting that the outbursts are caused by perturbations of stream filaments into an Earth intersecting orbit by Jupiter. If this is correct, then the behaviour of the April Lyrids does not represent a mystery either, though it does illustrate that relative outbursts can occurr for reasons other than a passage of the Earth through a dense cloud of recently ejected meteoroids.

3.3. THE LEONIDS

The history of the developement of meteor stream science is peppered with observations of strong outbursts associated with the Leonids. Indeed, it is often claimed that it was the Leonid displays in 1799 and 1833 that gave birth to the study of meteor streams. at first sight, the Leonids display all the characteristics of a young stream that we described earlier, namely very strong displays whenever the parent comet is close to perihelion but very weak otherwise. Indeed, most modellers (*eg* Wu and Williams 1996, Brown and Jones 1996) assume models based on this notion to try to predict the behaviour of the Leonids at the turn of millenium.

Unfortunately, the Leonids have been observed for a long time (well over 1000 years) and can hardly be regarded as young. Also the change from an outburs to a non-outburst is very sharp, a gradual decline would be expected. Williams (1998) has suggested that perturbations due to Uranus are responsible for clearing meteoroids out of the stream in most parts of the orbit away from the parent. Hence, even in a well studied stream there are still surprises to be found.

4. Conclusions

Meteor showers have now been observed for two centuries and the general physics which governs their behaviour is generally understood. However a number of aspects of their developement and evolution requires further studies. Paramount amongst these are the details of the ejection process from a cometary nucleus and the details of the evolution of individual streams that gives rise to spectacular outbursts.

References

1. Adams J. C., 1867, On the orbit of the November meteors,*Mon. Not. R. astr. Soc*,**27**, 247-252
2. Arter T.R. & Williams I.P., 1997, Periodic behaviour of the April Lyrids, *Mon. Not. R. astr. Soc.*, **286**, 163-172

3. Benzenberg J.F. & Brandes H.W.,1800, Versuch die entfernung, die geschwindigkeit und die bahn der sternschnuppen zu bestimmen, *Annalen der Phys*,**6**, 224-232
4. Brown P. & Jones J., 1996, Dynamics of the Leonid Meteoroid Stream :a Numerical Approach, in *Physics, Chemistry and Dynamics of Interplanetary Dust* Eds Gustafson B.A.S & Hanner M.S., ASP Conf.Ser, 113-116
5. Brown P. & Rendtel J., 1996, The Perseid Meteoroid stream: Characterization of Recent Activity from Visual Observations, *Icarus*, **124**, 414-428
6. Cook A.F., 1973, A working list of Meteor Streams, in Evolutionary and physical properties of Meteoroids, Eds Hemenway, C. L., P. M. Millman and A.F. Cook, NASA SP-319, Washington DC, 183-191
7. Gustafson B. A. S., 1989, Comet ejection and dynamics of nonspherical dust particles and meteoroids, *Astrophys. Jl.*,**337**, 945-949
8. Hasegawa I.,1993, Historical records of meteor showers, in *Meteoroids and their parent bodies*, Eds Stohl, J. & I. P. Williams, Slovak Academy of Sciences, Bratislava, 209-223
9. Harris N.W. Yau K.K. & Hughes D.W., 1995, The True extent of the nodal distribution of the Perseid meteoroid stream, *Mon. Not. R. astr . Soc.*, **273**, 999-1015
10. Herrick E. C.,1837, On the shooting stars of August 9th and 10th 1837, and on the probability of the annual occurrence of a meteoric shower in August, *American Jl.Sci.*,**33**, 176-180
11. Herrick E. C.,1838, Further proof of an annual Meteoric Shower in August, with remarks on Shooting Stars in general,*American Jl. Sci.*,**33**, 354-364
12. Hughes D.W., 1995, *Earth Moon and Planets*,**86**, 311,
13. Le Verrier U. J. J., 1867, Sur les etoiles filantes de 13 Novembre et du 10 Aout, *Comptes Rendus*, **64**, 94-99
14. Linblad B.A. & Porubcan V., 1992, Activity of the Lyrid Meteoroid stream, in *Asteroids Comets Meteors 91*, Eds Harris A.W & Bowell E., Lunar and Planetary Institute, Tucson, 367-370
15. Olmstead D.,1834, Observations on the meteors of 13 Nov.1833, *American Jl. Sci.*, **25**, 354-411
16. Schiaparelli G. V.,1867, Sur la relation qui existe entre les cometes et les etoiles filantes, *Astronomische Nachrichten*,**68**, 331-332
17. Steel D., 1998, The Leonid Meteor showers and the genesis of the *Ancient Mariner*, *A & G*,**13** 20-23
18. Twining A. C., 1834, Investigations respecting the meteors of Nov.13th, 1833,*American Jl. Sci.*,**26**, 320-352
19. Whipple F. L.,1951, A comet model II. Physical relations for comets and meteors,*Astrophys. Jl.*,**113**, 464-474
20. Williams I. P.,1996, What can meteoroid streams tell us about the ejection velocities of dust from comets, *Earth Moon and Planets*,**72**, 321-326
21. Williams I.P., 1998, The Leonid Meteor shower: why are there storms but no regular annual activity?, *Mon. Not. R. astr. Soc*, **292**, L37-L40
22. Williams I.P. & Wu Z.,1994, The current Perseid meteor shower, *Mon. Not.R. astr. Soc*, **269**, 524-528
23. Wu, Z. & Williams I.P.,1996, Leonid meteor storms, *Mon. Not. R. astr. Soc*, **264**, 980-990
24. Yeomans, D. K.,1991, Comets: a Chronological History of Observations, Science, Myth and Folklore, J. Wiley and Son, New York

PLANETARY CLOSE ENCOUNTERS: THE ENGINE OF COMETARY ORBITAL EVOLUTION

G.B. VALSECCHI
I.A.S. – Planetologia, Area di Ricerca C.N.R.
via Fosso del Cavaliere, I-00133 Roma, Italy

Abstract. The dynamical evolution of comets from their source regions to the observed orbits is mainly due to close encounters with the giant planets. A very important parameter for these encounters is the planetocentric velocity; fast encounters, taking place on hyperbolic planetocentric orbits, are in general effective only if very deep, whereas slow ones, during which temporary satellite captures may occur, can greatly modify cometary orbits even if they are rather shallow. Especially in the case of slow encounters, the outcomes can be extremely sensitive to initial conditions.

1. Introduction

Comets can be found anywhere in the Solar System, from the surface of the Sun to the outskirts of the region in which solar gravity dominates over other perturbations. There are, however, at least two reasons why observed comets cannot have been for very long in the orbits in which we see them: on those orbits physical aging goes on very quickly, and the dynamical lifetime against ejection on a hyperbolic trajectory is very short compared to the age of the Solar System.

Two main reservoirs have been identified that should be able to supply observable comets at the observed rate: the Oort cloud, for the long-period comets (Oort 1950) and the short-period ones of Halley type (see, e.g., Carusi and Valsecchi 1992), and the Edgeworth-Kuiper belt, for the Jupiter family of comets (Fernández 1980). The Oort cloud is further subdivided into a more massive, tightly bound, inner core, surrounded by the outer part of the cloud, that actually corresponds to what Oort (1950) originally proposed.

B.A. Steves and A.E. Roy (eds.), The Dynamics of Small Bodies in the Solar System, 187–196.

The situation is sketched in Fig. 1, where dots represent 291 original orbits of long-period comets, as well as the orbits of all short-period comets, taken from Marsden and Williams (1996), whereas small circles represent the orbits of the trans-neptunian objects, of the Centaurs, and of the asteroids with $Q > 4.5$ AU and $e > 0.4$. The Figure is a log-log plot of aphelion distance Q vs. perihelion distance q, both in AU; the area in which $q > Q$ is by definition a forbidden one, and orbits with $q < 0.0465$ AU lead to collision with the Sun. The position of the major planets (except Mercury and Pluto) is indicated by large dots along the $q = Q$ condition.

Various regions are identified on the plot:

- region LP, containing the observed long-period comets ($P > 200$ yr);
- region HT, containing the Halley-type comets ($20 < P < 200$ yr);
- region JF, containing the Jupiter-family ($20 < P$ yr);
- region OOC, representing the classical, outer Oort cloud;
- region IOC, showing the inner core of the Oort cloud;
- region EKB, containing the members of the Edgeworth-Kuiper belt.

The dividing line based on P between LP, HT and JF comets is the classical one; a better criterion to distinguish between HT and JF comets has been proposed by Carusi et al. (1986) and Carusi and Valsecchi (1987), and is based on the Tisserand parameter (see below).

Note that many EKB members protrude out of the EKB region as marked on the plot, due to their orbital eccentricity: they are the so-called "plutinos", in 2/3 mean motion resonance with Neptune, like Pluto; moreover, we have (somewhat arbitrarily) put the border between IOC and EKB at $Q = 100$ AU, so that object 1996 TL_{66} ends up in the inner Oort cloud. Finally, note that several objects, the Centaurs, appear to be "on the way" between the EKB and the JF regions.

2. Planetary encounters

The transitions from regions LP to HT, and from region EKB to JF, take place because of encounters with the giant planets, that are the quickest means by which a cometary orbit can be deeply transformed; in fact, close encounters play the most important role in governing the dynamical evolution of comets when they move within the planetary region.

A very important aspect to be taken into account is the planetocentric velocity of the comet at encounter; its unperturbed value (i.e. that obtained ignoring the mass of the planet) is to a good approximation given by

$$U_p = \sqrt{3 - T_p} \tag{1}$$

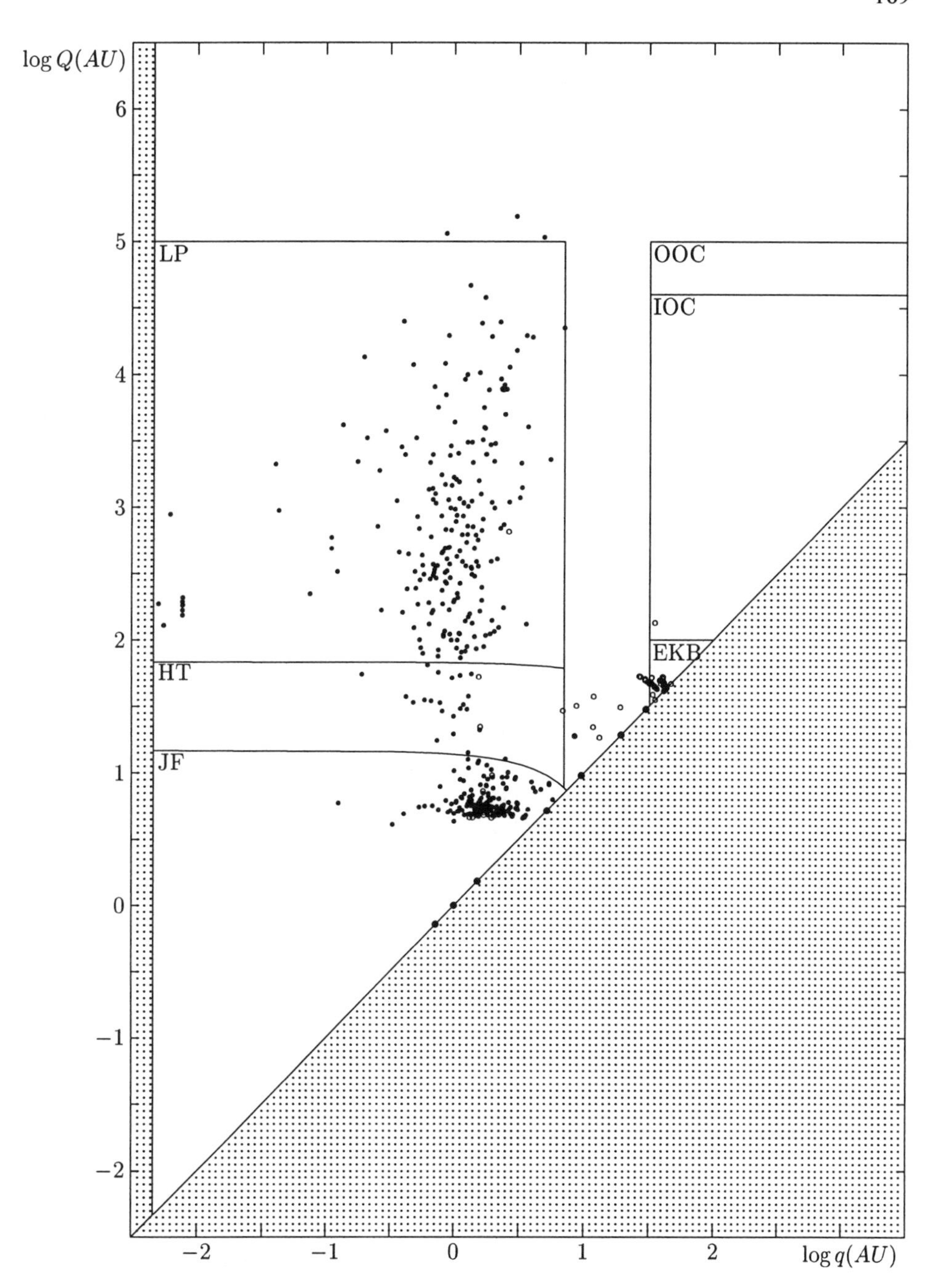

Figure 1. Log-log plot of aphelion distance Q versus perihelion distances q of the orbits of observed comets and of peculiar asteroids (see text).

where U_p is calculated from the Tisserand parameter

$$T_p = \frac{a_p}{a} + 2\sqrt{\frac{a}{a_p}(1-e^2)}\cos i \tag{2}$$

(a_p is the orbital semimajor axis of the planet encountered, and U_p is in units of the orbital velocity of the latter). T_p is a near-invariant for close encounters with the respective planet, so that its pre-encounter value is restored when the close interaction is over (for a quantitative discussion, see Carusi et al. 1995a).

As noted by Kresák (1972), almost all HT comets are characterized by $T_J < 2$ (the suffix J stands for Jupiter), so that for them $U_J > 1$, whereas U_J is smaller for the short-period comets belonging to the Jupiter family. This has been turned into a classification tool by Carusi et al. (1986) and Carusi and Valsecchi (1987), who proposed that periodic comets with $T_J < 2$ be considered of Halley type, and those with $T_J > 2$ as belonging to the Jupiter family. Therefore, encounters of JF comets with Jupiter are characterized by lower relative velocities, that make their orbits sensitive to planetary perturbations already at rather large distances.

Carusi and Valsecchi (1982) and Greenberg et al. (1988) have identified three phases in the time sequence of the approaching branch of a close encounter: a first phase where the motion is essentially heliocentric, with perturbations by the planet increasing with time but unable to produce qualitative changes in the orbit of the comet; a second phase where the perturbations by the Sun and the planet are of roughly the same size; and a third phase during which the comet is practically under the control of the planet, with the Sun as a perturber. The sequence is repeated, in opposite order, in the outgoing branch. The phase in which the perturbations of the Sun and of the planet are comparable is completely ignored in Öpik's theory of close encounters (Öpik, 1976), that in fact works well only for sufficiently fast encounters of objects in planet-crossing orbits (Greenberg et al. 1988).

In general, for encounters in which the planetocentric trajectory is hyperbolic, the deeper the encounter, the stronger its effects on the orbital parameters of the comet; a deviation from this simple rule of thumb takes place, for extremely close encounters, under the conditions described by Valsecchi (1992) and Valsecchi and Manara (1997). In short, if a comet passes extremely close to a massive planet, its velocity vector may be rotated by nearly 180°, and the resulting perturbations in orbital energy, that in this case are largely independent of the details of the encounter, may turn out to be smaller than those caused by a less deep encounter under appropriate circumstances.

On the other hand, for comets on orbits with $T_J \approx 3$, or even slightly larger, we can speak of slow encounters: here the relative velocity can be so low that the planetocentric energy is negative, and the comet undergoes a temporary satellite capture.

Many examples of this behaviour concerning real comets have been studied in recent times (Rickman 1979; Carusi and Valsecchi 1979, 1981; Carusi et al. 1981, 1985, 1995b; Tancredi et al. 1990; Benner and McKinnon 1995; Kary and Dones 1996); the best known cases are those of 39P/Oterma, 82P/Gehrels 3, 111P/Helin-Roman-Crockett and D/Shoemaker-Levy 9. In particular, numerical studies of motion of the latter comet have shown that the comet presumably underwent a very long satellite capture before the collision with Jupiter.

Carusi et al. (1981, 1982a) studied the orbital evolution 39P/Oterma after 1930 in a restricted elliptic Sun-Jupiter-comet 3-body problem, and found that it is extremely sensitive to small variations in the initial conditions. By varying only the starting mean anomaly of the comet, they showed that the jovicentric pattern during the encounter lasting from 1935 to 1941 changed in a peculiar way: for a given range of initial values they found a pattern, that abruptly changed into a totally different one at a certain value of the initial anomaly, remaining then qualitatively unchanged for another interval in anomaly, and so on. At the transitions between different patterns, very long satellite captures could take place.

These results can be reproduced also in the simple restricted circular Sun-Jupiter-comet 3-body problem. Using the integrator RADAU (Everhart 1985), we have integrated for 20 000 d a fictitious comet with orbital parameters similar to those of 39P/Oterma in 1933 ($a = 6.9224$ AU, $e = 0.1641$, $i = 3.08°$, $\omega = 243.05°$, $\Omega = 35.52°$, $M = 256.36°$); Jupiter is on a circular orbit of radius 5.203 AU, of null inclination, with initial $M_J = 124°$, and its inverse mass is 1047.355.

Figure 2 shows the results. In the top left frame it is possible to see that our fictitious object, like the real comet, is put, after the first encounter, in a 3/2 mean motion resonance with Jupiter, and after the shortest possible stay there (three revolutions), encounters Jupiter again and is sent into an orbit rather similar to the initial one.

If we vary the initial anomaly by $-0.02°$, setting $M = 256.34°$, the jovicentric pattern is similar during the first encounter, but the comet is now out of the exact resonance, and the second encounter does not take place within the time span considered (Fig. 3).

If we go in the other direction, varying the anomaly by $0.02°$, ($M = 256.38°$), we find that also in this case the jovicentric pattern remains similar during the first encounter, but the comet is again slightly out of the

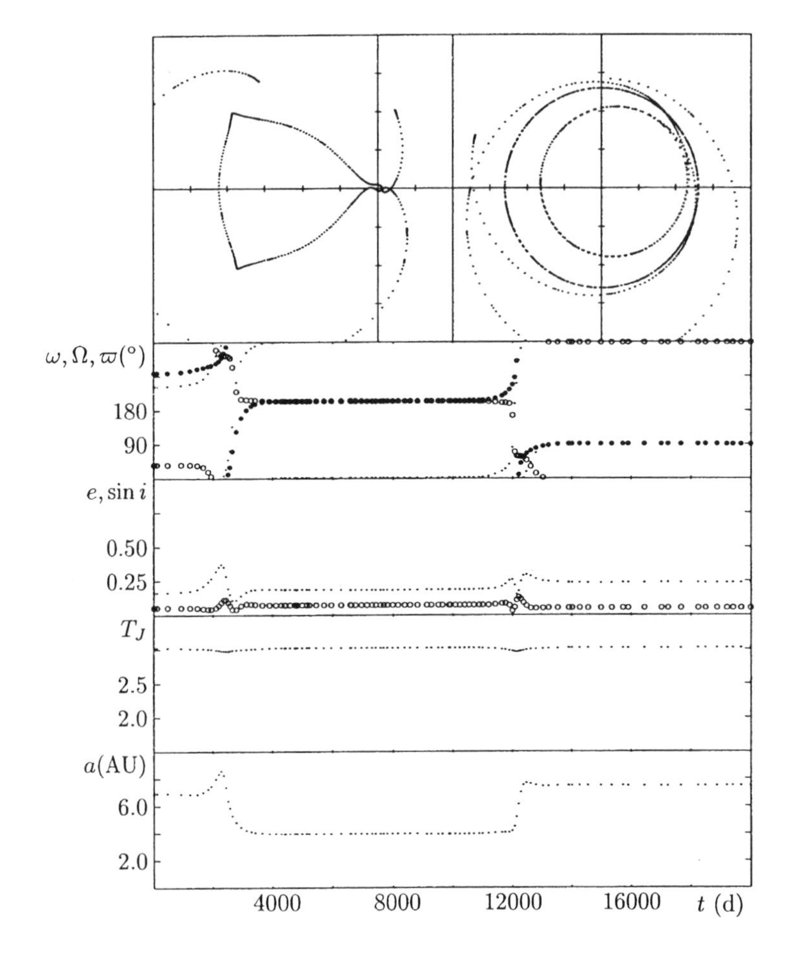

Figure 2. Evolution in the jovicentric rotating frame (top left) and in the heliocentric sidereal frame (top right) of a fictitious comet with initial orbital parameters similar to those of 39P/Oterma in 1933, integrated in a restricted circular Sun-Jupiter-comet 3-body problem for 20 000 d. In the remaining frames are given the time behaviours of, from bottom to top, a, T_J, e and $\sin i$ (small circles), and finally ω (small dots), Ω (small circles), and ϖ (large dots).

exact resonance; this time the second encounter takes place, but leaves the comet in an interior orbit (Fig. 4).

If we continue in this game, we can reach the situation in which there is an abrupt change of orbital pattern, and long temporary satellite captures become possible; this is obtained for the initial anomaly $M = 257.6°$. Figure 5 shows the result: a very long satellite capture (note that this integration has been prolonged to 30 000 d), lasting for many tens of years, qualitatively similar to that of D/Shoemaker-Levy 9.

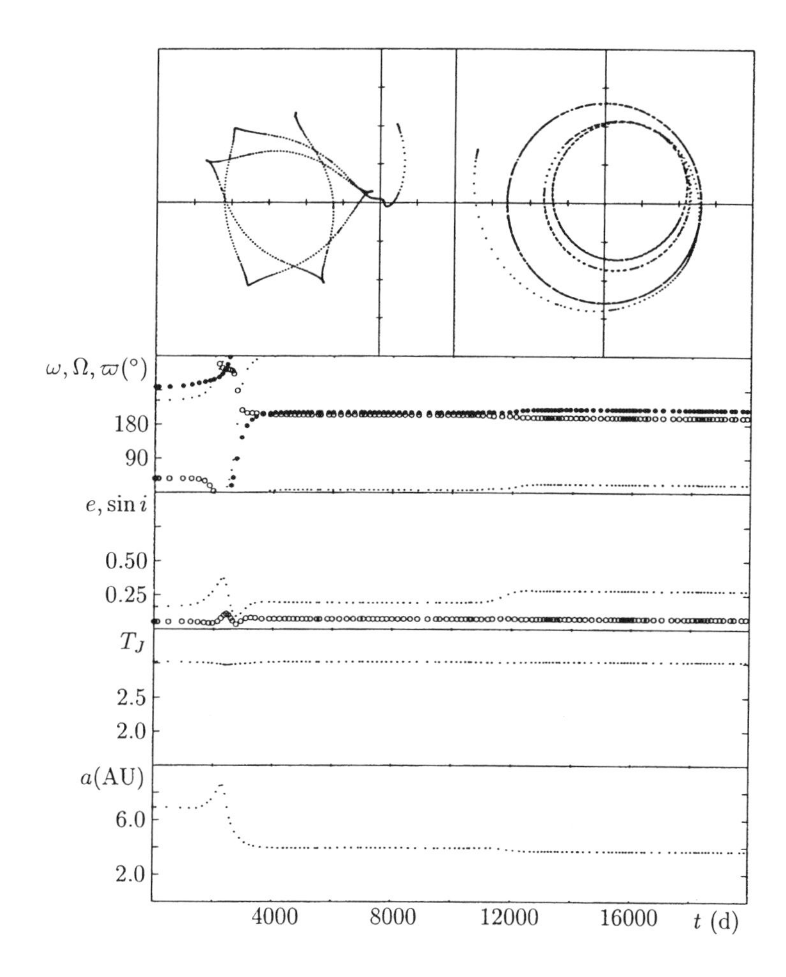

Figure 3. Same as Fig. 2, for a different initial mean anomaly (see text).

By continuing to change the initial anomaly, one can find the entire suite of orbital patterns explored by Carusi et al. (1982a).

3. Conclusions

Close encounters with the giant planets are the quickest way to displace the orbits of comets throughout the planetary region; in particular, encounters with Jupiter affect strongly the evolution of JF comets.

If the jovicentric orbit at an encounter is hyperbolic, the planetocentric pattern changes in a rather simple way if the initial conditions of the encounter are varied, as illustrated by Carusi et al. (1982b) for the 1779 encounter of D/Lexell with Jupiter.

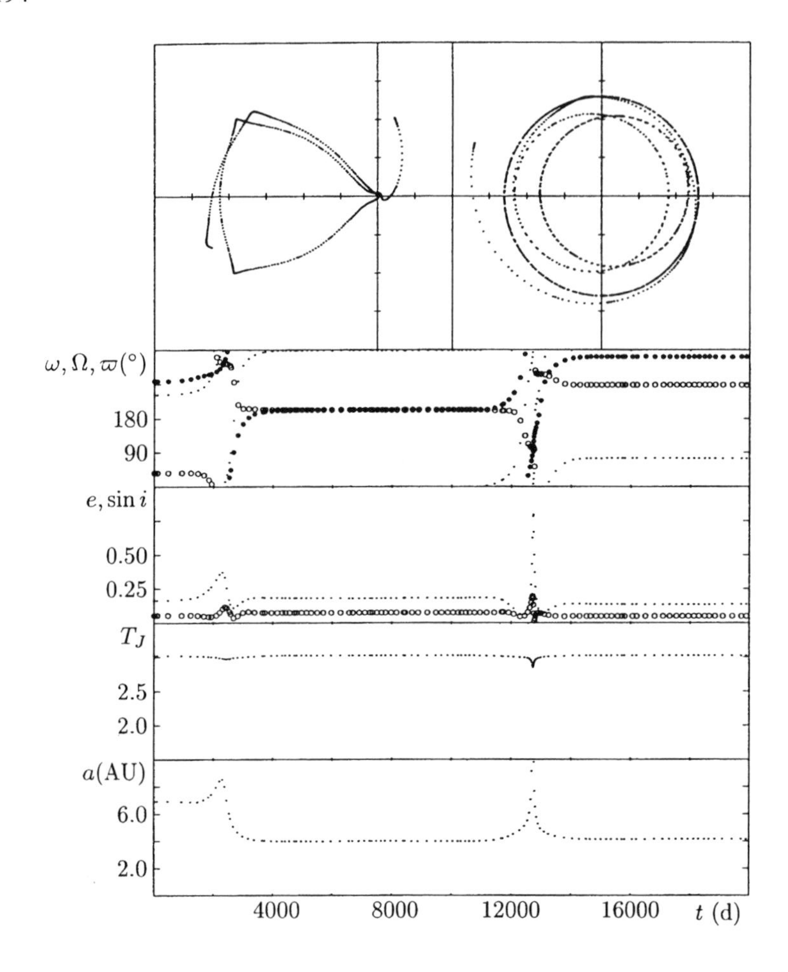

Figure 4. Same as Figs. 2 and 3, for a different initial mean anomaly (see text).

If, on the other hand, the jovicentric orbit turns out to be elliptic, the pattern at the encounter can become complicated, and different families of trajectories are obtained for slightly different initial conditions, with very long temporary satellite captures taking place with a non negligible frequency, as suggested by the recent case of comet D/Shoemaker-Levy 9.

Systematic exploration of these families of trajectories have been carried out in the planar Hill problem by Petit and Hénon (1986), but still have to be carried out in either the three-dimensional Hill's problem or the restricted circular 3-body problem.

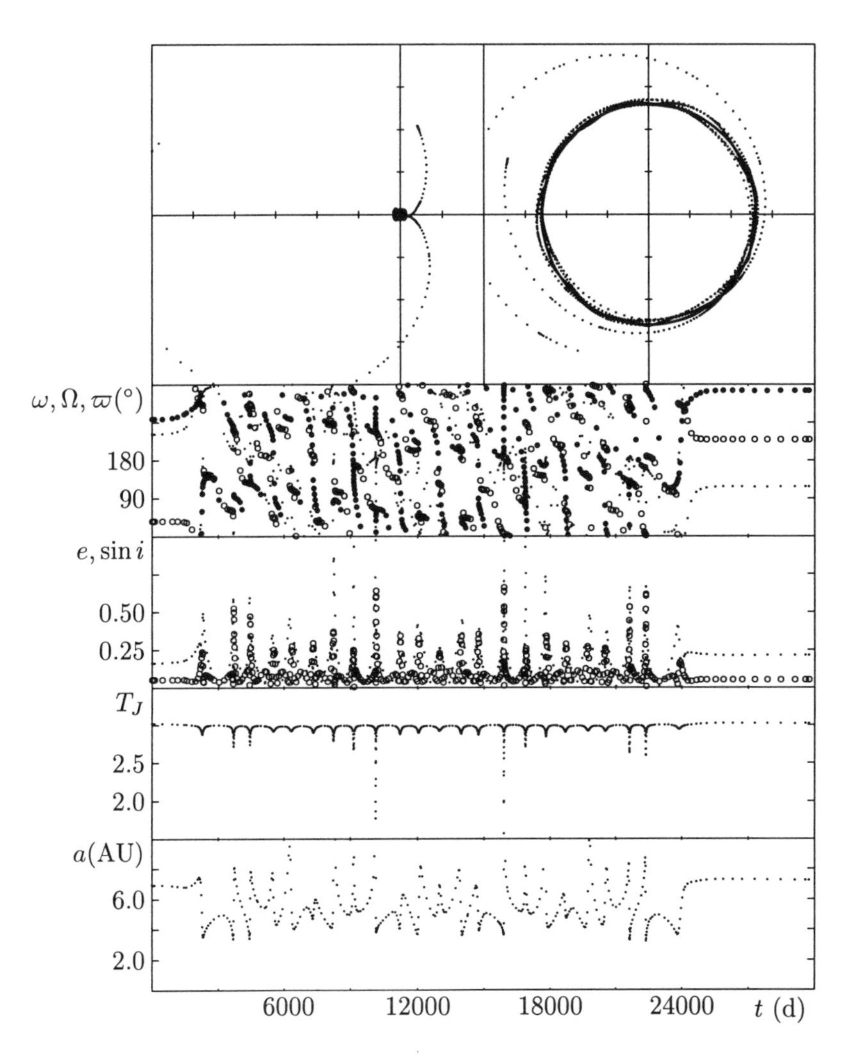

Figure 5. Same as Figs. 2, 3 and 4, for a different initial mean anomaly (see text). Note the longer duration of this integration.

References

1. Benner, L.A.M. and McKinnon, W.B. (1995), On the Orbital Evolution and Origin of Comet Shoemaker-Levy 9, *Icarus*, **118**, pp. 155-168.
2. Carusi, A. and Valsecchi, G.B. (1979), Numerical Simulations of Close Encounters between Jupiter and Minor Bodies, in *Asteroids* (T. Gehrels ed.), Univ. Arizona Press, Tucson, USA, pp. 391-416.
3. Carusi, A. and Valsecchi, G.B. (1981), Temporary Satellite Captures of Comets by Jupiter, *Astron. Astrophys.* **94**, pp. 226-228.
4. Carusi, A. and Valsecchi, G.B. (1982), Strong perturbations at close encounters with Jupiter, in *Sun and Planetary System* (W. Fricke and G. Teleki eds.), D. Reidel, Dordrecht, Holland, pp. 379-384.
5. Carusi, A. and Valsecchi, G.B. (1987), Dynamical evolution of short-period comets,

in *Interplanetary matter* (Z. Ceplecha and P. Pecina eds.), Czechoslov. Acad. of Sci. Publ., Prague, Czechoslovakia, pp. 21-28.
6. Carusi, A. and Valsecchi, G.B. (1992), Dynamics of comets, in *Chaos, Resonance and Collective Dynamical Phenomena in the Solar System* (S. Ferraz-Mello ed.), Kluwer Acad. Publ., Dordrecht, Holland, pp. 255-268.
7. Carusi, A., Kresák, Ľ. and Valsecchi, G.B. (1981), Perturbations by Jupiter of a chain of objects moving in the orbit of Comet Oterma, *Astron. Astrophys.* **99**, pp. 262-269.
8. Carusi, A., Kresák, Ľ. and Valsecchi, G.B. (1982a), Orbital patterns of interplanetary objects at close encounters with Jupiter, *Bull. Astron. Inst. Czechoslov.* **33**, 141-150.
9. Carusi, A., Kresáková, M. and Valsecchi, G.B. (1982b), Perturbations by Jupiter of the particles ejected from Comet Lexell, *Astron. Astrophys.* **116**, 201-209.
10. Carusi, A., Kresák, Ľ., Perozzi, E. and Valsecchi, G.B. (1985), *Long-Term Evolution of Short-Period Comets*, Adam Hilger, Bristol.
11. Carusi, A., Kresák, Ľ., Perozzi, E. and Valsecchi, G.B. (1986), Some General Features of the Dynamics of Halley-Type Comets, in *Proceedings of the 20th ESLAB Symposium on the Exploration of Halley's Comet*, ESA SP-250, **II**, pp. 413-418.
12. Carusi, A., Kresák, Ľ. and Valsecchi, G.B. (1995a), Conservation of the Tisserand Parameter at Close Encounters of Interplanetary Objects with Jupiter, *Earth, Moon and Planets* **68**, 71-94.
13. Carusi, A., Kresák, Ľ. and Valsecchi, G.B. (1995b), *Electronic Atlas of Dynamical Evolutions of Short-Period Comets*, available on the World Wide Web at URL `http://www.ias.rm.cnr.it/ias-home/comet/catalog.html`.
14. Everhart, E. (1985), An Efficient Integrator that Uses Gauss-Radau Spacings, in *Dynamics of Comets: Their Origin and Evolution* (A. Carusi and G.B. Valsecchi eds.), D. Reidel, Dordrecht, Holland, pp. 185-202.
15. Fernández, J.A. (1980), On the existence of a comet belt beyond Neptune, *Mon. Not. R. Astron. Soc.* **192**, 481-491.
16. Greenberg, R., Carusi, A. and Valsecchi, G.B. (1988), Outcomes of planetary close encounters: a systematic comparison of methodologies, *Icarus* **75**, 1-29.
17. Kary, D.M. and Dones, L. (1996), Capture Statistics of Short-Period Comets: Implications for Comet D/Shoemaker-Levy 9, *Icarus*, **121**, pp. 207-224.
18. Kresák, Ľ. (1972), Jacobian integral as a classificational and evolutionary parameter of interplanetary bodies, *Bull. Astron. Inst. Czechoslov.* **23**, 1-34.
19. Oort, J.H. (1950), The structure of the cloud of comets surrounding the Solar System and a hypothesis concerning its origin, *Bull. Astron. Inst. Neth.* **11**, 91-110.
20. Öpik, E.J. (1976), *Interplanetary Encounters*, Elsevier, New York, USA.
21. Petit, J.-M. and Hénon, M. (1986), Satellite encounters, *Icarus*, **66**, pp. 536-555.
22. Rickman, H. (1979), Recent Dynamical History of the Six Short-Period Comets Discovered in 1975, in *Dynamics of the Solar System* (R.L. Duncombe ed.), D. Reidel, Dordrecht, Holland, pp. 293-298.
23. Tancredi, G., Lindgren, M. and Rickman, H. (1990), Temporary Satellite Capture and Orbital Evolution of Comet P/Helin-Roman-Crockett, *Astron. Astrophys.*, **239**, pp. 375-380.
24. Valsecchi, G.B. (1992), Close Encounters, Planetary Masses and the Evolution of Cometary Orbits, in *Periodic Comets*, (J.A. Fernández and H. Rickman eds.), Univ. de la República, Montevideo, Uruguay, pp. 81-96.
25. Valsecchi, G.B. and Manara, A. (1997), Dynamics of comets in the outer planetary region. II Enhanced planetary masses and orbital evolutionary paths, *Astron. Astrophys.* **323**, 986-998.

GALACTIC PERTURBATIONS IN THE MOTION OF COMETS

H. PRĘTKA
Astronomical Observatory, A. Mickiewicz University
ul. Słoneczna 36, 60-286 Poznań, Poland
e-mail: pretka@phys.amu.edu.pl

Abstract. In the last several years it has been recognized that the galactic tidal force plays a dominant role in the dynamical evolution of comets in the Oort cloud. In particular, galactic perturbations are considered as one of the most important effects producing observable comets from the Oort cloud and from the interstellar medium. In this work we examine numerically the effect of the galactic potential on the long–term evolution of comets, concentrating on detailed analysis of individual comets from the observable population and their statistical characteristic. We compare also the influence of different galactic force components on the motion of a comet, raising a question what are the qualitative and statistical results of using a simple model of the galactic disk potential instead of a more realistic but more complicated model of a Galaxy as a whole.

1. Introduction

It was proposed by Oort in 1950 that the solar system is currently surrounded by a huge, roughly spherical cloud of comets. These comets have large perihelion distances and need some mechanism to be brought to the region of observability. Oort argued that stochastic action of passing stars may change angular momenta of the comets and direct them into orbits with perihelia small enough to make these comets observable (Oort 1950). However, it has been recognized that the galactic tidal force may be a more efficient mechanism of changing perihelion distances of long–period comets (Byl 1986, Heisler and Tremaine 1986, Yabushita 1989). Taking into account a dominant role of the galactic tidal field greatly changed our view on the structure of the Oort cloud as well as the estimation of the outer limit of the cloud (Heisler and Tremaine 1986) and the number of comets in it (Yabushita and Tsuzii 1989). In particular, it has been found that the systematical tidal action of the galactic disk is an order of magnitude

B.A. Steves and A.E. Roy (eds.), The Dynamics of Small Bodies in the Solar System, 197–202.

stronger than the influence of the galactic centre and two times stronger than the randomizing effect of stellar passages (Heisler and Tremaine 1986).

2. Galactic disk model

According to this results the dynamics of comets from the Oort cloud in many papers (Heisler and Tremaine 1986, Matese and Whitman 1992, Dybczyński and Prętka 1997, Breiter et al. 1996, Maciejewski and Prętka 1998, Wiegert and Tremaine 1997) was studied with approximation of the galactic perturbations by the gravitational force coming from a flat, infinite disk with a smooth matter distribution. In this case Newton's equations of the motion of a comet can be written in the following form (Heisler and Tremaine, 1986):

$$\ddot{x} = -\frac{\mu}{r^3}x\,, \qquad \ddot{y} = -\frac{\mu}{r^3}y\,, \qquad \ddot{z} = -\frac{\mu}{r^3}z - 4\pi G\rho z \tag{1}$$

where x, y, z are heliocentric rectangular coordinates of a comet in the galactic reference frame, ρ is the matter density in the solar neighbourhood and G is the gravitational constant. In this model, the motion of the Sun around the galactic centre and its vertical oscillations through the galactic plane are neglected. The detailed description of the dynamics of comets in this model may be found in (Prętka and Dybczyński 1994, Maciejewski and Prętka 1998). Massive Monte Carlo simulations done by Dybczyński and Prętka (1996, 1997) pointed out that the population of comets coming from the Oort cloud to the region of observability only due to the galactic disk perturbations should have very particular statistical characteristics. The purpose of such statistical investigations is usually to compare the simulated observable sub–population with the population of one–apparition comets observed in the vicinity of the Sun. It was shown that only comets which reside in the Oort cloud on orbits with very high inclination to the galactic plane have a chance to change their perihelia enough to become observable. This result may change significantly an estimation of the number of comets in the Oort cloud: only a very small sub–population of comets may be visible when we assume that the major mechanism perturbing the motion of comets is the galactic disk. Thus the number of comets in the cloud may be bigger than the earlier estimate. The qualitative study of the secularly averaged problem has been done by Matese and Whitman (1989) and Breiter et al. (1996). Applicability of the averaged equations of the cometary motion was discussed by Dybczyński and Prętka (1997). It was pointed out that it is not always well justified, especially in the case of the chaotic motion (see Prętka 1997, Maciejewski and Prętka 1998) or if the position of a comet in its orbit is required.

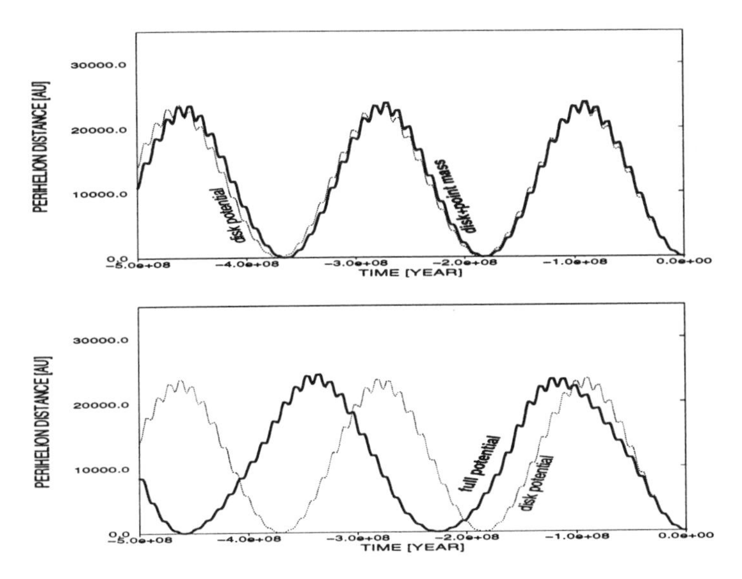

Figure 1. Long–term evolution curves of perihelion distance of a comet due to three different galactic potentials.

3. Application of other galactic models

The simple approximation of the galactic tidal perturbations may be insufficient in the case when we consider the motion of a comet on the edge of the Oort cloud or far beyond it. To make a practical comparison how different galactic force components perturb the motion of comets in the Oort cloud we consider three models of galactic perturbations: the model of a flat disk described above ('disk potential'), the second one containing the galactic disk term and the galactic centre term added as a point mass interaction ('disk+point mass potential'–with the Sun moving around the galactic centre on the circula orbit strictly in the galactic plane), and the third one (Dauphole, Colin, et al., 1996) having three components coming from the central part (so called bulge term), disk and halo terms respectively (and called here 'full galactic potential'–including near-circular motion of the Sun in the galactic plane and its vertical oscillations above and below the galatctic disk):

$$\Phi(\varpi, z) = \Phi_b(r) + \Phi_d(\varpi, z) + \Phi_h(r) \tag{2}$$

where ϖ and z are cylindric galactic coordinates of a body and $r^2 = \varpi^2 + z^2$. The first two simpler models, after averaging, lead to solutions resembling those proposed by Kozai (1962) for asteroids perturbed by Jupiter. More detailed description of all models may be found in (Prętka 1998). As in (Prętka 1998) we investigated both the long–term evolution of an individ-

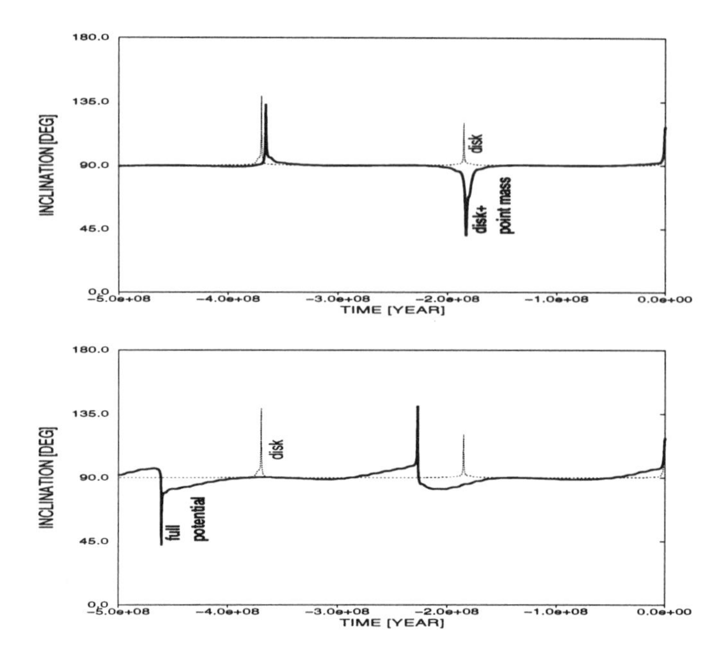

Figure 2. Long–term evolution of inclination of the cometary orbit to the galactic disk plane due to three different galactic potentials.

ual comet and the statistical characteristics of the observable comet population. Fig. 1 and Fig. 2 show an example of typical evolution of orbital elements of a comet due to the three different galactic potential models. The most signifficant differences appear in inclination and in perihelion distance. In contrast to a simple 'disk potential' model it is possible to observe changes of the orbit of a comet from prograde to retrograde when we use a more complicated model of the galactic perturbations (see Fig. 2). In a typical case amplitudes of curves representing the evolution of perihelion distance (Fig. 1) differ with each other in tens or even hundreds of astronomical units. These facts may have a significant meaning, especially when we study the motion of the individual comet (e.g. if we want to simulate the observable comet) in the galactic tidal field. From the same initial conditions we may obtain in one galactic model a comet passing at the distance of a few astronomical units from the Sun, but usage of a different galactic model may result in a comet which misses the planetary system even at a distance of hundreds of astronomical units and never be visible. In this case the approximation of the galactic perturbations by the gravitational influence of a flat, infinite disk with constant value of matter density is not sufficient enough. To show what are the statistical consequences of these differences we simulated observable population of comets using the three described models. The simulation was similar to that presented in (Prętka

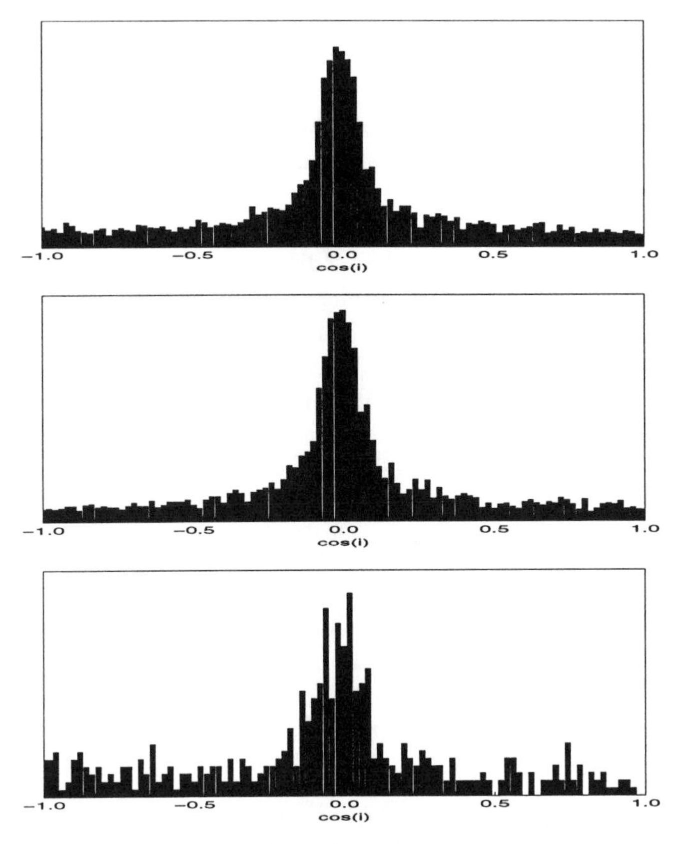

Figure 3. Distributions of the initial values of inclination of the orbit for three different galactic models: 'disk potential', 'disk+point mass potential' and 'full galactic potential' respectively.

1998): we chose randomly orbital elements of a comet and then traced its motion numerically to the moment when the comet entered the planetary region (defined as a sphere with a radius of 50 AU centered in the Sun). In the mentioned paper we fixed the semimajor axis of a comet to 50000 AU to study the statistical influence of different models for comets in the outer part of the Oort cloud and make these studies more efficient. In this paper we choose initial values of the semimajor axis and the eccentricity from distributions obtained by (Duncan et al. 1987) to check this influence also on a population with smaller semimajor axes. Results of this work show that the sub–population of observable comets has very similar characteristics independently on the model of the Galaxy in use. In Fig. 3 we present distributions of initial values of inclination obtained from the simulation for observable comets. It may be seen that almost all comets in these populations have very high inclinations for most of the time they spend in the Oort cloud. There are not signifficant differences between distributions obtained for any orbital element in different models of the galactic perturbations. Moreover, in all cases populations of observable comets are equal in size (they contain only 6.5% of a whole initial population taken to the simulation). Results of a previous simulation with a fixed, larger semimajor axis (Prętka 1997), equal to 50000 AU, show that the observable population of comets in more complicated model ('full galactic potential') was more numerous than in simpler models ('disk potential' and 'disk+point mass potential'). It means that the simpler model including only the component coming from the galactic disk is suitable to discuss the statistical properties of comets with semimajor axes typical for the Oort cloud region. But in the case where we study the motion of an individual comet or consider the statistical picture of more distant comets more realistic model of the galactic perturbations should be used. This fact has some practical meaning as the integration based on more complicated models is time–consuming and therefore less efficient.

4. Acknowledgements

I am very grateful to Dr P.A. Dybczyński and Dr S. Breiter for helpful disscusion and advice. I also want to thank organizers of the NATO ASI for supporting my participation on this meeting. This research was supported by the KBN grant No 2.PO3D.001.11.

References

Byl, J.: 1986, *Earth, Moon, and Planets*, **36**, 263.
Breiter, S., Dybczyński, P.A., and Elipe, A.: 1996, *Astron. Astrophys.*, **315**, 618.
Dauphole, B., Colin, J., and Geffert, M., Odenkirchen, M., and Tucholke, H.-J.: 1996, in *Unsolved Problems of the Milky Way* (L. Blitz and P. Teuben, Eds), p. 697.
Duncan, M., Quinn, T., and Tremaine, S.: 1987, *Astr. J*, **94**, 1330.
Dybczyński, P.A., and Prętka, H.: 1996, *Earth, Moon and Planets*, **72**, 13.
Dybczyński, P.A. and Prętka, H.: 1997, in: *Proceedings of the Conference 'Dynamics and Astrometry of Natural and Artificial Celestial Bodies'* (I. Wytrzyszczak et al., Eds), Poznań, p. 149.
Heisler, J. and Tremaine, S.: 1986, *Icarus* **65**, 13.
Kozai, Y.: 1962, *Astr. J*, **67**, 591.
Maciejewski, A.J., and Prętka, H.: 1998, in preparation to *Astron. Astrophys.*.
Matese, J.J., Whitman, P.G.: 1989, *Icarus*, **82**, 389.
Matese, J.J., Whitman, P.G.: 1992, *Celest. Mech. & Dyn. Astron.*, **54**, 13.
Oort, J.H.: 1950, *Bull. astr. Insts Neth.*, **11**, 91.
Prętka, H., and Dybczyński, P.A.: 1994, in: *Proceedings of the Conference 'Dynamics and Astrometry of Natural and Artificial Celestial Bodies'* (K. Kurzyńska et al., Eds), Poznań, p. 299.
Prętka, H.: 1997, in: *Proceedings of the Conference 'Dynamics and Astrometry of Natural and Artificial Celestial Bodies'* (I. Wytrzyszczak et al., Eds), Poznań, p. 155.
Prętka, H.: 1998: to be published in *Proceedings of the 'IV International Workshop on Positional Astronomy and Celestial Mechanics'*, Peniscola.
Wiegert, P., and Tremaine, S.: 1997 *private communication.*
Yabushita, S., and Tsuzii, T.: 1989, *Mon.Not.R.astr.Soc.* **241**, 59.
Yabushita, S.: 1989, *Astr. J.* **97**, 262.

SECTION THREE:
NATURAL AND ARTIFICIAL SATELLITES

INTRODUCTION

C.D. MURRAY
Astronomy Unit, Queen Mary and Westfield College,
Mile End Rd, London E1 4NS, U.K.

Natural and artificial satellites constitute an important class of the small bodies of the solar system. *Satellites* (from the Latin for "attendants") was the word chosen by Kepler in 1611 to describe Galileo's discovery of secondary planets orbiting Jupiter. Currently 63 natural planetary satellites are known to exist: Earth and Pluto have 1 each, Mars 2, Neptune 8, Jupiter 16, Uranus 17, and Saturn 18; there are good dynamical grounds for believing that more remain to be discovered. Indeed, the discovery of Caliban and Sycorax (Gladman *et al.* 1997), two outer moons of Uranus moving in retrograde orbits, was made within six months of the Maratea meeting. With the discovery of a moon orbiting the asteroid (243) Ida (Belton *et al.* 1995), only the second asteroid to have been seen close-up, we also have to consider the possibility that natural satellites of asteroids are commonplace. As well as the natural satellites there are several hundred active artificial satellites and probably several thousand inactive ones.

Broadly speaking the natural satellites can be classified into three groups:

- The medium to large (radii > 100 km) satellites have near-circular, regular orbits close to their planet's equatorial plane; these are probably primordial satellites which have survived in some form since the planet's formation.
- The small, inner satellites on regular orbits close to the planet; these satellites can play a role in ring dynamics and are probably the collision products of larger objects.
- The small, outer satellites on eccentric, inclined and even retrograde orbits; these are likely to be captured objects.

With sizes ranging from chunks of rock a few kilometres across to objects larger than the planet Mercury, the natural satellites exhibit a wide

B.A. Steves and A.E. Roy (eds.), The Dynamics of Small Bodies in the Solar System, 203–206.

variety of dynamical phenomena either as perturbed or perturbing bodies. The review articles by Greenberg (1977) and Peale (1986) and the respective books in which they appear provide comprehensive overviews of the dynamics and physical properties of planetary satellites.

In some cases the dynamical processes are directly responsible for the physical properties of satellites. For example, the active vulcanism on Io (see the recent review by Spohn 1997), one of the Galilean satellites of Jupiter, is a direct consequence of its orbital resonance with the adjacent satellite, Europa. Indeed, observations of Europa by NASA's *Galileo* space probe suggest that there is good evidence that the corresponding tidal heating of Europa may have resulted in an ocean of liquid water beneath its icy crust. Therefore the dynamical interaction of natural satellites can have important consequences for the existence of life.

The fact that the solar system exhibits a distinct preference for simple numerical relationships between pairs of objects was first shown by Roy & Ovenden (1954). Most of the commensurabilities discussed by Roy & Ovenden occur between natural satellites and the Saturn system has some of the best examples with Mimas, Enceladus, Tethys, Dione, Titan and Hyperion all involved in resonant relationships. At least two of these, Mimas and Titan, are also known to be responsible for some of the resonant features in Saturn's spectacular ring system. The article by Message reviews the dynamics of regular, resonant satellites in the Saturn system while an analytical approach to the specific dynamics of the unusual Mimas-Tethys inclination resonance is undertaken by Pro[illegible]in. The role of small, close satellites in producing resonant features in planetary ring systems is reviewed in the article by Murray.

Satellites which share orbits are referred to as "coorbital". In the solar system there are numerous examples of coorbital objects, the best known being the Trojan asteroids librating in "tadpole" orbits about the stable leading (L_4) and trailing (L_5) Lagrangian equilibrium points in the Sun–Jupiter system. However, the saturnian system contains examples of satellites in both "tadpole" and "horseshoe" orbits. Janus and Epimetheus are often referred to as *the* coorbital satellites and their unusual "horseshoe" configuration provides a beautiful example of the richness of solutions to the three-body problem. The long-term dynamics of such coorbital satellites are examined in detail in the article by Waldvogel while Morais & Murray report on the perturbations of coorbital satellites due to resonances with other satellites. Curiously, there may even be a connection between coorbital dynamics and the origin of the irregular, retrograde satellites of the outer planets: Namouni (1998) has recently shown that it is possible for objects in eccentric orbits coorbital with a planet to become retrograde planetary satellites, at least on a temporary basis.

Why do most of the outer planets possess large retinues of satellites while others in the inner solar system have none? Although solar tides are usually invoked to explain the difference, the role of purely gravitational effects and chaos is addressed in the article by Yokoyama. The gravitational interactions of large numbers of satellites can usually be studied by using secular perturbation theory. However, the presence of resonances, or even near-resonances requires a secular theory to higher order in the masses. Such a method is investigated by Christou & Murray and applied to the uranian satellite system.

A fundamental difference between dynamical studies of satellite orbits and those of planetary orbits is the form of the central potential. Although the Sun can be taken to be a spherical object and treated as a point mass, a planet is subjected to rotational and tidal distortions. This requires any study of satellite dynamics to take account of the precessional effects of a non-spherically symmetric central potential. For example, a small inner satellite of Saturn orbiting just beyond the main rings would have a mean motion of $\sim 600°\mathrm{d}^{-1}$ and a precession rate due the planet's oblateness of $\sim 0.5\%$ of this value. The article by Sansaturio, Vigo-Aguiar & Ferrándiz examines the whole question of the integrability of the motion of a test particle under various approximations to the planet's potential. It is also important to realise that the interesting dynamics of a satellite (or planet) are not always confined to its orbital motion— its rotational propoerties can also provide a rich source of diverse phenomena, with Hyperion providing the best paradigm in this respect (see, for example, Wisdom *et al.* 1984 and Black *et al.* 1995). The article by Maciejewski examines the problems associated with modelling rotational motion.

The advent of the space age has resulted in a situation where the number of artificial satellites is at least an order of magnitude larger than the number of planetary satellites. Although subject to the same law of gravity and therefore amenable to the same analytical techniques, the orbits of artificial satellites are pre-determined by the role they must play. In this sense planners can pick and choose the orbit that is the most suitable for the mission. For example, a geosynchronous orbit may be required for a communications satellite while a polar orbit might be more appropriate for an astronomical survey satellite such as IRAS. For studies of the Earth's magnetosphere and its interaction with the solar wind satellites on orbits with large semi-major axis and eccentricity are required; these are then subject to the complicating effects of lunar and solar perturbations. The dynamics of such satellites with "multi-day" periods are described in the article by Érdi. Another unusual class of satellite orbits, the so-called "halo" orbits, is examined in the article by Andreu & Simó. Although such orbits are bizarre in comparison with the more mundane world of planetary satellites, they are just as physically real and are proving to be extremely useful in obtaining specific scientific observations.

The study of artificial satellite dynamics can allow several approximations to be made. For example, the problem of two fixed centres, investigated in the article by Irigoyen, was of use in planning the Apollo missions to the Moon. With current plans for placing satellites in orbit around asteroids and comets, the need to model ever more bizarre potentials will arise. The article by Zafiropoulos & Stavliotis examines the motion of a satellite in orbit around a hemispherical shell.

Research into all types of satellite motion has grown in recent years. This is partly because the data set of natural satellites has doubled in the last forty years thanks to spacecraft flybys of the outer planets and improved sensitivity of ground-based telescopes. There are already hints as to where currently hidden satellites may lie, with Saturn's rings thought harboring several candidates. Similarly the numbers of artificial satellites have increased dramatically with future plans for global personal communications satellite networks and a plethora of robotic space missions set to orbit a variety of planets and minor bodies in the solar system. In proceedings dedicated to the memory of Victor Szebehely, it is appropriate that the three-body problem, so beloved of Victor, lies at the heart of our current understanding of so many aspects of satellite dynamics.

References

1. Belton, M.J.S., Chapman, C.R., Thomas, P.C., Davies, M.E., Greenberg, R., Klaasen, K., Byrnes, D., D'Amario, L., Synnott, S., Johnson, T.V., McEwen, A., Merline, W., Davis, D.R., Petit, J.-M., Storrs, A., Veverka, J. and Zellner, B. (1995) Bulk density of asteroid 243-Ida from the orbit of its satellite Dactyl, *Nature* **374**, 785–788.
2. Black, G.J., Nicholson, P.D., and Thomas, P.C. (1995) Hyperion: Rotational dynamics, *Icarus* **117**, 149–171.
3. Gladman, B.J., Nicholson, P.D., Burns, J.A., Kavelaars, J.J., Marsden, B.G., Williams, G.V. and Offutt, W.B. (1997) Discovery of two distant irregular moons of Uranus. *Nature*, **392**, 897–899.
4. Greenberg, R. (1977). Orbit-orbit resonances among natural satellites, in *Planetary Satellites*, ed. J.A. Burns (University of Arizona Press, Tucson).
5. Namouni, F. (1998) Secular interactions of coorbiting objects. *Icarus* (in press).
6. Peale, S.J. (1986) Orbital resonances, unusual configurations and exotic rotation states among planetary satellites, in *Satellites*, ed. J.A. Burns and M.S. Matthews (University of Arizona Press, Tucson).
7. Roy, A.E. and Ovenden, M.W. (1954) On the occurrence of commensurable mean motions in the solar system, *Mon. Not. R. Astr. Soc.* **114**, 232–241.
8. Spohn, T. (1997) Tides of Io, in *Tidal Phenomena*, eds. H. Wilhelm, W. Zurn and H.-G. Wenzel (Springer, Berlin).
9. Wisdom, J., Peale, S.J. and Mignard, F. (1984) The chaotic rotation of Hyperion, *Icarus* **58**, 137–152.

ORBITS OF SATURN'S SATELLITES: SOME ASPECTS OF COMMENSURABILITIES AND PERIODIC ORBITS

P.J. MESSAGE
University of Liverpool,
L69 3BX, U.K.
sx20@liverpool.ac.uk

Abstract. These lectures introduce some of the features of the impact of near-commensurability of orbital period on the motions of some of the larger satellites of Saturn; also how periodic solutions of the gravitational problem of three bodies arise in near-commensurability situations, and something of how this relates to the orbital motion of the satellites.

1. Introduction to the satellites of Saturn.

Let us begin by listing those satellites of Saturn which were discovered before the end of the last century, in order of their discovery, with their mean opposition magnitudes:

Year	Satellite	Discoverer	mn. oppn. mag.
1655	Titan	C. Huygens	8.4
1671	Iapetus	G. D. Cassini	10.9 to 11.9
1672	Rhea	G. D. Cassini	9.7
1684	Tethys	G. D. Cassini	10.3
1684	Dione	G. D. Cassini	10.4
1789	Enceladus	Wm. Herschel	11.8
1789	Mimas	Wm. Herschel	12.9
1848	Hyperion	Bond/Lassell	14.2
1898	Phoebe	Pickering	16.5

B.A. Steves and A.E. Roy (eds.), The Dynamics of Small Bodies in the Solar System, 207–225.

It is seen that these satellites were discovered, over more than two centuries, almost exactly in order of diminishing apparent brightness. Now let us list all those satellites currently known, in order of mean distance from Saturn, with their major semi-axes *(a)*, in thousands of kilometres, their orbital periods *(P)*, in days, and their mean opposition magnitudes *(m)* (the data are from the *Handbook of the British Astronomical Association, 1997*):

Satellite	a	P	m.
Pan	133.6	0.575	20
Atlas	137.6	0.602	18
Prometheus	139.4	0.613	16.5
(F Ring)	140		
Pandora	141.7	0.629	16
Epimetheus	151.4	0.694	15.5
Janus	151.5	0.695	14.5
Mimas	185.5	0.942	12.9
Enceladus	238.0	1.37	11.8
Tethys	294.7	1.888	10.3
Telesto	294.7	1.888	19
Calypso	294.7	1.888	18.5
Dione	377.4	2.737	10.4
Helene	377.4	2.737	18.5
Rhea	527.0	4.518	9.7
Titan	1221.9	15.945	8.4
Hyperion	1481.0	21.278	14.2
Iapetus	3561.3	79.331	10.2/11.9
Phoebe	12952	550.3*(R)*	16.5

(*R* indicates that the orbital motion of Phoebe is retrograde). The incidence of small-integer near-commensurabilities of orbital period between pairs of the major satellites can be seen: Mimas and Tethys *1:2*, Enceladus and Dione *1:2*, and Titan and Hyperion *3:4*. Also many of the smaller satellites follow orbits in *1:1* resonance with larger ones: the motion of each of Telesto and Calypso approximates to a Lagrange-type motion forming a perpetual equilateral triangle with Tethys and Saturn, and Helene likewise with Dione and Saturn. Also the "co-orbital" satellites, Epimetheus and Janus, follow motions in which they exchange orbits each time the faster catches up with the slower: see the lectures by Dr. Waldvőgel at this Advanced Study Institute, and previous ones, for the theory of this type of motion. Dr. Murray's lectures to this Institute will include more about the smaller satellites and the rings, so let us now consider some aspects of the motions of the larger satellites. The following table sets out, for each of these, the mean angular orbital motion *(n)*, in degrees per day, in more precision, and, where it is known, the mass *(m)*, in units of a millionth of the mass of Saturn:

Satellite	n	m
Mimas	381.9945121*	0.0648*
Enceladus	262.731901*	0.225*
Tethys	190.697913*	1.068*
Dione	131.534932*	1.992*
Rhea	79.690047*	3.2♯
Titan	22.577015¶	232.2♭
Hyperion	16.919989♭	
Iapetus	4.537951*	3
Phoebe	-0.655511†	

(∗ from Harper and Taylor *(1993)*; ♯ from Jeffreys *(1953)*; ¶ from Kozai *(1957)*; ♭ from Message *(1993)*; † from Rose *(1979)*). As we consider the effect of the near-commensurabilities of orbital period on the motions of these satellites, let us keep in mind that, in the expressions for the perturbations by one satellite on the orbit of another, as given by first-order theory, linear combinations of the mean motions appear as denominators, while the mass of the perturbing body appears of course as a factor. Thus near-commensurabilities will be associated with larger co-efficients of the corresponding terms in the perturbations. From the values just given, we see that the mean motion of Mimas, less twice that of Tethys, gives a difference of 0.598686 degrees per day, which is 0.0015673 times the mean motion of Mimas. This ratio is some 1400 times the ratio of the mass of Tethys to that of Saturn. Also, the mean motion of Enceladus, less twice that of Dione, gives -0.337963 degrees per day, which is -0.0012863 times the mean motion of Enceladus. This ratio is some 640 times the ratio of the mass of Dione to that of Saturn. These commensurabilities are sufficiently close to ensure that the nature of the motion is significantly different from motion in the absence of small-integer commensurability, so that the perturbation expressions given by Poisson's method do not even necessarily even provide an approximation to the actual perturbations, that is, these are "deep resonances" in the sense of Garfinkel *(Garfinkel, 1966, see also Garfinkel, Jupp, and Williams, 1971, and Jupp, 1973)*. Nevertheless, these ratios, 1400 and 640, are sufficiently large to put these two pairs into the "moderately close commensurability" case, as designated in an earlier paper *(Message 1966)*. On the other hand, three times the mean motion of Titan, less four times that of Hyperion, gives 0.0510496 degrees per day, which is 0.0022611 times the mean motion of Titan, and this ratio is less than ten times the mass of Titan in terms of that of Saturn, so that the effect of this commensurability on the perturbations of the motion of Hyperion by Titan is of a yet greater order, leading to a yet further change in the nature of the motion, putting this pair into the what, in the earlier paper, was designated the "very close commensurability" case. We shall see later some of the implications of this distinction between these two levels of closeness of commensurability.

2. The Equations of Motion, and how periodic orbits can arise in cases of near-commensurability.

Now let us begin to consider the effect of near commensurabilities on the motion of the satellites, setting out the equations for the perturbations of the motion of one satellite by another, at first in a fairly simple form. Suppose a satellite, P, of mass m, is perturbed by another satellite, P', of mass m'. Let the position vectors of P and P', referred to the position of Saturn, S, be $\mathbf{r}$ and $\mathbf{r}'$, respectively. Then the equation of motion for P is

$$\frac{\mathrm{d}^2\mathbf{r}}{\mathrm{d}t^2} = -\frac{\mu\mathbf{r}}{r^3} + \mathbf{grad}R, \tag{1}$$

where

$$\mu = G(M+m),$$

G being the gravitational constant,
and M the mass of Saturn.
Also R, the "disturbing function", is given by

$$R = Gm'\left\{\frac{1}{\Delta} - \frac{r\cos\chi}{r'^2}\right\},$$

Δ being the distance PP',
and χ the angle subtended by P and P' at S.
The Fourier expansion of Δ^{-1},

$$\frac{1}{\Delta} = \frac{1}{r}\left\{\frac{1}{2}B_0(\rho) + \sum_{i=1}^{\infty} B_i(\rho)\cos i\chi\right\},$$

introduces the Laplace co-efficient, $B_i(\rho) = B_i^{(1/2)}(\rho)$, which is expressible as a power series in $\rho = r/r'$. Suppose now that the motions of P and P' are coplanar. Then equation (1), expressed in terms of the plane polar co-ordinates, (r, θ), of P, gives

$$\ddot{r} - r\dot{\theta}^2 = -\frac{\mu}{r^2} + \frac{\partial R}{\partial r}, \qquad (2)$$

and

$$r\ddot{\theta} + 2\dot{r}\dot{\theta} = \frac{1}{r}\frac{\partial R}{\partial r}, \qquad (3)$$

Now of course $\chi = \theta - \theta'$, where θ' is the polar angle of P', and we may write

$$R = Gm' \sum_{n=0}^{\infty} B_i^* \cos i\chi,$$

where B_i^* depends on r and r'. (For $i > 1$, it is $(1/r)B_i(r/r')$ if $r < r'$, while, if $r > r'$, it is $(1/r')B_i(r'/r)$. If $i = 1$, from this must be subtracted r/r'^2, and if $i = 0$, the factor $1/2$ must be applied.) Then equations (2) and (3) become

$$\ddot{r} - r\dot{\theta}^2 = -\frac{\mu}{r^2} + Gm\prime \sum_{n=0}^{\infty} \frac{\partial B_i^*}{\partial r} \cos i(\theta - \theta'),$$

and

$$r\ddot{\theta} + 2\dot{r}\dot{\theta} = -\frac{Gm\prime}{r} \sum_{n=1}^{\infty} B_i^* i \sin i(\theta - \theta'),$$

respectively. Now let us suppose that the motion of P is close to circular, and put

$$r = a + \delta r,$$

and

$$\theta = nt + \epsilon + \delta\theta,$$

where a, n, and ϵ are constants, and work to first order in δr, $\delta\theta$, and m'. Then, to this precision,

$$\delta\ddot{r} - 3n^2\delta r - 2na\delta\dot{\theta} = Gm' \sum_{n=0}^{\infty} \frac{\partial B_i^*}{\partial r} \cos i(\theta - \theta'), \qquad (4)$$

and

$$a\delta\ddot{\theta} + 2n\delta\dot{r} = -\frac{Gm'}{a} \sum_{n=1}^{\infty} B_i^* i \sin i(\theta - \theta'), \qquad (5)$$

and the constants a and n must be related so that $n^2a^3 = \mu$. Then integrating equation (5) gives

$$a\delta\dot{\theta} + 2n\delta r = \frac{Gm'}{a(n-n')}\sum_{n=1}^{\infty} B_i^* \cos i(\theta - \theta'), \tag{6}$$

where $n' = \dot{\theta}'$, and we have omitted the constant of integration which corresponds to a shift in a. Elimination of $\delta\dot{\theta}$ with use of equation (4) gives

$$\delta\ddot{r} + n^2\delta r = Gm'\sum_{n=1}^{\infty} B_i^{**} \cos i(\theta - \theta') \tag{7}$$

where

$$B_i^{**} = \frac{\partial B_i^*}{\partial r} + \frac{2n}{a(n-n')}B_i^*.$$

The solution of equation (7) is

$$\delta r = c\cos(nt+\epsilon) + \sum_{n=1}^{\infty} c_i \cos i(\theta - \theta'), \tag{8}$$

where

$$\left\{n^2 - i^2(n-n')^2\right\} c_i = Gm'B_i^{**}, \tag{9}$$

and c and ϵ are constants of integration. Substitution of this into equation (6) and integrating gives

$$\delta\theta = -2\frac{c}{a}\sin(nt+\epsilon) + \sum_{n=1}^{\infty} d_i \sin i(\theta - \theta'), \tag{10}$$

where

$$i(n-n')^2 a d_i = \frac{Gm'}{a}B_i^* - 2n(n-n')c_i, \tag{11}$$

Now if c is larger than all of the c_i and ad_i, then the term $c\cos(nt+\epsilon)$ in δr corresponds to a free eccentricity, e, with $c = -ae$. But equation (9) shows that if, for some integer i, $n'/n \approx (i \pm 1)/i$, then c_i will be large. Suppose then that we have a small-integer near-commensurability of the type

$$(p+1)n - pn' = \nu,$$

for some integer p, where $|\nu|$ is very small compared to both n and n'. Then

$$c_p \approx -G\frac{m'}{2n\nu}B_p^{**}, \text{ and}$$
$$d_p \approx -2\frac{c_p}{a},$$

so that both $|c_p|$ and $|d_p|$ are large. If in fact $|c_p|$ is larger than $|c|$, so that the $i = p$ term in the summation in equation (8) is the largest periodic term in r, then this term will correspond to the eccentricity, e, of the orbit, which will therefore take a forced value such that, to first order in $|c_p/a|$, $c_p = -ae$. Then the mean anomaly, $\ell = \lambda - \varpi$, will always be equal to $\pi - p(\theta - \theta')$. (Here λ is the mean longitude, differing here from θ by small periodic terms.) So the apse longitude, ϖ, will be always equal to $\lambda + p(\theta - \theta') - \pi$, from which it is seen that ϖ will have the forced angular motion ν. This is the case of a periodic motion of Poincaré's *première sorte*, in which the critical argument, $(p+1)\lambda - p\lambda' - \varpi$, of the disturbing function, remains constant (in this case equal to π), apart from short-period terms with the period $2/(|n' - n|)$, which will be the period of the relative motion of S, P, and P'. (Here λ' is the mean longitude of P'.)

3. Formulation of the Gravitational Problem of Three Bodies in Jacobi's Form

Consider a system consisting of Saturn, S, of mass m_S, and two satellites, P_1 and P_2, of masses m_1 and m_2, respectively. We will use Jacobi's system of relative position vectors, that is, $\mathbf{r}_1$ is the position vector of P_1 referred to S, and $\mathbf{r}_2$ is the position vector of P_2 referred to G_1, where G_1 is the mass-centre of S and P_1. In terms of these, the kinetic energy, T, of the system may be expressed as a sum of squares:

$$2T = M\dot{\bar{\mathbf{r}}}^2 + \varepsilon\left\{\beta_1\dot{\mathbf{r}}_1^2 + \beta_2\dot{\mathbf{r}}_2^2\right\},$$

where $\bar{\mathbf{r}}$ is the position vector of the mass-centre of the whole system, referred to an inertial origin, $M = m_S + m_1 + m_2$ is the mass of the whole system, also

$$\varepsilon\beta_1 = \frac{m_S m_1}{m_S + m_1},$$

and

$$\varepsilon\beta_2 = \frac{m_2(m_S + m_1)}{m_S + m_1 + m_2},$$

where we take ε to be the greater of the ratios m_1/m_S and m_2/m_S, and so it is small when the ratios of the masses of the satellites to that of Saturn are. Also the potential energy,

$$V = -G\left\{\frac{m_S m_1}{SP_1} + \frac{m_S m_2}{SP_2} + \frac{m_1 m_2}{P_1 P_2}\right\},$$

in which G is here the gravitational constant, may be expressed as

$$V = -\varepsilon\left\{\frac{\mu_1\beta_1}{|\mathbf{r}_1|} - \frac{\mu_2\beta_2}{|\mathbf{r}_2|} - \mathcal{R}\right\},$$

where each μ_j is chosen so that $\varepsilon\mu_j\beta_j = Gm_S m_j$, and

$$\mathcal{R} = \varepsilon G\beta_1\beta_2\left\{\frac{1}{|\mathbf{r}_1 - \mathbf{r}_2|} - \frac{\mathbf{r}_1 \cdot \mathbf{r}_2}{|\mathbf{r}_2|^3}\right\} + O(\varepsilon^3).$$

The equations of motion are, after cancelling the factor ε,

$$\beta_j\frac{\mathrm{d}^2\mathbf{r}_j}{\mathrm{d}t^2} = -\frac{\mu_j\beta_j\mathbf{r_j}}{|\mathbf{r_j}|^3} + \frac{\partial R}{\partial \mathbf{r}_j}, \quad (j = 1, 2).$$

In the limit $\varepsilon \longrightarrow 0$, we have $\mathcal{R} \longrightarrow 0$, and we have two independent Kepler inverse square central force problems. Let us regard this as the "unperturbed" situation, and we develop perturbation methods to study the actual, $\varepsilon \neq 0$, situation.

The equations may be put into Hamiltonian form by taking $\mathbf{p}_j = \beta_j\dot{\mathbf{r}}_j$ as the momentum conjugate to $\mathbf{r}_j$, and then the Hamiltonian function is

$$\mathcal{H}(\mathbf{r}_1, \mathbf{r}_2, \mathbf{p}_1, \mathbf{p}_2) = \sum_{j=1}^{2}\left\{\frac{\mathbf{p}_j^2}{2\beta_j} - \frac{\mu_j\beta_j}{|\mathbf{r}_j|}\right\} - \mathcal{R}.$$

Let us now take, for each vector $\mathbf{r}_j$ and the velocity vector $\dot{\mathbf{r}}_j$ corresponding to it, the three canonical co-ordinates λ_j (the mean longitude), ϖ_j (the apse longitude), and Ω_j (the longitude of the ascending node) which, with the three other Keplerian elements a_j (the major semi-axis), e_j (the eccentricity of the orbit), and I_j (the inclination of the orbit plane to the reference plane), prescribe the instantaneous Kepler orbit corresponding to the position vector $\mathbf{r}_j$ and velocity vector $\dot{\mathbf{r}}_j$. The momenta conjugate to λ_j, ϖ_j, and Ω_j are then Λ_j, Π_j, and $\mathcal{N}_j$, respectively, which are given in terms of a_j, e_j, and I_j by the following equations:

$$\Lambda_j = \beta_j\sqrt{\mu_j a_j},$$

$$\Pi_j = \beta_j\sqrt{\mu_j a_j}\left\{\sqrt{(1-e^2)} - 1\right\},$$

and

$$\mathcal{N}_j = \beta_j\sqrt{\mu_j a_j}\sqrt{(1-e^2)}(\cos i - 1), \quad (j = 1, 2).$$

Introduce ε_j so that $\Pi_j = -\frac{1}{2}\Lambda_j\varepsilon_j^2$, for $j = 1$ and 2. Then

$$\varepsilon_j = e_j - \frac{1}{8}e_j^3 - \frac{1}{128}e_j^5 - \frac{1}{1024}e_j^7 + O(e_j^9).$$

Also

$$\mathcal{N}_j = -2(\Lambda_j + \Pi_j)\sin^2(i_j/2).$$

The Hamiltonian function giving the equations of motion in terms of the six co-ordinates λ_1, λ_2, ϖ_1, ϖ_2, Ω_1, and Ω_2, and their conjugate momenta Λ_1, Λ_2, Π_1, Π_2, $\mathcal{N}_1$, and $\mathcal{N}_2$, is

$$\mathcal{H}' = \mathcal{H}_0 - \mathcal{R},$$

where

$$\mathcal{H}_0 = -\sum_{j=1}^{2} \frac{\beta_j^3\mu_j^2}{2\Lambda_j^2}.$$

In the unperturbed case, with $\mathcal{R}$ identically zero, it is clear that all the ϖ_j, Ω_j, Λ_j, Π_j, and $\mathcal{N}_j$, are constants, and $\lambda_j = n_j t + \epsilon_j$, where the ϵ_j are constants, as then are the mean motions $n_j = \beta_j^3\mu_j^2/\Lambda_j^3 = \mu_j^{(1/2)}a_j^{(-3/2)}$. Since R is a periodic function of each of the co-ordinates, it may be written

$$\mathcal{R} = \sum_k K_k \cos P_k, \qquad (12)$$

where k denotes the six-vector with integer components:

$$k = (k_1, k_2, k_3, k_4, k_5, k_6),$$

and we have put

$$P_k = k_1\lambda_1 + k_2\lambda_2 + k_3\varpi_1 + k_4\varpi_2 + k_5\Omega_1 + k_6\Omega_2,$$

the summation in (12) being over all sets k with $k_1 \geq 0$ and with $k_5 + k_6$ even, and, since $\mathcal{R}$ depends only on the relative positions of the bodies, which is not affected by a change of origin of the longitudes, with only those terms for which

$$k_1 + k_2 + k_3 + k_4 + k_5 + k_6 = 0, \qquad (13)$$

(Sine terms do not appear because the lengths and scalar products of the vectors $\mathbf{r}_j$ are not altered if every one of the co-ordinates λ_j, ϖ_j, Ω_j is changed in sign, since this simply reflects in the line of origin of longitudes the figure formed by S and the two satellites P_j. Also K_k is a function of the Λ_j, Π_j, and $\mathcal{N}_j$, having the d'Alembert property, that if it is expressed in terms of the Λ_j, ε_j, and i_j, it has the it has the factors $\varepsilon_1^{|k_3|}$, $\varepsilon_2^{|k_4|}$, $\sin^{|k_5|}(i_1/2)$, $\sin^{|k_6|}(i_2/2)$, and otherwise involves the ε_j and i_j through ε_j^2 and $\sin^2(i_j/2)$. Equivalently, $\mathcal{R}$ may be expressed as a power series in $\varepsilon_j \cos \varpi_j$, $\varepsilon_j \sin \varpi_j$, $\sin(i_j/2)\cos\Omega_j$, and $\sin(i_j/2)\sin\Omega_j$, the co-efficients being functions of the λ_j and Λ_j only.

4. Conditions near a small-integer commensurability of orbital period

Suppose the mean motions of the two satellites are close to a small-integer commensurability:

$$(p+q)n_2 \approx pn_1,$$

where p and q are small integers, q being positive. Let us use the co-ordinates

$$\begin{aligned}
\theta_j &= \{(p+q)\lambda_2 - p\lambda_1\}/q - \varpi_j, \quad (j=1,2),\\
\theta_j^* &= \{(p+q)\lambda_2 - p\lambda_1\}/q - \Omega_j, \quad (j=1,2),\\
\phi &= \lambda_1 - \lambda_2,
\end{aligned}$$

and

$$\chi = (\lambda_1 + \lambda_2)/2,$$

to which the conjugate momenta are, respectively,

$$\begin{aligned}
\Theta_j &= -\Pi_j\\
&= (\Lambda_j \varepsilon_j^2)/2, \quad (j=1,2),\\
\Theta_j^* &= -\mathcal{N}_j\\
&= 2(\Lambda_j + \Pi_j)\sin^2(i_j/2), \quad (j=1,2),\\
\Phi &= (\Lambda_1 - \Lambda_2)/2 - (2p+q)/(2q)(\Pi_1 + \Pi_2 + \mathcal{N}_1 + \mathcal{N}_2),\\
X &= \Lambda_1 + \Lambda_2 + \Pi_1 + \Pi_2 + \mathcal{N}_1 + \mathcal{N}_2.
\end{aligned}$$

The general angular argument in $\mathcal{R}$ becomes

$$P_k = -\{p(k_1+k_2)/q + k_1\}\phi + k_3\theta_1 + k_4\theta_2 + k_5\theta_1^* + k_6\theta_2^*.$$

We note that χ is ignorable, so that X is constant. This is the integral of total angular momentum. Now suppose that use of a Lie series or equivalent type of transformation (see, *e.g.* the lectures in a previous Advanced Study Institute in *Message (1987)*):

$$(\phi, \theta_j, \theta_j^*; \Phi, \Theta_j, \Theta_j^*) \longmapsto (\tilde{\phi}, \tilde{\theta}_j, \tilde{\theta}_j^*; \tilde{\Phi}, \tilde{\Theta}_j, \tilde{\Theta}_j^*),$$

brings us to a Hamiltonian function

$$\tilde{\mathcal{H}}(\tilde{\phi}, \tilde{\theta}_j, \tilde{\theta}_j^*; \tilde{\Phi}, \tilde{\Theta}_j, \tilde{\Theta}_j^*) \;=\; \tilde{\mathcal{H}}_0 - \tilde{\mathcal{R}}$$

in which there are no short-period terms, that is, no dependence on ϕ, the equations giving the transformation being of the type

$$\begin{aligned}
\phi &= \tilde{\phi} + \Sigma' L_{0,k} \sin \tilde{P}_k, \\
\theta_j &= \tilde{\theta}_j + \Sigma' L_{j,k} \sin \tilde{P}_k, \quad (j = 1, 2), \\
\theta_j^* &= \tilde{\theta}_j^* + \Sigma' L_{j+2,k} \sin \tilde{P}_k, \quad (j = 1, 2), \\
\Phi &= \tilde{\Phi} + \Sigma' M_{0,k} \cos \tilde{P}_k, \\
\Theta_j &= \tilde{\Theta}_j + \Sigma' M_{j,k} \cos \tilde{P}_k, \quad (j = 1, 2), \\
\Theta_j^* &= \tilde{\Theta}_j^* + \Sigma' M_{j+2,k} \cos \tilde{P}_k, \quad (j = 1, 2),
\end{aligned}$$

with

$$\tilde{P}_k \;=\; -\{p(k_1 + k_2)/q + k_1\}\tilde{\phi} + k_3\tilde{\theta}_1 + k_4\tilde{\theta}_2 + k_5\tilde{\theta}_1^* + k_6\tilde{\theta}_2^*,$$

the summations here being in each case over those sets k of co-efficients k_j for which the co-efficient of $\tilde{\phi}$ in $\tilde{P}_k$ is *not* zero. The co-efficients $L_{j,k}$ and $M_{j,k}$ are functions of the momenta $\tilde{\Phi}$, $\tilde{\Theta}_j$, and $\tilde{\Theta}_j^*$. The transformation is chosen so that $\tilde{\mathcal{R}}$ has the form:

$$\tilde{\mathcal{R}} \;=\; \Sigma'' \tilde{K}_k \cos \tilde{P}_k, \tag{14}$$

this summation being over those sets k of co-efficients k_j for which the co-efficient of $\tilde{\phi}$ in $\tilde{P}_k$ *is* zero, that is, $p(k_1 + k_2)/q + k_1 = 0$, so that $k_1 + k_2 = mq$ for some integer m, and then $k_1 = -mp$. Now only those terms appear for which $k_5 + k_6$ is even, equal say to $2j$ for some integer j. Thus the conditions governing which sets of co-efficients appear in the terms of the series expansion (14), recalling equation (13), reduce to $mq + k_3 + k_4 + 2j = 0$, so that

$$\tilde{P}_k \;=\; k_3\tilde{\theta}_1 - \{mq + k_3 + 2j\}\tilde{\theta}_2 + k_5\tilde{\theta}_1^* + (2j - k_5)\tilde{\theta}_2^*.$$

5. Conditions for a periodic solution

Let us now seek solutions of the equations of motion in which the co-ordinates $\tilde{\theta}_j$ and $\tilde{\theta}_j^*$, and the momenta $\tilde{\Theta}_j$ and $\tilde{\Theta}_j^*$, are constant. Then the untransformed co-ordinates and momenta will be functions of time having the period of the co-ordinate $\tilde{\phi}$, which will have the constant rate of change $n_1^\sharp - n_2^\sharp$, where $n_j^\sharp$ is the rate of change of $\tilde{\lambda}_j$ (the counterpart of λ_j in the system of transformed co-ordinates), and so is equal to

$$\begin{aligned} n_j^\sharp &= \frac{\partial \tilde{\mathcal{H}}}{\partial \tilde{\Lambda}_j} \\ &= \frac{\beta_j^3 \mu_j^2}{\tilde{\Lambda}_j^3} - \frac{\partial \tilde{\mathcal{R}}}{\partial \tilde{\Lambda}_j} \\ &= \mu_j^{\frac{1}{2}} \tilde{a}_j^{-\frac{3}{2}} - \frac{2}{\beta_j \tilde{n}_j \tilde{a}_j} \frac{\partial \tilde{\mathcal{R}}}{\partial \tilde{a}_j} + \frac{\tilde{b}_j(\tilde{a}_j - \tilde{b}_j)}{\beta_j \tilde{n}_j \tilde{a}_j^4 \tilde{e}_j} \frac{\partial \tilde{\mathcal{R}}}{\partial \tilde{e}_j} + \frac{\tan(\tilde{i}_j/2)}{\beta_j \tilde{n}_j \tilde{a}_j \tilde{e}_j} \frac{\partial \tilde{\mathcal{R}}}{\partial \tilde{i}_j}, \end{aligned}$$

where $\tilde{a}_j$, $\tilde{e}_j$, $\tilde{i}_j$, $\tilde{n}_j$, and $\tilde{b}_j$ are the same functions of $\tilde{\Phi}$, $\tilde{\Theta}_j$ and $\tilde{\Theta}_j^*$ that a_j, e_j, i_j, n_j, and b_j are of Φ_j, Θ_j, and Θ_j^*, respectively, and all quantities here are evaluated with the transformed co-ordinates and momenta, so that the $n_j^\sharp$ will be constant in such a motion. Thus the quantities describing the relative positions of S, P_1, and P_2 will also be periodic with this rate of change, so that the relative configuration of the three bodies will be periodic with the period $2\pi/(n_1^\sharp - n_2^\sharp)$. Let us now examine how these conditions may be met. For $\tilde{\Theta}_j$ to be constant in the motion we must have $\frac{\partial \tilde{\mathcal{H}}}{\partial \tilde{\theta}_j}$ equal to zero, that is,

$$\Sigma'' k_{j+2} \tilde{K}_k \sin \tilde{P}_k = 0,$$

and for $\tilde{\Theta}_j^*$ to be constant in the motion we must have $\frac{\partial \tilde{\mathcal{H}}}{\partial \tilde{\theta}_j^*}$ equal to zero, that is,

$$\Sigma'' k_{j+4} \tilde{K}_k \sin \tilde{P}_k = 0.$$

These are satisfied if

$$\sin \tilde{P}_k = 0$$

for all those sets k in the Σ'' summation. This is certainly true if each of $\tilde{\theta}_1$, $\tilde{\theta}_2$, $\tilde{\theta}_1^*$, and $\tilde{\theta}_2^*$ is either zero or π. For $\tilde{\theta}_j$ to be constant requires $\frac{\partial \tilde{\mathcal{H}}}{\partial \tilde{\Theta}_j}$ to be zero, which in turn requires that

$$\nu^\sharp = \frac{\tilde{b}_j}{\beta_j \tilde{n}_j \tilde{a}_j^4 \tilde{e}_j} \frac{\partial \tilde{\mathcal{R}}}{\partial \tilde{e}_j} + \frac{\tan(\tilde{i}_j/2)}{\beta_j \tilde{n}_j \tilde{a}_j \tilde{e}_j} \frac{\partial \tilde{\mathcal{R}}}{\partial \tilde{i}_j},$$

where

$$\nu^\sharp = \left((p+q)n_2^\sharp - pn_1^\sharp\right)/q.$$

Also for $\tilde{\theta}_j^*$ to be constant requires $\frac{\partial\tilde{\mathcal{H}}}{\partial\tilde{\Theta}_j^*}$ to be zero, which in turn requires that

$$\nu^\sharp = \frac{1}{\beta_j\tilde{n}_j\tilde{a}_j\tilde{b}_j\sin\tilde{i}_j}\frac{\partial\tilde{\mathcal{R}}}{\partial\tilde{i}_j},$$

Now $\tilde{\mathcal{R}}$ always contains terms quadratic in the $\tilde{e}_j$ and the $\tilde{i}_j$, of the form

$$\frac{\varepsilon G\beta_1\beta_2}{\tilde{a}_2}\{A-B\},$$

where

$$A = K_2\{\tilde{e}_1^2+\tilde{e}_2^2-\sin^2\tilde{i}_1-\sin^2\tilde{i}_2+\sin\tilde{i}_1\sin\tilde{i}_2\cos(\tilde{\theta}_1^*-\tilde{\theta}_2^*)\}$$

and

$$B = K_2'\tilde{e}_1\tilde{e}_2\cos(\tilde{\theta}_1-\tilde{\theta}_2),$$

but, in the case $q=1$, there are also critical terms linear in the $\tilde{e}_j$, say

$$\frac{\varepsilon G\beta_1\beta_2}{\tilde{a}_2}\left\{K_{1,0}\tilde{e}_1\cos\tilde{\theta}_1+K_{0,1}\tilde{e}_2\cos\tilde{\theta}_2\right\},$$

so that in this case the conditions for a periodic solution are

$$\nu^\sharp = \frac{\varepsilon G\beta_2}{\tilde{n}_1\tilde{a}_1^2\tilde{a}_2}\left\{K_{1,0}\frac{1}{\tilde{e}_1}\cos\tilde{\theta}_1+K_2-K_2'\frac{\tilde{e}_2}{\tilde{e}_1}\cos(\tilde{\theta}_1-\tilde{\theta}_2)+\dots\right\},$$

and

$$\nu^\sharp = \frac{\varepsilon G\beta_1}{\tilde{n}_2\tilde{a}_2^3}\left\{K_{0,1}\frac{1}{\tilde{e}_2}\cos(\tilde{\theta}_2)+K_2-K_2'\frac{\tilde{e}_1}{\tilde{e}_2}\cos(\tilde{\theta}_1-\tilde{\theta}_2)+\dots\right\},$$

so that

$$\tilde{e}_1 = \frac{\tilde{n}_1}{\nu^\sharp}\varepsilon\beta_2\frac{\tilde{a}_1}{\tilde{a}_2}K_{1,0}\cos\tilde{\theta}_1+O\{(\varepsilon\beta_j\tilde{n}_j/\nu^\sharp)^2\},$$

and

$$\tilde{e}_2 = \frac{\tilde{n}_2}{\nu^\sharp}\varepsilon\beta_1K_{0,1}\cos\tilde{\theta}_2+O\{(\varepsilon\beta_j\tilde{n}_j/\nu^\sharp)^2\}.$$

This shows how, in the case of moderately close commensurability, in the sense mentioned in section 1 above, that is, when $\varepsilon\beta_2\tilde{n}_1/\nu^\sharp$ and $\varepsilon\beta_1\tilde{n}_2/\nu^\sharp$ are both small, periodic solutions of Poincaré's *première sorte* occur in the case $q=1$, with $\tilde{e}_1$ approximately proportional to $\varepsilon\beta_2\tilde{n}_1/\nu^\sharp$, and with $\tilde{e}_2$ approximately proportional to $\varepsilon\beta_1\tilde{n}_2/\nu^\sharp$. This confirms the indications of section 2, including the fact that the forced eccentricities increase as the commensurability becomes closer, that is, as $\nu^\natural$ approaches zero. In fact, when the very close commensurability case is reached, so that $\varepsilon\beta_2/\tilde{n}_1\nu^\natural$ and $\varepsilon\beta_1\tilde{n}_2/\nu^\natural$ are not both small, the forced eccentricities will usually be large compared with the ratios of the perturbing masses to the mass of the primary. In the very close commensurability case, periodic solutions are possible for values of q other than 1. Periodic solutions in the very close commensurability case are either of Poincaré's *deuxième sorte*, in which the solutions are planar, or of the *troisième sorte*, in which the solutions are three-dimensional. In the case $q = 1$, there is a continuous transition from the *première sorte* to the *deuxième sorte* on the approach to exact commensurability, and transition is made from the moderately close commensurability case to the very close commensurability case.

6. Periodic solutions in the case $q = 1$: the case of Enceladus and Dione

The critical arguments may be written

$$\theta_1 = \ell_1 - (p+1)\phi,$$

and

$$\theta_2 = \ell_2 - p\phi$$

(where ℓ_j is the mean anomaly $\lambda_j - \varpi_j$), so, since in a periodic solution of the type we have been considering, $\tilde{\theta}_1$ and $\tilde{\theta}_2$ are each constant, it follows that, while ϕ is making a complete revolution, ℓ_1 makes $p+1$ revolutions, and ℓ_2 makes p revolutions. Now in the case of Enceladus ($j = 1$), and Dione ($j = 2$), we have $p = 1$, so that $\theta_1 = \ell_1 - 2\phi$, and $\theta_2 = \ell_2 - \phi$, and these satellites are in fact very close to such a periodic solution, in which $\tilde{\theta}_1 = 0$, and $\tilde{\theta}_2 = \pi$. If they were exactly following such a periodic solution, then, at a conjunction of the two satellites, that is, when $\phi = 0$, we would have $\ell_1 = 0$, that is, Enceladus, which is on the inner of the two orbits, would be at perisaturnium, and also $\ell_2 = \pi$, so that Dione, which is on the outer of the two orbits, would be at aposaturnium. Therefore the two satellites would be as far apart as is possible at a conjunction. This is a *stable* periodic solution, as was shown in lectures at an earlier Advanced Study Institute (see *Message 1982a; section 3.1*). When the satellites have reached opposition, that is, when ϕ has increased to π, ℓ_1 will have increased by a complete revolution, so that Enceladus will once again be at perisaturnium, whereas ℓ_2 will have increased by half a revolution to π, so that Dione will now also be at perisaturnium. For a more general discussion of the possible patterns of occurrence of apses and syzygies in periodic solutions of this nature see *Message (1985).* Now in the actual motion, $\tilde{\theta}_1$ librates about its mean value of zero, and the eccentricity of Enceladus librates about a forced mean value, which, determined from observation, enabled H. Struve *(1916)*, and afterwards G. Struve *(1930)*, Jeffreys *(1953)*, and Kozai *(1955)* to make estimates of the mass of Dione. Also $\tilde{\theta}_2$ circulates, that is, it changes always in the same sense, so that the mean value of the eccentricity of Dione is a free quantity, and so provides no means of directly estimating the mass of Enceladus, which Kozai derived from the mass of Dione and the ratio of the amplitudes of of those periodic terms, in the mean longitudes of the two satellites, which are associated with the libration.

7. The effect of Saturn's oblateness on these solutions.

The oblateness of Saturn is sufficiently large to play a significant part in the perturbations of the orbits of the inner satellites, in fact it would be the dominant part if it were not for the linking of most of their orbital periods into deep resonance pairs. Let us examine the effect of the oblateness on the aspects of orbital motions which we have been considering. We may express the gravitational potential due to Saturn, at a point P, whose distance from Saturn's centre is r, and whose latitude from Saturn's equator is β, by

$$V(P) = -\frac{Gm_S}{r} + G\frac{(C-A)}{r^3}P_2(\sin\beta),$$

where we are neglecting terms of order $1/r^5$. Here C is the moment of inertia of Saturn about its polar axis (supposed here to be an axis of symmetry), and A is its moment of inertia about any axis in its equator. We are using the Legendre poynomial

$$P_2(\sin\beta) = (3\sin^2\beta - 1)/2.$$

If we write

$$V(P) = -\frac{Gm_S}{r} - \mathcal{R}_J,$$

then, to the order to which we are working,

$$\mathcal{R}_J = -\frac{Gm_S J_2 r_S^2}{r^3}P_2(\sin\beta)$$

is the disturbing function for the action of the oblateness of Saturn. (Here $J_2 = (C-A)/(m_S r_S^2)$, and r_S is the radius of Saturn's equator.) The mean value of $\mathcal{R}_J$, averaged over an orbit of P, is, in terms of the elements of that orbit,

$$\tilde{\mathcal{R}}_J = -\frac{\tilde{n}^2\tilde{a}^3 J_2 \tilde{r}_S^2}{4\tilde{b}^3}(3\sin^2\tilde{i} - 2). \tag{15}$$

¿From this we derive the contribution of the oblateness perturbation to the mean motion of the apse:

$$\begin{aligned} \dot{\tilde{\varpi}}_J &= \frac{\tilde{b}}{\tilde{n}\tilde{a}^2\tilde{e}}\frac{\partial\tilde{\mathcal{R}}_J}{\partial\tilde{e}} + \frac{\tan(\tilde{i}/2)}{\tilde{n}\tilde{a}\tilde{b}}\frac{\partial\tilde{\mathcal{R}}_J}{\partial\tilde{i}} \\ &= \frac{3\tilde{n}\tilde{a}^2 J_2 \tilde{r}_S^2}{4\tilde{b}^4}(5\cos^2\tilde{i} - 2\cos\tilde{i} - 1). \end{aligned}$$

Take $J_2 = 0.01667$, $r_S = 59670$ *km.*, and for Enceladus $a = 238020$ *km.*, giving, for the contribution of Saturn's oblateness to the mean motion of the apse of Enceladus, $\dot{\varpi}_{JEn} = 0.4129$ degrees per day. (Here the suffix "*En*" indicates elements of the orbit of Enceladus, while "*Di*" indicates those of Dione.) If we include this contribution in the estimate of the mean motion of the critical argument θ_1, before taking into account the effect of the near-commensurability, that is, before including any effect of the mutual perturbation on the apse motion, we obtain

$$n^\sharp_{En} - 2n^\sharp_{Di} + \dot{\varpi}_{JEn} = 0.0749$$

in degrees per day, or 0.000569 times the mean motion of Dione. This shows that, as far as this critical argument is concerned, we are still in the "moderately close" case. Turning to Dione, for which $a = 377400$ *km.*, we then find that $\dot{\varpi}_{JEn} = 0.0822$ degrees per day. Carrying out the corresponding calculation for the critical argument θ_2, we obtain

$$n^\sharp_{En} - 2n^\sharp_{Di} + \dot{\varpi}_{JDi} = -0.2558$$

in degrees per day, which is -0.001945 times the mean motion of Dione, so this is still well within the "moderately close" case. Thus the relevant periodic solutions are those of the *première sorte*, despite the effect of Saturn's oblateness.

8. The non-planar case, and the motion of Mimas and Tethys.

Let us now turn to the case of non-planar motions, which is relevant to the case of Mimas and Tethys. The periodic solutions of the general three-body problem which are relevant here are those of Poincaré's *troisième sorte.* (For some discussion of such solutions, the nature of successive conjunctions, oppositions, and node crossings, and a demonstration of their existence in certain cases, based on Roy and Ovenden's mirror theorem *(1955)*, see *Message (1982b).)* But for these satellites we must not neglect the effect of the oblateness on the motion of the nodes. From equation (15) we find that the contribution of the main oblateness term to the mean motion of the node is

$$\begin{aligned}\dot{\tilde{\Omega}}_J &= \frac{1}{\tilde{n}\tilde{a}\tilde{b}\sin\tilde{i}}\frac{\partial\tilde{R}}{\partial\tilde{i}} \\ &= -(3\tilde{n}\tilde{a}^2 J_2 r_S^2)/(2\tilde{b}^4)\cos\tilde{i}.\end{aligned}$$

For Mimas, we have $a = 185520$ *km.*, and so the contribution from Saturn's oblateness to the mean motion of the node of Mimas is $\dot{\tilde{\Omega}}_{JMi} = -0.988$ degrees per day. (Here the suffix "Mi" indicates elements of the orbit of Mimas, and "Te" will indicate those of Tethys.) Modifying the calculation of the mean motion of the critical argument $\theta_1^* = \lambda_{Mi} - 2\lambda_{Te} + \Omega_{Mi}$ to include this contribution, but without yet including the effect of mutual perturbations, we obtain

$$n_{Mi}^{\sharp} - 2n_{Te}^{\sharp} + \dot{\tilde{\Omega}}_{JMi} = -0.389$$

in degrees per day, which is -0.00204 times the mean motion of Tethys. For Tethys, we have $a = 294660$ *km.*, and so the contribution from Saturn's oblateness to the mean motion of the node of Tethys is $\dot{\tilde{\Omega}}_{JTe} = -0.196$ degrees per day, and hence the corresponding calculation for the critical argument $\theta_2^* = \lambda_{Mi} - 2\lambda_{Te} + \Omega_{Te}$, we find

$$n_{Mi}^{\sharp} - 2n_{Te}^{\sharp} + \dot{\tilde{\Omega}}_{JTe} = 0.402$$

in degrees per day, which is 0.00211 times the mean motion of Tethys. Each of these is well within the "moderately close" case, and, since the solutions of the *troisième sorte* only occur within the "very close commensurability" case, this indicates that the motion of Mimas and Tethys cannot approximate to a periodic solution of the *troisième sorte.* This is borne out by considering the actual mean motions of the relevant arguments. That of ϕ is 191.296599 degrees per day. The mean motion of the node of Mimas, Ω_{Mi} is -365.025 degrees per year (*Harper and Taylor, 1993*), or -0.999405 degrees per day, so that the mean motion of $\lambda_{Mi} - \Omega_{Mi}$ is 382.993917 degrees per day. The mean motion of the node of Tethys, Ω_{Te}, is -72.245 degrees per year (*Harper and Taylor, 1993*), or -0.197800 degrees per day, so that the mean motion of $\lambda_{Te} - \Omega_{Te}$ is 190.895713 degrees per day. Thus between successive conjunctions of the two satellites, that is, while ϕ is increasing by 2π, the argument $\lambda_{Mi} - \Omega_{Mi}$

increases by 4.0041895π, and $\lambda_{Te} - \Omega_{Te}$ increases by 1.995809π, so that the satellites do not return exactly to their nodes in this interval, which would be necessary to satisfy the conditions for a periodic solution of the *troisième sorte*. But if we consider the critical angle

$$\theta^* = \theta_1^* + \theta_2^* = 2\lambda_{Mi} - 4\lambda_{Te}^* + \Omega_{Mi} + \Omega_{Te},$$

which has the mean rate of change 0.013 degrees per day, which is 0.000035 times the mean motion of Mimas (about 30 times the mass of Mimas in terms of that of Saturn), we see that we are closer to the "very close commensurability" case than we are for the angles θ_1^* and θ_2^* themselves. In fact θ^* is indeed a librating angle in the Mimas and Tethys motion; it is librating about zero with an amplitude of about 94 degrees.

References

Garfinkel, B., 1966 *Astronomical Journal* **71**, 657-669.

Garfinkel, B., Jupp, A.H., and Williams, C., 1971 *Astronomical Journal* **76**, 157-166.

Harper, D., and Taylor, D.B., 1993 *Astronomy and Astrophysics* **268**, 326-349.

Jeffreys, H., 1953 *Monthly Notices of the Royal Astronomical Society* **113**, 81-96.

Jupp, A.H., 1973 *Celestial Mechanics* **7**, 347-355.

Kozai, Y., 1955 *Publications of the Astronomical Society of Japan* **7, No.4**, 176-189.

Kozai, Y., 1957 *Annals of the Tokio Astronomical Observatory* Series **2, 5**, No.**2**.

Message, P.J., 1966 "On nearly-commensurable periods in the restricted problem of three bodies, with calculations of the long-period variations in the interior 2:1 case", in *Proceedings of Symposium No.25 of the International Astronomical Union* 197-222.

Message, P.J., 1982a "Some aspects of Motion in the General Planar Problem of Three Bodies; in particular in the vicinity of periodic solutions associated with near small-integer commensurabilities of Orbital Period" in *"Applications of Modern Dynamics to Celestial Mechanics and Astrodynamics"* ed. V Szebehely, Reidel, 77-101.

Message, P.J., 1982b, *Celestial Mechanics* **28**, 107-118.

Message, P.J., 1985 "Some Results of Resonance and Periodic Motions" in *"Stability of the Solar System and its minor natural and artificial bodies"* ed. V. Szebehely, Reidel, 193-199.

Message, P.J., 1987 "Planetary Perturbation Theory from Lie Series, including Resonance and Critical arguments" in *"Long-term Dynamical*

Behaviour of natural and Artificial N-Body Systems" ed. A.E. Roy, Kluwer, 47-72.
Message, P.J., 1993 *Celestial Mechanics* **56**, 277-284.
Rose, L.E., 1979 *Astronomical Journal* **84**, 1067-1071.
Roy, A.E., and Ovenden, M.W., 1955 *Monthly Notices of the Royal Astronomical Society* **115** 296-309.
Struve, G., 1930 *Veröff. Univ.-Sternwarte, Berlin-Babelsberg* **6**, Heft**4**.
Struve, H., 1916 *Sitzungber. Preuss. Akad. Wiss.* 1098-1110.

HOW IDEAL IS THE MIMAS-TETHYS RESONANCE?

J. PROBIN
University of Liverpool,
Liverpool, UK

Introduction

Mimas and Tethys orbit Saturn with a near 2:1 resonance in their orbital periods. Certain properties of this system are of interest to dynamical astronomers, for instance the unusually small eccentricities of the satellites. In this paper we shall firstly examine the previous approach to the system developed by H. Struve (presented in [1]), and then employ the aforementioned circular orbit property to tackle the system via a different method that we believe will improve upon Struve's solution.

Struve's Formulation of the Problem

Struve began by using basic geometry in the Mimas-Tethys-Saturn system to express the distance between the satellites, Δ, in terms of Cartesian variables. He then expresses this as functions of the major semi-axis a_j, the mean longitude λ_j, the inclination i_j and the longitude of ascending node Ω_j, of the appropriate satellite; the inclination appearing through the variable

$$\gamma_j = \sin\frac{i_j}{2} \tag{1}$$

By removing some less significant terms, expanding to second order in γ_1 and γ_2 and substituting an infinite Laplace-type series in place of one particular expression, Struve eventually derives the following

$$\frac{1}{\Delta} = P_i^0 + P_i^1$$

where

$$P_i^0 = -\frac{1}{4}\gamma_1\gamma_2\sum B^{(i)}\cos\left[(i+1)\lambda_2 - (i-1)\lambda_1 - \Omega_1 - \Omega_2\right] \tag{2}$$

B.A. Steves and A.E. Roy (eds.), The Dynamics of Small Bodies in the Solar System, 227–232.

$$P_i^1 = -\frac{1}{8}\left(\gamma_1^2 + \gamma_2^2\right) \sum B^{(i)} \cos(i-1)(\lambda_2 - \lambda_1) \tag{3}$$

where the subscripts 1 and 2 refer to Mimas and Tethys respectively, and the $B^{(i)}$s are multiples of the Laplace coefficients, with dimension $(length)^{-1}$.

From analysis of his data, Struve had noted that the particular combination of variables he denoted by

$$W = 4\lambda_2 - 2\lambda_1 - \Omega_1 - \Omega_2 \tag{4}$$

moved at a rate of 3 ° per year, two orders of magnitude smaller than any other angular combination (as is to be expected given the 2:1 resonance of the system). Thus he argued that the $i = 3$ term would dominate the P_i^0 expression, with the $i = 1$ term dominating the P_i^1 expression as the only non-periodic term, hence providing the approximation

$$\frac{1}{\Delta} = -\frac{1}{4}B^{(3)}\gamma_1\gamma_2 \cos W - \frac{1}{8}B^{(1)}\left(\gamma_1^2 + \gamma_2^2\right) + \text{short period terms} \tag{5}$$

This of course leads immediately to the Disturbing Function for the system and hence provides Lagrange's Equations. Thus we now have analytic expressions for the rates of change of $a_j, \gamma_j, \Omega_j, n_j and \lambda_j$ for each satellite, all of these expressions being in terms of trigonometric functions of the critical angle W.

This work to this stage was hardly new, and indeed these same ten equations had previously been integrated years earlier. However Struve tried a different approach. Lagrange's Equations lead to

$$\frac{dW}{dt} = 4n_2 - 2n_1 + K\cos W \tag{6}$$

where K is appropriately defined, and hence

$$\frac{d^2W}{dt^2} + h^2 \sin W + s \sin W \frac{dW}{dt} = 0 \tag{7}$$

where h^2 and s are defined in [1].

By substituting numerical values Struve showed that

$$s\frac{dW}{dt} << h^2 \tag{8}$$

thus he approximated (7) by

$$\frac{d^2W}{dt^2} + h^2 \sin W = 0 \tag{9}$$

Of course (9) is just the simple pendulum equation.

Hence after using the available data to satisfy himself that the motion of the system was librational, Struve could then solve for the critical angle in terms of Jacobi elliptic functions and provide an analytic solution for the system, as seen in [1].

Subsequently his son G. Struve, and later Y. Kozai, used this solution as the mathematical basis for extensive analysis of their contemporary observational data to determine the parameters of the solution and the orbits.

The Ideal Resonance Problem

The Ideal Resonance Problem (IRP) is defined to have the following Hamiltonian

$$-F(x,y) = B(x) + 2\mu^2 A(x)\sin^2 y \tag{10}$$

y : generalised co-ordinate

x : generalised momentum

μ : small parameter

Applying Hamilton's Equations with the right hand sides set to zero provides the Centre of Libration:

$$y_0 = n\pi \tag{11a}$$

$$B'(x_0) = 0 \tag{11b}$$

The phase plane of the IRP (see [2]) is identical in form to that of the simple pendulum. In fact the Hamiltonian of the simple pendulum can be generated from (10) by setting

$$B(x) = x^2 \tag{12a}$$

$$A(x) = \text{constant} \tag{12b}$$

Hence from Struve's work we suspect that the Mimas-Tethys system could be modelled as an IRP, and we search for a change of co-ordinates which would transform the Hamiltonian of the system into the IRP form. Then the algorithm found in [2] would furnish an analytic solution in terms of elliptic functions to any order in μ.

The Mimas-Tethys Hamiltonian

The Mimas-Tethys Hamiltonian can be obtained from the $N = 2$ case of the general Hamiltonian for N secondaries orbiting a primary (See [3] for definition of the parameters)

$$H\left(\underline{r}_j,\underline{p}_j\right)=\sum_{j=1}^{N}\left\{\frac{\underline{p}_j^2}{2\beta_j}-\frac{\mu_j\beta_j}{|\underline{r}_j|}\right\}-R \tag{13}$$

where

$$R=\epsilon\sum_{j=1}^{N-1}\sum_{k=j+1}^{N}\beta_j\beta_k\left(\frac{1}{|\underline{r}_j-\underline{r}_k|}-\frac{\underline{r}_j\cdot\underline{r}_k}{|\underline{r}_k|^3}\right)+O(\epsilon^2) \tag{14}$$

In this case

$$R=\frac{\epsilon\beta_1\beta_2}{\Delta} \tag{15}$$

because the second, indirect portion of R does not contain the critical angle but only short period terms which can be averaged out. (Here $\epsilon^{-1}=70,000$). Thus

$$H=-\frac{\beta_1^3\mu_1^2}{2\Lambda_1^2}-\frac{\beta_2^3\mu_2^2}{2\Lambda_2^2}-\frac{\epsilon\beta_1\beta_2}{\Delta} \tag{16}$$

It must be noted that at this stage the Hamiltonian is not expressed in canonical co-ordinates and so must only be considered a general function.

The Change of Variables

We now wish to find a transformation to combine our two three-dimensional systems into a single six-dimensional system. However it can be seen from (10) that the Ideal Resonance Problem is expressed only in one dimension, and thus if we are to achieve our aim of expressing the Mimas-Tethys Hamiltonian in the form of (10) the transformation must produce one active variable (the critical angle) and five ignorable co-ordinates.

In light of this we make the transformation from

$$(\lambda_1,\varpi_1,\Omega_1,\lambda_2,\varpi_2,\Omega_2,\Lambda_1,\Pi_1,N_1,\Lambda_2,\Pi_2,N_2)$$

to

$$(\theta,\phi,\sigma,\psi,\theta^*,\theta^{**},\Theta,\Phi,\Sigma,\Psi,\Theta^*,\Theta^{**})$$

where

$$\theta = 4\lambda_2-2\lambda_1-\Omega_1-\Omega_2 \tag{17a}$$
$$\phi = \lambda_1-\lambda_2 \tag{17b}$$
$$\sigma = \Omega_1-\Omega_2 \tag{17c}$$
$$\psi = \lambda_1+\lambda_2+\varpi_1+\varpi_2 \tag{17d}$$
$$\theta^* = 2\lambda_2-\lambda_1-\varpi_1 \tag{17e}$$
$$\theta^{**} = 2\lambda_2-\lambda_1-\varpi_2 \tag{17f}$$

ϕ is the difference in mean longitudes and therefore a fast variable which can be averaged out over time.
σ is the difference in the longitudes of the nodes which is not quite so fast but can still be averaged out over time.

When the substitution (17) is made all the arguments in the disturbing function can be expressed without ψ , making it an ignorable and thus Ψ a constant (in fact a fraction of the total angular momentum of the system).

Now D'Alembert's Principle says that if any term in the disturbing function contains a multiple of ϖ_j in its angular argument then that term has as a factor the corresponding power of the eccentricity e_j. We approximate the almost circular orbits of Mimas and Tethys by setting both eccentricities to zero, and thus from (17e) and (17f), Θ^* and Θ^{**} will have a zero coefficient in the general disturbing function and can be treated as ignorable.

Thus we are left with one active variable, the critical angle θ , and five ignorable co-ordinates $\phi, \sigma, \psi, \theta^*, \theta^{**}$ as desired.

We find the conjugate momenta which makes this a canonical set through the use of a standard generating function which gives $\Lambda_1, \Pi_1, N_1, \Lambda_2, \Pi_2, N_2$ as a function of $\Theta, \Phi, \Sigma, \Psi, \Theta^*, \Theta^{**}$. The last five elements of the latter set, by virtue of being conjugate to ignorable co-ordinates, will all be constants of the motion. Of course setting the eccentricities to zero must reduce the number of degrees of freedom of any co-ordinate set by two, and here the effect is to render

$$\Psi = \Theta^* = \Theta^{**} \tag{18}$$

If we apply this new equality, and now define a new set of constants by

$$\begin{aligned} \alpha_1 &= \Sigma \\ -2\alpha_4 &= \Phi - \Psi \\ 4\alpha_5 &= -\Phi + 5\Psi \end{aligned} \tag{19}$$

then the old and new momenta for the Mimas-Tethys system are related by

$$\begin{aligned} \Lambda_1 &= -2(\Theta + \alpha_4) \\ \Lambda_2 &= 4(\Theta + \alpha_5) \\ \Pi_1 &= 0 \\ \Pi_2 &= 0 \\ N_1 &= -\Theta + \alpha_1 \\ N_2 &= -\Theta - \alpha_1 \end{aligned} \tag{20}$$

Using (20) we are now in a position to transform the Hamiltonian into our new co-ordinates via equations (1)-(3).

Let us define some new functions $g_1(\Theta), g_2(\Theta), h(\Theta)$, as follows

$$\gamma_1^2 = -\frac{1}{4}\left(\frac{\Theta-\alpha_1}{\Theta+\alpha_4}\right) = \frac{g_1(\Theta)}{B^{(1)}} \tag{21}$$

$$\gamma_2^2 = \frac{1}{8}\left(\frac{\Theta+\alpha_1}{\Theta+\alpha_5}\right) = \frac{g_2(\Theta)}{B^{(1)}} \tag{22}$$

$$\gamma_1\gamma_2 = -\frac{1}{4}\left[\frac{(\Theta-\alpha_1)}{(-2\Theta+\alpha_4)}\frac{(\Theta+\alpha_1)}{(\Theta+\alpha_5)}\right]^{\frac{1}{2}} = \frac{4}{B^{(3)}}h(\Theta) \tag{23}$$

Hence if we introduce the following notation

$$b(\Theta) = \frac{\beta_1^3\mu_1^2}{8(\Theta+\alpha_4)^2} + \frac{\beta_2^3\mu_2^2}{32(\Theta+\alpha_5)^2} - \frac{\epsilon\beta_1\beta_2}{8}\left[g_1(\Theta)+g_2(\Theta)\right] \tag{24}$$

$$A(\Theta) = \beta_1\beta_2 h(\Theta) \tag{25}$$

$$B(\Theta) = b(\Theta) - \epsilon A(\Theta) \tag{26}$$

and adopt the traditional notation of a Hamiltonian in an Ideal Resonance Problem so that instead of $H(\Theta)$, we write $F(\Theta)$, then we arrive at the final expression

$$-F(\theta,\Theta) = B(\Theta) + 2\epsilon A(\Theta)\sin^2\frac{\theta}{2} \tag{27}$$

Hence we have demonstrated that the Mimas-Tethys problem is an IRP.

Work is currently in progress to find the centre of libration by solving equation (11b) to second order in ϵ (ϵ^3being of order10^{-14}).We can then obtain a solution to the problem by applying the result to the algorithm found in paper [2].

We hope that future papers will show how by using the Extended Ideal Resonance Problem theory presented in [4] it is possible to generate a more accurate analytic solution than Struve.

References

1. Tisserand, F. (1896) *Traite de Mecanique Celeste (Tome IV)*, Gauthier-Villars, Paris.
2. Jupp, A.H. (1969) A Solution of the Ideal Resonance Problem for the Case of Libration,*Astronomical Journal* **74**, 35-43.
3. Message, P.J. (1993) Perturbation Theory: Techniques and Limitations, in A.E. Roy and B.A. Steves (eds.), *From Newton to Chaos*, Plenum Press, New York, pp 5-19.
4. Garfinkel, B. (1975) An Extended Ideal Resonance Problem, *Celestial Mechanics* **12**, 203-214.

THE DYNAMICS OF PLANETARY RINGS AND SMALL SATELLITES

C.D. MURRAY
Astronomy Unit, Queen Mary and Westfield College,
Mile End Rd, London E1 4NS, U.K.

1. Introduction

All the giant outer planets of the solar system possess ring systems, each with an accompanying retinue of small satellites. Although the ring systems of Jupiter, Saturn, Uranus and Neptune have their individual characteristics, one property they share is that small satellites have played a rôle in determining their dynamical structure. This has usually been achieved by the mechanism of resonance, whereby the gravitational effect of a perturbing satellite is enhanced at particular orbital locations.

The theory of spiral density waves, originally proposed to explain spiral structure in galaxies, was adapted to explain features such as the prominent Cassini Division in Saturn's rings (Goldreich & Tremaine 1979). Along with this new understanding came the terminology of galactic dynamics, often resulting in confusion in the discussion of resonances and their effects. One of the goals of this article is to approach the study of resonance in planetary rings from the classical direction (by means of the disturbing function) while following a parallel route using the new terminology.

We begin with a brief review of the physical and dynamical properties of each system of rings and satellites. From an analysis of the planetary disturbing function we classify the different types of resonances that can occur and proceed to show their equivalents using the terminology of galactic dynamics. We demonstrate how the location (in semimajor axis) of an exact resonance is calculated for the case of perturbed material orbiting an oblate planet. The particular dynamical problems associated with narrow rings are discussed, showing the need for some confinement mechanism. We demonstrate how the basics of the shepherding satellite mechanism can be understood by a simple application of the Tisserand relation. We conclude

B.A. Steves and A.E. Roy (eds.), The Dynamics of Small Bodies in the Solar System, 233–256.

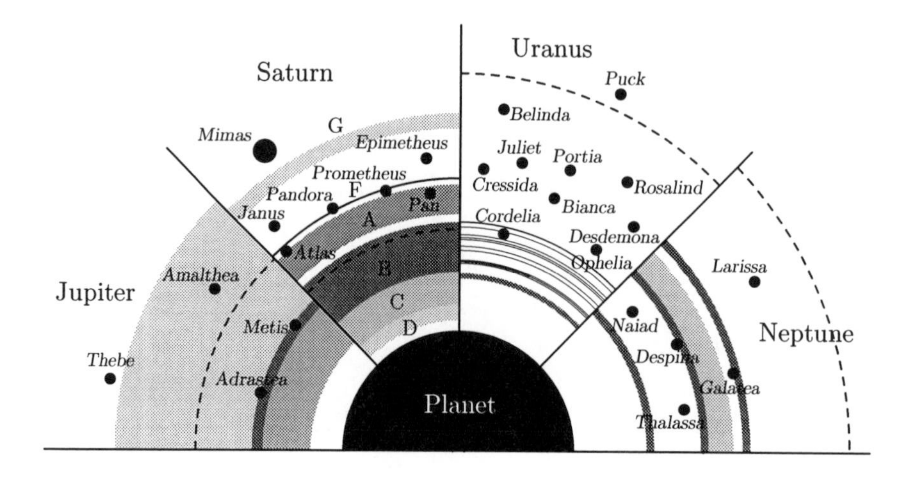

Figure 1. The ring systems and small satellites of the outer planets shown in relation to their respective planetary radii. The capital letters for Saturn denote the names of the major rings. The dashed curve denotes the location of the synchronous orbit for each planet. (Adapted from Nicholson & Dones, 1991).

with specific examples of ring-satellite interactions taking place in the ring systems of Saturn and Neptune.

2. Ring Systems of the Outer Planets

Ring systems are known to exist around the planets Jupiter, Saturn, Uranus and Neptune. Although Saturn's rings were first observed by Galileo Galilei in 1610, it was not until the serendipitous discovery of the rings of Uranus in 1977 (Elliot *et al,* 1977) that another ring system was detected. The Voyager spacecraft discovered the jovian ring in 1979, while an understanding of the true nature of Neptune's rings and ring arcs had to await the arrival of the Voyager 2 spacecraft in 1989. The book by Greenberg & Brahic (1984) provides a comprehensive summary of our knowledge of ring systems more than a decade ago. The review articles by Nicholson & Dones (1991) and Murray (1994) provide more recent surveys and include information about Neptune's ring system. A more complete discussion of ring dynamics and the effects of resonance is given in Murray & Dermott (1998).

Figure 1 follows the example of Nicholson & Dones (1991) and shows the rings and accompanying satellites on a uniform scale of planetary radii. Dermott *et al.* (1979) suggested that it is more appropriate to consider the concept of the Roche zone rather than a unique Roche limit, as defining the location where a satellite would be tidally disrupted. They showed that such a zone should occur in the range 1.44–2.42 planetary radii. This is in good agreement with the location of the planetary ring systems (see Fig. 1). It is clear from this plot that the small satellites are found close to the ring systems. Whether or not the satellites have played an important rôle in the origin of planetary rings, it is clear that the dynamical interactions between rings and satellites can be important because of their close proximity.

3. Classification of Resonances

A necessary condition for the existence of a resonance between two gravitationally interacting bodies is that their mean motions should satisfy the relationship

$$\frac{n'}{n} \approx \frac{p}{p+q} \tag{1}$$

where n and n' denote the mean motions of the inner and outer bodies respectively and p and q are positive integers.

In the context of planetary rings a more precise definition can be found by considering one of two approaches. The first involves an examination of the properties of the planetary disturbing function while the second involves an understanding of the various frequencies that arise when studying motion around an oblate planet. Here we develop the theory for an exterior satellite perturbing an interior ring particle, but it can easily be extended to consider the converse situation.

3.1. THE DISTURBING FUNCTION

Consider a test particle moving in an orbit with semimajor axis a, eccentricity e, inclination I, longitude of pericentre ϖ, longitude of ascending node Ω and mean longitude λ. A perturbing satellite moves in an exterior orbit with elements a', e', I', ϖ', Ω' and λ'. The gravitational potential experienced by the test particle with position vector $\mathbf{r}$ due to the exterior satellite of mass m' and position vector $\mathbf{r}'$ is

$$\mathcal{R} = \frac{\mathcal{G}m'}{|\mathbf{r}' - \mathbf{r}|} - \mathcal{G}m' \frac{\mathbf{r}' \cdot \mathbf{r}}{r'^3} \tag{2}$$

where the first and second terms on the right-hand side are referred to as the direct and indirect contributions, respectively. The disturbing function $\mathcal{R}$ can be expanded as an infinite series in both sets of orbital elements. It has the general form

$$\mathcal{R} = \sum S \cos \varphi. \tag{3}$$

The angle φ is a linear combination of longitudes of the form

$$\varphi = j_1\lambda' + j_2\lambda + j_3\varpi' + j_4\varpi + j_5\Omega' + j_6\Omega \tag{4}$$

where the j_i are all integers satisfying the d'Alembert property that $\sum_i j_i = 0$ with the additional constraint that $|j_5+j_6|$ must be even (see, for example, Murray & Harper 1993 and Hamilton 1994). To lowest order, S has the form

$$S = \frac{\mathcal{G}m'}{a'} f(\alpha) e'^{|j_3|} e^{|j_4|} s'^{|j_5|} s^{|j_6|} \tag{5}$$

where $\alpha = a/a' < 1$, $s = \sin I/2$ and $s' = \sin I'/2$. Terms with $j_1 = j_2 = 0$ are referred to as *secular* and their effect can be treated separately from the resonant terms (see Murray & Dermott 1998). The function $f(\alpha)$ can be written as

$$f(\alpha) = f_{\mathrm{d}}(\alpha) + f_{\mathrm{e}}(\alpha) \tag{6}$$

where the f_{d} and f_{e} terms arise from the direct and indirect terms for an external perturber respectively. Explicit expressions for $f_{\mathrm{d}}(\alpha)$ for those arguments where $|j_1 + j_2| = 1$ (first-order resonances) or $|j_1 + j_2| = 2$ (second-order resonances) are given in Table 1. These are expressed in terms of Laplace coefficients, $b_s^{(j)}$ defined by

$$\frac{1}{2} b_s^{(j)}(\alpha) = \frac{1}{2\pi} \int_0^{2\pi} \frac{\cos j\psi \, \mathrm{d}\psi}{(1 - 2\alpha\cos\psi + \alpha^2)^s}. \tag{7}$$

For our purposes the only indirect terms we need to consider are $f_{\mathrm{e}}(\alpha) = -2\alpha$ for $\varphi = 2\lambda' - \lambda - \varpi'$, and $f_{\mathrm{e}}(\alpha) = -(27/8)\alpha$ for $\varphi = 3\lambda' - \lambda - 2\varpi'$.

A resonance occurs at a semimajor axis where the time derivative of a specific linear combination φ is zero. In other words, the semimajor axis of the resonance is the solution of the non-linear equation

$$\dot{\varphi} = j_1 n' + j_2 n + j_3\dot{\varpi}' + j_4\dot{\varpi} + j_5\dot{\Omega}' + j_6\dot{\Omega} \tag{8}$$

where we have neglected the variation of the mean longitude at epoch and the quantities $\dot{\varpi}'$, $\dot{\varpi}$, $\dot{\Omega}'$ and $\dot{\Omega}$ are all functions of a and a' and can be determined from Lagrange's planetary equations. When the central planet is oblate there will be additional contributions to the disturbing function which will also affect the rates of change of the pericentres and the nodes (see Eqs. (16–17) below). In fact, for motion near the planet these rates can be dominated by the oblateness.

The first- and second-order resonances arising from the general arguments listed in Table 1 can be classified according to the eccentricity or inclination term associated with each argument (Table 2). Note from Eq. (5)

TABLE 1. Direct terms associated with first- and second-order resonant terms in the expansion of the disturbing function. The symbol D denotes $\mathrm{d}/\mathrm{d}\alpha$.

j_1	j_2	j_3	j_4	j_5	j_6	$f_{\mathrm{d}}(\alpha)$
j	$1-j$	0	-1	0	0	$\frac{1}{2}\left[-2j-\alpha D\right] b_{1/2}^{(j)}$
j	$1-j$	-1	0	0	0	$\frac{1}{2}\left[-1+2j+\alpha D\right] b_{1/2}^{(j-1)}$
j	$2-j$	0	-2	0	0	$\frac{1}{8}\left[-5j+4j^2-2\alpha D+4j\alpha D+\alpha^2 D^2\right] b_{1/2}^{(j)}$
j	$2-j$	-1	-1	0	0	$\frac{1}{4}\left[-2+6j-4j^2+2\alpha D-4j\alpha D-\alpha^2 D^2\right] b_{1/2}^{(j-1)}$
j	$2-j$	-2	0	0	0	$\frac{1}{8}\left[2-7j+4j^2-2\alpha D+4j\alpha D+\alpha^2 D^2\right] b_{1/2}^{(j-2)}$
j	$2-j$	0	0	0	-2	$\frac{1}{2}\alpha b_{3/2}^{(j-1)}$
j	$2-j$	0	0	-1	-1	$-\alpha b_{3/2}^{(j-1)}$
j	$2-j$	0	0	-2	0	$\frac{1}{2}\alpha b_{3/2}^{(j-1)}$

TABLE 2. The types of first- and second-order resonances based on the resonant argument.

Resonant Argument	Resonance Type
$j\lambda'+(1-j)\lambda-\varpi'$	e'
$j\lambda'+(1-j)\lambda-\varpi$	e
$j\lambda'+(2-j)\lambda-2\varpi'$	e'^2
$j\lambda'+(2-j)\lambda-\varpi'-\varpi$	ee'
$j\lambda'+(2-j)\lambda-2\varpi$	e^2
$j\lambda'+(2-j)\lambda-2\Omega'$	I'^2
$j\lambda'+(2-j)\lambda-\Omega'-\Omega$	II'
$j\lambda'+(2-j)\lambda-2\Omega$	I^2

that the power of the eccentricity or inclination is the absolute value of the coefficient of the longitude of pericentre or ascending node.

3.2. THE EFFECT OF OBLATENESS

A ring particle orbiting an oblate planet experiences the effect of the zonal harmonic coefficients J_2, J_4, etc., in addition to the standard $1/r$ potential. These give rise to three separate frequencies, n (the modified mean motion), κ (the epicyclic frequency) and ν (the vertical frequency) given by

$$n^2 = \frac{\mathcal{G}m_{\mathrm{p}}}{r^3}\left[1+\sum_{i=1}^{\infty} A_{2i} J_{2i}\left(\frac{R_{\mathrm{p}}}{r}\right)^{2i}\right] \tag{9}$$

$$\kappa^2 = \frac{\mathcal{G}m_{\rm p}}{r^3}\left[1+\sum_{i=1}^{\infty} B_{2i}J_{2i}\left(\frac{R_{\rm p}}{r}\right)^{2i}\right] \tag{10}$$

$$\nu^2 = \frac{\mathcal{G}m_{\rm p}}{r^3}\left[1+\sum_{i=1}^{\infty} C_{2i}J_{2i}\left(\frac{R_{\rm p}}{r}\right)^{2i}\right] \tag{11}$$

where $m_{\rm p}$ is the mass of the planet, $R_{\rm p}$ is the radius of the planet, r is the radius of the particle's orbit and the coefficients A_{2i}, B_{2i} and C_{2i} are given in Table 3 (taken from Nicholson & Porco, 1988).

TABLE 3. The coefficients in the expansions of the orbital frequencies

i	A_{2i}	B_{2i}	C_{2i}
1	+3/2	−3/2	+9/2
2	−15/8	+45/8	−75/8
3	+35/16	−175/16	+245/16
4	−315/128	+2205/128	−2835/128
5	+693/256	−6237/256	+7623/256
6	−3003/1024	+33033/1024	−39039/1024

Note that in the case of a perfectly spherical, homogeneous planet $n = \kappa = \nu = n_0$ where

$$n_0^2 = \frac{\mathcal{G}m_{\rm p}}{r^3} \tag{12}$$

is the usual definition of the mean motion of a test particle in a near-circular orbit ($r \approx a$) about a spherical planet. It is clear that each of the frequencies n, κ and ν is close to n_0 for small J_{2n}; it is the small differences from n_0 that play a crucial rôle in determining the location of ring resonances. For example, if we include terms up to $\mathcal{O}(R_{\rm p}/r)^4$, the actual mean motion is given by

$$n = n_0\left[1+\frac{3}{4}J_2\left(\frac{R_{\rm p}}{r}\right)^2 - \left(\frac{9}{32}J_2^2+\frac{15}{16}J_4\right)\left(\frac{R_{\rm p}}{r}\right)^4\right]. \tag{13}$$

For distant satellites $r \gg R_{\rm p}$ and $n \approx n_0$. However, for close satellites the higher order terms make significant contributions to the mean motion and must be included.

The resulting expressions for the rates of change of the particle's longitude of pericentre and longitude of ascending node are

$$\dot{\varpi} = n - \kappa \tag{14}$$

$$\dot{\Omega} = n - \nu. \tag{15}$$

Therefore, including terms up to $\mathcal{O}(R_\mathrm{p}/r)^4$ for n, κ and ν, the pericentre and node rates can be written as

$$\dot{\varpi} = +n_0\left[\frac{3}{2}J_2\left(\frac{R_\mathrm{p}}{r}\right)^2 - \frac{15}{4}J_4\left(\frac{R_\mathrm{p}}{r}\right)^4\right] \tag{16}$$

$$\dot{\Omega} = -n_0\left[\frac{3}{2}J_2\left(\frac{R_\mathrm{p}}{r}\right)^2 - \left(\frac{9}{4}J_2^2 + \frac{15}{4}J_4\right)\left(\frac{R_\mathrm{p}}{r}\right)^4\right]. \tag{17}$$

Note that to lowest order $\dot{\varpi}$ and $\dot{\Omega}$ are equal in magnitude and opposite in sign.

A perturbing satellite will have its own set of frequencies n', κ' and ν' given by Eqs. (9–11) with r replaced by r'. The *pattern speed,* Ω_ps, of the satellite's perturbing potential is the angular frequency of a reference frame in which this potential is stationary. This depends on the exact combination of frequencies under consideration and may be written as

$$m\Omega_\mathrm{ps} = mn' + k\kappa' + p\nu' = (m+k+p)n' - k\dot{\varpi}' - p\dot{\Omega}' \tag{18}$$

where m, k and p are integers and m is non-negative. Resonances occur when an integer multiple of the difference between n and Ω_ps is equal to zero (for *corotation resonances*), or the natural frequency of the radial or vertical oscillations of the ring particle (for *eccentric* or *Lindblad resonances*, and *vertical resonances* respectively). Note that there is no fundamental difference between these resonances and those described from a disturbing function approach.

3.3. COROTATION RESONANCES

A corotation resonance occurs where the pattern speed of the perturbing potential matches the orbital frequency of the particle. In this case

$$m(n - \Omega_\mathrm{ps}) = 0 \tag{19}$$

and the resonance condition becomes

$$(m+k+p)n' - mn - k\dot{\varpi}' - p\dot{\Omega}' = 0. \tag{20}$$

If we ignore the variation of the mean longitude of epoch, this in turn can be considered as setting the condition $\dot{\varphi}_\mathrm{cr} = 0$ where φ_cr is the resonant angle given by

$$\varphi_\mathrm{cr} = (m+k+p)\lambda' - m\lambda - k\varpi' - p\Omega'. \tag{21}$$

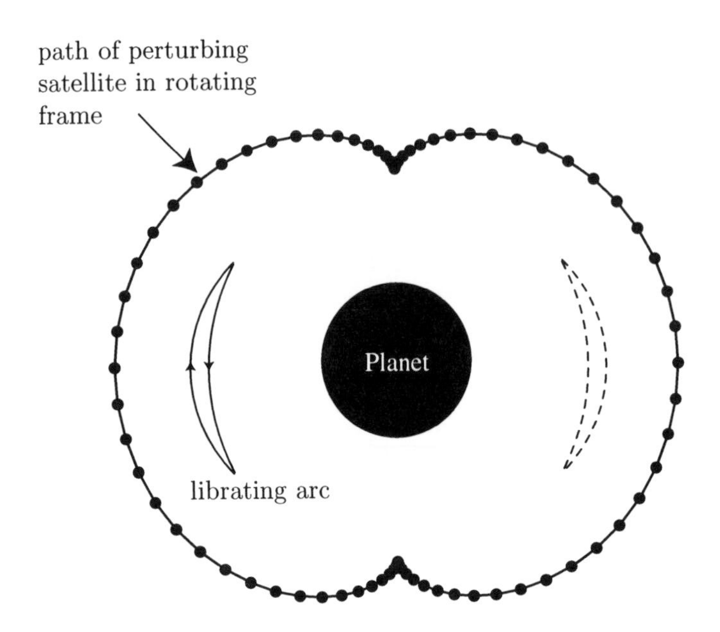

Figure 2. The geometry of the 3:2 corotation resonance ($j = 3$, $k = 1$, $p = 0$) for the case where the eccentricity of the perturbing satellite is 0.19. Points are plotted on the path of the satellite at equal time intervals. Note that there are two possible equilibrium positions, each of which could contain librating arcs of material.

If we adopt the convention that the resonant angle is always written such that the coefficient of λ' in φ is always j then

$$\varphi_{\rm cr} = j\lambda' + (k + p - j)\lambda - k\varpi' - p\Omega' \qquad (22)$$

where $j = m+k+p$. We can see that this argument satisfies the d'Alembert relation and we already know that for this to be a valid argument $|p|$ must be even; this is because the inclinations always occur in even powers and hence the sum of the coefficients of the nodes must always be even. Furthermore this is the argument of a resonance of order $|k + p|$. It is clear that the 1:1 (or coorbital) resonance, where $p = k = 0$, is a special case of a corotation resonance.

The geometry of a simple corotation resonance is shown in Fig. 2 for the case of a 3:2 resonance with $j = 3$, $k = 1$, $p = 0$ and $e' = 0.19$. In this case there are two possible libration points in the frame rotating with the pattern speed of the perturbing satellite. In this frame the path of the satellite is closed. Because the particle is close to, but not at the exact resonance (i.e. it is not located at one of the two equilibrium points) it will librate about the equilibrium point.

The maximum width (in semimajor axis) of a corotation resonance can be calculated by deriving an expression for $\ddot{\varphi}_{\rm cr}$ and making use of a pendulum model (Dermott 1984). The relevant part of the disturbing function (cf. Eq. (5)) is

$$\mathcal{R} = \frac{\mathcal{G}m'}{a'} f_{\rm d}(\alpha) e'^{|k|} s'^{|p|} \cos \varphi_{\rm cr}. \tag{23}$$

The exact form of $f_{\rm d}(\alpha)$ depends on the resonance in question. For example, in the case of the 3:2 resonance discussed above we have $k = 1$ and $p = 0$ which corresponds to the second argument Table 1 with $j = 3$. Using the pendulum approach the width, $W_{\rm cr}$, of a general corotation resonance can be written as a function of the magnitude of $\mathcal{R}$ as (Dermott 1984)

$$W_{\rm cr} = 8 \left(\frac{a\,|\mathcal{R}|}{3\mathcal{G}m_{\rm p}} \right)^{1/2} a. \tag{24}$$

3.4. LINDBLAD RESONANCES

A Lindblad resonance occurs where the pattern speed of the perturbing potential matches the epicyclic or radial frequency of the particle. In this case

$$m(n - \Omega_{\rm ps}) = \pm\kappa \tag{25}$$

where the upper and lower signs correspond to the inner (ILR) and outer (OLR) Lindblad resonance respectively. The use of $\pm$ permits us to consider a ring particle that is orbiting inside or outside the orbit of the perturbing satellite. The resonance condition can also be written as

$$(m \mp 1)n \pm \dot{\varpi} - m\Omega_{\rm ps} = 0 \tag{26}$$

or

$$(m + k + p)n' - (m \mp 1)n - k\dot{\varpi}' \mp \dot{\varpi} - p\dot{\Omega}' = 0. \tag{27}$$

In terms of the resonant angle, $\varphi_{\rm lr}$, we have

$$\varphi_{\rm lr} = (m + k + p)\lambda' - (m \mp 1)\lambda - k\varpi' \mp \varpi - p\Omega'. \tag{28}$$

Using our standard notation we can write this as

$$\varphi_{\rm lr} = j\lambda' + (k + p \pm 1 - j)\lambda - k\varpi' \mp \varpi - p\Omega' \tag{29}$$

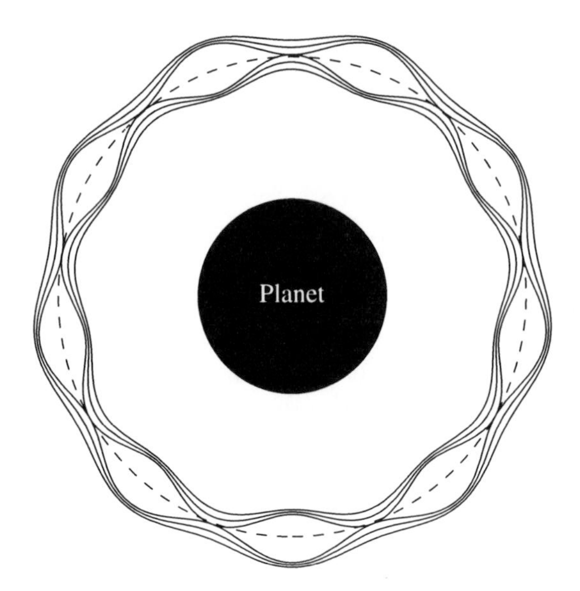

Figure 3. Schematic representation of the streamlines arising from the 9:8 inner Lindblad resonance. The dashed curve denotes exact resonance. The forced eccentricity, and hence the amplitude of the streamline wave, decreases away from the resonance.

where, as before, $j = m + k + p$. Note that this is a resonant argument of order $|k + p \pm 1|$. Also, since the sum of the coefficients of Ω and Ω' must be even, p has to be even for all vertical resonances.

In order to calculate the maximum width of a Lindblad resonance it is important to remember that we are dealing with a collection of ring particles, each of which is responding to the perturbing potential. The Lindblad resonance induces a forced eccentricity on the ring particles such that, at a given semimajor axis, the particles move in streamline motion. The resulting pattern in the rotating frame is a "wavy" ring. The size of the forced eccentricity will decrease as the distance from the exact resonance increases, with a phase change of 180° on opposite sides of the resonance. The width of the resonance is determined by the separation from the exact resonance such that the value of the forced eccentricity is just sufficient for the outer streamline to intersect the inner one. The mechanism is discussed in Porco & Nicholson (1987) and we use their notation in what follows. The resulting paths are illustrated in Fig. 3 for the case of the 9:8 Lindblad resonance with $k = p = 0$.

To lowest order Lagrange's planetary equations give the time variation of the particle's eccentricity, e as

$$\frac{\mathrm{d}e}{\mathrm{d}t} = -\frac{1}{na^2 e}\frac{\partial \mathcal{R}}{\partial \varpi}. \tag{30}$$

We can use this equation to calculate a general expression for the forced eccentricity, e_{f}, at the strong, first-order Lindblad resonances given by $k = p = 0$.

For $k = p = 0$ we have

$$\varphi_{\mathrm{lr}} = j\lambda' + (1-j)\lambda - \varpi \tag{31}$$

corresponding to a first-order, e-type, ILR. In this case the associated part of the disturbing function is

$$\mathcal{R} = \frac{\mathcal{G}m'}{a'}\, e\, \frac{1}{2}\mathcal{A}_0 \cos\varphi_{\mathrm{lr}} \tag{32}$$

where, from the first term in Table 1,

$$\mathcal{A}_0 = [-2j - \alpha D]\, b_{1/2}^{(j)}. \tag{33}$$

Hence

$$\frac{\mathrm{d}e}{\mathrm{d}t} = -\frac{1}{2} n\alpha(m'/m_{\mathrm{p}})\mathcal{A}_0 \sin\varphi_{\mathrm{lr}} \tag{34}$$

and if we write the approximate solution as $e \approx e_0 + e_{\mathrm{f}} \cos\varphi_{\mathrm{lr}}$ where $e_0 \approx 0$ and e_{f} is the forced eccentricity due to the resonance, we have

$$e_{\mathrm{f}} = \frac{n\alpha(m'/m_{\mathrm{p}})\mathcal{A}_0}{2[jn' - (j-1)n]} \tag{35}$$

If we write $a = a_{\mathrm{res}} + \Delta a$ (where $\Delta a \ll a_{\mathrm{res}}$) for the semimajor axis of the particle we can make use of Kepler's third law and a series expansion in $\Delta a/a_{\mathrm{res}}$ to write the amplitude of the resulting forced wave as

$$ae_{\mathrm{f}} = \frac{\alpha a^2 (m'/m_{\mathrm{p}})|\mathcal{A}_0|}{3(j-1)\,|a - a_{\mathrm{res}}|} = \frac{A_{\mathrm{lr},0}}{|a - a_{\mathrm{res}}|} \tag{36}$$

for the case $k = p = 0$ where

$$A_{\mathrm{lr},0} = \frac{\alpha a^2 (m'/m_{\mathrm{p}})|\mathcal{A}_0|}{3(j-1)}. \tag{37}$$

For a critical value of the semimajor axis, a_{crit}, the amplitude of the ring wave equals the separation in semimajor axis from the exact resonance. For this value we have

$$|a_{\mathrm{crit}} - a_{\mathrm{res}}|^2 = A_{\mathrm{lr},0} = (\Delta a_{\max})^2 = (W_{\mathrm{lr},0}/4)^2 \tag{38}$$

where $W_{\mathrm{lr},0}$ is the full width of the $k = 0$ Lindblad resonance. Hence

$$W_{\mathrm{lr},0} = 4a \left[\frac{\alpha (m'/m_{\mathrm{p}}) |\mathcal{A}_0|}{3(j-1)} \right]^{1/2} . \tag{39}$$

These expressions can be obtained from those given by Porco & Nicholson (1987) using the substitution $j = m$.

The values of $\alpha = a/a'$ and the corresponding Laplace coefficients in the definition of $\mathcal{A}_0$ can be calculated for any particular value of j. Provided j is sufficiently large, it can be shown that $\alpha \mathcal{A}_0/(j-1) \approx 1.6$ and hence, from our expression for $W_{\mathrm{lr},0}$ we can write

$$W_{\mathrm{lr},0} \approx 4a \left[\frac{1.6}{3} (m'/m_{\mathrm{p}}) \right]^{1/2} \approx 2.9 (m'/m_{\mathrm{p}})^{1/2} a. \tag{40}$$

This gives us a useful expression for the width of the general $k = p = 0$ ILR. Note that this width is approximately the same for all first-order resonances.

3.5. VERTICAL RESONANCES

A vertical resonance occurs where the pattern speed of the perturbing potential matches the vertical frequency of the particle. In this case

$$m(n - \Omega_{\mathrm{ps}}) = \pm \nu \tag{41}$$

where the upper and lower signs correspond to the inner (IVR) and outer (OVR) vertical resonances respectively. The resonance condition can also be written as

$$(m \mp 1)n \pm \dot{\Omega} - m\Omega_{\mathrm{p}} = 0 \tag{42}$$

or

$$(m + k + p)n' - (m \mp 1)n - k\dot{\varpi}' \mp \dot{\Omega} - p\dot{\Omega}' = 0. \tag{43}$$

In terms of the resonant angle, φ_{vr}, we have

$$\varphi_{\mathrm{vr}} = (m + k + p)\lambda' - (m \mp 1)\lambda - k\varpi' \mp \Omega - p\Omega'. \tag{44}$$

Using our standard notation we can write this as

$$\varphi_{\mathrm{vr}} = j\lambda' + (k + p \pm 1 - j)\lambda - k\varpi' - p\Omega' \mp \Omega \tag{45}$$

where, as before, $j = m + k + p$. Note that this is a resonant argument of order $|k + p \pm 1|$. Also, since the sum of the coefficients of Ω and Ω' must be even, p has to be odd for all vertical resonances.

We can develop a theory for the variation of the forced inclination induced by the vertical resonance as a function of the distance from exact resonance using a theory similar to the one already used above for the Lindblad resonances. In this case we can write

$$\frac{\mathrm{d}I}{\mathrm{d}t} = -\frac{1}{na^2 \sin I}\frac{\partial \mathcal{R}}{\partial \Omega}. \tag{46}$$

Consider the case of an IVR with $k = 0$, $p = 1$ and a resulting resonant argument given by

$$\varphi_{\mathrm{vr}} = j\lambda' + (2 - j)\lambda - \Omega' - \Omega. \tag{47}$$

This is the lowest order resonant argument which satisfies the condition that p has to be odd. This corresponds to a second order, II' resonance. The associated part of the disturbing function is

$$\mathcal{R} = -\frac{\mathcal{G}m'}{a'} ss' \mathcal{A} \cos \varphi_{\mathrm{vr}} \tag{48}$$

where, from Table 1, $\mathcal{A} = \alpha b_{3/2}^{(j-1)}$. Hence

$$\frac{\mathrm{d}I}{\mathrm{d}t} = -\frac{1}{4} n\alpha (m'/m_{\mathrm{p}}) \sin I' \mathcal{A} \sin \varphi_{\mathrm{vr}}. \tag{49}$$

The resulting value of the forced inclination, I_{f} can be written as

$$aI_{\mathrm{f}} = \frac{A_{\mathrm{vr}}}{|a - a_{\mathrm{res}}|} \tag{50}$$

where

$$A_{\mathrm{vr}} = \frac{\alpha a^2 (m'/m_{\mathrm{p}}) \sin I' \, |\mathcal{A}|}{6(j-2)}. \tag{51}$$

In the case of Lindblad resonances the width of the resonance was determined by calculating the minimum separation from exact resonance where the radial variations just managed to produce intersection of streamlines. There is no analagous mechanism in the case of vertical resonances in a coplanar ring system. However, although there is still a forced inclination given by Eq. (50), it is important to remember that in the real situation this does not rise to infinity as exact resonance is approached. A more careful analysis shows that the small divisor approach breaks down close to resonance and we must resort to the pendulum-like equation of motion to understand the resonance mechanism.

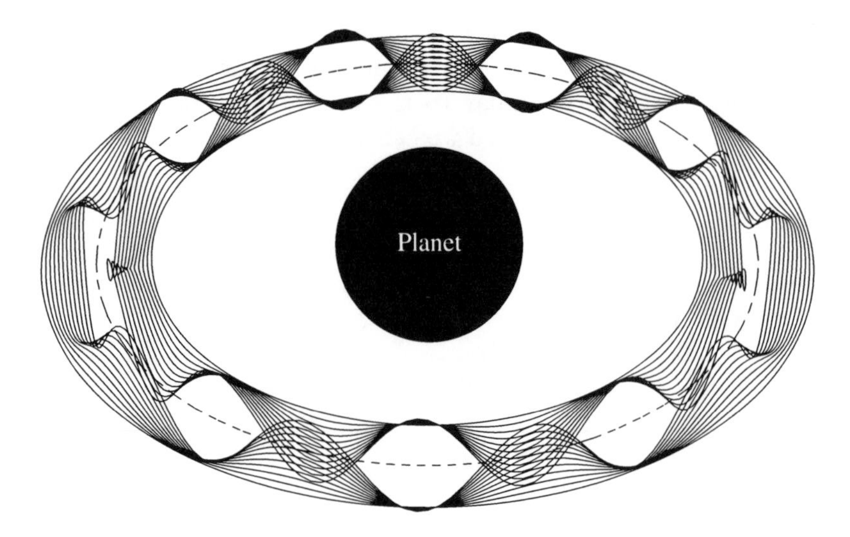

Figure 4. Schematic representation of the streamlines arising from a vertical resonance. The dashed curve denotes exact resonance. The forced inclination, and hence the amplitude of the streamline wave, decreases away from the resonance.

Figure 4 shows an illustration of the effect of the forced inclination on either side of an exact vertical resonance (denoted by the dashed line). Note that according to Eq. (50) the magnitude of the forced inclination decreases as the distance from the exact resonance increases. At sufficiently large distances there is effectively no forced inclination and the ring particles remain in the equatorial plane. Note also that there is a phase shift of the longitude of ascending node of 180° on either side of the exact resonance. This is similar to the 180° shift in the longitude of pericentre on either side of the Lindblad resonance (see Fig. 3).

4. Resonant Locations

Using Kepler's third law, the approximate semimajor axis of an internal $p+q:p$ resonance is given by $a = [p/(p+q)]^{2/3}a'$, where a' is the semimajor axis of the perturbing object. However, the semimajor axis of the *exact* resonance depends on the particular resonant argument under consideration. Consider a general resonant argument of the form given in Eq. (4). This argument is associated with the general $|j_1| : |j_2|$ resonance. As stated above, if we ignore the variation of the mean longitude at epoch, the location of the exact resonance is that given by the semimajor axis which satisfies the equation $\dot{\varphi} = 0$.

In the case of resonances between satellites and ring particles the effect of the planet's oblateness usually dominates the motion of the particle's pericentre and node, especially in regions close to the planet. The contribution of the planet's oblateness to n, $\dot{\varpi}$ and $\dot{\Omega}$ is a function of the orbital radius (see Eqs. (13) and (16–17)). Unfortunately n, $\dot{\varpi}$ and $\dot{\Omega}$ depend on r (or a) in a non-linear manner so that a numerical method is required in order to find the location of the exact resonance for a given argument.

Table 4 shows the results of such calculations for the second-order resonances associated with the 5:3 Mimas commensurability. Here there are six possible resonant arguments if we include terms up to order 2 in the eccentricities and inclinations. In order to derive the locations of the resonances we have taken the mean motion, perichrone rate and node rate of Mimas to be $381.9945°\mathrm{d}^{-1}$, $1.0008°\mathrm{d}^{-1}$ and $-0.9995°\mathrm{d}^{-1}$ respectively (Harper & Taylor, 1993).

TABLE 4. The resonances associated with the 5:3 commensurability of the saturnian satellite, Mimas. In the classification column ILR, IVR, CER, and CIR denote inner Lindblad resonance, inner vertical resonance, corotation eccentricity resonance and corotation inclination resonance respectively. All the resonances lie in Saturn's A ring.

i	φ_i	Type	Class.	j	m	k	p	n ($°\mathrm{d}^{-1}$)	a (km)
1	$5\lambda' - 3\lambda - 2\Omega$	I^2	—	5	–	–	–	638.886	131793
2	$5\lambda' - 3\lambda - \Omega - \Omega'$	II'	IVR	5	4	0	1	638.102	131900
3	$5\lambda' - 3\lambda - 2\Omega'$	I'^2	CIR	5	3	0	2	637.324	132007
4	$5\lambda' - 3\lambda - 2\varpi'$	e'^2	CER	5	3	2	0	635.990	132191
5	$5\lambda' - 3\lambda - \varpi - \varpi'$	ee'	ILR	5	4	1	0	635.219	132298
6	$5\lambda' - 3\lambda - 2\varpi$	e^2	—	5	–	–	–	634.454	132404

In Table 4 we indicate the "type" of the resonance, using the terminology of the disturbing function approach, as well as the "classification" of the resonance (denoted by "Class." in the table) using the terminology of ring dynamics. The effect of the oblateness has caused the resonances to be spread out over more than 600 km in semimajor axis. Note that the e^2 and I^2 resonances are not classified under the standard terminology of ring dynamics.

5. Spiral Waves and Bending Waves

A satellite resonance in a ring of sufficiently high surface density introduces an azimuthal variation in the gravitational potential. We have already seen that a $p+1:p$ inner Lindblad resonance gives rise to a $p+1$-lobed pattern in the motion of the particle in the frame rotating with the mean motion of

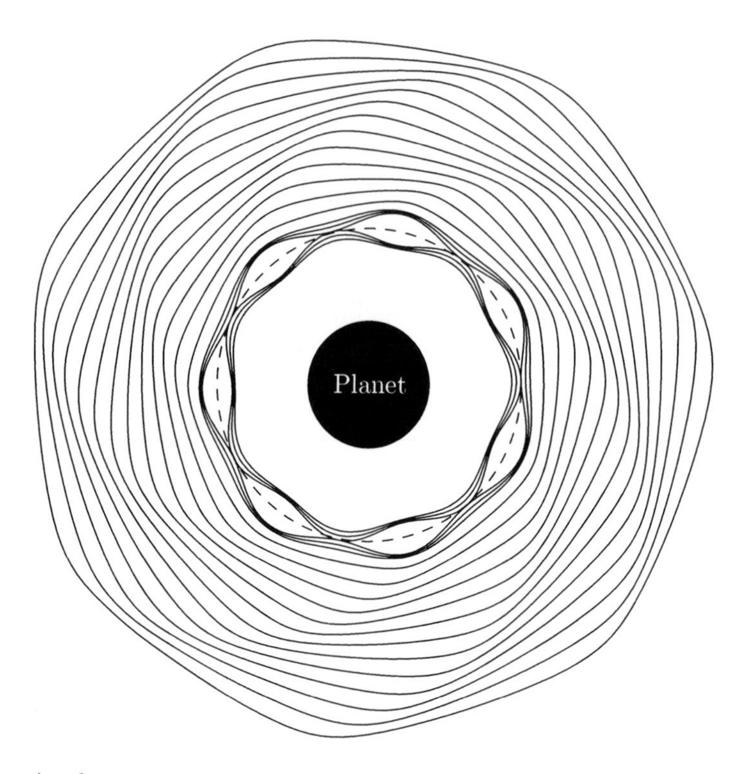

Figure 5. A schematic representation of the seven-armed spiral structure associated with the 7:6 inner Lindblad resonance (ILR). The dashed line denotes the location of the exact resonance.

the perturber (see Fig. 3). A study of the effect of this modified potential on surrounding material shows that it results a trailing pattern with $p+1$ spiral arms (see, for example, Fig. 5 for the case of the 7:6 ILR). This is an example of a *spiral density wave.* In practice the spiral is very tightly wound but a radial profile would always show the characteristic decrease in wavelength with increasing distance from exact resonance. There is an equivalent phenomenon due to vertical resonances leading to the formation of *spiral bending waves.* For an inner vertical resonance the result is trailing bending waves that propagate inwards from the exact resonance.

Goldreich & Tremaine (1978) adapted the standard galactic spiral density wave theory and used it to explain the formation of the Cassini division in Saturn's rings. Shu (1984) provides a review of the basic theory. In Table 4 we used the example of the Mimas 5:3 commensurability to calculate the exact location of the resulting resonances. Mimas has an inclination of 1.5° and its vertical resonances are comparable in strength to its Lindblad resonances. However, because ring eccentricities are small the corotation resonances are much weaker. Table 4 shows that Mimas's 5:3 ILR lies 398 km beyond its 5:3 IVR. The resulting effect on the rings is shown in Fig. 6,

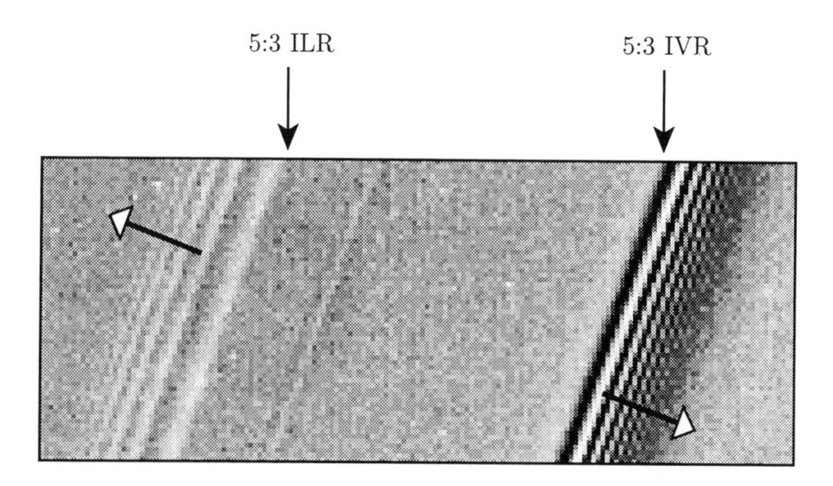

Figure 6. A Voyager image showing the outward-propagating, trailing spiral density wave resulting from the 5:3 Mimas inner Lindblad resonance and the inward propagating, trailing bending wave resulting from the 5:3 Mimas inner vertical resonance. The superimposed arrows denote the direction of propagation of each wave. The resonances are separated by approximately 400 km

part of a Voyager 2 image of Saturn's A ring. The spiral density wave due to the 5:3 ILR at the left is propagating outwards while at the right of the image is the spiral bending wave due to the 5:3 IVR propagating inwards. The additional radial feature just interior to the 5:3 density wave is due to an ILR with Prometheus. Note that the bending wave produces vertical motion of the particles that produces localised warping of the ring plane. This explains the different contrast between the two wave features.

6. Narrow Rings and Shepherding

Narrow planetary rings should spread due to the combined effects of two major processes. For a ring composed of particles of size d, collisions between ring particles should cause spreading of a time scale $t_{\rm coll} \propto d^{-2}$ while the time scale for spreading due to the effects of Poynting-Roberston light drag is $t_{\rm pr} \propto d$ (Dermott 1984). For example, for the parameters appropriate to the rings of Uranus, the maximum lifetime is $\sim 10^7$ y, or only 0.2% of the age of the solar system. This suggests that either such rings are relatively young or that there is some confining mechanism. Goldreich & Tremaine (1979) proposed that each narrow ring is prevented from spreading due to the gravitational influence of a small satellite orbiting on either side of the ring. Each satellite acts to "shepherd" ring particles and the net result is a torque from each satellite leading to ring confinement.

The basics of the shepherding mechanism can be explained by consider-

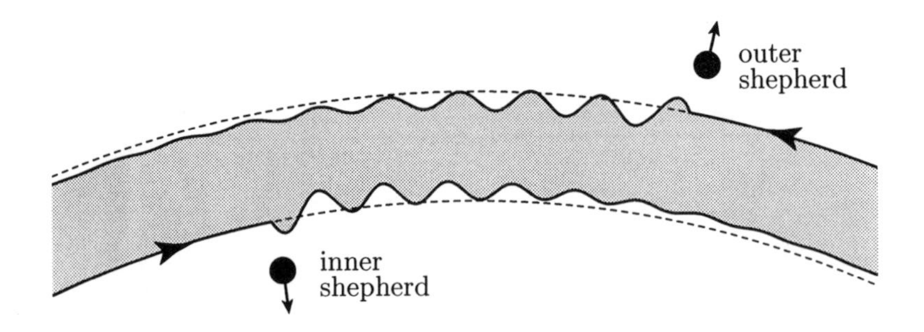

Figure 7. Schematic representation of the shepherding mechanism. In a frame moving with the outer shepherd the ring material (on initially circular orbits) interior to it passes by and a wave is created due to the eccentricity the ring particles receive from the satellite. The wave propagates downstream from the satellite. After the wave has damped due to collisions the common semimajor axes of the ring particles has been reduced because of an exchange of angular momentum with the satellite. The net result is an apparent "repulsion" between the ring and the satellite with the satellite moving outwards. For the inner shepherd the wave propagates upstream and the satellite's orbit decreases. The actions of both satellites provide a confining mechanism for the ring but also cause secular changes in the semimajor axes of both satellites denoted by the small arrows. (Adapted from Dermott 1984.)

ing an encounter between a satellite and ring particle initially on a circular orbit. A simple understanding of such an encounter shows how a satellite can produce a wave on a ring edge and as well as an apparent "repulsion" of the two orbits. Let a and a' denote the semimajor axes of the particle and satellite respectively, where we take $a' > a$ and a' is assumed to be fixed. Let m' and $m_{\rm p}$ denote the masses of the satellite and planet respectively, with $m' \ll m_{\rm p}$. We assume that the satellite is moving on a circular orbit exterior to the particle (upper part of Fig. 7). At encounter the particle receives an impulse from the satellite which leads to changes in the orbital elements of the particle. At this stage we can ignore the effect of the particle on the satellite.

The angular momentum per unit mass of the particle is

$$h = \sqrt{\mathcal{G} m_{\rm p} a(1 - e^2)} \tag{52}$$

and hence the changes in the orbital elements are related to the change in angular momentum by

$$2\frac{\delta h}{h} = \frac{\delta a}{a} - 2e^2 \tag{53}$$

where, since the orbit is initially circular, we have taken $\delta e = e$.

It can be shown (see Dermott 1984) that $\delta a/a \gg e^2$ and so δh is determined by δa. Curiously, $\delta a/a$ is $\mathcal{O}(m'/m_{\rm p})^2$ and can be calculated by deriving δe, a quantity of $\mathcal{O}(m'/m_{\rm p})$, from the Jacobi constant or the Tisserand relation for the planet-satellite-particle three-body system. In the case of zero inclination where a' is not taken to be unity, the Tisserand relation (see Roy 1988) can be written as

$$\frac{a'}{a} + 2\left(\frac{a}{a'}\right)^{1/2}\left(1 - e^2\right)^{1/2} = C \tag{54}$$

where C is a constant up to $\mathcal{O}(m'/m_{\rm p})$. If $\Delta a = a - a'$ denotes the small separation of the semimajor axes in our model, Eq. (54) can be expanded to give

$$\frac{3}{4}\left(\frac{\Delta a_n}{a'}\right)^2 - e_n^2 \approx C - 3 \tag{55}$$

where we have now generalised the equation such that Δa_n and e_n are the values of Δa and e after the n-th encounter with the satellite (see Dermott & Murray 1981a). Note that there are no linear terms in Δa_n in Eq. (55).

The eccentricity that the particle receives can be calculated using Gauss's form of the planetary perturbation equations (see Roy 1988). For small e the variation is given by

$$\frac{\mathrm{d}e}{\mathrm{d}t} \approx \frac{1}{na}\left(\bar{R}\sin f + 2\bar{T}\cos f\right) \tag{56}$$

where $\bar{R}$ and $\bar{T}$ denote the radial and tangential components of the perturbing force (per unit mass) and f is the true anomaly of the particle. Treating the particle's interaction with the satellite as a radial impulse lasting a time Δt, from Eq. (56), the resulting eccentricity is

$$e \approx \frac{1}{na}\bar{R}\,\Delta t\,\sin f. \tag{57}$$

If we consider an impulse that lasts for a time interval $\Delta t \approx 0.2P$, where $P = 2\pi/n$ is the orbital period of the particle then the relative angular velocity of the particle with respect to the satellite is

$$U = \frac{3}{2}\frac{n}{a}\Delta a_0 \tag{58}$$

and hence

$$\Delta t = \frac{2\Delta a_0}{Ua}. \tag{59}$$

If we assume that the particle has a true anomaly $f \approx \pi/2$ (i.e. the particle is close to quadrature) then $\sin f \approx 1$ and the resulting eccentricity is

$$e \approx \frac{4}{3}\frac{m'}{m_{\rm p}}\left(\frac{a}{\Delta a_0}\right)^2 \tag{60}$$

(cf. Lin & Papaloizou 1979). Julian & Toomre (1966) showed that the coefficient in this equation is 2.24 rather than 4/3 once the tangential force on the particle is included.

Because we assume that all particles encountering the satellite move on initially circular orbits, every particle at the edge of the ring will receive exactly the same eccentricity from the satellite. Furthermore, since the perturbation is applied over such a short time interval, the orbits of the post-encounter particles will be unperturbed keplerian ellipses of identical a and ϵ but with systematic differences in phase. In a rotating reference frame the path of an individual particle moving on an elliptical orbit resembles a sinusoidal curve, as the particle moves from its pericentre to apocentre over the course of an orbital period. The following particle will follow the same path and so on for all subsequent particles. The resulting effect is that in the rotating frame the satellite appears to produce a standing wave on the ring's edge. This wave propagates upstream of an outer satellite and downstream of an inner satellite (see Fig. 7). In the non-rotating frame the wave will be seen to move at the same rate as the mean motion of the satellite. Because the wave is a consequence of the particles' new-found eccentricity, the amplitude, A, of the wave is simply half the difference between the apocentric and pericentric distances. Hence $2A = a(1+e) - a(1-e) = ae$ and

$$A = 2.24 \frac{m'}{m_{\mathrm{p}}} \left(\frac{a}{\Delta a_0} \right)^2 a. \tag{61}$$

The particle's mean motion is n and that of the satellite is $n+U$. Therefore the number of orbits between encounters is n/U. In that time the particle will have covered a distance of $2\pi a$ with respect to the satellite. Hence the wavelength is

$$\ell = \frac{2\pi a}{n/U} = 3\pi\, \Delta a_0. \tag{62}$$

Note that the amplitude of the wave is a function of Δa_0 and m', while the wavelength is a function of Δa_0 alone. Therefore observations of an edge wave allow determination of the mass and radial distance of the satellite that produced it, even though the satellite may not have been detected directly.

It is the change in semimajor axis that is at the heart of the shepherding mechanism for narrow rings. Figure 7 shows that it is the combined effect of a satellite on either side of the ring that leads to its confinement. When the Voyager 2 spacecraft reached Uranus in January 1986 two small satellites, Cordelia and Ophelia, were detected on either side of the outermost ϵ ring of the planet. The outer edge of the ϵ ring is located at the 14:13 ILR with Ophelia (the exterior satellite) while the inner edge is located at the 24:25 OLR with Cordelia (the inner satellite) (Porco & Goldreich 1987). Additional shepherding moons between the narrow rings of Uranus have not been seen, although Cordelia could be involved in the confinement (Murray & Thompson 1990). The F ring of Saturn has two satellites, Prometheus and Pandora orbiting on either side of it and this is thought to be another example of the shepherding mechanism.

7. The A ring of Saturn

The proximity of satellites such as the F ring shepherds Prometheus and Pandora, and the coorbital pair Janus and Epimetheus to the main rings of Saturn ensures that their gravitational influence is felt by the rings even though their masses are small compared with the larger but more distant moons of Saturn. In fact, apart from a few distinct features due to density and bending waves at Mimas resonances, virtually all of the structure in Saturn's A ring can be understood as arising from the effects of Prometheus, Pandora and an embedded satellite.

Figure 8 shows part of a Voyager images of the A ring with the location of Prometheus and Pandora ILRs indicated. The correspondence is excellent. Although density waves are created at each resonance the resulting wavelength is small, leading to just a single feature at each ILR. Two additional features in the image are due to a density wave and a bending wave caused by the Mimas 8:5 commensurability. The only other feature is the 30 km Keeler Gap near the outer part of the ring; there is some evidence for the presence of a perturbing satellite in the gap (Cooke 1991). The A ring edge is located close to the 7:6 ILR with Janus, one of the coorbital satellites, and so a 7-lobed pattern would be expected as in Fig. 5. Indeed, the pattern observed by Porco *et al.* (1984) is consistent with that of a 7-lobed figure. It is interesting to note that the ± 10 km shift of Janus's semimajor axis every 4 y due to its perturbations from Epimetheus (Dermott & Murray 1981b, Nicholson *et al.* 1992) means that the edge of the A ring should also change location by a similar amount with the same period.

Wavy edges were detected in Voyager images of the 325 km-wide Encke gap in Saturn's A ring by Cuzzi & Scargle (1985). Measurements of the wavelength and amplitude of the features allowed estimates of the mass and distance of the satellite assumed to be causing the waves (see Sect. 6 above). The satellite, now named Pan, was finally discovered by Showalter (1991) using low resolution images from the Voyager archive (see Fig. 9). Although Pan was unresolved in the images, its motion was entirely consistent with that of an object with a semimajor axis of 133582.8 ± 0.8 km, placing it in the centre of the Encke gap (Showalter 1991).

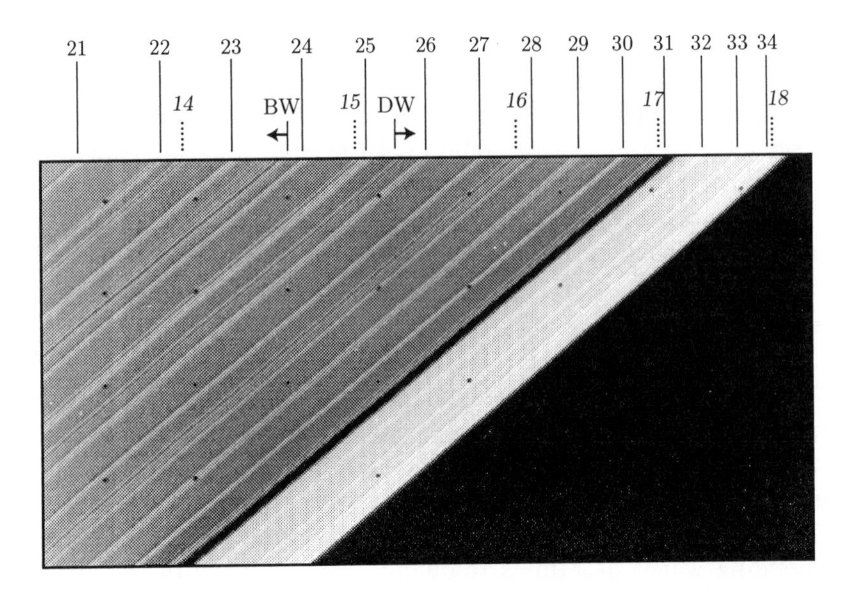

Figure 8. A Voyager image of the outer part of Saturn's A ring showing the structure created by the $p+1:p$ resonances with Prometheus (solid lines and number) and Pandora (dotted lines and italicised numbers). BW and DW denote the locations of the bending wave and density wave due to the Mimas 8:5 resonance, with the arrows denoting their direction of propagation. The Keeler gap near the outer edge is 30 km wide.

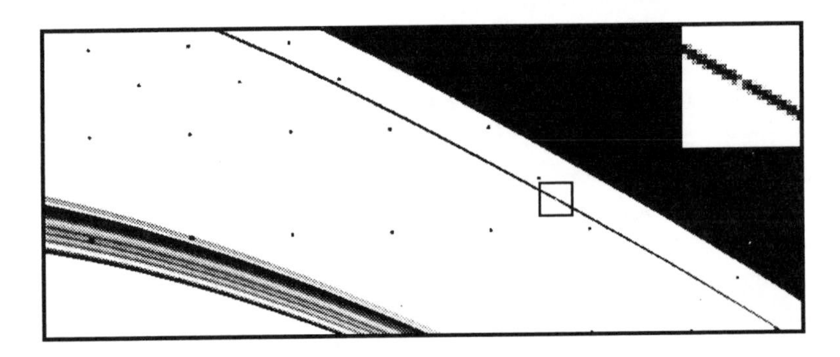

Figure 9. An enhanced Voyager 2 image of Saturn's A ring showing an unresolved detection of the satellite Pan (highlighted by square and enlarged on inset at upper right) in the Encke gap in Saturn's A ring. A sequence of such images taken approximately five minutes apart demonstrate movement of Pan consistent with keplerian motion at this location.

8. The Adams Ring of Neptune

The accidental discovery of the rings of Uranus by occultation in 1977 led to a number of attempts to detect a ring system around Neptune. However, only about 10% of all observations showed an occultation event and

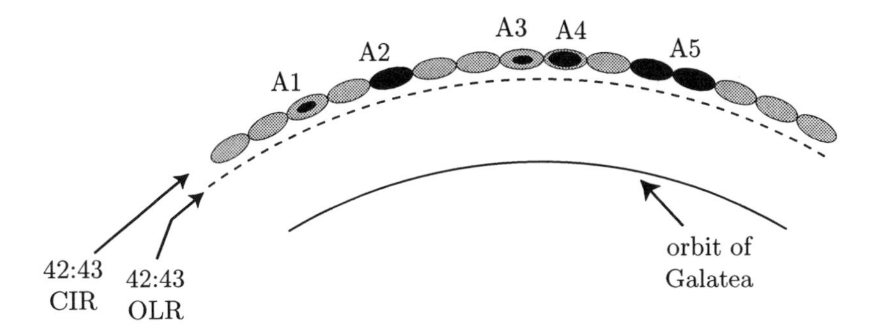

Figure 10. Schematic illustration of the resonant mechanisms thought to be responsible for the azimuthal structure of the Adams ring of Neptune. The small ovals denote some of the 84 equilibrium points associated with the 42:43 corotation inclination resonance with the satellite Galatea. Some of these contain optically thicker material (dark ovals). The dashed line denotes the location of Galatea's 42:43 outer Lindblad resonance.

those only on one side of the planet suggesting incomplete rings or arcs of material. The observations were finally understood when Voyager 2 images during the 1989 flyby showed that Neptune had an optically thin ring system but that one of its rings, the outermost Adams ring, contained several arcs of optically thicker ($\tau \sim 0.04$) material; these were responsible for producing the occultation events. The narrow arcs had to be confined in radius and longitude if they were to have lifetimes of more than a few decades.

Porco (1991) suggested that the arcs were maintained by the gravitational effects of a small satellite, Galatea, orbiting $\sim$ 900 km inside the Adams ring. Galatea's 42:43 outer corotation inclination resonance provides 86 equilibrium sites (the small ovals in Fig. 10), only some of which are filled, or partially filled by the optically thicker material. Each equilibrium site would have an angular extent of 4° and have a radial width of 600 m. While the 42:43 outer CIR provided azimuthal confinement, the 42:43 OLR, located $\sim$ 1.5 km interior to the ring, provided the radial confinement.

9. Summary

The ring systems of the outer planets provide many examples of resonant phenomena. Although features such as density waves require an understanding of group effects, the locations of satellite resonances and the basic mechanisms of satellite perturbations can be calculated for a wide variety of systems. In the case of the Encke gap this has led to the successful prediction of a new satellite, Pan, while most of the radial structure of Saturn's A ring and the azimuthal structure of Neptune's ring arc can also be explained using resonances.

References

Cooke, M.L. (1991) Saturn's rings: radial variation in the Keeler gap and C ring photometry *Ph.D. dissertation* Cornell University, Ithaca.
Cuzzi, J.N. and Scargle, J.D. (1985) Wavy edges suggest moonlet in Saturn's Encke gap *Astrophys. J.* **292**, 276–290.
Dermott, S.F. (1984) Dynamics of narrow rings. In *Planetary Rings* (Eds. Greenberg, R. and Brahic, A.) pp. 589–637, University of Arizona Press, Tucson.
Dermott, S.F., Gold, T. and Sinclair, A.T. (1979) The rings of Uranus: nature and origin. *Astron. J.* **84**, 1225–1234.
Dermott, S.F. and Murray, C.D. (1981a) The dynamics of tadpole and horseshoe orbits. I. Theory *Icarus* **48**, 1–11.
Dermott, S.F. and Murray, C.D. (1981b) The dynamics of tadpole and horseshoe orbits. II. The coorbital satellites of Saturn *Icarus* **48**, 12–22.
Elliot, J.L., Dunham, E.W. and Mink, D.J. (1977) The rings of Uranus. *Nature* **267**, 328–330.
Goldreich, P. and Tremaine, S. (1978) The formation of the Cassini Division in Saturn's rings. *Icarus* **34**, 240–253.
Goldreich, P. and Tremaine, S. (1979) Towards a theory for the Uranian rings. *Nature* **277**, 97–99.
Greenberg, R. and Brahic, A. (1984) *Planetary Rings.* University of Arizona Press, Tucson.
Hamilton, D.P. (1994) A comparison of Lorentz, planetary gravitational, and satellite gravitational resonances. *Icarus* **109**, 221–240.
Harper, D. and Taylor, D.B. (1993) The orbits of the major satellites of Saturn. *Astron. Astrophys.* **268**, 326–349.
Julian, W.H. and Toomre, A. (1966) Non-axisymmetric responses of differentially rotating disks of stars. *Astrophys. J.* **146**, 810–830.
Lin, D.N.C. and Papaloizou, J. (1979) Tidal torques on accretion discs in binary systems with extreme mass ratios. *Mon. Not. R. astr. Soc.* **186**, 799–812.
Murray, C.D. (1994) Planetary ring dynamics. *Phil. Trans. R. Soc. Lond.* A **349**, 335–344.
Murray, C.D. and Dermott, S.F. (1998) *Solar System Dynamics.* Cambridge University Press, Cambridge (In press).
Murray, C.D. and Harper, D. (1993) Expansion of the planetary disturbing function to eighth order in the individual orbital elements. *QMW Maths Notes* No. 15, Queen Mary and Westfield College, London.
Murray, C.D. and Thompson, R.P. (1990) Orbits of shepherd satellites deduced from the structure of the rings of Uranus. *Nature* **348**, 499–502.
Nicholson, P.D. and Dones, L. (1991) Planetary rings. *Rev. Geophys. Supp.* **29**, 313–327.
Nicholson, P.D., Hamilton, D.P., Matthews, K. and Yoder, C.F. (1992) New observations of Saturn's coorbital satellites. *Icarus* **100**, 464–484.
Nicholson, P.D. and Porco, C.C. (1988) A new constraint on Saturn's zonal gravity harmonics supplied by Voyager observations of an eccentric ringlet. *J. Geophys. Res.* **93**, 10209–10224.
Porco, C.C. (1991) An explanation for Neptune's ring arcs. *Science* **253**, 995–1001.
Porco, C.C. and Goldreich, P. (1987) Shepherding of the Uranian rings. I. Kinematics. *Astron. J.* **93**, 724–729.
Roy, A.E. (1988) *Orbital Motion.* Adam Hilger, Bristol.
Showalter, M. (1991) Visual detection of 1981S13, Saturn's eighteenth satellite, and its role in the Encke gap. *Nature* **351**, 709–713.
Shu, F. (1984) Waves in planetary rings. In *Planetary Rings* (Eds. Greenberg, R. and Brahic, A.) pp. 513–561, University of Arizona Press, Tucson.

Long-Term Evolution of Coorbital Motion

Jörg Waldvogel
Applied Mathematics
Swiss Federal Institute of Technology ETH
CH-8092 Zurich, Switzerland

Abstract

In these lectures the planar problem of three bodies with masses m_0, m_1, m_2 will be used as a model of coorbital motion, thus leaving the analysis of three-dimensional effects to later work. For theoretical as well as for numerical studies the choice of appropriate variables is essential. Here Jacobian coordinates and a rotating frame of reference will be used. The application of the Hamiltonian formalism in connection with complex notation will greatly simplify the differential equations of motion.

The results obtained are partially of experimental nature, based on reliable numerical integration. Obviously, chaos plays an important role. An orderly behaviour occurs for small mass ratios $\epsilon := (m_1 + m_2)/m_0$; however, the typical phenomena persist even for mass ratios as large as 0.01. In particular, proper coorbital motion seems to be chaotic, but stable for very long periods of time. The interaction of the satellites, as they approach each other, is qualitatively described by Hill's lunar problem. Temporary capture between independently revolving satellites is delicate and can only happen when close encounters are involved. It seems to be able to persist for very long times, though, even for mass ratios as large as $\epsilon = 0.1$.

1 Introduction

Coorbital motion is a particular case of three-body motion: the motion of two small satellites about a central body. For simplicity, only the planar case will be considered. Although coorbital motion constitutes a simplified model of three-body motion it displays surprisingly complex dynamics. In earlier work mainly a single close encounter in the motion of satellites in close circular orbits was considered, see, e.g. [4, 9].

In this study the succession of many close encounters is considered. Dermott and Murray [2] observed that coorbital motion may eventually decay, and they gave an estimate of the lifetime. The basic situation consists of a large central planet m_0 and two small satellites m_1, m_2 revolving about m_0 in the same plane.

B.A. Steves and A.E. Roy (eds.), The Dynamics of Small Bodies in the Solar System, 257–276.

Pairs of small satellites revolving about a central planet are called coorbital if their orbits are close in an appropriate sense. Three types of motion may be distinguished, all of which actually occur in the Saturnian ring system: (1) one-to-tone resonance or *proper coorbital motion*, such as the motion of Janus and Epimetheus, (2) *stable revolution*, such as the motion of the F ring shepherds Pandora and Prometheus, (3) *temporary capture*, such as the motion of two neighbouring ring particles who remain in bound state for an extended period of time.

In order to handle the long-term evolution of coorbital motion it is important to use appropriate coordinates. For describing a single close encounter coordinates which lead to Hill's lunar problem in the limiting case of small satellite masses seem to be appropriate [9]. Therefore we will basically introduce Hill's coordinates; however, the time intervals of weakly perturbed Kepler motion between close encounters will have to be considered as well.

Since we restrict ourselves to the planar case we will use complex notation for convenience. First, we will collect the tools of complex notation, complex gradients and canonical transformations. Jacobian coordinates and Levi-Civita regularization will be summarized for later use.

Then, the behaviour at a single close encounter will be summarized, following earlier work by Hénon and Petit [4, 6], and by Spirig and Waldvogel [9, 12]. Various coordinate systems with the above mentioned properties will be used with the goal of adequately describing coorbital motion. Numerical experiments suggest that coorbital pairs of type (1) have a finite lifetime that can be very long for small mass ratios and favourable initial conditions.

2 Tools

2.1 Complex Notation

For discussing the planar three-body problem we use the Hamiltonian formalism and complex notation. The basic equations will be collected together with the notation to be used.

Let $\vec{z} = (x, y)^T$ be the Cartesian coordinates of a point mass, and introduce the complex coordinate z and its conjugate $\bar{z}$ according to

$$z = x + iy \in \mathbb{C}, \quad \bar{z} = x - iy, \quad i = \sqrt{-1} \,. \tag{1}$$

Consider the function $H(x, y)$, e.g. a Hamiltonian with the dependence on the momenta suppressed, and write it in terms of z and $\bar{z}$,

$$H(x, y) = \widetilde{H}(z, \bar{z}) \,. \tag{2}$$

Then the complex gradient

$$\operatorname{grd} H := \frac{\partial H}{\partial x} + i\,\frac{\partial H}{\partial y} \tag{3}$$

becomes

$$\operatorname{grd} H := 2\,\frac{\partial \widetilde{H}}{\partial \bar{z}}\,, \tag{4}$$

as is seen by differentiating (2) by means of (1).

Next, we consider a Hamiltonian depending on several complex coordinates z_k $(k = 1, \ldots, N)$, their canonically conjugated momenta p_k, and the corresponding complex conjugated variables $\bar{z}_k$, $\bar{p}_k$ $(k = 1, \ldots, N)$. The typical transformation of variables to be considered is a conformal map in each coordinate. Therefore, with Z_k, P_k $(k = 1, \ldots, N)$ being the new coordinates and momenta, we define the coordinate transformation by the functions

$$z_k = z_k(Z_1, Z_2, \ldots, Z_N),\ k = 1, \ldots, N\,, \tag{5}$$

analytic in every variable, and we are looking for the definition of the canonically conjugated momenta P_k. Using the well-known technique of the generating function (see, e.g. [8]) we proceed as follows. From the generating function

$$W(Z_k, p_k) := Re \sum_{k=1}^{N} z_k(Z) \cdot \bar{p}_k \tag{6}$$

the new momenta are obtained as the partial derivatives of W with respect to the new coordinates, in complex notation:

$$P_k = 2\,\frac{\partial W}{\partial \overline{Z}_k}\,, \quad k = 1, \ldots, N\,. \tag{7}$$

2.2 The Hamiltonian

The technique discussed in the previous section conveniently allows to set up the equations of motion in various coordinate systems. Consider now the three point masses m_0, m_1, m_2 at the inertial positions $x_0, x_1, x_2 \in \mathbb{C}$, respectively, with the center of mass at rest, $\Sigma\, m_j\, x_j = 0$. We will formulate the equations of motion in terms of the relative coordinates with respect to x_0,

$$z_1 := x_1 - x_0, \quad z_2 := x_2 - x_0\,. \tag{8}$$

The position x_0 of the reference body m_0, in turn, may be recovered from

$$(m_0 + m_1 + m_2)\, x_0 = -m_1\, z_1 - m_2\, z_2\,.$$

It is easily seen that the canonically conjugated momenta in complex notation are

$$p_j = m_j\,\frac{dx_j}{dt}\,, \quad j = 1, 2,\,.$$

The Hamiltonian H and angular momentum C of the planar three-body problem are then given by

$$\begin{aligned} H &= \frac{|p_1+p_2|^2}{2m_0} + \frac{|p_1|^2}{2m_1} + \frac{|p_2|^2}{2m_2} - \frac{m_0\, m_1}{|z_1|} - \frac{m_0\, m_2}{|z_2|} - \frac{m_1\, m_2}{|z_2-z_1|}\,, \\ C &= Im(\bar{z}_1\, p_1 + \bar{z}_2\, p_2)\,. \end{aligned} \tag{9}$$

In this study we will consider coorbital configurations with m_0 as the central body and m_1, m_2 as the possibly interacting satellites. Therefore, it is useful to introduce Jacobi coordinates R, D with respect to m_0 and the center of mass of m_1 and m_2:

$$R = \mu_1\, z_1 + \mu_2\, z_2, \quad D = z_2 - z_1\,, \tag{10}$$

where

$$\mu_j = \frac{m_j}{m_1+m_2}\,, \quad j = 1, 2 \tag{11}$$

are the relative masses of the satellites. This yields the transformation

$$z_1 = z_1(R,D) = R - \mu_2\, D, \quad z_2 = z_2(R,D) = R + \mu_1\, D\,. \tag{12}$$

With the generating function according to (6),

$$W(R, D, p_1, p_2) = Re\Big(z_1(R,D)\,\bar{p}_1 + z_2(R,D)\,\bar{p}_2\Big)\,,$$

we obtain the definition of the canonically conjugated momenta P_R, P_D from Equ. (7):

$$\begin{aligned} P_R &= 2\,\frac{\partial W}{\partial \overline{R}} = p_1 + p_2 \\ P_D &= 2\,\frac{\partial W}{\partial \overline{D}} = \mu_1\, p_2 - \mu_2\, p_1\,. \end{aligned}$$

Solving this for the old momenta yields

$$p_1 = \mu_1\, P_R - P_D, \; p_2 = \mu_2\, P_R + P_D\,, \tag{13}$$

and the Hamiltonian and angular momentum (9) become

$$\begin{aligned} H &= \frac{|P_R|^2}{2m_0} + \frac{|\mu_1\, P_R - P_D|^2}{2m_1} + \frac{|\mu_2\, P_R + P_D|^2}{2m_2} \\ &\quad - \frac{m_0\, m_1}{|R - \mu_2\, D|} - \frac{m_0\, m_2}{|R + \mu_1\, D|} - \frac{m_1\, m_2}{|D|}\,, \\ C &= Im(\overline{R}\, P_R + \overline{D}\, P_D)\,. \end{aligned} \tag{14}$$

2.3 Regularization

Since repeated close encounters of m_1 and m_2 are involved in coorbital motion it may be necessary to regularize the respective binary collisions. A review of Levi-Civita's regularizing transformation [5], written in complex notation, is given in this section.

(i) A single binary collision between a particle at position z with momentum p and a central body at the origin is regularized by introducing new coordinates and momenta Z, P according to Levi-Civita's conformal canonical transformation

$$z = Z^2, \quad p = \frac{P}{2\bar{Z}} \,. \tag{15}$$

The second equation follows from (7) with the generating function $W = \frac{1}{2}\,(Z^2\bar{p} + \bar{Z}^2 p)$. Regularization is achieved by introducing the fictitious time s and the new Hamiltonian K according to

$$\begin{aligned} dt &= r \cdot ds, \quad r = |z| = |Z|^2 \\ K &= r(H - h)\,, \end{aligned} \tag{16}$$

where h is the fixed value of H on the orbit under consideration.

(ii) Hill's lunar problem has a special significance for coorbital motion, see Section 3. It approximates coorbital motion in a rotating (and pulsating) coordinate system during the close encounters of the satellites [9]. The model is given by the Hamiltonian [12]

$$H = \frac{1}{2}\,|p|^2 + Im(\bar{p}\,z) - \frac{3}{8}\,(z^2 + \bar{z}^2) - \frac{1}{4}\,z\bar{z} - \frac{1}{|z|}\,. \tag{17}$$

Using the transformations (15), (16) yields the new Hamiltonian

$$K = \frac{P\bar{P}}{8} + |Z|^2 \left(\frac{1}{2}\,Im(\bar{P}Z) - \frac{3}{8}\,(Z^4 + \bar{Z}^4) - \frac{1}{4}\,Z^2\bar{Z}^2 - h\right) - 1\,, \tag{18}$$

a polynomial of degree 2 in P and 6 in Z. For completeness the equations of motion are given here:

$$\frac{dZ}{ds} = 2\,\frac{\partial K}{\partial \bar{P}}\,, \quad \frac{dP}{ds} = -2\,\frac{\partial K}{\partial \bar{Z}}\,, \quad \frac{dt}{ds} = Z\bar{Z}\,.$$

(iii) As a by-product, we mention the possibility of simultaneously regularizing all binary collisions in the planar three-body problem [11]. The goal is to introduce *two* new complex coordinates Z_1, Z_2 such that the three complex relative positions $z_1, z_2, z_2 - z_1$ appear as complete squares, $z_1 = \zeta_1^2$, $z_2 = \zeta_2^2$, $z_2 - z_1 = \zeta_0^2$. Since this implies

$$\zeta_2^2 = \zeta_0^2 + \zeta_1^2\,, \tag{19}$$

it suffices to parameterize the relation (19) (formally Pythagoras' theorem) by 2 new variables Z_1, Z_2. We choose the well-known relations using quadratic polynomials for ζ_1, ζ_2,

$$z_1 = \Big(\frac{Z_1^2 - Z_2^2}{2}\Big)^2, \quad z_2 = \Big(\frac{Z_1^2 + Z_2^2}{2}\Big)^2, \quad z_2 - z_1 = Z_1^2\, Z_2^2 \,.$$

Equ. (7), together with the generating function

$$W = \frac{1}{2}\,(z_1\,\bar{p}_1 + \bar{z}_1\,p_1 + z_2\,\bar{p}_2 + \bar{z}_2\,p_2)$$

provides the definition of the conjugated momenta:

$$\begin{pmatrix} P_1 \\ P_2 \end{pmatrix} = \begin{pmatrix} \overline{Z}_1 & \overline{Z}_1 \\ -\overline{Z}_2 & \overline{Z}_2 \end{pmatrix} \begin{pmatrix} (\overline{Z}_1^2 - \overline{Z}_2^2)\,p_1 \\ (\overline{Z}_1^2 + \overline{Z}_2^2)\,p_2 \end{pmatrix} .$$

Solving for p_1, p_2 yields

$$\begin{aligned} p_1 &= \frac{\overline{Z}_2\,P_1 - \overline{Z}_1\,P_2}{2\,\overline{Z}_1\,\overline{Z}_2(\overline{Z}_1^2 - \overline{Z}_2^2)} \\ p_2 &= \frac{\overline{Z}_2\,P_1 + \overline{Z}_1\,P_2}{2\,\overline{Z}_1\,\overline{Z}_2(\overline{Z}_1^2 + \overline{Z}_2^2)} \,. \end{aligned}$$

Finally, the time transformation

$$dt = r_0\,r_1\,r_2 \cdot ds, \quad r_j = |z_j|$$

transforms the Hamiltonian (9) into the regularized

$$K = K(Z_1, Z_2, P_1, P_2) = r_0\,r_1\,r_2(H - h) \;,$$

which turns out to be a polynomial of degree 2 in P, and of degree 12 in Z. A recent account of implementing this regularization is given in [3].

3 A Single Close Encounter

We restrict ourselves to summarizing results obtained with the circular model described in [9]. Approximative data from two known coorbital pairs in the Saturnian ring system will be given.

Let

$$\epsilon := (m_1 + m_2)/m_0 << 1 \tag{20}$$

be the ratio of the total satellite mass $m_1 + m_2$ and the central mass m_0, and assume the satellites m_1, m_2 to initially move on nearly identical circles of radii R_1, R_2, respectively, about m_0. Following the definitions in Equ. (10) we introduce

$$R_{12} := \mu_1 R_1 + \mu_2 R_2, \quad \Delta := |R_2 - R_1| \tag{21}$$

as the "mean" orbital radius (more precisely, the orbital radius of the satellites' common center of mass at a close encounter) and the orbital separation.

Furthermore, we introduce the relative separation

$$\delta := \Delta / R_{12} \tag{22}$$

and the impact parameter

$$c := \delta \cdot \epsilon^{-1/3} . \tag{23}$$

In [9] it is established that during a close encounter of the satellites coorbital motion behaves like a particular solution of Hill's lunar problem. This is seen by using Jacobi coordinates R, D according to (10) and introducing scaled rotating and pulsating coordinates z according to

$$D = \epsilon^{1/3} Rz, \quad z = x + iy . \tag{24}$$

If the coorbital configuration is near-circular, i.e. if

$$R = R_0 \, e^{i\omega t} \text{ with } \omega^2 R_0^3 = m_0 ,$$

then the limit $\epsilon \to 0$ results in Hill's lunar equations

$$\begin{aligned} \ddot{x} - 2\dot{y} - 3x \; + \; x/r^3 \; &= \; 0, \; r = \sqrt{x^2 + y^2} \\ \ddot{y} + 2\dot{x} \qquad \; + \; y/r^3 \; &= \; 0 \end{aligned} \tag{25}$$

with the energy integral

$$\frac{1}{2} (\dot{x}^2 + \dot{y}^2) - \frac{3}{2} x^2 - \frac{1}{r} = h . \tag{26}$$

As a consequence of the definition (23), a single close encounter is described by the uniquely determined orbit satisfying Hill's lunar equations (25) and

$$\lim_{t \to -\infty} x(t) = c > 0 . \tag{27}$$

The main result on circular coorbital motion is based on the limiting case $c \to 0$. Introducing scaled coordinates

$$\xi = c^{-1} x, \quad \eta = c^2 y, \quad \tau = c^3 t \tag{28}$$

into (25), (26) yields in zeroth order

$$\begin{aligned} - 2\eta' - 3\xi &= 0 \\ \eta'' + 2\xi' + \frac{1}{\eta|\eta|} &= 0 \\ \frac{1}{2}\eta'^2 - \frac{3}{2}\xi^2 - \frac{1}{|\eta|} &= c^{-2}h \, , \end{aligned} \tag{29}$$

where primes denote derivatives with respect to τ. The orbit in the limit $c \to 0$ exists if $h \to \infty$ such that $c^{-2}h$ has a finite limit. Elimination of η' from the first and the third equation of (29) yields

$$-\frac{3}{8}\xi^2 - \frac{1}{|\eta|} = c^{-2}h \, .$$

Since the above limit corresponds to $\xi \to 1$, $\eta \to \infty$ we obtain $c^{-2}h \to -\frac{3}{8}$, and the possible limiting orbits are given by

$$\eta = \frac{\pm 8/3}{1-\xi^2} \quad \text{for} \quad |\xi| < 1 \, . \tag{30}$$

In particular, the closest approach occurs for $\xi = 0$ which implies

$$|\eta_{\min}| = \frac{8}{3} \, . \tag{31}$$

In order to characterize a close encounter in coorbital motion it suffices to consider solutions of Hill's lunar problem for which the limit (27) exists, referred to as *non-oscillating orbits*. These orbits approximate the motion of coorbital satellites (described in rotating Jacobi coordinates) during a close encounter in the limit of small mass ratios $\epsilon \to 0$.

In Fig. 1 (overview) and Fig. 2 (details) a few typical non-oscillating orbits corresponding to various values of the impact parameter $c > 0$ are plotted. They all have their incoming branch in the first quadrant. Three types of orbits may be distinguished according to the value of c compared to the separating values

$$c_1 = 1.33611\,71883, \qquad c_2 = 1.71877\,99380 \, . \tag{32}$$

These types will be briefly discussed in the following, using the same nomenclature as for the types of coorbital motion introduced in Section 1.

Orbits of Type 1 are obtained if the impact parameter satisfies $c \in (0, c_1)$.

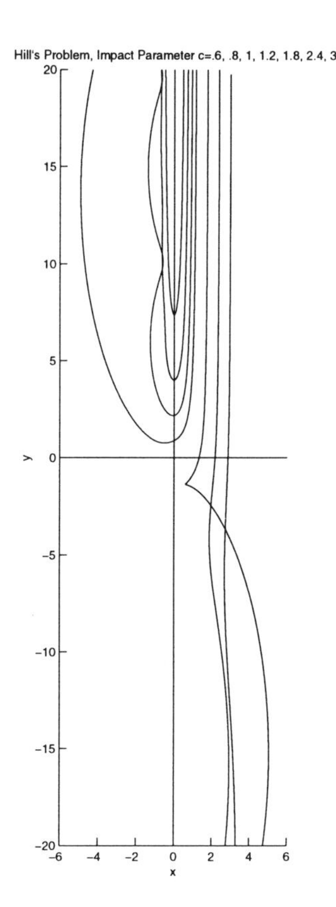

Fig. 1 The family of non-oscillating orbits in Hill's lunar problem. The incoming branch is in the first quadrant. From left to right, the values of the impact parameter are $c = 0.6, 0.8, 1.0, 1.2$ (Type 1), 1.8, 2.4, 3.0 (Type 2).

In the entire interval the orbit persistently escapes in the second quadrant as $t \to +\infty$. For sufficiently small values of c (in practice for $c < 0.8$) the orbit is well approximated by Equ. (30). In this interval the entire orbit depends continuously on the parameter c, and it is bounded away from the collision singularity at the origin. The limiting orbit for $c = c_1$ is asymptotic to a periodic orbit of Hill's problem with energy $h = -.375c_1^2$ and never escapes. Accordingly, the close encounters of coorbital motion corresponding to these values of c have the following properties:

- The leading body is the same before and after the close encounter.
- The upper and lower bodies exchange their roles such that the center of mass of the satellites remains at the same distance from m_0.

- For $c << 1$ (in practice $c < 0.8$) the minimum distance of the satellites during the close encounter is approximated by

$$D_{\min} = \epsilon^{1/3}\, R \cdot \frac{8}{3}\, c^{-2} = \frac{8}{3}\, R\epsilon\delta^{-2}\,, \tag{33}$$

as follows from (24), (28), (31), and (23). Here ϵ and δ are defined by (20) and (22), respectively.

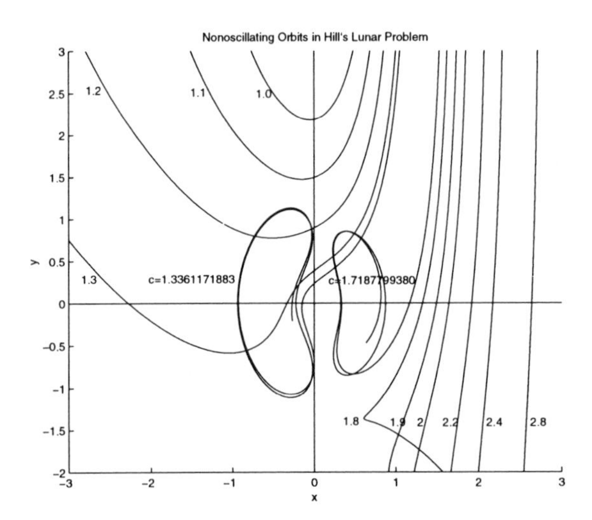

Fig. 2 Detail of Fig. 1. The limiting cases of orbits of Types 1 and 2 corresponding to $c = c_1$ and $c = c_2$, respectively (see Equ. (32)) are asymptotic to kidney-shaped periodic orbits of Hill's problem. Due to the hyperbolicity of these orbits (as fixed points of the Poincaré map) only a few revolutions can be traced numerically.

In the Saturnian ring system the satellites Janus and Epimetheus are currently on orbits of this kind. In Section 1 they were referred to as orbits of Type 1 or proper coorbital motion. Other notions are one-to-one resonance or horseshoe orbits, due to the horseshoe-like shape of the relative orbit in a rotating coordinate system. In Table 1 below the approximate orbital data of this pair of satellites are collected.

Close encounters of Type 2 are characterized by a sufficiently large value of the impact parameter c, more precisely by the condition $c > c_2$. Fig. 1 and Fig. 2 show that non-oscillating orbits of this type end in an escape in the fourth quadrant for $t \to \infty$. In the whole interval the entire orbit again depends continuously on c and is bounded away from the collision singularity at the origin. The limiting orbit for $c = c_2$ is again asymptotic to a periodic orbit of Hill's problem and never escapes. The close encounters of the corresponding coorbital motion have the properties

- The leading satellite becomes trailing and vice-versa.
- The upper satellite remains in the upper orbit, the lower satellite remains in the lower orbit.

Therefore, the lower (and faster) satellite merely passes the upper (and slower) one, without much interaction. In a circular configuration this *stable revolution* generally persists for a long time if $c > 3$. In the Saturnian ring system the F ring shepherds Pandora and Prometheus are an example of this situation (Table 1).

Non-oscillating orbits of Type 3 correspond to an impact parameter in the so-called transition region defined by $c \in (c_1, c_2)$. They depend on c in a very complicated, discontinuous way. The family has fractal structure with infinitely many orbits asymptotic to periodic orbits. Short intervals of continuous dependence on c are always followed by discontinuities. The orbits of the family cannot be bounded away from the collision singularity at the origin; the family contains infinitely many collision orbits. The orbits may generically end in an escape in the second, third or fourth quadrant (see Fig. 3). Therefore the corresponding close encounter in coorbital motion

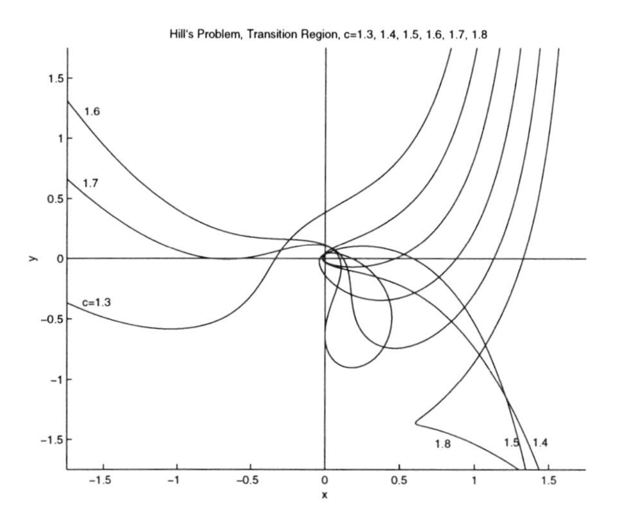

Fig. 3 Non-oscillating orbits from the transition region corresponding to the values $c = 1.4, 1.5, 1.6, 1.7$ of the impact parameter (from left to right on the incoming branch in the first quadrant). Every orbit involves close approaches with the singularity at the origin. The orbits corresponding to $c = 1.3$ and $c = 1.8$ outside the transition region are added for clarity.

may ultimately be of Type 1 or Type 2, but only after possible collisions of the satellites. Actual celestial bodies would thus not survive close encounters in the transition region.

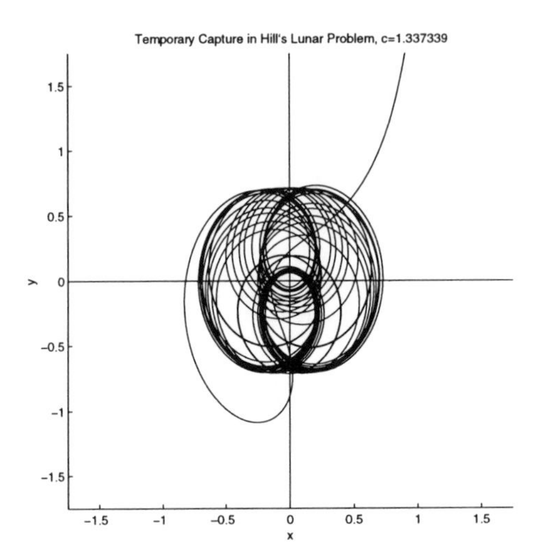

Fig. 4 Initial segment of the non-oscillating orbit with impact parameter $c = 1.337339$. The orbit is close to a quasiperiodic solution (torus); it escapes only after hundreds of revolutions (not shown).

An interesting phenomenon of non-oscillating orbits in the transition region is the existence of orbits close to quasiperiodic solutions (*tori*) of Hill's problem. These orbits carry out hundreds of revolutions about the origin before escaping (Fig. 4). In coorbital motion this corresponds to temporary capture between the two satellites in orbit about the central body. The experiments suggest that temporary capture can be achieved for an arbitrarily long time span. However, near-collisions between the satellites seem to be necessary.

In Table 1 below approximate orbital data of two actual coorbital pairs in the Saturnian ring system are listed. For the central body, Saturn, a normalized mass (mass times gravitational constant) of $m_0 = 3.8 \cdot 10^7$ km^3 sec^{-2} is assumed. The quantities ϵ, R_{12} and $\Delta, \delta, c, D_{\min}$ are defined in (20), (21), (22), (23), (33), respectively. The periods of revolution T_k follow from the third Keplerian law,

$$T_k = 2\pi\sqrt{R_k^3/m_0}\,, \quad k = 1, 2\,,$$

and the synodic period T_{syn} (the time between two consecutive close encounters) is given by

$$T_{\mathrm{syn}} = \frac{1}{T_1^{-1} - T_2^{-1}} \doteq \frac{2}{3} T_1 \, \delta^{-1} \, . \tag{34}$$

In the last line of Table 1 we give a measure T_{enc} of the duration of a close encounter of coorbital satellites. We say that the interaction begins when the trailing satellite "sees" the leading one in the tangential direction of its own orbit. The interaction is said to end in the inverse constellation. Simple geometry yields

$$T_{\mathrm{enc}} = \frac{L}{2\pi R_{12}} \cdot T_{\mathrm{syn}} \, ,$$

where

$$L \doteq \sqrt{8 R_{12} \Delta}$$

is the length of the chord of the outer orbit which is tangential to the inner one. Combining these equations yields

$$T_{\mathrm{enc}} \doteq \frac{2\sqrt{2}}{3\pi} T_1 \cdot \delta^{-1/2} \, . \tag{35}$$

	Type 1 Janus-Epimetheus	Type 2 Pandora-Prometheus
ϵ	$8 \cdot 10^{-9}$	$1.64 \cdot 10^{-9}$
R_{12}	151 460 km	140 270 km
Δ	50 km	2350 km
δ	0.00033	0.0167
c	0.165	14.2
D_{min}	29700 km	2350 km
T_1, T_2	16.68 h	14.7 h, 15.1 h
T_{syn}	1404 days	24.8 days
T_{enc}	275 h = 16.5 revol.	2.3 revolutions

Table 1

4 Long-Term Evolution

The long-term evolution of a coorbital pair is determined by the close encounters between the satellites. It strongly depends on the successive types of the close encounters as they are discussed in Section 3. We will first define coordinate systems appropriate for describing all phases of coorbital motion:

the close encounters, the intervening time intervals of perturbed Kepler motions, and the intervals of temporary capture between the satellites.

The starting point is the system of Jacobi coordinates R, D and their canonically conjugated momenta P_R, P_D (see Equs. (10), (13)), resulting in the Hamiltonian (14). In order to describe the relative motion of m_2 with respect to m_1 in a simple way we introduce a rotating and pulsating coordinate system. One way to do this is to measure the relative coordinate D with respect to the center-of-mass coordinate R, i.e. to introduce the complex ratio

$$Q := D/R \tag{36}$$

as a new coordinate. In order to carry out this transformation within the canonical formalism we augment (36) by the identical transformation $S := R$, thus denoting the new (unchanged) center-of-mass coordinate by S. Hence, the coordinate transformation becomes

$$\begin{aligned} R &= S \\ D &= S \cdot Q\,. \end{aligned} \tag{37}$$

Following the procedures of Section 2.1 we obtain the canonically conjugated momenta P_S, P_Q from the generating function

$$W = \frac{1}{2}\,(S\,\overline{P}_R + \overline{S}\,P_R + SQ\,\overline{P}_D + \overline{S}\,\overline{Q}\,P_D)$$

as

$$P_S = 2\,\frac{\partial W}{\partial \overline{S}}\,, \quad P_Q = 2\,\frac{\partial W}{\partial \overline{Q}}\,.$$

Solving these equations for the old momenta yields

$$P_R = P_S - \frac{\overline{Q}}{\overline{S}}\,P_Q, \quad P_D = \frac{P_Q}{\overline{S}}\,. \tag{38}$$

Substituting (37), (38) into the Hamiltonian (14) yields

$$\begin{aligned} H = \; & \frac{1}{2|S|^2}\left(\Big(\frac{1}{m_0} + \frac{1}{m_1 + m_2}\Big)\left|\overline{S}\,P_S - \overline{Q}\,P_Q\right|^2 \right. \\ & \left. + \Big(\frac{1}{m_0} + \frac{1}{m_2}\Big)\left|\,P_Q\right|^2\right) \\ & - \frac{1}{|S|}\left(\frac{m_0 m_1}{|1-\mu_2 Q|} + \frac{m_0 m_2}{|1+\mu_1 Q|} + \frac{m_1 m_2}{|Q|}\right). \end{aligned} \tag{39}$$

Obviously, this coordinate system generates singularities when the center of mass of m_1 and m_2 passes through the origin, $S = 0$. This cannot happen during a close encounter of m_1, m_2; however, it may happen if the satellites occupy nearly opposite positions on their orbits about m_0.

The following consideration involving two small satellites moving on independent circular Keplerian orbits

$$z_k = r_k \, e^{i\omega_k t}, \; (k = 1, 2), \; \omega_1^2 r_1^3 = \omega_2^2 r_2^3$$

will indicate the behavior of Q near this singularity. Equs. (10), (37) imply

$$Q = \frac{z_2 - z_1}{\mu_2 z_2 + \mu_1 z_1} \, .$$

Simplifying by z_1 reveals the orbit of the variable Q as the image of the circle $z = z_2/z_1 = r_2/r_1 \cdot e^{i(\omega_2 - \omega_1)t}$ under a Möbius transformation, i.e. again a circle. If S passes through the origin, Q passes through infinity, i.e. its orbit is a straight line (a circle with infinite radius).

On the other hand, during a close encounter of m_1 and m_2 the coordinate S nearly travels on a Keplerian orbit; therefore Q describes the relative motion of z_2 with respect to z_1 in the rotating-pulsating coordinates defined by the Kepler motion of S. Hence the quotient Q behaves as Hill's coordinates of Section 3 in every close encounter (compare Fig. 1 with Fig. 6 below). These segments are connected by near-circular arcs (see Fig. 5 below). In the variable Q the types of motion discussed earlier can easily be distinguished:

Type 1: Proper coorbital motion is characterized by a long sequence of close encounters, each one characterized by a single near-perpendicular intersection with the imaginary axis. The orbit of Q has the shape of a horseshoe. For $m_1 = m_2$ the connecting near-circular arcs may pass through the right or the left half-plane and seem to change in a chaotic way.

Type 2: Stable revolution is characterized by an orbit of Q consisting of perturbed circles without intersections with the imaginary axis.

Type 3: Temporary capture: Q enters the vicinity of the origin and stays there for a long time.

The Hamiltonian (39) may finally be regularized in two steps: By applying Levi-Civita's transformation to the variable Q the collisions between m_1 and m_2 are regularized. With q, P_q being the new coordinates and momenta, and τ_1, K_1 being the new time and Hamiltonian, we transform

$$Q = q^2, \quad P_Q = \frac{P_q}{2\overline{q}}, \quad dt = |q|^2 \, d\tau_1 \quad K_1 = |q|^2 (H - h) \, , \tag{40}$$

where h is the value of the Hamiltonian on the orbit under consideration.

The second step is to transform the variable S in a similar way:

$$S = s^2, \quad P_S = \frac{P_s}{2\overline{s}}, \quad d\tau = |s|^4 \, d\tau_1, \quad K = |s|^4 \, K_1 \, . \tag{41}$$

Here s, P_s, τ, K are again the new coordinates, new momenta, independent variable, and Hamiltonian. In addition to being analogous to the transformation (40), (41) "linearizes" the unperturbed equations of motion of the center of mass [10] and introduces the true anomaly of this motion as a new independent variable. These variables were used by Scheibner [7] in the elliptic restricted three-body problem. The transformed Hamiltonian becomes

$$\begin{aligned} K = \; & \frac{1}{8}\left(\frac{1}{m_0} + \frac{1}{m_1+m_2}\right)|q|^2\,|\bar{s}\,P_s - \bar{q}\,P_q|^2 + \frac{1}{8}\left(\frac{1}{m_1}+\frac{1}{m_2}\right)|P_q|^2 \\ & - |s|^2\left(\frac{m_0 m_1\,|q|^2}{|1-\mu_2 q^2|} + \frac{m_0 m_2\,|q|^2}{|1+\mu_1\,q^2|} + m_1 m_2\right) - h\,|q|^2\,|s|^4\,, \end{aligned} \tag{42}$$

and the equations of motion are

$$\begin{aligned} \frac{dq}{d\tau} &= 2\,\frac{\partial K}{\overline{P}_q}\,, \quad \frac{dP_q}{d\tau} = -2\,\frac{\partial K}{\partial \bar{q}} \\ \frac{ds}{d\tau} &= 2\,\frac{\partial K}{\overline{P}_s}\,, \quad \frac{dP_s}{d\tau} = -2\,\frac{\partial K}{\partial \bar{s}}\,, \quad \frac{dt}{d\tau} = |qs^2|^2\,. \end{aligned} \tag{43}$$

In the types of motion and time intervals considered in this work the collisions of the satellites with m_0 are excluded (i.e. the denominators in (42) are > 0); hence Equ. (43) are expected to be well suited for studying planar coorbital motion.

5 Results and Conclusions

For generating orbits of coorbital pairs we use roughly opposite initial positions and circular initial velocities. The initial positions will be defined according to Equ. (8), i.e. by

$$x_1 = x_0 + z_1, \quad x_2 = x_0 + z_2\,,$$

where

$$x_0 = -\frac{m_1 z_1 + m_2 z_2}{M}\,, \quad M = m_0 + m_1 + m_2\,.$$

Nearly opposite values for z_1 and z_2, e.g.

$$z_2 = -1, \quad z_1 = 1+\delta, \quad |\delta| << 1 \tag{44}$$

were chosen in all the experiments. According to Kepler's laws the circular initial velocities are given by the complex momenta

$$p_1 = i m_1\sqrt{\frac{M}{x_1}}\,, \quad p_2 = -i m_2\sqrt{\frac{M}{-x_2}}\,. \tag{45}$$

A large number of orbits of coorbital pairs of all three types were computed numerically, using initial data according to (44), (45) with δ and ϵ (Equ. (20)) in the range $10^{-1} \dots 10^{-3}$. For sufficiently small mass ratios ϵ motion of Type 1 is obtained initially if $|\delta| < c \cdot \epsilon^{1/3}$ and $c < 1.33$. In this way it was possible to observe a sufficient number of revolutions of coorbital motion as well as the process of its decay. The representation of the relative motion of m_2 with respect to m_1 in rotating-pulsating coordinates Q (Equs (36), (10)) proves to be an excellent means for visualizing the long-term behavior of the coorbital pair.

In all the experiments it was possible to continue the computation until the coorbital pair broke up. Similar results were obtained by G. Auner and R. Dvorak [1] in a numerical study with small mass ratios $\epsilon < 10^{-4}$. In most of their experiments with $\delta > 0.7 \cdot \epsilon^{1/3}$ break-up of the coorbital pair within 10'000 close encounters was observed. Therefore it may be conjectured that coorbital pairs have a *finite lifetime* (which may be very long, though).

We restrict ourselves to visualize the long-term evolution and decay of coorbital motion by means of a typical example.

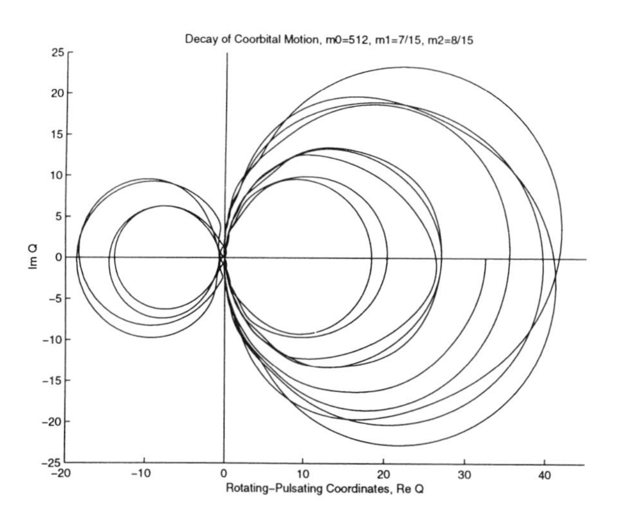

Fig. 5 Relative motion in a coorbital pair, shown in rotating-pulsating coordinates. Masses: $m_0 = 512$, $m_1 = 7/15$, $m_2 = 8/15$. Circular initial positions: $z_1 = -1$, $z_2 = 1.01$. The pair breaks up after 10 close encounters; 4 more close encounters of the subsequent Type-2 motion are shown.

Figure 5, giving an overall view of the relative motion in rotating-pulsating coordinates, clearly displays 10 near-circular connecting arcs representing 10 intervals of weakly perturbed Kepler motion between 10 close encounters.

These circles are slightly deformed since the Kepler orbits have small eccentricities. The succession of connecting arcs is generally non-monotonic, in fact is seems to be totally irregular and impredictable. After the break-up of the coorbital pair the motion of this system transforms into Type-2 motion (stable revolution); 4 more close encounters are shown in the figure.

The behavior near the close encounters is seen in Fig. 6, a mere close-up view of the neighbourhood of the origin of Fig. 5. The 10 close encounters of Type-1 motion are represented by the U-shaped branches entering the picture from above and from below. These encounters slowly "deteriorate" in an irregular but systematic way: the minimum distance decreases until the pair breaks up in a near-collision. The branches on the left side of the figure represent the first 4 close encounters in the subsequent Type-2 motion.

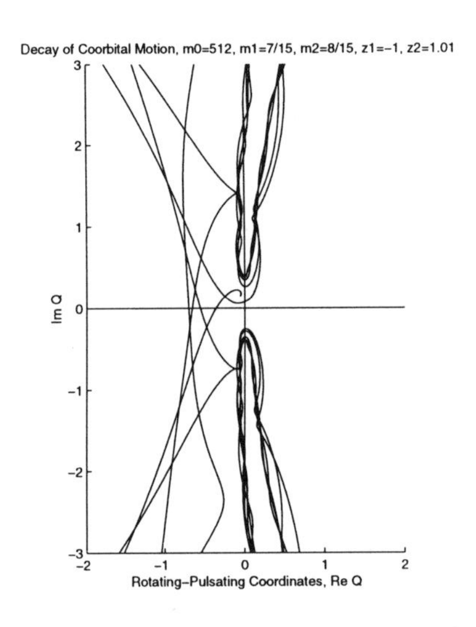

Fig. 6 Detail of Fig. 5 (neighbourhood of the origin). The coorbital pair breaks up with the tenth close encounter.

Finally, we mention a surprising observation concerning temporary capture between two satellites in nearly identical orbits about a central body. In Section 3 (Figure 4) we gave an example of long temporary capture in Hill's lunar problem, which is the limiting case $\epsilon \to 0$ of coorbital motion. Our experiments suggest that the phenomenon of temporary capture, between coorbital satellites possibly for arbitrarily long times, exists also for finite mass ratios $\epsilon > 0$. Fig. 7 shows the initial section of an orbit corresponding to $\epsilon = 1/343$ which remains in bound state for more than 100 revolutions. Similar orbits have been found for mass ratios as large as $\epsilon = 0.1$. Note the striking similarity between Fig. 7 and Fig. 4.

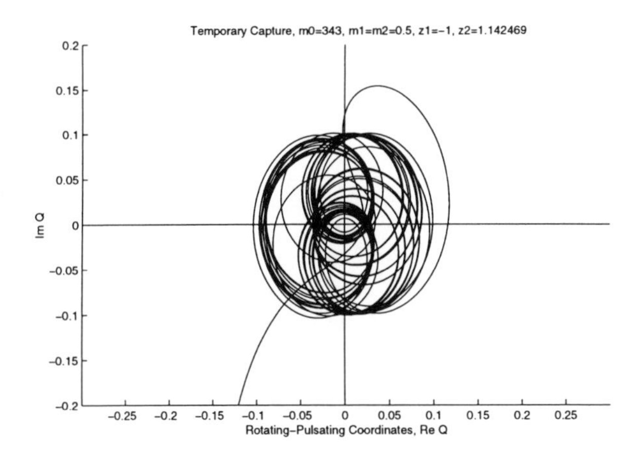

Fig. 7 Temporary capture between two satellites in orbit. Masses: $m_0 = 343$, $m_1 = m_2 = 1/2$. Circular initial positions: $z_1 = -1$, $z_2 = 1.142469$. The bound state breaks up only after more than 100 revolutions (not shown here). Note the similarity with Fig. 4.

The analysis of some of the phenomena of this section by means of appropriate surfaces-of-section is deferred to a later paper.

References

[1] Auner, G. and Dvorak, R.: *Close planetary orbits and their cosmological importance.* In: J. Hagel, M. Cunha, R. Dvorak (eds), Perturbation Theory and Chaos in Nonlinear Dynamics with Emphasis to Celestial Mechanics. Universität Wien, 1-17, (1995).

[2] Dermott, S.F. and Murray, C.D.: The dynamics of tadpole and horseshoe orbits, I, II. *Icarus*, **48**, 1-22, (1981).

[3] Gruntz, D. and Waldvogel, J.: *Orbits in the planar three-body problem.* In: W. Gander, J. Hrebicek (eds), Solving Problems in Scientific Computing using Maple and Matlab. Springer-Verlag, 37-57, (1991).

[4] Hénon, M. and Petit, J.M.: Series expansions for encounter-type solutions of Hill's problem. *Celest. Mech.*, **38**, 67-100, (1986).

[5] Levi-Civita, T.: Sur la régularisation du problème des trois corps. *Acta Math.*, **42**, 99-144, (1920).

[6] Petit, J.M. and Hénon, M.: Satellite encounters. *Icarus*, **66**, 536-555, (1986).

[7] Scheibner, W.: Satz aus der Störungstheorie. *J. Reine Angew. Math.*, **65**, 291, (1866).

[8] Siegel, C.L., and Moser, J.: *Lectures on Celestial Mechanics.* Springer-Verlag, 290 pp. , (1971).

[9] Spirig, F. and Waldvogel, J.: *The three-body problem with two small masses: A singular-perturbation approach to the problem of Saturn's coorbiting satellites.* In: V. Szebehely (ed), Stability of the Solar System and its Minor Natural and Artificial Bodies. Reidel, 53-63, (1985).

[10] Stiefel, E. and Scheifele, G.: *Linear and Regular Celestial Mechanics.* Springer-Verlag, 301 pp., (1971).

[11] Waldvogel, J.: A new regularization of the planar problem of three bodies. *Celest. Mech.*, **6**, 221-231, (1972).

[12] Waldvogel, J. and Spirig, F.: *Chaotic motion in Hill's lunar problem.* In: A.E. Roy and B.A. Steves (eds), From Newton to Chaos: Modern Techniques for Understanding and Coping with Chaos in N-Body Dynamical Systems. Plenum Press, 217-230, (1995).

STABILITY OF PERTURBED COORBITAL SATELLITES

M.H.M. MORAIS AND C.D. MURRAY
Astronomy Unit, Queen Mary and Westfield College,
Mile End Rd, London E1 4NS, U.K.

1. Introduction

Coorbital configurations are special solutions of the three-body problem. In the restricted case (when the third body is a massless particle), it is well known that for small displacements from the triangular equilibrium points L_4 and L_5, the particle moves in a linearly stable 1:1 libration, provided the mass ratio is less than a critical value. There are two types of coorbital motion: "tadpole" orbits that enclose one of the triangular points and "horseshoe" orbits that enclose L_3, L_4 and L_5 (Dermott & Murray, 1981).

Several examples of coorbital configurations exist in the solar system. Jupiter has the Trojan asteroids librating about its triangular points with the Sun. Asteroid (5261) Eureka orbits near Mars' L_5 point (Mikkola *et al.*, 1994) and asteroid (3753) 1986TO is in a horseshoe orbit with the Earth (Wiegert *et al.*, 1997). However, all the known coorbital *satellites* occur in the saturnian system. Janus and Epimetheus are in a horseshoe configuration, while examples of tadpole orbits exist about the L_4 point of Dione (Helene) and the L_4 and L_5 points of Tethys (Telesto and Calypso, respectively). There may also be coorbital companions of Mimas (Gordon *et al.*, 1996) and Enceladus (Baum *et al.*, 1981).

In this paper we report on preliminary work on the dynamics of companions of Mimas, Enceladus, Tethys and Dione and try to understand the major dynamical differences between them as potential coorbital systems. According to Sinclair (1984) the existence of the Enceladus–Dione and Mimas–Tethys 2:1 mean motion resonances has to be taken into account in any coorbital model. If two satellites are involved in a resonance, then a coorbital of either satellite can traverse that resonance, during each 1:1 libration. The strength of the resulting perturbation depends on the

B.A. Steves and A.E. Roy (eds.), The Dynamics of Small Bodies in the Solar System, 277–282.

mass of the perturber, the closeness to the resonance and the time spent in or near it.

2. The dynamical model

Consider a system of two satellites orbiting an oblate planet where one of the satellites has a coorbital companion. In what follows unprimed variables (r, a, e, n, λ, ϖ: radial distance, semimajor axis, eccentricity, mean motion, mean longitude, longitude of apse) refer to the massless coorbital, primed variables to the inner satellite (mass m') and double primed variables to the outer satellite (mass m''). The central planet has unit mass, equatorial radius R and dominant zonal harmonic J_2.

Message (1966) showed that the secular solution for a zero amplitude L_4 coorbital of, for instance, the inner satellite is:

$$\begin{aligned} h &= e \sin \varpi = e' \sin(\varpi' + 60°) + c \sin(\gamma t + \chi) \\ k &= e \cos \varpi = e' \cos(\varpi' + 60°) + c \cos(\gamma t + \chi) \end{aligned} \tag{1}$$

where $\gamma = (27/8)(m'/(1+m'))n$ is the secular frequency and c (proper eccentricity) and χ are constants determined by the initial conditions (e_0, ϖ_0). The stationary configuration has $e = e'$ and $\varpi - \varpi' = 60°$.

Érdi (1979) suggested that the solution given above is better seen in a plot of e against $\varpi - \varpi'$. If $|c| < e'$ then $\varpi - \varpi'$ librates around 60° with frequency γ; if $|c| \geq e'$ then it circulates. Maximum eccentricity $e_{max} = c + e'$ occurs at $\varpi - \varpi' = 60°$. Equation (1) can also be generalized for non-zero amplitude tadpoles: the secular period and the longitude at which maximum eccentricity occurs decrease with increasing tadpole amplitude.

With the inclusion of the oblateness potential (to second order in the eccentricities) we obtain $\varpi' = \beta t + \varpi_0'$ (where $\beta = (3/2)J_2(R/a')^2 n'$) and the solution becomes:

$$\begin{aligned} h &= e' \sin(\beta t + \varpi_0' + 60°) + c \sin((\gamma + \beta)t + \chi) \\ k &= e' \cos(\beta t + \varpi_0' + 60°) + c \cos((\gamma + \beta)t + \chi) \end{aligned} \tag{2}$$

hence

$$e^2 = e'^2 + c^2 + 2ce' \cos(\gamma t + \chi - \varpi_0' - 60°)$$

and

$$\tan(\varpi - \varpi') = \frac{e' \sin 60° + c \sin(\gamma t + \chi - \varpi_0')}{e' \cos 60° + c \cos(\gamma t + \chi - \varpi_0')}$$

Thus, to this order the oblateness does not affect the solution in the $(\varpi - \varpi', e)$ diagram. However, Eq. (2) breaks down if the eccentricities become large because higher-order terms need to be included in the oblateness potential.

Now we analyse the effect of a 2:1 resonance on the dynamics. The Hamiltonian for a coorbital of the interior satellite is:

$$H = -\frac{G(1+m')}{r} + \mathcal{R}_{1:1} + \mathcal{R}_{\text{obl}} + \mathcal{R}_{2:1}. \tag{3}$$

Taking the 1:1 averaged disturbing function $\mathcal{R}_{1:1}$ from Message (1966) and expanding the oblateness potential $\mathcal{R}_{\text{obl}}$ to second order and the 2:1 resonance potential $\mathcal{R}_{2:1}$ to first order in the eccentricities, Sinclair (1984) obtained the (h, k) solution for a particle at the triangular points. He found that at L_4

$$\begin{aligned} h \;\approx\;& c\sin((\gamma+\beta)t+\chi) + e'_0\sin(\beta t + \varpi'_0 + 60^\circ) \\ &+ e'_{\text{f}}\sin(2\lambda'' - \lambda' - 60^\circ) \end{aligned} \tag{4}$$

where e'_0 and e'_{f} are, respectively, the proper and the forced eccentricities of the interior satellite. Combining with the equivalent expression for k gives

$$\begin{aligned} e^2 \;\approx\;& c^2 + {e'_0}^2 + {e'_{\text{f}}}^2 + 2ce'_0\cos(\gamma t + \chi - \varpi'_0 - 60^\circ) \\ &+2ce'_{\text{f}}\cos(2\lambda'' - \lambda' - \varpi' - \gamma t - \chi + \varpi'_0 - 60^\circ) \\ &+2e'_0e'_{\text{f}}\cos(2\lambda'' - \lambda' - \varpi' - 120^\circ). \end{aligned} \tag{5}$$

However, Sinclair's derivation assumes that the 2:1 resonance does not destroy the 1:1 equilibrium. We believe that this assumption is not always justified and in these circumstances there are important consequences for the long-term behaviour of the coorbital.

Coorbitals with a significant amplitude can evolve in and out of the 2:1 resonance during each 1:1 libration, and hence the resonant perturbations will be non-coherent. The net effect could be a random walk of the orbital elements. A likely mechanism for instability in this kind of coorbital systems are collisions caused by large and/or irregular variations in e. Particles in horseshoe orbits can collide with their parent satellite but a tadpole coorbital is more likely to collide with existing coorbital material. A satellite of mass m has a coorbital region of radial half-width $\Delta \approx 2(m/(3(m+1)))^{\frac{1}{3}}$. If a tadpole's e is comparable to Δ, then it traverses this entire region and the simultaneous existence of tadpole and horseshoe orbits is unlikely (cf. Sinclair, 1984).

3. Numerical results

To simulate a massless coorbital of a saturnian satellite under the effect of an oblate planet and a 2:1 resonance with another satellite, we numerically integrated this restricted 4-body problem. We assumed that all the

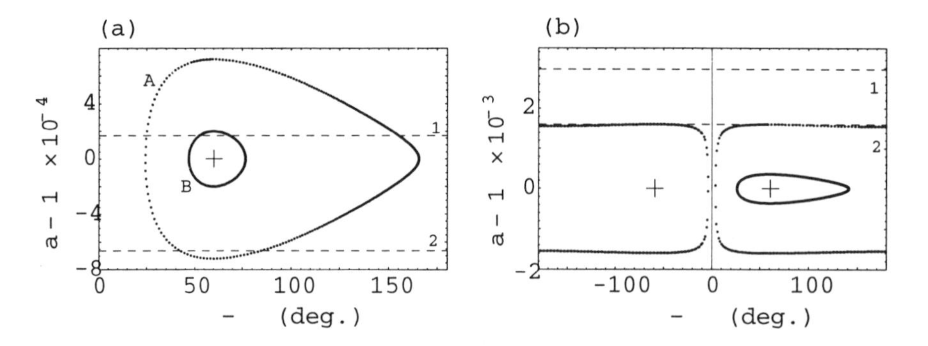

Figure 1. Variation of semimajor axis with longitude difference for unperturbed sample coorbitals of (a) Enceladus and (b) Mimas. The locations of the resonances associated with the arguments θ_1 and θ_2 are indicated by the dashed lines.

motion occurs in Saturn's equatorial plane since the actual inclinations of the selected satellites are all small. This gives two resonant arguments, $\theta_1 = 2\lambda'' - \lambda' - \varpi'$ and $\theta_2 = 2\lambda'' - \lambda' - \varpi''$, associated with the 2:1 resonance.

In all the integrations the inner satellite was started at pericentre and the outer satellite at apocentre of ellipses with aligned apses. Semimajor axes and eccentricities were taken from Harper & Taylor (1993). The coorbital was started with a small radial displacement δa from L_4. The semimajor axis and mean motion of the coorbital parent satellite were taken to be unity.

3.1. ENCELADUS AND DIONE

Enceladus and Dione are involved in a 2:1 eccentricity resonance with critical argument θ_1. Figure 1a shows the location of these resonances and the a variation of two L_4 coorbitals of Enceladus: one with large ($\sim 140°$) amplitude and the other with smaller ($\sim 30°$) amplitude, labelled A and B, respectively.

Figure 2a shows that coorbital A has irregular behaviour. Although it starts as an L_4 coorbital, its amplitude suddenly increases and it becomes an L_5 coorbital. The eccentricity in Fig. 2b also behaves in an irregular fashion with $e > 0.01$ at times. Our numerical studies have shown that coorbital B has a much more regular behaviour. It has librating critical argument $2\lambda'' - \lambda - \varpi$ and its eccentricity exhibits a frequency of $\sim \dot{\theta}_1$ in agreement with Eq. (6). The mass of Enceladus ($m' = 0.2 \times 10^{-6}$) is smaller than that of Dione ($m'' = 1.9 \times 10^{-6}$) and so we expect the 2:1 resonance to have a larger effect on coorbitals of Enceladus. In fact, our integrations of tadpole coorbitals of Dione showed that these conform to Érdi's secular solution.

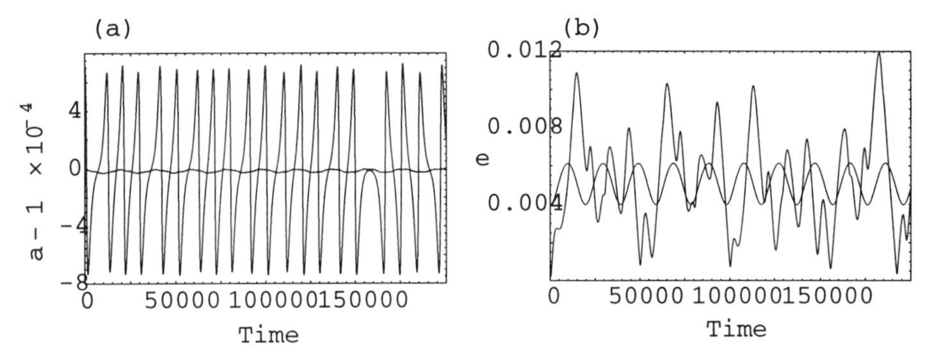

Figure 2. Evolution of (a) semimajor axis and (b) eccentricity for coorbital A in Fig. 1a. The regular, smaller amplitude curves show the variation of a' and e'. Time is in units such that the orbital period of Enceladus is 2π.

3.2. MIMAS AND TETHYS

Mimas and Tethys are in an inclination resonance with critical argument $4\lambda'' - 2\lambda' - \Omega'' - \Omega'$. This cannot be reproduced with our model as we only consider planar motion. However, this implies that the satellites and their respective tadpole coorbitals will not be significantly affected by the eccentricity resonances as these are far away (see Fig. 1b). We found that tadpole coorbitals of Tethys evolve according to Érdi's theory, further proof that the effect of the resonance is small.

Figure 3 shows the secular evolution of a particle started at Mimas' L_4 point. Its maximum eccentricity $e_{\max} \approx 0.005$ disagrees with the value of 0.04 predicted by Eq. (1) (dashed curve in Fig. 3). This is due to the fact that terms up to fourth order in e' have to be taken into account in the oblateness potential for Mimas' motion.

4. Conclusions

The 2:1 resonance with Dione should have a strong effect on coorbitals of Enceladus. Large amplitude tadpoles have irregular orbits and are probably short-lived due to collisions. Sinclair (1984) suggests also that this erratic nature of the orbits may have prevented the accretion of material to form companion satellites.

Coorbital companions of Dione and Tethys are not significantly affected by the 2:1 resonance. Their coorbital regions have half widths $\Delta \sim 10^{-2}$, thus only tadpoles with large c (i.e. $c \geq c_{\text{crit}}$ where $c_{\text{crit}} = \Delta - e'$) will potentially collide with horseshoe material. Mimas is the less massive satellite and thus it has the smallest coorbital region ($\sim 5 \times 10^{-3}$). The coorbital shown in our integration is able to cross the whole horseshoe zone.

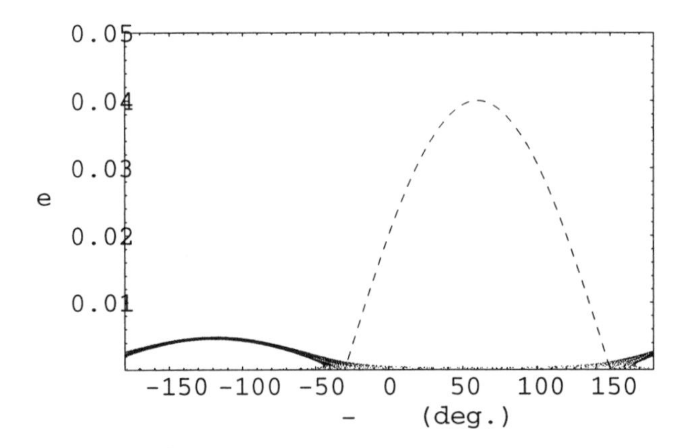

Figure 3. Actual variation of e and $\varpi - \varpi'$ for a Mimas coorbital perturbed by Tethys (points) together with the unperturbed variation (dashed curve).

In the case of the Mimas–Tethys system the inclusion of inclinations in our model would allow coorbitals to approach the actual resonance. By analogy with the Enceladus–Dione case, we would still expect coorbitals of Tethys ($m'' = 1.1 \times 10^{-6}$) to be less affected by the resonance than those of Mimas ($m' = 0.06 \times 10^{-6}$).

Our preliminary work verifies Sinclair's (1984) conclusions concerning the irregular behaviour of material coorbital with Enceladus. We also agree that horseshoe and tadpole coorbitals of Dione and Tethys can co-exist. However, our results on Mimas' coorbitals disagree with Sinclair's analysis, although a more detailed study is needed before any firm conclusions can be reached about their stability.

M.H.M. Morais is supported by grant BD/5072/95 - Program PRAXIS XXI from JNICT, Portugal.

References

Baum, W.A., Kreidl, T., Westphal, J.A., Danielson, G.E., Seidelmann, P.K., Pascu, D., Currie, D.G. *Icarus* **47**, 84–96 (1981).
Dermott, S.F. and Murray, C.D. *Icarus* **48**, 1–11 (1981) .
Érdi, B. *Celest. Mech.* **20**, 59–67 (1979).
Gordon, M.K., Murray, C.D. and Beurle, K. *Icarus* **121**, 114–125 (1996).
Harper, D. and Taylor, D.B. *Astron. Astrophys* **268**, 326–349 (1993).
Message, P.J. *Lectures in Applied Mathematics* **6**, 70–78 (1966).
Mikkola, S., Innanen, K.A., Muinonen, K. and Bowell, E. *Celest. Mech.* **58**, 53–64 (1994).
Sinclair, A.T. *Astron. Astrophys* **136**, 161–166 (1984).
Wiegert, P.A., Innanen, K.A., Mikkola, S. *Nature* **387**, 685–686 (1997).

INSTABILITY AND CHAOTIC MOTION FOR SOME FICTITIOUS SATELLITES OF VENUS AND MARS

TADASHI YOKOYAMA
Universidade Estadual Paulista
C. Postal 178 CEP 13.500-970
Rio Claro
Brasil

1. Introduction

Among the remarkable differences between inner and outer planets, we can mention the striking scenario of their satellite system : while in general, the giant planets have a rich satellite system, the inner planets are very poor. In order to explain this situation, many authors have invoked the effects of the tides which could have been decisive to provoke the disappearance or the fall of such satellites. Usually, it is well agreed that the time scale for tides is not so short. In this work, just considering gravitational effects (solar perturbation and the oblateness of the mother planet), it is possible to show the existence of strong instability and chaotic motion that may occur in a very short time (some thousand of years). The conditions under which this occurs is dictated by some special high values of the obliquity of the ecliptic (ϵ) and the distance of the satellite from the planet. As shown by Laskar, Touma and Wisdom, the obliquity of the ecliptic of the! inner planets is not primordial. According to them, ϵ might have varied up to very large inclinations. Therefore considering some high ϵ and provided that the distance of the satellite is usually in the neighbourhood of a critical value, a strong chaotic motion with big increase of eccentricity and inclination can occur in a very short time.

2. Averaged Hamiltonian

The main part of the averaged disturbing force due to the Sun on a satellite is (Kinoshita and Nakai)[4]:

B.A. Steves and A.E. Roy (eds.), The Dynamics of Small Bodies in the Solar System, 283–288.

$$\{R_\odot\}_s = \frac{M_\odot}{M+M_\odot} \frac{n_\odot^2 a^2}{(1-e_\odot^2)^{3/2}}$$
$$\left[\frac{1}{8}\left(1+\frac{3}{2}e^2\right)\left(3\cos^2 I-1\right)+\frac{15}{16}e^2\sin^2 I\cos 2g\right] \quad (1)$$

where k^2 is the gravity constant, M is planet's mass. The semimajor axis, eccentricity, inclination, argument of the pericenter, node and mean motion of the satellite are indicated by a, e, I, g, h, n, respectively. For the corresponding elements of the Sun we use the index $(\odot)$. The coordinate system is fixed at the center of the gravity of the planet while its orbital plane is taken as the reference plane.

For the oblateness, usually the equator of the planet is used as the reference plane, and the secular part of the disturbing function (Kinoshita and Nakai)[4]is:

$$\{R_{J_2}\}_s = \frac{1}{4}n^2 a_p^2 \frac{1}{(1-e^2)^{3/2}} J_2\left(3\cos^2 i-1\right) \quad (2)$$

where a_p is the planet's equatorial radius, J_2 is the oblateness coeficient, i is the inclination referred to the equator. Writting $\{R_{J_2}\}_s$ with respect to orbital plane, we have:

$$\{R_{J_2}\}_s = \frac{1}{8}n^2 a_p^2 \frac{1}{(1-e^2)^{3/2}} J_2$$
$$\left[\left(3\cos^2\epsilon-1\right)\left(3\cos^2 I-1\right)-3\sin 2\epsilon\sin 2I\cos h+\right.$$
$$\left.3\sin^2\epsilon\sin^2 I\cos 2h\right] \quad (3)$$

The equations of motion are now easily obtained from the Hamiltonian:

$$H^* = \{R_{J_2}\}_s + \{R_\odot\}_s \quad (4)$$

which can be trivially written in terms of the classical Delaunay canonical variables (G, H, g, h).

In the case when the equator of the planet is choosen as the reference plane, the oblateness part is given by (2) while the solar part needs to be rewritten :

$$\{R_\odot\}_s = \frac{k^2 M_\odot Q_\odot a^2}{2a_\odot^3}\left[\frac{P}{4}\left(1-3\cos^2 I-3\cos^2\epsilon+9\cos^2 I\cos^2\epsilon\right)+\right.$$

$$\frac{3}{4}Z\sin^2 I\left(-1+3\cos^2\epsilon\right)\cos 2g+$$
$$\frac{3}{4}P\sin^2 I\sin^2\epsilon\cos\left(2h-2h_\odot\right)+$$
$$\frac{3}{8}Z\left(1+\cos I\right)^2\sin^2\epsilon\cos\left(2g+2h-2h_\odot\right)-$$
$$\frac{3}{2}Z\sin I\sin\epsilon\left(1+\cos\epsilon\right)\cos\epsilon\cos\left(2g+h-h_\odot\right)+$$
$$3P\sin I\cos I\sin\epsilon\cos\epsilon\cos\left(h-h_\odot\right)+$$
$$\frac{3}{8}Z\left(1-cosI\right)^2\sin^2\epsilon\cos\left(2g-2h+2h_\odot\right)+$$
$$\left.\frac{3}{2}Z\sin I\left(1-\cos I\right)\sin\epsilon\cos\epsilon\cos\left(2g-h+h_\odot\right)\right] \tag{5}$$

where:

$$Q_\odot=\frac{L_\odot^3}{2G_\odot^3} \qquad P=1+\frac{3}{2}e^2 \qquad Z=\frac{5}{2}e^2$$

For satellites close to the planet, the oblateness dominates and $R_\odot$ given by (5) is just a perturbation.

3. Large Variation Of The Eccentricity

In the integration of Hamiltonian (4), we have to fix some values for the semimajor axis a and ϵ. In order to do that, first of all, let's define a special value of a. Depending on the mutual distance between planet-satellite-Sun, either $\{R_\odot\}_s$or $\{R_{J_2}\}_s$ will be dominant (Goldreich, Kinoshita and Nakai)[2,4]. A rough estimate of the critical semimajor axis such that the magnitude of both perturbations are almost equal can be obtained equating (1) to (2):

$$a_{crit}=\left[2\frac{M}{M_\odot}J_2 a_\odot^3 a_p^2\left(1-e_\odot^2\right)^{3/2}\right]^{1/5} \tag{6}$$

As for ϵ, the obliquities of the inner planets could have varied in a large range (Laskar, Touma and Wisdom)[5,7], depending on their precession constant (α). For instance, due to planetary perturbations, Venus' ϵ can vary in the interval 0^0 to 90^0, if α is about 20"/year. According to Laskar [5], this value of α might have occurred when the period of the planet was nearly 20 hours.

If the hydrostatic equilibrium theory is assumed (Jeffreys)[3], we can estimate the corresponding value of J_2 (Ward)[8]. In doing that, let's see the behaviour of the eccentricity for several values of ϵ. Numerous integrations were tested always starting from equatorial and almost circular orbits.

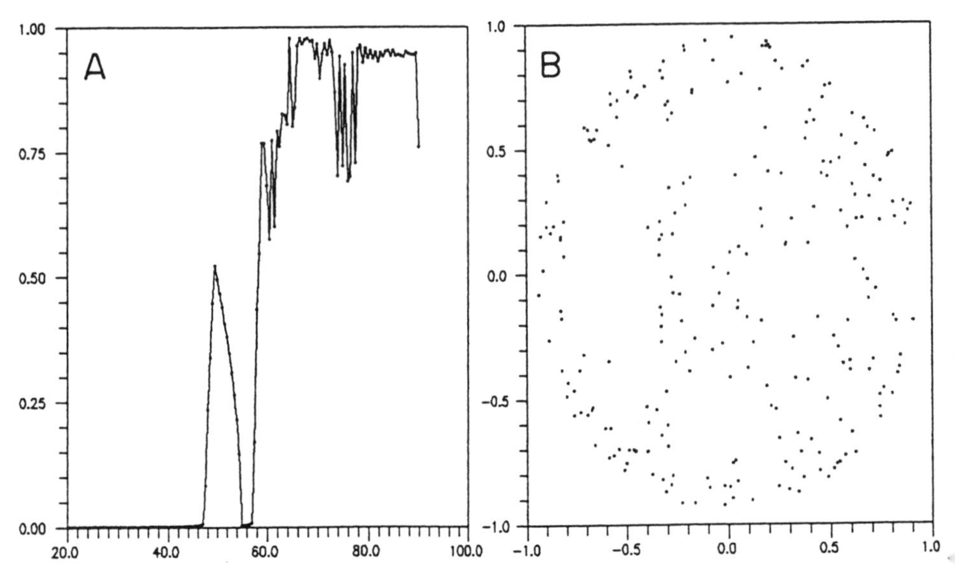

Figure 1. **A**: x-axis: inclination of the orbital plane of Venus with respect to its equator. y-axis: maximum eccentricity of the satellite attained in an integration over 5000 years. **B**: x-axis=$e\cos g$. y-axis=$e\sin g$. Surface section for a satellite of Venus. All the points are due to one single orbit. See the text.

Usually when the semimajor axis a is much smaller than a_{crit}, these orbits are very stable and no significant variation of the eccentricity occurs. However when a approaches a_{crit} and ϵ is sufficiently large, eccentricity can increase up to critical values. In the Figure 1A, we show the maximum value of e attained during an integration over 5000 years for each ϵ in the range 20-90 degrees. All the orbits started with initial $e = 0.001$, on the equatorial plane. The semimajor axis adopted was $a = 6.2a_p$, which corresponds to a_{crit} when $J_2 = 0.321E-3$. In this Figure when ϵ is greater than 60^0, the eccentricity can easily goes to prohibitive values that may cause collision with Venus, or with some internal orbits. This occurs in a few thousand of years while the time scale for the variation of ϵ is of the order of myears. To see more details about the trajectories when a is near a_{crit}, let's show som! e surface section curves. For a sa tellite initially placed on the equator, with $a = 6.2a_p$, $e = 0.001$, and $\epsilon = 80^0$, the coordinates $x = e\cos g$, $y = e\sin g$ are plotted at each $h = \pi$ (Figure 1B). All the points are due to one single orbit and we can see that the motion is completely chaotic. Even for $a = 10a_p$ and $\epsilon \geq 55^0$, we have still found strong irregular motion , which shows that the chaotic region is not so small.

Mars is an interesting planet since its obliquity is still in chaotic motion, and it is varying from 0^0 to 60^0 (Laskar, Touma and Wisdom)[5,7]. Therefore, taking the present parameters of Mars and fixing $a = 13.2a_p$, we obtain figures very similar to Figure 1A and 1B. As before, when ϵ approaches 60^0, in a very short time the eccentricity increase to high values and averaged equations show chaotic motion.

Up to now we have taken only values of semimajor axis in the neighbourhood of a_{crit}. For $a \ll a_{crit}$ (inner satellites) or $a \gg a_{crit}$ (outer satellites) many other numerical experiments have confirmed the results obtained by (Kinoshita and Nakai)[4] in the case of Uranus' system: for outer satellites (always placed on the equator of the planet, in almost circular orbit) if ϵ is high, then in a very short time, the eccentricity may increase to critical values, whereas for some very inner satellites the eccentricity always remains small, no matter the value of ϵ. In this sense, Phobos and Deimos (inner satellites of Mars) are completely safe, in spite of the chaotic evolution of the obliquity of Mars. However, even in this case ($a \ll a_{crit}$), depending on the inclination of the satellite, some interesting situations exist due to some commensurabilities between the argument of pericenter and the node of the satellite. Therefore some v! ery big increase of eccentricities may still occur. The possibilities of several resonances are clear when the Hamiltonian is written using the equator as the reference plane (section 2). It is easy to see that these resonance conditions depend mostly on the inclination i of the satellite with respect to the equator. On the other hand, some particular values of ϵ may increase drastically the effect of the solar perturbation.

4. Conclusion

Depending on the obliquity a strong chaotic motion can appear in the neighbourhood of a_{crit}. Since the value of this critical semimajor depends on J_2 (which in its turn depends on the planet's spin), in the past this chaotic motion might have been important in the dynamics of possible satellites of the inner planets.

As we said before, here we are presenting only some brief results and other resonance relations (e.g.inclusion of the secular variation of the elements of the planet), should be reported elsewhere.

5. Acknowledgements

We thank Research Foundation of São Paulo (FAPESP)for very important financial support and also all the members of Group of Solar System of the Astronomy Department of São Paulo University, due to several and stimulating discussions.

References

1. Burns J.A.(1973) Where are the satellites of the inner planets ? *Nature Physical Science*, **242**, 23-25.
2. Goldreich P.(1966) History of the Lunar orbit, *Rev. of Geophysics*, **4**, 411-434.
3. Jeffreys H.(1959) *The Earth*, Cambridge Univ.Press, London.
4. Kinoshita H. and Nakai H. (1991) Secular perturbations of fictitious satellites of Uranus, *Cel. Mech. and Dyn. Astron.*, **52**, 293-303.
5. Laskar J. and Robutel P.(1993) The chaotic obliquity of the planets, *Nature*, **361**, 608-612.
6. Laskar J. et al (1993) Stabilization of Earth's obliquity by the Moon, *Nature*, **361**, 615-617.
7. Touma J. and Wisdom J.(1993) The chaotic obliquity of Mars, *Science*, **259**, 1294-1297.
8. Ward W. et al (1979) Past obliquity of Mars: The role of the Tharsis uplift, *Journal of Geoph. Research*, **84**, 243-259.

APPLICATIONS OF A HIGH ORDER SECULAR PERTURBATION THEORY

The Uranian satellites and beyond

A. A. CHRISTOU AND C. D. MURRAY
Astronomy Unit
Queen Mary and Westfield College
London E1 4NS
United Kingdom

Abstract. We have employed computer algebra to derive explicit expressions for the secular planetary theory of the second order in the masses described in Duriez (1979) and Laskar (1985). In applying this theory to the problem of the long-term motion of the Uranian satellites, we have achieved a marked improvement over the result of the analytical treatment by Malhotra *et al.* (1989). Other possible applications of this work are also discussed.

1. Introduction

In stark contrast to the satellite systems of the other gas giants, the classical Uranian satellite system is free of dynamical intricacies such as mean motion resonances and unusually large eccentricities or inclinations, innermost Miranda being a possible exception. For these reasons, secular theory may be invoked to account for and even quantify the dominant variations in their long period motion.

Originally, the application of first order Laplace-Lagrange theory by Dermott & Nicholson (1986) showed that long-term gravitational interactions, especially between Titania and Oberon, are significant compared to the effects of planetary oblateness and should not be omitted from any theory that attempts to model their motion.

After the Voyager 2 Uranus encounter in January 1986, Laskar & Jacobson (1987) attempted to fit the theory of Laskar (1986) into spacecraft- and Earth-based observations. They eventually succeeded in deriving values for the masses comparable to the Voyager radio and optical tracking estimates

B.A. Steves and A.E. Roy (eds.), The Dynamics of Small Bodies in the Solar System, 289–294.

made by Tyler *et al.* (1986). However, it was found that the mere proximity of the system to mean motion resonances was sufficient to alter its fundamental precession eigenfrequencies by a maximum of 15%. For that reason, the theory that was used to fit the observations contained empirical correction factors derived from numerical testing.

Malhotra *et al.* (1989) (hereafter referred to as MFMN) were able to treat these discrepancies analytically by extending the secular theory used to second order in the satellites' masses. Thus it became apparent that the main cause for the poor agreement between their numerical and analytical results could be traced to the effects of the 2:1 Umbriel-Titania and 3:2 Titania-Oberon near mean motion resonances (NMMRs). However, there remained significant residuals ($\sim 2\%$) that could probably be explained by the effect of other, weaker near-resonances.

The question that our work has tried to answer is how close to the actual dynamics can one bring the analytical solution for the Uranian satellites by including *all* significant NMMRs? This investigation might also be of interest in light of the recent attempt by Taylor (1998) to fit a large number of observations spanning almost 200 years into a numerical model of the satellites' motion.

2. Method and results

In our effort to model the secular effect of near-resonances we have adopted the theory described by Duriez (1979) and Laskar (1985). The use of the *Mathematica* computer algebra package (Wolfram, 1991) has allowed us to present the output in a form which is explicit in terms of the system parameters. It can thus be readily employed for the study of *individual* near-resonances. For the reader who is unfamiliar with the theoretical background we present the principal results below. The details can be found in the references already cited and the work by Christou & Murray (1997).

We use the set of variables $\{p_i, q_i, z_i, \bar{z}_i, \zeta_i, \bar{\zeta}_i\}_{i=1,2,\cdots,n}$ defined as

$$z_i = e_i \exp \mathrm{i}\varpi_i \ , \quad \zeta_i = \sin(I_i/2) \exp \mathrm{i}\Omega_i, \tag{1}$$

$$a_i = A_i(1+p_i)^{-2/3} \ , \quad \lambda_i = N_i t - \mathrm{i} q_i \tag{2}$$

where $\{a_i, e_i, I_i, \varpi_i, \Omega_i, \lambda_i\}_{i=1,2,\cdots,n}$ denote the semimajor axis, eccentricity, inclination, longitude of pericentre, longitude of ascending node and mean longitude of the ith satellite, bars denote the complex conjugates and $\mathrm{i}^2 = -1$. The symbols A and N are used to denote the *proper* semimajor axis and mean motion of each body. These are considered to be constants of the motion.

Lagrange's planetary equations may be written in the compact form:

$$\frac{\mathrm{d}\mathbf{V}}{\mathrm{d}t} = \Lambda(\mathbf{V}, t) \tag{3}$$

where $\mathbf{V}$ denotes our variable vector. We are searching for a near-identity variable transformation $\mathbf{V} \rightarrow \mathbf{V_o}$ of the form

$$\mathbf{V} = \mathbf{V_o} + \Delta\mathbf{V}(\mathbf{V_o}, t) \tag{4}$$

such that $\mathbf{V_o}$ satisfies the second order, longitude-averaged equation

$$\frac{\mathrm{d}\mathbf{V_o}}{\mathrm{d}t} = \langle\Lambda(\mathbf{V_o}, t)\rangle + \left\langle \frac{\partial\Lambda}{\partial\mathbf{V_o}}(\mathbf{V_o}, t)\Delta\mathbf{V}(\mathbf{V_o}, t)\right\rangle \tag{5}$$

where the quantities contained in $\langle\ \rangle$ have been averaged over, and are hence independent of, the longitudes. It is subsequently found that an expression for $\Delta\mathbf{V}(\mathbf{V_o}, t)$ which uniquely defines the transformation is

$$\frac{\partial\Delta\mathbf{V}}{\partial t}(\mathbf{V_o}, t) = \{\Lambda(\mathbf{V_o}, t)\} \tag{6}$$

where we use $\{\ \}$ to denote those terms in $\Lambda(\mathbf{V_o}, t)$ which are longitude-dependent. By employing our knowledge of the d'Alembert rules for each of the six planetary equations we can generate the coefficient of any term present in the second order part of Eq. (5) as a function of the masses and the semimajor axes of the associated satellites.

In the case of the Uranian satellite system, we shall only consider the linear terms up to the second order in the masses. Since the eccentricities and inclinations of the satellites are small, the contribution of nonlinear terms is negligible even though relatively large values of the semimajor axes ratio (up to 0.75) are found in the system.

To order 1 in the masses, the resulting equations can be written as

$$\dot{\mathbf{z}} = \mathrm{i}C\mathbf{z}\ , \quad \dot{\boldsymbol{\zeta}} = \mathrm{i}D\boldsymbol{\zeta} \tag{7}$$

where C and D are matrices which depend only on the semimajor axes and masses of the five satellites. Note that in the linear regime the eccentricity and the inclination problems are separated.

Here we concentrate on the eccentricity solution for Eqs. (7). This is of the general form

$$z_i(t) = \sum_{j=1}^{n} C_j^{(i)} \exp\left[\mathrm{i}(g_j t + \beta_j)\right] \tag{8}$$

where C_j is the real scaled amplitude vector corresponding to eigenfrequency g_j and phase β_j. The above expressions are commonly referred to as the Laplace-Lagrange solution.

Our representation of the second order contribution can be incorporated into the linear problem in the form of a correction matrix δC. This matrix will simply be added to the Laplace-Lagrange matrix C and this new problem will be solved in the same manner as before.

TABLE 1. Initial configuration of the Uranian satellite system used in this work

Satellite	m/M_p ($\times 10^5$)	a (km)	e	I (deg.)
Miranda	0.1	129742	0.002564	4.2805
Ariel	1.8	190798	0.003330	0.3105
Umbriel	1.1	265824	0.005342	0.3602
Titania	3.2	436030	0.001347	0.1421
Oberon	3.4	583078	0.001331	0.1009

Physical parameters: $R_p = 26200$ km, $J_2 = 0.003345$, $J_4 = -0.0000321$, $GM_p = 5.784184 \times 10^6$ km^{3}sec^{-2}

Special care should be taken when choosing the set of initial conditions to be used in such a theory since osculating elements have to be deconvolved from their short-periodic component before they can be used in a secular perturbation model. For this reason we have used the averaged orbital element set (Table 1) which was produced by MFMN by means of digital filtering of a numerical simulation. It has also been necessary to integrate numerically the equations of motion of the five principal Uranian satellites using the same *osculating* elements as MFMN with the SWIFT package written by Levison & Duncan (1994) in order to fit values for their proper mean motions and semimajor axes which were not used in that

TABLE 2. Comparison of eigenfrequency values derived by numerical and analytical methods

Mode number	MFMN (Num.)	L-L Theory	MFMN (Theor.)	This work
1	20.299	20.283	20.289	20.291
2	6.000	5.961	5.965	5.995
3	2.909	2.855	2.874	2.906
4	1.924	1.608	1.874	1.933
5	0.367	0.352	0.367	0.367

paper. Finally, the oblateness-induced component of the orbital precession was incorporated into the diagonal elements of matrix C. An added complication was due to the fact that the values of the coefficients present in these expressions depend on using either the *instantaneous osculating* mean motions or the *geometric* mean motions (see Greenberg, 1981).

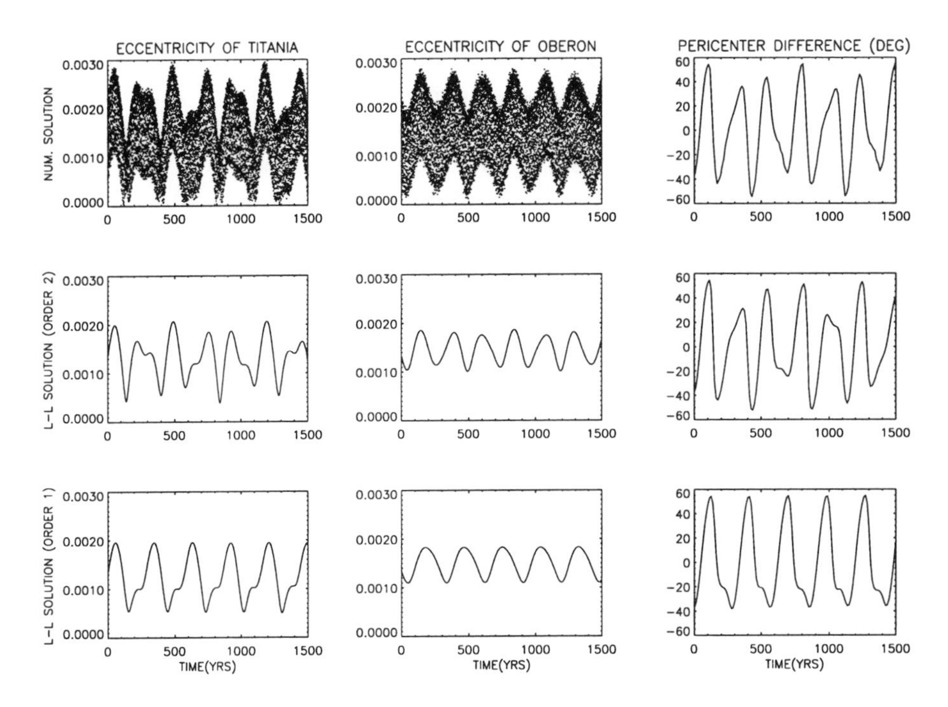

Figure 1. Comparison of results for Titania and Oberon. The upper right plot has been generated by averaging the numerical output every 12.5 years.

It is evident from the results presented in Table 2 that a marked improvement over MFMN has been achieved at least in terms of the maximum discrepancy between the theoretically and numerically computed eigenfrequencies. Figure 1 reveals that the actual numerical solution is closely followed by our order 2 model whereas the Laplace-Lagrange solution is not able to reproduce basic features present in the numerical data like, for example, the number of periodic oscillations of the eccentricity over the given timespan. Note that, as originally suggested by Dermott & Nicholson (1986), the pericentres of the outer two satellites might actually be locked depending on their specific physical and orbital properties.

The relatively large masses of Titania and Oberon can probably account for the large number of arguments ($\sim$ 20) that has to be included in the model in order to achieve the required precision. Fortunately, for sufficently high order terms this can be done by using a geometric series approximation.

3. Future directions

The relative simplicity of the Uranian satellite system has provided us with an opportunity to gauge the extent to which it can be modelled using a linear, second order, secular perturbation theory.

Although the satellites of the other gas giants do not share this system's unique characteristics, it would seem profitable to apply this theory to the study of other (extrasolar) planetary systems where, at present, the available observations provide poor, if any, constraints on the existence of "typical" configurations. Dynamical studies of such systems are at such an early stage that even the study of relatively simple configurations are likely to provide us with useful insight on what to expect from future observational results.

References

Christou, A.A., Murray, C.D. (1997), *Astron. Astrophys.* **327**, 416
Dermott, S.F., Nicholson, P.D. (1986), *Nature* **319**, 115
Duriez, L. (1979), *Première Thèse*, Lille
Greenberg, R. (1981), *Astron. J.* **86**, 912
Laskar, J. (1985), *Astron. Astrophys.* **144**, 133
Laskar, J. (1986), *Astron. Astrophys.* **166**, 349
Laskar, J., Jacobson, R.A. (1987), *Astron. Astrophys.* **188**, 212
Levison, H. F., Duncan, M. J. (1994), *Icarus* **108**, 18
Malhotra, R., Fox, K., Murray, C.D., Nicholson, P.D. (1989), *Astron. Astrophys.* **221**, 348
Taylor, D.B. (1998), *Astron. Astrophys.* **330**, 362
Tyler, G.L., Sweetnam, D.N., Anderson, J.D., Campbell, J.K., Eshleman, V.R., Hinson, D.P., Levy, G.S., Lindal, G.F., Marouf, E.A., Simpson, R.A. (1986), *Science* **233**, 79
Wolfram, S. (1991), *Mathematica*, Addison-Wesley, Second Edition

NON–INTEGRABILITY OF THE MOTION OF A POINT MASS AROUND A PLANET OF ARBITRARY SHAPE

M.E. SANSATURIO
E.T.S. de Ingenieros Industriales,
Paseo del Cauce s/n, E-47011 Valladolid, Spain
e-mail: marsan@wmatem.eis.uva.es

AND

I. VIGO-AGUIAR AND J.M. FERRÁNDIZ
Dept. de Análisis Matemático y Matemática Aplicada,
Escuela Politécnica Superior E-03080 Alicante, Spain

Abstract. In this paper, several results on the non–integrability of the problem of a satellite under different assumptions on the mass distribution of the primary are collected. Among them, we pay special attention to the non–integrability through meromorphic integrals of any truncation of the zonal satellite problem, as well as to that of the $(J_2 + J_{22})$–problem or 'general main satellite problem'.

1. Introduction

More than one century ago, the theory of orbits went through an important qualitative change when Bruns and then Poincaré (see [10]) proved by means of several methods and hypotheses that the three body problem has no more first integrals apart from the ten classical ones. Since then, great progress has been made in the knowledge of the behaviour of such non–integrable systems (including chaos), as well as in the methods which allow us to investigate the non–integrability. A panorama of the evolution of the ideas and results concerning the study of problems of several bodies can be found in the usual treatises – for instance, see [9] or [5].

However, questions about non–integrability of problems of satellites of external bodies remained open till this decade. Irigoyen and Simó [3] published the first result along this line, proving that the main problem of the

B.A. Steves and A.E. Roy (eds.), The Dynamics of Small Bodies in the Solar System, 295–302.

satellite of an oblate primary, or J_2–problem, was not completely integrable through meromorphic integrals. The proof was based on some theorems by Ziglin [14] and Yoshida [11], which will be commented on briefly in section 2. Later, an analogous method allowed Ferrándiz and Sansaturio [1] to establish that the J_{22}–problem did not have a complete system of first integrals. Such integrals were supposed to be rational since, in three dimensions, the non–integrability according to Ziglin depends on a condition of rational independence of the eigenvalues of certain matrices which does not allow us to use the continuity arguments as happens in dimension 2.

In [2], the present authors set up the non–integrability through meromorphic integrals of the truncation of the zonal satellite problem at any order. Then in [6], we solved the case in which the perturbation was an arbitrary sectorial harmonic and established that the main satellite problem, or $(J_2 + J_{22})$–problem, is not completely integrable either.

In this paper we gather together the above mentioned results, either with much briefer proofs or by generalizing them.

2. Brief Summary of Yoshida's Theorem

Very broadly speaking, Ziglin's method [14] takes into consideration an analytic Hamiltonian for which a particular solution describing a curve non–homologous to zero on a certain Riemann surface is selected. Then, the non–integrability depends on the properties of the monodromy group associated to the variational equations along such a solution, and particularly on certain non–resonance conditions.

In the case of a Hamiltonian

$$H = \frac{1}{2}\mathbf{p}^2 + V(\mathbf{q}),$$

where $V(\mathbf{q})$ is homogeneous of an integer degree $k \neq 0, \pm 2$, Yoshida [11] obtained very useful criteria, based on Ziglin's results, by making use of the existence of a straight–line solution of the form

$$\mathbf{q} = \mathbf{c}\phi(t), \qquad \mathbf{p} = \mathbf{c}\dot{\phi}(t),$$

where $\phi(t)$ is a solution of $\ddot{\phi}(t) + \phi^{k-1} = 0$ and the constant vector $\mathbf{c}$ is a solution to the algebraic equation $\mathbf{c} = V_{\mathbf{q}}(\mathbf{c})$.

In a standard way, the variational equations can be transformed into a Gauss hypergeometric equation, with well known monodromy group and finally, the non–integrability depends on the eigenvalues of $V_{\mathbf{qq}}(\mathbf{c})$, one of them always being equal to $k-1$.

In the planar case, $n = 2$, the problem turns out to be non–integrable if a coefficient referred to as the integrability coefficient, $\lambda = \mathrm{Tr}V_{\mathbf{qq}}(\mathbf{c}) - (k-1)$, lies in certain non–integrability regions S_k ([11]; p. 128, eq. (1.3)), which are derived from the fact that the trace of some monodromy matrices is > 2. For $n > 2$, as shown in [13], non–integrability follows if the parameters

$$\Delta\rho_i = \sqrt{1 + \frac{8k\lambda_i}{(k-2)^2}}$$

are rationally independent, where the λ_i are the eigenvalues of $V_{\mathbf{qq}}(\mathbf{c})$.

The reformulation of these criteria into polar or spherical coordinates presented in [6] has turned out to be useful in dealing with orbital problems, since it allows us to simplify the calculations, sometimes very drastically. In the planar case, the integrability coefficient in polar coordinates is easily computed through

$$\lambda_p = \frac{W_{\theta\theta}(\theta_0)}{kW(\theta_0)},$$

where $V(r,\theta) = r^k W(\theta)$ is the potential function, k being an integer but $k \neq 0, \pm 2$, and θ_0 is a solution to $W_\theta(\theta) = 0$ such that $W(\theta_0) \neq 0$. This coefficient relates to Yoshida's integrability coefficient through the equation $\lambda_p = \lambda - 1$.

3. Non–integrability of the Truncated Zonal Satellite Problem

In this section we present some results about this problem recently obtained by the authors [2], although the proof has been considerably shortened.

By using cylindrical coordinates $(\rho, \varphi, z; p_\rho, p_\varphi, p_z)$, the homogeneous Hamiltonian for this problem can be written as

$$H = \frac{1}{2}\left(p_\rho^2 + p_z^2\right) + \frac{p_\varphi^2}{2\rho^2} - \frac{\mu}{r} + \sum_{k=2}^{n} \frac{V_k}{r^{k+1}} + p_0\,, \tag{1}$$

where $r = \sqrt{\rho^2 + z^2}$, $V_k = \varepsilon_k P_k(z/r)$, ε_k are constant and $P_k(x)$ is the Legendre polynomial of order k.

From the fact that p_φ is a first integral of (1) it follows that, if this Hamiltonian admits a meromorphic integral independent from the other two and in involution with them, such an integral is analytic with respect to p_φ. Consequently, the integrability in the Liouville sense of (1) implies that of the Hamiltonian

$$H = \frac{1}{2}\left(p_\rho^2 + p_z^2\right) - \frac{\mu}{r} + \sum_{k=2}^{n} \frac{V_k}{r^{k+1}} + p_0\,, \tag{2}$$

which is obtained by making $p_\varphi = 0$ in (1). It is immediate that (2) admits a particular solution with $\rho = p_\rho = 0$. However, since an explicit expression for the associate monodromy matrices which are required by Ziglin's method is not known, it is convenient to introduce a suitable limit problem.

The change of scale defined by

$$t = \beta \bar{t}\,, \qquad \rho = \beta^{\frac{2}{n+3}}\,\bar{\rho}\,, \qquad z = \beta^{\frac{2}{n+3}}\,\bar{z}\,,$$
$$p_0 = \beta^{-\frac{2(n+1)}{n+3}}\,\bar{p}_0\,, \qquad p_\rho = \beta^{-\frac{n+1}{n+3}}\,\bar{p}_\rho\,, \qquad p_z = \beta^{-\frac{n+1}{n+3}}\,\bar{p}_z\,,$$

gives a new homogeneous Hamiltonian, which after multiplying by $\beta^{\frac{2(n+1)}{n+3}}$ and taking limits as $\beta \to 0$, reduces to

$$K = \frac{1}{2}\left(p_\rho^2 + p_z^2\right) + V + p_0\,, \quad \text{where} \quad V = \frac{\varepsilon_n}{r^{n+1}} P_n\left(\frac{z}{r}\right) \tag{3}$$

and the bar in the new variables has been removed to simplify the notation.

The Hamiltonian (3) is of the form required by Yoshida's theorem [11], since the potential V is homogeneous of degree $k = -n-1$. As is well known, the Hessian of V computed on the vector $\mathbf{c}$ quoted in section 2 always has the eigenvalue $k-1$, which in our case turns out to be $-n-2$. It is easy to establish that this is precisely the value of $\partial^2 V/\partial z^2$, so that the integrability coefficient reduces to $\partial^2 V/\partial \rho^2$. Moreover, due to the symmetry with respect to the OZ axis, it is obvious that

$$\frac{\partial^2 V}{\partial \rho^2} = \frac{\partial^2 V}{\partial x^2} = \frac{\partial^2 V}{\partial y^2} \quad \text{in} \quad \rho = x = y = 0\,,\ z = c\,.$$

Since V verifies Laplace equation, we get

$$\left.\frac{\partial^2 V}{\partial \rho^2}\right|_{\substack{\rho = 0 \\ z = c}} = -\frac{1}{2}\left.\frac{\partial^2 V}{\partial z^2}\right|_{\substack{\rho = 0 \\ z = c}} = 1 + \frac{n}{2}\,,$$

a value which lies in the non–integrability region S_{-n-1} and coincides with that previously obtained by the authors in [2] but with rather more tedious calculations.

Once we have established that the limit problem (3) does not have a complete system of meromorphic integrals, it follows that neither does the problem (2) in a neighbourhood of $\beta = 0$.

Now, it is enough to apply the usual continuity argument – see [12] or [3] –, since the non–integrability regions correspond to those for which the trace of the monodromy matrices is > 2.

However, notice that such an argument is more complicated than simply checking the aforementioned inequality. In fact, it is necessary to verify that the Riemann surface, the poles and the loops which are used to calculate the monodromy matrices do not suffer significant qualitative changes for any problem close to the limit problem, i.e., in a neighbourhood of $\beta = 0$

after performing the change of scale. Nevertheless, this can be verified in an immediate way.

To this end it is enough to realize that the basic solution can be considered of the form $\rho = 0$, $z = \psi(t)$, where ψ is a solution of the differential equation

$$\ddot{\psi} + \frac{\partial V}{\partial z}(0, \psi) = 0\,,$$

which admits as a first integral

$$\frac{1}{2}(\dot{\psi})^2 + V(0, \psi) = h\,,$$

and that the Riemann surface involved in Ziglin's theorem is defined by the integral

$$t = \int \frac{dw}{\sqrt{p(w)}}\,, \qquad \text{with } p(w) = 2(h - V(0, w))\,.$$

The monodromy group of the normal variation equation is then built by taking closed circuits around the branch points corresponding to the roots of $p(w) = 0$.

In the limit problem, we can take $h = \varepsilon_n\, P_n(\psi/\sqrt{\psi^2})$ and $n+1$ simple poles, roots of $w^{n+1} = 1$, appear. If the potential is a finite sum of zonal harmonics it is obvious that the solution $\rho = 0$ still exists and the equation $w^{n+1} = 1$ would be replaced by another one with the same leading and independent terms and with lower powers of w multiplied by factors which are positive powers of $\beta^{\frac{2}{n+3}}$ and hence it goes on having $n+1$ simple roots provided β is small enough.

Notice that this result includes as a particular case the non–integrability through meromorphic integrals of the truncated two fixed centres problem in the asymmetric case. The non–integrability of such a problem in the symmetric case was first established by Irigoyen [4] by following a scheme similar to that used for the J_2–problem in [3]. We would like to point out that the starting solution used in Irigoyen's work is not suitable for the study of odd harmonics.

On the other hand, although in (1) only the harmonics with negative powers have been included, it is easy to realize that if we add up a finite number of zonal harmonics of the form $r^k V_k$ to (1), the conclusion still holds. Such terms could be envisaged as coming from the truncation of the development of a perturbation due to a third body by simply assuming some suitable conditions (e.g. the Keplerian orbit is circular and equatorial).

4. The General Main Satellite Problem

In this section we slightly generalize some results previously obtained by the authors [6] in the sense that we succeed in proving that the (J_2+J_{22})–problem is not completely integrable through meromorphic integrals.

Assuming that the primary rotates with constant angular velocity ω around the Oz axis and using Cartesian coordinates in the rotating system $Oxyz$ attached to the primary, the Hamiltonian is of the form

$$H=\frac{1}{2}(p_x^2+p_y^2+p_z^2)-\omega(xp_y-yp_x)-\frac{\mu}{r}+V_2+V_{22}\,, \tag{4}$$

with

$$V_2=\frac{\varepsilon_2}{r^3}P_2\left(\frac{z}{r}\right), \qquad V_{22}=-\frac{\varepsilon_{22}}{r^5}(x^2-y^2)$$

and ε_2, ε_{22} are constants which are proportional to the small parameters J_2, J_{22}, respectively. Notice that whenever the moments of inertia verify $A<B<C$, as in the case of the Earth, ε_2 and ε_{22} are positive.

Since any solution with initial conditions $z=p_z=0$ always remains in the plane $z=0$, by making $z=p_z=0$ in (4), we obtain a new Hamiltonian H_0 whose flow is included in that of (4).

In polar coordinates such a Hamiltonian H_0 is expressed as

$$H_0=\frac{1}{2}\left(p_r^2+\frac{p_\theta^2}{r^2}\right)-\omega p_\theta-\frac{\mu}{r}-\frac{\varepsilon_2}{2r^3}-\frac{\varepsilon_{22}}{r^3}\cos 2\theta\,. \tag{5}$$

It is immediate that if (4) has a first integral which is analytic in $z=0$, $p_z=0$ and is independent from H, it will give rise to another first integral which is independent from H_0 for the subproblem (5). On the other hand, if this integral was analytic in ω, at least in $\omega=0$, it would also provide an integral of the auxiliary problem

$$K_0=\frac{1}{2}\left(p_r^2+\frac{p_\theta^2}{r^2}\right)-\frac{\mu}{r}-\frac{\varepsilon_2}{2r^3}-\frac{\varepsilon_{22}}{r^3}\cos 2\theta\,, \tag{6}$$

corresponding to a non–rotating primary.

The problem (6) can be treated in a very similar way to that of the zonal harmonics. A suitable change of scale allows us to go to a limit problem, in which the Keplerian term disappears, that is

$$K_1=\frac{1}{2}\left(p_r^2+\frac{p_\theta^2}{r^2}\right)+\frac{1}{r^3}W(\theta)\,, \quad \text{with} \quad W(\theta)=-\frac{\varepsilon_2}{2}-\varepsilon_{22}\cos 2\theta\,. \tag{7}$$

The integrability coefficient in polar coordinates is extremely easy to calculate. According to what is indicated in section 2, it is enough to consider the solution $\theta_0 = 0$ to $W_\theta(\theta) = 0$. Therefore,

$$\lambda_p = \frac{W_{\theta\theta}(0)}{(-3)W(0)} = \frac{4\varepsilon_{22}}{3(\varepsilon_2/2 + \varepsilon_{22})} > 0\,.$$

This coefficient lies in one of the non–integrability regions given in [6], so that (7) does not have any additional meromorphic integral.

The usual continuity argument proves that (6) also verifies this property. This allows us to conclude that the original problem (4) does not have any global first integral which is meromorphic and also analytic in $w = 0$, and in $z = p_z = 0$. This result generalizes those previously obtained by the authors ([2], [6]), where only the non–existence of additional rational integrals was proved, either for the case $\varepsilon_2 = 0$ or for the general case.

5. Final comments

There are several questions of practical interest which could arise in connection with the non–integrability, such as the size of the chaotic zones (as it was established by Simó [8] in the J_2–problem) and, in general, its relevance from a practical point of view. Most of these questions are still open and the answers do not seem to be trivial. Some considerations about the physical meaning of the orbits required by Ziglin's method, as well as some numerical simulations can be seen in [7], where it is shown the appearance of chaos (hard or soft) for moderate values of J_{22} in the neighbourhood of the 1:1 resonant orbit.

6. Acknowledgements

This work has been partially supported by the Spanish CICYT Grant number ESP97-1816-C04-02. Helpful comments by an unknown referee are also greatly acknowledged.

References

1. Ferrándiz, J.M. and Sansaturio, M.E. (1995) Non–existence of rational integrals in the J_{22}–problem, *Phys. Lett.* A, **207**, 180–84.
2. Ferrándiz, J.M., Sansaturio, M.E. and Vigo–Aguiar, I. (1996) Non integrability of the truncated zonal satellite Hamiltonian at any order, *Phys. Lett.* A, **221**, 153–57.
3. Irigoyen, M. and Simó, C. (1993) Non integrability of the J_2–problem, *Celest. Mech.*, **55**, 281–87.
4. Irigoyen, M. (1996) Analytical non–integrability of the truncated two fixed centres problem in the symmetric case, *J. of Diff. Eq.*, **131**, 267–76.
5. Marchal, C. (1990) *The Three Body Problem*, Elsevier, Amsterdam.

6. Sansaturio, M. E., Vigo–Aguiar I. and Ferrándiz J.M. (1997) Non–integrability of some Hamiltonian systems in polar coordinates, *J. Phys. A: Math. Gen.*, **30**, 5869–5876.
7. Sansaturio, M. E., Vigo–Aguiar I. and Ferrándiz J.M. (in press) Non–integrability and chaos in satellite orbital dynamics, Paper AAS98-131, Advances in the Astronautical Sciences Vol. 99 I & II, Robert H. Jacobs Series Editor.
8. Simó, C. (1991) Measuring the lack of integrability of the J_2-problem for Earth's satellites, in *Predictability, Stability and Chaos in N-Body Dynamical Systems*, A.E. Roy, Ed., NATO ASI series B272, Plenum Publishing Corporation, 305–309.
9. Szebehely, V.G, (1967) *Theory of Orbits. The Restricted Problem of Three Bodies*, Academic Press, New York.
10. Whittaker, E.T. (1904) *A Treatise on the Analytical Dynamics of Particles and Rigid Bodies*, Reprinted by Cambridge University Press, New York (1965).
11. Yoshida, H. (1987) A criterion for the non–existence of an additional integral in Hamiltonian systems with a homogeneous potential, *Physica* D, **29**, 128–42.
12. Yoshida, H. (1988) Non integrability of the truncated Toda lattice Hamiltonian at any order, *Commun. Math. Phys.*, **116**, 529–38.
13. Yoshida, H. (1989) A criterion for the non–existence of an additional analytic in Hamiltonian systems with n degrees of freedom, *Phys. Lett.* A, **141**, 108–12.
14. Ziglin, S.L. (1983) Branching of solutions ad non–existence of first integrals in Hamiltonian mechanics, *Funct. Anal. Appl.*, **16**, 181–99.

DYNAMICS OF SATELLITES WITH MULTI-DAY PERIODS

B. ÉRDI
Department of Astronomy
Eötvös University, Budapest, Hungary

Abstract. Lunar perturbations of the Earth's artificial satellites with large semi-major axis are investigated. Long-term evolution of the orbits are studied through double averaging of the perturbing function. A first order secular solution is derived from Lagrange's planetary equations. Types of perturbations are discussed and secular resonances between the perturbed orbits of the satellites and the Moon are shown.

1. Introduction

For the investigation of the Earth's magnetosphere and the interplanetary space outside of it, artificial satellites with orbits of large semi-major axis and large eccentricity are used. An example is the recently launched Interball satellite (semi-major axis a=102 387 km, eccentricity e=0.92, period T=3.8 days, epoch: August 3, 1995). Since the orbits of these satellites are very elongated, the Moon and the Sun can perturb them significantly. For the long time evolution of the orbits it is important to know the effects of luni-solar perturbations. For example, due to luni-solar perturbations, the perigee height can decrease to a level where air drag might become a decisive factor on the dynamics of these satellites.

Luni-solar perturbations can be treated in several ways (see for example [1], [2], [3], [4], [5], [6]), analytical approaches, based on series expansions, methods of averaging, and numerical techniques are used. Szebehely and Zare [7] demonstrated that important indirect perturbations of several years period occur in the eccentricity and inclination of the satellite orbits with multi-day periods, originating from the perturbed motion of the Moon. It is the purpose of this paper to make an analytical approach to this problem in order to derive approximate analytical expressions through double aver-

B.A. Steves and A.E. Roy (eds.), The Dynamics of Small Bodies in the Solar System, 303–307.

aging, from which the indirect perturbations of the Moon on the satellites can be determined.

2. Averaging of the Lunar Perturbing Function

The perturbations of the orbital elements of an Earth's satellite coming from the perturbing effects of the Moon can be determined from Lagrange's planetary equations. The lunar perturbing function can be expressed as

$$R = \frac{Gm'}{r'} \sum_{l \geq 2} \left(\frac{r}{r'}\right)^l P_l(\cos\psi), \tag{1}$$

where G is the gravitational constant, m' is the mass of the Moon, r and r' are the geocentric distances of the satellite and the Moon, P_l are Legendre polynomials, and ψ is the geocentric elongation of the satellite from the Moon.

When considering the motion of a satellite, it is natural to refer its orbit to an equatorial frame. However, as noted by Kozai [8], lunar perturbations are more conveniently obtained if the coordinates of the Moon are referred to the ecliptic instead of the equator. In the ecliptic frame the Moon's orbital inclination is nearly constant and the longitude of its ascending node can be well approximated by a linear function of time. Under these assumptions, the development of the lunar perturbing function according to the orbital elements is given in [3] and [5].

To study the long time variations of the satellite orbit, the short periodic terms of the perturbing function, depending on the mean anomalies (through the true anomalies) of both the satellite and the Moon can be averaged out. If there is no resonance between the satellite and the Moon, the elimination of the short periodic terms can be performed by making use of Hansen's expansion ([9]). Performing the double averaging on the development of the lunar perturbing function given in [5], we obtain:

$$\begin{aligned} \bar{R} &= Gm' \sum_{l \geq 2} \sum_{m=0}^{l} \sum_{s=0}^{l} \sum_{p=0}^{l} \sum_{q=0}^{l} (-1)^{k_1} \frac{\varepsilon_m \varepsilon_s (l-s)!}{2a'(l+m)!} \left(\frac{a}{a'}\right)^l \\ & \quad \times F_{lmp}(i) F_{lsq}(i') X_0^{l,l-2p}(e) X_0^{-(l+1),l-2q}(e') \\ & \quad \times \left\{ (-1)^{k_2} U_l^{m,-s} \cos(\bar{\theta}_{lmp} + \bar{\theta}'_{lsq} - y_s \pi) \right. \\ & \qquad \left. + (-1)^{k_3} U_l^{m,s} \cos(\bar{\theta}_{lmp} - \bar{\theta}'_{lsq} - y_s \pi), \right\}, \end{aligned} \tag{2}$$

where

$$\bar{\theta}_{lmp} = (l-2p)\omega + m\Omega,$$

$$\bar{\theta}'_{lsq} = (l-2q)\omega' + s(\Omega' - \frac{\pi}{2}),$$

a, e, i, ω, Ω are the orbital elements of the satellite (the inclination i, the argument of the perigee ω, and the longitude of the ascending node Ω referred to the equator), a', e', i', ω', Ω' are the orbital elements of the Moon (i', ω', Ω' referred to the ecliptic). $F_{lmp}(i)$, $F_{lsq}(i')$ are Kaula's inclination functions [10]. $X_0^{l,l-2p}(e)$, $X_0^{-(l+1),l-2q}(e')$ are Hansen's coefficients which can be expressed as

$$X_0^{l,l-2p}(e) = (1+\beta^2)^{-l-1} X_{0,0}^{l,l-2p}(\beta),$$

where

$$\beta = \frac{e}{1+\sqrt{1-e^2}}.$$

For $2p - l > 0$

$$X_{0,0}^{l,l-2p}(\beta) = (-\beta)^{2p-l} \begin{pmatrix} 2p+1 \\ 2p-l \end{pmatrix} F(-l-1, 2p-2l-1, 2p-l+1; \beta^2).$$

For $2p - l \leq 0$

$$X_{0,0}^{l,l-2p}(\beta) = (-\beta)^{l-2p} \begin{pmatrix} 2l-2p+1 \\ l-2p \end{pmatrix} F(-l-1, -2p-1, l-2p+1; \beta^2).$$

F denotes the hypergeometric series. Since $l \geq 2$, the first argument in F is negative and thus the series becomes a polynomial in β^2. Analogous expressions give the coefficients $X_0^{-(l+1),l-2q}(e')$.

The functions $U_l^{m,s}$ take care of the fact that the Moon's orbit is referred to the ecliptic

$$U_l^{m,s}(\varepsilon) = (-1)^{m-s} \sum_{r=r_1}^{r_2} (-1)^{l-m-r} \begin{pmatrix} l+m \\ m+s+r \end{pmatrix} \begin{pmatrix} l-m \\ r \end{pmatrix} \times \cos^{m+s+2r}\left(\frac{\varepsilon}{2}\right) \sin^{-m-s+2l-2r}\left(\frac{\varepsilon}{2}\right),$$

where $r_1 = \max[0, -(m+s)]$, $r_2 = \min[l-s, l-m]$, and ε is the obliquity of the ecliptic to the equator.

Finally, $\varepsilon_m = 1$ if $m = 0$, and $\varepsilon_m = 2$ if $m \neq 0$; $\varepsilon_s = 1$ if $s = 0$, and $\varepsilon_s = 2$ if $s \neq 0$; k_1 is the integer part of $m/2$; $k_2 = t(m+s-1)+1$, $k_3 = t(m+s)$, $t = 0$ if $l-1$ is even, and $t = 1$ if $l-1$ is odd; $y_s = 0$ if s is even, and $y_s = 1/2$ if s is odd.

Note, that the expansion (2) is valid for arbitrary inclination, and also for any eccentricity $e < 1$, since Hansen's trigonometric expansion is convergent for $e < 1$ and moreover the coefficients $X_0^{l,l-2p}$ are polynomials in β.

3. First Order Secular Solution

The long-term evolution of a satellite orbit under the effect of the Moon can be determined from Lagrange's planetary equations by making use of the averaged perturbing function (2). In a linear approximation, a, e, i can be taken constant on the right-hand side of the equations, while ω and Ω change linearly with time due to the oblateness of the Earth: $\omega = \dot{\omega} + \omega_0$, $\Omega = \dot{\Omega} + \Omega_0$, where

$$\begin{aligned} \dot{\omega} &= \frac{3}{4} J_2 n \left(\frac{r_0}{a}\right)^2 \frac{5\cos^2 i - 1}{(1-e^2)^2}, \\ \dot{\Omega} &= -\frac{3}{2} J_2 n \left(\frac{r_0}{a}\right)^2 \frac{\cos i}{(1-e^2)^2}, \end{aligned} \tag{3}$$

and J_2 is the second zonal harmonic coefficients of the Earth's potential, r_0 is the mean equatorial radius of the Earth, n is the mean motion of the satellite.

The argument of the perigee ω' and the longitude of the ascending node Ω' of the Moon's orbit can also be approximated as a linear function of time, due to the perturbations of the Sun: $\omega' = \dot{\omega}' + \omega_0'$, $\Omega' = \dot{\Omega}' + \Omega_0'$, while a', e', i' are constant. The solution of Lagrange's equations under these conditions result in a first order secular theory for the satellite. The integration of the equations depends on the argument

$$\bar{\theta}_{lmp} \pm \bar{\theta}'_{lsq} = \gamma^{\pm}_{lmpsq} t + \delta_{lmpsq}$$

with

$$\gamma^{\pm}_{lmpsq} = (l-2p)\dot{\omega} + m\dot{\Omega} \pm [(l-2q)\dot{\omega}' + s\dot{\Omega}']$$

and δ_{lmpsq} constant.

Since the short periodic terms have been averaged out, no secular or long periodic perturbations occur in the semi-major axis. Secular perturbations are related to terms with $l-2p=0$, $m=0$, $l-2q=0$, $s=0$, that is when $\gamma^{\pm}_{lmpsq}=0$. It can be easily seen that there are no secular terms in the eccentricity e and in the inclination i, however, ω, Ω change secularly.

Long periodic perturbations are connected with $\gamma^{\pm}_{lmpsq} \neq 0$. These perturbations occur in every orbital elements, except a. For example, the long periodic perturbations in the eccentricity are:

$$\begin{aligned} \delta e &= Gm' \frac{\sqrt{1-e^2}}{na^2 e} \sum_{l \geq 2} \sum_{m=0}^{l} \sum_{s=0}^{l} \sum_{p=0}^{l} \sum_{q=0}^{l} (-1)^{k_1} \frac{\varepsilon_m \varepsilon_s (l-s)!}{2a'(l+m)!} \left(\frac{a}{a'}\right)^l (l-2p) \\ &\quad \times F_{lmp}(i) F_{lsq}(i') X_0^{l,l-2p}(e) X_0^{-(l+1),l-2q}(e') \end{aligned}$$

$$\times \left\{ (-1)^{k_2} \frac{U_l^{m,-s}}{\gamma_{lmpsq}^{+}} \cos(\bar{\theta}_{lmp} + \bar{\theta}'_{lsq} - y_s \pi) \right.$$

$$\left. +(-1)^{k_3} \frac{U_l^{m,s}}{\gamma_{lmpsq}^{-}} \cos(\bar{\theta}_{lmp} - \bar{\theta}'_{lsq} - y_s \pi) \right\} .$$

When $\gamma_{lmpsq}^{\pm} \approx 0$, secular resonances between the frequencies $\dot{\omega}$, $\dot{\omega}'$ and $\dot{\Omega}$, $\dot{\Omega}'$ may give rise to large perturbations in the eccentricity and in the inclination. The equations $\dot{\omega} = \dot{\omega}'$, $\dot{\Omega} = \dot{\Omega}'$ with $\dot{\omega}' = 401.0547$"/day, $\dot{\Omega}' = -190.6341$"/day and $\dot{\omega}$, $\dot{\Omega}$ given by equations (3) define surfaces of secular resonances in the space of the orbital elements a, e, i. An analysis of these equations shows that the first type of secular resonance occurs only under the critical inclination $i = 63^o 26'$, and in both resonances values of large a and e are connected with small inclinations. For any given value of the eccentricity, the two resonances do not overlap.

Further details of the solution will be given in another paper.

Acknowledgements

Grants from the NATO and from the Hungarian Foundation for Scientific Research (OTKA T019409) are gratefully acknowledged.

References

1. Cook, G. E. (1972) Basic theory for PROD, a program for computing the development of satellite orbits, *Celestial Mechanics* **7**, 301-314.
2. Roth, E. A. (1973) Fast computation of high eccentricity orbits by the stroboscopic method, *Celestial Mechanics* **8**, 245-249.
3. Giacaglia, G. E. O. (1974) Lunar perturbations of artificial satellites of the Earth, *Celestial Mechanics* **9**, 239-267.
4. Janin, G. (1974) Accurate computation of highly eccentric satellite orbits, *Celestial Mechanics* **10**, 451-467.
5. Lane. M. T. (1989) On analytic modeling of lunar perturbations of artificial satellites of the Earth, *Celestial Mechanics* **46**, 287-305.
6. Zare, K., Szebehely, V., and Liu, J. J. F. (1996) A set of regular and uniform elements for satellites of high eccentricity and multiday period, *Spaceflight Mechanics* **93**, 589-599.
7. Szebehely, V. and Zare, K. (1997) Regions of chaoticity of multi-day satellite orbits, *NATO ASI* Aquafredda di Maratea.
8. Kozai Y. (1973) A new method to compute lunisolar perturbations in satellite motions, *SAO Special Report* **349**.
9. Plummer, H. C. (1918) *An Introductory Treatise on Dynamical Astronomy*, Cambridge University Press.
10. Kaula, W. M. (1966) *Theory of Satellite Geodesy*, Blaisdell Publishing Co.

TRANSLUNAR HALO ORBITS IN THE QUASIBICIRCULAR PROBLEM

M.A. ANDREU AND C. SIMÓ
Dept. Matemàtica Aplicada i Anàlisi, Univ. Barcelona
Gran Via 585, 08007 Barcelona, Spain

Abstract. The Quasibicircular Problem (QBCP) is a four body problem where three masses are revolving in a quasibicircular motion (that is, a coherent motion close to bicircular), the fourth mass being small and not influencing the motion of the three primaries. The Earth-Moon-Sun-Spacecraft case is considered and the Hamiltonian which governs the motion of the fourth mass is derived. That is a Hamiltonian with three degrees of freedom depending periodically on time. One application of this model has been the computation of quasiperiodic translunar Halo orbits. These orbits have been obtained using a continuation method starting at the Halo orbits of the RTBP. The Halo orbits of the QBCP are refined to Halo orbits of the Solar System using a Parallel Shooting method and JPL ephemeris.

1. Quasibicircular solution of the three body problem

We are going to derive a planar quasibicircular solution of the general three body problem (3BP), using the masses of the Moon, Earth and Sun, which are denoted by μ, $1-\mu$ and m_s, respectively. The units of mass, time and length are taken in such a way that the gravitational constant, the Earth-Moon distance and their mean motion are equal to 1.

We consider the Jacobi formulation of the 3BP using the relative position vectors $\vec{r}$ and $\vec{R}$ (see figure 1). The equations of the motion can be written in complex notation, using z and Z instead of $\vec{r}$ and $\vec{R}$ (see [4]). They are

$$\ddot{z} = -\frac{z}{r^3} + m_s\left(\frac{Z-\mu z}{r_{es}^3} - \frac{Z+(1-\mu)z}{r_{ms}^3}\right), \qquad (1)$$

B.A. Steves and A.E. Roy (eds.), The Dynamics of Small Bodies in the Solar System, 309–314.

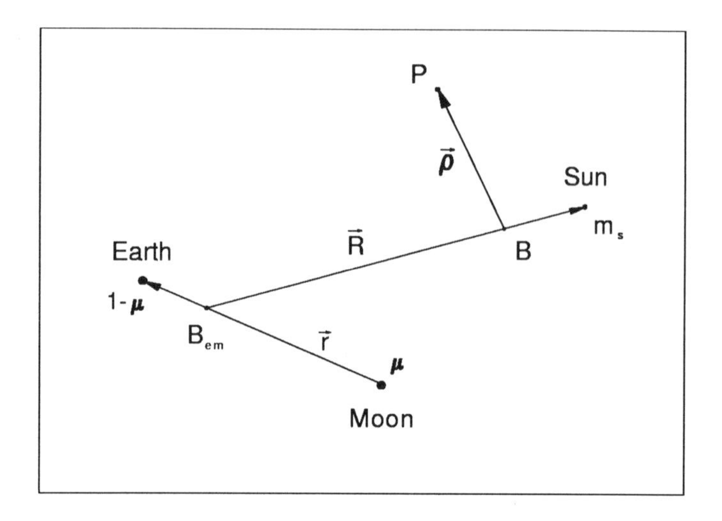

Figure 1. Quasibicircular problem in an inertial frame of reference.

$$\ddot{Z} = -(1+m_s)\left(\mu \frac{Z+(1-\mu)z}{r_{ms}^3} + (1-\mu)\frac{Z-\mu z}{r_{es}^3}\right), \qquad (2)$$

where $r = |z|$, $r_{ms} = |Z + (1-\mu)z|$, $r_{es} = |Z - \mu z|$. We want a solution

$$z \simeq e^{it}, \qquad Z \simeq a_s e^{in_s t},$$

where n_s is the mean motion of the Sun and a_s the mean distance between Sun and Earth-Moon barycenter. It is possible to find such a solution as

$$z = e^{it} \sum_{j=-\infty}^{+\infty} b_j e^{ij\omega_s t}, \qquad Z = a_s e^{in_s t} \sum_{j=-\infty}^{+\infty} c_j e^{ij\omega_s t}, \qquad (3)$$

being $\omega_s = 1 - n_s$ and $b_j, c_j \in \mathbb{R}$. For this purpose, we introduce the small parameter $\epsilon = 1/a_s \simeq 1/389$, and look for a solution of the form

$$z = e^{it}(1 + \sum_{k\geq 1} \epsilon^k u_k), \qquad Z = a_s e^{in_s t}(1 + \sum_{k\geq 1} \epsilon^k v_k),$$

where u_k and v_k are trigonometric polynomials with basic frequency ω_s. Substituting into the differential equations 1 and 2 and equating terms of the same order in ϵ, the polynomials u_k and v_k can be obtained recursively. To compute each polynomial, a linear differential equation with constant coefficients is solved. Finally, adding all the terms in powers of ϵ, we get the coefficients b_j and c_j of the expressions 3.

2. Quasibicircular Problem

It is a four body problem where three masses are revolving in a quasibicircular motion, the fourth mass being small and not influencing the motion of the three primaries. In our case, the three primaries are Sun, Earth and Moon, so the vectors $\vec{r}(t)$ and $\vec{R}(t)$ are the functions computed in the previous section. The fourth particle can be, for instance, an spacecraft.

The figure 1 shows the QBCP in an inertial reference system. To express the Hamiltonian of the mass P in a suitable way, we perform a change of coordinates (translation + rotation + homothetic transformation) such that Earth and Moon are located at the fixed points $(\mu, 0, 0)$ and $(\mu - 1, 0, 0)$, respectively, as is usual in RTBP. Then, the Hamiltonian is

$$\begin{aligned} H &= \frac{1}{2}\alpha_1(p_x^2 + p_y^2 + p_z^2) + \alpha_2(p_x x + p_y y + p_z z) + \alpha_3(p_x y - p_y x) + \\ &+ \alpha_4 x + \alpha_5 y - \alpha_6 \left(\frac{1-\mu}{q_{pe}} + \frac{\mu}{q_{pm}} + \frac{m_s}{q_{ps}} \right) , \end{aligned}$$

where

$$q_{pe}^2 = (x-\mu)^2 + y^2 + z^2 , \qquad q_{pm}^2 = (x - \mu + 1)^2 + y^2 + z^2 ,$$
$$q_{ps}^2 = (x - \alpha_7)^2 + (y - \alpha_8)^2 + z^2 ,$$

and α_k are of the form

$$\begin{aligned} \alpha_k(t) &= \alpha_{k0} + \sum_{j\geq 1} \alpha_{kj} \cos(j\omega_s t) , \qquad k = 1, 3, 4, 6, 7, \\ \alpha_k(t) &= \sum_{j\geq 1} \alpha_{kj} \sin(j\omega_s t) , \qquad k = 2, 5, 8. \end{aligned}$$

The α_k functions can be computed from $\vec{r}(t)$ and $\vec{R}(t)$ and its derivatives. The table 1 shows some Fourier coefficients of these functions.

Notice that the Hamiltonian of the QBCP can be considered as a perturbation of the RTBP with Earth-Moon mass parameter. The main difference is that the QBCP in non autonomous. It depens periodically on time with frequency ω_s. The period is denoted by $T = 2\pi/\omega_s$.

3. Quasiperiodic Halo Orbits in the QBCP

We want to obtain an orbit of the form

$$x = \sum_{k\in Z^2} x_k \gamma_1^{k_1} \gamma_2^{k_2} , \qquad y = i \sum_{k\in Z^2} y_k \gamma_1^{k_1} \gamma_2^{k_2} , \qquad z = \sum_{k\in Z^2} z_k \gamma_1^{k_1} \gamma_2^{k_2} ,$$

TABLE 1. Greatest Fourier coefficients of α_k functions. See more in [1].

j	α_1 c	α_2 s	α_3 c	α_4 c
0	1.00184161e+00		1.00000000e−00	−9.75524233e−04
1	5.76751773e−04	−2.64437603e−04	5.63412600e−04	2.15476436e+00
2	1.43877703e−02	−1.32868690e−02	1.88968744e−02	3.65748447e−04
3	−2.63036297e−06	9.38609321e−06	−9.91175880e−06	3.29567338e−03
4	1.17627836e−04	−1.21850906e−04	1.56870814e−04	3.30103140e−07
5	−8.06858139e−08	1.52212760e−07	−1.70776258e−07	1.27884069e−05
6	9.84324977e−07	−1.07210266e−06	1.31961368e−06	−2.62379795e−09
7	−1.17205439e−09	1.88937126e−09	−2.13655004e−09	6.53380551e−08
8	8.31190597e−09	−9.32498503e−09	1.11716892e−08	−3.89172071e−11
j	α_5 s	α_6 c	α_7 c	α_8 s
0		1.00090746e+00	−6.31406957e−02	
1	−2.19257075e+00	2.87092175e−04	3.88563862e+02	−3.89743726e+02
2	−3.33721049e−04	7.18717800e−03	1.73691020e−01	−1.73427917e−01
3	−3.29500143e−03	−2.35118315e−06	3.38290807e+00	−3.38569649e+00
4	−3.10063505e−07	4.58575897e−05	1.57483757e−04	−1.55588663e−04
5	−1.27777734e−05	−3.84868362e−08	2.93636049e−02	−2.93758267e−02
6	2.65280641e−09	3.27067750e−07	−1.22443455e−05	1.22585121e−05
7	−6.52847925e−08	−4.40696648e−10	2.53893543e−04	−2.53959689e−04
8	3.89172071e−11	2.45260066e−09	−2.27892904e−07	2.28002922e−07

where x_k, y_k, $z_k \in \mathbb{R}$, k denotes the multiindex (k_1, k_2) and

$$\gamma_1 = \exp(i(\omega_0 t + \varphi_0)) , \qquad \gamma_2 = \exp(i\omega_s t) ,$$

where ω_0 is a prefixed frequency. The phase φ_0 is arbitrary. Moreover, we have symmetry conditions $x_k = x_{-k}$, $y_k = -y_{-k}$ and $z_k = z_{-k}$.

The second order differential equations of the QBCP are

$$-\ddot{x} + 2\dot{y} + (1 + 2c_2)x + \sum_{n\geq 2} c_{n+1}(n+1)T_n + \alpha_{10}\ddot{x} + \alpha_{11}\dot{x} + \alpha_{12}\dot{y} +$$

$$+\alpha_{13}x + \alpha_{14}y + \alpha_{15} + \alpha_{16}\sum_{n\geq 2} c_{n+1}(n+1)T_n + \alpha_{20}\sum_{n\geq 2}\frac{\partial G_n}{\partial x} = 0 , \quad (4)$$

$$-\ddot{y} - 2\dot{x} - (c_2 - 1)y + y\sum_{n\geq 2} c_{n+1}R_{n-1} + \alpha_{10}\ddot{y} - \alpha_{12}\dot{x} + \alpha_{11}\dot{y} -$$

$$-\alpha_{14}x + \alpha_{17}y + \alpha_{18} + \alpha_{16}y\sum_{n\geq 2} c_{n+1}R_{n-1} + \alpha_{20}\sum_{n\geq 2}\frac{\partial G_n}{\partial y} = 0 , \quad (5)$$

$$-\ddot{z} - c_2 z + z\sum_{n\geq 2} c_{n+1}R_{n-1} + \alpha_{10}\ddot{z} + \alpha_{11}\dot{z} + \alpha_{19}z +$$

$$+\alpha_{16}z\sum_{n\geq 2} c_{n+1}R_{n-1} + \alpha_{20}\sum_{n\geq 2}\frac{\partial G_n}{\partial z} = 0 . \quad (6)$$

The functions α_k for $k = 10, \ldots, 20$ are T-periodic and they can be computed from the previous α_k. The terms of the summations come from the expansion of the potential in Legendre polynomials. The coefficients c_n were introduced in [5] for the RTBP. T_n, R_n and G_n are polynomials in x, y and z, but G_n is also T-periodic on time. These functions are computed recursively (see [3] and [1]).

Let $\mathcal{F}$ be the space of real Fourier functions with two basic frequencies ω_0 and ω_s, $\mathcal{F}_0$ be the subspace of $\mathcal{F}$ of the even functions and $\mathcal{F}_1$ be the subspace of $\mathcal{F}$ of the odd functions. The equations 4, 5 and 6 allow to define

$$F : \mathcal{F}_0 \times \mathcal{F}_1 \times \mathcal{F}_0 \longrightarrow \mathcal{F}_0 \times \mathcal{F}_1 \times \mathcal{F}_0 \,.$$

So, the problem is reduced to solve a non linear equation $F(X) = 0$. It can be done using Newton's method

$$X^{(j+1)} = X^{(j)} - DF^{-1}(X^{(j)})F(X^{(j)}) \,,$$

with $X^{(0)}$ the Fourier coefficients of a Halo orbit of the RTBP. As the equation is infinite dimensional, it is necessary a truncation and the use of an adaptative method. Let $M_j = \|F(X^{(j)})\|_\infty$ and $\kappa \in \mathbb{R}$, $0 < \kappa \ll 1$, say $\kappa = 10^{-5}$. In each iteration of the Newton's method, the selected frequencies are the ones such that their coefficients in $F(X^{(j)})$ are larger than κM_j. We remark that some steps of a continuation method are required.

The figure 2 shows the amplitude of the orbits computed. We realize the gap corresponding the low order resonance $\omega_0 = 2\omega_s$. The dynamics of this zone has not been studied yet.

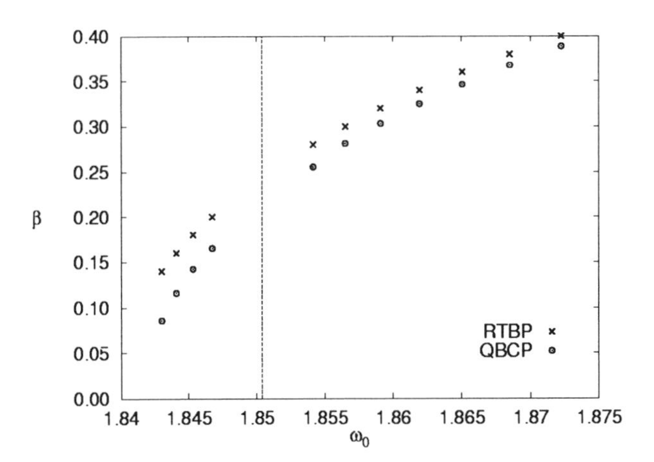

Figure 2. Frequency ω_0 and z-amplitude of the Halo orbits around L_2. The vertical dashed line corresponds to the frequency $\omega_0 = 2\omega_s$.

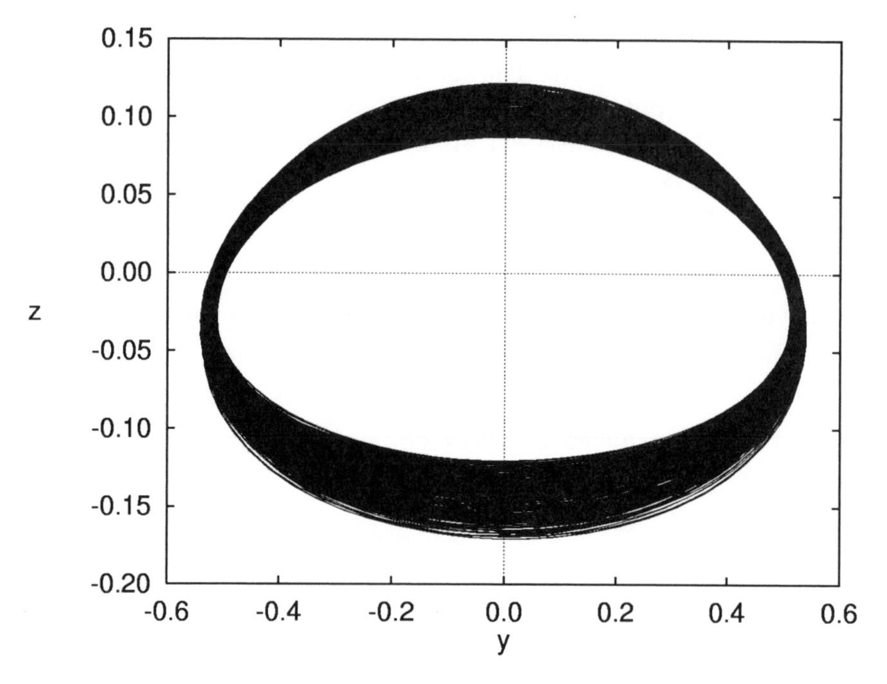

Figure 3. yz projection of a Halo orbit in the Solar System for a time span about 19 years. The unit of length is the distance between Moon and L_2 (about 60000 km).

4. Translunar Halo orbits in the real Solar System

The motivation to study the translunar Halo orbits is the application to spacecraft missions. These orbits were proposed by Farquhar in [2] as a means to stablish a continuous communications link between the Earth and the far side of the Moon. However, it was difficult to compute this kind of orbits for a long time interval. Presently, Halo orbits of the QBCP can be refined to Halo orbits of the Solar System for a long time span. The figure 3 shows one of these orbits for a time span of 19 years. To produce this orbit a Parallel Shooting method and JPL ephemeris are used (see [1]).

References

1. Andreu, M.A. (in preparation) *The Quasibicircular Problem,* Thesis, Dept. Matemàtica Aplicada i Anàlisi, Universitat de Barcelona.
2. Farquhar, R.W. (1971) *The Utilization of Halo Orbits in Advanced Lunar Operations,* NASA TN D-6365.
3. Gómez G., Jorba A., Masdemont J. and Simó C. (1991) *Study refinement of semi-analytical halo orbit theory,* ESOC contract 8625/89/D/MD(SC), Final Report.
4. Marchal, C. (1990) *The Three-Body Problem,* Elsevier.
5. Richardson, D.L. (1980) A note on a Lagragian formulation for motion about the collinear points, *Celestial Mechanics* **22**, 231–236.

PROBLEMS CONNECTED WITH THE ROTATIONAL DYNAMICS OF CELESTIAL BODIES

ANDRZEJ J. MACIEJEWSKI
Toruń Centre for Astronomy, N. Copernicus University, PL-87–100 Toruń, Gagarina 11, Poland

1. Introduction

Celestial bodies in the Solar System are not point masses, they are extended and deformable bodies. When this fact must be taken into account, two extreme approaches are widely used: the object is considered either as a perfectly rigid body or perfect self-gravitating fluid. Here, I discuss only these models that assume that bodies are perfectly rigid. This approach is adequate for small bodies in the Solar System.

For a system consisting of a finite number of rigid bodies interacting mutually according to Newton's law of universal gravitation, we can formulate two classes of problems: restricted and non-restricted ones. In a restricted problem, we assume that orbits of mass centers of the bodies are known and we study only rotational motion. In an unrestricted problem we reject this simplification. Even in the simplest non-trivial case of an unrestricted problem of a point mass and single rigid body, we meet fundamental and unsolved problems.

The aim of this paper is to point out two very important questions concerning the dynamics of 'simple' restricted and non-restricted models of the rotational motion.

2. Rigid satellite in circular orbit

Let us consider the restricted problem of rotational motion of a rigid satellite $\mathcal{B}$ moving in a circular orbit around a fixed gravitational center O (see Fig. 1). In this model the gravitational torque acting on the satellite is approximated by the first non-trivial term. Equations of the rotational motion

B.A. Steves and A.E. Roy (eds.), The Dynamics of Small Bodies in the Solar System, 315–320.

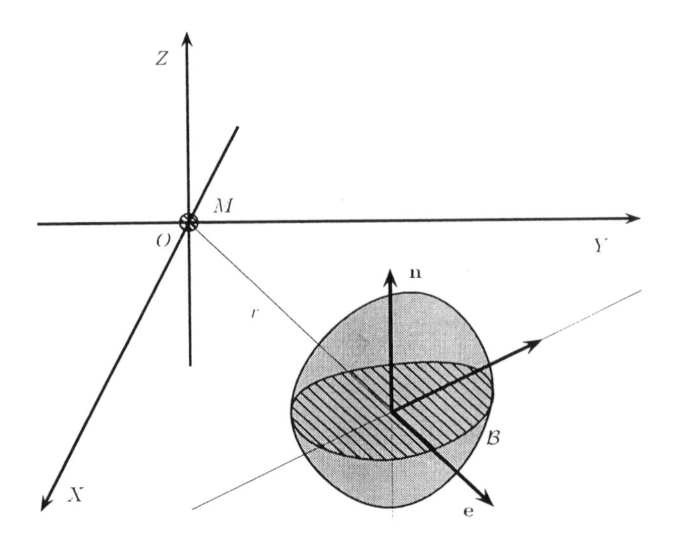

Figure 1. A rigid satellite in a circular orbit around a gravitational center

of the satellite can be written in the following form (see [5, 6])

$$\frac{d}{dt}\mathbf{M} = \mathbf{M} \times \mathbf{\Omega} + 3\mathbf{e} \times \mathbf{I}\mathbf{e}, \quad \frac{d}{dt}\mathbf{e} = \mathbf{e} \times (\mathbf{\Omega} - \mathbf{n}), \quad \frac{d}{dt}\mathbf{n} = \mathbf{n} \times \mathbf{\Omega}, \qquad (1)$$

where $\mathbf{\Omega}$, $\mathbf{M} = \mathbf{I}\mathbf{\Omega}$, $\mathbf{I} = \mathrm{diag}(A, B, C)$ are the angular velocity, the angular momentum and the inertia tensor respectively; $\mathbf{e}$ is the unit vector in the direction of the radius vector of the center of mass of the satellite, and $\mathbf{n}$ is the unit vector normal to the orbital plane. All vectors are taken with respect to the principal axes reference frame and units are chosen in such a way that the Kepler orbital frequency is equal to one. Equations (1) possess the energy integral

$$H_1 = H(\mathbf{M}, \mathbf{e}, \mathbf{n}) = \frac{1}{2}\langle \mathbf{M}, \mathbf{I}^{-1}\mathbf{M}\rangle - \langle \mathbf{M}, \mathbf{n}\rangle + \frac{3}{2}\langle \mathbf{e}, \mathbf{I}\mathbf{e}\rangle, \qquad (2)$$

and the following geometric integrals

$$H_2 = \langle \mathbf{e}, \mathbf{e}\rangle, \qquad H_3 = \langle \mathbf{n}, \mathbf{n}\rangle, \qquad H_4 = \langle \mathbf{e}, \mathbf{n}\rangle. \qquad (3)$$

One can consider Equations (1) as the Euler equations with the Hamiltonian (2) on the dual to a certain nine dimensional Lie algebra $\mathcal{L}_9$ (see [6] for details). From general theory it follows that on a common level of integrals (3) system (1) is Hamiltonian. For its complete integrability we have to know two additional first integrals. Thus, one can formulate at least two interesting questions: for which values of parameters (A, B, C) system possesses one, or two additional first integrals. To the best of the author's knowledge the answer to the first problem, except for the simple case of a symmetric satellite (two of the principal moments of inertia are equal), is unknown. What concerns the second question, only the case of a spherically symmetric satellite is known to be completely integrable.

There exist some numerical results showing that system (1) is not integrable and chaotic for a wide range of parameters [18, 19].

When searching parameters values for which the system can be integrable one can use the approach of S. V. Kowalevski [9, 10] with extensions made by A. M. Lyapunov [13] (see books [12, 11] for a detailed discussion). Omitting details, we can summarize the results of application of the Kowalevski-Lyapunov method to system (1) in the following.

Theorem 1 *Except for the case of a spherically symmetric body system (1) has a multi-valued solution.*

This result suggests that we cannot expect that, for some parameters value, the system possesses two additional algebraic first integrals. An amazing fact connected with the application of the Kowalevski-Lyapunov method to system (1) is the following: in all cases all the Kowalevski exponents are integers, but the Kovalewski matrix is not semi-simple.

One can suspect that more conclusive and stronger results can be obtained if it is possible to apply the Ziglin theorem [20, 21]. In fact, when this theorem is successfully applied then it strictly proves the non-existence of an additional meromorphic first integral. In order to apply this theorem we have to know a solution of the system whose maximal analytic continuation gives rise to a non-trivial Riemannian surface. For the problem considered one can take a family of pendulum like solutions filling an invariant manifold $\mathcal{M} = \{(\mathbf{M}, \mathbf{e}, \mathbf{n}) \in R^9 \mid M_1 = M_2 = n_1 = n_2 = e_3 = 0, n_3 = 1\}$. However, to apply the Ziglin theorem, we have to determine the monodromy group of normal variational equations (see cited papers of Ziglin). In the case considered, we obtain a system of four linear equations with double-periodic coefficients. It seems that for this system there is no way to obtain necessary information about the monodromy group by means of analytical methods. The proof of Theorem 1 and the description of a numerical application of the Ziglin theorem can be found in [16].

3. Stationary motions of an unrestricted problem

Let us consider a system consisting of two mutually attracting bodies one of which is a point mass $\mathcal{P}$ with mass M, the other is a rigid body $\mathcal{B}$ with an axis of symmetry S and mass m (see Fig. 2). The system is conservative, and, besides, the energy, the total linear and angular momentum of the system, and the projection of total angular momentum of the body onto the symmetry axis are conserved. Using these integrals, we can make

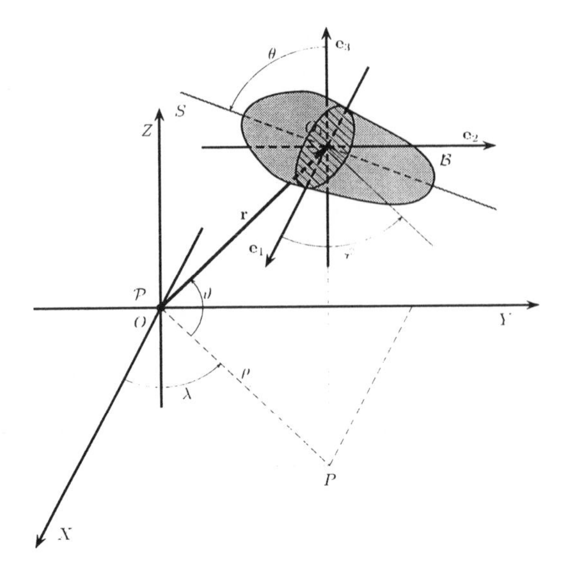

Figure 2. Geometry of unrestricted problem of a point mass and symmetric rigid body

a standard reduction in the system. As the result of this reduction, we obtain a Hamiltonian system with four degrees of freedom. The equilibria of this reduced Hamiltonian system are called stationary motions, or relative equilibria.

According to [4], the most detailed and complete analysis of the described unrestricted problem of a point mass and a symmetric rigid body was done by [8]. The amazing fact is that in [8] only the so called Lagrangean stationary solutions were found. For a Lagrangean solution orbits of both bodies lie in one plane. As it is well understood now, there can exist non-Lagrangean stationary solutions for which orbital planes of the bodies are parallel but do not coincide [3, 15]. Thus, we can formulate the following problems for the system described above: (i) do non-Lagrangean stationary solutions exist? (ii) how many stationary solutions exist in the problem and how they bifurcate? Here we give simple arguments showing that the answer to the first question is positive.

The Hamiltonian of the problem can be written in the following form

$$H = \frac{1}{2\mu}\left(p_1^2 + \frac{(p_5 - p_3)^2}{q_1^2} + p_2^2\right) + \frac{1}{2A}\left(\frac{p_3 - p_6\cos q_4}{\sin q_4}\right)^2 + \frac{1}{2A}p_4^2 + V, \quad (4)$$

where $(q_1, q_2, q_3, q_4, q_5, q_6) = (\rho, z, \varphi - \lambda, \theta, \lambda, \psi)$ are the generalized coordinates and p_i, for $i = 1, \ldots, 6$ denotes the respective moments, μ is the reduced mass of the body; (ρ, z, λ) are cylindric coordinates of the mass center of the body $\mathcal{B}$ in $OXYZ$ reference frame (see Fig. 2) with axes parallel to the axes of an inertial reference frame; the Euler angles (φ, θ, ψ)

of the type 3–1–3 describe the orientation of the principal axes reference frame with respect to $OXYZ$ frame. Potential $V = V(r, \hat{x}_3)$ has the form

$$V(r,\hat{x}_3) = -GM\left\{\frac{m}{r} + \frac{V_2(r,\hat{x}_3)}{r^3} + \frac{1}{r^3}\sum_{n=3}^{\infty} A_n \left(\frac{a_0}{r}\right)^{n-2} P_n(\hat{x}_3)\right\}, \quad (5)$$

where $r = |\mathbf{r}|$ (see Fig. 2); A_n are coefficients that depend only on the shape and mass distribution of the body, P_n are Legendre polynomials, and a_0 is the mean radius of the body, and

$$V_2(r,\hat{x}_3) = \frac{1}{2}(C - A)[1 - 3\hat{x}_3^2], \quad \hat{x}_3 = -\frac{\rho}{r}\sin\theta\sin(\varphi - \lambda) - \frac{z}{r}\cos\theta, \quad (6)$$

where A, $B = A$, C are the principal moments of inertia of the body (see[7]). Direct inspection shows that Hamiltonian (4) does not depend on q_5, and q_6, thus p_5 and p_6 are first integrals and they can be treated as parameters of the problem. We denote $p_5 = \delta$ and $p_6 = \gamma$. The stationary solutions are the equilibria of Hamilton's equations of motion, thus they are solutions of the following system of nonlinear equations:

$$\frac{\partial H}{\partial q_i} = 0, \qquad \frac{\partial H}{\partial p_i} = 0, \quad \text{for} \quad i = 1, 2, 3, 4. \quad (7)$$

To prove that there exist non-Lagrangean stationary solutions first we solve system (7) with $V = V_2$. There exist three families of Lagrangean solution for such an approximation. For a Lagrangean solution we have $q_2 = z = 0$. Next, we can assume that the dimensions of the body are small with respect to the radius of orbit. This allows us to consider $\epsilon = a_0/r$ in expansion (5) as a small parameter. It can be shown that the families of solutions we found for $\epsilon = 0$ generically have analytical continuation for $\epsilon > 0$. For a solution obtained in this way we have $q_2 = \epsilon a + \cdots$, where $a \neq 0$ is proportional to the third harmonic A_3. This approach was applied in [1, 2, 3, 14, 15] for proving the existence of stationary solutions in different kinds of restricted and unrestricted problems.

We show that there exists a non-Lagrangean stationary solution in a case when ϵ is arbitrary. For $\gamma = 0$ (i.e. the body does not rotate around its axis of symmetry) system (7) has the following solution: $p_1 = p_2 = P_4 = q_2 = 0$, $q_3 = q_4 = \pi/2$, $p_3 = \delta/(1 + \mu q_{10}^2/A)$, $q_1 = q_{10}$, where q_{10} is a solution of the following equation

$$\frac{\mu\delta^2 q_{10}}{A + \mu q_{10}^2} = \frac{\partial V}{\partial r}(q_{10}, -1).$$

It can be shown that this equation has a solution for a wide class of potential functions V. Starting from this solution, we are able to prove that there exist its analytic continuation for $|\gamma| > 0$. For such continuation we obtain that $z = \gamma b + \cdots$, where $b \neq 0$, i.e., this is a family of non-Lagrangean stationary solutions. Details and proofs of all statements given above will be published elsewhere [17]

References

1. M. Z. Aboelnaga. Regular rotational-translational motion of two arbitrary rigid bodies and their stability. *Astronom. Zh.*, 66(3):604–611, 1989.
2. M. Z. Aboelnaga. Stationary motions of an axi-symmetric satellite in a circular orbit. *Kosm. Issled.*, 27(2):176–179, 1989.
3. Yu. V. Barkin. 'Oblique' regular motion of a satellite and some small effects in the motion of the Moon and Phobos. *Kosm. Issled.*, 15(1):26–36, 1985.
4. Yu. V. Barkin and V. G. Demin. *Translatory-Rotatory Motion of Celestial Bodies*, volume 20 of *Itogi Nauki i Tekhniki, Seria Astronomia.* VINITI, Moscow, 1982.
5. V. V. Beletskii. *Motion of a Satellite about its Mass Center.* Nauka, Moscow, 1965. In Russian.
6. O. I. Bogoyavlenskii. *Overturning solitons. Nonlinear integrable equations.* Nauka, Moscow, 1991. in Russian.
7. G. N. Duboshin. *Celestial Mechanics. Fundamental Problems and Methods.* Nauka, Moscow, second, revised and enlarged edition, 1968. In Russian.
8. H. Kinoshita. Stability motions of an axisymmetric body around a spherical body and their stability. *Publ. Astron. Soc. Jpn.*, 22:383–403, 1970.
9. S. Kowalevski. Sur le problème de la rotation d'un corps solide autour d'un poit fixe. *Acta Math.*, 12:177–232, 1888.
10. S. Kowalevski. Sur une propriété du systéme d'équations différentielles qui définit la rotation d'un corps solide autour d'un poit fixe. *Acta Math.*, 14:81–93, 1890.
11. V. V. Kozlov and S. D. Furta. *Asymptotic of Solutions of Strongly Nonlinear Differential Equations.* Moscow University Press, Moscow, 1996. In Russian.
12. Valerii V. Kozlov. *Symmetries, Topology and Resonances in Hamiltonian Mechanics.* Springer-Verlag, Berlin, 1996.
13. A. M. Lyapunov. On a certain property of the differential equations of the problem of motion of a heavy rigid body, having a fixed point. *Soobshch. Khar'kovsk. Mat. Obshch.*, 4(3):123–140, 1894. In Russian.
14. A. J. Maciejewski. Regular precessions in the restricted problem of the rotational motion. *Acta Astronomica*, 44:301–316, 1994.
15. A. J. Maciejewski. Reduction, relative equilibria and potential in the two rigid bodies problem. *Celest. Mech.*, 63:1–28, 1995.
16. A. J. Maciejewski. Certain property of equations of the rotational motion of a satellite in circular orbit. 1997. in preparation.
17. A. J. Maciejewski. Stationary solution in the unrestricted problem of the rotational motion. 1997. in preparation.
18. A. J. Maciejewski and K. Goździewski. Nonintegrability, separatrices crossing and homoclinic orbits in the problem of rotational motion of a satellite. In T. Bountis, editor, *Chaotic Dynamics. Theory and Practice*, volume 298 of *NATO ASI Series, Series B: Physics*, pages 145–159, New York and London, 1992. Plenum Press.
19. A. J. Maciejewski and K. Goździewski. Solutions homoclinic to regular precessions of a symmetric satellite. *Celest. Mech.*, 61:347–368, 1995.
20. S. L. Ziglin. Branching of solutions and non-existence of first integrals in Hamiltonian mechanics. I. *Functional Anal. Appl.*, 16:181–189, 1982.
21. S. L. Ziglin. Branching of solutions and non-existence of first integrals in Hamiltonian mechanics. II. *Functional Anal. Appl.*, 17:6–17, 1983.

CLOSED FORM EXPRESSIONS FOR SOME GRAVITATIONAL POTENTIALS

Triangle, Rectangle, Pyramid and Polyhedron

ROGER A. BROUCKE
Department of Aerospace Engineering and Engineering Mechanics
The University of Texas at Austin
Austin, TX 78712

Abstract. We give elementary derivations of the gravitational Newtonian potential created by different mass distributions. We begin with several two-dimensional objects such as rectangles, triangles and polygons. We emphasize the presence of two kinds of terms: logarithms and Arc-Tangents. We also show the connection with the well-known logarithmic potential of the wire segment. All these potentials are expressed in closed form.

We also give a few theorems which allow us to reduce several potentials to the potential of a more simple but equivalent mass distribution. For instance, the potential at the vertex of a triangle relates to the potential created by the wire segment at the side opposite to this vertex. On the other hand, the potential at the apex of a pyramid, created by the whole massive pyramid relates to the potential created by an equivalent mass distribution at the base of the pyramid. We use these properties to derive closed-form expressions for the potential created by a polyhedron at inner as well as outer points.

1. Introduction

The standard method for computing the gravitational field of the earth or any other irregular-shaped or non-homogeneous body consists in making expansions in spherical harmonics and determining the C_{nm} and S_{nm} coefficients of degree n and order m. High values of n and m (> 100) are now typically used in Industry. It is well known not only that these expansions

B.A. Steves and A.E. Roy (eds.), The Dynamics of Small Bodies in the Solar System, 321–340.

become computationally very intensive but also that the series practically do not converge. We are therefore facing some kind of a crisis now, with regard to the expansion in spherical harmonics of the potential of the Earth. This problem appears actually even more serious if one thinks of the other bodies in the solar system, especially some of the asteroids such as Gaspra and Ida which have extremely irregular shapes.

For these reasons, we find it necessary to initiate new research on alternate forms of solution of the Laplace Equation. The object of the present article is to explore the practicality of using the closed-form expressions for the potential of some homogeneous bodies with well defined geometric shapes.

Our exposition will go from the particular to the general, in the sense that we begin with simple shapes and end with the more complex ones. To begin with we will review the potential of a segment of straight line, mainly because the well known logarithmic expression plays a rather important role in relation to two- and three-dimensional bodies as well. So, the next few sections will deal with two-dimensional bodies, (plates). We review the rectangular plate and we give some generalizations of the exercises that are found in Kellog. Then we treat triangles in great detail and we show how the generalization to arbitrary polygons can be made. Finally, we will give closed-form expressions for the potential of three-dimensional solids, especially pyramids and irregular polyhedra with an arbitrary number of planar faces.

Our approach will intentionally be kept as elementary as possible. We will not use the Gauss Divergence theorem or Green's theorem. All our derivations are based on a short table of elementary integrals given in Appendix A at the end of the article. All the expressions are in closed form, as well for the potential as for the components of the acceleration. The results are essentially for homogeneous bodies only, but this restriction could be relaxed by imbedding bodies of different density.

In a sense it is a little surprising that no attempts have been made during the space era, to model the gravitation field of celestial bodies with the present type of methods, because there is actually quite a bit of literature on the subject. The approach that we are following here to obtain the potential of the polyhedron is really a detailed implementation of the scheme that was proposed by Tisserand in a paragraph in his famous "Mecanique Celeste," (1890, Vol. 2, page 68). The Russian literature also has information on work done by Meller and Sludsky in the 1860's, as is mentioned in a more modern article on the gravity field of homogeneous polyhedra by Strakhov and Lapina (1990).

We need to note that the potential of the rectangular parallelipiped was derived in detail in MacMillan's book (1930, pages 74-79), as well as in several later articles, such as Nagy (1966) and Bannerjee (1977). MacMillan's formula spans a page and a half and has 12 Logarithms and 24 Arc-Tangents. Among the recent work, there also are two articles by J. Waldvogel, on the potential of a cube (1976) and on the polyhedron (1981).

Most of the literature on these subjects can be found in the Geophysics Journals. For instance, some of the articles on this problem are by Nagy (1966) and Bannerjee and Gupta (1977), who give a direct integration in rectangular coordinates. They derive the z-component of the acceleration instead of the potential. Both have the usual logarithmic terms. Nagee also has Arc-sine terms, while Bannerjee and Gupta have the Arc-Tangent terms instead. The same problem was also solved by Sorokin (1951) and Haaz (1953), also with Log and Arc-Tangent terms.

Okabe (1979) derived the gravitational potential as well as it's first and second derivatives, for an arbitrary polyhedral body, using the Gauss Divergence theorem. Another detailed derivation based on the Gauss theorem was recently published by R. Werner (1994). The key to the use of the Divergence Theorem is in a short identity that allows us to write the inverse distance $(1/r)$ as the Divergence of half the unit vector along r. This formula is in some of the textbooks, such as MacMillan (1930, page 80) or Levallois (1970, Vol. 3, page 79).

Cady (1980) derives equations for the vertical gravity field of "homogeneous polygonal bodies", using direct integration in rectangular coordinates. Two of the integrations are done analytically and the third numerically. Pohanka (1988) gives another derivation of the gravity field of a homogeneous polyhedral body, again in terms of logarithms and Arc-Tangents.

In the present article, potential always means "Potential Function". The constant of Gravity is always represented by the symbol G.

2. The Potential of a Straight Wire Segment

It is a rather elementary exercise in potential theory, to derive the expression for the potential of a thin homogenous massive rod. It is given in most textbooks (Kellog, 1929, page 56), (MacMillan, 1930, pages 42-43). These derivations are usually given in terms of a rather special configuration of the rod (on the x-axis or z-axis for instance), in order to simplify the algebra. The result is then given in terms of coordinates but it can be transformed to the intrinsic form, in terms of distances only, by using what we call MacMillan's device, described in his book (MacMillan, 1930, page 74). An example of an early derivation is found in one of G. Green's (1828, page 40) papers. It is in a non-intrinsic form, explicitly in terms of coordinates.

We want to give here another version of this derivation, valid for any arbitrary orientation of the rod in a 3-D space, in order to emphasize the importance of the intrinsic representation of the basic results in potential theory.

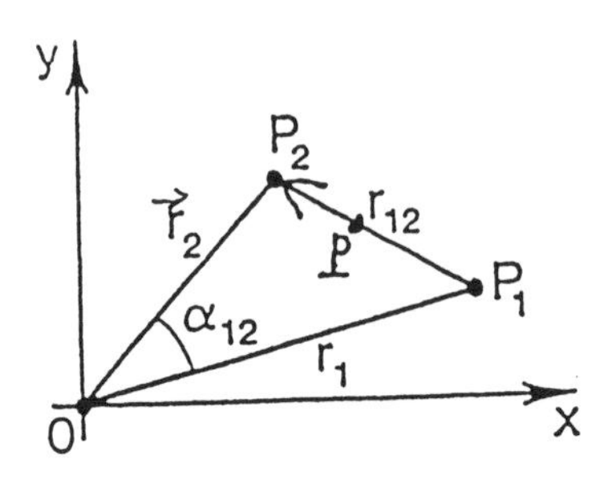

Fig. 1

Let the rod or segment of straight wire, with linear density σ, extend in space from point P_1 to point P_2, (fig. 1) with position Vectors $\vec{r}_1(x_1y_1z_1)$ and $\vec{r}_2(x_2y_2z_2)$. An arbitrary point P on the rod can be represented by the vector

$$\vec{r} = \vec{r}_1 + \lambda(\vec{r}_2 - \vec{r}_1) = \vec{r}_1 + \lambda\vec{r}_{12} \; , \tag{2.1}$$

where λ is a scalar parameter, varying from 0 to 1. The total mass of the rod is $M = \sigma r_{12}$. We will find the expression for the potential function U at the origin 0. The distance to an arbitrary point is given by

$$d^2 = \vec{r}^{\,2} = (x^2 + y^2 + z^2) = \lambda^2 r_{12}^2 + \lambda(r_2^2 - r_1^2 - r_{12}^2) + r_1^2 \; . \tag{2.2}$$

Here r_{12} is the length of rod and r_1, r_2 are the distances of the origin 0 to the two endpoints. The line-element ds on the rod is obtained by differentiating (2.1) with respect to λ:

$$ds = |d\vec{r}| = r_{12} d\lambda \; . \tag{2.3}$$

The total potential at 0 is thus the definite integral in λ:

$$U = G\sigma \int_{\lambda=0}^{\lambda=1} \frac{r_{12} d\lambda}{d} = GM \int_0^1 \frac{d\lambda}{d} \; . \tag{2.4}$$

The denominator d is a square root of a quadratic function in λ. We have thus a simple integral that is found in the tables.

$$U = \frac{GM}{r_{12}} \log \left[\frac{r_1 + r_2 + r_{12}}{r_1 + r_2 - r_{12}} \right] \; . \tag{2.5}$$

We can write this important result in a few different forms. For instance, we can use a hyperbolic inverse tangent:

$$U = \frac{GM}{2r_{12}} th^{-1}\left(\frac{r_{12}}{r_1 + r_2}\right) . \tag{2.6}$$

One of the remarkable properties of the straight wire segment is that it's attraction is equivalent to that of a properly defined tangent circular wire segment. We refer the reader to MacMillan, (1930, page 5) for this. We will not use this property in the present article.

We also mention that the potential is well defined on the extensions of the wire, in both directions of the wire. The force of attraction becomes infinite however on the wire itself. The equipotential surfaces around the wire are ellipsoids of revolution. Orbits have been computed in the field the wire segment by P. Halamek (1988).

3. The Potential of a Triangle at it's Vertex

We will first derive an expression for the potential of a triangle P_1OP_2 at it's vertex 0, the origin of the coordinate system Oxy, (fig. 2). This will show us an interesting connection with the logarithmic potential of a straight segment of a wire.

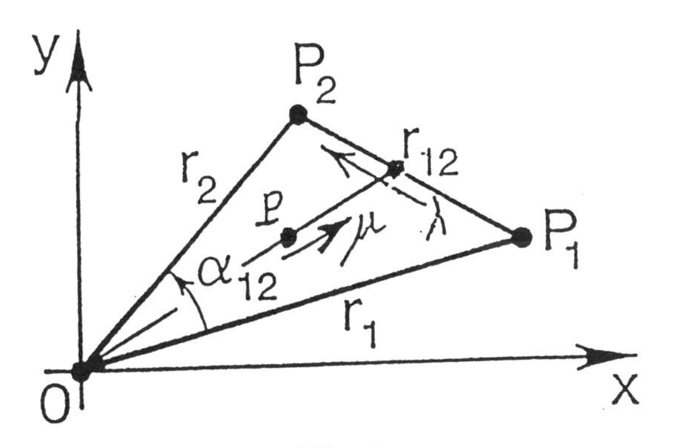

Fig. 2

As is shown in figure 2, the triangle can be defined by the position vectors $\vec{r}_1, \vec{r}_2$ of P_1 and P_2 with coordinates (x_1, y_1) and (x_2, y_2). The total area of the triangle is $A = |\vec{r}_1 \times \vec{r}_2|/2 = (x_1y_2 - y_1x_2)/2$.

We first parameterize the side $\overline{AB} = \vec{r}_{12}$ with a parameter λ, (0 to 1), such that an arbitrary point Q on this side has the position vector

$$\vec{r}_Q = +\vec{r}_1 + \lambda(\vec{r}_2 - \vec{r}_1) = \vec{r}_1 + \lambda\vec{r}_{12} .$$

Then we introduce a second parameter μ, (0 to 1) to define an arbitrary point P on the line $\overline{OQ}$: This point P is thus defined by the two variables λ and μ:

$$\vec{r}_P = \mu\vec{r}_Q = \mu\vec{r}_1 + \lambda\mu(\vec{r}_2 - \vec{r}_1) \ . \tag{3.1}$$

The area element dA is given by the relation

$$dA = (x_1y_2 - y_1x_2)\mu \ d\mu \ d\lambda \ , \tag{3.2}$$

while the distance $\overline{OP}$ is defined by the formula:

$$r_Q^2 = \mu^2\vec{r}_Q^2 = \mu^2 \left[\lambda^2 r_{12}^2 + \lambda(r_2^2 - r_1^2 - r_{12}^2) + r_1^2\right] \ . \tag{3.3}$$

The total potential at 0 is then obtained by a double integration, in λ and μ:

$$U = G\sigma \int_0^1 \int_0^1 \frac{dA}{\mu r_Q} = G\sigma(x_1y_2 - y_1x_2) \int_0^1 \int_0^1 \frac{d\mu d\lambda}{r_Q} \ . \tag{3.4}$$

The integrand does not contain the variable μ, so that we have immediately:

$$U = G\sigma(x_1y_2 - y_1x_2) \int_0^1 \frac{d\lambda}{r_Q} = 2GM_{\mathrm{TR}} \int_0^1 \frac{d\lambda}{r_Q} \ . \tag{3.5}$$

We see that we gave reduced the problem of the triangle to the problem of the wire segment, seen in eq. (2.4). The wire segment is here the opposite side $\vec{r}_{12}$ if the triangle. However, the exact comparison of the equations shows that the mass of the wire must be equal to *twice* the mass of the triangle in order to have the same potential.

The result of the integration is thus the usual logarithmic term. This result can now be generalized to the potential of more general triangles or even planar polygons, (in their plane). The result will be a sum of logarithmic terms. For instance, (figure 3), the potential of the triangle ABC at the interior point D can be obtained by adding three triangles. If D is an exterior point, we must subtract the triangle DAC, (figure 4). We have thus the general formula:

$$U_{ABC} = U_{DAB} + U_{DBC} \pm U_{DAC} \ . \tag{3.6}$$

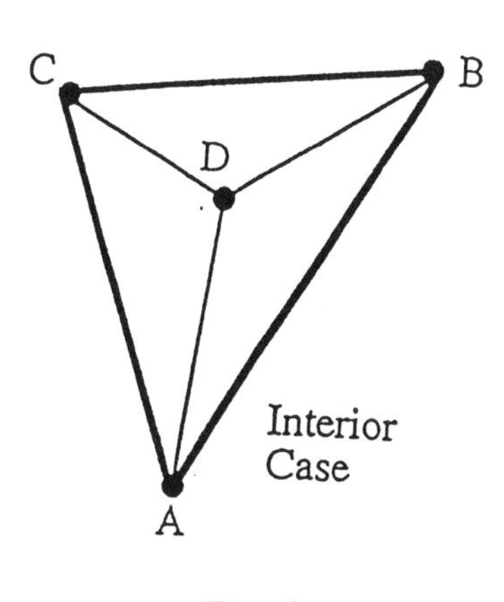

Fig. 3

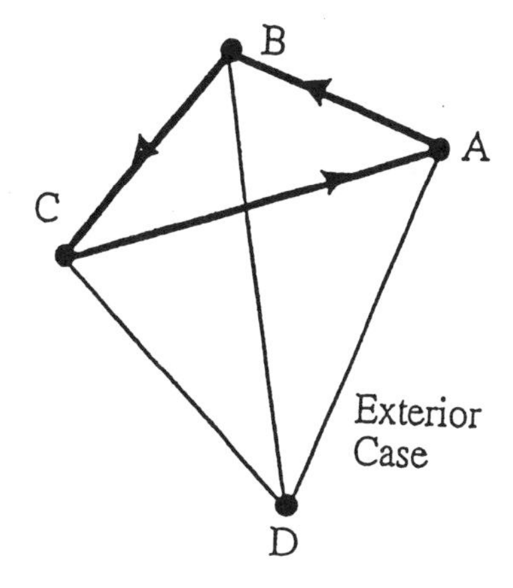

Fig. 4

This will be the sum of three logarithms. The potential is equivalent to the total potential of three wire segments, each with the appropriate mass:

$$U_{ABC} = U_{AB} + U_{BC} \pm U_{AC} \ . \tag{3.7}$$

In order to remove the sign ambiguity, the triangle ABC should be oriented, say in the direct sense (= counterclockwise). The + sign is taken if D is seen on the left-side by a point traveling over the sides of the triangle.

As was said above, all this applies to polygons with n sides: the potential will be the algebraic sum of n logarithmic terms, all similar to those for the wire segment potential.

Corollary: Assume a rectangle with sides parallel to the coordintate axes and with two opposite vertices at (0,0) and (x_1, y_1). Then, the potential at a corner of the homogenous rectangular plate is the same as the potential created by the two homogeneous wire segments at the sides opposite to this corner, if each side has a mass equal to the total mass M of the rectangle.

In other words, the potential at 0 is given by the formula

$$U = \frac{GM}{x_1} \log \left[\frac{d + x_1}{y_1} \right] + \frac{GM}{y_1} \log \left[\frac{d + y_1}{x_1} \right] , \tag{3.8}$$

where d is the length of the diagonal of the triangle. Each of the two terms is the potential of a wire segment.

In the following sections, we will find the expressions for the potential of some special and then some more general triangles. We will not restrict ourselves to two dimensions any longer. We will however assume that the triangles are in the xy-plane and we will find their potential at a point which is not in the plane of the triangle. Initially we will assume that this point is on the z-axis, but we will later generalize to any arbitrary point (x, y, z). We may also note that the literature on potential theory does not contain much on triangular plates. An exception is the exercise 9, in MacMillan, (1930, pages 22-23), which gives the acceleration due to a plate, in the form of an isosceles triangle, at a point outside it's plane.

4. The Potential of the Special Triangular Plate T_1

We will first give the potential, at a point $P(0, 0, z)$ on the z-axis, created by the right triangle shown in figure 5. It is located in the xy-plane. The side $\overline{P_oP_1}$ is horizontal. The problem is completely defined by the three constants (x_1, y_1, z). We use the abbreviation $\beta_1 = x_1/y_1$. We also define the distances:

$$d_Y^2 = y_1^2 + z^2 \quad ; \quad d_1^2 = x_1^2 + y_1^2 + z^2 \quad ; \quad d^2 = x^2 + y^2 + z^2 \; . \tag{4.1}$$

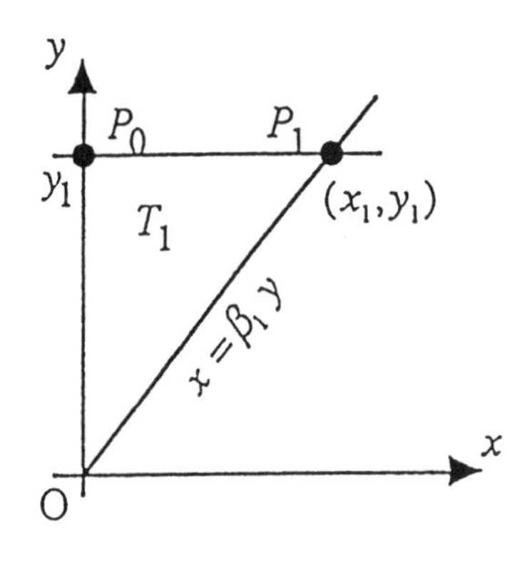

Fig. 5

The potential at P is then obtained by a double integration in x and y. We will integrate in x first:

$$\begin{aligned} U_1 &= G\sigma \int_{y=0}^{y_1} dy \int_{x=0}^{x=\beta_1 y} \frac{dx}{d} \\ &= G\sigma \int_{y=0}^{y_1} dy \left[\log(x + \sqrt{x^2 + y^2 + z^2}) \right]_0^{\beta_1 y} \; . \end{aligned} \tag{4.2}$$

We used integral 1. from our table in Appendix A. Next we substitute the limits 0 and $\beta_1 y$ for x and we integrate in y. Using the integrals 3 and 4 from Table A gives us the final result:

$$U_1 = G\sigma \left\{ y_1 \log \left[\frac{x_2 + d_1}{d_Y} \right] - z \tan^{-1} \left(\frac{d_1}{\beta_1 z} \right) + z \tan^{-1} \left(\frac{|z|}{\beta_1 z} \right) \right\} \; . \tag{4.3}$$

The same result is of course obtained if the order of the integrations is reversed. The result holds if $z = 0$, but in that case, we have only the logarithmic term.

5. The Potential of the Special Triangular Plate T_2

This is basically only given because it will allow us to recover a result in Kellog's book. We take another right triangle T_2, (fig. 6). It has a vertex $P_1(x_1y_1)$ and a vertex at the origin 0. It is bordered by the x-axis, a vertical line $x = x_1$ and the oblique line $x = \beta_1 y$. The notations are the same as in the previous sections. We define the distance d_X by:

$$d_X^2 = x_1^2 + z^2 \, . \tag{5.1}$$

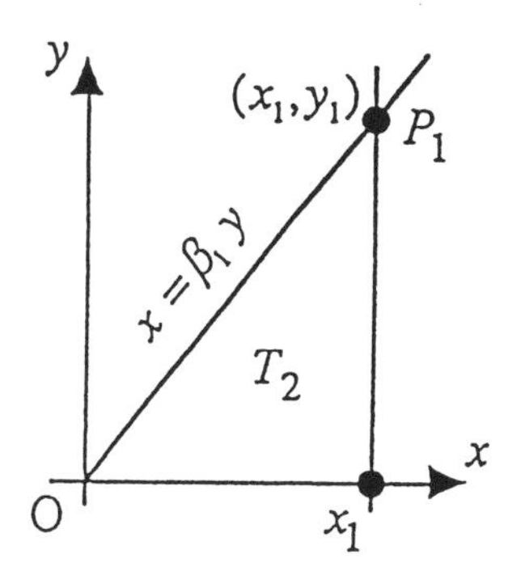

Fig. 6

We integrate again in x first:

$$U_2 = G\sigma \int_{y=0}^{y_1} dy \left[\log \left(x + \sqrt{x^2 + y^2 + z^2} \right) \right]_{x=\beta_1 y}^{x=x_1} \, . \tag{5.2}$$

The results are slightly more complicated than in the previous section, because both of the integration limits are non-zero (in x). But we will use the integrals 5 and 6 from our table A. This gives us the final result:

$$\begin{aligned} U_2 &= G\sigma \left\{ x_1 \log \left(\frac{d_1 + y_1}{d_X} \right) - z \tan^{-1} \left(\frac{x_1 y_1}{z \, d_1} \right) \right. \\ &\left. + z \tan^{-1} \left(\frac{d_1}{\beta_1 z} \right) - z \tan^{-1} \left(\frac{|z|}{\beta_1 z} \right) \right\} . \end{aligned} \tag{5.3}$$

We see that the above formula contains three Inverse Tangent-terms. They could be combined with the use of the standard Trigonometric formulas. We will later comment on this type of simplifications.

6. The Potential of the Special Rectangular Plate

The results of the two previous sections can be added up to form the potential of the rectangle at a point on the z-axis, (the line perpendicular to the rectangle at one of it's corners). A close inspection of the two potentials U_1 and U_2 shows that they have several of the Inverse Tangent terms in common, with opposite signs, so that the sum of the two expressions is a rather simple result:

$$\begin{aligned} U_2 &= G\sigma\left\{x_1 \log\left(\frac{d_1+y_1}{d_X}\right) + y_1 \log\left(\frac{d_1+x_1}{d_Y}\right)\right. \\ &\left. -z\ \tan^{-1}\left(\frac{x_1 y_1}{z d_1}\right)\right\} . \end{aligned} \tag{6.1}$$

This result for the rectangular plate was given by Kellog (1929, page 57) as an exercise using integral 5 in our Appendix A. Our results agree with Kellog, (except for the notations of course). In another exercise (page 12), Kellog gives the three components of the acceleration caused by the gravitational potential of the rectangular plate. Pierce (1902, page 28) computes the z-component of the acceleration due to a rectangular plate, at a point P on the perpendicular through it's center.

7. The Potential of the General Rectangular Plate

It is a trivial exercise to generalize the problem of the previous section to a more general configuration of the rectangular plate, as shown in figure 7. The corners have arbitrary locations, but the sides remain parallel to the coordinate axes.

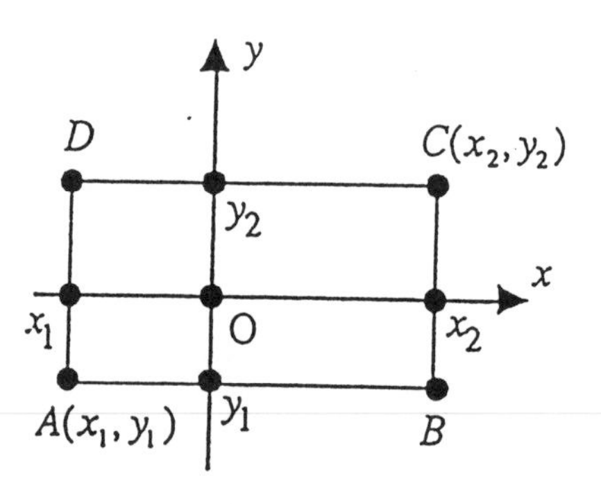

Fig. 7

The first integration gives the standard term of the form log $(r + x)$, (Integral 1 in A) while the next integration uses integral 5 of A (which is really Kellog's formula). When the limits of integration are properly applied, we finally get a sum of eight terms, four logarithms and four Inverse tangents. They involve the distances of the arbitrary point $P(0,0,z)$ on the z-axis to the four corners, such as for instance

$$d_A^2 = x_1^2 + y_1^2 + z^2 \; , \tag{7.1}$$

with similar expressions for d_B, d_C and d_D. We have then the following eight functions.

$$\begin{aligned}
G_1 &= \log[(x_2 + d_C)/(x_1 + d_D)] \quad ; \quad H_1 = \tan^{-1}[x_1 y_2/(z d_D)] \; , \\
G_2 &= \log[(y_2 + d_C)/(y_1 + d_B)] \quad ; \quad H_2 = \tan^{-1}[x_2 y_2/(z d_C)] \; , \\
G_3 &= \log[(y_1 + d_A)/(y_2 + d_D)] \quad ; \quad H_3 = \tan^{-1}[x_2 y_1/(z d_B)] \; , \\
G_4 &= \log[(x_1 + d_A)/(x_2 + d_B)] \quad ; \quad H_4 = \tan^{-1}[x_1 y_1/(z d_A)] \; .
\end{aligned}$$

The log-terms can basically be associated with the sides of the rectangle, while the Inverse tangents are associated with it's four angles. The potential function is then given by the following expression:

$$U = G\sigma \left\{ y_2 G_1 + x_2 G_2 + x_1 G_3 + y_1 G_4 + z(H_1 - H_2 + H_3 - H_4) \right\} \; . \tag{7.2}$$

Now, we want to make a rather important generalization of this formula, to the case where the Point P has arbitrary coordinates (x, y, z), instead of being on the z-axis. This task is rather easy, because the four vertices A, B, C, D have completely arbitrary locations. Consequently we may translate the whole system by arbitrary amounts (x, y) along the two axes.

To obtain the new and more general result, it is sufficient to replace in all the above formulas, (x_1, y_1, x_2, y_2) respectively by $(x_1 - x, y_1 - y, x_2 - x, y_2 - y)$. This gives us the new expression of the potential U as a function of the three variables (x, y, z). As a consequence we can now also compute the components of the acceleration: the gradient of U. We find

$$\left\{ \begin{aligned}
F_x &= G\sigma(-G_2 - G_3) \; , \\
F_y &= G\sigma(-G_1 - G_4) \; , \\
F_z &= G\sigma(+H_1 - H_2 + H_3 - H_4) \; .
\end{aligned} \right. \tag{7.3}$$

It is important to note that in taking the partial derivatives of $U(x, y, z)$, the arguments of the logarithms and Inverse Tangents may be treated as constants (MacMillan, 1930, pages 79-81). This simplifies the work considerably. Similar but less general expressions for the accelerations were given by Kellog (1929, page 12).

The availability of the above general expressions for the acceleration gives us the possibility, for instance to compute and study the properties of orbits around a rectangular plate.

8. The Potential of a General Triangular Plate

In this section we summarize the list of formulas that give the potential at $P(0,0,z)$ of an arbitrary triangle. We first need to compute the three sides of triangle (r_{12}, r_{23}, r_{31}) followed by the three distances of the vertices to the origin, (r_1, r_2, r_3) and three distances to P, (d_1, d_2, d_3). We also compute the three dot-and cross-products (D_{12}, D_{23}, D_{31}) and (C_{12}, C_{23}, C_{31}).

Next we compute the three logarithms:

$$\begin{aligned} L_{12} &= \log\left[(d_1 + d_2 + r_{12})/(d_1 + d_2 - r_{12})\right] , \\ L_{23} &= \log\left[(d_2 + d_3 + r_{23})/(d_2 + d_3 - r_{23})\right] , \\ L_{31} &= \log\left[(d_3 + d_1 + r_{31})/(d_3 + d_1 - r_{31})\right] . \end{aligned} \tag{8.1}$$

As for the three arc-tangents, we first compute a numerator

$$N = -z(C_{12} + C_{23} + C_{31}) , \tag{8.2}$$

followed by three denominators:

$$\begin{aligned} D_1 &= z^2(r_1^2 + D_{23} - D_{31} - D_{12}) - C_{12}C_{31} , \\ D_2 &= z^2(r_2^3 + D_{31} - D_{12} - D_{23}) - C_{23}C_{12} , \\ D_3 &= z^2(r_3^2 + D_{12} - D_{23} - D_{31}) - C_{31}C_{23} . \end{aligned} \tag{8.3}$$

This allows us to compute the following sum of four angles:

$$\Sigma = \tan^{-1}\left(\frac{Nd_1}{D_1}\right) + \tan^{-1}\left(\frac{Nd_2}{D_2}\right) + \tan^{-1}\left(\frac{Nd_3}{D_3}\right) + S_z\pi . \tag{8.4}$$

The potential function is then given by

$$U = G\sigma\left\{\frac{C_{12}}{r_{12}}L_{12} + \frac{C_{23}}{r_{23}}L_{23} + \frac{C_{31}}{r_{31}}L_{31} + z\Sigma\right\} . \tag{8.5}$$

This expression is of course invariant with respect to an arbitrary rotation around the z-axis. The symbol S_z stands for sign (z), the sign of the variable z.

9. The Potential of a General Quadrilateral

As we have said in the previous sections, it is now easy to generalize our problem to any polygon with n sides. There will be n logarithms and n arc-tangents. The notations which have been introduced in the previous sections are such that many terms can be obtained by circular permutation of the indices. As an illustration we will give here the detailed set of formulas for an irregular quadrilateral defined by it's four vertices (x_i, y_i).

Again, we first compute the four sides as well as the four distances to 0 and to $P(0,0,z)$. Next we need to compute several cross-products and dot-products of the 4 vectors $\vec{r}_i$. More precisely, we need eight products of each kind:

$$D_{12}, D_{23}, D_{34}, D_{41}, D_{13}, D_{24}, D_{31}, D_{42} ,$$
$$C_{12}, C_{23}, C_{34}, C_{41}, C_{13}, C_{24}, C_{31}, C_{42} .$$

Strictly speaking, only 6 of each need to be computed because of the symmetry properties such as $D_{42} = D_{24}$ and $C_{42} = -C_{24}$.

The four logarithmic terms $(L_{12}, L_{23}, L_{34}, L_{41})$ are defined similarly to the same terms for the triangle. On the other hand, for the four arc-tangents we will here write out explicitly the denominators D_i and the numerators N_i:

$$\begin{aligned} N_1 &= -zd_1(C_{12} + C_{41} - C_{42}) , \\ N_2 &= -zd_2(C_{23} + C_{12} - C_{13}) , \\ N_3 &= -zd_3(C_{34} + C_{23} - C_{24}) , \\ N_4 &= -zd_4(C_{41} + C_{34} - C_{31}) , \end{aligned} \tag{9.1}$$

$$\begin{aligned} D_1 &= z^2(d_1^2 + D_{42} - D_{41} - D_{12}) - C_{12}C_{41} , \\ D_2 &= z^2(d_2^2 + D_{13} - D_{12} - D_{23}) - C_{23}C_{12} , \\ D_3 &= z^2(d_3^2 + D_{24} - D_{23} - D_{34}) - C_{34}C_{23} , \\ D_4 &= z^2(d_4^2 + D_{31} - D_{34} - D_{41}) - C_{41}C_{34} . \end{aligned} \tag{9.2}$$

So we have the following sum of five angles:

$$\begin{aligned} \Sigma &= \tan^{-1}\left(\frac{N_1}{D_1}\right) + \tan^{-1}\left(\frac{N_2}{D_2}\right) + \tan^{-1}\left(\frac{N_3}{D_3}\right) \\ &\quad + \tan^{-1}\left(\frac{N_4}{D_4}\right) + 2S_z\pi . \end{aligned} \tag{9.3}$$

The complete potential function is then

$$U = G\sigma \left\{ \frac{C_{12}}{r_{12}} L_{12} + \frac{C_{23}}{r_{23}} L_{23} + \frac{C_{34}}{r_{34}} L_{34} + \frac{C_{41}}{C_{41}} L_{41} + z\Sigma \right\} . \tag{9.4}$$

The result is rotationally invariant, around the z-axis. This result also remains valid when z tends to zero: the angle Σ has a well-defined constant limit, so that the term $z\Sigma$ vanishes. Only the four logarithmic terms survive, which result is consistent with the fact that the attraction is now equivalent to the attraction of the perimeter of the quadrilateral, with the appropriate homogeneous mass distribution on each of the sides.

In a practical application in three dimensions, it should of course be verified that the four points (x_i, y_i, z_i) forming the quadrilateral lie in a plane. This can be done for instance by checking that the following determinant is zero:

$$\det \begin{bmatrix} x_2 - x_1 & y_2 - y_1 & z_2 - z_1 \\ x_3 - x_1 & y_3 - y_1 & z_3 - z_1 \\ x_4 - x_1 & y_4 - y_1 & z_4 - z_1 \end{bmatrix} = 0 \,. \tag{9.5}$$

10. The Components of the Acceleration

In this section we will give a generalization of the results of the two previous sections, the triangle and the quadrilateral. In these two sections we got an expression for the potential at a special point $P(0, 0, z)$ on the z-axis. We will now generalize these results for an arbitrary point $P(x, y, z)$. At the same time, we will also show how to get the gradient of the potential. This will actually all be very similar to what we did for the rectangular plate.

Both in the case of the triangle or the quadrilateral, we use the fact that the vertices are now arbitrary points in the plane, so that we may translate the coordinate system by an arbitrary amount (x, y) along the two coordinate axes.

The more general result, giving us the potential function $U(x, y, z)$, at an arbitrary point P, is obtained by replacing in the expressions for the potential all the vertex-coordinates (x_i, y_i) by the new quantities $(x_i - x, y_i - y)$. This gives the new formulas for the potential. No other changes are required.

The next important application is now to compute the gradient of $U(x, y, z)$, giving us the components of the acceleration and the equations of motion of a particle. We also mention again that in taking the partial derivatives of $U(x, y, z)$, the arguments of the logarithms and Inverse tangents may be treated as constants (MacMillan, 1930, pages 70-81).

As for the triangular plate, the three components of the acceleration are then:

$$\left\{ \begin{array}{lcl} F_x & = & -G\sigma[(y_2 - y_1)L_{12}/r_{12} + (y_3 - y_2)L_{23}/r_{23} + (y_1 - y_3)L_{31}/r_{31}] \,, \\ F_y & = & +G\sigma[(x_2 - x_1)L_{12}/r_{12} + (x_3 - x_2)L_{23}/r_{23} + (x_1 - x_3)L_{31}/r_{31}] \,, \\ F_z & = & +G\sigma \sum \,. \end{array} \right. \tag{10.1}$$

As for the quadrilateral plate, the three components of the acceleration are then given by completely similar formulas:

$$\begin{cases} F_x &= -G\sigma[(y_2 - y_1)L_{12}/r_{12} + (y_3 - y_2)L_{23}/r_{23} \\ &\quad +(y_4 - y_3)L_{34}/r_{34} + (y_1 - y_4)L_{41}/r_{41}] \ , \\ F_y &= -G\sigma[(x_2 - x_1)L_{12}/r_{12} + (x_3 - x_2)L_{23}/r_{23} \\ &\quad +(x_4 - x_3)L_{34}/r_{34} + (x_1 - x_4)L_{41}/r_{41}] \ , \\ F_z &= +G\sigma\Sigma \ . \end{cases} \tag{10.2}$$

11. The Completely Intrinsic Formulation

In the previous sections, we have given general formulas for the potential created by a polygon located in the xy-plane. The formulas are general in the sense that the vertices have arbitrary coordinates of the form $(x_i, y_i, 0)$. The fact that the polygon is in the xy-plane could still be considered as a lack of generality. However it is possible, with only some minor changes, to obtain a completely intrinsic formulation where the polygon is allowed to have a completely arbitrary orientation in 3D-space.

It is sufficient to replace the z-coordinate of the attracted unit mass P by it's signed distance to the polygon. This distance can be written in an intrinsic form with the use of the mixed triple product of vectors. Let the unit mass P have the position vector $\vec{r}(xyz)$. We define the vector $\vec{v}_1 = \vec{V}_1 - \vec{r}$ from P to any vertex of the polygon, say $\vec{V}_1$. Let also $\hat{n}$ be the unit-vector normal to the polygon. The signed distance Z from P is then the projection of $\vec{v}_1$ on $\hat{n}$: $Z_F = \vec{v}_1 \cdot \hat{n}$.

Actually the unit-vector $\hat{n}$ can easily be written in terms of the cross-product of any two vectors in the plane of the polygon. If we take the edge vectors between the vertices 1, 2, and 3, ($\vec{V}_{12} = \vec{V}_2 - \vec{V}_1$; $\vec{V}_{23} = \vec{V}_3 - \vec{V}_2$), with sides r_{12} and r_{23}, the signed distance from P can be expressed by the general formula:

$$Z = \vec{v}_1 \cdot (\vec{V}_{12} \times \vec{V}_{23})/r_{12}r_{23} \ . \tag{11.1}$$

This expression should be substituted for the z-coordinate of P in the formulas of the previous sections.

12. The Attraction of a Massive Pyramid at it's Apex

The formulas given in this section are inspired from MacMillan's perspectivity theorem (MacMillan, 1930, page 9).

We have thus a thin horizontal plate, (figure 8), $P_1P_2P_3P_4$, located at a height h above the origin O. We want to give a formula for it's attraction on a unit-mass at the origin O. The area of the plate is B, its surface density is σ and it's mass is $M_B = \sigma B$. The Newtonian force of attraction is of course inversely proportion to h^2. Let us consider only one component F, to simplify the notations:

$$F = G\alpha \frac{\sigma B}{h^2} = G\alpha \frac{M_B}{h^2}\,, \tag{12.1}$$

where α is a direction cosine, independent of the height h of the plate, (a dimensionless quantity).

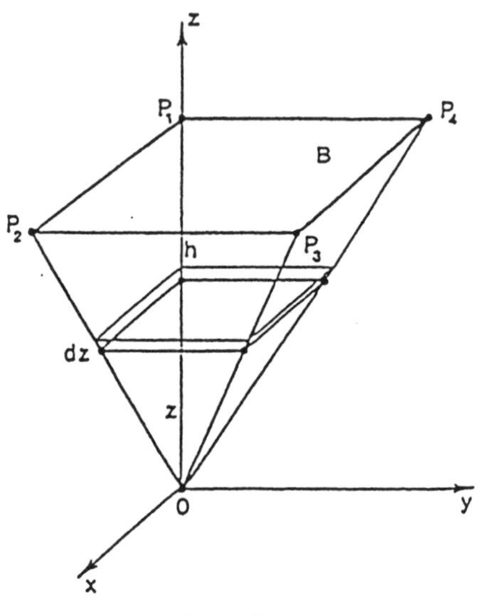

Fig. 8

In fact, what the perspectivity theorem really says is that F is independent of h, (when we replace the plate by another one, parallel and in perspective with it). This is because the mass is then proportional to h^2, assuming that we compare plates of the same surface density.

Now we refer to the same figure 8, but we will consider the attraction at O, due to the massive pyramid $OP_1P_2P_3P_4$, with constant volume density ρ, height h and base B. To find the total force of attraction at O, we first consider a thin horizontal section, at height z and with thickness dz. The area of this section is $B \cdot (z/h)^2$. The volume of the layer is $B \cdot (z/h)^2 \cdot dz$. The attraction at O, due to this layer is then

$$dF = G\alpha\rho \left(\frac{z}{h}\right)^2 B\frac{dz}{z^2} = G\alpha\rho B \cdot \frac{dz}{h^2}\,, \tag{12.2}$$

where the constant α is as defined in the previous section. We note that dF is independent of the height z of the layer.

The total attraction of the complete pyramid is now the integral over z, from O to h of the previous expression (55):

$$F = \frac{G\alpha\rho Bh}{h^2} = \frac{G\alpha}{h^2}(3M_P) , \tag{12.3}$$

where $M_P = \rho Bh/3$ is the total mass of the pyramid.

Consequently, the expression for the force F is similar to the expression (15.1) that was given above for the thin plate at the base of the pyramid. This is the famous theorem:

> "*The attraction due to a pyramid, at its apex O, is the same as the attraction due to a plate at the base of the pyramid, with three times the mass of the pyramid. The direction of the attraction is the same*".

This theorem is given at several places in the literature, (MacMillan, 1930, page 8), (Tisserand, 1890, Vol. 2, page 68), but the derivation is usually done for a narrow pyramid with infinitesimal base dS instead of our finite area B. The fact that we draw a rectangular base in figure 12 plays of course no role in the derivation of the result.

We consider this as an important theorem, as it will allow us to reduce the gravity of three-dimensional volumes to the gravity created by two-dimensional surfaces, (usually at the boundaries of the volume), (Levallois, 1960, Vol. 3, page 82).

A similar reasoning could be done for the potential, rather than the accelerations, the result will be slightly different however: the factor will be 3/2 instead of 3.

Referring to the same figure 8, the potential at 0, created by the top plate $P_1P_2P_3P_4$ will be proportional to its mass and inversely proportional to the distance h:

$$U_B = G\beta\frac{\sigma B}{h} = G\beta\frac{M_B}{h} , \tag{12.4}$$

where β is a dimensionless proportionality factor, independent of h and σ is the surface density.

Let us also consider the potential (created at 0), by a thin layer (thickness dz) at a height $z(< h)$; the proportionality factor β is the same:

$$dU = G\beta\frac{dM}{z} . \tag{12.5}$$

The area of this layer is $B(z/h)^2$ so that its mass, (for a given volume density ρ) and potential are:

$$dM = \frac{\rho Bz^2dz}{h^2} \quad ; \quad dU = \frac{G\beta\rho Bzdz}{h^2} . \tag{12.6}$$

An integration in z from 0 to h gives the total potential at the apex 0 created by the pyramid:

$$U_P = \frac{G\beta\rho B}{h^2} \cdot \frac{h^2}{2} = \frac{G\beta\rho B}{2} . \tag{12.7}$$

If we introduce the volume and mass of the pyramid, the above expression may be written in the form

$$U_P = \frac{G\beta}{2h}(3M_P) = \frac{G\beta}{h} \cdot \left(\frac{3M_P}{2}\right) . \tag{12.8}$$

We have thus the following theorem resulting from the comparison of (15.4) and (15.8):

> "*The potential due to a pyramid, at its apex 0, is the same as the potential of a thin plate at the base, with 3/2 times the mass of the pyramid.*"

13. The Potential Created by a Polyhedron

We have seen in the previous section that the potential of (three-dimensional) pyramids can be obtained with the formulas for the potential of two-dimensional masses (the bases of the pyramids).

Now we consider convex homogeneous polyhedra, with an arbitrary number of planar faces of polygonal shape. Let us first consider the potential at an exterior point A. We first construct all the pyramids from A to all the faces of the polyhedron. There will be two classes of pyramids: the first class contains all the pyramids constructed on the visible faces (from A) and the second class all the pyramids on the hidden faces.

The total potential at A, due to the polyhedron is then the algebraic sum of the potentials of all the pyramids, assuming a + sign for a hidden face and a – sign for a visible face. Actually, the distinction between hidden and visible faces is computationally quite automatic. We need to take the unit normal to each face. Then we take its dot product with the line of sight and this immediately gives us the required sign: + for a visible face and – for a hidden face.

The situation is even more simple in the case of a Point A which is inside the polyhedron. We again construct all the pyramids from A to the faces of the polyhedron and we compute the separate potentials of these inner pyramids. All we need to do now is simply add all these potentials (all of them with the same + sign).

This approach go getting the potential of a polyhedron was proposed by Tisserand (1890, Volume 2, page 68).

Appendix A – A Table of Integrals

In the present appendix we collect some of the most important indefinite integrals which have been used in this article. The first three integrals are taken from the tables. The first one is the most fundamental one, showing how the logarithm function appears in this work as the integral of $1/r$, (where $r^2 = x^2 + y^2 + z^2$). In the second integral we again obtain a logarithm as a result. The remaining four integrals operate on this logarithm. It is seen that, in all cases, arc-tangents appear in the results. The integral 5 is taken from Kellog's book (1929, page 57). Note that Kellog's integral has a misprint in the last of the five terms: the x in the numerator should be changed into a ζ. All of these integrals can be easily checked by differentiation or by integration by parts.

In integral 2, we used the abreviations:

$$X = cx^2 + bx + a \quad ; \quad q = 4ac - b^2$$

In the integral 6, the constant C depends on A: $C^2 = A^2 + 1$.

The six integrals are as follows:

1. $$\int \frac{dx}{r} = \log(x + r) = th^{-1}\left(\frac{x}{r}\right) .$$
2. $$\int \frac{dx}{\sqrt{X}} = \frac{1}{\sqrt{c}} \log\left[\frac{2cx + b}{\sqrt{q}} + \frac{2\sqrt{c}}{\sqrt{q}}\sqrt{X}\right] .$$
3. $$\int \log\sqrt{a^2 + x^2} \cdot dx = x \log\sqrt{a^2 + x^2} - x + a \tan^{-1}\left(\frac{x}{a}\right) .$$
4. $$\int \log[Ax + r]dx = x \log[Ax + r] - x - y \tan^{-1}\left[\frac{r}{Ay}\right] + y \tan^{-1}\left[\frac{x}{y}\right] .$$
5. $$\int \log(y + r)dx = x \log(r + y) + y \log(r + x) - x + z \tan^{-1}\left(\frac{x}{z}\right) - z \tan^{-1}\left(\frac{xy}{zr}\right) .$$
6. $$\int \log(Ax + \sqrt{C^2x^2 + a^2})dx = x \log(Ax + \sqrt{C^2x^2 + a^2}) - x + a \tan^{-1}\left(\frac{x}{a}\right) + a \tan^{-1}\left(\frac{aA}{\sqrt{C^2x^2 + a^2}}\right) .$$

Acknowledgements

We had several conversations with R. Werner, who worked out a detailed derivation of the potential of the polyhedron, (1994), based on the Gauss Divergence theorem and the standard formula, (Macmillan, 1930 or Levallois, 1970), for the inverse distance in terms of the Divergence of a unit vector.

References

1. Banerjee, B. and Das Gupta, S. P., 1977, "Gravitational Attraction of a Rectangular Parallelepiped," *Geophysics*, Vol. 42, pages 1053-1055.
2. Cady, J. W., 1980, "Calculation of gravity and magnetic anomalies of finite-length right polygonal prisms," *Geophysics*, Vol. 45, pages 1507-1512.
3. Golizdra, G. Ya., 1981,"Calculation of the gravitational field of a Polyhedron," Izvestiya, *Earth Physics*, Vol. 17, pages 625-628.
4. Green, G., 1828, "An essay on the Application of Mathematial Analysis to the Theories of Electricity and Magnetism", Nottingham.
5. Halamek, P., 1988, Ph.D. Dissertation, University of Texas, Austin, Texas.
6. Kellog, O. D., 1929, "Foundations of Potential Theory," Dover, N.Y.
7. Levallois, J. J., 1970, "Geodesie Generale," Vol. 3, Eyrolles, Paris.
8. MacMillan, W. D., 1930, "The theory of the Potential," Dover, N.Y.
9. Nagy, D.,1966, "The gravitational attraction of a right rectangular prism," Geophysics, Vol. 31, pages 362-374.
10. Okabe, M., 1979, "Analytical expressions for the gravity of polyhedral bodies", *Geophysics*, Vol. 44, pages 730-741.
11. Pierce, B. O., 1902, "The Newtonian Potential Function", Boston.
12. Pohanka, V., 1988, "Optimum Expression for Computation of the Gravity field of a Homogeneous Polyhedral Body," *Geophysical Prospecting*, Vol. 36, pages 733-751.
13. Strakhof, V. N. and Lapina, M. I., 1990, "Direct Gravimetric and Magnetometric problems for Homogeneous Polyhedrons," *Geophysical Journal*, Vol. 8, pages 740-756.
14. Tisserand, F., 1890, "Traite de Mecanique Celeste," Vol. 2, Gauthier - Villars, Paris.
15. Waldvogel, J., 1976, "The Newtonian Potential of a Homogeneous Cube," *ZAMP*, Vol. 27, pages 867-871.
16. Waldvogel, J., 1979, "The Newtonian Potential of Homogeneous Polyhedra," *ZAMP*, Vol. 30, pages 388-398.
17. Werner, R., 1994, "The Potential of a Polyhedron", Celestial Mechanics, Vol. 59, Pages 253-278.

ORBITAL ELEMENTS OF A SATELLITE MOVING IN THE POTENTIAL OF A HEMISPHERICAL SHELL

B. ZAFIROPOULOS AND CH. STAVLIOTIS
Department of Physics
University of Patras
26110 Patras, Greece

Abstract. This paper gives the perturbed elements of a satellite moving in the potential of a hemispherical shell. In the first section of this investigation the potential of the shell is calculated explicitly in terms of spherical harmonics. The disturbing accelerations produced by the above potential have been substituted into the Gauss-Lagrange planetary equations. The orbital elements produced, after integrating these equations, have been expressed by means of summations and the zero-order Hansen coefficients. The results include any order of odd spherical harmonics.

1. INTRODUCTION

The potential of a hemispherical surface can be equivalently represented by the sum of the potentials of an infinitive number of circular rings. The potential of a ring of radius A, located at $z = 0$, at any point on the positive z-axis is

$$U_r = -\mu_r \left(A^2 + z^2\right)^{-\frac{1}{2}}, \tag{1}$$

where μ_r is the product of the ring mass times the gravitational constant.

We consider sections of the shell at distances ζ from the plane $z = 0$. By means of Equation (1), we obtain for the potential of the hemispherical shell

$$\begin{aligned} U &= -\mu_r \int_0^A \left[A^2 - \zeta^2 + (z-\zeta)^2\right]^{-\frac{1}{2}} d\zeta \\ &= -\mu_r \left(\frac{1}{z}\right) \left[(A^2 + z^2)^{\frac{1}{2}} - (z - A)\right]. \end{aligned} \tag{2}$$

B.A. Steves and A.E. Roy (eds.), The Dynamics of Small Bodies in the Solar System, 341–348.

Developing the above expression in descending powers of z we have

$$U = -\frac{\mu}{z}\left[1+\frac{A}{2z}-\frac{1}{2^3}\frac{A^3}{z^3}+\frac{1}{2^4}\frac{A^5}{z^5}-...\right], \; z > A, \tag{3}$$

where $\mu = \mu_r A$ is the product of the hemispherical shell mass times the gravitational constant.

Equation (3) gives the potential of the shell for any point on the z-axis of symmetry, with $z > A$. According to a theorem stated by Legendre, the potential of the body can be estimated at all points with distance $r > A$, in terms of Legendre polynomials P_n. Substituting in the last equation $P_{n-1}r^{-n}$ for z^{-n} we obtain the general expression for the potential of a hemispherical shell (cf. e.g. MacRobert, 1967, p. 142), for $r > A$

$$U = -\frac{\mu}{A}\left[\frac{A}{r}+\frac{1}{2}\left(\frac{A}{r}\right)^2 P_1(cos\vartheta)-\frac{1}{2^3}\left(\frac{A}{r}\right)^4 P_3(cos\vartheta)+\frac{1}{2^4}\left(\frac{A}{r}\right)^6 P_5(cos\vartheta)-...\right], \tag{4}$$

where ϑ is the co-latitude. The above series contains only odd spherical harmonics.

Equation (4) can be presented in the compact form of summation as follows

$$U = -\frac{\mu}{r}\left[1+\sum_{\nu=1}^{\infty}\frac{(-1)^{\nu+1}(2\nu-3)!!}{(2\nu)!!}\left(\frac{A}{r}\right)^{2\nu-1} P_{2\nu-1}(\cos\vartheta)\right], \; r > A, \tag{5}$$

where

$$(2\nu-3)!! = (2\nu-3)\cdot(2\nu-5)\cdot\ldots\cdot 1\,, \qquad (2\nu)!! = (2\nu)\cdot(2\nu-2)\cdot\ldots\cdot 2 \tag{6}$$

and we define $(2\nu-3)!! = 1$, for $\nu = 1$. In the same way we produce similar expressions for various other axi-symetric potentials in terms of Legendre polynomials. As an example we give below the potentials due to a disc and a homogeneous hemisphere.

For the expressions given below, r represents the distance from the center of the body and ϑ stands for the co-latitude.

Disk Potential. The gravitational potential of a circular disk of radius A is (Ramsey, 1981)

$$U = -\frac{2\mu}{r}\left[\frac{1}{2}+\sum_{\nu=1}^{\infty}\frac{(-1)^{\nu}(2\nu-1)!!}{(2\nu+2)!!}\left(\frac{A}{r}\right)^{2\nu} P_{2\nu}(\cos\vartheta)\right], \; r > A. \tag{7}$$

Potential of a Homogeneous Hemisphere. The relative expression is (Zafiropoulos and Stavliotis, 1996)

$$U = -\frac{\mu}{r}\left[1+3\sum_{\nu=0}^{\infty}\frac{(-1)^{\nu}(2\nu-1)!!}{(2\nu+4)!!}\left(\frac{A}{r}\right)^{2\nu+1} P_{2\nu+1}(\cos\vartheta)\right], r > A. \tag{8}$$

where we define $(2\nu - 1)!! = 1$, for $\nu = 0$. The potentials of all the above bodies possess an axis of symmetry. Similar expressions can be obtained for the potential of a ring, an oblate spheroid, a magnet, etc.

2. EQUATIONS OF MOTION

The orbit of the satellite is specified by six independent parameters known as the elements of the orbit. We choose these elements to be: the longitude of the ascending node Ω, the inclination of the orbital plane ι, the semi-major axis α, the eccentricity of the orbit e, the argument of the pericenter ω, and the mean anomaly χ at time $t = 0$.

The motion of a small mass satellite around a hemispherical shell is governed by the equation

$$\frac{d^2 \vec{r}}{dt^2} = -\nabla U\,, \tag{9}$$

where $\vec{r}$ is the position vector of the satellite and the potential U is given by Equation (5).

The system of the three second-order differential equations defined by (9) can be solved in a closed form only in the presence of a spherical potential of the form $U_0 = -\frac{\mu}{r}$. It is useful, when a precise solution cannot be obtained, to work out analytic approximations. We assume that the orbit of the satellite does not depart appreciably from a Keplerian one and employ the osculating elements to describe the motion of the satellite.

By means of Equation (5), we have, for the disturbing potential

$$V = U - U_0 = -\frac{\mu}{r} \sum_{\nu=1}^{\infty} \frac{(-1)^{\nu+1}(2\nu-3)!!}{(2\nu)!!} \left(\frac{A}{r}\right)^{2\nu-1} P_{2\nu-1}(\cos\vartheta)\,. \tag{10}$$

Using simple spherical trigonometry we get, for the angle ϑ,

$$\cos\vartheta = \sin\iota \, \sin u\,. \tag{11}$$

The relation (9) is equivalent to a system of six first-order differential equations which are known as the Gauss-Lagrange equations. Using as independent variable the true anomaly υ measured from the pericenter of the orbit, we have (cf. e.g. Smart 1953)

$$\frac{d\Omega}{d\upsilon} = \frac{r^3 \sin u}{H^2 \sin\iota} W\,, \tag{12}$$

$$\frac{d\iota}{d\upsilon} = \frac{r^3 \cos u}{H^2} W\,, \tag{13}$$

$$\frac{d\alpha}{dv} = \frac{2\,\alpha^2\, r}{H^2}\{ e\, r\, sin\, v\, R + p\, S\}, \tag{14}$$

$$\frac{de}{dv} = \frac{r^2}{H^2}\{ p\, sin\, v\, R + (p + r)\, cos\, v\, S + e\, r\, S\}, \tag{15}$$

$$\frac{d\omega}{dv} = \frac{r^2}{e\, H^2}\{ -p\, cos\, v\, R + (p + r)\, sin\, v\, S\} - cos\, \iota\, \frac{d\Omega}{dv}, \tag{16}$$

$$\frac{d\chi}{dv} = \frac{r^2\sqrt{1 - e^2}}{e\, H^2}\{ (p\, cos\, v - 2\, e\, r)\, R - (p + r)\, sin\, v\, S\}, \tag{17}$$

where R, S and W are the components of the disturbing accelerations, u and v denote the true anomalies measured from the ascending node and the periastron, respectively, H is the angular momentum of the satellite about the centre of the force, given by $H = \sqrt{\mu p}$, p is the semi-latus rectum and r represents the distance of the satellite specified by

$$r = \frac{\alpha\,(1 - e^2)}{1 + e\, cos\, v} = \frac{p}{1 + e\, cos\, v}. \tag{18}$$

3. THE PERTURBING ACCELERATIONS R, S AND W

In order to employ the system of equations (12) to (17) we need to estimate the components of the disturbing accelerations. These are defined as the rates of change of the perturbing potential V. Thus we have:

$$R = \frac{\partial V}{\partial r}, \tag{19}$$

$$S = -\frac{cos\, u\, sin\, \iota}{r\, sin\, \vartheta}\frac{\partial V}{\partial \vartheta}, \tag{20}$$

$$W = -\frac{cos\, \iota}{r\, sin\, \vartheta}\frac{\partial V}{\partial \vartheta}. \tag{21}$$

We use the explicit formulae for the various order spherical harmonics and insert $cos\,\vartheta = sin\,\iota\, sin\, u$. f stands for $sin^2 \iota$. The method employed is general and we are able to incorporate any order of odd spherical harmonics.

Putting (Zafiropoulos, 1987)

$$\rho_{1,0} = 1, \quad \rho_{3,0} = \frac{3}{4}(1 - \frac{5}{4} f), \quad \rho_{3,1} = \frac{5\, f}{16}, \ etc. \tag{22}$$

we obtain, for the component R,

$$R = \frac{\mu\, sin\, \iota}{r^2} \sum_{\nu=1}^{\infty} \left(\frac{A}{r}\right)^{2\nu - 1} \sum_{j=0}^{\nu - 1} \rho_{2\nu - 1, j}\, sin\, [\,(2j + 1)\, u\,]. \tag{23}$$

Similarly, for the component S, we make the following substitutions

$$\sigma_{1,0} = \frac{1}{2}, \quad \sigma_{3,0} = \frac{3}{16}\left(1 - \frac{5f}{2}\right), \quad \sigma_{3,1} = \frac{15f}{32}, \quad etc. \tag{24}$$

and, consequently, we write

$$S = -\frac{\mu \sin\iota \cos u}{r^2} \sum_{\nu=1}^{\infty} \left(\frac{A}{r}\right)^{2\nu-1} \sum_{j=0}^{\nu-1} \sigma_{2\nu-1,j} \cos(2ju). \tag{25}$$

Finally, the component W, in the compact form of summation is

$$W = -\frac{\mu}{r^2} \cos\iota \sum_{\nu=1}^{\infty} \left(\frac{A}{r}\right)^{2\nu-1} \sum_{j=0}^{\nu-1} \sigma_{2\nu-1,j} \cos(2j\,u). \tag{26}$$

4. THE ORBITAL ELEMENTS

The perturbed elements of the satellite can be presented in the compact form of summations by means of the zero-order Hansen coefficients. The relation which permits us to introduce the zero-order Hansen coefficients into the expressions for the perturbed elements of the orbit, has the form

$$\left(\frac{\alpha}{r}\right)^n = \sqrt{1-e^2} \sum_{k=0}^{n} (2 - \delta_{0,k}) X_0^{-(n+2),k} \cos kv, \tag{27}$$

where the symbol $\delta_{0,k}$ denotes the Kronecker's delta and $X_0^{n,m}$ are the Hansen's coefficients (Zafiropoulos and Kopal 1982 and Zafiropoulos and Zafiropoulos 1982).

Replacing R, S, W and $\frac{a}{r}$, in Equations (12) to (17), by means of Equations (23), (25), (26) and (27), respectively, we obtain the following expressions for the perturbed elements

$$\delta\Omega = \Omega - \Omega_0 = \frac{\cot\iota}{4\sqrt{1-e^2}} \sum_{\nu=1}^{\infty} \left(\frac{A}{\alpha}\right)^{2\nu-1} \sum_{j=0}^{\nu-1} \sigma_{2\nu-1,j} \times$$
$$\times \sum_{k=0}^{2\nu-2} (2 - \delta_{0,k}) X_0^{-2\nu,k} C_{jk}^1(v,\omega) \Big|_{v=0}^{v=v}, \tag{28}$$

where

$$C_{jk}^1(v,\omega) = \frac{\cos[(2j+k+1)v + (2j+1)\omega]}{(2j+k+1)} + \frac{\cos[(2j-k+1)v + (2j+1)\omega]}{(2j-k+1)} -$$
$$- \frac{\cos[(2j+k-1)v + (2j-1)\omega]}{(2j+k-1)} - \frac{\cos[(2j-k-1)v + (2j-1)\omega]}{(2j-k-1)}.$$

$$\delta\,\iota \;=\; \iota - \iota_0 = -\,\frac{\cos\iota}{4\sqrt{1-e^2}}\sum_{\nu=1}^{\infty}\left(\frac{A}{\alpha}\right)^{2\nu-1}\sum_{j=0}^{\nu-1}\sigma_{2\nu-1,j}\;\times$$

$$\times\sum_{k=0}^{2\nu-2}\left(2-\delta_{0,k}\right)X_0^{-2\nu,k}\,S^1_{jk}(\upsilon,\omega)\,\Big|_{\upsilon=0}^{\upsilon=\upsilon}\,. \tag{29}$$

where

$$S^1_{jk}(\upsilon,\omega) \;=\; \frac{\sin[(2j+k+1)\upsilon+(2j+1)\omega]}{(2j+k+1)}+\frac{\sin[(2j+k-1)\upsilon+(2j-1)\omega]}{(2j+k-1)}+$$

$$+\;\frac{\sin[(2j-k+1)\upsilon+(2j+1)\omega]}{(2j-k+1)}+\frac{\sin[(2j-k-1)\upsilon+(2j-1)\omega]}{(2j-k-1)},$$

$$\delta\,\alpha \;=\; \alpha-\alpha_0 = -\,\frac{\alpha\sqrt{1-e^2}\,\sin\iota}{2}\sum_{\nu=1}^{\infty}\left(\frac{A}{\alpha}\right)^{2\nu-1}\times$$

$$\times\Big\{\sum_{j=0}^{\nu-1}\sigma_{2\nu-1,j}\sum_{k=0}^{2\nu}\left(2-\delta_{0,k}\right)X_0^{-(2\nu+2),k}\,S^1_{jk}(\upsilon,\omega)\;-$$

$$-\;\frac{e\,\alpha}{p}\sum_{j=0}^{\nu-1}\rho_{2\nu-1,j}\sum_{k=0}^{2\nu-1}\left(2-\delta_{0,k}\right)X_0^{-(2\nu+1),k}\,S^2_{jk}(\upsilon,\omega)\Big\}\,\Big|_{\upsilon=0}^{\upsilon=\upsilon} \tag{30}$$

where

$$S^2_{jk}(\upsilon,\omega) \;=\; \frac{\sin[(2j+k)\upsilon+(2j+1)\omega]}{(2j+k)}+\frac{\sin[(2j-k)\upsilon+(2j+1)\omega]}{(2j-k)}-$$

$$-\;\frac{\sin[(2j+k+2)\upsilon+(2j+1)\omega]}{(2j+k+2)}-\frac{\sin[(2j-k+2)\upsilon+(2j+1)\omega]}{(2j-k+2)}\,.$$

$$\delta\,e \;=\; e-e_0 = \frac{\sin\iota\sqrt{1-e^2}}{8}\sum_{\nu=1}^{\infty}\left(\frac{A}{\alpha}\right)^{2\nu-1}\Big\{\sum_{k=0}^{2\nu-1}\left(2-\delta_{0,k}\right)\times$$

$$\times\,X_0^{-(2\nu+1),k}\Big[2\sum_{j=0}^{\nu-1}\rho_{2\nu-1,j}\,S^2_{jk}(\upsilon,\omega)\;-\;\sum_{j=0}^{\nu-1}\sigma_{2\nu-1,j}\,S^3_{jk}(\upsilon,\omega)\Big]\;-$$

$$-\frac{\alpha}{p}\sum_{k=0}^{2\nu-2}\left(2-\delta_{0,k}\right)X_0^{-(2\nu),k}\Big[\sum_{j=0}^{\nu-1}\sigma_{2\nu-1,j}\,S^3_{jk}(\upsilon,\omega)\;+$$

$$+ 2\,\varepsilon \sum_{j=0}^{\nu-1} \sigma_{2\nu-1,j}\, S^1_{jk}(\upsilon,\omega)\,\Big]\Big\}\Big|_{\upsilon=0}^{\upsilon=\upsilon}, \tag{31}$$

where

$$\begin{aligned} S^3_{jk}(\upsilon,\omega) &= \frac{sin[(2j+k)\upsilon+(2j+1)\omega]}{(2j+k)} + \frac{sin[(2j+k)\upsilon+(2j-1)\omega]}{(2j+k)} + \\ &+ \frac{sin[(2j-k)\upsilon+(2j+1)\omega]}{(2j-k)} + \frac{sin[(2j-k)\upsilon+(2j-1)\omega]}{(2j-k)} + \\ &+ \frac{sin[(2j+k+2)\upsilon+(2j+1)\omega]}{(2j+k+2)} + \frac{sin[(2j+k-2)\upsilon+(2j-1)\omega]}{(2j+k-2)} + \\ &+ \frac{sin[(2j-k+2)\upsilon+(2j+1)\omega]}{(2j-k+2)} + \frac{sin[(2j-k-2)\upsilon+(2j-1)\omega]}{(2j-k-2)}. \end{aligned}$$

$$\begin{aligned} \delta\,\omega &= \omega-\omega_0 = \frac{\sin\iota\,\sqrt{1-e^2}}{8\,e} \sum_{\nu=1}^{\infty} \Big(\frac{A}{\alpha}\Big)^{2\nu-1} \Big\{ \sum_{k=0}^{2\nu-1} (2-\delta_{0,k}) X_0^{-(2\nu+1),k} \times \\ &\times \Big[2\sum_{j=0}^{\nu-1} \rho_{2\nu-1,j}\, C^2_{jk}(\upsilon,\omega) + \sum_{j=0}^{\nu-1} \sigma_{2\nu-1,j}\, C^3_{jk}(\upsilon,\omega)\Big] + \frac{\alpha}{\rho} \sum_{k=0}^{2\nu-2} (2-\delta_{0,k}) \times \\ &\times X_0^{-2\nu,k} \sum_{j=0}^{\nu-1} \sigma_{2\nu-1,j}\, C^3_{jk}(\upsilon,\omega)\Big\}\Big|_{\upsilon=0}^{\upsilon=\upsilon} - cos\,\iota\;\delta\,\Omega\,, \end{aligned} \tag{32}$$

where

$$\begin{aligned} C^2_{jk}(\upsilon,\omega) &= \frac{cos[(2j+k)\upsilon+(2j+1)\omega]}{(2j+k)} + \frac{cos[(2j-k)\upsilon+(2j+1)\omega]}{(2j-k)} + \\ &+ \frac{cos[(2j+k+2)\upsilon+(2j+1)\omega]}{(2j+k+2)} + \frac{cos[(2j-k+2)\upsilon+(2j+1)\omega]}{(2j-k+2)}, \end{aligned}$$

$$\begin{aligned} C^3_{jk}(\upsilon,\omega) &= \frac{cos[(2j+k)\upsilon+(2j-1)\omega]}{(2j+k)} - \frac{cos[(2j+k)\upsilon+(2j+1)\omega]}{(2j+k)} - \\ &- \frac{cos[(2j-k)\upsilon+(2j+1)\omega]}{(2j-k)} + \frac{cos[(2j-k)\upsilon+(2j-1)\omega]}{(2j-k)} + \\ &+ \frac{cos[(2j+k+2)\upsilon+(2j+1)\omega]}{(2j+k+2)} - \frac{cos[(2j+k-2)\upsilon+(2j-1)\omega]}{(2j+k-2)} + \\ &+ \frac{cos[(2j-k+2)\upsilon+(2j+1)\omega]}{(2j-k+2)} - \frac{cos[(2j-k-2)\upsilon+(2j-1)\omega]}{(2j-k-2)}. \end{aligned}$$

$$
\begin{aligned}
\delta\chi &= \chi-\chi_0= \\
&= \sin\iota \sum_{\nu=1}^{\infty}\left(\frac{A}{\alpha}\right)^{2\nu-1}\Big\{\frac{-p}{4\,\alpha\, e}\sum_{k=0}^{2\nu-1}(2-\delta_{0,k})\,X_0^{-(2\nu+1),k}\Big[\sum_{j=0}^{\nu-1}\rho_{2\nu-1,j}\,C_{jk}^2\,(v,\omega) \\
&\quad +\frac{1}{2}\sum_{j=0}^{\nu-1}\sigma_{2\nu-1,j}\,C_{jk}^3(v,\omega)\Big] + \sum_{k=0}^{2\nu-2}(2-\delta_{0,k})\,X_0^{-2\nu,k} \\
&\quad \Big[\sum_{j=0}^{\nu-1}\rho_{2\nu-1,j}\,C_{jk}^4(v,\omega) - \frac{1}{8\,e}\sum_{j=0}^{\nu-1}\sigma_{2\nu-1,j}\,C_{jk}^3\,(v,\omega)\Big]\Big\}\Big|_{v=0}^{v=v}, \qquad (33)
\end{aligned}
$$

where

$$
\begin{aligned}
C_{jk}^4(v,\omega) = \frac{cos[(2j+k+1)v+(2j+1)\omega]}{(2j+k+1)} &+ \\
+\frac{cos[(2j-k+1)v+(2j+1)\omega]}{(2j-k+1)}&.
\end{aligned}
$$

The previous equations (28) to (33) provide the orbital elements of a satellite moving in the potential of a hemispherical shell. The elements are presented in the most general form by means of summations and the zero-order Hansen coefficients.

Tests have been performed that prove the correctness of the calculated elements. We can separate secular from periodic terms. The procedure is easily applied to similar forms of potentials (Zafiropoulos and Stavliotis 1995).

REFERENCES

MacRobert, T.M., 1967, *Spherical Harmonics*, Pergamon Press Ltd, 3rd Ed.

Ramsey, A.S., 1981, *Newtonian Attraction*, Cambridge Univ. Press, Cambridge.

Smart, W.M., 1953, *Celestial Mechanics*, Longmans, London.

Zafiropoulos, B. and Kopal, Z., 1982, Astrophys. Space Sci. **88**, 355.

Zafiropoulos, B. and Zafiropoulos, F., 1982, Astrophys. Space Sci. **88**, 401.

Zafiropoulos, B., 1987, Astrophys. Space Sci. **139**, 353.

Zafiropoulos, B. and Stavliotis, Ch., 1995, in "From Newton to Chaos", ed. Roy, A.E. and Steves, B.A., Plenum Press, p.235.

Zafiropoulos, B. and Stavliotis, Ch., 1996, in Proceedings of "Second Hellenic Astronomical Conference", ed. Contadakis, M.E., et al. p.602.

SECTION FOUR: FEW BODY SYSTEMS

INTRODUCTION

A.E. ROY
Dept. of Physics and Astronomy,
University of Glasgow, Glasgow, U.K.

AND

B.A. STEVES
Dept. of Mathematics,
Glasgow Caledonian University, Glasgow, U.K.

Traditionally, the many-body point mass gravitational problem has been divided into the two-body problem, the few-body problem and the many-body problem. Dynamical situations encountered in the solar system lie within the province of the few-body problem where orbits have to be calculated precisely, in contrast to many body problems such as star clusters or galaxies where statistical or hydrodynamical approaches can be applied.

Fortunately the majority of problems involving small bodies in the solar system fall into the category of disturbed Keplerian two-body motion. Until recently general perturbation analytical methods were used to obtain solutions to an arbitrary degree of accuracy. The modern solar system with its planet-crossing comets and asteroids and the space trajectories of interplanetary craft, has seen the development of many special perturbation approaches, such as the long-term numerical investigation of the orbits of the five outer planets and Uranus' system of satellites.

Many of the problems can be modelled by the restricted circular three-body problem. The Jacobi integral has been widely used in deriving information about the orbital stability of the particle by a study of the surfaces of zero velocity. In recent years, this method has been extended to the general three-body problem using the c^2H stability criterion, where $\mathbf{c}$ is the angular momentum and H is the total energy of the system.

Unfortunately there does not seem to be any analytical stability criterion for four or more bodies. Isolated three-body systems do not exist in

B.A. Steves and A.E. Roy (eds.), The Dynamics of Small Bodies in the Solar System, 349–350.

the solar system, so that the use of the c^2H criterion, like the restricted three-body Jacobi integral, gives only an approximation to the truth. For this reason a number of recent approaches to the problem of the stability of few-body dynamical systems adopt a kind of actuarial point of view. Just as actuarial data on the lifestyles of human beings enable predictions to be made about their survival rates, so the study of chaotic and resonant processes enables ordered and chaotic regions of the phase space to be mapped. An example of this is the study of the rates of depletion of certain regions in the asteroid belt and the study of slow chaos or 'stickiness' near ordered regions. Indeed many of the papers in this volume are excellent examples of the present lines of investigation into the global behaviour of three or four body systems.

In the present section are more examples of the diversity of ways in which our understanding of the behaviour of bodies in the few-body problem has grown. Steves, Barnett and Roy show how the Steves-Roy Finite-Time Stability Method can be applied to planets of binary stars to obtain minimum lifetimes comparable to the age of our solar system, while Kiseleva and Eggleton examine the effect of tidal friction in triple systems as a means of producing close stellar and planetary orbits. They apply their method to cases such as the triple system β Per and some recently discovered extra-Solar planets. A single formula for the stability of hierarchical triples is derived by Mardling and Aarseth, as a result of their study of the correspondence between the binary-tides problem in astro-physics and the three-body problem.

Taidakova and Gor'Kavyi propose the use of a new numerical integrator, giving examples of its use in the case of the nonconservative dynamics of particles around a star. Gozdziewski and Maciejewski study the dynamics and stability of a system of two bodies interacting gravitationally, where one body is spherically symmetric and the other is axially symmetric. In a paper by Chesley and Zare, the authors study a Poincaré map in the planar isosceles three-body problem which leads to global information on the dynamical behaviour for different mass ratios.

Steves, Roy and Bell model a restricted four-body system of equal masses moving in coplanar, initially circular orbits. They show how certain initial configuration solutions to the four-body equal mass problem evolve to the five Lagrangian solutions of the Copenhagen problem as two of the masses are reduced to zero. Gomatam takes their analysis further by developing a method for studying the stability of four-body configurations and applying it to the four body equal mass configuration solutions.

There is no doubt that the few-body problem is alive and well and enjoying the attention of most of the celestial mechanics community.

SOME SPECIAL SOLUTIONS OF THE FOUR-BODY PROBLEM

B.A. STEVES
Dept. of Mathematics,
Glasgow Caledonian University, Glasgow, U.K.

A.E. ROY
Dept. of Physics and Astronomy,
University of Glasgow, Glasgow, U.K.

AND

M. BELL
Dept. of Physics and Astronomy,
University of Glasgow, Glasgow, U.K.

1. Introduction

The restricted three-body problem with two unequal finite point masses and a test particle of infinitesimal mass moving in their gravitational fields has been the subject of many studies since its formulation (See for example Szebehely 1967). To minimise the number of variables and initial conditions, the problem is usually further restricted by constraining the two finite masses to move in circular orbits about their common centre of mass. This allows the so-called Jacobi integral to be obtained which has been widely used from the days of G. W. Hill in deriving information about the orbital stability of the particle by a study of the surfaces of zero-velocity (Szebehely, 1967) given by the Jacobi integral and the initial conditions.

A further restriction confines the particle to the orbital plane of the two finite masses reducing the problem to fourth order or three when the Jacobi integral is taken into account. This circular restricted coplanar model has also inspired a large literature of papers in which the problems of finding families of periodic orbits and studying their stability have been explored. Among the most productive researcher in this field was Henri Poincaré.

Practically, it is only since the advent of modern high-speed computers that celestial mechanicians have been able to study in an exhaustive manner

B.A. Steves and A.E. Roy (eds.), The Dynamics of Small Bodies in the Solar System, 351–372.

the restricted, three-dimensional, three-body problem, a study that retains its relevance today in the problem of the stability of planets of double stars (Steves, Barnett and Roy, 1998). A century ago, however, computing power was in its infancy and every effort was made to introduce as many restrictions as possible to minimise the computational labours involved in seeking a global understanding of the test particle's orbital behaviour.

At that time, the Copenhagen school of celestial mechanicians gave their name to a coplanar restricted three-body point mass system they set up, consisting of two finite equal masses moving in a circle about their common centre of mass and a test particle of infinitesimal mass moving in their orbital plane. This further restriction brought the restricted three-body problem within the grasp of the Copenhagen School's computational facilities. The Copenhagen model's simplicity enabled a global picture of its families of periodic orbits to be obtained.

Two of the present authors (Steves and Roy, 1998) have recently modelled a restricted four-body dynamical system which minimises significantly the number of variables and boundary condition parameters, yet remains useful as a method of studying the stability of real hierarchical four-body stellar systems. Definition of this model - hereafter called the Caledonian problem - and a summary of its useful properties will be the subject of the first part of this paper. The second part will be devoted to an exposition of how certain initial configurations of the four equal masses in the Caledonian problem can be reduced to the five Lagrangian solutions of the Copenhagen problem. In so doing, the existence of families of special solutions of the four-body problem will be established forming bridges between the Caledonian problem and the Copenhagen problem. Finally, a further reduction of the four-body problem will produce a solution relevant to the Trojan asteroid problem.

2. The Caledonian Problem

2.1. PRACTICALITY

It is well-known that about two thirds of the stars in the galaxy belong to binary systems, their periods of revolution ranging from hours to centuries. About one-fifth of these binaries are found to be triple systems and a rough estimate suggests that about one-fifth of these triples are quadruples or more. For example, Castor, one of the twins, is really a sextuple stellar system. Taking the number of stars in the galaxy to be 10^{11} and using the above estimates, the number of quadruple stellar systems in our galaxy may be of order 10^{9}.

The sun is a star of around average stellar mass with fifteen percent of stellar masses lying within a few percent of solar mass. Thus about

5×10^4 of the quadruple stellar systems in the galaxy may well consist of stars of almost solar mass. The number of quadruple stellar systems with all members within a few percent of the sun's mass is therefore of order one million. A four-body stellar system with all four stars of equal mass is therefore of practical dynamical interest.

2.2. HIERARCHICAL QUADRUPLE SYSTEMS

Just as in the solar system, orbital motion in quadruple stellar systems is invariably found to consist of perturbed two-body motions. All four-body stellar systems are found to exist in two different hierarchies, the *double binary* and the *linear*. In the former, the four stars form two binaries, the centres of mass of each binary being much more distant from each other than the distances separating the component stars in each binary. In fact this type of hierarchy is better described as a triple binary, the binaries being formed by (i) stars one and two, (ii) stars three and four, (iii) the centre of mass of stars one and two and the centre of mass of stars three and four. In the linear hierarchy there are again three disturbed Keplerian orbits, (i) stars one and two, (ii) star three orbiting the centre of mass of stars one and two and (iii) star four orbiting the centre of mass of the first three stars.

Multiple systems of more than four components can also be arranged in a hierarchical manner of binaries moving in disturbed Keplerian orbits, the binary components consisting of bodies and/or centres of mass. Such systems have been described using generalised Jacobi coordinates (Walker, 1983; Roy, 1988).

The definition of *hierarchical* can therefore be given as follows: *A dynamical system of n bodies is said to be hierarchical if at a given epoch it can be defined to exist as a clearly identified number of disturbed two-body motions, where the two bodies in a pair may be made up of masses, a mass and a centre of mass, or two centres of mass.*

The hierarchy is said to be changed if any of the designated disturbed two-body systems is disrupted. A measure of the stability of the hierarchy is given by the time duration during which the hierarchy remains unchanged. It goes without saying that parts of a hierarchy may exhibit relatively quick changes while other parts show long-term stability. In the solar system, planet-crossing asteroids provide an example of the former kind while the planetary system itself is an example of the latter.

2.3. STABILITY OF QUADRUPLE SYSTEMS

In dynamical systems of n bodies operating under Newton's law of gravitation, the ten integrals of the motion exist for all values of $n \geq 2$, namely the six integrals concerning the motion of the system's centre of mass, the three angular momentum integrals and the energy integral. If $\bar{c}$ is the total angular momentum and H is the system's energy, then their combination c^2H in the case $n = 3$ provides a stability criterion c^2H_{crit}, the value of c^2H_{crit} being found from the appropriate Lagrange collinear arrangement of the three masses m_1, m_2, m_3. If the actual value of $c^2H < c^2H_{crit}$, then if m_1 and m_2 form a binary sub-system with m_3 orbiting the centre of mass of m_1 and m_2 at a distance greater than the separation of the binary components, the triple system is stable in the sense that m_3 cannot disrupt the binary, though m_3 may recede to infinity.

The c^2H criterion has been used in many studies of the general three-body problem (Marchal and Saari, 1975; Szebehely and Zare, 1977; Zare, 1977; Roy, 1979; Roy et al, 1984; Valsecchi et al, 1984). In the restricted circular three body problem, the corresponding quantity J (the Jacobi constant) has likewise been used in many stability studies of that problem.

It is well-known, however, that for $n \geq 4$, there is no analytical stability criterion valid for all time. It would appear, therefore, that stability studies in such few-body problems, $n \geq 4$, must be restricted to the search for a finite time criterion of an *actuarial* kind. By 'actuarial', we imply that for a family of n-body systems whose parameters satisfy the criterion, we cannot say that any particular n-body system in the family will remain stable (ie hierarchy unchanged) for the time duration given by the criterion but we can say that half the members of the family will remain stable for that time duration.

Some work on the stability of four-body systems has been done by Milani and Nobili and by Roy, Walker and MacDonald. Milani and Nobili (1983) considered the four-body linear hierarchical system Sun-Mercury-Venus- Jupiter and obtained estimates of the time it would take for such a hierarchy to be broken in different ways. Roy, Walker and MacDonald (1985) considered a number of fictitious coplanar four-body systems with various distributions of masses and separations. By numerical integration they found how long it took for such systems to change their hierarchy. They also considered how such 'stability' durations were related to the c^2H criterion given by the triple sub-sets of bodies within the quadruple systems.

The studies referred to above, while valuable, in no way gave a global perspective to the problem of the stability of four-body dynamical systems. The work of Roy, Walker and MacDonald in particular highlighted the need for a suitably restricted four-body model that would retain the majority of the properties of four-body systems while reducing the number of variables and initial conditions to the point where a global perspective might be obtainable.

2.4. COMPARISON OF THE GENERAL FOUR-BODY PROBLEM AND THE CALEDONIAN PROBLEM

2.4.1. *The General Four-Body Problem*

The general three-dimensional four-body problem is of 24th order. With four unequal masses the initial parameter values at $t = 0$ consist of

$$(x_i, y_i, z_i; \dot{x}_i, \dot{y}_i, \dot{z}_i; m_i)_{t=0}, i = 1, 2, 3, 4$$

giving a total of 28 values.

Ten classical integrals exist in the general problem of four bodies. The elimination of the node and the change from time as the independent variable can be made, effectively reducing the order of the problem from 24 to 12. It cannot be reduced further.

2.4.2. *The Restricted Coplanar Initially Circular Equal Mass Four-Body Problem (The Caledonian Problem)*

The Caledonian Problem (see Figure 1) has the following properties:

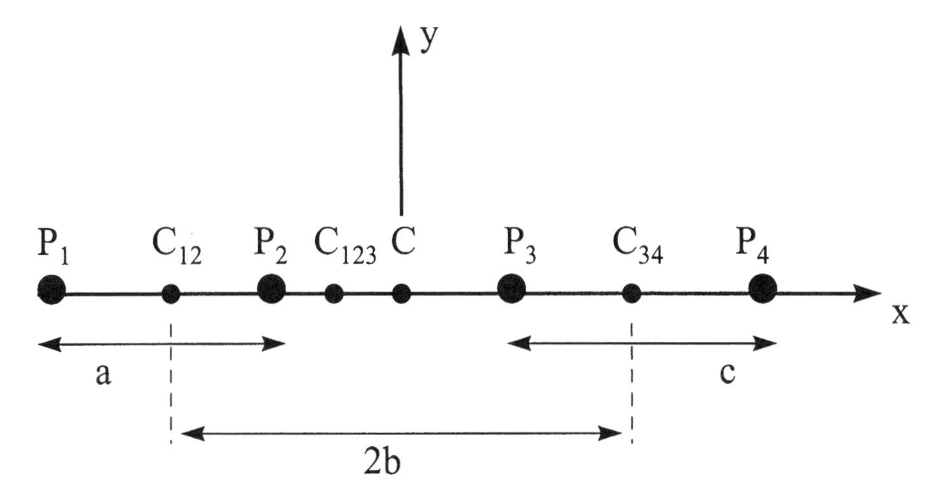

Figure 1. The initial configuration of the Caledonian problem.

A. The four bodies P_1, P_2, P_3 and P_4 are taken to be equal in mass and point-sized.

B. All four move in coplanar orbits about the centre of mass of the system.

C. At $t = 0$, the bodies are collinear and are moving co-rotationally with velocity vectors perpendicular to the line they are on. By condition B, the velocity vectors are coplanar.

D. The velocity magnitudes at $t = 0$ are chosen so that the hierarchy is linear or double binary.

For the *linear hierarchy*:
P_1 and P_2 are initially moving in a circular orbit about their centre of mass C_{12}.
P_3 and C_{12} are initially moving in circular orbits about C_{123}, the centre of mass of P_1, P_2 and P_3.
P_4 and C_{123} are initially moving in circular orbits about C, the centre of mass of P_1, P_2, P_3 and P_4.

For the *double binary hierarchy*:
P_1 and P_2 are initially moving in a circular orbit about their centre of mass C_{12}.
P_3 and P_4 are initially moving in a circular orbit about C_{34}.
C_{12} and C_{34} are initially moving in a circular orbit about C.

E. In the double binary hierarchy, a further simplification can be made if the initial separation of P_1 and P_2 is taken to be equal to the initial separation of P_3 and P_4. The distances $C_{12}C$ and CC_{34} are also equal in both hierarchies.

2.4.3. *Computational Advantages of the Caledonian Problem over the Four-Body Problem*

1. The general problem was of 24th order with 28 initial parameter values if all four masses were different. In the restricted model, properties A, C and D reduce the problem to 16th order with the number of independent initial parameter values reduced to four, viz a, b, c and m (the bodies' mass). If the double binary hierarchy is adopted initially and property E is used, the number of initial parameter values in that model is reduced to three viz, a, b and m.
2. Properties B and C ensure that the system is in a mirror configuration at $t = 0$ with all four bodies collinear, the velocity vectors being perpendicular to the line of bodies. By the Roy-Ovenden mirror theorem (Roy & Ovenden 1955) the dynamical behaviour of the system after $t = 0$ is a mirror image of its behaviour before $t = 0$. The situation is advantageous in studies of the evolution of the system and in any search for periodic orbits, the condition for a periodic system being to find a second mirror configuration.
3. If property E is used in the double binary case, a further symmetry is provided in that model. If $c = a$,

$$x_{10} = -x_{40} \; ; \qquad \dot{y}_{10} = -\dot{y}_{40}$$

$$x_{20} = -x_{30}\ ; \qquad \dot{y}_{20} = -\dot{y}_{30}.$$

In that model, therefore, the behaviour of P_4 at any time is a mirror of the behaviour of P_1, while the behaviour of P_3 at any time is a mirror of the behaviour of P_2 at that time.

4. In the general four-body problem there exist the classical ten integrals of energy: (1), of angular momentum (3), and of centre of mass (6). The elimination of the node and the change from time as the independent variable can be made, effectively reducing the order of the problem from 24 to 12.

   In the present restricted problem the classical integrals are 6 in number, the energy (1), the angular momentum (1) and the centre of mass (4). The elimination of the node and the replacement of time as the independent variable further reduces the order by 2, so that the problem, of order 16, reduces to 8. In the double binary model, with $c = a$, advantage 3 above increases the centre of mass integrals effectively from 4 to 8, since at any time

$$\begin{array}{llll} x_4 = -x_1\ ; & y_4 = -y_1\ ; & \dot{x}_4 = -\dot{x}_1\ ; & \dot{y}_4 = -\dot{y}_1 \\ x_3 = -x_2\ ; & y_3 = -y_2\ ; & \dot{x}_3 = -\dot{x}_2\ ; & \dot{y}_4 = -\dot{y}_1, \end{array}$$

   For that particular model, therefore, the order may be effectively reduced to 4.

2.5. EQUATIONS OF MOTION IN THE CALEDONIAN PROBLEM

We now consider the equations of motion, taking account of the two different hierarchies' properties.

Let the four bodies P_i have masses m_i, $i = 1, 2, 3, 4$, where the masses are not equal at this stage. Let C be the centre of mass of the four bodies. Let $\mathbf{r_i}$ be the radius vector from C to body P_i, $i = 1, 2, 3, 4$. Let the units chosen be such that the constant of gravitation G is of unit value.

Then, putting $\boldsymbol{\rho_{ij}} = \boldsymbol{r_{ij}}/r_{ij}^3$,

$$\begin{array}{rcl} m_1\ddot{\mathbf{r}}_1 & = & m_1(m_2\boldsymbol{\rho}_{12} + m_3\boldsymbol{\rho}_{13} + m_4\boldsymbol{\rho}_{14}) \\ m_2\ddot{\mathbf{r}}_2 & = & m_2(-m_1\boldsymbol{\rho}_{12} + m_3\boldsymbol{\rho}_{23} + m_4\boldsymbol{\rho}_{24}) \\ m_3\ddot{\mathbf{r}}_3 & = & m_3(-m_1\boldsymbol{\rho}_{13} - m_2\boldsymbol{\rho}_{23} + m_4\boldsymbol{\rho}_{34}) \\ m_4\ddot{\mathbf{r}}_4 & = & m_4(-m_1\boldsymbol{\rho}_{14} - m_2\boldsymbol{\rho}_{24} - m_3\boldsymbol{\rho}_{34}) \end{array} \qquad (1)$$

Then $\sum_{i=1}^{4} m_i\ddot{\mathbf{r}}_i = \mathbf{0}$, so that $\sum_{i=1}^{4} m_i\dot{\mathbf{r}}_i = \mathbf{p}$ and $\sum_{i=1}^{4} m_i\mathbf{r}_i = \mathbf{p}t + \mathbf{q}$, where $\mathbf{p}$ and $\mathbf{q}$ are constant vectors. If C, the centre of mass of the system is the origin and at rest, $\mathbf{p} = \mathbf{q} = 0$.

In the Caledonian Problem all four masses are equal. Let their mass be M. Then Equations (1) reduce to:

$$\begin{aligned} \ddot{\mathbf{r}}_1 &= M(\boldsymbol{\rho}_{12} + \boldsymbol{\rho}_{13} + \boldsymbol{\rho}_{14}) \\ \ddot{\mathbf{r}}_2 &= M(-\boldsymbol{\rho}_{12} + \boldsymbol{\rho}_{23} + \boldsymbol{\rho}_{24}) \\ \ddot{\mathbf{r}}_3 &= M(-\boldsymbol{\rho}_{13} - \boldsymbol{\rho}_{23} + \boldsymbol{\rho}_{34}) \\ \ddot{\mathbf{r}}_4 &= M(-\boldsymbol{\rho}_{14} - \boldsymbol{\rho}_{24} - \boldsymbol{\rho}_{34}) \end{aligned} \tag{2}$$

Now consider the two types of hierarchy.

(a) The linear hierarchy (See Figure 2)

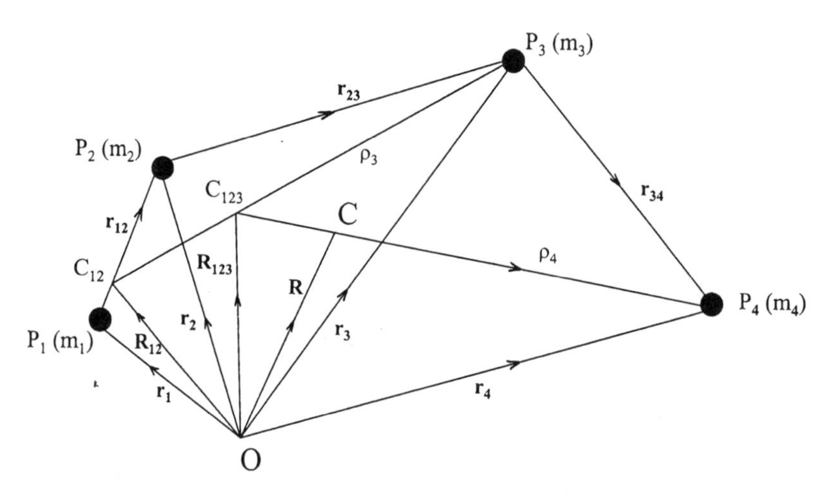

Figure 2. A four body linear hierarchy.

In this hierarchy we want equations in $\mathbf{r}_{12}$, $\boldsymbol{\rho}_3 = C_{12}P_3$, $\boldsymbol{\rho}_4 = C_{123}P_4$. After a little reduction (Steves and Roy, 1998), it is found that

$$\begin{aligned} \ddot{\mathbf{r}}_{12} &= M\left[-2\boldsymbol{\rho}_{12} + (\boldsymbol{\rho}_{23} - \boldsymbol{\rho}_{13}) + (\boldsymbol{\rho}_{24} - \boldsymbol{\rho}_{14})\right] \\ 2\ddot{\boldsymbol{\rho}}_3 &= M\left[-3\,(\boldsymbol{\rho}_{13} + \boldsymbol{\rho}_{23}) + 2\boldsymbol{\rho}_{34} - \boldsymbol{\rho}_{14} - \boldsymbol{\rho}_{24}\right] \\ 3\ddot{\boldsymbol{\rho}}_4 &= -4M(\boldsymbol{\rho}_{14} + \boldsymbol{\rho}_{24} + \boldsymbol{\rho}_{34}) \end{aligned}$$

(b) The double binary hierarchy. (See Figure 3)
We want equations in

$$\mathbf{r}_{12} = P_1P_2, \ \boldsymbol{\rho} = C_{12}C_{34}, \ \mathbf{r}_{34} = P_3P_4$$

It is then found that

$$\begin{aligned} \ddot{\mathbf{r}}_{12} &= M\left[-2\boldsymbol{\rho}_{12} + (\boldsymbol{\rho}_{23} - \boldsymbol{\rho}_{13}) + (\boldsymbol{\rho}_{24} - \boldsymbol{\rho}_{14})\right] \\ \ddot{\mathbf{r}}_{34} &= M\left[-2\boldsymbol{\rho}_{34} + (\boldsymbol{\rho}_{23} - \boldsymbol{\rho}_{24}) + (\boldsymbol{\rho}_{13} - \boldsymbol{\rho}_{14})\right] \\ \ddot{\boldsymbol{\rho}} &= -M\left[(\boldsymbol{\rho}_{13} + \boldsymbol{\rho}_{14}) + (\boldsymbol{\rho}_{23} + \boldsymbol{\rho}_{24})\right] \end{aligned}$$

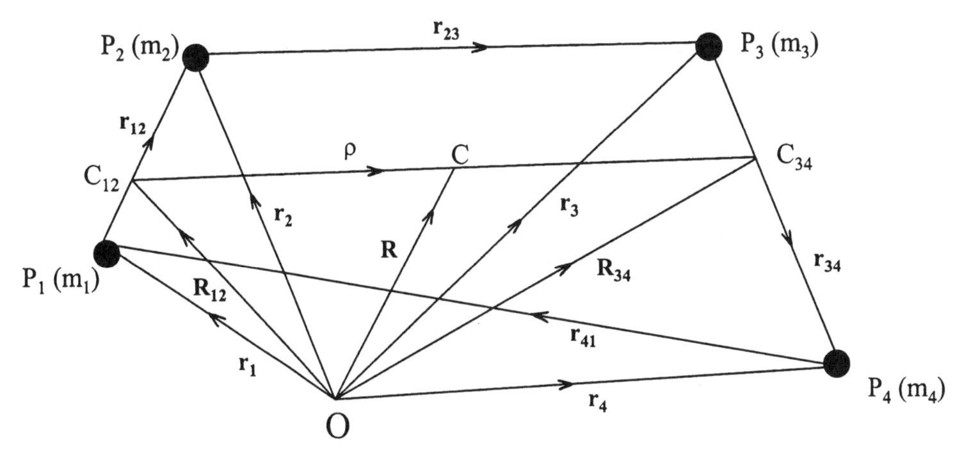

Figure 3. A four body double binary hierarchy.

The choice of units such that G is unity is easily achieved by using as the units of distance, mass and time, the astronomical unit, the solar mass and a new unit of time, the *yix*, defined to be one *year* divided by 2π, of duration roughly one-*sixth* of a year.

2.6. THE INITIAL CONDITIONS IN THE HIERARCHIES

(a) Linear hierarchy (See Figure 1)

Recalling that all masses are equal, it is clear from Figure 1 that for any reasonable duration of the hierarchy we must have, at $t = 0$, $P_3C_{12} > P_1C_{12} = P_2C_{12}$; $\quad P_4C_{123} > C_{123}P_3$. At $t = 0$,

$$P_1P_2 = a; \quad C_{12}C = CC_{34} = b; \quad P_3P_4 = c$$

Let

$$P_1P_2 = a_2 = a; \quad P_3C_{12} = a_3 = 2b - \frac{c}{2}; \quad P_4C_{123} = a_4 = \frac{4}{3}b + \frac{2}{3}c.$$

Then

$$\alpha_1 = \frac{a_2}{a_3} = \frac{a}{2b - \frac{c}{2}}; \quad \alpha_2 = \frac{a_3}{a_4} = \frac{2b - \frac{c}{2}}{\frac{4}{3}b + \frac{2}{3}c}$$

are important parameters in the stability of the system. At $t = 0$, the system is collinear, with all velocity vectors at right angles to the line $P_1P_2P_3P_4$. P_1 and P_2 are initially in circular orbits about C_{12}; P_3 and C_{12} are initially in circular orbits about C_{123}; P_4 and C_{123} are initially in circular orbits about C. Assume C to be at rest and take inertial rectangular axes Cx and Cy through C. Co-rotational orbits are assumed, though are not necessary. Then, at $t = 0$, if x_{i0}, y_{i0} are the coordinates of P_i, $i = 1, 2, 3, 4$ it is easily shown (Steves and Roy, 1998) that :

$$
\begin{aligned}
&\dot{x}_{i0} = y_{i0} = 0. \\
&x_{10} = -(b + \tfrac{a}{2}); \quad \dot{y}_{10} = \sqrt{M}\left(-\frac{1}{\sqrt{2a_2}} - \frac{1}{\sqrt{3a_3}} - \frac{1}{\sqrt{4a_4}}\right) \\
&x_{20} = -(b - \tfrac{a}{2}); \quad \dot{y}_{20} = \sqrt{M}\left(\frac{1}{\sqrt{2a_2}} - \frac{1}{\sqrt{3a_3}} - \frac{1}{\sqrt{4a_4}}\right) \\
&x_{30} = b - \tfrac{c}{2}; \qquad \dot{y}_{30} = \sqrt{M}\left(\frac{2}{\sqrt{3a_3}} - \frac{1}{\sqrt{4a_4}}\right) \\
&x_{40} = b + \tfrac{c}{2}; \qquad \dot{y}_{40} = \sqrt{M}\left(\frac{3}{\sqrt{4a_4}}\right),
\end{aligned}
$$

$\dot{y}_{i0}$ are obtained using the circular two-body relation $n^2a^3 = \mu = G(m_1 + m_2)$ and thus

$$\dot{y} = \sqrt{\frac{G(m_1 + m_2)}{a}}$$

(b) Double binary hierarchy (See Figure 1)
As before, at $t = 0$,

$$y_{i0} = \dot{x}_{i0} = 0, \; i = 1, 2, 3, 4.$$

and

$$x_{10} = -(b + \frac{a}{2}), \; x_{20} = -(b - \frac{a}{2}), \; x_{30} = b - \frac{c}{2}, \; x_{40} = b + \frac{c}{2}$$

Then

$$
\begin{aligned}
\dot{y}_{10} &= \sqrt{M}\left(-\frac{1}{\sqrt{2a}} - \frac{1}{\sqrt{2b}}\right) \\
\dot{y}_{20} &= \sqrt{M}\left(\frac{1}{\sqrt{2a}} - \frac{1}{\sqrt{2b}}\right) \\
\dot{y}_{30} &= \sqrt{M}\left(-\frac{1}{\sqrt{2c}} + \frac{1}{\sqrt{2b}}\right) \\
\dot{y}_{40} &= \sqrt{M}\left(\frac{1}{\sqrt{2c}} + \frac{1}{\sqrt{2b}}\right)
\end{aligned}
$$

2.7. STRATEGIES IN EXPLORING THE CALEDONIAN PROBLEM

Although the Caledonian Problem does not admit of a Jacobi integral, almost every other strategy used in the Copenhagen problem can be applied in studying the evolution and stability of the trajectories in the former problem. The small number of starting parameters should enable a global study to be carried out using present-day computing power.

Among these methods are:

(a) To search for families of periodic orbits and to examine their linear stability. Some work along these lines has been done by Hadjidemetriou (1978) in four-body cases in the solar system.

(b) Direct long-term numerical integrations. While being the most direct method of following the trajectories of the bodies in the quadruple system, such an approach is time-consuming and subject to inevitable error-accumulation. Nevertheless, the accumulation of a large data-base of orbits, close encounters, cross-overs, escapes, break-ups of the initial hierarchies after known durations is invaluable in providing a data-base against which the reliability of other methods as predictors of short or long term stability can be tested.

(c) The use of the Roy-Walker empirical stability parameters. Roy(1979) introduced empirical stability parameters, the epsilons, that give insight into the duration of a general three-body system's hierarchical stability. In a series of papers, Roy, Walker, Carusi, Valsecchi, Emslie and MacDonald (Roy, 1979; Walker et al, 1980; Walker and Roy, 1983a, 1983b; Valsecchi et al, 1984; Roy et al, 1985) showed that in the general three-body problem the c^2H stability criterion is unnecessarily restrictive as a measure of hierarchical stability. They also demonstrated that there is a range $q = c^2H - (c^2H)_{crit} > 0$ in which, for sufficiently small q, the hierarchy will last for very long times compared to the system's synodic period.

3. From Caledonia to Copenhagen

3.1. THE LAGRANGIAN SOLUTIONS OF THE RESTRICTED THREE-BODY PROBLEM (RTBP)

It is well-known that in the general three-body problem there exist five special equilibrium configurations. Either the three masses m_1, m_2, m_3 are collinear or occupy the vertices of an equilateral triangle. In the former case, the positions of m_1, m_2 and m_3 are given by the solution of Lagrange's quintic equation

$$\begin{aligned}(m_1 + m_2)X^5 + (3m_1 + 2m_2)X^4 + (3m_1 + m_2)X^3 \\ -(m_2 + 3m_3)X^2 - (2m_2 + 3m_3)X - (m_2 + m_3) = 0\end{aligned}$$

where

$$X = \frac{x_2 - x_3}{x_1 - x_2},$$

x_1, x_2 and x_3 being the distances of m_1, m_2 and m_3 from the centre of mass. By Descartes' Rule of Signs there is only one positive root of X, which defines uniquely the distribution of the three particles in the order m_1, m_2, m_3. The other two orders (namely 231 and 213) give two more distinct straight-line solutions

In the equilateral triangular solution, it is easy to show that such a configuration is an equilibrium one. Letting $\mathbf{r_i}$ be the radius vector of m_i, $i = 1, 2, 3$ and letting $\boldsymbol{\rho_{ij}} = \boldsymbol{r_{ij}}/r_{ij}^3$, where $\boldsymbol{r_{ij}} = \boldsymbol{r_j} - \boldsymbol{r_i}$, the equations of motion of the three particles are given by

$$\begin{aligned} m_1\ddot{\mathbf{r}}_1 &= m_1(m_2\boldsymbol{\rho}_{12} + m_3\boldsymbol{\rho}_{13}) \\ m_2\ddot{\mathbf{r}}_2 &= m_2(-m_1\boldsymbol{\rho}_{12} + m_3\boldsymbol{\rho}_{23}) \\ m_3\ddot{\mathbf{r}}_3 &= m_3(-m_1\boldsymbol{\rho}_{13} - m_2\boldsymbol{\rho}_{23}) \end{aligned}$$

Then

$$\sum_{i=1}^{3} m_i\ddot{\mathbf{r}}_i = \sum_{i=1}^{3} m_i\dot{\mathbf{r}}_i = \sum_{i=1}^{3} m_i\mathbf{r}_i = 0 \tag{3}$$

taking the centre of mass to be the origin and at rest.

Assuming, by hypothesis, that $r_{12} = r_{23} = r_{31} = r$, we have

$$\ddot{\mathbf{r}}_1 = r^{-3}\left[m_2\left(\mathbf{r}_2 - \mathbf{r}_1\right) + m_3\left(\mathbf{r}_3 - \mathbf{r}_1\right)\right],$$

or, using (3),

$$\ddot{\mathbf{r}}_1 = -Mr^{-3}\mathbf{r}_1,$$

where $M = \sum_{i=1}^{3} m_i$. Similarly,

$$\ddot{\mathbf{r}}_2 = -Mr^{-3}\mathbf{r}_2;\ \ddot{\mathbf{r}}_3 = -Mr^{-3}\mathbf{r}_3.$$

If circular motion of the triangle is desired, let

$$\mathbf{r_i} \cdot \dot{\mathbf{r}}_\mathbf{i} = 0,\ |\dot{\mathbf{r}}_\mathbf{i}| = n|\mathbf{r_i}|,\ n^2 = M/r^3.$$

Then

$$\mathbf{r_i} = \mathbf{r_{i_0}} \cos nt + \frac{\dot{\mathbf{r}}_{\mathbf{i_0}}}{n} \sin nt,$$

where $\mathbf{r_{i_0}}$ and $\dot{\mathbf{r}}_{\mathbf{i_0}}$ are the radius vector and velocity vector of P_i at $t = 0$.

Thus, if we consider only circular motion, the bodies' radius vectors $\mathbf{r_i}$ obey differential equations of the Simple Harmonic Motion (SHM) form

$$\ddot{\mathbf{r}}_\mathbf{i} = -n^2\mathbf{r_i}$$

where the bodies' relative geometry rotates with no change in scale and with a constant angular velocity n. Thus, in principle, for systems of two or more bodies, geometric configurations that reduce the equations of motion of the bodies to the SHM form, with the same angular velocity for each body, will give a rotating configuration solution with no change of scale or angular velocity.

The RTBP has five corresponding special equilibrium solutions where the particle of infinitesimal mass resides at one of five points, usually denoted by the letters L_1, L_2, L_3, L_4, and L_5. The first three remain collinear with the line P_1P_2 where P_1 and P_2 are the two finite masses. The others, L_4 and L_5, form equilateral triangles with P_1 and P_2 (Figure 4).

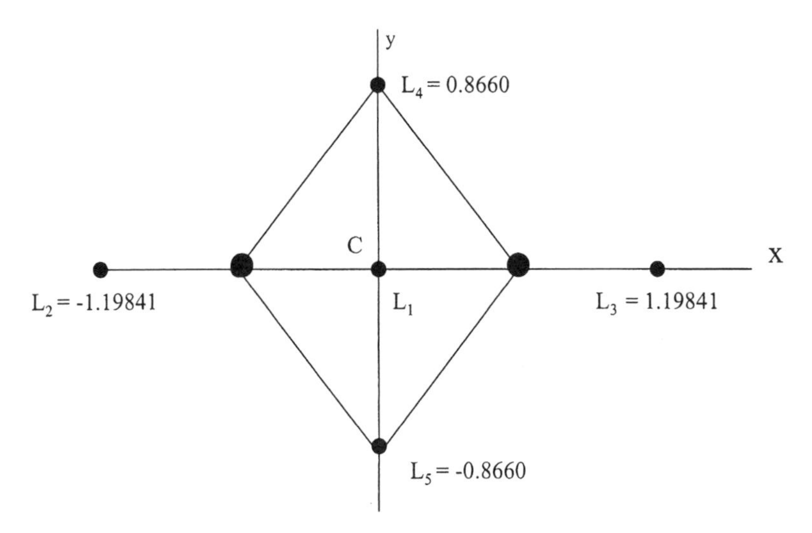

Figure 4. The Lagrange equilibrium solutions to the Copenhagen problem

In the Copenhagen Problem (Figure 4), with $m_1 = m_2 = m$, if a frame of coordinates rotating with constant angular velocity n is centred on the centre of mass C of the two finite mass bodies and P_1 and P_2 are placed at $(-0.5, 0)$ and $(+0.5, 0)$ respectively, the Lagrange solutions in the rotating frame become $L_3 = -L_2 = (1.19841, 0)$; $L_1 = (0, 0)$; $L_4 = -L_5 = (0, \sqrt{\frac{3}{4}}) = (0, 0.8660)$ (See any standard text on the Lagrange solutions).

The system can rotate with constant angular velocity n about C, the centre of mass of P_1 and P_2 with $n = 2\pi/T$ where

$$T = 2\pi\sqrt{\frac{a^3}{2Gm}},$$

$a/2$ being the radius of P_1 and P_2's orbit about C.

Indeed, it has long been known that special solutions of the n-body problem exist where equilibrium solutions appear for particular geometrical configurations, for example where n masses are placed at the vertices of an n-gon of equal sides. Moulton (1910) gave straight-line solutions of the problem of n-bodies, while Palmore (1973, 1975a, 1975b, 1976) classified relative equilibrium solutions in the four-body equal mass problem. Simo (1978), in his paper on relative equilibrium solutions in the four-body problem, considered arbitrary mass four-body configurations and also solutions involving three masses and a particle. He presented a survey of the solutions which exist for arbitrary masses, using arguments involving the counting of the number of bifurcation sets and different invariant manifolds.

In their recent paper (1998), Roy and Steves used the method of reducing the equations of motion to the SHM form with the same mean motion to derive three well-known equilibrium geometries where the four equal masses, initially placed and given appropriate initial velocities, will circulate thereafter at constant angular velocity with no change in system geometry or scale. The configurations A. a square, B. an equilateral triangle, and C. a straight line are shown in Figure 5.

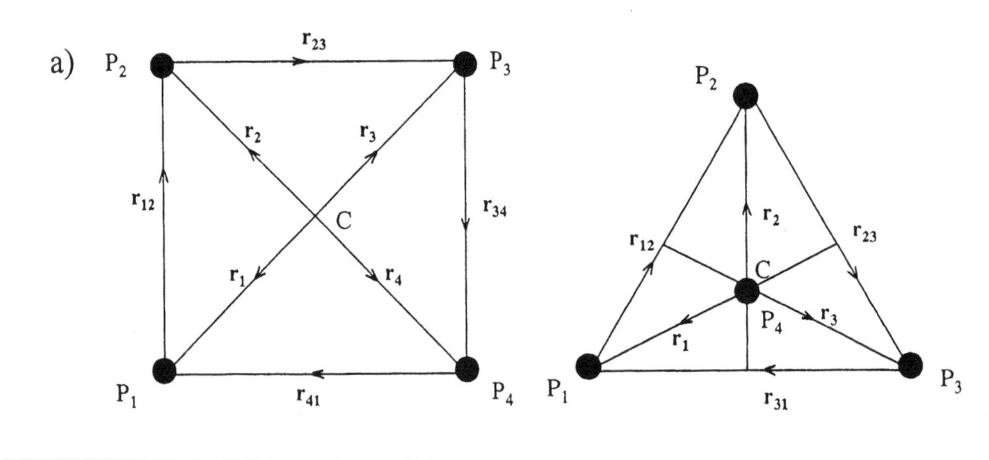

Figure 5. The four body equal mass equilibrium configurations a) Case A: square, b) Case B: equilateral triangle and c) Case C: straight line.

The derivation of the solution for Case A is given below. Let the mass of each body be M, the centre of mass be C at the origin, the x-axis pass through P_1P_3 and the y-axis pass through P_2P_4. In this case, if a solution is found where the geometry is invariant,

$$\begin{aligned} \mathbf{r}_4 &= -\mathbf{r}_2; \quad \mathbf{r}_3 &= -\mathbf{r}_1 \\ \mathbf{r}_{34} &= -\mathbf{r}_{12}; \quad \mathbf{r}_{41} &= -\mathbf{r}_{23} \end{aligned}$$

Also,

$$\begin{aligned} r_{12} &= r_{23} = r_{34} = r_{41} = a, \text{ say} \\ r_{13} &= r_{24} = b = \sqrt{2}a. \end{aligned} \tag{4}$$

Hence

$$\ddot{\mathbf{r}}_i = -R\mathbf{r}_i, \quad i = 1, 2, 3, 4.$$

where

$$R = Ma^{-3}(2 + \frac{1}{\sqrt{2}}).$$

But a is constant by hypothesis, so that, writing $n^2 = R$, the solution is

$$\mathbf{r}_i = \mathbf{r}_{i_0} \cos nt + \frac{\dot{\mathbf{r}}_{i_0}}{n} \sin nt,$$

where $\mathbf{r}_{\mathbf{i}_0}$ and $\dot{\mathbf{r}}_{\mathbf{i}_0}$ are the radius vector and velocity vector of P_i at $t = 0$. For circular orbits, we put

$$\mathbf{r}_i \cdot \dot{\mathbf{r}}_i = 0 \text{ and } |\dot{\mathbf{r}}_\mathbf{i}| = n|\mathbf{r}_i|.$$

A method for studying the stability of Cases A, B and C, which can be extended to four-body problems of unequal masses, is given in this volume by Gomatam, Steves and Roy (1998). Cases A, B and C are well-known to be unstable.

In a similar fashion, invariant geometries can be demonstrated for four-body systems where one pair of bodies each has a mass equal to $m = \mu M$, where M is the mass of each of the other pair of bodies. By gradually reducing the masses $m = \mu M$ of two of the four bodies, keeping them equal in mass m as μ approaches 0, it can be shown that most of the special solutions of the equal mass four-body problem, including the Caledonian problem can be reduced to solutions of the Copenhagen problem. In so doing, families of equilibrium solutions for all values of μ from $1 \geq \mu \geq 0$ can be traced. Each of the four bodies can be shown to perform circular orbits with constant angular velocity about the centre of mass of the system.

Table 1 below shows the different Lagrange equilibrium points reached by keeping two of the bodies' masses constant at M, while the masses $m = \mu M$ of the other two remain equal to each other but are gradually reduced to zero. Details of the derivation and evolution of these families in the range $0 < \mu < 1$ are given in Roy and Steves (1998).

Cases	Masses Reduced	Lagrange points reached by the reduced masses
A1	P_2, P_3	$P_2, P_3 \rightarrow L_4$
A2	P_2, P_4	$P_2 \rightarrow L_4; P_4 \rightarrow L_5;$
B1	P_2, P_4	Still to be explored
B2	P_1, P_3	Still to be explored
C1	P_1, P_4	$P_1 \rightarrow L_2; P_4 \rightarrow L_3;$
C2	P_2, P_3	$P_2, P_3 \rightarrow L_1$
C3	P_1, P_3	$P_1 \rightarrow L_2; P_3 \rightarrow L_1;$
C4	P_3, P_4	$P_3, P_4 \rightarrow L_3$

Table 1. The evolution of four-body equilibrium configurations to the Copenhagen equilibrium configurations, as two masses are reduced equally from M to zero. See Figure 5.

4. The Trojan Asteroid Problem

The Trojan asteroids form two groups of asteroids that oscillate about the L_4 and L_5 equilateral triangle points with the Sun and Jupiter forming the other corners of the triangle. We now apply the methodology of Roy and Steves (1998) to obtain a model of relevance to the Trojan case.

Let the four-body coplanar problem have masses m_1, m_2, m_3 and m_4 where m_1 has mass M, with $m_2 = m_4 = \mu M$ and $m_3 = \mu_1 M$, with $0 < \mu < 1$ and $0 < \mu_1 < 1$.

We try for a kite shaped equilibrium configuration with all four bodies moving in circular orbits of the same period about the centre of mass C which is at rest. For the centre of mass to be at the origin and at rest:

$$\mathbf{r_1} + \mu_1 \mathbf{r_3} = -\mu(\mathbf{r_2} + \mathbf{r_4}) \tag{5}$$

A kite shaped solution will have one line of symmetry passing through P_1P_3. Let this line be the x-axis, producing the following relations:

$$\begin{aligned} r_{A2} &= r_{A4}; \quad \mathbf{r_{A2}} &= -\mathbf{r_{A4}} \\ r_{12} &= r_{14}; \quad r_{23} &= r_{34} \\ r_2 &= r_4 \end{aligned}$$

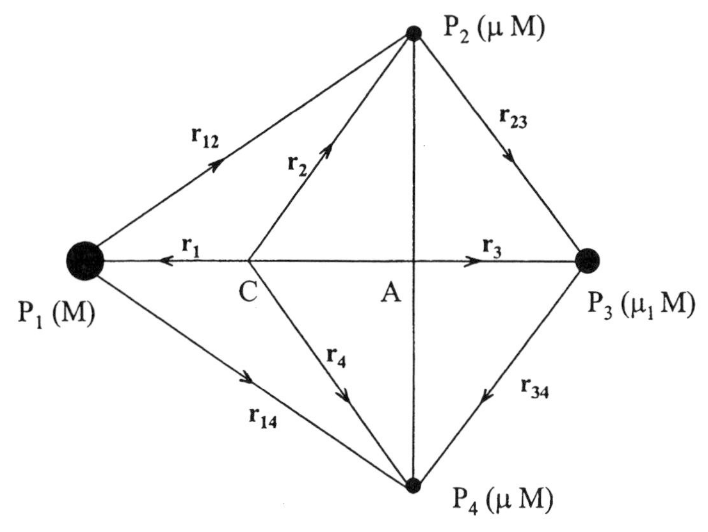

Figure 6. A kite shaped equilibrium configuration solution where bodies P_2 and P_4 are of equal mass.

where the notation r_{Ai} is used to denote the distance AP_i, where $i = 1, 2, 3$ or 4. x components of the centre of mass relation (5) also produce the equations:

$$\begin{aligned} \mathbf{r_1} + \mu_1 \mathbf{r_3} &= -2\mu \mathbf{r_A} \\ r_1 &= 2\mu r_A + \mu_1 r_3 \end{aligned} \tag{6}$$

Due to the symmetry, we need only two parameters to describe the kite shape completely. Let these parameters be α and β such that:

$$\begin{aligned} r_{13} &= \alpha r_{1A}; & \mathbf{r_{13}} &= \alpha \mathbf{r_{1A}} \\ r_{24} &= \beta r_{1A}; & \mathbf{r_{24}} &= \mathbf{r_4} - \mathbf{r_2} \end{aligned}$$

where r_{1A} will be used as the basic unit of scale.

If we now choose as directions for our orthogonal system of coordinates:

$$\hat{x} = \frac{\mathbf{r_{1A}}}{r_{1A}}$$

$$\hat{y} = -\frac{\mathbf{r_{24}}}{r_{24}}$$

the distances and vectors in the kite shape configuration can now be written as functions of α, β and r_{1A} and the vectors $\mathbf{r_{1A}}$ and $\mathbf{r_{24}}$.

For example, using Pythagoras' theorem:

$$\begin{aligned} r_{12} = r_{14} &= c r_{1A} \\ r_{23} = r_{34} &= d r_{1A} \\ r_{3A} &= (\alpha - 1) r_{1A} \end{aligned} \tag{7}$$

where

$$c = [1 + \tfrac{\beta^2}{4}]^{\frac{1}{2}} \qquad d = [(\alpha-1)^2 + \tfrac{\beta^2}{4}]^{\frac{1}{2}} \tag{8}$$

Also using the centre of mass relation (6), we get:

$$\begin{aligned} r_1 &= \gamma r_{1A} \\ r_A &= (1-\gamma) r_{1A} \\ r_3 &= (\alpha-\gamma) r_{1A} \\ r_2 = r_4 &= [\tfrac{\beta^2}{4} + (1-\gamma)^2]^{\frac{1}{2}} r_{1A} \end{aligned} \tag{9}$$

where

$$\gamma = \frac{2\mu + \alpha\mu_1}{1 + 2\mu + \mu_1} \tag{10}$$

Likewise the vectors become:

$$\begin{aligned} \mathbf{r_A} &= (1-\gamma)\mathbf{r_{1A}}; & \mathbf{r_{A2}} &= -\tfrac{\mathbf{r_{24}}}{2} \\ \mathbf{r_1} &= -\gamma\mathbf{r_{1A}}; & \mathbf{r_{12}} &= \mathbf{r_{1A}} - \tfrac{\mathbf{r_{24}}}{2} \\ \mathbf{r_3} &= (\alpha-\gamma)\mathbf{r_{1A}}; & \mathbf{r_{14}} &= \mathbf{r_{1A}} + \tfrac{\mathbf{r_{24}}}{2} \\ \mathbf{r_2} &= (1-\gamma)\mathbf{r_{1A}} - \tfrac{1}{2}\mathbf{r_{24}}; & \mathbf{r_{23}} &= (\alpha-1)\mathbf{r_{1A}} + \tfrac{\mathbf{r_{24}}}{2} \\ \mathbf{r_4} &= (1-\gamma)\mathbf{r_{1A}} + \tfrac{1}{2}\mathbf{r_{24}}; & \mathbf{r_{34}} &= -(\alpha-1)\mathbf{r_{1A}} + \tfrac{\mathbf{r_{24}}}{2} \end{aligned} \tag{11}$$

Then as before, letting $\boldsymbol{\rho_{ij}} = \boldsymbol{r_{ij}}/r_{ij}^3$, the equations of motion are:

$$\begin{aligned} \ddot{\mathbf{r}}_1 &= M\,[\mu\boldsymbol{\rho_{12}} + \mu_1\boldsymbol{\rho_{13}} + \mu\boldsymbol{\rho_{14}}] \\ \ddot{\mathbf{r}}_2 &= M\,[-\boldsymbol{\rho_{12}} + \mu_1\boldsymbol{\rho_{23}} + \mu\boldsymbol{\rho_{24}}] \\ \ddot{\mathbf{r}}_3 &= M\,[-\boldsymbol{\rho_{13}} - \mu\boldsymbol{\rho_{23}} + \mu\boldsymbol{\rho_{34}}] \\ \ddot{\mathbf{r}}_4 &= M\,[-\boldsymbol{\rho_{14}} - \mu\boldsymbol{\rho_{24}} - \mu_1\boldsymbol{\rho_{34}}] \end{aligned} \tag{12}$$

Using equations (7) to (11), the equations of motion can be written solely in terms of α, β, r_{1A}, $\mathbf{r_{1A}}$ and $\mathbf{r_{24}}$. Thus $\ddot{\mathbf{r}}_1$ becomes

$$\ddot{\mathbf{r}}_{\mathbf{1A}} = -\frac{M}{r_{1A}^3}\left[\frac{1}{\gamma}\left(\frac{2\mu}{c^3} + \frac{\mu_1}{\alpha^2}\right)\right]\mathbf{r_{1A}} \tag{13}$$

and $\ddot{\mathbf{r}}_3$ becomes:

$$\ddot{\mathbf{r}}_{\mathbf{1A}} = -\frac{M}{r_{1A}^3}\left[\frac{1}{(\alpha-\gamma)}\left(\frac{1}{\alpha^2} + \frac{2\mu(\alpha-1)}{d^3}\right)\right]\mathbf{r_{1A}} \tag{14}$$

By simplifying the equations $(\ddot{\mathbf{r}}_4 - \ddot{\mathbf{r}}_2)$ and $(\ddot{\mathbf{r}}_4 + \ddot{\mathbf{r}}_2)$ from (12), we get also:

$$\ddot{\mathbf{r}}_{\mathbf{24}} = -\frac{M}{r_{1A}^3}\left[\frac{1}{c^3} + \frac{2\mu}{\beta^3} + \frac{\mu_1}{d^3})\right]\mathbf{r_{24}} \tag{15}$$

$$\ddot{\mathbf{r}}_{\mathbf{1A}} = -\frac{M}{r_{1A}^3}\left[\frac{1}{(1-\gamma)}(\frac{1}{c^3} - \frac{\mu_1(\alpha-1)}{d^3})\right]\mathbf{r}_{\mathbf{1A}} \tag{16}$$

Note that Equation (16) is redundant, as it is a linear combination of Equations (13) and (14) by the centre of mass relation (5). For an equilibrium solution, the coefficients of the vectors on the right hand side of Equations (13) to (16) must be equal and negative. Therefore, equating the coefficients of Equations (13) and (14) gives:

$$F_1 = -\frac{1}{(\alpha-\gamma)\alpha^2} + 2\mu\left[\frac{1}{\gamma c^3} - \frac{\alpha-1}{(\alpha-\gamma)d^3}\right] + \frac{\mu_1}{\gamma\alpha^2} = 0 \tag{17}$$

Similarly, using (13) and (15) we obtain:

$$F_2 = -\frac{1}{c^3} + 2\mu\left[\frac{1}{\gamma c^3} - \frac{1}{\beta^3}\right] + \mu_1\left[\frac{1}{\gamma\alpha^2} - \frac{1}{d^3}\right] = 0 \tag{18}$$

Given values of μ and μ_1, equations (17) and (18) provide two equations with two unknowns α and β. Solutions for α and β then stipulate the equilibrium positions of the four bodies according to the equations (11) for $\mathbf{r_i}(\alpha, \beta, r_{1A})$ where $\mathbf{r_{1A}} = r_{1A}\hat{\mathbf{x}}$ and $\mathbf{r_{24}} = \beta r_{1A}\hat{\mathbf{y}}$. Figures 7 a) to c) show the family of equilibrium configurations for $\mu_1 = 1$, 0.5 and 0.001, where μ is decreased from 1 to 0 in steps of 0.1 for a) and b), and for c) in steps of 0.1 from 1 to 0.1, steps of 0.01 from 0.1 to 0.01 and steps of 0.001 from 0.01 to 0.

The configurations in each family are scaled such that r_{13} is always 1 with the origin being the midpoint between the two bodies P_1 and P_3. Hence P_1 and P_3 are always located at $(-0.5, 0)$ and $(0.5, 0)$, respectively to make for easier comparison with the Lagrange equilibrium solutions of the Restricted Three-Body Problem. The location of the centre of mass in this coordinate system varies as μ is decreased. For each value of μ_1, the equilibrium positions of P_2 and P_4 approach the L_4 and L_5 points as μ is reduced to zero. Figure 7 c) where $\mu_1 = 0.001$ represents a family of equilibrium solutions in a kite-shaped geometry: beginning with $\mu = 1$, a system of three suns and a Jupiter! and ending with $\mu = 0$, a system of a sun, Jupiter and two Trojan asteroids of infinitesimal mass. Between these two solutions exist an infinity of equilibrium configurations of different kite-shapes. Thus there exists a series of equilibrium configurations for the system, a sun, Jupiter and two Trojan asteroids of finite mass.

Thus, if the two asteroids represented by μM, though very small in mass, produce even minor gravitational attractions on each other, then the equilibrium configuration of the four bodies is not a rotating double equilateral triangle, but a slightly different kite shape. It is interesting to speculate that the centres of mass of the two Trojan groups of asteroids

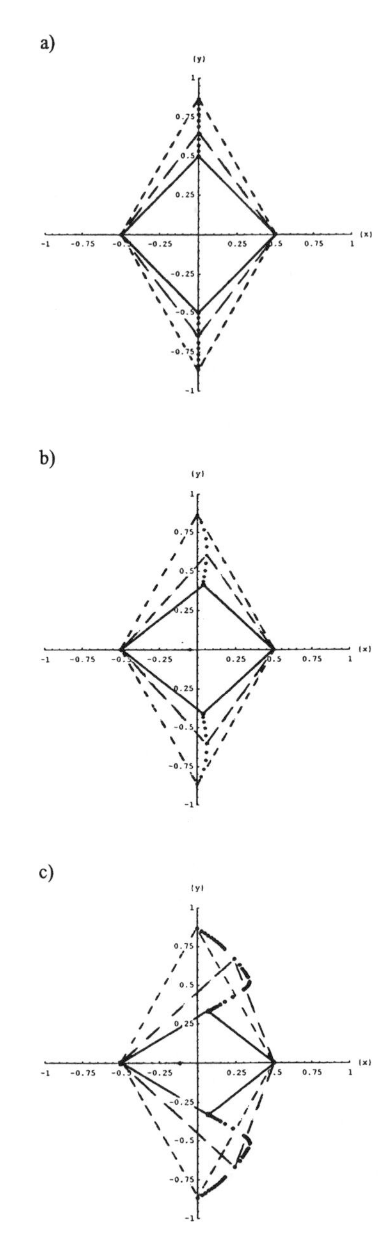

Figure 7. Families of kite shaped equilibrium configurations for a) $\mu_1 = 1$, b) $\mu_1 = 0.5$ and c) $\mu_1 = 0.001$.

in the Sun-Jupiter case may lie nearer the slightly distorted form of the double equilateral triangular configuration given above than precisely at the apexes of the double equilateral triangle itself. Nevertheless the deviation is small, as can be seen if we take $\mu = 10^{-8}$ and $\mu_1 = 0.001$, where μ_1 then corresponds to the mass of Jupiter and μ represents one third the mass of the moon which is of course much more massive than any asteroid. In this case the departure of the 'asteroids' from L_4 and L_5 is only of the order 10^3 kilometers. For typical masses of asteroids this deviation would be much smaller. The question of the stability of this slightly distorted configuration will be treated elsewhere.

References

1. Gomatam, J., Steves, B.A. and Roy, A.E. (1998) Some equal mass four-body equilibrium configurations: Linear stability analysis. *The Dynamics of Small Bodies in the Solar System*, eds. B.A. Steves and A.E. Roy, Kluwer, Dordrecht, Section 4.
2. Hadjidemetriou, J.D. (1978) Instabilities in periodic planetary-type orbits. *Instabilities in Dynamical Systems* ed. V. Szebehely, Reidel, Dordrecht.
3. Marchal, C. and Saari, D.G. (1975) Hill regions for the general three-body problem. *Celest. Mech.* **12**, 115.
4. Milani, A. and Nobili, A.M. (1983) On the stability of hierarchical four-body systems. *Celest. Mech.* **31**, 241.
5. Moulton, F.R. (1910) The Straight Line Solutions of the Problem of N Bodies. *Ann. Math.* **12**, 1.
6. Palmore, J.I. (1973) Classifying Relative Equilibria. I. *Bull. Amer. Math. Soc.* **79** No.5, 904.
7. Palmore, J.I. (1975a) Classifying Relative Equilibria. II. *Bull. Amer. Math. Soc.* **81** No.2, 489.
8. Palmore, J.I. (1975b) Classifying Relative Equilibria. III. *Lett. Math. Phys.* **1**, 71.
9. Palmore, J.I. (1976) Measure of Degenerate Relative Equilibria. I. *Ann. of Math.* **104**, 421.
10. Roy, A.E. (1979) Empirical stability criteria in the many- body problem. *Instabilities in Dynamical Systems* ed V. Szebehely, Reidel, Dordrecht.
11. Roy, A.E. (1988) *Orbital Motion*, 3rd ed., Adam Hilger, Bristol.
12. Roy, A.E., Carusi, A., Valsecchi, G. and Walker, I.W. (1984) The use of the energy and angular momentum integrals to obtain a stability criterion in the general hierarchical three-body problem. *Astron. Astrophys.* **141**, 25.
13. Roy, A.E. and Ovenden, M.W. (1955) On the occurrence of commensu-

rable mean motions in the Solar System II The mirror theorem. *Mon. Not. Roy. Astron. Soc.* **115**, 296.
14. Roy, A.E. and Steves, B.A. (1998) Some Special Restricted Four-Body Problems: II. From Caledonia to Copenhagen. *Planet. Space Sci.*, **46**, No. 5.
15. Roy, A.E., Walker, I.W. and McDonald, A.J.C. (1985) Studies in the stability of hierarchical dynamical systems. *Stability of the Solar System and its Minor Natural and Artificial Bodies* ed V. Szebehely, Reidel, Dordrecht.
16. Simo, C. (1978) Relative Equilibrium Solutions in the Four-Body Problem. *Celest. Mech.* **18**, 165.
17. Steves, B.A., Barnett, A.D. and Roy, A.E. (1998) The finite-time stability method applied to planets of binary systems. *The Dynamics of Small Bodies in the Solar System*, eds. B.A. Steves and A.E. Roy, Kluwer, Dordrecht, Section 4.
18. Steves, B.A. and Roy, A.E. (1998) Some Special Restricted Four-Body Problems: I. Modelling the Caledonian Problem. *Planet. Space Sci.*, **46**, No. 5.
19. Szebehely, V. (1967) *Theory of Orbits*, New York, Academic.
20. Szebehely, V. and Zare, K. (1977) Stability of classical triplets and of their hierarchy. *Astron. Astrophys.* **58**, 145.
21. Valsecchi, G., Carusi, A. and Roy, A.E. (1984) The effect of orbital eccentricities on the shape of the Hill-type analytical stability surfaces in the general three-body problem. *Celest. Mech.* **32**, 217.
22. Walker, I.W. (1983) Stability criteria in many-body systems IV Empirical stability parameters for general hierarchical dynamical systems. *Celest. Mech.* **29**, 149.
23. Walker, I.W., Emslie, A.G. and Roy, A.E. (1980) Stability criteria in many-body systems I An empirical stability criterion for co-rotational three-body systems. *Celest. Mech.* **22**, 371.
24. Walker, I.W. and Roy, A.E. (1983a) Stability criteria in many-body systems III Empirical stability regions for corotational, coplanar, hierarchical three-body systems. *Celest. Mech.* **29**, 117.
25. Walker, I.W. and Roy, A.E. (1983b) Stability criteria in many-body systems V On the totality of possible hierarchical general four-body systems. *Celest. Mech.* **29**, 267.
26. Zare, K. (1977) Bifurcation points in the planar problem of three bodies. *Celest. Mech.* **16**, 35.

SOME EQUAL MASS FOUR-BODY EQUILIBRIUM CONFIGURATIONS: LINEAR STABILITY ANALYSIS

J. GOMATAM AND B.A. STEVES
Department of Mathematics,
Glasgow Caledonian University,
Glasgow, U.K.

AND

A.E. ROY
Department of Physics and Astronomy,
Glasgow University,
Glasgow, U.K.

Abstract. A linear stability analysis of the coplanar, rigid motion of 4 equal masses about their centre of mass is presented. We consider cases where particles are located at (A) the centre of a square, (B) the vertices and centroid of an equilateral triangle and (C) designated points on a line. The variational equations are analysed in the rotating frame. These equations decompose into invariant subspaces of perturbation evolution in the orbital plane and in the normal direction. All the three cases, A, B and C are linearly unstable. It is worth pointing out that, for cases A and B, perturbations normal to the orbital plane have amplitudes which grow as a power of time t.

1. Introduction

The recondite nature of the n-body problem in celestial mechanics has emphasised the need for analysing various special cases of symmetric geometry and mass equality (Moulton, 1910; Palmore, 1975; Steves, Roy and Bell, 1998). A linear stability analysis of the relative equilibrium solution of the 4-body problem for perturbation in the orbital plane was presented by Simo (1978) for arbitrary and restricted masses; this led to a prescription of mass values which ensured linear stability. The objective of this paper

B.A. Steves and A.E. Roy (eds.), The Dynamics of Small Bodies in the Solar System, 373–378.

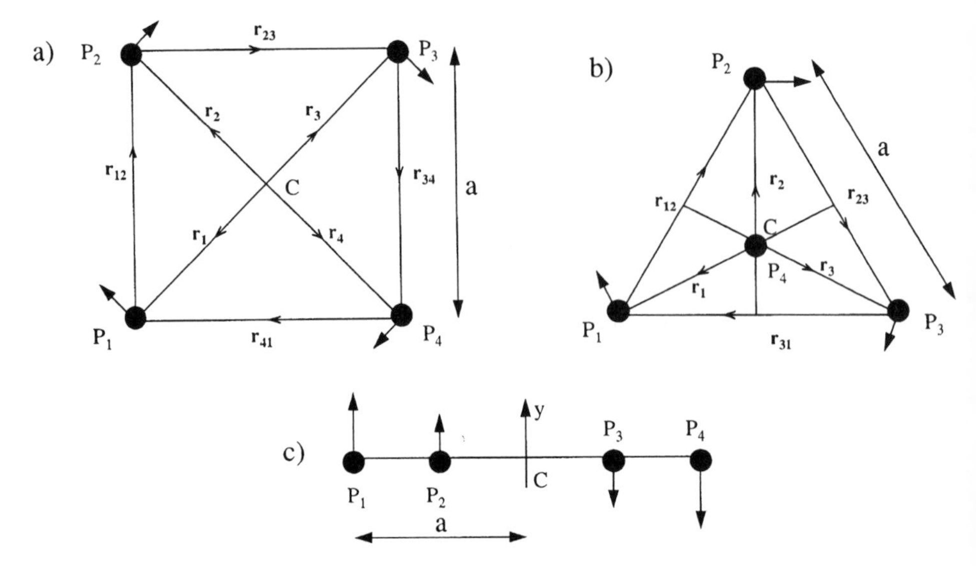

Figure 1. a) Case A: Four equal masses at the corners of a square of side a.
b) Case B: Three equal masses at the vertices of an equilateral triangle, with an equal mass at the centroid.
c) Case C: Four equal masses at P_1, P_2, P_3, and P_4.

is to tailor the linear stability analysis to the highly symmetric cases of the Caledonian problem: equal mass square (Case A), equilateral triangle (Case B) and collinear (Case C) configurations. We offer an analysis of the eigenvalues of the associated Jacobian in terms of the ratio m/a^3, where m is the mass and a is a linear dimension characterising the geometric configuration. All these special solutions of the equal mass four-body problem turn out to be linearly unstable both to perturbations in the orbital plane as well as in the normal direction. In the general 4-body problem (Simo 1978) perturbations normal to the orbital plane lead to linear stability. In our investigations, in Cases A and B, due to high degree of symmetry imposed by the equality of masses, perturbations normal to the orbital plane lead to a Jacobian with multiple imaginary roots. This implies that these perturbations, while oscillatory, have amplitudes which grow as a power of time t. The procedure presented here can, in principle, be applied to an equal mass n-body problem, so long as a rigidly rotating, planar geometric configuration obtains as a special solution.

2. Variational Equations

In this section, we provide a brief introduction to the equations of motion of n bodies of mass m_i, $i = 1, ..., n$, moving under the influence of mutual gravitational forces and the corresponding variational equations for the evolution of perturbations of any given solution, presumed to be known. With $\mathbf{r_i}$ denoting the position vector of m_i with respect to the centre of mass and $G = 1$, we have:

$$\ddot{\mathbf{r}}_\mathbf{i} \equiv \frac{d^2\mathbf{r_i}}{dt^2} = \sum_{j=1,j\neq i}^{n} m_j \boldsymbol{\rho_{ij}}, \quad i = 1, ..., n \tag{1}$$

$$\boldsymbol{\rho_{ij}} \equiv \frac{\boldsymbol{r_{ij}}}{r_{ij}^3} \quad i, j = 1, ..., n; \quad i \neq j \tag{2}$$

$$\mathbf{r_{ij}} \equiv \mathbf{r_j} - \mathbf{r_i} \tag{3}$$

$$r_{ij} \equiv |\mathbf{r_{ij}}| \, . \tag{4}$$

Let $\mathbf{r_i}^*$ be a known solution to equation (1) and $\mathbf{p_i}$, a small (to be defined in what follows) perturbation. Therefore, let

$$\mathbf{r_i} = \mathbf{r_i^*} + \mathbf{p_i} \tag{5}$$

$$\mathbf{r_{ij}} = \mathbf{r_{ij}^*} + \mathbf{p_{ij}} \tag{6}$$

where

$$\mathbf{p_{ij}} \equiv \mathbf{p_i} - \mathbf{p_j} \tag{7}$$

$$p_{ij} \equiv |\mathbf{p_{ij}}| \tag{8}$$

Assume that $p_{ij} << r_{ij}^*$ and expand $\boldsymbol{\rho_{ij}}$ to first order in p_{ij}.

$$\boldsymbol{\rho_{ij}} \approx \frac{\mathbf{r_{ij}^*}}{r_{ij}^{*3}} + \left(\frac{1}{r_{ij}^{*3}}\right) \left[\mathbf{p_{ij}} - 3\mathbf{r_{ij}^*} \frac{\left(\mathbf{r_{ij}^*} \cdot \mathbf{p_{ij}}\right)}{r_{ij}^{*2}}\right] + O(p_{ij}^2) \tag{9}$$

Substituting equation (9) in equation (1) and using the fact that the $\mathbf{r_i^*}$ satisfy the equations of motion (1), we arrive at the variational equations for $\mathbf{p_i}$:

$$\ddot{\mathbf{p}}_\mathbf{i} = {\sum_{j=1}^{n}}' m_j \frac{\mathbf{p_{ij}}}{r_{ij}^{*3}} - 3{\sum_{j=1}^{n}}' m_j \frac{\mathbf{r_{ij}^*}}{r_{ij}^{*5}} \left(\mathbf{r_{ij}^*} \cdot \mathbf{p_{ij}}\right) \tag{10}$$

Here ${\sum_{j=1}^{n}}' \equiv \sum_{j=1,j\neq i}^{n}$

Equations (10) are in general non-autonomous, depending on the functional forms of $\mathbf{r_{ij}^*}$. Solutions $\mathbf{r_i^*}$ are said to be linearly, asymptotically stable if $\lim_{t\to\infty} \mathbf{p_i} \to \mathbf{0}$.

We confine the analysis of equations (1) and (10) to coplanar, equal mass four-body problems where point-like bodies located in symmetric configurations move rigidly about their centre of mass (Steves and Roy, 1998). The general procedure is to assume a particular geometric configuration and implement the following criteria for further simplification:

$$\ddot{\mathbf{r}}_i^* = -n^2 \mathbf{r}_i^* \; ; \tag{11}$$

at t=0,

$$\begin{aligned} \mathbf{r}_i^*(0).\dot{\mathbf{r}}_i^*(0) &= 0 \qquad \text{(circular motion)} & (12)\\ \dot{r}_i^*(0) &= n r_i^*(0) \quad \text{(rigid motion)}. & (13) \end{aligned}$$

Here the constant n depends on m, the mass of the body and a length parameter characterising the geometry of the rigidly rotating configuration. Since $\mathbf{r}_i^*$'s are sinusoidal in t, coefficients of $\mathbf{p_{ij}}$ in equation (10) contain terms periodic in time. However transforming equation (10) to a frame of reference, rotating with the rigid configuration (synodical frame) leads to a system of differential equations with constant coefficients. The problem then becomes essentially algebraic.

3. Method of Analysis: Case A as an example

The required transformations are

$$\mathbf{p_i} = C_{i1}\mathbf{r}_1^* + C_{i2}\mathbf{r}_2^* + C_{i3}\hat{\mathbf{k}} \; , \quad i = 1, 2, 3, 4; \tag{14}$$

where $\mathbf{r}_1^*$ and $\mathbf{r}_2^*$ are known solutions and $\hat{\mathbf{k}}$ is the unit vector perpendicular to the orbital plane. Henceforth, we will denote $\mathbf{r}_1^*$ and $\mathbf{r}_2^*$ by $\mathbf{r_1}$ and $\mathbf{r_2}$ respectively. In what follows, we will illustrate the method of calculation for the Case A, where four equal masses are located at the corners of a square of side a units.

Consider

$$\begin{aligned} \ddot{\mathbf{p}}_i &= m\left[\frac{\mathbf{p_{12}}}{r_{12}^3} + \frac{\mathbf{p_{13}}}{r_{13}^3} + \frac{\mathbf{p_{14}}}{r_{14}^3}\right] \\ &\quad -3m\left[\frac{\mathbf{r_{12}}}{r_{12}^5}(\mathbf{r_{12}}.\mathbf{p_{12}}) + \frac{\mathbf{r_{13}}}{r_{13}^5}(\mathbf{r_{13}}.\mathbf{p_{13}}) + \frac{\mathbf{r_{14}}}{r_{14}^5}(\mathbf{r_{14}}.\mathbf{p_{14}})\right] \end{aligned} \tag{15}$$

For the case A:

$$r_{12} = r_{14} = a \; , \quad r_{13} = a\sqrt{2} \quad \text{and } \mathbf{r_1}.\mathbf{r_2} = 0 \; . \tag{16}$$

Express $\mathbf{p_1}$ in terms of the triplet $\{\mathbf{r_1}, \mathbf{r_2}, \hat{\mathbf{k}}\}$:

$$\mathbf{p_1} = C_{11}\mathbf{r_1} + C_{12}\mathbf{r_2} + C_{13}\hat{\mathbf{k}} \tag{17}$$

with similar expressions for $\mathbf{p_2}, \mathbf{p_3}$ and $\mathbf{p_4}$.

Now we can calculate various scalar products occurring in equation (15), while exploiting simplifying features of Case A. For instance

$$\mathbf{r_{12}}.\mathbf{p_{12}} = (C_{22} - C_{11} - C_{21} + C_{11})\, a^2 . \tag{18}$$

Since $\dot{\mathbf{r}}_1 = n\mathbf{r_2}, \dot{\mathbf{r}}_2 = -n\mathbf{r_1}$, we obtain

$$\ddot{\mathbf{p}}_1 = \left(\ddot{C}_{11} - 2\dot{C}_{12}n - C_{11}n^2\right)\mathbf{r_1} + \left(\ddot{C}_{12} + 2\dot{C}_{11}n - C_{12}n^2\right)\mathbf{r_2} + \ddot{C}_{13}\hat{\mathbf{k}}. \tag{19}$$

In all these calculations, $n = \sqrt{\frac{m}{a^3}\left(2 + \frac{1}{\sqrt{2}}\right)}$.

Substituting equations (19) and (18) etc., in (15) and equating coefficients of $\mathbf{r_1}, \mathbf{r_2}$ and $\hat{\mathbf{k}}$ on both sides, we arrive at a set of linear, autonomous, second-order differential equations for the C_{ij}'s. A further simplification obtains, because of the decoupling of perturbations in the direction $\hat{\mathbf{k}}$ from those confined to the orbital plane. More explicitly, with

$$V \equiv (C_{11}, C_{21}, C_{31}, C_{41}, C_{12}, C_{22}, C_{32}, C_{42},)^T \in \mathbf{R}^8 \tag{20}$$

$$U \equiv (C_{13}, C_{23}, C_{33}, C_{43})^T \in \mathbf{R}^4 \tag{21}$$

the variational equations for $\mathbf{p_i}$, $i = 1, 2, 3, 4$ reduce to

$$M_{||}\ddot{V} + N_{||}\dot{V} + P_{||}V = 0 \tag{22}$$

$$M_{\perp}\ddot{U} + N_{\perp}\dot{U} + P_{\perp}U = 0 \tag{23}$$

in the rotating frame. M, N and P's are matrices of appropriate dimensionalities, with constant entries depending on m/a^3. Assuming that

$$C_{ij} = C_{ij}(0)e^{\sigma t} \qquad i = 1, 2, 3, 4;\ j = 1, 2, 3 \tag{24}$$

the stability of the motion is determined by the following polynomial equations for σ:

$$\text{Determinant}\left[\sigma^2 M_{||} + \sigma N_{||} + P_{||}\right] \equiv f_{||}\left(\sigma, \frac{m}{a^3}\right) = 0 \tag{25}$$

$$\text{Determinant}\left[\sigma^2 M_{\perp} + \sigma N_{\perp} + P_{\perp}\right] \equiv f_{\perp}\left(\sigma, \frac{m}{a^3}\right) = 0 \tag{26}$$

For the Case A, equation (25) is of 16th degree in σ while equation (26) is an 8th degree polynomial.

All algebraic and numerical calculations have been performed with the aid of Mathematica, Version 2.3. Instability is indicated by the presence of any σ such that Re $\sigma > 0$. Here Re σ stands for Real part of σ. We have solved equations (25) and (26) numerically for $m/a^3 = 10^{-2}$, $10^{-1}, 1, 10$ and 10^2. The results are summarised in Table 1.

Case	Nature of roots of $f_{\perp}(\sigma, m/a^3) = 0$	Nature of roots of $f_{\|\|}(\sigma, m/a^3) = 0$
A	All roots are imaginary; two double roots imply instability	All roots have $\text{Re}\,\sigma > 0$
B	$\sigma = 0$ is a double root; rest of the roots are imaginary with two double roots	All roots have $\text{Re}\,\sigma > 0$
C	One root with $\text{Re}\,\sigma > 0$; the rest of the roots imaginary	All roots have $\text{Re}\,\sigma > 0$

Table 1. Nature of roots of $f_{\perp} = 0$ *and* $f_{||} = 0$ *for* $10^{-2} < m/a^3 < 10^2$.

4. Conclusions

All the three cases examined turn out to be linearly unstable to perturbations in the orbital plane as well as in the orthogonal direction. In the latter situation the occurrence of multiple imaginary roots (Cases A and B) imply oscillatory perturbations which grow as a power of t. In the Case C, perturbations in the direction of $\hat{\mathbf{k}}$ grow exponentially due to the occurrence of a root σ, with $\text{Re}\,\sigma > 0$. The procedure presented here can be applied to any rigidly rotating, planar solution of the n-body problem with equal masses.

References

1. Moulton, F.R. (1910) The Straight Line Solutions of the Problem of N Bodies. *Ann. Math.*, **12**, 1.
2. Palmore, J.I. (1975) Classifying Relative Equilibria III. *Lett. Maths. Phys.*, **1**, 71.
3. Steves, B.A., Roy, A.E. and Bell M. (1998) Some Special Solutions of the Four-Body Problem: I. Modelling the Caledonian Problem. *The Dynamics of Small Bodies in the Solar System*, eds. B.A. Steves and A.E. Roy, Kluwer Academic Publishers, Dordrecht, Section 4.
4. Roy, A.E. and Steves, B.A. (1998) Some Special Restricted Four-Body Problems: II. From Caledonia to Copenhagen. *Planet. Space. Sci.*, **46**, No.5.
5. Simo, C. (1978), Relative Equilibrium Solutions in the Four-Body Problem. *Celest. Mech.*, **18**, 165.

THE FINITE-TIME STABILITY METHOD APPLIED TO PLANETS OF BINARY STARS

B.A. STEVES AND A.D. BARNETT
Dept. of Mathematics,
Glasgow Caledonian University, Glasgow, U.K.

AND

A.E. ROY
Dept. of Physics and Astronomy,
University of Glasgow, Glasgow, U.K.

1. Introduction

In recent years, the search for extrasolar planets and planetary systems has produced some remarkable results. Since the first detection in 1995 (Mayor and Queloz, 1995) of an object half the mass of Jupiter orbiting the star 51 Pegasi, nineteen other discoveries of extrasolar planets/brown dwarves around main sequence stars (Schneider, 1998) have been confirmed.

To date, the search for extrasolar planets has been primarily concentrated on single stars, partly because of the lack of knowledge of the stability of planets in multiple star systems. Yet 65 % of stars are to be found in gravitationally bound systems of two or more stars.

The long-term stability of fictitious planets of binary systems has been explored by a number of authors using numerical integrative methods over fixed time spans (See for example, Harrington, 1977; Donnington and Mikulskis, 1992, 1995; Hale, 1994). Due to limitations on the computation time, these time spans fall far short of the accepted lifetime of our solar system of 5 x 10^9 years.

The stability of the binary star-planet problem can also be studied analytically using the c^2H stability criterion for the general three-body problem. See Steves, Roy and Bell (1998) in this volume for a short description of this method of analysis.

In this paper, the stability of planet-binary star systems is studied by applying the Finite-Time Stability Criterion Method (FTSCM) developed

B.A. Steves and A.E. Roy (eds.), The Dynamics of Small Bodies in the Solar System, 379–384.

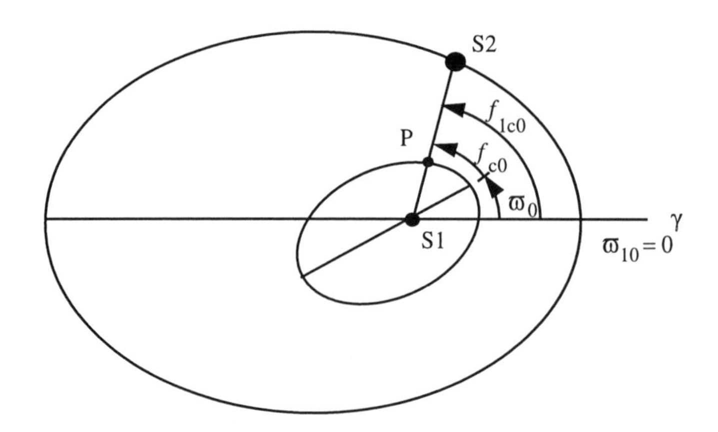

Figure 1. The initial configuration of the elliptical restricted three body problem. S1, P, and S2 denote Star 1, the planet, and Star 2 respectively. At the initial time $t = 0, \varpi_{10}$ is arbitrarily set to 0, without loss of generality. In the circular case ϖ_0 is arbitrarily set to 0, so therefore $f_{1c0} = f_{c0}$

by Steves and Roy (Steves, 1990; Steves and Roy, 1995) to investigate the stability of solar system satellites against solar perturbations. Their approach combined characteristics of both analytical and numerical methods to produce a stability criterion valid not for all time, but for known finite durations. In their application of the FTSCM to sun-perturbed satellites, Steves and Roy (1995) showed that the main satellites of Jupiter, Saturn, Uranus and Mars had minimum lifetimes ranging from 10^6 to 10^{11} years, whereas almost every real planetary satellite perturbed by the sun had no guarantee of stability under the c^2H stability criterion.

2. The Finite-Time Stability Criterion Method (FTSCM)

The model developed by Steves and Roy (1995) for a planet-satellite system perturbed by the Sun involves a three-body hierarchical system where the satellite and planet are assumed to move in disturbed Keplerian orbits about their centre of mass, while the Sun moves in a larger Keplerian orbit about the centre of mass of the planet and satellite. Hereafter this hierarchy is called a (planet-satellite)-Sun hierarchy. The problem is further reduced to an elliptic restricted coplanar three-body problem, where the mass of the satellite is assumed negligible compared to the masses of the planet and Sun. The Sun is therefore assumed to move in a fixed elliptical orbit about the planet.

This model is eminently suitable for studying the minimum lifetimes of planets in binary systems which have a (Star 1-planet)-Star 2 type hierarchy.

Figure 1 shows the initial configuration of the two stars and planet for this model, where S1, P and S2 denote Star 1, the planet and Star 2 respectively. Let the osculating orbital elements of the planet's orbit be the semi-major axis a, the eccentricity e, the longitude of pericentre ϖ, and the mean longitude of the epoch ϵ. n and f_0 are the mean motion and the initial true anomaly of the planet. The corresponding quantities for Star 2's orbit are denoted with a subscript 1, ie a_1, e_1, ϖ_1, ϵ_1, n_1 and f_{10}. Let the masses of Star 1, the planet and Star 2 be M, m and m_1, with m negligible compared to M and m_1. Star 2 is therefore assumed to move in a fixed elliptical orbit about Star 1.

The FTSCM involves applying successively to the three-body hierarchical problem, (body 1- body 2)-body 3, a series of increasingly less pessimistic stability criteria which are valid for finite lengths of time. The successive levels of stability criteria are based on the natural periodic cycles found in the hierarchical three-body system. Estimates of these finite times during which the three body system is clearly stable can then be used as minimum durations of the orbits of the inner pair of bodies about their centre of mass against perturbations of the outermost third body. If the minimum lifetime obtained from the first most pessimistic stability criterion is long enough to allow the next less pessimistic stability criterion to take effect, then the minimum duration of the system using the second stability criterion can be found. If this second minimum duration is long enough to enable the third stability criterion to be applied, a third, even longer, minimum duration can be calculated, and so on. In this manner, the minimum lifetime of the inner pair system can be extended until all the possible stability criteria have been invoked.

In the elliptic restricted coplanar three-body model, the stability criteria are based on the natural periodic cycles: the *synodic*, the *conjunction* and the *mirror* cycles, where the cycles are defined as follows for the (Star 1-planet)-Star 2 hierarchy:

- *synodic* cycle: the revolution of the planet beginning and ending at an opposition with Star 2 and therefore centred on a conjunction with Star 2.
- *conjunction* cycle: the revolution of the conjunction line through approximately 2π radians to end at the conjunction closest to the initial conjunction with respect to the moving apse of the planet's orbit (Steves and Roy, 1995).
- *mirror configuration* cycle: the revolution of the conjunction line, beginning with the conjunction lying nearest to the planet's pericentre, through approximately 2π radians to end at the next closest conjunction to the planet's pericentre (Steves and Roy, 1995).

For the stability criterion itself, the eccentricity of the satellite in the (planet-satellite)-Sun problem or the planet in the (Star 1-planet)-star 2 problem was found to be the best orbital parameter to indicate that the system is approaching an unstable situation (Steves and Roy, 1995). As the planet's eccentricity increases towards a value of one, the system approaches an unstable situation where the planet will either collide with Star 1 or escape from its gravitational influence.

The changes in the planet's orbital elements e, a, ϖ and ϵ over one synodic period are found analytically to third order by expanding Lagrange's planetary equations about the small parameters e, e_1, $\alpha = \frac{a}{a_1}$, $\nu = \frac{n}{n_1}$. For an exhaustive description of the analysis of the FTSCM see Steves and Roy (1995). At each stage in the FTSCM, the stability criterion takes the most pessimistic viewpoint. It first assumes that the worst possible change in the planet's orbital eccentricity over one synodic cycle is added on to the planet's eccentricity every synodic cycle. This procedure is highly pessimistic since changes in the eccentricity over a synodic period vary in a cyclic manner about zero and depend on the position of the initial planet-Star 2 conjunction with respect to the apsidal lines of the two orbits. The eccentricity is allowed to accumulate in this manner until some arbitrarily chosen upper limit for the eccentricity is reached. At this point the system is taken to be approaching an unstable situation. We choose the limit e_u to be 0.5, since the analytical theory for the changes in the orbital elements of the planet over one synodic period is accurate to 5% for values of e up to 0.5. The time taken to reach the arbitrarily chosen upper limit of the eccentricity provides a measurable minimum lifetime $T^{(1)}_{min}$ for the planet system at the first level of the FTSCM.

If this minimum lifetime is long enough for many conjunction cycles to pass, then the longer second level minimum duration of the system using the conjunction cycle can be found. Here it is assumed that the worst possible change in the planet's orbital eccentricity over the conjunction cycle is added on to the planet's eccentricity every conjunction cycle. Again this procedure is highly pessimistic as the change in eccentricity over a conjunction cycle also varies in a cyclic manner about zero. The time taken for the eccentricity to accumulate to e_u is then taken to be the minimum duration $T^{(2)}_{min}$ of the planet system at the second level of the FTSCM. The third level minimum duration $T^{(3)}_{min}$ uses the same principles but is based on the mirror configuration cycle.

3. Results

The minimum lifetimes at the three stability levels were determined for a variety of the initial parameters α, e, e_1 and μ. Figure 2 shows typical surfaces of minimum lifetimes at the third level plotted in $\alpha - e$ space, for $e_1 = 0.05$, and $\mu = 10$ and 100. The minimum lifetimes are given as the

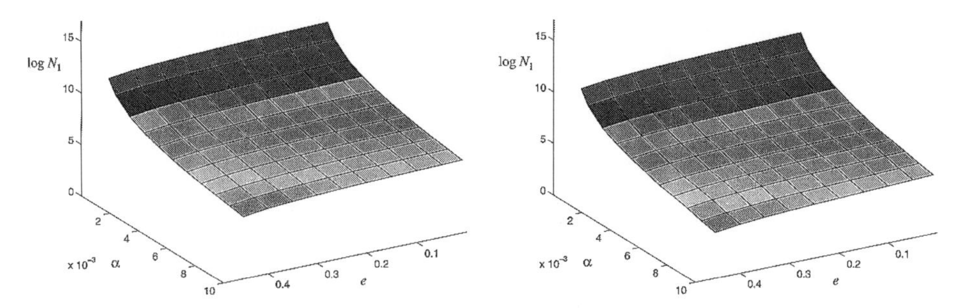

Figure 2. Surface of $\log N_1$ in $\alpha - e$ space for the third level elliptical case, where $e_1 = 0.05$, α: 0.0005 - 0.01, e: 0.01 - 0.45 and (a) $\mu = 10$, (b) $\mu = 100$.

logarithm of minimum numbers of synodic periods N_1 endured.

Generally, as expected, the minimum lifetimes decrease with increased e, α and μ. Larger initial values of e will mean that less time is required to increase the eccentricity to the upper limit. Larger initial values of α indicate Star 2 has been moved closer to the planet and hence its perturbations on the planet's orbit will be larger and the planet will endure for shorter times. And finally larger values of μ indicate the mass of Star 2 has increased, which also would increase the size of the perturbations on the planet's orbit. The minimum lifetimes at the third level for $\mu = 10$ and 100, range from 10^2 to 10^{10} synodic periods for a range in α of 0.0005 to 0.01 and in e of 0.01 to 0.45.

To gain a physical understanding of the magnitudes of the minimum durations achieved by the FTSCM, the method was also applied to an Earth-type planet orbiting within a binary star system of two solar mass stars, with $e = 0.01$, $a = 1$ AU and $\mu = 1$. Star 2 is assumed to move in a fixed circular orbit about Star 1, ie $e_1 = 0$, with radii ranging from 20 AU to 1000 AU. The minimum durations in years are given in Table 1.

The third level shows that the minimum duration at a binary star separation of 50 AU is of the order of 10^8 years, which is about two percent of the age of the solar system. Thus, under the assumptions made in the model, Earth-type planetary orbits which exist for near the age of the solar system are possible in binary star systems with separations larger than about 50 AU. It should be noted that the assumptions made in the model take the most pessimistic view, even at the third level and therefore stable

Earth-type planetary orbits may be possible at lower separations.

	Minimum Duration of Stability (yrs)		
Separation	$T_{min}^{(1)}$	$T_{min}^{(2)}$	$T_{min}^{(3)}$
20 AU	$7.37. \times 10^{2}$	1.31×10^{5}	1.09×10^{6}
30 AU	2.75×10^{3}	3.95×10^{5}	6.98×10^{6}
40 AU	6.68×10^{3}	1.69×10^{6}	4.31×10^{7}
50 AU	1.34×10^{4}	3.31×10^{6}	1.06×10^{8}
100 AU	1.13×10^{5}	1.14×10^{8}	1.03×10^{10}
500 AU	1.48×10^{7}	2.50×10^{11}	2.26×10^{14}
1000 AU	1.19×10^{8}	3.07×10^{12}	7.99×10^{62}

Table 1: The minimum durations of stability for an Earth-type planet orbiting a star which is part of a binary star system of two equal solar masses.

References

1. Donnison, J.R. and Mikulskis, D.F. (1992) Three-body orbital stability criteria for circular orbits. *Mon. Not. Roy. Astr. Soc.* **254**, 21.
2. Donnison, J.R. and Mikulskis, D.F. (1995) The effect of eccentricity on 3-body orbital stability-criteria and its importance for triple star systems. *Mon. Not. Roy. Astr. Soc.* **272**, 1.
3. Hale A., (1994) Orbital coplanarity in solar-type binary systems: Implications for planetary system formation and detection. *Astron. Jour.* **107**, 306.
4. Harrington, R.S. (1977) Planetary orbits in binary stars. *Astron. Jour.* **82**, 753.
5. Mayor, M. and Queloz, D. (1995), *Nature*, **378**, 355.
6. Schneider, J. (1998) Extrasolar planets catalog, http://www.usr.obspm.fr/departement/darc/planets/catalog.htm
7. Steves, B.A. (1990) Finite-time stability criteria for sun-perturbed planetary satellites *PhD Thesis*, Univ. Glasgow.
8. Steves, B.A. and Roy, A.E. (1995) Finite-time stability criteria for sun-perturbed planetary satellites *Amer. Astron. Soc. Advances in the Astronaut. Sci.* **89** Univelt Publishers, San-Diego, California.
9. Steves, B.A., Roy, A.E. and Bell, M. (1998) Some special solutions of the four-body problem *The Dynamics of Small Bodies in the Solar System*, eds B.A. Steves and A.E. Roy, Kluwer, Dordrecht.

DYNAMICS AND STABILITY OF THREE-BODY SYSTEMS

ROSEMARY MARDLING
Department of Mathematics, Monash University,
Clayton, Victoria, Australia

AND

SVERRE AARSETH
Institute of Astronomy, Cambridge University,
Madingley Road, Cambridge

Abstract. We discuss the importance of the three-body problem in astrophysics, and present a summary of some new results which use a correspondence between the binary-tides problem and the three-body problem to derive a single formula for the stability of hierarchical triples.

1. Introduction

The simplicity and complexity of the three-body problem has allured many people to its study, some devoting their lives to uncovering its secrets. Our interest in the three-body problem is of astrophysical origin, with application in several distinct areas:

1. The formation and evolution of temporarily and permanently bound sub-systems of three or more stars in star clusters, be they clusters of proto-stars in star forming regions, open (galactic) clusters consisting of up to several tens of thousands of stars, or globular clusters of up to a million stars. A typical interaction involves the temporary capture of a single star by a binary. Such a system will inevitably disintegrate by ejecting one of the stars, although some systems formed this way can be very long lived. The fact that triples formed via the capture of a single object by a binary are unstable is intimately related to the subject of this contribution. A bound system can only form if the interaction involves some kind of energy loss. For instance, if it involves a fourth

B.A. Steves and A.E. Roy (eds.), The Dynamics of Small Bodies in the Solar System, 385–392.

star, it is possible for a permanently bound triple to be formed, with one of the four stars escaping the system with the excess energy.

2. Capture interactions are of fundamental importance in the theory of planet formation. It is now believed that rocky planets, as well as the cores of gas giants, are formed via a runaway accretion process which involves the gravitational capture of planetesimals by ever more massive proto-planets. This is a three-body process with the central star as the third body. Understanding how energy and angular momentum are exchanged in these interactions allows us to address questions such as how the planets acquired their various spins, both in magnitude and direction.
3. The existence of small bodies with binary companions such as the asteroid Ida and its companion Dactyl (Chapman et al. 1995) are testimony to the process of capture followed by removal of energy by a fourth body (or indeed several other bodies).
4. Predicting which configurations of stars are possible, from triples to quadruples to quintuples and beyond, requires an understanding of the nature of stability in the three-body problem, in particular for arbitrary orbital configurations.
5. In the realm of extra-solar planetary systems, we may use our knowledge of stability in the three-body problem to predict whether planets can exist in stable orbits in particular binary configurations.

2. The binary-tides problem and the three-body problem

Having attempted to convince the reader of the importance of the three-body problem in astrophysics, we now present a summary of our results, the details of which will appear elsewhere. We have mentioned above the process of capture of a single object by a binary. One starts with what are essentially *two* objects - a single star (or planetesimal) and a binary - whose relative motion is unbound. As the third object approaches the binary, it begins to experience its extended nature, whereupon energy and angular momentum are exchanged. This interaction may lead to the binary becoming more or less tightly bound. In the former case, it is possible for the binary to absorb more than the excess energy of unbound motion, resulting in the *temporary* capture of the third body; it is inevitable that one of the objects will escape the system. The motion is chaotic in the sense that miniscule changes in the initial configuration lead to an entirely different evolution, although the probability is one that the final configuration will consist of a single star and a binary.

The process of triple capture is intimately related to another capture process: that of *two-body tidal capture* in which the tides of a star absorb the excess energy of unbound motion of a second star as it passes within a few stellar radii of the first (Mardling 1995a,b). In normal stellar environments this process is extremely unlikely, but in regions where the stellar density is high, such as the cores of globular clusters, the process can occur frequently enough for it to be important in the creation of various exotic objects such as blue stragglers (in which the final outcome is coalescence of the two stars), low-mass X-ray binaries and millisecond pulsars.

The evolution of the orbit of a tidal capture binary is chaotic, that is, the direction in which energy flows between the orbit and the tides, as well as the amount of energy exchanged, is entirely unpredictable. This is in contrast to the way energy is exchanged in a normal binary which is close enough to interact tidally. Were the tidal motions unable to dissipate some of the energy acquired from the orbit, the system would, again with probability one, unbind at some time in the future, usually quite soon after capture. However, any amount of damping more likely than not allows such a binary to survive to become permanently bound. It will continue to dissipate tidal energy until it reaches a point where the tidal interaction is no longer chaotic, after which evolution proceeds as in a normal binary. Figure 1 illustrates the difference in behaviour for chaotic and normal tidal evolution.

One can write down an approximate formula which describes the boundary between chaotic and normal behaviour in a tidally interacting binary (figure 2a). Such a boundary is, in reality, not well defined in the sense that one finds regions of stability and chaos intertwined in its vicinity. However, for practical purposes it is useful to have such a formula. If the mass ratio of the stars is q, a binary with an initial eccentricity e will be chaotic if its initial periastron separation (in units of radius r_* of the tidally active star) is less than R_p/r_*, given quite accurately by (Mardling & Aarseth 1998)

$$R_p/r_* = C\left[\frac{1+q}{\omega^2}\frac{(1+e)^{1.2}}{(1-e)^{1/2}}\right]^{2/5}. \tag{1}$$

Here $C = 1.836$ and ω is the (scaled) frequency of the most energetic mode of oscillation of the tides. This expression was obtained using similarity arguments, with the constant C being the only empirical parameter.

It is the analogy between two-body tidal capture and three-body point mass capture that led us to devise a formula which predicts the stability or otherwise of hierarchical triples. The way energy and angular momentum are exchanged between the inner and outer orbits of a stable hierarchy is very similar to the way they are exchanged in a binary undergoing normal tidal evolution. It occurs in a quasi-periodic fashion, with (generally)

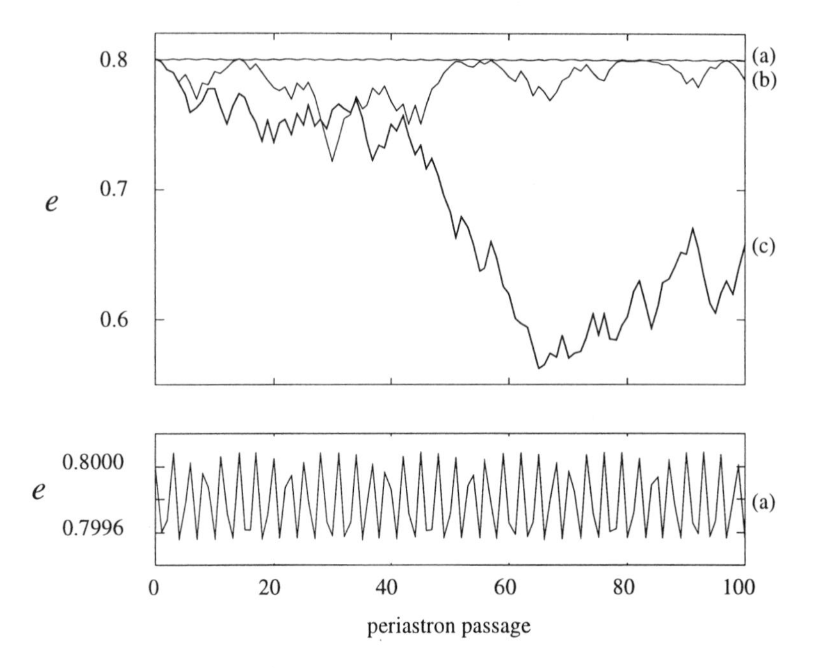

Figure 1. Chaotic vs. normal tidal interaction. Curve (a) (shown magnified at the bottom of the figure) is typical of normal tidal evolution. The eccentricity varies little from orbit to orbit so that the energy in the tides is small. In contrast, curves (b) and (c) display the extreme sensitivity to initial conditions typical of chaotic systems, with the eccentricity varying over a wide range and the tidal energy becoming very large. The initial eccentricity for each example was 0.8, while the initial periastron separation (in units of the stellar radius) was 3.1, 2.9 and 2.90001 for cases (a), (b) and (c), respectively.

small amounts of energy and angular momentum moving in and out of the inner binary. This is consistent with the well known result that the inner semimajor axis is an adiabatic invariant (its average value is constant). In the absence of energy removal from the system, the average behaviour of both a tidal binary and a three-body system is invariant and the system is stable. Figures 3c and 3d illustrate stable behaviour. Two nearby orbits are plotted here with only small deviations. This should be compared to figure 1a. Note that, unlike in the binary-tides case, the eccentricity can vary considerably in a stable triple, although initially close orbits will remain close.

The way energy and angular momentum are exchanged between the inner and outer orbits of an *unstable* triple is very similar to the way they are exchanged in a binary whose tide-orbit interaction is chaotic. This is particularly obvious if one plots the eccentricity of the outer orbit against the number of outer orbits (figures 3a,b).

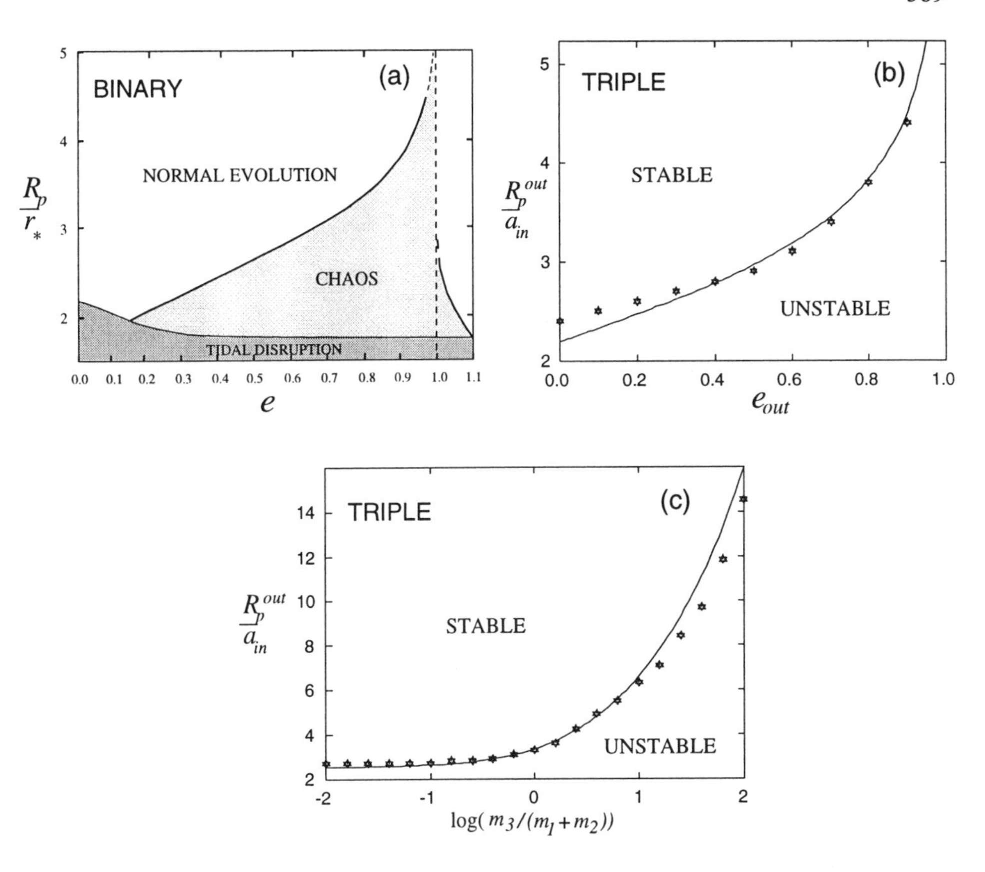

Figure 2. Stability boundaries for (a) tidally interacting binaries with equal mass components and density profiles appropriate for fully convective stars. R_p/r_* is the initial periastron separation in units of the radius of the tidally active star and e is the orbital eccentricity. The boundary moves down for more centrally condensed stars and up if the tidally active star is less massive than its companion. (b) hierarchical triples with equal mass components and initially circular inner binaries. R_p^{out}/a_{in} is the periastron separation of the outer binary in units of the semimajor axis of the inner binary, and e_{out} is the outer orbital eccentricity. Here the outer orbit of a hierarchy is defined as the orbit of the most distant object about the centre of mass of the other two. The stars represent an empirically determined boundary, while the curve was calculated using equation (1). (c) hierarchical triples with $e_{in} = e_{out} = 0.5$, with e_{in} the inner eccentricity. R_p^{out}/a_{in} is plotted against the mass ratio of the outer binary, $m_3/(m_1 + m_2)$.

A formula very similar to (1) predicts the stability or otherwise of triples with arbitrary inner and outer eccentricities and masses. At this stage, however, it assumes that the orbits are coplanar, which represents an upper limit to the stability of inclined systems. We obtained this formula by setting up a correspondence between the variables in the binary-tides problem

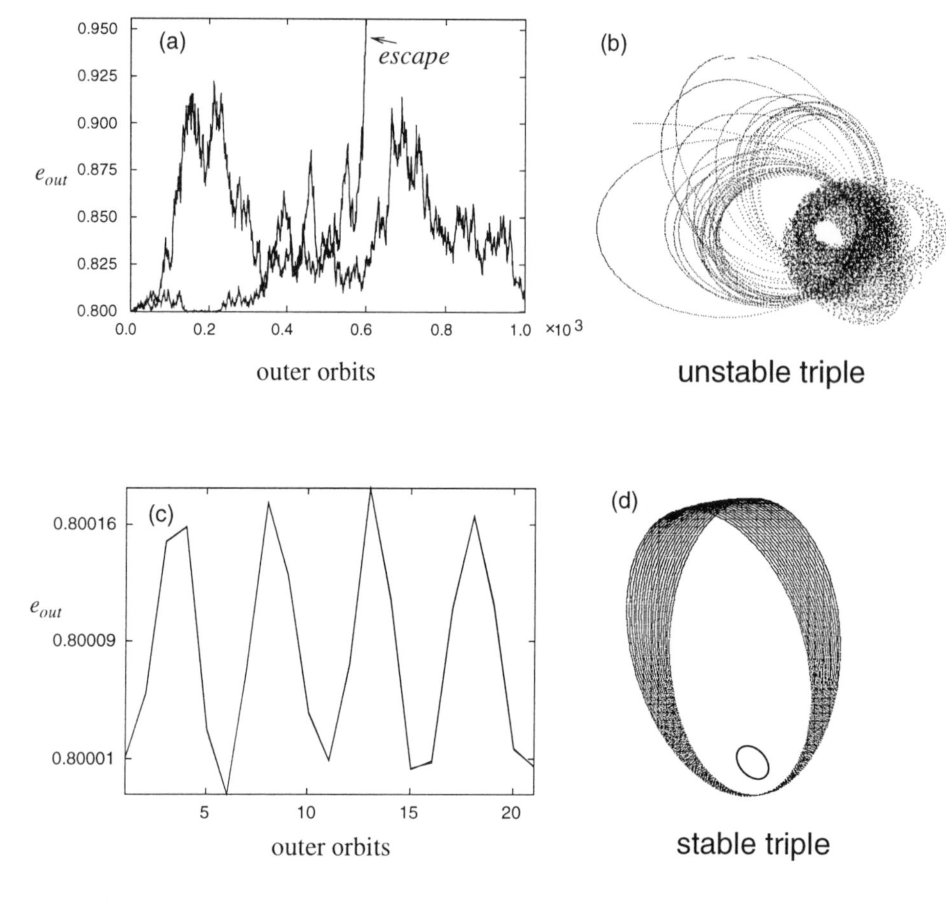

Figure 3. Chaos and stability in triples. (a) shows the chaotic evolution of two initially close orbits. The third body escapes in one case. (b) shows the orbits of an unstable triple. The smaller orbit to the right is the inner binary. Extreme apsidal advance is evident, as well as large variations in the eccentricity of both the inner and outer binaries. (c) illustrates stable behaviour, with two initially close orbits remaining close. (d) shows the outer orbit of a stable triple with regular apsidal advance and (in this case) little variation in eccentricity. The behaviour is very similar to the binary-tides case.

and the hierarchical triple problem. It is given by

$$R_p^{out}/a_{in} = C\left[(1+q_{out})\frac{(1+e_{out})}{(1-e_{out})^{1/2}}\right]^{2/5}, \tag{2}$$

where now $C = 2.8$, e_{out} is the eccentricity of the outer orbit and $q_{out} = m_3/(m_1+m_2)$ is the mass ratio of the outer 'binary'. Note that the inner eccentricity does not appear in this expression. Figures 2a,b illustrate the accuracy of this formula by comparing it with empirically determined data, obtained using a three-body regularization code (Aarseth & Zare 1974). A system was deemed stable if two orbits, initially differing by 1 part in 10^5 in the eccentricity, remained close after 100 orbits. Although in cases very close to the stability boundary it is possible that such apparently stable orbits are in fact unstable, we have found this stability criterion adequate for our purposes. One can examine the stability boundary in much more detail if one is interested in its fine structure, for example, its resonant behaviour (Kiseleva et al 1994). However, for practical purposes such as its use in star cluster simulations, the resolution equation (2) provides is invaluable and of greater validity than criteria based on numerical fitting (Eggleton & Kiseleva 1995).

3. Star cluster simulations

The stability criterion has proved extremely useful in N-body simulations of star clusters containing a significant proportion of primordial binaries (Aarseth 1997, Mardling & Aarseth 1998). A number of hierarchical systems form during the cluster evolution and their long-lived stability requires special consideration because of the short inner binary period. Consequently, we replace any such binary with its c.m. motion once stability has been ascertained, whereupon the two-body solution is advanced by KS regularization until the outer pericentre distance exceeds the permitted value. It is fairly common to find several stable subsystems at the same time, especially during later stages; hence there is a real need for special treatment since otherwise the calculation would become prohibitively expensive.

We demonstrate the importance of recognizing stable hierarchies by illustrating a recent episode. However, first it should be mentioned that strongly interacting subsystems of 3 - 5 members are studied by the chain regularization method (Mikkola & Aarseth 1993) which allows for arbitrarily close encounters of the members. The event in question was initiated by a single particle impacting an energetic binary, thus forming a temporary compact triple suitable for chain treatment. During this interaction, a super-hard binary approached so closely that it formed a natural extension of the chain and was 'absorbed' into the system. Soon thereafter one of the original triple components escaped, thereby reducing the membership to four. The subsequent integration ceased to advance, with no resolution of the outcome even in the presence of a small external perturbation.

The situation was only clarified by including a procedure for checking the stability of a *quadruple* system in a similar manner to that used for triple systems. If the quadruple consists of a pair of binaries (2+2), one identifies the most compact binary and applies the stability formula taking this sub-system to be the outer object. If the quadruple is a hierarchy inside a hierarchy (3+1), again one identifies the closest pair but takes this to be a single object which forms the inner binary with the next closest object. Analysis of the four-body configuration showed that the stability condition above was satisfied, with the outer pericentre distance of a super-hard binary falling 30% outside the critical value after making an empirical correction (Mardling & Aarseth 1998) for the large inclination of 112°. Hence the chain procedure was terminated at once and the quadruple accepted for standard hierarchical treatment with KS regularization.

Inspection of this model calculation with $N = 10^4$ single stars and 500 hard primordial binaries reveal nine examples of chain termination due to the three-body stability test. It is highly significant that each of these events were preceded by a strong interaction involving two binaries, thereby justifying the three-body procedure which has been in use for some time.

Reference to figure 2b shows that when the outer binary is highly eccentric, it is possible to have an unstable triple in which energy exchange between the inner and outer orbits is very small, so that the 'disintegration timescale' is correspondingly very long. Practical calculations occasionally provide such examples which do not satisfy the stability criterion. Such border-line cases may prove extremely time-consuming for direct integrations, especially when large period ratios are involved. In one recent example, the outer orbit had an eccentricity exceeding 0.99 and the period ratio was 1.8×10^4; yet the pericentre was still some 20% inside the critical value for stability after making an additional allowance for the inclination (134°). The parameters for this triple were supplied as initial conditions to the three-body regularization code for further study. Although the inner eccentricity exhibited 26 cyclical variations in the range [0.05, 0.58], the semi-major axis showed little evidence of any secular evolution over 7×10^7 orbits. Likewise, the outer pericentre was conserved to high accuracy, with oscillations of relative size $\simeq 1 \times 10^{-3}$. This example highlights the need for distinguishing between absolute and practical stability, where a formally unstable system may still be considered stable over a significant time interval.

References

1. Aarseth, S.J. 1997 (in preparation)
2. Aarseth, S.J. & Zare, K. 1974, Celest. Mech. 10, 185
3. Chapman, C. R. et al. 1995, Nature 374, 783
4. Eggleton, P.P. & Kiseleva, L.G. 1995, Astrophys. J. 455, 640
5. Kiseleva, L.G., Eggleton, P.P. & Anosova, J.P. 1994, MNRAS 267, 161
6. Mardling, R.A. 1995a, Astrophys. J. 450, 722
7. Mardling, R.A. 1995b, Astrophys. J. 450, 732
8. Mardling, R.A. & Aarseth, S.J. 1998 (in preparation)
9. Mikkola, S. & Aarseth, S.J. 1993, Celest. Mech. Dyn. Astron. 57, 439

NEW NUMERICAL METHOD FOR NON-CONSERVATIVE SYSTEMS

T.A. TAIDAKOVA
Crimean Astrophysical Observatory
334242, Simeiz, Crimea, Ukraine

AND

N.N. GOR'KAVYI
Crimean Astrophysical Observatory
334242, Simeiz, Crimea, Ukraine

Abstract. We discuss the efficiency of the implicit second-order integrator for an investigation of the nonconservative dynamics of particles around a star in a co-rotating coordinate system. A big advantage of this numerical integrator is its stability for conservative systems: the error of the semi-major axis and the eccentricity does not accumulate with an increasing number of the time steps. We tested this method for two dissipative systems: one including the Poynting-Robertson drag and the other including the PR-drag with a perturbing planet.

1. Introduction

The dynamics and evolution of interplanetary particles are determined by several effects which include: (i) the Poynting-Robertson (P-R) drag; (ii) resonance effects by planets; (iii) gravitational encounters with these planets. The effect of solar radiation on interplanetary dust particles' trajectories (the Poynting-Robertson effect) has been studied by a number of authors numerically and analytically (Gor'kavyi et.al., 1997). Meanwhile not much is known about some important features of gravitational scattering of particles by planets and dust flows near planetary resonances. In this paper we discuss an implicit second-order numerical integrator (Potter,1973; Taidakova,1990; Taidakova,1997). For the nondissipative case the discretization errors in the energy, the semi-major axis and the eccentric-

B.A. Steves and A.E. Roy (eds.), The Dynamics of Small Bodies in the Solar System, 393–398.

ity by the impicit second-order intergrator show only periodic changes and do not grow with an increasing number of time steps. We used this integrator for an investigation of the dynamics of non-elastic particles in the Neptunian arcs (Gor'kavyi, Taidakova, 1993), gravitational scattering of comets near Beta Pictoris (Gor'kavyi, Taidakova, 1995a) and an origin of retrograde satellites of outer planets (Gor'kavyi, Taidakova, 1995b). Based on our numerical model "Saturn-2" (Gor'kavyi, Taidakova, 1995b), we surmise that the outermost group of not yet discovered retrograde satellites with semimajor axes of the orbits in the range $(24-31)10^6 km$ (with $(25-26)10^6 km$ being the most probable value) may exist near Saturn. By their sizes, orbital properties, and origin, they must be analogs of Jupiter's and Uranus's retrograde satellites. (The remarkable discovery of two new retrograde satellites of Uranus by Philip Nicholson, Joseph Burns, Brett Gladman and J.J. Kavelaars indicates that the list of minor retrograde satellites of our Solar system is not yet complete.)

2. Implicit Integrator

The equations of motion of a particle in the gravitational field of the Sun and the planet with mass m_{pl} in the corotating coordinate system take the form (Taidakova 1990, 1997):

$$\begin{aligned} \ddot{x} &= 2\dot{y} + x + F_x \\ \ddot{y} &= -2\dot{x} + y + F_y \\ \ddot{z} &= F_z \quad , \end{aligned} \tag{1}$$

where:

$$\begin{aligned} F_x &= -(1-\beta)\frac{m_{st}(x+m_{pl})}{R^3} - \frac{m_{pl}(x-m_{st})}{r^3} + P_x \\ F_y &= -(1-\beta)\frac{m_{st}y}{R^3} - \frac{m_{pl}y}{r^3} + P_y \\ F_z &= -(1-\beta)\frac{m_{st}z}{R^3} - \frac{m_{pl}z}{r^3} + P_z \\ P_x &= (1+sw)\alpha[\frac{[(x+m_{pl})\dot{x} + (\dot{y}-m_{pl})y + \dot{z}z](x+m_{pl})}{R^2} + (\dot{x}-y)] \\ P_y &= (1+sw)\alpha[\frac{[(x+m_{pl})\dot{x} + (\dot{y}-m_{pl})y + \dot{z}z]y}{R^2} + (\dot{y}+x)] \\ P_z &= (1+sw)\alpha[\frac{[(x+m_{pl})\dot{x} + (\dot{y}-m_{pl})y + \dot{z}z]z}{R^2} + \dot{z}] \quad , \end{aligned}$$

where:

$$\begin{aligned}
\alpha &= \frac{-\beta m_{st}}{cR^2} \\
R &= \left((x+m_{pl})^2 + y^2 + z^2\right)^{1/2} \\
r &= \left((x-m_{st})^2 + y^2 + z^2\right)^{1/2} \\
x_{st} &= -m_{pl} \quad , \qquad x_{pl} = m_{st}, \\
m_{st} + m_{pl} &= 1 \quad , \qquad x_{pl} - x_{st} = 1 \quad .
\end{aligned}$$

Here the total mass of the Sun and the planet is taken as the unit of mass, and the distance between the planet and the Sun is taken as the unit of length. The unit of time is chosen in such a way that the angular velocity of orbital motion of the planet is equal to unity, and, hence, its orbital period is 2π. Let $v^* = v_x + i\,v_y$, $x^* = x + i\,y$ *and* $F^* = F_x + i\,F_y$. We obtain rather than (1):

$$\begin{aligned}
\frac{dz}{dt} &= v_z \\
\frac{dv_z}{dt} &= F_z \\
\frac{dx^*}{dt} &= v^* \\
\frac{dv^*}{dt} &= -2\,i\,v^* + x^* + F^* \quad .
\end{aligned} \tag{2}$$

We may solve the equation $\frac{dU(t)}{dt} + R(U(t),t) = 0$ with initial conditions $U(t_0) = U_0$ by the implicit second-order integrator described in Potter (1973):

$$v^{[n+1]} = v^{[n]} - \frac{1}{2}\left(R^{[n]} + R^{[n+1]}\right)\Delta t \quad . \tag{3}$$

In our equations (2) R is the function of x, y, z . We will calculate this function R in space-time points $n + 1/2$. ¿From (2) by the use of Eq.(3) we derive the equations for new integrator (Taidakova 1990, 1997):

$$\begin{aligned}
v_x^{[n+1]} &= \frac{v_x^{[n]}(1-\Delta^2 t) + (2v_y^{[n]} + x^{[n+\frac{1}{2}]} + F_x^{[n+\frac{1}{2}]})\Delta t + (y^{[n+\frac{1}{2}]} + F_y^{[n+\frac{1}{2}]})\Delta^2 t}{1+\Delta^2 t} \\
v_y^{[n+1]} &= \frac{v_y^{[n]}(1-\Delta^2 t) - (2v_x^{[n]} - y^{[n+\frac{1}{2}]} - F_y^{[n+\frac{1}{2}]})\Delta t + (x^{[n+\frac{1}{2}]} + F_x^{[n+\frac{1}{2}]})\Delta^2 t}{1+\Delta^2 t}
\end{aligned}$$

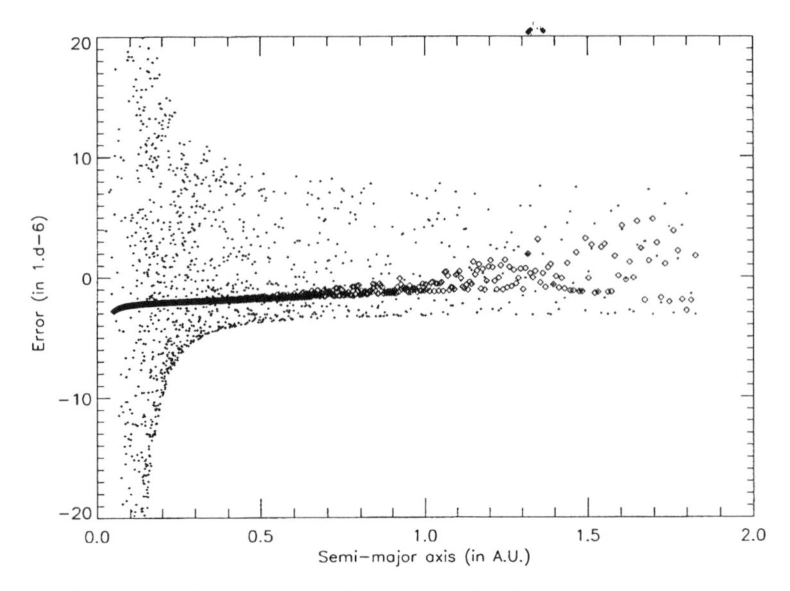

Figure 1. Evolution of the numerical errors in the first integral of motion with 2-th order Potter integrator. Crosses show results for a constant step. Diamonds show results for steps, which decrease spasmodic after every 3 periods of the test particle.

$$\begin{aligned} v_z^{[n+1]} &= v_z^{[n]} + F_z^{[n+\frac{1}{2}]}\Delta t \\ x^{[n+1]} &= x^{[n]} + (v_x^{[n+1]} + v_x^{[n]})\Delta t/2 \\ y^{[n+1]} &= y^{[n]} + (v_y^{[n+1]} + v_y^{[n]})\Delta t/2 \\ z^{[n+1]} &= z^{[n]} + (v_z^{[n+1]} + v_z^{[n]})\Delta t/2 \quad , \end{aligned} \tag{4}$$

$$\text{where: } x^{[n+\frac{1}{2}]} = x^{[n]} + v_x^{[n]}\Delta t/2\ ;\quad y^{[n+\frac{1}{2}]} = y^{[n]} + v_y^{[n]}\Delta t/2;$$
$$z^{[n+\frac{1}{2}]} = z^{[n]} + v_z^{[n]}\Delta t/2\ ;\quad F_{x,\,y,\,z}^{[n+\frac{1}{2}]} = F\left(x^{[n+\frac{1}{2}]}, y^{[n+\frac{1}{2}]}, z^{[n+\frac{1}{2}]}, t^{[n+\frac{1}{2}]}\right) .$$

3. Application to Nonconservative Systems

We consider a nonconservative systems including the Sun and a test particle. The particle's initial semimajor axis is 1.82665 and its initial eccentricity is 0.88. For Poynting-Robertson drag we take $\beta = 0.03$. The integration step length for the integration is 0.001 (orbital period of the planet or the coordinate system is 2π). The accuracy of our integrator is represented by the difference in the values of a first integral of motion $\delta C = C(t) - C(0)$, where $C = \frac{a(1-e^2)}{e^{4/5}}$ (see, for example, Gor'kavyi et.al., 1997).

In Figure 1 we show an evolution of an error of the first integral of motion during a PR-drift to the Sun. Results are better if stepsize of the integrator will be decreasing $\propto a^{3/2}$ interruptedly. When the timestep changes steadily, the error increases very fast. Finally we consider test particles sub-

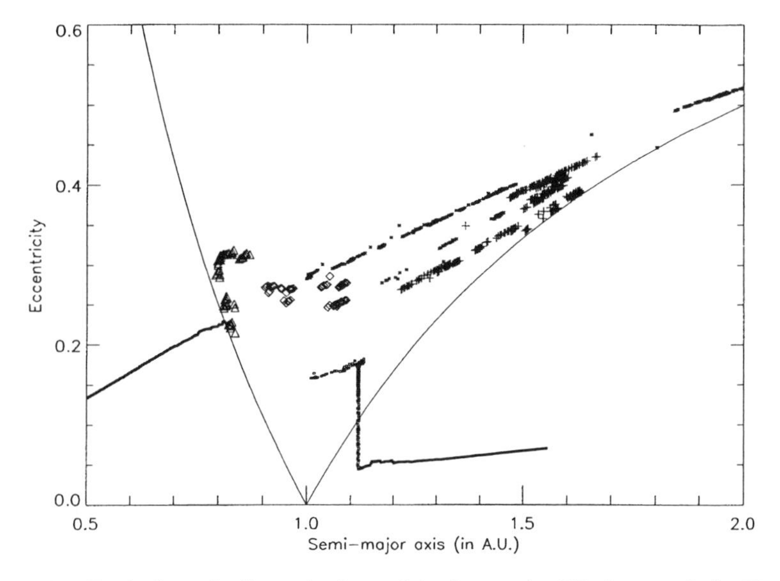

Figure 2. Evolution of a "martian" particle due to the PR-drag and the Earth

ject to PR-effect and the Earth as a perturbing planet in a circular coplanar orbits around the central body. The particle's initial semimajor axis is 1.55 and its initial eccentricity is 0.07. The particle's parameters are typical for martian micrometeorites. In Figure 2 we show the evolution of semimajor axis and eccentricity of this particle. For correct investigation of the evolution of the particle during close encounters with the Earth, it is better if the timestep can be decreased near a planet. For the stability of the error, a total time of this decreasing must be small in comparison with an orbital period of the particle.

4. Conclusions

We presented an integrator that may be applied to nonconservative systems. By comparisons with another method (Cordeiro et.al., 1997), our integrator showed good results in terms of the stability of the error. An important advantage is that our method may be applied for close approaches with the planets, which can frequently occur for dust particles that spiral toward the star due to PR-drag.

5. Acknowledgments

T.T. and N.G. acknowledge the NATO Advanced Study Institute, Director of the NATO ASI Prof. Archie E. Roy and Dr. Bonnie Steves for hospitality and financial support. T.T. has been supported by a Small Research Grant

TABLE 1. **CV by a Martian Micrometeorite**

Time (in 1000 years)	(a,e)-position	Comments
T=0	a=1.55, e=0.07	Birthday (escape from Mars?)
0¡T¡11.25	1.117¡a¡1.55, 0.044¡e¡0.07	Slow PR-drift
11.25¡T¡17.35	a=1.117, 0.044¡e¡0.18	Capture in resonance 6:5 and resonant heating
17.35¡T¡19.31	1.018¡a¡1.117, 0.158¡e¡0.18	Escape from resonance and slow PR-drift
19.31¡T¡29.24	1.21¡a¡1.67, 0.27¡e¡0.44	Gravitational scatterings by the Earth + PR-drift
29.24¡T¡31.01	0.91¡a¡1.09, 0.25¡e¡0.29	Gravitational scatterings by the Earth
31.01¡T¡45.67	1.00¡a¡2.66, 0.28¡e¡0.62	Strong gravitational scatterings + PR-drift
45.67¡T¡47.00	0.80¡a¡0.87, 0.22¡e¡0.32	Gravitational scatterings for $a < 1$ + PR drift
47.00¡T¡53.03	0¡a¡0.82, 0¡e¡0.23	Escape from the Earth' zone and PR-drift to Sun

of the AAS from the Gaposchkin's Research Fund.

References

1. Cordeiro, R.R., Gomes,R.S., Vieira Martins, R. (1997) A Mapping for Nonconservative Systems, *Celestial Mechanics and Dynamical Astronomy*, **65**, p.407
2. Gor'kavyi, N., Ozernoy, L., Mather, J., Taidakova,T. (1997) Quasi-Stationary States of Dust Flows Under Poynting-Robertson Drag: New Analytical And Numerical Solutions, *ApJ*, **488**, Oct.10. p.268
3. Gor'kavyi, N.N., Taidakova, T.A. (1993) Theory of the Neptunian arcs. A multicomponent epiton and Galatea, *Astronomy Lett.*, **19(2)**, p.142
4. Gor'kavyi, N.N., Taidakova, T.A. (1995a), Beta Pictoris and Numerical Study of the Giant Planets Hypothesis, in *Circumstellar Dust Disks and Planet Formation*, ed. R.Ferlet, A.Vidal-Madjar, Editions Frontieres, Gif sur Yvette Cedex - France, p.99
5. Gor'kavyi, N.N., Taidakova, T.A. (1995b) The Model for Formation of Jupiter, Saturn and Neptune Satellite Systems, *Astronomy Lett.*, **21(6)**, p.939
6. Potter, D. (1973) *Computational Physics.* John Wiley &Sons Ltd., London-New York-Sydney-Toronto.
7. Taidakova, T. (1990) The Numerical Analysis of the Dynamics of Particles About a Planet. I. Four-Body Problem. *Nauch.Inform.Astrosoveta Akademii Nauk SSSR*, Riga, Zinatne, **68**, p.72
8. Taidakova, T. (1997) A New Stable Method for Long-Time Integration in an N-Body Problem, in *Astronomical Data Analyses, Software and Systems VI, ASP Conf. Ser. 125*, ed. G. Hunt & H.E.Payne, San Francisco: ASP, p. 174

TIDAL FRICTION IN TRIPLE SYSTEMS: A MEANS OF PRODUCING CLOSE STELLAR AND PLANETARY ORBITS

L. G. KISELEVA AND P. P. EGGLETON
Institute of Astronomy, Madingley Road, Cambridge CB3 0HA, UK
e-mail: lgk,ppe@ast.cam.ac.uk

Abstract. In hierarchical triple systems the combination of tidal friction (TF) with fluctuations of eccentricity due to the third body can lead to potentially large but slow changes in the inner orbit, especially if the two orbits have high (40° or more) relative inclination. We model the dynamical evolution of triple systems using a force law which includes a combination of point-mass gravity, quadrupolar distortion (QD) of each body by the other two, and a dissipative TF term. In hypothetical cases of triple systems with relative orbital inclination $i = 100^{\circ}$ (as in the well-known triple stellar system β Per), the effect of the third star is periodically to increase the inner eccentricity up to nearly unity, provided we neglect the effects of QD and TF. The combined effect of QD and TF may reduce the fluctuations of the inner eccentricity, and in some cases the binary orbit may shrink quite drastically after a suitably long interval of time. These results can be applied to systems where all three components are of stellar mass, and also to triple systems with one binary component, or even two components including the distant one, being Jupiter-like planets. This is potentially important for the long-term evolution of such systems and can probably explain the origin of very short-period orbits for some recently discovered extra-Solar planets, such as τ Boo – which has both a Jupiter-like companion in a 3d orbit and an M2V companion in a $\sim$2000 yr orbit.

1. Introduction

In binary stars, TF dissipates a fraction of the orbital energy at constant angular momentum and will circularise binary orbits on a rather short timescale compared with the nuclear timescale, provided that at least one

B.A. Steves and A.E. Roy (eds.), The Dynamics of Small Bodies in the Solar System, 399–406.

star of the binary has a radius comparable to the separation between binary components. This dissipation effectively ceases once the orbit is circularised. In a hierarchical triple system such dissipation cannot cease entirely, as neither inner nor outer orbit can become exactly circular because of the perturbation of the third star. Thus in a hierarchical triple such as λ Tau or β Per (Algol) TF can lead to a steady secular decrease of the inner semimajor axis, accompanied by transfer of angular momentum from the inner to the outer pair, persisting over the whole nuclear lifetime of the system. The situation can be even more dramatic if two orbits have high relative inclination. It can be shown analytically and numerically (Kozai 1962, Mazeh & Shaham 1979, Marchal 1990, Heggie 1996, Kiseleva 1996) that for non-coplanar triple systems there is a quasi-periodic change of the inner eccentricity (on a timescale $\sim P_{\rm out}^2/P_{\rm in}$) during which it reaches a maximum value $e_{\rm in}^{\rm max}$. This value only depends on the inclination i between the two orbital planes; other parameters affect only the timescale. If $i \approx 90^o$, $e_{\rm in}^{\rm max} \approx 1$ and the two components of the close pair may collide or suffer a very strong tidal interaction. The combined influence of TF and of the third component on the binary orbit may prevent collision but produce other interesting and even dramatic results, such as for example a severe shrinking of the orbit. Recent discoveries of a number of extra-Solar Jupiter-mass planets with very short-period orbits of only a few days around their parent stars, such as 51 Peg (Mayor & Queloz 1995, Marcy et al. 1997), τ Boo and υ And (Butler et al. 1997), make this potential mechanism of production of short period orbits particularly interesting for investigation.

2. Models and Results

For the coplanar system λ Tau ($m_1 = 1.9\,M_\odot, m_2 = 7.2\,M_\odot, P_{\rm in} = 3.95$d; $m_3 = 0.7\,M_\odot, P_{\rm out} = 33$d, $X = 8.3$; Fekel & Tomkin 1982) we find that, because of the third body, $\overline{e}_{\rm in} = 0.007$, with $e_{\rm in}$ fluctuating between 0.0 and 0.014 on timescales which range from less than an inner orbit to $\sim$400 outer orbits or $\sim$35yr. This eccentricity is worryingly large, since the inner pair is semidetached; we should expect that the mass transfer from $*1$ to $*2$ will be very strongly modulated, on the timescale of the inner orbit, since the pressure scale height in the atmosphere of $*1$ is $\sim 0.001 a_{\rm in}$, i.e. small compared with the expected fluctuations in separation. Tidal friction can be expected to reduce the fluctuations, though not to zero, and at least partly stabilise the situation. But too much TF will lead to a secular shrinkage of the inner orbit on a timescale that can be short compared with the nuclear timescale ($\sim 1.2 \times 10^8$ yr). If we require that the orbit should not shrink that rapidly then we can set an upper limit to the strength of TF that this interesting system permits.

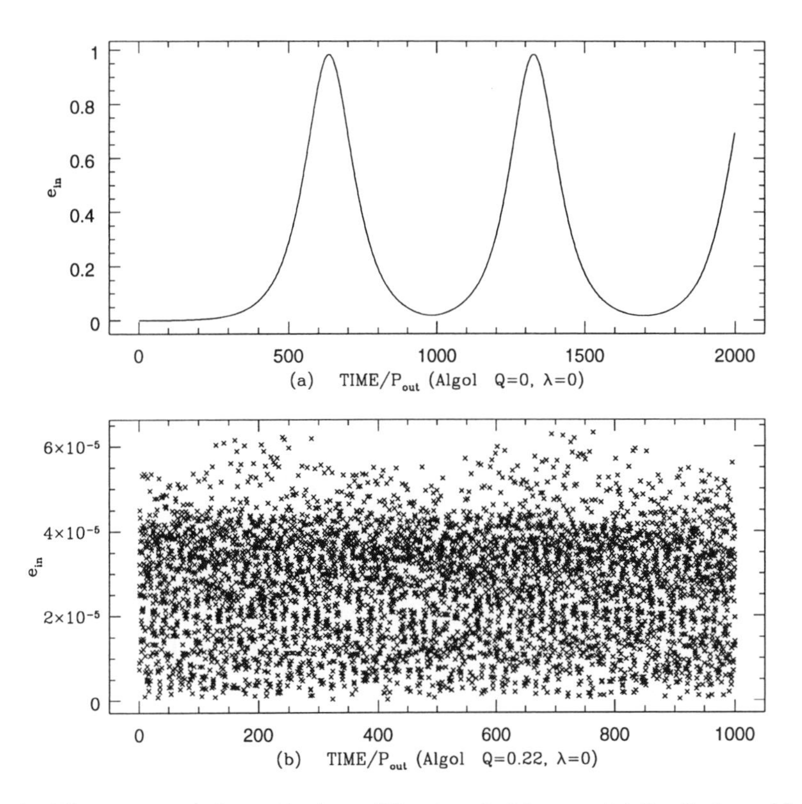

Figure 1. The top panel shows the large 'Kozai cycles' in eccentricity that would occur in Algol, where the outer orbit is inclined at 100° to the inner, *provided* that the 3 stars are treated as point masses. e_{in} fluctuates between 0 and 0.985. The lower panel shows that if the stars are allowed to adopt the QD expected, the cycles are virtually suppressed and e_{in} (sampled once per outer orbit) fluctuates about a very small mean, $\sim 3.5 \times 10^{-5}$.

Fig 1a shows that e_{in} can fluctuate by a much larger amount if the outer orbit is nearly perpedicular to the inner, as in Algol (β Per) ($m_1 = 0.8\,M_\odot, m_2 = 3.7\,M_\odot, P_{in} = 2.87\,\mathrm{d};\ m_3 = 1.7\,M_\odot, P_{out} = 1.86\,\mathrm{yr},\ e_{out} = 0.23;\ i = 100°,\ X = 237$; Tomkin & Lambert 1978, Lestrade *et al.* 1993). If we approximate the system (rather too crudely – see below) as three point masses, then e_{in} cycles rather smoothly between 0 and 0.985, while i fluctuates between 100° and 140°. Such 'Kozai cycles' (Kozai 1962) presumably do not actually occur in this semidetached system. They can be damped to a small value by TF, but in fact they are also strongly reduced by the non-dissipative effect of the quadrupole moments of the two stars in the inner pair (Fig 1b). This effect produces apsidal motion which is much more rapid than the apsidal motion due to $*3$, and so prevents the Kozai cycles from operating; e_{in} never exceeds $\sim 10^{-4}$ if QD is included.

We approximate the equations of motion as follows. For the force between $*1$ and $*2$ only, we write (Eggleton *et al.* 1997):

$$m_1\ddot{\mathbf{r}}_1 = -m_2\ddot{\mathbf{r}}_2 =$$

$$-\mathbf{r}_{12}\left[\frac{Gm_1m_2}{r_{12}^3}+\frac{6G(m_2^2A_1+m_1^2A_2)}{r_{12}^8}+\frac{27}{2}\frac{\sigma_1m_2^2A_1^2+\sigma_2m_1^2A_2^2}{r_{12}^{10}}\mathbf{r}_{12}.\dot{\mathbf{r}}_{12}\right] \quad (1)$$

where

$$A_i = \frac{R_i^5Q_i}{1-Q_i} \quad , \quad \sigma_i = \frac{\lambda_i}{Q_i^2m_1R_i^2}\sqrt{\frac{Gm_i}{R_i^3}} \quad , \quad \mathbf{r}_{12}\equiv\mathbf{r}_1-\mathbf{r}_2\,. \quad (2)$$

Here R_i is the stellar radius, and Q_i is a version of the apsidal motion constant, a dimensionless measure of the quadrupolar distortability which gives $Q\sim0.22-0.03$ for polytropes of index $n\sim1.5-3$. The second force term in Eq. 1 is a conservative term due to QD, and the third force term is an approximation to the dissipative effect of TF. The coefficient λ_i is a dimensionless tidal-friction dissipation rate, with $\lambda\sim1$ implying that a free quadrupolar distortion would be damped on something like the oscillation period of the distortion. Eq. 1 is easily generalised to 3 (or more) stars. The method used in the numerical work was the regularized CHAIN method (Mikkola & Aarseth 1993) with perturbations.

We apply Eq. 1 to the following situations:
(a) For λ Tau, we investigated a range of λ from 10^{-5} to 10. Unless $\lambda\lesssim10^{-4}$ the timescale of shrinkage was found to be very much less than the nuclear timescale, $\sim1.2\times10^8$ yr, which is also the timescale on which the orbit is expected to expand due to mass transfer. Such a result seems unlikely, and so we suggest that λ Tau has reached a situation of transient equilibrium, in which its period tries to increase due to mass transfer, but is prevented from increasing, at least at the same rate, by the fact that the third-body perturbation is causing the inner pair to lose angular momentum (to the outer pair) via TF at a balancing rate. This implies a value of λ in this system of about 10^{-5}, a value which we feel is reasonable on physical grounds.
(b) In Algol the effect of the QD term in Equn (1) is much more marked than in λ Tau. The main effect is to introduce an extra term into the apsidal motion. If this is larger than the three-body term already present, then it in effect randomises the phase, and so prevents the gradual accumulation of eccentricity in a Kozai cycle (Fig 1). Thus we do not in fact need the TF term to kill the cycle in present-day Algol. Unfortunately this also means that we do not get an estimate for λ in Algol, as we did in λ Tau.
(c) However, the situation may have been different different at an early stage in Algol's evolution. Fig 2 shows what can happen in 'proto-Algol' – a hypothetical model of β Per at age zero. We take the inner masses

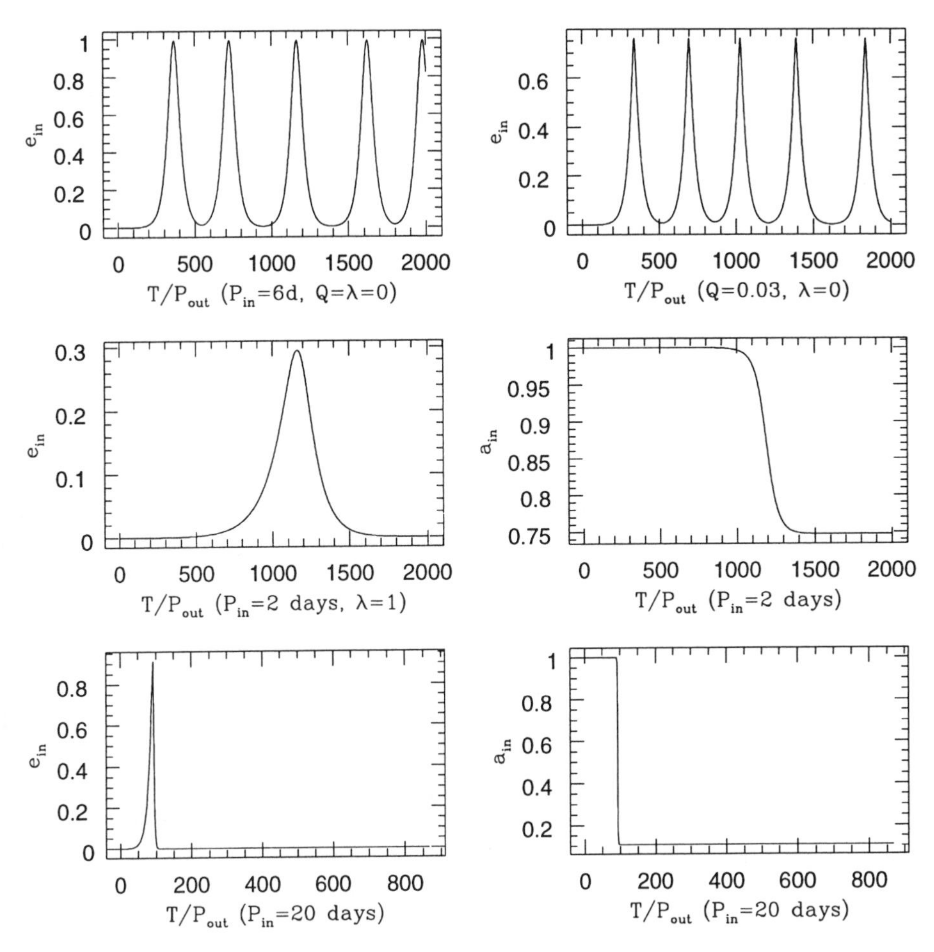

Figure 2. Top two panels: Kozai cycles ($e_{\rm in}$ as a function of time, measured in initial outer orbits) in 'proto-Algol', a hypothetical state of Algol before nuclear evolution and mass transfer began. We assume smaller (ZAMS) components, a wider initial inner orbit (6d), and zero TF. The effect of including QD (right panel) or not (left panel) only reduces the amplitude of the Kozai cycles slightly (cf. Fig. 2a). Centre panels: $e_{\rm in}$ and $a_{\rm in}$ as functions of time when TF is included (for a somewhat shorter initial period, 2d, than above). During the first (and only) Kozai cycle, TF becomes so strong at periastron, once $e_{\rm in} \sim 0.3$, that it circularises the orbit to a new smaller separation. Bottom panels: with a somewhat longer initial period (20d), the effect is much more drastic. The Kozai cycle has to go further before tidal friction takes over, so that the semimajor axis is shrunk by a much larger factor.

to be 2.5 and $2\,M_\odot$, with ZAMS radii, and consider a variety of initial values of $P_{\rm in}$, but otherwise take the same parameters for the outer orbit as for present-day Algol (above). Here the Kozai cycles are not damped by the QD term (Fig 2), as in actual Algol (Fig 1b), because (i) the stars are smaller at zero age, and (ii) the inner period is presumed longer. The

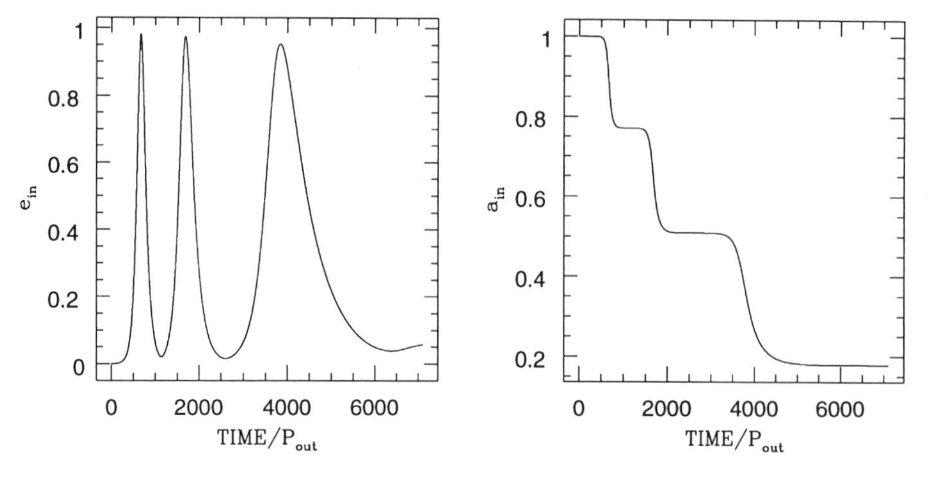

Figure 3. For some combinations of QD and TF, the Kozai cycles can persist for some time but be progressively damped. This is a hypothetical system with parameters like Algol but with stars that are ten times smaller relative to their Roche radii.

middle panels include TF as well as QD for $P_{\rm in} = 2\,{\rm d}$, and show that after only one Kozai cycle, which brings the inner pair of stars close together at periastron at the peak of the cycle, the semimajor axis is rapidly decreased by 25%. The bottom two panels show a much more drastic shrinkage of the orbit, if it starts with longer period ($P_{\rm in} = 20\,{\rm d}$). However, in all cases the orbit only shrinks to the size which allows TF to kill the next Kozai cycle and to circularise the orbit. After that the orbit remains nearly of the same size with $e_{\rm in}$ close to zero.

Fig 3 shows how the combination of Kozai cycle and TF may affect the orbit of the inner binary in an 'Algol-like' stellar system if both close-binary components were 10 times smaller than they are now; alternatively we can think of both periods as longer by $10^{1.5}$. In this case every time the stars approach each other very closely at the peak of the Kozai cycle TF becomes strong and shrinks the orbit, until after three approaches it is close enough for further cycles to be entirely suppressed by TF.

Fig 4 shows the evolution of the orbit of a Jupiter-mass ($M \sim 10^{-3} M_\odot$) planet around a solar-type companion in a stellar system with a third body of $1 M_\odot$ (top panels), $0.1 M_\odot$ (middle panels), and $10^{-3} M_\odot$ (bottom panels). The initial period ratio was 30. The inclination of the planetary orbital plane with respect to the outer orbit is taken to be the same as in β Per, i.e. $100°$. Such a high inclination does not seem very improbable (e.g. Holman et al. 1997). If the third body was captured by a dynamical encounter during the early stellar formation phase of a dense star-forming region, we expect a more-or-less uniform distribution of i over solid angle, i.e. $n(i) \propto \sin i$. For each mass of third body considered, we see that a single cycle takes

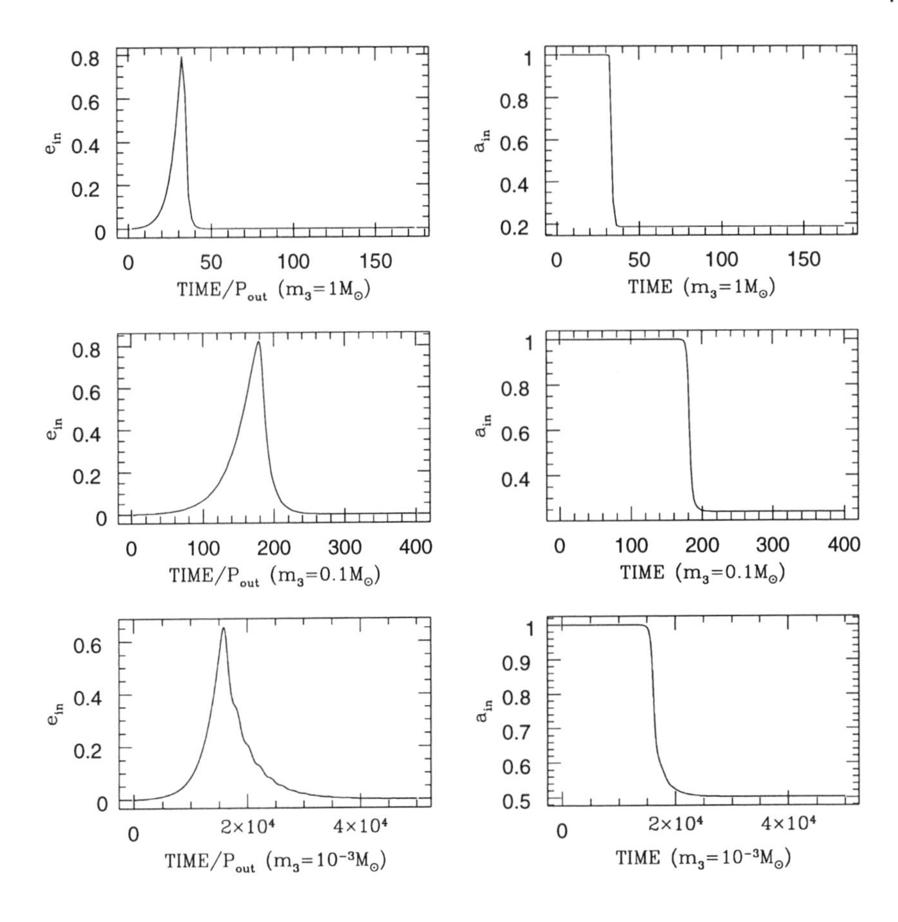

Figure 4. A Kozai cycle, and its destruction by TF, in a 'Sun-Jupiter' system with a third body of $1M_\odot$ (top panels), $0.1M_\odot$ (middle panels) and $10^{-3}M_\odot$. The orbits are inclined at 100°. The cycle developed to the much the same peak of eccentricity in all three cases (left panels), but more slowly with the lower-mass third body. Tidal friction shrank the orbit drastically at the peak of the cycle (right panels).

place which brings the 'Sun-Jupiter' pair close together and allows TF to recircularise the orbit at a considerably shorter period. We were restrained from doing calculations involving more realistic systems, where for example the initial period of 'Jupiter' is years rather than days, and the period ratio is thousands rather than tens, because they are expensive of computer time. But we believe that this is a possible mechanism for putting Jupiter-mass planets into short-period ($\sim$few day) orbits about stars. Note that the mass of the third body mainly affects the timescale, and had only a minor effect on the Kozai cycle.

We conclude that, although the combination of third body (which might perhaps be of very low, even planetary, mass) and TF may only be important within a limited range of parameters, it may produce – particularly in clusters, since triples with high inclination are rather readily formed in clusters (Kiseleva 1996, Kiseleva et al. 1996) – some fraction of *close* binary-star or star-planet systems, such as 51 Peg or τ Boo . The F7V dwarf τ Boo has a companion of $\gtrsim 4$ Jupiter masses in a 3.3d circular orbit (Butler et al. 1997), and also a companion M2V star at about $5''$ with a very tentative orbit (Hale 1994) of ~ 2000 yr and $e \sim 0.9$. We have not yet attempted to model these specific parameters, but the combination of Kozai cycles and TF could have worked, as above, to shrink the Jupiter orbit from an initial period of say $1 - 10$ yrs to the present short period, provided that the Jupiter orbit and the M2V orbit are highly inclined to each other – probably nearer 90° than 100° or 80°.

Acknowledgements

The authors thank NATO for the Collaborative Research Grant CRG 941288, and are also grateful grateful to NATO for financial support to attend the meeting.

References

1. Butler, P. R., Marcy, G. W., Williams, E., Hauser, H. & Shirts, P. 1997 ApJ, 474, L115
2. Eggleton, P. P., Kiseleva, L. G. & Hut, P. 1997, ApJ, in press
3. Eggleton, P. P. & Kiseleva, L. G. 1995, ApJ, 455, 640
4. Fekel, F. C. & Tomkin, J. 1982, ApJ, 263, 289
5. Hale, A. 1994, AJ, 107, 306
6. Heggie, D. C. 1996, private communication
7. Holman, M., Touma, J. & Tremaine, S. 1997, Nature, 386,254
8. Kiseleva, L. G., Eggleton, P. P. & Anosova, J. P. 1994, MN, 267, 161
9. Kiseleva, L. G. 1996, in *Dynamical Evolution of Star Clusters*, eds. P. Hut & J. Makino, p233
10. Kiseleva, L. G., Aarseth, S., de la Fuente Marcos, R. & Eggleton P. P. 1996, in *Origins, Evolution and Destinies of Binaries in Clusters* eds. E. F. Milone & J.-C. Mermilliod, ASP Conf series, **90**, p433
11. Kozai, Y. 1962, AJ, 67, 591.
12. Lestrade, J.-F., Phillips, R. B., Hodges, M. W. & Preston, R.A. 1993, ApJ, 410, 808
13. Marchal, C. 1990. *The Three-Body Problem*; Elsevier, Amsterdam
14. Mazeh, T. & Shaham, J. 1979, A&A, 77, 145
15. Mikkola, S. & Aarseth, S.J. 1993, Cel. Mech., 57, 439.
16. Tomkin, J. & Lambert, D. L. 1978, ApJ, 222, L119

SPECIAL VERSION OF THE THREE BODY PROBLEM

KRZYSZTOF GOŹDZIEWSKI AND ANDRZEJ J. MACIEJEWSKI
Toruń Centre for Astronomy, N. Copernicus University, Poland

1. Introduction

In this paper, we consider a system of two bodies interacting gravitationally. One of them is spherically symmetric and the other is axially symmetric. In the full non-linear settings, this problem permits planar motion when the mass center of the spherically symmetric body moves in a plane containing the axis of symmetry of the second body. Such planar unrestricted problem was studied by Kokoriev and Kirpichnikov [7] with one additional specification. The authors assumed that the gravity field of the axially symmetric body is well approximated by the gravity field of two point masses lying in the axis of symmetry of the body. Such formulation gives rise to an amazingly interesting problem for analytical and numerical studies. For example, it gives completely different ideas about the number of relative equilibria in the problem of two rigid bodies. We will present a detailed discussion of the problem of Kokoriev and Kirpichnikov (for short, it will be called KK problem from hereafter) in our forthcoming paper [4] .

The KK problem is an adequate model for several interesting astronomical situations. In order to mention just a few, let us denote the rigid body by $\mathcal{B}$, and let $\mathcal{S}$ represent the sphere. Let the masses of the bodies be M an m, respectively. The case $m \gg M$ represents the motion of a dumb-bell satellite in the gravity field of a spherical planet. In the case $m \ll M$ we have the problem of a spherical satellite moving in a polar orbit, in the field of a prolate planet. In the limiting case $m \to 0$ we get the Chermnykh problem [3], i.e., the problem of motion of a probe mass in the gravitational field of a dumb-bell, rotating with a constant angular velocity ω. The case $\omega = 1$ corresponds to the restricted three body problem, and $\omega = 0$ to the Euler problem of two fixed centers.

The KK problem can be also generalized in a number of ways. Here we propose two approaches. Approximating the gravitational field of the rigid body by the field of a *complex* dumb–bell, we obtain a version of the generalized Euler problem. We describe this model in [10]. Also, the model of motion of an irregular

B.A. Steves and A.E. Roy (eds.), The Dynamics of Small Bodies in the Solar System, 407–412.

satellite moving in the equatorial plane of a spheroidal planet may be considered a possible generalization of the basic problem [5].

2. The problem and equations of motion

We developed the idea of Kokoriev an Kirpichnikov starting from the very beginning. It is a very fruitful approach because of two reasons: we can show the source of the problem, and ways of its generalization.

First, we considered an isolated system consisting of a rigid body $\mathcal{B}$ and a sphere S, which interact gravitationally. Using results of Maciejewski [8], who considered a problem of two rigid bodies, we wrote down Newton-Euler equations of motion of the system in a very general form. It may be shown [4] that, after reductions, the equations may be written in a simple, vectorial form in the rigid body fixed principal axes frame. They have the same structure as the equations of motion of a rigid body moving in the gravitational field of a fixed center, considered in [11]. We also showed that the equations have a Hamiltonian structure with respect to a non-canonical Poisson bracket. The details may be found in [4].

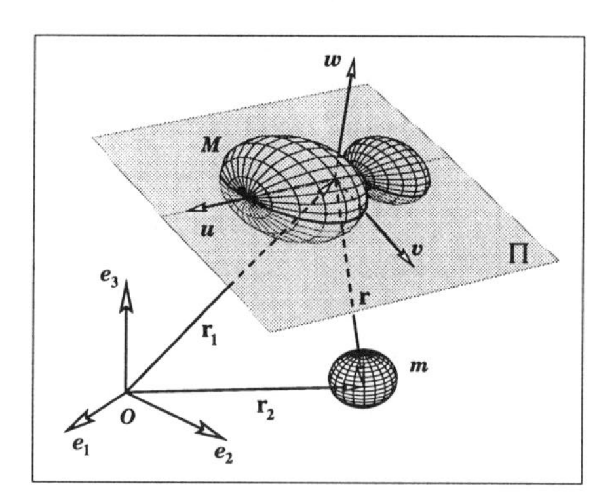

Figure 1. Geometry of the problem of Kokoriev and Kirpichnikov.

The system of equations of motion says little about the dynamics till the potential of mutual interaction of the bodies is not specified. At this point, it can be done in a number of ways. Having in mind the original KK problem, we assumed that the rigid body has a plane Π of dynamic symmetry (Fig 1). It is easy to prove [4] that in this situation there exists an invariant manifold of planar orbital motion of the system. After restriction of the system to this invariant manifold and reduction with respect to the integral of total angular momentum, we obtain a system with two degrees of freedom given by the following Hamiltonian

$$H = \frac{1}{2}(p_1^2 + p_2^2) + \frac{1}{2\kappa}(\gamma - q_1 p_2 + q_2 p_1)^2 + V(q_1, q_2), \tag{1}$$

where $V(q_1,q_2)$ is the gravitational potential. Canonical variables (q_1,q_2) denote coordinates of the spherical body taken with respect to the principal axes frame of $\mathcal{B}$. Due to our choice of physical units, the variables are non-dimensional. This also leads to non-dimensional parameters: $\kappa \in (0,\infty)$ denotes the inertia moment of $\mathcal{B}$ with respect to the axis perpendicular to the plane Π, $\gamma \in [0,\infty)$ is the module of total angular momentum.

At this stage, the problem is still quite general, because the Hamiltonian contains unspecified potential V. For this reason, Eq. 1 defines *a whole class* of dynamical systems. Of course, the KK problem belongs to this class. Let us assume that the gravitational potential of $\mathcal{B}$ is approximated by the potential of a dumb-bell, i.e., two point masses m_A and m_B, located in a constant distance, such that $M = m_A + m_B$. For such a case the potential has the following form

$$V = -\frac{1-\mu}{\sqrt{(q_1+\mu)^2+q_2^2}} - \frac{\mu}{\sqrt{(q_1-1+\mu)^2+q_2^2}},$$

where the mass parameter $\mu = m_B/M$. For reasons of symmetry we assume that $\mu \in [0,1/2]$. Thus the problem is described by three parameter family of Hamiltonian systems. Its importance flows from the fact that it is a nontrivial generalization of the classical problems of Celestial Mechanics [7]. Let us underline that the problem is also *unrestricted*, i.e., the relative orbital motion and the rotational motion are coupled.

As another example let us consider an approximation of the potential of $\mathcal{B}$ by the potential of *a complex dumb-bell* [10], i.e., $m_{A,B} = \frac{1}{2}M(1 \pm \mathrm{i}\delta)$, $c, \delta \in \mathbb{R}$, located in the imaginary q_1 axis, at coordinates $z_\pm = c(\delta \pm \mathrm{i})$, where $c = \mathrm{const.}$ Moreover, this leads to a *real* potential function of the form

$$V = -\frac{\mu}{\sqrt{(q_1+\mathrm{i}\mu)^2+q_2^2}} - \frac{1-\mu}{\sqrt{(q_1-\mathrm{i}(1-\mu))^2+q_2^2}}.$$

From the physical point of view it approximates the gravitational potential of a flattened, axially symmetric body.

3. Triangular libration points

According to Poincaré, we should look for families of solutions originating in a dynamical system. Generally, stationary solutions are the simplest for the analysis. The analysis consists of finding conditions of existence of the equilibria, as well as their stability in a linear and nonlinear approximation.

In analogy to the restricted problem of three bodies, the equilibria are classified as collinear (lying in the symmetry axis of $\mathcal{B}$) and triangular (outside this axis). They correspond to relative circular orbits of the rigid body and the sphere.

Moreover, the number of the equilibria may be quite different from the restricted problem.

The conditions of existence, as well as phase coordinates of the triangular equilibria may be found analytically. For that purpose, following the general idea in [7], we introduce a set of parameters $\{\sigma, \kappa, \omega_0\}$ through relations

$$\sigma = \mu(1-\mu), \quad \gamma = \omega_0(\kappa - \sigma) + \omega_0^{-1/3}, \quad \mu_0 = \frac{\sigma}{\kappa - \sigma}.$$

Here ω_0 denotes the angular velocity of $\mathcal{B}$ in the stationary motion. Then, we find, that the triangular equilibria exist only for $\omega_0 \leq \omega_k \equiv 2\sqrt{2}$. The number of stationary solutions is equal to the number of solutions of the following equation

$$\phi(\omega_0) = \gamma, \quad \text{where} \quad \phi(\omega_0) = \omega_0(\kappa - \sigma) + \omega_0^{-1/3}, \quad \text{and} \quad \gamma \in \mathbb{R}_+.$$

Two cases should be considered here. If the parameter $\mu_0 \geq 0$ then $\phi(\omega_0)$ possesses a global minimum at $\omega_0 = \omega_*$. Further, if $\omega_* \geq \omega_k$ then, for every $\gamma \geq \gamma_k = \phi(\omega_k)$ there exists a pair of triangular libration points, located in the vertices of two isosceles triangles, which have a common base. This is illustrated in the left panel of Fig. 2. If the minimum of $\phi(\omega_0)$ lies in the interval $(0, \omega_k)$ then, for every value $\phi(\omega_0) \in (\gamma_*, \gamma_k)$, there exist *two* pairs of triangular libration points, and, for $\phi(\omega_0) \geq \gamma_k$, there exists one pair of equilibria. As it may be seen in the bottom panel of Fig. 2, as ω_0 grows up to ω_k, in the case when only a pair of equilibria exists, they tend to L_2 located in the symmetry axis of the dumb-bell. If $\omega_0 \to 0$ the points escape to infinity. In the case when two pairs of equilibria exist, with $\omega_0 \to \omega_k$ the internal pair of the equilibria tends to L_2. The other two equilibria move outside the bifurcation point, corresponding to $\phi(\omega_0) = \gamma_*$.

The case $\mu_0 < 0$ was not considered by Kokoriev and Kirpichnikov. For this case the function is a monotonic function. Then, for every $\gamma > \gamma_k$ there exists one pair of equilibria. The conclusions presented here were obtained basically following the ideas of [7], moreover, we show in [4] that they are also the result of a different, quite elementary analysis.

The results, concerning the number of equilibria, differ qualitatively from the conclusions obtained by a perturbative approach (see for instance [1], [2]).

4. Stability analysis

The linear stability analysis shows that the triangular equilibria are unstable for $\mu_0 < 0$, or for $\mu_0 \geq 3$. For fixed values of μ_0, the region of linear stability may be described simply on the plane (σ, ω_0). Such choice of parameters leads to an analytical representation of the curves limiting the stability region in the form of function graphs $\sigma_{k_1:k_2}(\omega_0)$, $k_1, k_2 \geq 0$. They are $(k_1 : k_2)$ resonance curves. Fig. 3 shows the diagram of linear stability analysis for a generic value of the parameter μ_0.

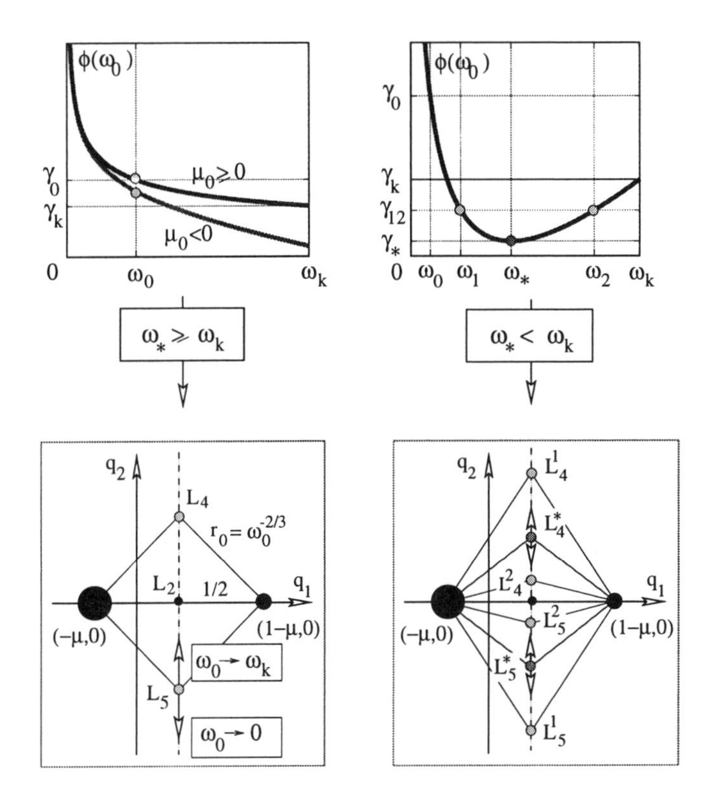

Figure 2. Conditions of existence and geometry of triangular libration points in the problem of Kokoriev and Kirpichnikov.

We performed the nonlinear stability analysis in the resonance cases of the first, the second and the third order. Details can be found in the paper [4]. Here we present them shortly.

Theorem *The stability analysis of equilibria in the KK problem shows that*

- *in the case of null eigenvalues the triangular points are unstable,*
- *the equilibria are generically unstable in the first order resonance, except for some limiting cases,*
- *at the second order resonance they are stable,*
- *at the third order resonance the equilibria are unstable, except for three sets of the parameters values; for them the coefficients of the resonant normal form, which decide on the stability, vanish.*

The latest case is related to the method of isolating the values of parameters, for which the system *might* be integrable. This is based on an application of the Kozlov theorem [9]. We performed such analysis, but, unfortunately, the "suspected" sets of parameters do not correspond to integrability.

We performed the analysis of stability to the very end in the limiting problem of Chermnykh (as $\mu_0 = 0$), the result are to be published in [6].

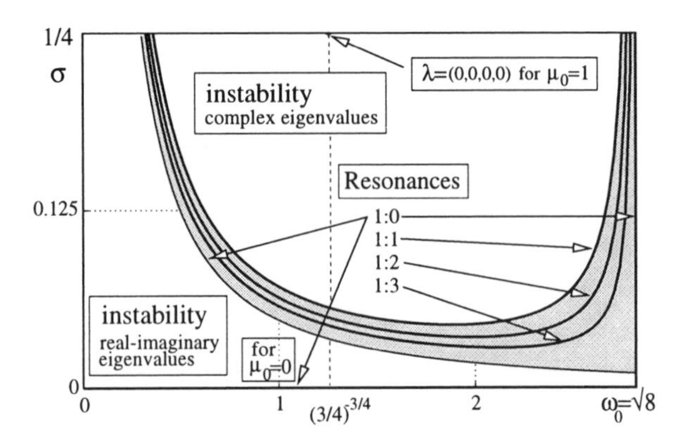

Figure 3. Diagram of linear stability for triangular libration points in the KK problem.

5. Acknowledgements

The work of K. Goździewski was supported by Research Grant No. 2P03C.008.12 by the Polish Committee of Scientific Research. We thank Zbroja for corrections of the manuscript.

References

1. Aboelnaga, M.Z. (1989) Regular rotational-translational motion of two arbitrary rigid bodies and their stability, *Astron. Zh.*, **66**(3).
2. Barkin, Y.W. (1975) Equations of motion of perturbed rotational motion of a rigid body about its mass center *Vest. Moscow Univ., Phys. Series*, **16**(1), pp. 46-53.
3. Chermnykh, S.V. (1987) On the stability of libration points in a certain gravitational field, *Vest. Leningrad Univ.*, **2**(**8**), pp. 10–13.
4. Goździewski K., and Maciejewski, A.J. (1997a) Equations of motion and relative equilibria in the special version of three body problem, *in preparation.*
5. Goździewski K., and Maciejewski, A.J. (1997b) The motion of a rigid satellite in the field of a spheroidal planet, *in preparation.*
6. Goździewski K., and Maciejewski, A.J. (1997c) Nonlinear stability of triangular libration points in the problem of Chermnykh, *Celest. Mech., submitted.*
7. Kokoriev, A.A., and Kirpichnikov, S.N., (1988) On the stability of triangular Lagrangean solutions of a system of two graviting bodies: an axi-symmetric and spherically symmetric, *Vest. Leningrad Univ.*, **1**(1), pp. 75–84.
8. Maciejewski, A.J. (1996) Reduction, relative equilibria and potential in the two rigid bodies problem, *Celest. Mech.*, **63**, pp. 1-28.
9. Kozlov, V.V. (1983) Integrability and non-integrability in Hamiltonian mechanics, *Russian Math. Surveys*, **38**(1), pp. 1–75.
10. Maciejewski, A.J., and Goździewski, K. (1997) On a generalized problem of Euler, *Celest. Mech., in preparation.*
11. Wang, L., Maddocks, J., and Krishnaprasad, P. (1992) Steady rigid-body motions in a central gravitational field, *J. Astron. Sci.*, **40**(4), pp. 449–478.

BIFURCATIONS IN THE MASS RATIO OF THE PLANAR ISOSCELES THREE-BODY PROBLEM

S. CHESLEY and K. ZARE
Department of Aerospace Engineering and Engineering Mechanics
University of Texas at Austin
Austin, Texas 78712

Abstract. We study a Poincaré map in the planar isosceles three-body problem, but we emphasize the mapping of areas in phase space over few iterations rather than single points over many iterations. This map, with a complementary symbolic dynamics, leads to global information. We identify a stable fixed point of the mapping with associated quasi-periodic motion for smaller mass ratios. The invariant KAM region around this fixed point vanishes at an inverse period doubling bifurcation at $m_3/m_1 \cong 2.581$. We also find a set on which a horseshoe map completely describes the motion. This simply chaotic set is destroyed at mass ratio $m_3/m_1 \cong 2.662$ leading to an interesting global bifurcation. Ranges of the mass ratio are identified on which the dynamics is qualitatively similar in a *global* sense. We also study the motion at the limiting values of the mass ratio.

1. Introduction

Poincaré surface of section plots obtained numerically by many iterations of the corresponding function and for many initial points have been very popular in the literature. This representation of Poincaré maps has two important disadvantages: the need for many iterations of the map, and the inability of the method to shed light on the structure of the chaotic domain. In a previous paper [7] (hereafter referred to as Paper I), we introduced a global representation of a Poincaré map in the planar isosceles three-body problem with three equal masses. This was accomplished by considering the problem as a mapping of areas rather than points. This point of view leads to an elegant geometric description of the dynamics in the chaotic regions, and to a nearly complete global description of the system. The description also reveals the existence (or non-existence) of certain classes of dynamical behavior and their corresponding measure in the map domain.

This Poincaré map f_α is a one parameter family of mappings depending on a mass parameter α. The case of equal masses ($\alpha = \frac{1}{3}$) has been studied in considerable detail in Paper I. The global problem (i.e., $0 < \alpha \leq 1$) is considered in this paper. In particular, we pay close attention to the bifurcation sets of the topological classifications with α as the bifurcation parameter.

B.A. Steves and A.E. Roy (eds.), The Dynamics of Small Bodies in the Solar System, 413–424.

Most studies of this problem have relied substantially on an analysis of the *triple collision manifold* [3,5,6]. These studies are quite limited in their ability to describe dynamics which occur far from triple close approach, and therefore they do not lead to a complete global description.

In what follows we present a brief description of the system in Section 2, and a short review of the relevant results from Paper I for $\alpha = \frac{1}{3}$ in Section 3. This review is necessary to discuss the main topic, the bifurcation problem, which we present in Section 4.

2. A Poincaré map for the planar isosceles problem

The planar isosceles three-body problem is a special case of the general problem. Here we provide only a brief development of the relevant equations and refer the reader to the literature (for example Paper I, Devaney [3], or Zare and Szebehely [8]) for details. For this problem certain symmetry conditions permit the reduction to only two degrees-of-freedom. The governing differential equations are

$$\frac{d^2x}{dt^2} = -\frac{Gm_1}{4x^2} - Gm_3\frac{x}{r^3}, \quad \frac{d^2y}{dt^2} = -G\left(2m_1 + m_3\right)\frac{y}{r^3} \tag{1}$$

where $r^2 = x^2 + y^2$, and x and y are respectively the Cartesian coordinates of m_2 and m_3 with respect to the center of mass of the equal-mass binary system m_1 and m_2. Equations (1) admit the energy integral

$$h = \frac{1}{2}\left(\frac{dx}{dt}^2 + \alpha\frac{dy}{dt}^2\right) - \frac{Gm_1}{4x} - \frac{Gm_3}{r} \tag{2}$$

where $\alpha = m_3/\left(m_1 + m_2 + m_3\right),\ 0 \leq \alpha \leq 1$.

Physically, the system describes the motion of a pair of equal mass primaries – the binary – moving rectilinearly on the x-axis, and a secondary mass m_3 moving rectilinearly on the y-axis. For the case of negative energy all orbits have an infinite number of binary collisions ($x = 0$). The motion of m_3 on the y-axis leads to either a finite or an infinite number of syzygy crossings ($y = 0$). The former signifies an escape or triple collision after a finite number of oscillations and the latter case represents a forever oscillatory motion. For $h < 0$ the binary separation is bounded, but the motion of m_3 is unbounded in general.

Equations of motion (1) are singular at the binary collision between m_1 and m_2 ($x = 0,\ y \neq 0$) and at the triple collision ($x = y = 0$). However, the singularity at the binary collision is not essential and it can be eliminated by various regularization techniques. A regularization in the extended phase space leads to the following Hamiltonian equations.

$$\frac{dQ_i}{d\tau} = \frac{\partial\Gamma}{\partial P_i}, \quad \frac{dP_i}{d\tau} = -\frac{\partial\Gamma}{\partial Q_i}, \quad i = 0, 1, 2, \tag{3}$$

with canonical variables

$$\begin{aligned} &Q_0 = t, && Q_1 = \sqrt{x}, && Q_2 = y, \\ &P_0 = -h, && P_1 = 2Q_1\, dx/dt, && P_2 = \alpha\, dy/dt, \end{aligned}$$

and Hamiltonian function

$$\Gamma = \frac{1}{2}\left(\frac{P_1^2}{4} + \frac{P_2^2 Q_1^2}{\alpha}\right) - \frac{Gm_1}{4} - \frac{Gm_3 Q_1^2}{\sqrt{Q_1^4 + Q_2^2}} + P_0 Q_1^2 \equiv 0. \tag{4}$$

We note that the present Hamiltonian formulation does not permit the analysis of the restricted isosceles problem ($\alpha = 0$).

2.1 REDUCTION TO A TWO-DIMENSIONAL MAP

This investigation relies on a one parameter family of Poincaré mappings f_α which we define here. (We refer the reader to Paper I for a more thorough exposition.) We use as a surface of section the plane defined by $Q_2 = 0$, and we eliminate P_2 by use of the energy equation (4). After selecting $G = 1$, $2(m_1 + 4m_3) = 1$, and $-8h = 1$ for normalization of units we have

$$P_2 = \pm\frac{\sqrt{\alpha}}{2Q_1}\left(1 - Q_1^2 - P_1^2\right)^{1/2} \tag{5}$$

on our surface of section. It is immediately clear that $Q_1^2 + P_1^2 \leq 1$ defines the admissible region of initial conditions. With $Q_1 \geq 0$ this is a closed semicircular region of unit radius. Every point (Q_1, P_1) in this region represents a unique initial condition with $Q_2 = 0$ and $P_2 \geq 0$ computed from (5).

We define the set A to be the interior of the closed semicircular region defined above. This is an excellent surface of section because almost every isosceles trajectory intersects A, with the only exceptions being certain isolated triple collision orbits. The boundary of A for which $Q_1^2 + P_1^2 = 1$ corresponds to the collinear homothetic solution. The boundary of A for which $Q_1 = 0$ corresponds to states of triple collision.

A solution started by a point in A must return to another point in A at the next syzygy crossing unless m_3 escapes or goes to triple collision. This may be viewed as a two-dimensional map $f_\alpha : \bar{A} \to A$,where $\bar{A} \subset A$ is the set of points which neither escape nor lead to triple collision. We further define $A_e \subset A$ as the subset corresponding to escape of m_3, and $A^* \subset A$ as the subset of points which lead to triple collision. Then A is the union of the three disjoint sets, $A = \bar{A} \cup A_e \cup A^*$, but f_α is only defined on $\bar{A}$. So for a given binary state $p_0 = (Q_1, P_1)_0 \in \bar{A}$, we define $f_\alpha(p_0)$ to be the binary state $p_1 = (Q_1, P_1)_1 \in A$ taken at the next syzygy crossing. It is well known that such Poincaré maps of Hamiltonian flows are area-preserving and that the eigenvalues of the Jacobian Df_α form reciprocal pairs (i.e., $\lambda_1\lambda_2 = 1$).

Unfortunately, in general, as in this case, Poincaré maps are not defined in closed form, so they must be obtained by numerical integration of the equations of motion (3). To study the properties of f_α as a global mapping we wish to consider how areas are transformed, but from a computational perspective we can only determine the precise behavior of a single point. To describe the global properties we have studied a square lattice of 62605 points in A (grid spacing of 0.005). At each lattice point we perform a single numerical integration. For further computational details we refer the reader to Paper I.

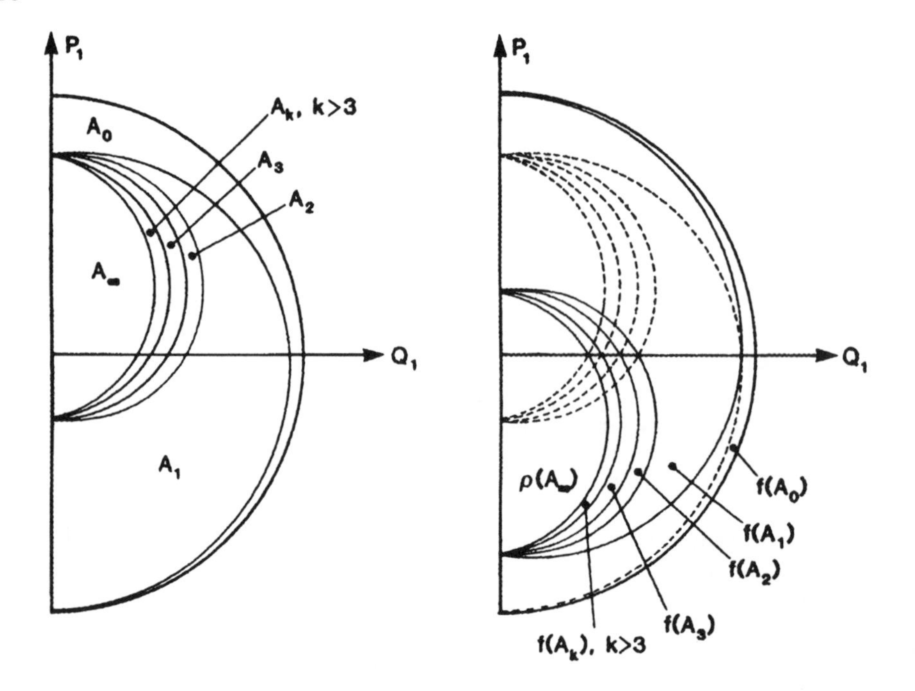

Figure 1. The geometrical description of f_α with $\alpha = \frac{1}{3}$. The diagram on the left depicts the partitioning of A into A_k, $(k = 0, 1, 2, \ldots)$. The diagram on the right depicts the images of the A_k. Here the preimages are shown with dashed lines to emphasize the intersections discussed in the text.

2.2 GLOBAL PROPERTIES OF THE POINCARÉ MAP

The global description of f_α, which is given in greater detail in Paper I, proceeds as follows. Since each point in $\bar{A}$ must return for a subsequent syzygy crossing, we may assign to it a non-negative integer k representing the number of binary collisions which ensue before the next syzygy crossing. This will partition A into open subsets A_k where all points in each subset have the same assigned k (see Figure 1). Additionally, we denote the points in A_e which lead to hyperbolic escape of m_3 by A_∞. (However we reserve the notation A_k to imply $k < \infty$.) Separating A_k and A_{k+1} is a locus of points B_k which lead to triple collision after k binary collisions. Notice that for $k \geq 2$, the B_k's accumulate to the boundary of A_∞ indicating that the area of $A_k \to 0$ as $k \to \infty$. In Paper I we established empirically that the areas of the A_k decay with increasing k according to a power law. The boundary of A_∞ which we denote by B_∞ corresponds to a set of points leading to parabolic escape of m_3. Under this notation we have the following relations

$$\bar{A} = \bigcup_{k=0}^{\infty} A_k, \quad A_e = A_\infty \cup B_\infty, \quad A^* = \bigcup_{k=0}^{\infty} B_k.$$

Due to the conservative nature of this system and the associated time reversibility we have a useful symmetry. Consider the reflection ρ about the Q_1-axis in the (Q_1, P_1)-plane defined by

$$(Q_1, P_1) \underset{\rho}{\longrightarrow} (Q_1, -P_1).$$

Now ρ can be considered as a reversal of the velocity, hence it could also be considered a reversal of the arrow of time. Thus the backward orbit of the reflection of p is the same as the reflection of the forward orbit of p and it has the same number of binary collisions as the forward orbit of p. This leads to the identity $\rho f_\alpha = f_\alpha^{-1}\rho$. It follows that for any $p \in A_k$, we have $\rho f_\alpha(p) \in A_k$ or $f_\alpha(p) \in \rho(A_k)$. That is

$$f_\alpha(A_k) = \rho(A_k). \tag{6}$$

Furthermore, if the orbit of p corresponds to the escape of m_3 ($p \in A_e$) the orbit of $\rho(p)$ corresponds to the capture of m_3. If the orbit of p terminates with a triple collision ($p \in A^*$) the orbit of $\rho(p)$ initiates at a triple ejection. The identity (6) hold only for the A_k and is not true for other sets in general, however it provides a convenient means of obtaining the forward images $f_\alpha(A_k)$. The asymmetry and the intersections of these subregions A_k with the Q_1-axis are the necessary ingredients for the intersections $A_j \cap \rho(A_k)$ illustrated in Figure 1. These intersections allow the existence of both the invariant region and the communication between the subregions which will be discussed in the next section.

The mapping on A_k, $k \geq 2$, is easy to describe. One may imagine each of these regions to be a crescent with an upper and lower tip, and an inner and outer edge. With this terminology the entire outer edge of A_k may be considered in a limiting sense to map into the upper tip of $f_\alpha(A_k)$. Similarly the inner edge maps into the lower tip of the image, the upper tip maps into the inner edge, and the lower tip maps into the outer edge. The map is smooth in the interior of each region, so we conclude that f_α is a contraction in a direction parallel to the edges, and an expansion normal to the edges. This contraction and expansion, with the intersections of the image and the pre-image shown in Figure 1, is the classical recipe for a horseshoe map. This description does not apply to the regions A_0 and A_1, although for A_1 the description is somewhat similar. But because A_1 does not have the crescent shape, the simple action described for $k \geq 2$ is not valid here. For A_0 the map simply pushes points from top to bottom until they finally map into A_1 after a finite number of iterations.

This partitioning of A into A_k and B_k is particularly conducive to a symbolic sequence representation of an orbit. We define the nth character in the symbolic sequence to be k if $f_\alpha^n(p) \in A_k$, for all integer n. Thus a sequence of k's corresponds to the sequence of A_k's visited under f_α^n (in both directions of time). If $f_\alpha^n(p)$ is not defined due to hyperbolic escape (capture) then we terminate the sequence on the right (left) with the symbol "∞". If $f_\alpha^n(p) \in B_k$, ($k \geq \infty$) we terminate the sequence with the symbol "$*_k$" to symbolize the triple collision/ejection or parabolic escape. We will use these sequences and subsequences to help qualitatively categorize and characterize orbits.

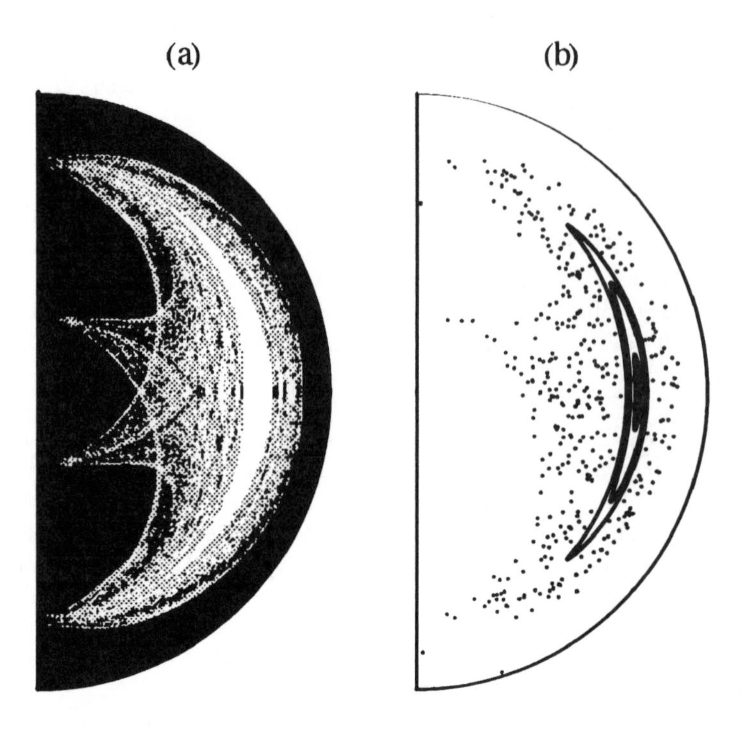

Figure 2. Different techniques for visualizing the dynamics. (a) depicts regions of capture-escape (black), and capture or escape (gray) after six iterations of f_α. (b) is the usual Poincaré surface of section plot. Observe that the techniques are complementary in the sense that (a) and (b) yield information on the structure in different regions. Both plots are for $\alpha = \frac{1}{3}$.

3. Review of results for three equal masses

Here we summarize several characteristics of the map f_α for the equal mass case which are important to the bifurcation analysis which follows. We again refer the reader to Paper I for extensive details.

For this mass ratio Figure 2(a) shows the spread of capture, escape, and capture-escape regions under six iterations of f_α. The capture region (colored gray) has been obtained from the union of successive forward iterates of the initial capture region $(f_\alpha^n \rho(A_\infty)\,,\ n = 0, 1, 2, \ldots, 6)$. The escape region (also colored gray) is the reflection of the capture region. The capture-escape region (colored black) is simply the intersection of the capture and escape regions. Clearly at this mass ratio, there are three distinct global regions in Figure 2(a): the solid black region, the solid white region, and the variegated region.

In the solid black region, every point represents a capture-escape solution with either zero or one binary collision between all successive syzygy crossings. Such orbits never enter A_k $(k \geq 2)$. Indeed, they escape immediately after entering A_1 (i.e., after the first binary collision) in either direction of time, so we may call it the *fast scattering region.* In the symbolic sequence representation these orbits are of the form "∞, ∞", "$\infty, 1, \infty$", "$\infty, 1, 1, \infty$", "$\infty, 1, 0, 1, \infty$", or "$\infty, 1, 0, ..., 0, 1, \infty$".

The white region converges to an invariant set for which every point corresponds to a bounded solution and a large subset form invariant KAM curves surrounding an elliptic fixed point p_1 in A_1. These invariant curves may be easily detected by scrutinizing a few initial conditions for many iterations as is usually done for Poincaré maps. (See Figure 2(b).) We call this region the *main invariant region* and the central elliptic fixed point p_1 the *main periodic orbit.* The main invariant region is bounded by a separatrix region formed by the intersecting invariant manifolds of a pair of hyperbolic periodic points of period-two. These points are also in A_1 and they are located near the tips of the largest invariant curve in Figure 2(b) (and near the tips of the solid white region in Figure 2(a)). The invariant manifolds intersect transversally, but with a very small splitting angle. In the symbolic dynamic representation, any orbit in the main invariant region is represented by a bi-infinite sequence of ones: "$\ldots 1, 1, 1, \ldots$".

The first stages of the Cantor set structure are readily apparent throughout the variegated region. We call this the *chaotic region.* It may be divided into orbits which "communicate" with A_1 and those which do not. The non-communicating set is easily understood in terms of the horseshoe mapping and its associated symbolic dynamics. In fact, in the non-communicating case almost every orbit (i.e., full measure) is capture-escape type with a finite number of syzygy crossings. We identify these orbits as *chaotic scattering.* Those orbits which are not capture, escape, or triple collision form a Cantor set and exhibit chaotic dynamics. The set of symbolic sequences associated with the non-communicating set is precisely given by the set of all sequences not including "0" or "1" or "$*_0$".

The communicating set is more complex, since in addition to the chaotic scattering and the chaotically bounded orbits, it has the additional possibility of stable periodic orbits with associated quai-periodic motion. This is so because, while the bounded chaotic orbits which do not communicate form a hyperbolic set, non-hyperbolicity *may* be introduced through communication with A_1. In Paper I we identify as an example a stable, communicating period-three orbit with symbol sequence "$\ldots, 5, 1, 1, \ldots$".

At this mass ratio all A_k intersect the Q_1-axis (see Figure 1), and as we mentioned before, this is one of the necessary ingredients for the existence of invariant sets (including fixed points). Indeed, in addition to the main periodic orbit p_1 in A_1, there exists a hyperbolic fixed point p_k on the Q_1-axis in each A_k $(k \geq 2)$.

4. Bifurcations

For the present study we wish to explore how the dynamics are affected by changes in the mass ratio. It is clear that the partitioning depicted in Figure 1 must change smoothly with α. Indeed we can consider the family of partitions under f_α as a partitioning of the half-cylinder $A \times (0, 1]$. Figure 3 depicts several cross-sections of this half-cylinder.

An important feature for $\alpha = \frac{1}{3}$ was the coexistence of two invariant sets, namely the main invariant region with stable behavior and the Cantor set with

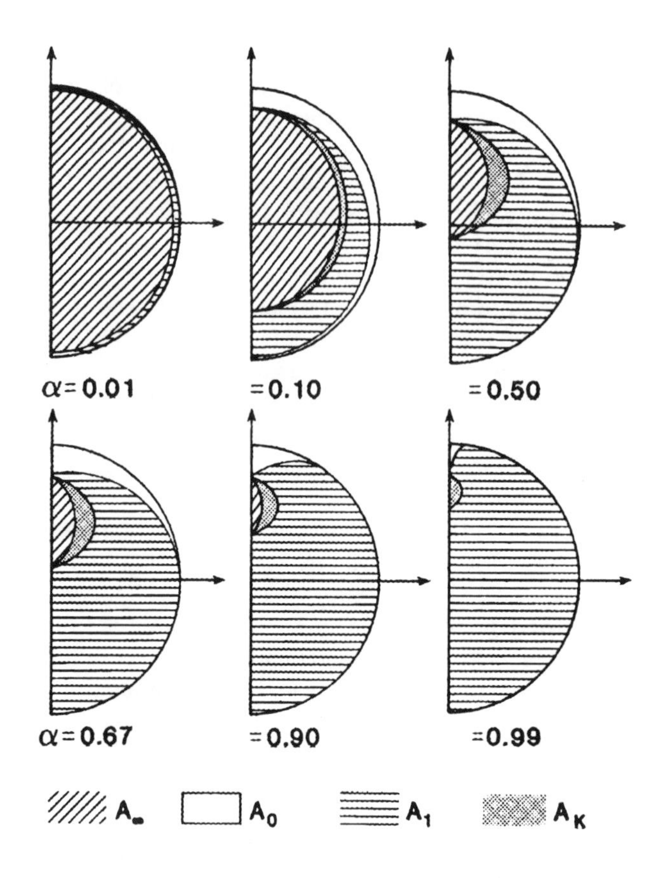

Figure 3. Partitioning of A at several values of α. In the legend A_k implies $k \geq 2$.

chaotic behavior. Here we are interested in finding the values of α which mark the creation (or destruction) of these sets. The existence of the main invariant region is connected to the stability of the main periodic orbit and we discuss this first in Section 4.1. The existence of the Cantor set structure is related to the intersections of the A_k with the Q_1-axis for $k \geq 2$. This can be readily observed from Figure 3 and it will be discussed in Section 4.2. Finally we summarize the results for $0 < \alpha \leq 1$ in Section 4.3.

4.1 INVERSE PERIOD DOUBLING BIFURCATION

As was mentioned earlier, the main periodic orbit is stable at $\alpha = \frac{1}{3}$. An inverse period doubling bifurcation of this periodic orbit occurs at $\alpha_{cr_1} = 0.5634101...$ This type of bifurcation also occurs in dissipative dynamical systems where it has been distinguished as one of the three generic types of intermittent transitions to chaos [4]. For $\alpha < \alpha_{cr_1}$ the stable periodic orbit, with the expected KAM tori, forms the main invariant region. This invariant region is bounded by a separatrix region formed from a pair of period-two unstable heteroclinic points. A standard

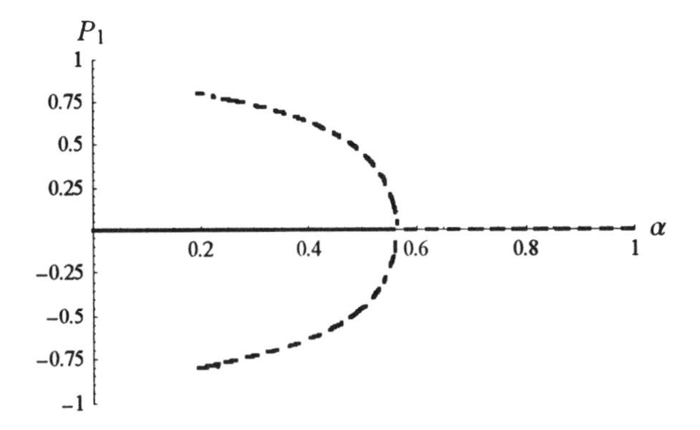

Figure 4. Bifurcation diagram of main periodic orbit. Dashes indicate instability.

bifurcation diagram is depicted in Figure 4. The locus of the period-one and -two points due to changing α is depicted in Figure 5. In this figure the dashed curve describes the path of the unstable period-two fixed points for $0.2 < \alpha < \alpha_{cr_1}$. We did not compute the path for values of α smaller than about 0.2 because the period-two orbit becomes strongly hyperbolic and our present computer algorithm is unable to follow the family further. Comparing Figures 3 and 5, we see that the upper locus is pushing into the very narrow upper tip of A_1 for small α. The solid curve on the Q_1-axis in Figure 5 is the path of the main periodic point on the stable range $0 < \alpha < \alpha_{cr_1}$, and the short dashed segment to the left of the bifurcation point is the path of the main periodic orbit on the unstable range $\alpha_{cr_1} < \alpha \leq 1$. The location of the bifurcation point in A is at (0.7335761...,0). At $\alpha = 1$ the orbit is at $p_c = \left(\sqrt{2}/2, 0\right)$. This is the circular Keplerian case which we discuss below.

4.2 GLOBAL BIFURCATION

In Figure 3 one can see that at $\alpha = \frac{1}{2}$ all A_k with $k \geq 2$ extend below the Q_1-axis, but in the slice at $\alpha = \frac{2}{3}$ they do not. This transition occurs at $\alpha_{cr_2} =$ 0.571... The rather elegant description of the dynamics presented in Paper I for $\alpha = \frac{1}{3}$ is drastically and suddenly altered as α increases above α_{cr_2}. Simó [5] has identified this critical value numerically as case of heteroclinic connection on the triple collision manifold, but his analysis could not reveal the global impact on the dynamics described here.

First observe that the very simple Cantor structure of orbits which never enter A_1 no longer exists for $\alpha > \alpha_{cr_2}$. In fact, since in this case every point in A_k $(k \geq 2)$ maps into A_1, all of the orbits which remain in A_k, $k \geq 2$, have been destroyed. This means that all orbits except the fast scattering are of the much more difficult communicating class.

We are also led to the rather obvious but striking conclusion that an infinite collection of periodic orbits are destroyed simultaneously as α increases above α_{cr_2}. As an example, the fixed points p_k (for $k \geq 2$) on the Q_1-axis move to the

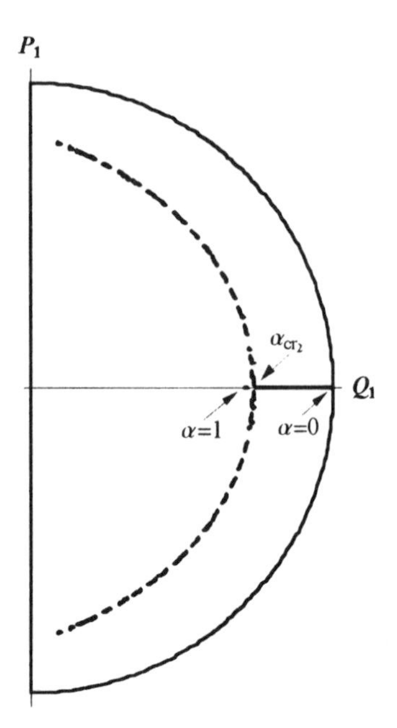

Figure 5. Loci of periodic points under varying α. Dashed lines indicate unstable orbits. The period-one points lie on the Q_1-axis, and the curves are the period-two points. These curves stop before $\alpha = 0$ because the locus for this periodic orbit becomes difficult to compute.

origin as $\alpha \to \alpha_{cr_2}$ and as they approach the critical value their stability index goes to infinity. Of course a similar behavior occurs for *all* periodic orbits which do not pass into A_1 at every other iteration. In terms of the symbolic dynamics described earlier, the only periodic orbits which survive this global bifurcation are those for which a "1" follows every integer $k \neq 1$. Finally we point out that almost every triple ejection-collision orbit with only one syzygy crossing (i.e., $B_j \cap \rho(B_k)$, $j, k \geq 1$) is destroyed for $\alpha > \alpha_{cr_2}$. The sole surviving one is located at $B_0 \cap \rho(B_0)$.

4.3 SUMMARY FOR $0 < \alpha \leq 1$

Now we are in a position to divide the bifurcation space into regions with similar behavior.

α near zero. From Figure 3 it is clear that as α becomes small, the region of hyperbolic escape (A_∞) covers almost all of A. At the same time, the A_k become very small, although they each continue to have positive measure for $\alpha > 0$. To explain this we point out that as α approaches zero, P_2 also approaches zero, according to (5). Then for very small α, *all* points $p \in A$ are near the collinear homothetic orbit. One would expect such orbits to lead to a triple close approach followed by hyperbolic escape as is indeed verified by Figure 3. We wish to emphasize that the full richness of dynamics described in Paper I for $\alpha = \frac{1}{3}$ persists at arbitrarily small $\alpha > 0$, but the interesting dynamics comprises a vanishingly small region in A.

$0 < \alpha < \alpha_{cr_1}$. Throughout this range the behavior is as described for the equal mass case. We note that Simó [5] has identified another critical value of $\alpha = .159...$ where we find some changes in the set of forbidden symbolic subsequences. But both the main invariant and Cantor regions exist, as well as the orbits which communicate between these regions. Regions of fast scattering are larger for small α, so this type of behavior is much more evident on this range than the others.

$\alpha_{cr_1} < \alpha < \alpha_{cr_2}$. On this (rather short) range the main periodic orbit is now unstable, hence there is no longer a main invariant set. However, there remains a small set for which the horseshoe map still applies, and there we still find the full richness of the Cantor structure. Very few orbits avoid A_1 altogether, so on this range most orbits are of the communicating class.

$\alpha_{cr_2} < \alpha < 1$. Now there is neither a KAM region nor a Cantor region. This means that all orbits must pass through A_1 and thus communication is dominant. The fast scattering orbits comprise smaller and smaller regions, vanishing as $\alpha \to 1$. Also as $\alpha \to 1$ the motion approaches integrable Keplerian motion (see below).We note that A_1 dominates and all of the other A_k become vanishingly small. However, using arguments based on the flow on the classical triple collision manifold, Simó [5] has shown that escape orbits exist for all values of $\alpha \in (0, 1)$. This means that for α arbitrarily close to unity all of the A_k persist.

It is well known that at $\alpha = \frac{55}{63} \cong 0.873$ there is a qualitative change in the dynamics near the homothetic collinear triple collision [3, 5]. At that value the collinear rest point on the triple collision manifold transitions from a focus to a node as α increases. The effects of this transition have been discussed elsewhere [1, 5, 6]. We point out that in our formulation the change to a node is manifested in Figure 3 by the shape of A_0. For $\alpha < \frac{55}{63}$ there is a small segment of A_0 touching the P_1-axis near $P_1 = -1$, but for larger α that segment vanishes. This result is based on simple arguments using the triple collision manifold. An open question for future study is the character of region A_0 for higher values of α. Does the region extend all the way down to the P_1-axis for all α, or does the lower limit of A_0 slide up along the boundary (collinear solution) as α increases above $\frac{55}{63}$?

$\alpha = 1$. Here, with all of the mass concentrated in the third body, the problem is integrable. This case has been called a double Kepler problem [2] because, in a barycentric frame, each of the binary masses describes an elliptic orbit if we imagine that the masses pass through each other unimpeded at the binary collision. Alternatively, if we consider the collision as the usual elastic bounce then the particles exchange orbits at the collision. At $\alpha = 1$ *all* points in A are parabolic-type fixed points and A_1 covers all of A. When α becomes less than unity all of these fixed points are destroyed, with the exception of the circular (main periodic) orbit p_c, which becomes hyperbolic. At $\alpha = 1$, with our normalization of units, we have $m_3 = \frac{1}{8}$ and the semi-major axis of the ellipse (in physical coordinates) is $a = \frac{1}{2}$. The circular orbit (the main periodic orbit) is at $p_c = \left(\sqrt{2}/2, 0\right)$. In Figure 6 level curves of eccentricity surround p_c, increasing to unit eccentricity on the boundary of A. As one moves around a level curve in Figure 6 the argument of perigee ω in the (x, y)-plane makes a complete revolution.

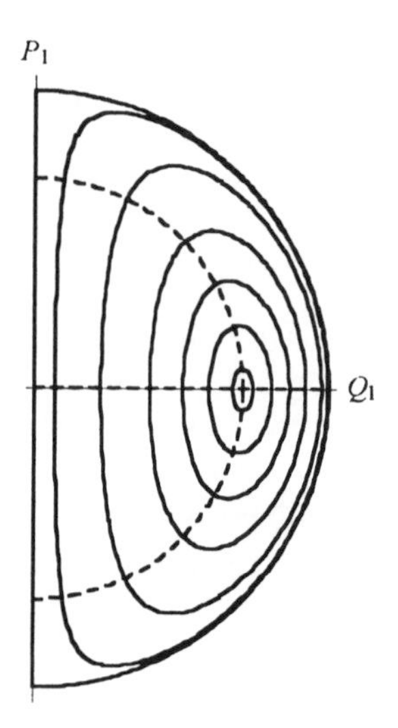

Figure 6. The characteristics of the Keplerian orbit for $\alpha = 1$. The central point is the circular orbit, p_c. Level curves of eccentricity surround p_c. The dashed curves radiating from p_c are curves of constant ω. The upper and lower dashed curves correspond to $\omega = 3\pi/2$ and $\pi/2$, respectively, and form a semicircle centered at the origin. The left and right dashed curves correspond to $\omega = 0$ and π, respectively.

Acknowledgments

The authors are indebted to the late Dr. Victor Szebehely for his encouragement and support. Illuminating discussions with Dr. Carles Simó and Dr. Rudolf Dvorak are gratefully acknowledged.

References

1. Pau Atela and Robert I. McLachlan. Global behavior of the charged isosceles three-body problem. *Internat. J. Bifur. Chaos Appl. Sci. Engrg.*, 4(4):865–884, 1994.
2. R. Broucke. On the isosceles triangle configuration in the planar general three-body problem. *Astron. Astrophys.*, 73:303–313, 1979.
3. Robert L. Devaney. Triple collision in the planar isosceles three body problem. *Invent. Math.*, 60:249–267, 1980.
4. Y. Pomeau and P. Manneville. Intermittent transition to turbulence in dissipative dynamical systems. *Comm. Math. Phys.*, 74:189, 1980.
5. Carles Simó. Analysis of triple collision in the isosceles problem. In *Classical Mechanics and Dynamical Systems*, pages 203–224, New York, 1981. Marcel Dekker.
6. Carles Simó and Regina Martínez. Qualitative study of the planar isosceles three-body problem. *Celestial Mech.*, 41(1-4):179–251, 1988.
7. K. Zare and S. Chesley. Order and chaos in the planar isosceles three-body problem. *CHAOS*, 8(2), 1998.
8. K. Zare and V. Szebehely. Order out of chaos in the three-body problem: Regions of escape. In A. E. Roy and B. A. Steves, editors, *From Newton to Chaos*, pages 299–313, New York, 1995. Plenum Press.

SECTION FIVE: STUDIES OF DYNAMICAL SYSTEMS

INTRODUCTION

G. CONTOPOULOS
Research Center for Astronomy,
Academy of Athens, Greece

AND

R. DVORAK
Institute for Astronomy,
University of Vienna, Austria

This chapter contains some new developments in the theory of dynamical systems.

In the first part the discovery and application of new fast methods to distinguish between ordered and chaotic orbits are described.

This subject started with the frequency analysis of orbits by Laskar [1,2] and Laskar, Froeschlé and Celletti [3]. The method is a refinement of the Fourier methods used in Physical Chemistry [4] and in Stellar Dynamics [5]. It can distinguish ordered and chaotic orbits already after 10^3 iterations, in cases where the use of Lyapunov Characteristic Exponents requires 10^6 iterations.

This method has been improved by Froeschlé, Froeschlé and Lohinger [6] and by Voglis and Contopoulos [7], who introduced the spectra of "stretching numbers" (or "Local Lyapunov Characteristic Numbers"). Similar information is provided by the directions of the deviations ξ from an orbit ("helicity angles" of Contopoulos and Voglis 1996). All these approaches allow a distinction of ordered and chaotic domains, by calculating orbits for only 10-20 iterations!

The most recent developments in this field are presented here in the papers of Contopoulos and Voglis; Efthymiopoulos, Voglis and Contopoulos; Froeschlé and Lega; and Froeschlé and Lohinger.

The first of these papers describes the use of the spectra for a fast separation of ordered and chaotic orbits in 2-D and 4-D systems. This

B.A. Steves and A.E. Roy (eds.), The Dynamics of Small Bodies in the Solar System, 425–428.

method is even faster than a systematic exploration of a Poincaré surface of section. It is applied to mappings and to Hamiltonian systems, e.g. models of galaxies.

This paper also emphasizes the role of the asymptotic curves of simple periodic orbits in explaining the main characteristics of the helicity spectra. Then it explains the transition from order to chaos by the destruction of invariant curves and the formation of cantori.

The second paper describes a method developed by Voglis [9] and Voglis and Efthymiopoulos [10] to derive the main frequencies of the orbits, by using "rotation angles" around a center (stable periodic orbit) and "twist angles" of the deviations ξ (i.e. the angles between ξ_{i+1} and ξ_i). This method gives both the main frequencies around the origin and the frequencies around the secondary islands. In this way the thin chaotic layers at the borders of the islands are clearly separated.

The paper by Froeschlé and Lega compares various criteria for distinguishing between order and chaos, namely the frequency map analysis of Laskar et al. [1,2,3], the sup-map method of Laskar [2] and Froeschle and Lega [11], the fast Lyapunov indicator method of Froeschlé et al. [12] and the twist angle method of Contopoulos and Voglis [13], extended by Froeschlé and Lega [14]. Their conclusion is that the various methods "are to be used as complementary tools".

Then they study 4-D maps and they apply the Morbidelli and Giorgilli theorem [15] in the neighbourhod of a KAM torus. Their calculations show an abundance of secondary KAM tori ("slave" tori) near a chief KAM torus, that are abruptly destroyed when the chief torus is destroyed. Finally a relationship is found between Lyapunov times and macroscopic instability times and a distinction is made between the "Nekhoroshev regime" and the "resonance overlapping regime".

The paper of Froeschlé and Lohinger explores the relationships between the "Local Lyapunov Characteristic Number", the "largest eigenvalue", and the "maximum stretching parameter" of an orbit. Only a loose connection exists between the distributions of these quantities.

Other indicators of chaos, based on the equation of geodesic deviation, have been considered in a paper by Cipriani and Di Bari. This equation appears in the framework of the Geometro-Dynamical Approach of Dynamical Systems. Of special interest is the use of the Finsler metric, that gives results compatible with those derived by using the Lyapunov Characteristic Numbers. An attempt to apply these methods to the Mixmaster Universe model is made.

Another problem of great current interest is the stickiness around islands of stability. This problem is reviewed here in detail by Dvorak and also to some degree by Contopoulos and Voglis. An orbit stays close to the last KAM curve surrounding an island for a long time, and then escapes to large distances. The stickiness is due to cantori with small holes surrounding the island. The time of stickiness increases exponentially when the distance from the island decreases, but very close to the island the increase is super-exponential. Additionally it has been shown that the structure of the stickiness zone is fractal. One can also locate the main cantorus responsible for the stickiness and find how the invariant curves of periodic orbits inside it cross the gaps of the cantorus.

Dvorak studies two problems in detail, the standard map and the Sitnikov problem. In the latter case stickiness was found even when the island of stability has disappeared completely.

A different problem is the distribution of periodic orbits in dynamical systems. Two Hamiltonian systems of 2 Degrees of freedom have been considered by Grousousakou and Contopoulos. In the first system the central periodic orbit is stable and becomes unstable when the perturbation increases. In the second case the central periodic orbit is unstable for small perturbations and undergoes an infinity of transitions to stability and instability for larger perturbations. The points representing the periodic orbits on a Poincaré surface of section form some characteristic lines. There are regular families of periodic orbits following a Farey tree, and irregular families, generated inside the lobes of the asymptotic curves of the unstable periodic orbits. The relation of these periodic orbits with the homoclinic tangle is explored.

This chapter ends with some papers on various dynamical problems:

The contribution of Harsoula and Voglis deals with the origin of counterrotating galaxies. In such galaxies the central part rotates in an opposite direction from the outer part. Most people try to explain these galaxies as produced by mergers of galaxies. In the present work an alternative, cosmological, method for producing counterrotation is explored.

The paper of Hermann introduces a fractal space-time, and derives Schroedinger's equation from Newton's equation for dynamical systems. A scale relativity for chaotic systems is derived. An application of this theory to the solar system is made, with emphasis on the asteroid belt and the satellites of Jupiter, Saturn and Uranus.

Aparicio and Floria deal with a Gylden-type system, i.e. a two-body problem with a Kepler parameter varying in time. A new set of variables is introduced that leads to a particular form of the equations of motion, similar to harmonic equations.

Finally the paper of Floria introduces a set of new variables that provide a uniform treatment of any perturbed two-body motion which are similar to the well known Delaunay variables.

References

1 Laskar, J.: 1988, *Astron. Astrophys.* **198**, 341.
2 Laskar, J.: 1990, *Icarus*, **88**, 266.
3 Laskar, J., Froeschlé, C. and Celletti, A.: 1992, *Physica D* **56**, 253.
4 Noid, D.W., Koszykowski, M.L. and Marcus, R.A.: 1977, *J. Chem. Phys.* **67**, 404.
5 Binney, J. and Spergel, D.: 1982, *Astrophys. J.* **252**, 308.
6 Froeschlé, C., Froeschlé, Ch. and Lohinger, E.: 1993, *Celest. Mech. Dyn. Astron*, **56**, 307.
7 Voglis, N. and Contopoulos, G.: 1994, *J. Phys.* **A27**, 4899.
8 Contopoulos, G. and Voglis, N.: 1996, *Celest. Mech. Dyn. Astron.* 64,1.
9 Voglis, N.: 1996, Oral Presentation, *Human Capital and Mobility Workshop*, Santorini, Greece.
10 Voglis, N. and Efthymiopoulos, C.: 1998, *J. Phys.* **A31**, 2913.
11 Froeschlé, C. and Lega, E.: 1996, *Celest. Mech. Dyn. Astron.* **64**, 21.
12 Froeschlé, C. Gonczi, R. and Lega, E.: 1997, *Planet. Space Sci.* **45**, 881.
13 Contopoulos, G. and Voglis, N.: 1997, *Astron. Astrophys.* **317**, 73.
14 Froeschlé, C. and Lega, E.: 1998, *Astron. Astrophys.* **334**, 355
15 Morbidelli, A. and Giorgilli, A.: 1995, *J. Stat. Phys.* **78**, 1607.

DYNAMICAL SPECTRA

G. CONTOPOULOS
Research Center for Astronomy, Academy of Athens
and
Department of Astronomy, University of Athens

AND

N. VOGLIS
Department of Astronomy, University of Athens

Abstract. We introduce the spectra of stretching numbers and helicity angles, that we call dynamical spectra. The main characteristics of the spectra of helicity angles are explained by calculating the asymptotic curves of simple periodic orbits.

A fast distinction between ordered and chaotic domains is made by calculating many orbits, along some lines in the phase space, for short intervals of time (e.g. 20 periods). This method is applied to 2-D and 4-D systems, both for mappings and realistic galactic models. We explain the phenomenon of stickiness near the outer edge of islands of stability which is due to the difficulty of crossing some cantori with small gaps surrounding the islands. We study the formation of cantori, and the crossing of cantori by the asymptotic curves of unstable periodic orbits.

1. Introduction

The usual distinction between order and chaos is by means of the (maximal) Lyapunov characteristic number (LCN)

$$LCN = \lim_{t \to \infty} \frac{\ln \mid \frac{\xi}{\xi_0} \mid}{t} \tag{1}$$

where ξ_0 is the initial infinitesimal deviation from an orbit and ξ is the deviation at time t (Fig. 1). The orbit is ordered if this limit is zero and chaotic if it is positive.

B.A. Steves and A.E. Roy (eds.), The Dynamics of Small Bodies in the Solar System, 429–454.

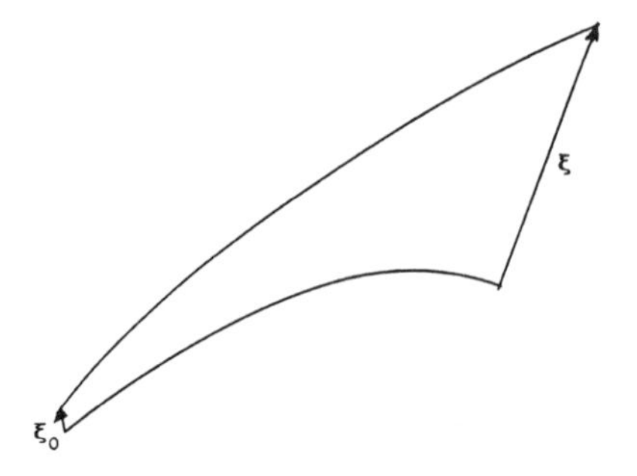

Figure 1. The infinitesimal deviation from an orbit at times 0 and t define the Lyanupov characteristic numbers in the limit

If the deviation $\mid \xi \mid$ is exponential

$$\mid \xi \mid = \mid \xi_0 \mid e^{at} \tag{2}$$

with constant $a > 0$, then

$$LCN = a. \tag{3}$$

On the other hand if the deviation $\mid \xi \mid$ is linear

$$\mid \xi \mid = \mid \xi_0 \mid t \tag{4}$$

then

$$LCN = \lim_{t \to \infty} \frac{lnt}{t} = 0 \tag{5}$$

Strictly speaking the above definition applies to bounded systems, because for orbits extending to infinity $LCN = 0$; in such cases we speak about chaotic or ordered scattering.

Furthermore in integrable systems, which are definitely nonchaotic, we may have unstable periodic orbits with positive LCN, but these are isolated and of measure zero. Therefore the definition (1) should apply to chaotic, or ordered domains. In fact a connected chaotic domain has a constant LCN.

In practice, in order to find numerically an approximate limit (1) we need a very long time, which is longer when the LCN is smaller. In galactic dynamics we usually need times of the order of 10^6 periods. But the age of the Universe is only about 10^2 periods. Thus it is customary to use "short time Lyapunov characteristic numbers " for fixed small t.

The first who used "short time LCN" were Nicolis et al (1983) and Fujisaka (1983), and many more authors followed (see references in Contopoulos and Voglis 1996). The shortest time in the case of maps is $t = 1$. This was introduced by Froeschlé et al (1993) and was called "local Lyapunov characteristic number".

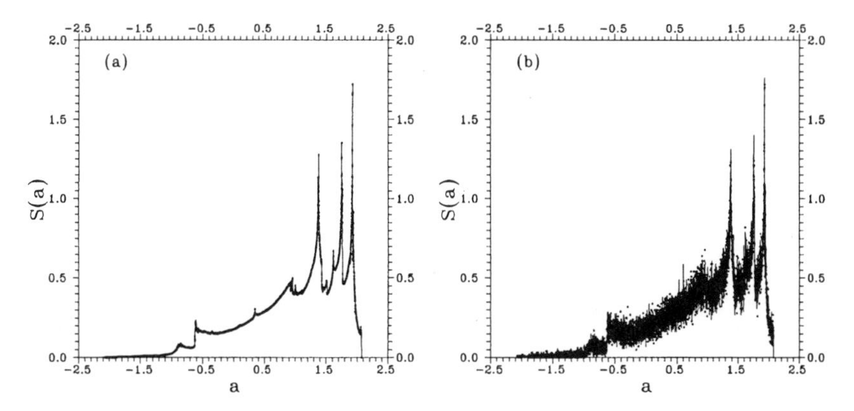

Figure 2. The spectrum of an orbit calculated for 10^N periods (solid line) and the next 10^N periods (dots) (a) $N = 8$, (b) $N = 5$.

We studied in detail this case (Voglis and Contopoulos 1994) and we called the quantity

$$a_i = \ln \mid \frac{\xi_{i+1}}{\xi_i} \mid \tag{6}$$

the "stretching number". The "spectrum of the stretching numbers" gives the number dN of values of a_i within an interval $(a, a + da)$ divided by da and by the total number of iterations N

$$S(a) = \frac{dN(a, a + da)}{N da} \tag{7}$$

In the same paper we have shown that $S(a)$ is invariant with respect to the initial point on an orbit. In Fig. 2a we show the spectrum of the standard map

$$\begin{aligned} x_{i+1} &= x_i + y_{i+1} \qquad\qquad (modulo 1) \\ y_{i+1} &= y_i + \frac{K}{2\pi} \sin 2\pi x_i \end{aligned} \tag{8}$$

for $K = 5$ and a number of iterations 10^8. The spectrum of the next 10^8 iterations falls exactly on the same curve. On the other hand the spectra calculated for two intervals of 10^5 iterations show some scatter (Fig. 2b).

In 2D maps the spectrum is also invariant with respect to the direction of the initial deviation x_0.

The chaotic spectra are further invariant with respect to initial conditions in the same connected chaotic domain. But different chaotic domains give different spectra. On the other hand the spectra of ordered orbits are

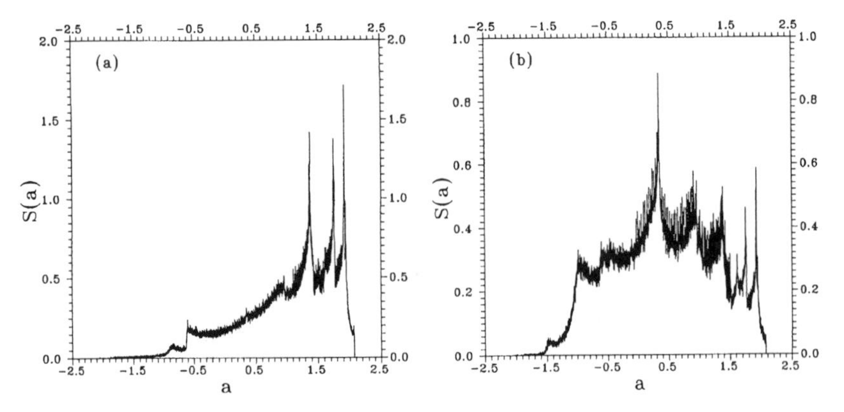

Figure 3. The composite spectrum of $10^2 \times 10^2$ orbits calculated for 10^2 periods each. (a) Initial conditions in a chaotic domain. (b) Part of the initial conditions in an ordered domain.

in general different, except if two orbits start on the same invariant curve (or invariant torus).

The invariance of the spectra with respect to initial conditions in the same chaotic domain allows the construction of an invariant spectrum by superimposing data from many orbits. In Fig. 3a we give a spectrum derived by calculating 10^4 orbits, with initial conditions in a grid of 10^2 points along x and y, for only 10^2 periods each. This spectrum is very similar to the spectrum 2a , but if part of the square of each initial conditions is in an ordered domain the composite spectrum is quite different (Fig.3b).

In the case of continuous systems we can define a stretching number for times even shorter than $t = 1$, e.g. for an intergration step Δt (Smith and Contopoulos 1996). If Δt is small the spectrum is independent of the value of Δt.

Another quantity of interest is the angle formed by the vector ξ with a fixed axis, say the x-axis (Fig. 4). This is called "helicity angle" Φ, while the difference of two seccessive helicity angles is called "twist angle" $\phi = \Phi_2 - \Phi_1$. In Fig. 4 we give also the well known "azimuthal angles" Θ and "rotation angles" $\theta = \Theta_2 - \Theta_1$ with respect to a center O.

In the case of closed invariant curves, and thin chaotic zones inside invariant curves, we can choose as center 0 the stable periodic orbit inside an island of stability. However in a completely chaotic domain (like fig. 5a) the point 0 is arbitrary.

In the same way as for stretching numbers we can define spectra of the helicity angles Φ, the twist angles ϕ, the azimuthal angles Θ, and the

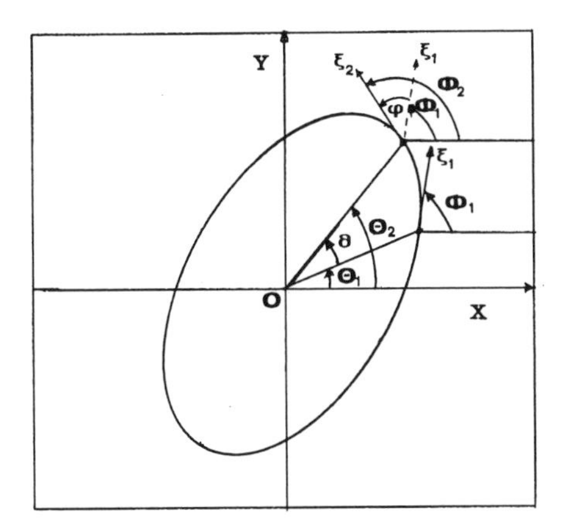

Figure 4. The helicity angles Φ_1,Φ_2 and the azimuthal angles Θ_1,Θ_2. The twist angle is $\phi = \Phi_2 - \Phi_1$ and the rotation angle is $\theta = \Theta_2 - \Theta_1$.

rotation angles θ. E.g. the spectrum of the twist angles is

$$S(\phi) = \frac{dN(\phi, \phi + d\phi)}{N d\phi} \tag{9}$$

2. The Structure of Chaotic Domains

The Lyapunov characteristic number is the average value of the stretching numbers

$$LCN = \int_{-\infty}^{\infty} aS(a)da \tag{10}$$

However two systems with the same LCN may have very different properties. In Fig. 5a, d we give the distribution of 10^4 successive iterates of the map (8) for $K = 10$, and of the conservative Hénon map

$$\begin{aligned} x_{i+1} &= 1 - K' x_i^2 - y_i \qquad (modulo 1) \\ y_{i+1} &= x_i \end{aligned} \tag{11}$$

for $K' = 7.407$. The two systems look equally chaotic and in fact the density of points in each case is constant all over the square $(0 \leq x < 1)$,$(0 \leq y < 1)$. Furthermore the Lyapunov characteristic numbers for these values of K and K' are equal, $LCN = 1.62$. However the spectra of stretching numbers (Fig. 5b, e) and of helicity angles (Fig. 5c, f) are quite different.

Nicolis et al (1983) emphasized that the dispersion of the values of the short time Lyapunov exponents around the mean is equally important as the mean itself. But the dispersions are insufficient to define the spectra, which have some characteristic peaks at particular values of a and of Φ.

The best way to understand the structure of phase space and the forms of the spectra is by calculating the asymptotic curves of the periodic orbits. In Fig. 6 we give the unstable asymptotic curve of the simplest unstable periodic orbit $(0,0)$. This curve starts like an almost straight line with an inclination of $\Phi = 42^o$. As it reaches the line $x = 1$, it continues from $x = 0$ because of the modulo 1, reaches the boundary $y = 1$, then continues from $y = 0$ and so on.

If we start an orbit on the asymptotic curve very close the origin its images are always on the same asymptotic curve.

A deviation ξ after a few iterations becomes almost tangent to the asymptotic curve. Therefore the helicity angles are almost equal to the inclinations of the asymptotic curve at the successive iterations of the initial point.

We notice that most segments of the asymptotic curve are parallel to the direction $\Phi = 42^o$. This explains the first large maximum of the spectrum of the helicity angles (Fig.5c). But according to a classical theorem the asymptotic curve comes arbitrarily close to any segment of it, without ever interesting itself, both in the forward and in the backward direction. This explains the symmetry of the spectrum $S(\Phi)$ with respect to a translation by 180°. In particular the first maximum of the curve of Fig. 5c to the right of -180° is a replica of the maximum of $\Phi = 42^o$, located at $\Phi = -138^o$.

In Fig. 6 we see another dominant direction of many lines near $\Phi = 55^o$. This gives the second maximum of the spectrum and its symmetric peak at $\Phi = -125^o$.

The other values of ϕ from -180° to $+180^o$ are all present, because the curves of Fig.6 turn continuously from one angle to another, but most of the angles are weakly populated, as it is seen in Fig. 5c.

The only problem that remains unsolved is why the form of the asymptotic curve is as it is. What we have found is derived only from numerical calculations supplemented by general theorems, e.g. that the asymptotic curve does not intersect itself (if one finds numerically intersections we know that the calculations are wrong). But a general theory of the forms of the asymptotic curves is lacking.

As a comparison Fig. 7 gives an unstable asymptotic curve of the simple unstable periodic orbit (x=y $= 0.256444$) in the Hénon map for $K' = 7.407$. We see that the arrangement of the asymtotic curve is very different than in Fig. 6. We can again explain the 180° periodicity of the spectrum of the helicity angles (Fig.5f), but the positions of the maxima are very different.

Another form of the asymptotic curve is given by the standard map for $K = 5$ (Fig.8). In this case there are two large islands that cannot be crossed by the asymptotic curve. A chaotic orbit fills the whole region outside the islands (Fig. 9a) but it may take a large time even to approach the islands

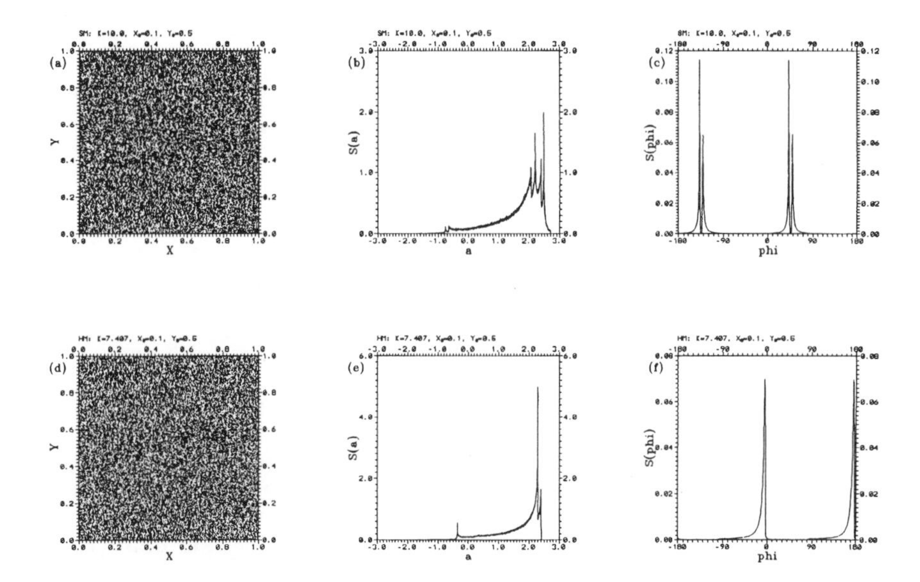

Figure 5. A comparison of the standard map for $K = 10$ (a), and its spectra of stretching numbers (b) and helicity angles (c), and of the Hénon map for $K' = 7.407$ (d), and its corresponding spectra (e, f).

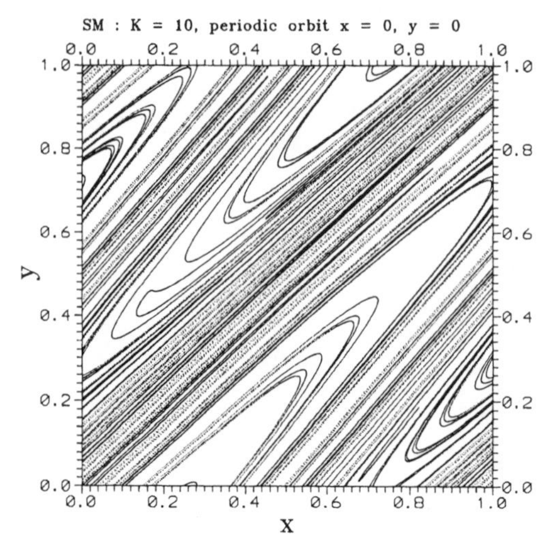

Figure 6. An unstable asymptotic curve from the unstable periodic orbit ($x = y = 0$) in the case of the standard map for $K = 10$.

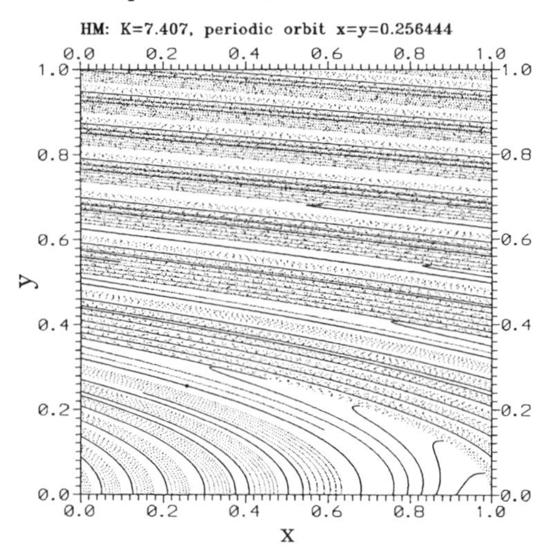

Figure 7. An unstable asymptotic curve from the unstable periodic orbit ($x = y = 0.256444$) in the case of the Hénon map for $K = 7.407$.

(this is seen in Fig.8, where the asymptotic curve has been drawn for a relatively short time). This phenomenon is related to the stickiness of the islands, a phenomenon that we will discuss in section 5.

The helicity spectrum of a chaotic orbit from Fig. 9a is given in Fig. 9b. This spectrum is similar to the spectrum of Fig. 5c. In particular it is periodic with respect to a translation by 180°.

In contrast the helicity spectrum of an ordered orbit does not have the 180° periodicity. An initial condition inside an island gives two closed invariant curves, one in each island. The iterates of the initial point alternate

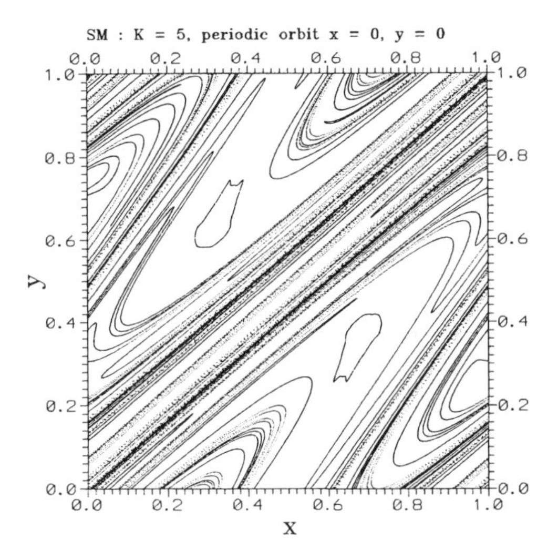

Figure 8. An unstable asymptotic curve from the unstable periodic orbit (0,0) in the standard map for $K = 5$. This curve does not enter the region of the islands.

from one island to the other. Thus we take only every second point, to concentrate our study on only one island (Fig.9c). We start a vector ξ_0 inwards from the invariant curve the corresponding helicity spectrum is the solid line of Fig. 9d that has no 180° symmetry. On the other hand if we start a vector ξ_0 outwards (exactly opposite to the above) we find a spectrum equal to the above but transposed by 180°. The explanation of this phenomenon is simple. An initial infinitesimal vector directed outwards has its endpoint on an invariant curve slightly outside the original invariant curve. But the successive points on this curve move clockwise at a slightly greater rate around the center than in the original curve. Thus the vector tends to become tangent to the invariant curve, turning clockwise. There is no reason to have an exact symmetry between the angles Φ and $\Phi + 180^o$, unless the invariant curve is small in size, in which case it is almost an exactly symmetric ellipse. On the other hand if the original ξ_0 is directed inwards its endpoint moves on an invariant curve slightly inside the original one and at at a slightly smaller rate. Thus ξ_0 tends to become tangent to the invariant curve but in the opposite direction. By taking the second x_0 exactly opposite to the first we have, by definition, for any angle Φ of the first vector x_0, an angle $\Phi + 180^o$ for the second (opposite) vector, thus the second spectrum is a transpose by 180° of the first.

We conclude that the helicity spectra $S(\Phi)$ allow a clear distinction between chaotic and ordered orbits. Chaotic orbits have a 180° - symmetric helicity spectrum, while ordered orbits do not have this symmetry except of some very particular cases (e.g. 180° - symmetric invariant curves).

Those exceptional cases can be easily distinguished, by considering one half of the invariant curve.

Furthermore one must take points on the same island; if one takes all values on both islands of Fig. 9c, one finds a 180° symmetric spectrum , because of the symmetry of the islands with respect to the center of Fig. 9c.

In the case of more degrees of freedom an alternative method based on the properties of the spectra of stretching numbers and helicity angles can be applied to distinguish between chaotic and ordered motion (Voglis et al 1997). In this case the possibility of distinction is more important because we cannot visualize islands of stability on a Poincaré surface of section.

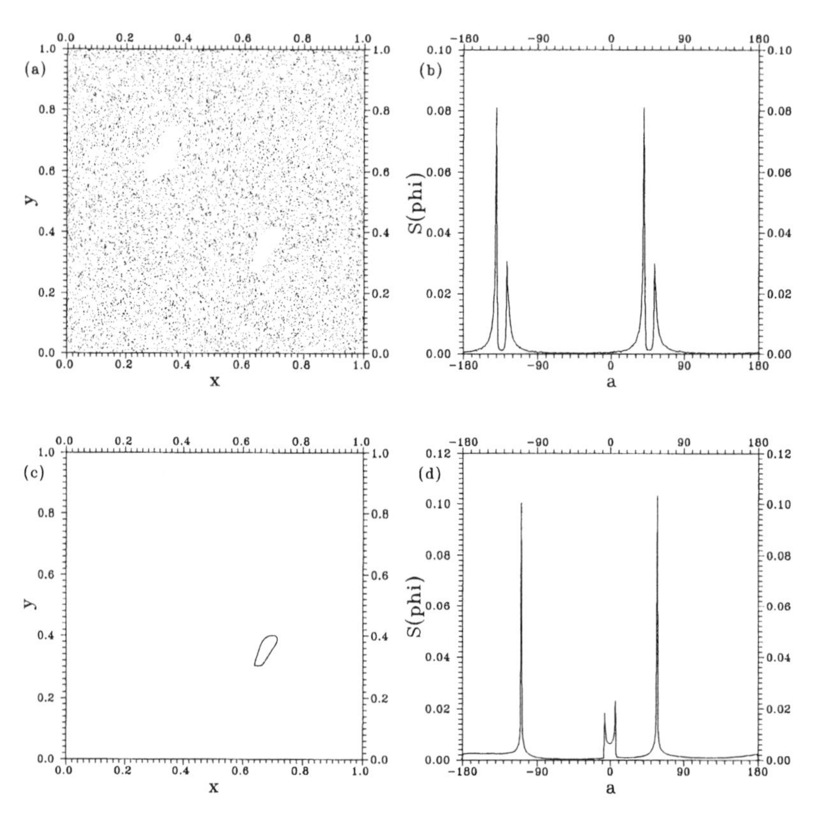

Figure 9. A chaotic and an ordered orbit in the standard map for $K = 5$ (a and c, respectively) and their helicity spectra (b and d).

3. Fast distinction between ordered and chaotic domains

As we have just explained, we can separate individual ordered and chaotic orbits by the form of their spectra. In order to do that, we need to have spectra more or less well defined. For this reason we need at least some

10^4 iterations. But there is a much faster method to distinguish order and chaotic domains in dynamical systems.

Our method is based on the fact that the spectra are invariant with respect to initial conditions in a chaotic domain, while they change gradually from one point to the next in an ordered domain. E.g. the average value of the helicity angle $< \Phi >$ is constant in the chaotic domain, while it varies smoothly in the ordered domain. The same is true for the average twist angle $< \phi >$ and stretching number $< a >$.

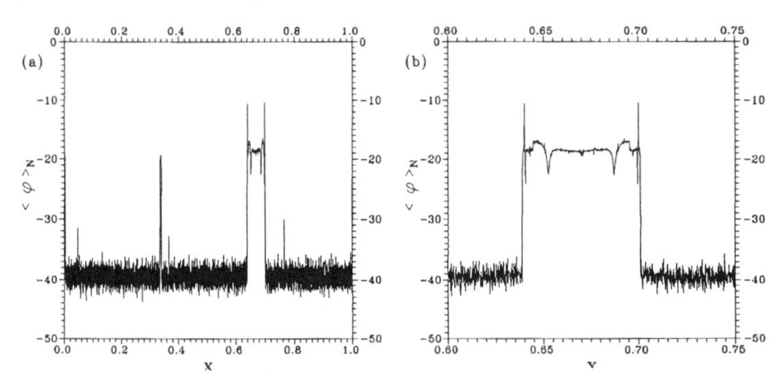

Figure 10. (a) Average values $< \phi >_N$ for orbits calculated for $N = 10^4$ periods for different x with a step $\Delta x = 10^{-4}$ and constant $y = 0.34$. (b) Magnification of the island region.

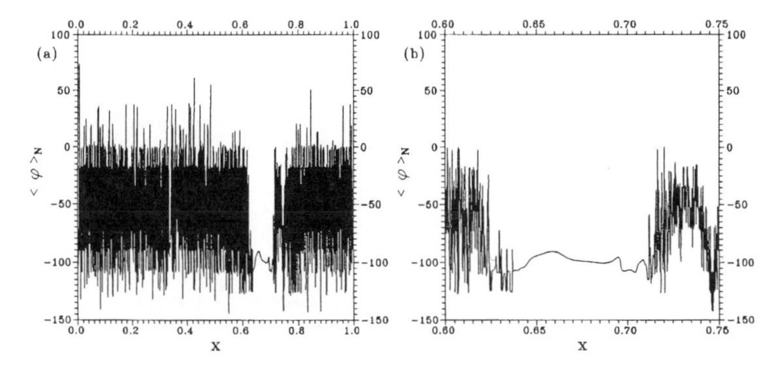

Figure 11. (a) Average values $< \phi >_N$ as in Fig. 10 but with $N = 20 - 10$, i.e. 10 iterations after the first 10 transients. (b) Magnification of the island region.

As an example we take 10^4 orbits along the horizontal line $y = 0.34$ that passes through both chaotic and ordered domains. For each orbit we calculate the value of $< \phi >_N$ after $N = 10^4$ iteractions. These values deviate from the limiting constant $< \phi >$, which is derived after $N \to \infty$ iterations, but the deviations are small (Fig.10a). On the other hand the values of $< \phi >_N$ vary smoothly inside an island of stability. More important is the fact that the change of $< \phi >_N$ at the borders of the island is abrupt. Thus even a rather small island can be identified (see the almost vertical line at about $x = 0.34$ in Fig. 10a).

The structure of the island can be seen in magnification in Fig. 10b. We see, first, that the deviations from a smooth line are extremely small in general. However we see some larger deviations near the borders of the island corresponding to secondary chaotic zones, and also some dips of the curve representing higher order islands.

The clear separation of the chaotic from the ordered domain allows us to do the same calculation (Contopoulos and Voglis 1997) for only 10 iterations after the first 10 transient points (Fig.11). The dispersion of the values of $< \phi >_N$ is now much larger, but we can again clearly distinguish between the chaotic and ordered domains. Even the small island near $x = 0.34$ can be distinguished.

Instead of the average values of $< \phi >_N$ one may use the average stretching numbers $< a >_N$, which are positive on the average for chaotic orbits and around zero for ordered orbits. The number of $N = 20 - 10$ iterations is again sufficient. However, with only 10 or 20 iterations one cannot say with certainty whether one particular orbit is chaotic or not. In fact individual chaotic orbits may have very discordant values of $< \Phi >_N$, $< \phi >_N$ or of $< a >_N$, for $N = 20 - 10$, that reach the range of values of $< \phi >_N$ or $< a >_N$ for ordered orbits, making their distinction practically impossible. Thus for individual orbits much larger values of N would be needed.

But as regards domains of chaotic and ordered orbits our method with $N = 20 - 10$ iterations is very effective. *It is even faster than a systematic calculation of Poincaré surfaces of section.*

Furthermore the calculation of 20 points from each orbit allows a rough delineation of the invariant curves, so that it is not necessary to use a dense grid of parallel lines, each with constant y.

The same method can be used also in systems of 3 degrees of freedom, or 4-D maps.

An example is the case of two coupled standard maps

$$\begin{aligned} x_1' &= x_1 + y_1' \\ y_1' &= y_1 + \frac{K}{2\pi}\sin 2\pi x_1 - \frac{\beta}{\pi}\sin 2\pi(x_2 - x_1) \\ x_2' &= x_2 + y_2' \qquad\qquad (mod 1) \\ y_2' &= y_2 + \frac{K}{2\pi}\sin 2\pi x_3 - \frac{\beta}{\pi}\sin 2\pi(x_1 - x_2), \end{aligned} \tag{12}$$

If $\beta = 0$ the two maps (x_1, y_1) and (x_2, y_2) are independent. However if $\beta \neq 0$ the two maps are coupled and a genuine 4-D map is formed. Such a map represents a Poincaré surface of section of a 3-D conservative dynamical system.

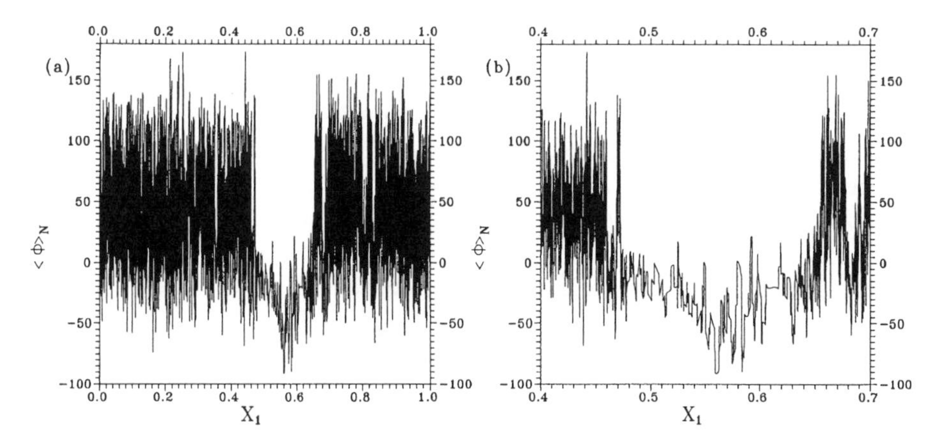

Figure 12. (a) Average values of $\phi_{x_1 x_2}$ for 10 iterations after the first 10 transients in the case of the 4-D map (12) with $K = 3$ and $\beta = 0.3$ along the line $y_1 = 0.1$, $x_2 = 0.62$, $y_2 = 0.2$. (b) Magnification of the island region.

In such a case, we define three helicity angles, namely $\Phi_{x_1x_2}$, $\Phi_{x_1y_1}$, $\Phi_{x_1y_2}$, giving the angles of the projections of the vectorξ on the planes x_1x_2, x_1y_1, x_1y_2 with the axis x_1.

The average values of $< \phi_{x_1x_2} >_N$ for $N = 20 - 10$ iterations (i.e. the average of 10 iterations after the first 10 transients) are given in Fig. 12 for 10^4 initial conditions along a line parallel to the x_1-axis. We see that the chaotic domains have the same average values of $< \phi >$, with large dispersions. But we can see also a big island of stability where the value of $< \phi >_N$ changes with x_1, with smaller dispersions. The separation of the island from the chaotic domain is again very clear.

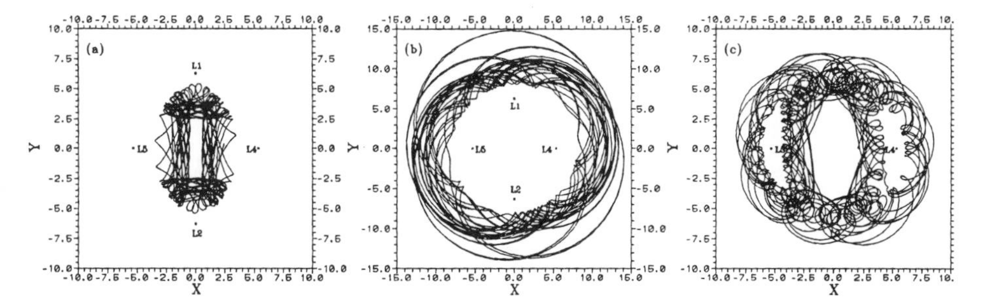

Figure 13. Three types of chaotic orbits in a barred galaxy: (a) inside corotation (b) outside corotation, and (c) both inside and outside corotation.

4. Application

The examples that we have seen up to now refer to simple maps, like the standard map and the Hénon map. But the same kinds of spectra appear in realistic models. In Fig. 13 we give types of chaotic orbits in a barred galaxy (inside corotation along the bar, outside corotation forming a ring around the bar, and orbits that go both inside and outside the bar). Such orbits stay along the spiral arms outside the bar and are important in constructing self consistent models of galaxies (Kaufmann and Contopoulos 1996).

Some spectra of streching numbers for galactic orbits are given in Fig. 14. Close to a stable periodic orbit the spectra are almost exactly symmetric around $a = 0$, terminating at two abrupt peaks for positive and negative a (Fig. 14a). Further away the spectra develop more peaks (Fig. 14b), while the chaotic spectra have a large number of peaks (Fig.14c) (Patsis et al 1997).

5. Stickiness

The transition between ordered and chaotic domains is a fuzzy zone, where the orbits stay for a long time but then diffuse into the chaotic domain. This phenomenon is known as stickiness (Contopoulos 1971, Shirts and Reinhardt 1982).

The sticky regions are limited by cantori with small holes that act as partial barriers for the orbits. However after long times the orbits pass through the gaps of the cantori, entering into the chaotic domain and may go very far from the ordered domain. But this process can be reversed, i.e. orbits in the chaotic domain may enter the sticky zone and stay there for a long time, before going out again.

An example of a sticky domain is shown in Fig. 15. The thin very dark

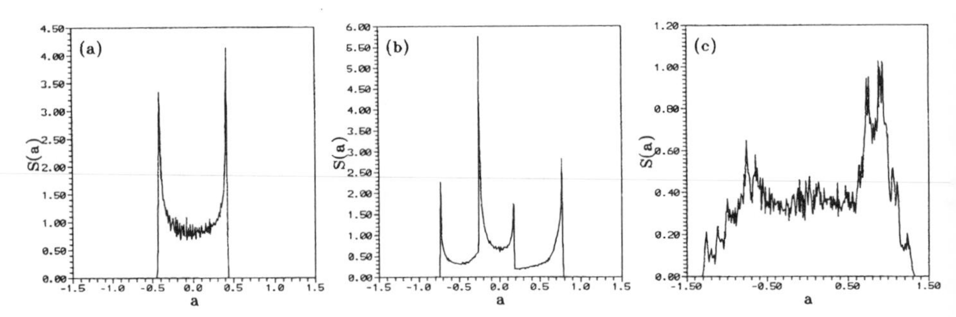

Figure 14. Spectra of orbits in a model of a barred galaxy : (a) a regular orbit close to a stable periodic orbit, (b) a regular orbit further away, (c) a chaotic orbit.

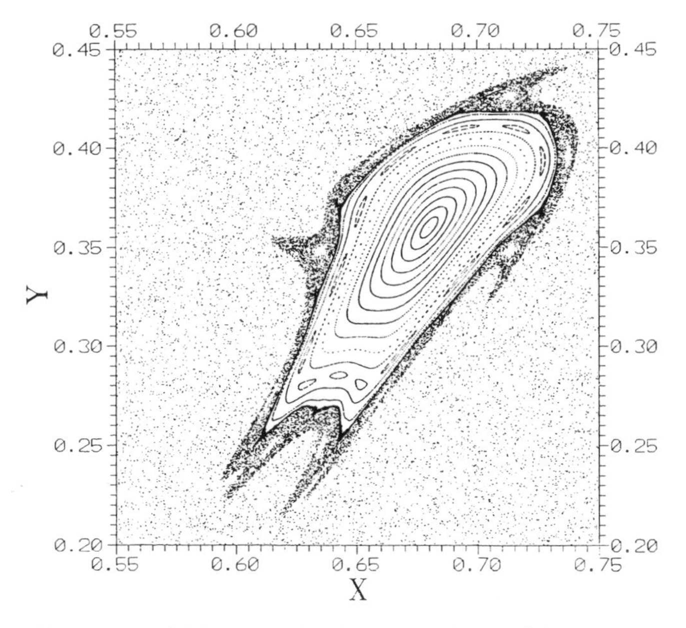

Figure 15. Sticky zones in the standard map (8) for $K = 5$.

region around the island represents the main sticky zone, while the less dark and thicker region represents a large number of zones, where stickiness lasts for relatively short times.

In practice the distinction between the various sticky zones is not quite clear. In Fig.16 we give the stickiness time as a function of x along a particular line $y = 0.36$, intersecting the main island of Fig. 15.

The stickiness time is also called "trapping time" or "escape time", and it is defined as the time required for an orbit to go out of an ellipse surrounding the sticky zones in Fig. 15.

The stickiness time undergoes large variations for very small changes of x. However Fig. 16 shows some characteristic trends. The stickiness time increases on the average exponentially as we approach the main island. This exponential dependence is shown as a straight line in Fig. 16. The right end of Fig. 16 (for $x \geq 0.64324$) is shown in more detail in Fig. 17. Individual stickiness times may deviate considerably from the average, but the exponential trend is clear. There are only three remarkable deviations from this exponential law.

(1) Very close to a big island at the right end of Figs. 16 and 17 the stickiness time is even larger than exponential. We call this region a superexponential regime, because it reminds us of the theoretical superexponential stability time very close to every invariant curve (Morbidelli and Giorgilli 1995).

(2) Inside the sticky zone, there are higher order islands where the trap-

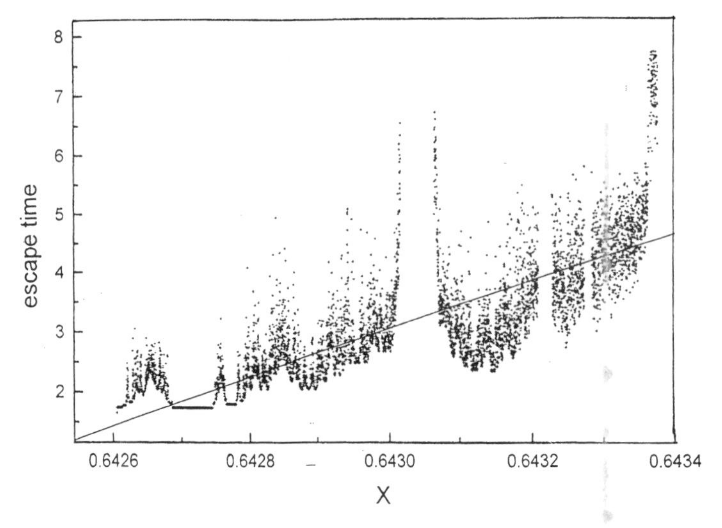

Figure 16. Stickiness time as a function of the distance from an island.

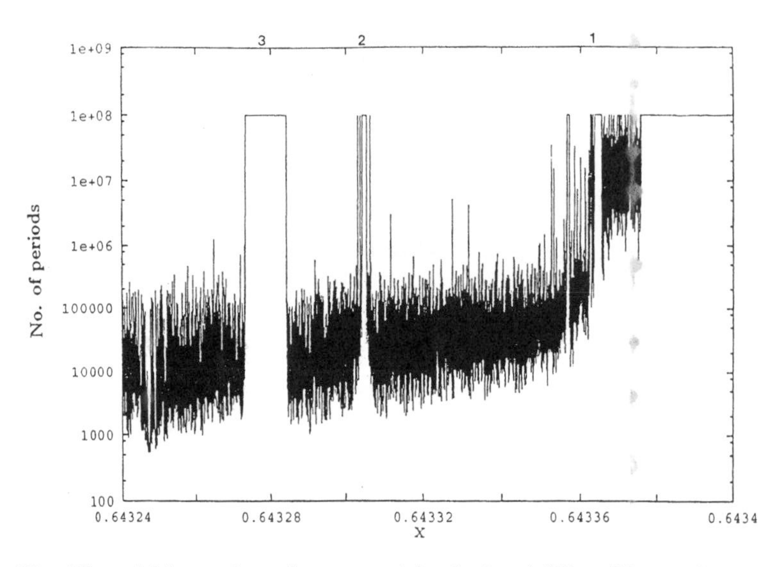

Figure 17. The stickiness time close to an island of stabilility. We see the exponential regime, the superexponential regime very close to the main island, and the islands of stability with infinite stickiness time.

ping lasts for ever. Near every such island the stickiness grows abruptly, and it seems that in all cases there is a local exponential and superexponential regime.

(3) Further away from the boundary of the main island the exponential law is not valid any more. Instead, there are localized peaks of high

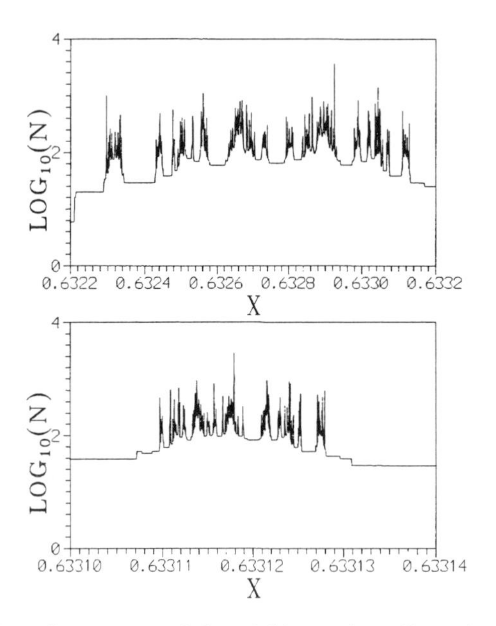

Figure 18. (a, b) Fractal structure of the stickiness time. Sucessive magnifications of a region beyond the left end of Fig.16 give the same forms for the succession of peaks and low stickiness intervals.

stickiness time separated by intervals where the stickiness time is low and almost constant (left end of Fig. 16).

The peaks of high stickiness time have a fractal structure. If we magnify a small region of Fig. 16 we find a similar structure (Fig.18a), and further magnifications show again a similar structure of peaks separated by intervals of low and almost constant stickiness time (Fig.18b). This is the main characteristic of fractal structure.

An explanation of this structure is based on the forms of the asymptotic curves of unstable periodic orbits in the sticky region. If the initial condition of an orbit is on the stable manifold (asymptotic curve) of an unstable periodic orbit, this orbit is trapped for ever, approaching the periodic orbit asymptotically.

An orbit which is close to a stable manifold, but not exactly on it, will eventually escape in general, following closely an unstable manifold of the same orbit. However the escape time may be very different if the deviation is on one or the other side of the stable manifold. The iterates of the original point approach the unstable periodic orbit from opposite sides, and then they follow two opposite unstable asymptotic curves which are not symmetric. In fact one asymptotic curve may lead fast to an escape out of the sticky zone, while the other may escape only after a very long time, making many oscillations before entering the chaotic zone. The close

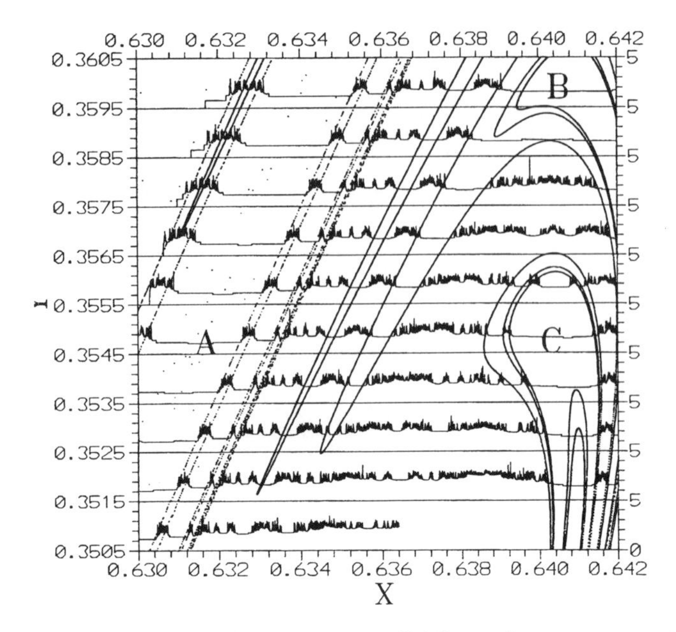

Figure 19. The stickiness times along many parallel lines of constant y, each separated from the next by $\Delta y = 0.001$, together with some stable asymptotic curves of an unstable periodic orbit. The domains A,B,C have low stickiness time.

relation between the peaks and the plateaus in the stickiness time on one hand, and the stable asymptotic manifold of an unstable periodic orbit on the other hand is shown in Fig.19 (Efthymiopoulos et al. 1997). We see that these lines follow closely the borders of some low stickiness plateaus like A,B,C.

These low stickiness plateaus are due to some escape channels (openings of the homoclinic tangle of the asymptotic curves) that lead to escape after an almost constant number of iterates. These can be seen clearly in Fig. 20, where we give the same asymptotic curves as in Fig. 19 over a larger area. Now the escape channels for the plateaus A,B, or C, allow an escape after a relatively low and constant time.

6. Formation of Cantori

The boundaries of the sticky zones are some cantori with small gaps that provide a barrier to orbits on both sides of them for long times. These cantori are formed by the destruction of "noble tori" (Greene 1979), that separate an outer and an inner chaotic domain. Noble tori are those with a "noble rotation number", i.e. a number that is represented as a continuous

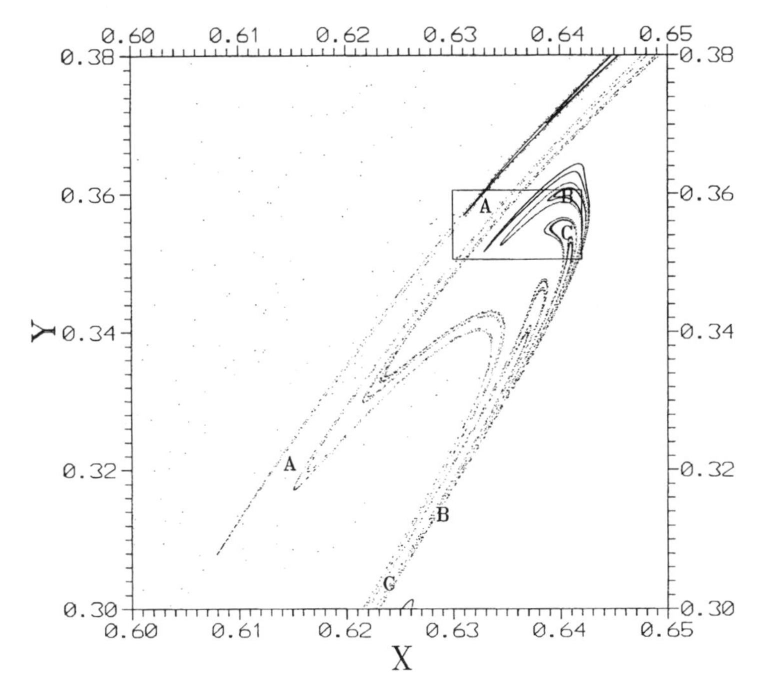

Figure 20. The asymptotic curves of Fig. 19 over a larger area. The domains A,B,C lead to escape channels.

fraction

$$a = \frac{1}{a_1 + \frac{1}{a_2 + \frac{1}{a_3 + \dots}}} \tag{13}$$

where all a_i above a certain order $i > N$ are equal to 1. This is written in the form $a = [a_1, a_2, a_3, ...]$. According to Greene's conjecture these tori are the last KAM tori (or "last KAM curves" in the case of 2 degrees of freedom) to be destroyed as a perturbation parameter K increases beyond a critical value K_c. The destroyed KAM curve is a cantorus. For small positive values of $(K - K_c)$ the gaps of the cantorus are small and the cantorus provides a partial barrier for the communication of the chaotic domains on both its sides. However for larger $(K - K_c)$ the gaps become large and no effective barrier is provided to the diffusion from one chaotic domain to another.

We follow the destruction of some noble tori and the formation of cantori in the standard map (8) when K is close to $K = 5$. In Fig. 21 we see a part of the boundary of the main island of the standard map for $K = 4.791$. We see some noble tori, like $[2,1,1,2,1,...]$, $[2,1,1,3,1,...]$, $[2,1,1,4,1,...]$ that separate completely the outer chaotic domain from an inner, localized, chaotic domain around the unstable orbit $2/5$.

However the noble numbers $[2,1,1,1,...]$ and $[2,1,1,5,1,...]$ correspond to cantori (i.e. destroyed tori).

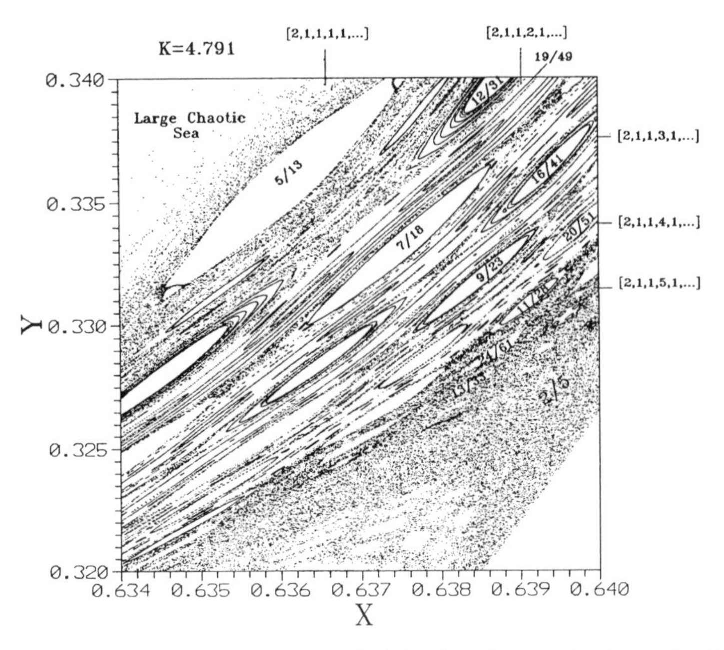

Figure 21. Part of the boundary of the main island of the standard map for $K = 4.791$, containing some noble tori and cantori.

As the perturbation K increases the noble tori are destroyed both from outside (where we have the large chaotic domain) and from inside (from the outer side of the chaotic domain around the unstable orbit 2/5). In Fig. 22 we see that a quite small increase of K from 4.791 to 4.793 has led to the destruction of the noble tori $[2, 1, 1, 2, 1, ...]$ and $[2, 1, 1, 4, 1, ...]$. However the noble torus $[2, 1, 1, 3, 1, ...]$ still exists. Finally a further increase of K to $K = 4.794$ leads to the destruction of this noble torus also, and to the communication of the inner chaotic domain around the unstable orbit 2/5 with the outer chaotic domain. At the critical value of $K = K_c$ the size of the main island decreases abruptly.

As K increases, the size of a torus with given rotation number increases. However at the same time the outermost tori are destroyed. Thus we have two competing influences affecting the size of the island. The size increases as the outer tori expand, but whenever a noble torus is destroyed the size decreases abruptly. The abrupt decrease is more conspicuous when a low order noble torus (i.e. one with $a_i = 1$, for $i > N$ with N=small) is destroyed. E.g. for $K = 4.794$ the last KAM curve is moving from beyond the 2/5 resonance to inside this resonant region.

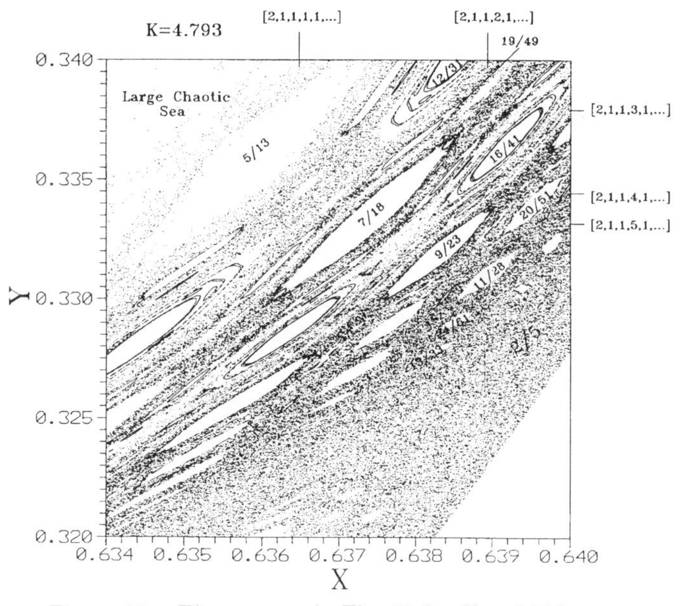

Figure 22. The same as in Fig. 21 for $K = 4.793$.

7. Crossing of a Cantorus

The position of a cantorus can be found by approaching it through periodic orbits. Namely, for each truncation of the continuous fraction representing a noble rotation number correspond a couple of periodic orbits, and the sequence of such periodic orbits tends to the cantorus. E.g. the noble number $a = [2, 4, 1, 1...]$ has as succesive truncations

$$\frac{1}{2}, \frac{4}{9}, \frac{5}{11}, \frac{9}{20}, \frac{14}{31}, \frac{23}{51}, \frac{37}{82}, \frac{60}{133}, \frac{97}{215}, \frac{157}{348}, \frac{254}{563}, \dots \tag{14}$$

These numbers are successively larger and smaller than the noble number a itself. For each truncation there are two periodic orbits. The first is unstable, while the second is stable for relatively small K, becoming unstable as K increases beyond a critical value $K_{n/m}$. Greene (1979) conjectured that the value of $K = K_c$, at which the last KAM curve is destoyed and a cantorus is formed, is such that all periodic orbits close to the cantorus have become unstable.

In a previous paper (Contopoulos et al 1982) we have used this conjecture in deriving the value of K_c by extrapolating the values of $K_{n/m}$ where n/m are the even or odd truncations of the noble number. Namely, we find that the values of $K_{n/m}$ for even tuncation, representing periodic orbits beyond the cantorus (in the present case the numbers $\frac{1}{2}$, $\frac{4}{9}$, $\frac{5}{11}$, $\frac{9}{20}$, $\frac{14}{31}$, $\frac{23}{51}$, $\frac{37}{82}$, approach the value of K_c becoming smaller at each successive approximation. Thus by extrapolating (linearly) the numbers $K_{n/m}$, we find an approximate value of K_c which is smaller than all $K_{n/m}$. The same limit is reached if we extrapolate the values of $K_{n/m}$ for successive odd truncations of the noble number, representing periodic orbits inside the cantorus (in the present case the numbers $\frac{4}{9}$, $\frac{5}{11}$, $\frac{9}{20}$, $\frac{14}{31}$, $\frac{23}{51}$. Again the limit K_c is smaller than all $K_{n/m}$.

For a value of K smaller than K_c all periodic orbits of type n/m are in couples (one stable-one unstable). For a value of K slightly larger than K_c the periodic orbits corresponding to high order truncations of a have become all unstable while orbits further away from the cantorus are still (partly) stable. As K increases the orbits in a larger and larger zone around it, both outside and inside, are only unstable.

In Fig. 23 we see many stable islands close to the cantorus $a = [2, 4, 1, 1, ...]$ for $K = 5$, like 9/20 and 23/51 outside the cantorus, and 14/31 inside the cantorus. As K increases some stable orbits become unstable and the corresponding islands disappear. For K=5 the higher order orbits 37/82, 60/133, 97/215 ... are already unstable.

In order to find the approximate position of a cantorus we find the position of the periodic orbits n/m inside and outside the cantorus. In particular we find the stable periodic orbits and when stable orbits do not exist (very close to the cantorus) we find the unstable periodic orbits that are the continuations of the stable orbits as K increases. In fact it is easier to find stable periodic orbits, than unstable ones, and once we have found them for a value of K smaller than K_c it is possible to continue them for larger K even if they become unstable.

In this way we approach the cantorus, not only as regards its position with respect to the main island but also as regards its gaps. This is seen in Fig. 23, where we see the positions of the points of the periodic orbit 97/215 (squares) inside the cantorus, and of the periodic orbit 157/348 (circles) outside the cantorus. These orbits are so close to each other that most circles overlap with some square. Of course the number and the positions of the circles are not exactly the same as those of the squares, but nevertheless their proximity is remarkable. The higher order approximations are in general close to these squares and circles (e.g. the stars corresponding to the periodic orbit 254/563).

Finally, we want to find how the crossing of a cantorus is realized. Many people have studied the crossing of cantori in maps, by calculating orbits starting on one side of the cantorus until one point of an orbit appears beyond the cantorus. This method is not accurate for two reasons: (1) because the jumps are discontinuous, and (2) because the successive iterates of an orbit go outside and inside the cantorus, but very close to it, for a

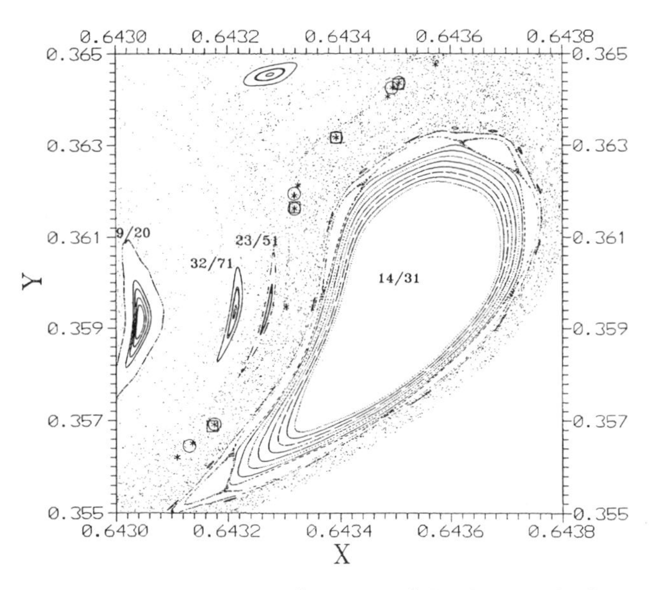

Figure 23. Part of the cantorus $a = [2, 4, 1, 1, ...]$ in the standard map for $K = 5$. The periodic orbits 97/215 (squares), 157/348 (circles) and 254/563 (stars) define the approximate position of the cantorus.

long time before going far away from the cantorus. Thus it is not clear each time whether a crossing has taken place or not.

Our method (Efthymiopoulos et al 1997) consitsts of calculating an unstable asymptotic curve of one unstable periodic orbit inside the cantorus and follow its continuation until it crosses the cantorus, and even farther , until it reaches any given distance outside the cantorus.

In Fig. 24 we show the unstable asymptotic curve starting from an unstable periodic orbit (point 0) as it crosses the cantorus $a = [2, 4, 1, 1...]$ and goes to large distances in the large chaotic domain outside the main island of stability.

The asymptotic curve starts downwards and to the left inside the cantorus (Fig.24a). After making some oscillations in this region (two oscillations in the present case), it moves outwards and crosses the cantorus just above and to the right of 0. Then it makes a number of oscillations, going inside and outside the cantorus, but it continues with longer and longer oscillations outside the cantorus.

As we continue the asymptotic curve further (Fig.24b) , we find that it enters the cantorus several times (see the oscilations close to the segment AB of Figs. 24a,b) but it goes out again to larger distances. A further constinuation of the asymptotic curve (Fig.24c) shows that it extends to so large distances outside the cantorus that it reaches the boundary of the square $(0 < x < 1, 0 < y < 1)$. Later on the asymptotic curve fills densely the whole square except the original island and a dual island symmetric to the first with respect to the center.

However the asymptotic curve comes back to the sticky zone, close and inside the cantorus, an infinite number of times. This eventually leads to an equalization of the density of points in the sticky zone and in the outer chaotic domain, but this equalization requires a very long time.

The present method shows in a continuous way the first crossing of a cantorus and the following crossings.

The form of the lobes of the asymptotic curve gives much information about the diffusion through cantori. First, we see that the lobes are for a long time almost parallel to the cantorus. In fact, the lobes become longer and longer, surrounding the whole island before going to large distances from the island.

Second, the crossing of a cantorus is done many times outwards and inwards before the asymptotic orbit goes very far, in which case the successive crossings become more and more rare.

Third, it is well known that an unstable asymptotic curve cannot cross itself, or other unstable asymptotic curves. Thus the unstable asymptotic curves from other periodic orbits in the same chaotic zone inside the cantorus must follow the oscillations of the above asymptotic curve, before going far away from the cantorus. This implies that different asymptotic curves are very close to each other close to the cantorus and cross the cantorus together. Orbits with initial conditions outside these asymptotic curves have their iterates between the lobes of the original asymptotic curve and similar lobes of other asymptotic curves, therefore these iterates move in the same way inwards and outwards, surrounding the island several times before going far from the cantorus.

As a consequence the phenomenon of diffusion through cantori can be understood correctly only by following the lobes of one unstable asymptotic curve as in Fig. 24. It is not a random diffusion of points outwards from the cantorus, but it is governed by the forms of the lobes of the unstable asymptotic curve.

The lobes, in turn, avoid all islands of stability, both inside and outside the cantorus. Thus, they can go further outside the cantorus only by passing between islands of stability, and at the same time avoiding the unstable asymptotic curves of the unstable periodic orbits outside the cantorus.

The sizes and general forms of the lobes in a homoclinic tangle have been studied numerically in a particular model by Contopoulos and Polymilis (1993). It was found that the low order intersections of the stable and unstable manifolds allow the prediction of the higher order intesections. Thus a quantitative description of the lobes can be made. A further quantitative study of the present case will be given in a future paper.

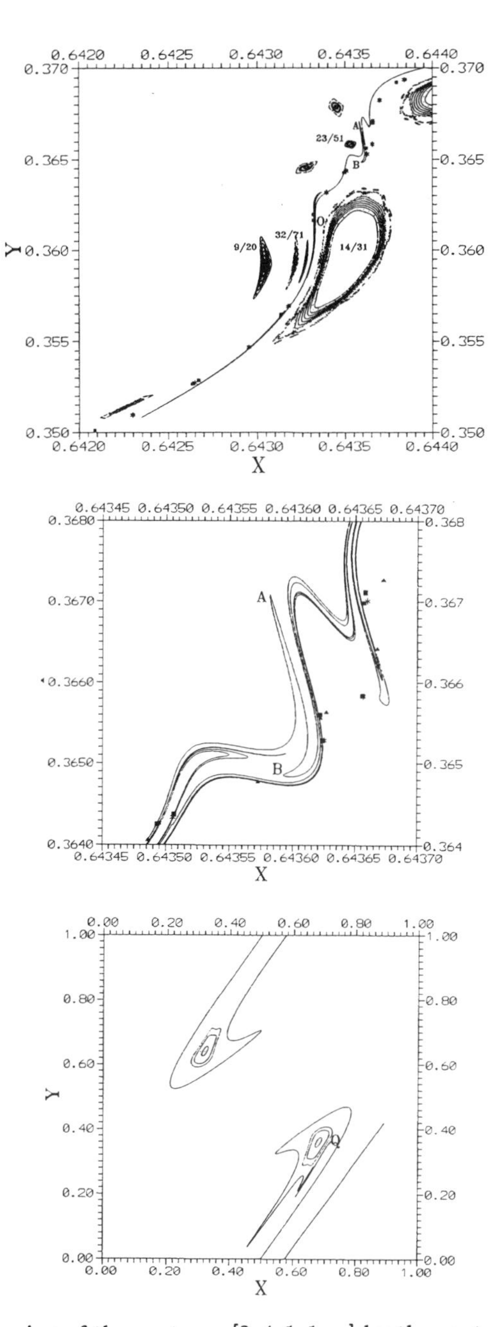

Figure 24. The crossing of the cantorus [2, 4, 1, 1, ...] by the unstable asymptotic curve of the periodic orbit 97/215. (a) The initial part of the asymptotic curve. (b) Oscillations of the asymptotic curve close to the segment AB. (c) Large oscillations of the asymptotic curve at a later time, leading to large distances from the island of stability.

Acknowledgement

This research was supported in part by the Greek National Secretariat for Research and Technology (PENED 293/1995).

References

1. Contopoulos, G. : 1971, *Astron. J.*, **76**, 147.
2. Contopoulos, G. and Polymilis, C. : 1993, *Phys. Rev. E*, **47**, 1546.
3. Contopoulos, G. and Voglis, N. : 1996, *Celest. Mech. Dyn. Astron.*, **64**, 1.
4. Contopoulos, G. and Voglis, N.: 1997, *Astron. Astrophys.*, **317**,73
5. Contopoulos, G., Varvoglis, H. and Barbanis, B. : 1982, *Astron. Astrophys.*, **172**, 55.
6. Efthymiopoulos, C., Contopoulos, G., Voglis, N. and Dvorak, R. : 1997 *J. Phys. A*, (in press).
7. Froeschlé, C., Froeschlé, Ch. and Lohinger, E. : 1993, *Celest. Mech. Dyn. Astron.*, **56**, 307.
8. Fujisaka, H. : 1983, *Prog. Theor. Phys.*, **70**, 1264.
9. Greene, J. M. : 1979, *J. Math. Phys.*, **20**, 1183.
10. Kaufmann, D. E. and Contopoulos, G. : 1996, *Astron. Astrophys.*, **309**, 381.
11. Morbidelli, A. and Giorgilli, A. : 1995, *J. Stat. Phys.*, **78**, 1607.
12. Nicolis, J. S., Meyer - Kress, G. and Haubs, G. : 1983, *Z. Naturforsh*, **38a**, 1157.
13. Patsis, P.A., Efthymiopoulos, C., Contopoulos, G. and Voglis, N. : 1997, *Astron. Astrophys.*, **326**, 493.
14. Shirts, R.B. and Reinhardt, W.P. : 1982, *J. Chem.Phys.*, **77**, 5204.
15. Smith, H. Jr. and Contopoulos G.: 1996, *Astron. Astrophys.*, **314**, 795.
16. Voglis, N. and Contopoulos, G. : 1994, *J. Phys. A*, **27**, 4899.
17. Voglis, N., Contopoulos, G. and Efthymiopoulos, C.: 1997, *Phys. Rev. E* **57**, 372.

ANGULAR DYNAMICAL SPECTRA AND THEIR APPLICATIONS

C. EFTHYMIOPOULOS[1,2], N. VOGLIS[1] AND G. CONTOPOULOS[1,2]
[1] *Department of Astronomy, University of Athens*
[2] *Center for Astronomy and Applied Mathematics, Academy of Athens*

Abstract. We define the invariant spectra of rotation angles and twist angles (angular dynamical spectra) and their moments (angular moments) in a 2D Poincaré map. The angular moments give the main frequencies around a stable fixed point of period 1 or higher. We give examples from the standard map. In particular we find the resonant structure (the location of islands, their frequencies, chaotic zones, and noble tori or cantori) near the last KAM curve.

1. Introduction

The analysis of the fundamental frequencies of orbits on a Poincaré map is an efficient method to obtain information on the overall resonant structure in phase space. An early example of frequency analysis for a 2D galactic Hamiltonian model was given by Contopoulos (1966). Fourier methods have been used for a frequency analysis of various dynamical systems (e.g. Noid et al. 1977, Binney and Spergel 1982). A recent method by Laskar (1990, 1993, Laskar et. al 1992), the NAFF method, has improved the accuracy considerably. Another recently applied method is "Hénon's method" (Léga and Froeschlé 1996).

Here we describe one more new frequency analysis method, applied in 2D maps: the method of the *angular dynamical spectra*. The invariant spectra of dynamical systems were introduced in two recent papers (Voglis & Contopoulos, 1994, Contopoulos & Voglis 1997). The method presented here is based on the spectra of *rotation angles* and *twist angles* (Voglis 1996). We give examples of its application and we compare it with pre-

B.A. Steves and A.E. Roy (eds.), The Dynamics of Small Bodies in the Solar System, 455–462.

vious methods. A detailed theoretical explanation of the method is given elsewhere (Voglis and Efthymiopoulos 1998, hereafter VE).

2. Description of the Method

Consider a two-dimensional area preserving map (e.g. a Poincaré map) of the form:

$$x_{i+1} = F(x_i, y_i, a) \qquad\qquad y_{i+1} = G(x_i, y_i, a) \tag{1}$$

were a is a non-linearity parameter. The corresponding linearized map is given by:

$$dx_{i+1} = \frac{\partial F}{\partial x_i}dx_i + \frac{\partial F}{\partial y_i}dy_i \qquad\qquad dy_{i+1} = \frac{\partial G}{\partial x_i}dx_i + \frac{\partial G}{\partial y_i}dy_i \quad . \tag{2}$$

Let (x_0, y_0) be a period-1 fixed point of the map (1). We can define the position vector of a point $\bar{R}_i \equiv (x_i - x_0, y_i - y_0)$ and the infinitesimal vector $\bar{\xi}_i \equiv (dx_i, dy_i)$. From $\bar{R}_i$ and $\bar{\xi}_i$ we can define two angles: the *rotation* angle $\theta_i = arcsin(\bar{R}_i \times \bar{R}_{i+1})$ and the *twist* angle $\phi_i = arcsin(\bar{\xi}_i \times \bar{\xi}_{i+1})$. The spectra of rotation angles and twist angles are the distributions of these angles after N iterations, namely:

$$S(\theta) = \frac{dN(\theta)}{Nd\theta} \qquad\qquad S(\phi) = \frac{dN(\phi)}{Nd\phi}. \tag{3}$$

The main property of the spectra $S(\theta)$ and $S(\phi)$ is that they are *invariant* along an orbit, regular or chaotic. Furthermore, the spectra of regular orbits are invariant with respect to the initial conditions on the same invariant curve and the spectra of chaotic orbits are invariant with respect to the initial conditions in the same chaotic domain. The invariance of the spectra was found by Voglis and Contopoulos 1994 and Contopoulos and Voglis 1997.

Given the invariance of the spectra $S(\theta)$ and $S(\phi)$ along an orbit, we can define the first spectral moments ν_θ and ν_ϕ as:

$$\nu_\theta = \frac{1}{2\pi}\oint S(\theta)\theta d\theta \qquad \nu_\phi = \frac{1}{2\pi}\oint S(\phi)\phi d\phi \tag{4}$$

The symbol of closed integration refers to the integration with respect to appropriate intervals of definition $[\theta_{min}, 2\pi + \theta_{min}]$ and $[\phi_{min}, 2\pi + \phi_{min}]$ of the angles θ and ϕ, so chosen that the spectra $S(\theta)$ and $S(\phi)$ have no discontinuities due to the 2π modulo.

The relation between ν_ϕ and ν_θ is explained in detail in VE. The vectors $\bar{\xi}_i$, except for the first few transient ones, tend to a unique sequence of slopes, independent of the initial slope. For a regular orbit on an invariant curve the vector $\bar{\xi}$ tends to become tangent to the invariant curve. For a chaotic orbit the vector $\bar{\xi}$ tends to become tangent to the unstable asymptotic curve of the simplest unstable periodic orbit (Contopoulos & Voglis, 1997). If $\bar{\xi}$ becomes tangent to an invariant curve around the center (x_0, y_0), then the number of revolutions of the vector $\bar{\xi}$ around its starting point is equal to the number of revolutions of the end point of the position vector $\bar{R}$ (starting point of $\bar{\xi}$) around the center (x_0, y_0). Thus we get:

$$\nu_\phi = \nu_\theta \quad . \tag{5}$$

On the other hand, the orbits of invariant curves inside higher order islands make an "epicyclic" motion, that is a revolution around (x_0, y_0) and at the same time a revolution around the stable periodic orbit at the center of each island, which is the "guiding center" of the epicyclic motion. In this case the number of revolutions of the vector $\bar{\xi}$ around its starting point is equal to the difference of the number of revolutions of the guiding center around the main center (x_0, y_0) minus the number of revolutions of the starting point of $\bar{\xi}$ around the guiding center. That is we have

$$\nu_\phi = \nu_\theta - \nu_\kappa, \qquad \nu_\theta = n/m \tag{6}$$

were ν_κ is the *epicyclic* frequency, i.e. the frequency of revolution around the guiding center and $\nu_\theta = n/m$ is the rotation number of the guiding center. Eq.(6) can be generalised even further if we have islands inside islands. In this case we obtain:

$$\nu_\phi = \nu_\theta - \nu_{\kappa 1} + \nu_{\kappa 2} - \nu_{\kappa 3} + \ldots \tag{7}$$

where $\nu_{\kappa i}$ are the frequencies of revolution around the islands of each successive order.

We give one example from the well studied standard map (Fig. 1, e.g. Laskar et al. 1992),

$$\begin{aligned} x_{i+1} &= x_i + a\sin(x_i + y_i) \qquad (x,y) \in (-\pi, \pi], \qquad mod(2\pi) \\ y_{i+1} &= x_i + y_i \qquad , \end{aligned} \tag{8}$$

for $a = -1.3$. In Figs. 2a,c we compare the curves $\nu_\theta(x)$ and $\nu_\phi(x)$ for a line of initial conditions joining the center $(0,0)$ to one of the stable fixed points of period 6, $x_6 = 0.64259$ $y_6 = 0.77707$. This line crosses also one of the islands 1/7 and 1/8 further out from the island 1/6. We perform $N = 16384$ iterations per orbit. In Fig. 4b we give also the curve $f(x)$ of the fundamental frequency analysis using an FFT for the same line of initial conditions.

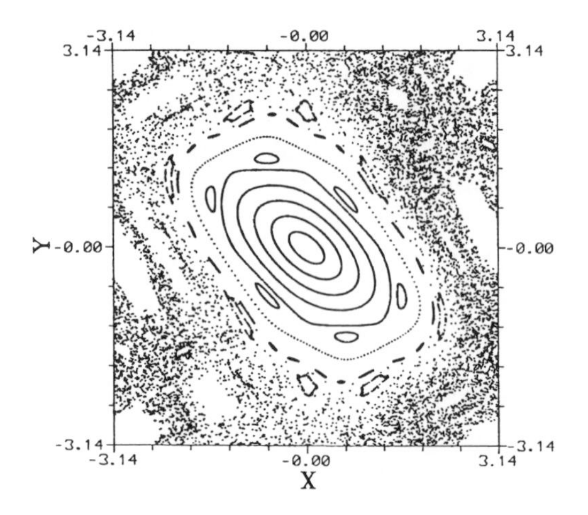

Figure 1. The phase portrait of the standard map for $a = -1.3$.

We observe that the curves $\nu_\theta(x)$ and $f(x)$ coincide, while the curve $\nu_\phi(x)$ coincides with them only in the regions of KAM curves around $(0, 0)$. Inside any island, the curve $\nu_\phi(x)$ has a U-shape. This is the shape of $\nu_\kappa(x)$ shifted by a constant $\nu_\theta = n/m$, the rotation number of the stable periodic orbit of the island.

An additional property of the curves $\nu_\phi(x)$ and $\nu_\theta(x)$ is that they tend to a constant level value in the region where the line of initial conditions crosses the large chaotic domain outside the KAM region of Fig. 1. This indicates that this is a *connected* chaotic domain. Namely, the fact that the spectra $S(\phi)$ and $S(\theta)$ are invariant with respect to the initial conditions in a chaotic domain means that the mean values ν_ϕ and ν_θ must take a constant value throughout the whole domain. However, the noise of ν_ϕ and ν_θ is much larger in the chaotic domain than in the regular domain.

The main advantages of the above method are: a) we can obtain the two frequencies, main and epicyclic, in a single run b) we do not need to specify the centers and multiplicities of particular islands (the stable periodic orbits) since in order to calculate the twist angles we only need to know the evolution of the infinitesimal vector $\bar{\xi}$. The fact that we can always find a value of ν_ϕ for every orbit, independently of the choice of center, is important because in some cases it may not be possible to define an obvious center, e.g. in many cases of chaotic orbits or of irregular periodic orbits which do not bifurcate from a center (Contopoulos 1970).

On the other hand, the accuracy of the above method is of order $O(1/N)$). This is not as high as the accuracy $O(1/N^3)$ of other methods (e.g. NAFF, or Hénon's method), but it is sufficient for many practical applications.

We use now the twist angles method in a difficult example, exploring the

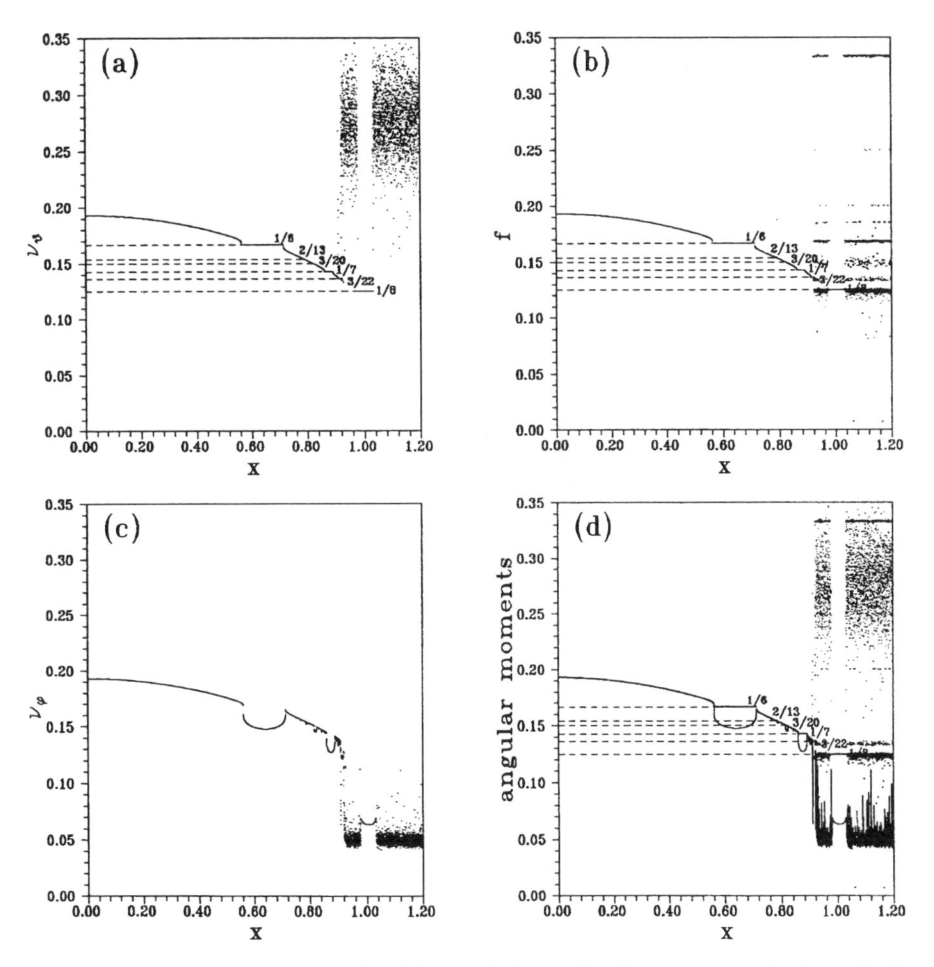

Figure 2. The curves a) $\nu_\theta(x)$, b) $f(x)$, c)$\nu_\phi(x)$ and d) a,b,c superposed, along the line $y = \lambda x$, where $\lambda = y_6/x_6$ is the slope of the line joining the center $(0,0)$ with one of the period 6 stable fixed points ($x_6 = 0.64259, y = 0.77707$).

phase space structure near the boundary of a KAM region. It is well known that the most robust tori, i.e. those which are less easily destroyed as we increase the perturbation a, are those with noble rotation numbers. These numbers can be given in the form of continued fraction approximations, namely:

$$[a_0, a_1, ...] \equiv \frac{1}{a_0 + \frac{1}{a_1 + \cdots}} \quad , \tag{9}$$

with $a_i = 1$ for all i beyond some j (called the order of the noble number). The rational truncations $[a_0, a_1, ..., a_i]$ of a noble rotation number correspond to the rotation numbers of periodic orbits which appear pairwise, stable and unstable, and approach the noble torus from both its sides. These

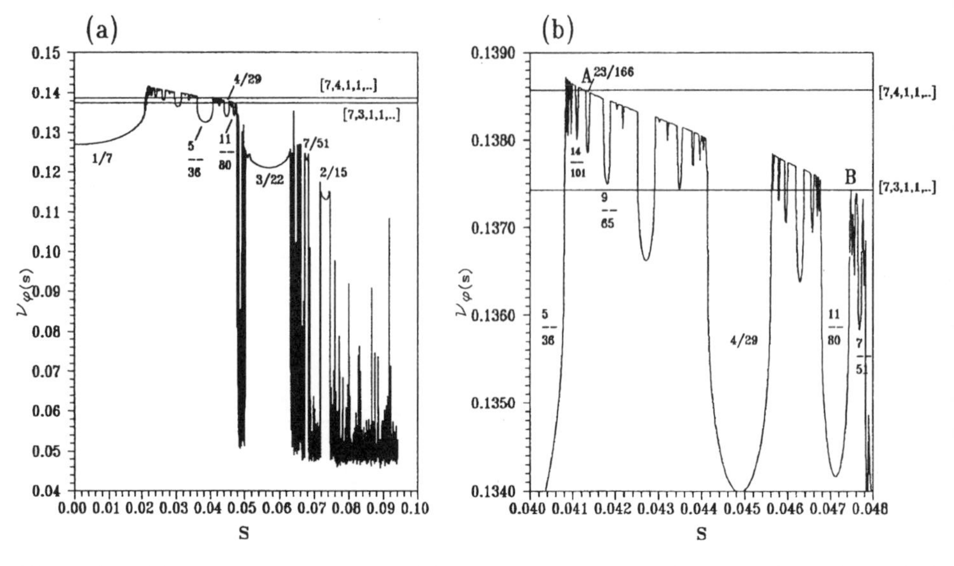

Figure 3. a) The curve ν_ϕ, for $a = -1.3$ as a function of the distance s from the period 7 stable fixed point ($x_{1/7} = 0.91287, y_{1/7} = 1.00521$) along the line joining this point with the period 8 stable fixed point ($x_{1/8} = 1.06497$, $y_{1/8} = 1.11658$). The rotation numbers of some islands are marked, b) a detail of Fig. 3a. The points A and B belong to KAM segments and give the positions of the noble tori [7, 4, 1, 1, ..] and [7, 3, 1, 1, ..] respectively.

periodic orbits create resonant zones which grow in size as the perturbation increases and eventually overlap. Then the noble torus is destroyed and a cantorus is formed. The cantorus has an infinite number of gaps, allowing communication between the various chaotic zones.

The question now is at which value of the perturbation does a given noble torus turn into a cantorus. To explore this question, we consider the boundary of the central KAM region of Fig. 1. We take a line of initial conditions passing through the centers of one of the 1/7 islands ($x = 0.91287$, $y = 1.00521$) and one of the 1/8 islands ($x = 1.06497, y = 1.11658$). For $a = -1.3$, the 1/7 island is inside the last KAM curve, while the 1/8 island is outside this KAM curve, in the large chaotic sea. In Fig. 3a we give the curve $\nu_\phi(s)$ (s is the distance from the fixed point $x_{1/7}, y_{1/7}$) along this line, for $N = 10^5$ iterations per orbit, and scanning step $ds = 10^{-5}$.

A large chaotic domain for s greater than $s \approx 0.047$ is distinguished in Fig. 3a. In this domain, the moment ν_ϕ tends to a value $\nu_\phi \approx 0.055$ constant throughout the whole domain, which indicates that this chaotic domain is connected . Some islands of stability, e.g. 3/22, 7/52 and 2/15 are embedded in this chaotic domain. On the other hand, for s smaller

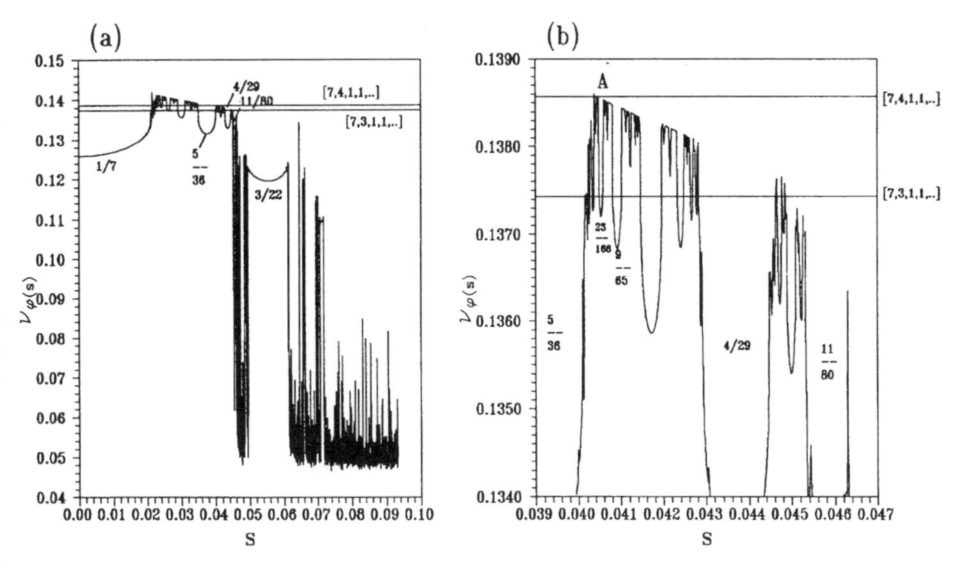

Figure 4. a) Same as Fig. 3a, for $a = -1.31$, b) Same as Fig.3b for $a = -1.31$. The line of the noble $[7, 3, 1, 1, ..]$ has no intersection with KAM segments of the curve ν_ϕ to the right of the island 11/80.

than about 0.047 the curve $\nu_\phi(s)$ consists of a) segments that belong to a smoothly decreasing curve, corresponding to KAM tori (KAM-segments), and b) U-shaped parts corresponding to islands. The rotation numbers of the islands form Farey sequences tending to noble numbers. For example, the islands 1/7, 4/29, 5/36, 9/65, 14/101, 23/166 (Fig. 3b) belong to the Farey sequence tending to $[7, 4, 1, 1, ...] = 0.1385705161099319$. Similarly, the islands 1/7, 3/22, 4/29, 7/51, 11/80, 18/131, belong to a sequence tending to $[7, 3, 1, 1, ...] = 0.1374307259386117$. The straight lines corresponding to the noble numbers $[7, 4, 1, 1, ...]$ and $[7, 3, 1, 1, ...]$ in Fig. 3b intersect the curve ν_ϕ at the points A and B belonging to KAM segments. These points mark the positions of the noble tori $[7, 4, 1, 1, ...]$ (point A) and $[7, 3, 1, 1, ...]$ (point B; a magnification of Fig. 3b shows that this point belongs to a small KAM segment). Both these tori exist for $a = -1.3$. But for $a = -1.31$ the connected chaotic domain has moved further to the left, including the island 11/80 (Fig. 4a). If the noble torus $[7, 3, 1, 1, ...]$ to the right of the island 11/80 existed, then it would not allow chaos to penetrate to the left of this island. Thus the noble torus $[7, 3, 1, 1, ...]$ is destroyed for $a = -1.31$. As a result, in Fig. 4b we see that there is no intersection of the line $[7, 3, 1, 1, ...]$ with a KAM segment.

In the same way we can find the consecutive destruction of all noble tori at higher values of the perturbation a.

Acknowledgements This research was supported in part by the Greek National Secretariat for Research and Technology (PENED 293/1995) and the Research Committee of the Academy of Athens under project 200/409. C.E. was supported by the Greek Foundation of State Scholarships (I.K.Y.).

References

1. Binney, J. and Spergel, D.: 1982, Astrophys. J. 252, 308.
2. Contopoulos, G.: 1966, in Nahon, F. and Hénon, M. (eds) "Les Nouvelles Méthodes de la Dynamique Stellaire" CNRS, ≡ Paris Bull. Astron. Ser. 3, 2, Fasc.1, 223.
3. Contopoulos, G.: 1970, Astron. J., 75, 96.
4. Contopoulos G., Voglis N., 1997, A&A 317,73.
5. Lega, E., and Froeschlé, C. : 1996, Physica D (in press)
6. Laskar, J.: 1990, Icarus 88, 266.
7. Laskar, J.: 1993, Physica D67, 257.
8. Laskar J., Froeschlé C., Celleti A., 1992, Physica D 56, 253.
9. Noid, D.W., Koszykowski, M. L. and Marcus, R.A.: 1977, J. Chem. Phys. 67, 404.
10. Voglis N., Contopoulos G., 1994, J.Phys. A 27, 4899.
11. Voglis N., 1996, Oral Presentation, Human Capital and Mobility Workshop, Santorini, October 17-20.
12. Voglis N., Efthymiopoulos C., 1998, "Angular Dynamical Spectra. A New Method For Determining Frequencies, Weak Chaos, and Cantori", J. Phys. A: Math. Gen. 31, 2913.

WEAK CHAOS AND DIFFUSION IN HAMILTONIAN SYSTEMS

From Nekhoroshev to Kirkwood.

CLAUDE FROESCHLÉ
Observatoire de la Cote d'Azur, CNRS/ UMR 6529,
Bd. de l'Observatoire, BP 4229,
06304 Nice, France

AND

ELENA LEGA
Observatoire de la Cote d'Azur, CNRS/ UMR 6529,
Bd. de l'Observatoire, BP 4229,
06304 Nice, France
CNRS-LATAPSES, 250 Rue A. Einstein,
06560 Valbonne, France

Abstract. Recent theoretical results have shown that the stability of a dynamical system is strictly related to the density of tori which foliate the phase space. Therefore, the numerical check of the dynamical behaviour of a given domain requires tools of analysis which should be as sensitive as the classical Lyapunov characteristic exponents but cheaper in computational time, in order to be applied to a large number of orbits. We define the different methods of analysis and we compare them using the two and four dimensional Standard map as a model problem. We apply such tools for a fine study of the vicinity of an invariant KAM torus. We then study standard map like mappings showing the relationship between the Lyapunov times and macroscopic instability times. The results obtained applied to the distribution of asteroids can help in solving some puzzling problems such that of "stable" chaos or that of the Kirkwood gaps.

1. Introduction

The now classical result of Nekhoroshev (1977), revisited by Morbidelli and Giorgilli (1995a), shows that the old and crucial question for the stability of a dynamical system is closely related to the structure and density of invariant tori which foliate the phase space. The motion in a given region of the phase space either exhibits a macroscopic diffusion due to overlapping

B.A. Steves and A.E. Roy (eds.), The Dynamics of Small Bodies in the Solar System, 463–502.

of resonances, i.e. Chirikov regime, or an exponentially slow diffusion which is very difficult to detect numerically. However, in this last case there must exist many invariant tori in the non resonant domain whose borders are nothing but the resonant lines which constitute the so called Arnold web. Conversely, for the Chirikov regime, the majority, if not the totality, of the invariant tori have disappeared.

In order to check in which regime is a given portion of the phase space, it is necessary to compute at least the largest Lyapunov exponent for a large number of orbits. But this could take a huge amount of time in particular for weakly chaotic orbits. Therefore it is interesting to define new methods of analysis which should be as sensitive to weak chaos as the LCEs but cheaper in computational time. Different methods have been developed at this purpose: the frequency map analysis (Laskar et al. 1992, Laskar 1993, Lega & Froeschlé 1996), the sup-map method (Laskar 1990, Froeschlé & Lega 1996), the fast Lyapunov indicator (Froeschlé et al. 1997a,1997b) and the twist angle method (Contopoulos and Voglis 1997).

We define and compare the different methods using the standard map as a model problem and we show how, the application of such tools, has allowed (Lega & Froeschlé 1996) to check numerically the properties outlined by the theoretical description given by the Morbidelli-Giorgilli theorem. We then study standard map like mappings showing the relationship between the Lyapunov times and macroscopic instability times (Morbidelli & Froeschlé 1996). The results obtained, applied to the distribution of asteroids, can help in solving some puzzling problems, such that of "stable" chaos (Milani & Nobili 1992) or that of the Kirkwood gaps. For instance, the numerical discovery of Milani and Nobili (1992) about asteroids exhibiting stable chaos in the main belt could be considered as indicating that the main belt is in a Nekhoroshev regime. In the same spirit, the 2/1 gap seems to correspond to a weak chaotic region and therefore it isn't in a Nekhoroshev regime.

In Section 2 we describe the set of tools developed for detecting strong (Section 2.1) and weak chaos (Section 2.2). In particular, in the case of weak chaos we give a simple example of application of the different tools using the two-dimensional Standard map as a model problem. The comparison between all the methods, and the test of their sensitivity, is made in Section 2.3 using the two and the four dimensional standard map. Section 3 is devoted to the applications. In Section 3.1 we have explored the fine structure of the phase space in the vicinity of the golden torus using the methods explained in Section 2.2, while in Section 3.2 we show the different dynamical behaviours which characterize an Hamiltonian system. In Section 3.3 an application to the asteroidal belt is given.

2. Tools for detection of chaos

2.1. THE CASE OF STRONG CHAOS

It is well known that for non-linear Hamiltonian systems there exist ordered regions with quasi-periodic orbits and regions with chaotic orbits. As far as strong chaos is concerned the Poincaré's surface of section is a performant tool to visualize the existence of chaotic orbits at least for systems with two degrees of freedom (Hénon and Heiles 1964, Hénon 1969). In this case, a measure of the divergence of nearby orbits allows to clearly distinguish between regular and chaotic motion. Such a measure is usually given by the Lyapunov Characteristic Exponents (hereafter LCEs). The reader can refer to Benettin et al. (1980) and to Froeschlé (1984) for a detailed review.

We recall in this Section the definition and the basic properties of both methods used for detecting strong chaos.

2.1.1. *Poincaré maps and mappings*

As it is well known, there are many reasons to use Poincaré maps to study dynamical systems instead of a full set of differential equations (Hénon 1981). First of all the study of an n-dimensional system is reduced to an $(n-1)$-dimensional space without loosing any of the properties of the full system. Then, when an explicit form of the mapping is known, a simple procedure of iteration of this form allows to get orbits, which, thanks to the speed of computation (in comparison to the integration of the system of differential equations) can be followed over a very long time. For the same reason a lot of orbits can be studied together and global properties of the system can be obtained up to very fine details. Therefore, mappings have been widely used to study fundamental issues in many domains like plasma physics, accelerator dynamics as well as celestial mechanics.

When the problem concerns general properties of dynamical systems, some well known maps are used like a real numerical laboratory. Let us just recall the standard map (Froeschlé 1970, Chirikov 1979), the Hénon's quadratic map (Hénon 1969) or the Hénon's non-symplectic map (Hénon 1976).

Since all the examples and results of this paper are based on the Standard map we recall its formulation. Let us consider the Hamiltonian pendulum:

$$H = \frac{x^2}{2} - \cos y \tag{1}$$

where x and y are conjugate variables. We can associate to the integrable system:

$$A = \left\{ \begin{array}{lclcl} \dot{y} & = & \frac{\partial H}{\partial x} & = & x \\ \dot{x} & = & -\frac{\partial H}{\partial y} & = & -\sin y \end{array} \right. \tag{2}$$

the mapping T_1:

$$T_1 = \begin{cases} y_1 &= y_0 + x_0 \Delta t \\ x_1 &= x_0 - \Delta t \sin y_0 \end{cases} \tag{3}$$

which is nothing but the Euler method to compute orbits of ordinary differential equations. If we consider (Froeschlé and Lega 1995) the phase space diagram obtained with T_1, even with a small value of Δt, the system appears slowly expanding. If we compute the determinant of the Jacobian matrix J we obtain: $|J| = 1 + \Delta t^2 \cos y_0$ which is not equal to one and therefore the mapping is not an area preserving one, as we would expect since the flow under study is Hamiltonian.

Let us make a slight change in the mapping and consider instead the mapping T_2:

$$T_2 = \begin{cases} y_1 &= y_0 + x_0 \Delta t \\ x_1 &= x_0 - \Delta t \sin y_1 \end{cases} \tag{4}$$

Again if $\Delta t \to 0$, $y_1 \to y_0$, $x_1 \to x_0$ and we obtain dividing by Δt the system of equations A i.e. $\lim \frac{y_1 - y_0}{\Delta t} = x_0$ gives $\dot{y} = x$ etc.

The determinant of the Jacobian matrix is now equal to 1 and even for large values of Δt ($\Delta t \simeq 0.5$) the phase space diagram is in agreement with that of the pendulum. The slight and powerful change of T_1 in order to obtain T_2 is the result of a search for a symplectic map. The Hamiltonian H can be considered as $H = H_1 + H_2$ with $H_1 = x^2/2$ and $H_2 = -\cos y$. The Euler method applied respectively to H_1 and H_2 gives rise to the area preserving mappings:

$$T_2' = \begin{cases} y_1 &= y_0 + x_0 \Delta t \\ x_1 &= x_0 \end{cases} \tag{5}$$

and:

$$T_2'' = \begin{cases} y_2 &= y_1 \\ x_2 &= x_1 - \sin y_1 \Delta t \end{cases} \tag{6}$$

The composite mapping $T_2 = T_2' \circ T_2''$ is also area preserving.

For large values of Δt ($\Delta t \simeq 1$) all the well known features of non-linear area preserving mapping (i.e. islands, chaotic zones etc.) appear. Therefore introducing the non-linearity parameter ϵ (Froeschlé 1970) we get the Standard map T:

$$T = \begin{cases} x_{i+1} &= x_i + \epsilon \sin(x_i + y_i) \quad (\mathrm{mod} 2\pi) \\ y_{i+1} &= x_i + y_i \quad (\mathrm{mod} 2\pi) \end{cases} \tag{7}$$

Another formulation of the Standard map is obtained considering the following Hamiltonian (Chirikov 1979):

$$H = \frac{x^2}{2} + K_0 \cos y + \sum_{n \neq 0} K_n(x) cos(y - nt) \tag{8}$$

If the constants K_n are small, then the Hamiltonian of the pendulum gives a good approximation to the system using the averaging principle. However, this procedure is no longer valid near the separatrix which is replaced by a narrow chaotic band when high-frequency terms are present. Chirikov modified the high-frequency terms considering a new Hamiltonian:

$$H_C = \frac{x^2}{2} + K_0 \cos y + K_0 \sum_{n \neq 0} cos(y - nt) \tag{9}$$

which can be considered closer to H than the Hamiltonian of the pendulum since the new high-frequency terms allow chaos. Using the Fourier transform of the Dirac δ function, H_C becomes:

$$H_C = \frac{x^2}{2} + K_0 \cos y \delta(t) \tag{10}$$

Then using the property that the delta function acts instantaneously the standard map is obtained by integration (see Lichtenberg and Lieberman 1983):

$$T = \left\{ \begin{array}{lcll} x_{i+1} & = & x_i + k_0 \sin(y_i) & (x \in \Re) \\ y_{i+1} & = & y_i + x_{i+1} & (\mathrm{mod} 2\pi) \end{array} \right. \tag{11}$$

2.1.2. *The Lyapunov characteristic exponents*

As already said the Lyapunov Characteristic Exponents measure the divergence of nearby orbits. We recall here the definition of the largest LCE in the case of a mapping. The reader can refer to Froeschlé (1984) to get the definition of the whole set of the LCEs. We consider the mapping T, in an m-dimensional phase space, and the corresponding tangent mapping defined as follows:

$$\left\{ \begin{array}{lcl} \vec{X}_{n+1} & = & T\vec{X}_n \\ \vec{V}_{n+1} & = & (\frac{\partial T}{\partial \vec{X}_n})\vec{V}_n \end{array} \right. \tag{12}$$

We iterate simultaneously these two mappings,taking as initial conditions a point $\vec{X}_0$ and a vector $\vec{V}_0$. The largest LCE of an initial point $\vec{X}_0$ is defined as:

$$\chi(\vec{X}_0) = \lim_{n \to \infty} \frac{1}{n} \sum_{j=1}^{n/h} \ln ||\vec{V}_j|| \tag{13}$$

where the vector $\vec{V}_j$ is renormalized to one at regular intervals of time $\Delta t = h$. In practice, the computation is done on a finite number of iterations and usually we call Lyapunov Characteristic Indicators (hereafter LCIs) the truncated values of the LCEs for a finite time n. In the case of strong chaos, it is not a problem to reveal the chaotic character of an orbit, since the largest LCI quickly stops to decrease and reaches a positive value. The real difficulty is to discriminate between a regular orbit and a weak chaotic one. In fact, in this case it can take a very long time to reveal the chaotic character and an arbitrary truncation of the computation can give misleading results. Therefore, although the theory of LCEs has a solid mathematical background, in order to overcome the computational difficulties which can arise, in particular for the case of weak chaos, the search of other indicators is still under development.

To this purpose, let us remark that the LCEs can be seen as the moment of order one of the distribution of the norms:

$$\chi_j = \ln(||\vec{V}_j||) \tag{14}$$

Froeschlé et al. (1993) have studied the properties of the set of χ_j's, called by them distribution of the Local Lyapunov Characteristic Indicators (LLCIs) (sometimes called stretching parameters, Contopoulos 1995).) They showed that the LLCIs allow to reveal the complex structure of the chaotic zone, recovering some of the informations that are lost in the computation of the LCIs.

2.2. THE CASE OF WEAK CHAOS

We summarize in this Section all the principal indicators used for a fine identification of large sets of orbits. In order to get the reader familiar with the different tools we will show for each one a simple "*cas d'école*" using the two dimensional Standard map as a model problem. We then compare the different methods and we test their sensitivity exploring the vicinity of an hyperbolic point associated to a very thin chaotic layer. An extension to the four dimensional standard map is given at the end of the Section.

2.2.1. *Fast Lyapunov indicators*

As we said in the previous section the computation of the LCEs may take a large amount of time in the case of weak chaos: practically much more than 6 to 10 times the Lyapunov time (the inverse of the LCE) which is considered as the time necessary to reach the LCE but not the time necessary to check the convergence.

Therefore, it is interesting to define new methods to detect weak chaos. Recently Froeschlé et al. (1997a,1997b) have defined an indicator strictly related to the largest LCE: the fast Lyapunov indicator (hereafter FLI). As explained in Section 2.1.2 when computing the LCEs a renormalization procedure is applied at given intervals of time on the tangential vectors evolving with the flow. This procedure is applied to avoid overflow of the lengths of the vectors in the case of a chaotic orbit. Curiously, a systematic study and use of the time at which such renormalization is necessary seems to have not been done yet.

In the present approach, the time necessary to reach a given value, either of the length of any vector or of the angle between vectors, is taken as an indicator of stochasticity.

It is well known that any vector will evolve under the action of the largest characteristic exponent. However, a transitory regime could occur if the initial vector is almost perpendicular to the expanding manifold. We have tested three different indicators (Froeschlé et al. 1996) and we have taken as fast Lyapunov indicator the one which is less dependent on the initial conditions. Starting with a p-dimensional basis $\vec{V}_p(0) = (\vec{v}_1(0), \vec{v}_2(0), ...\vec{v}_p(0))$, embedded in an n dimensional space, and with an initial condition $P(0) = (x_1(0), x_2(0), ...x_n(0))$, we take at each iteration the largest among the vectors of the evolving basis. We define the indicator ψ_3 as follows:

$$\psi_3(t) \quad = \quad \sup_j ||\vec{v}_j(t))|| \quad j = 1, ...p \tag{15}$$

Let us remark that for a high dimensional system the computations of all vectors can take a large amount of time and the method would no longer be a cheap one. In fact, we can find the supremum between all the vectors within a short time, for example 10^2 iterations, and than just follow its evolution as we have found that it remains the greatest vector.

Fig.1 shows a set of orbits of the standard map for $\epsilon = 0.8$. For a cross section of the x-axis at $y = 0$ we have computed the time $N(\gamma)$ necessary for ψ_3 to reach a threshold value of $\gamma = 10^{10}$. The total number of iterations is $N = 20\,000$. Of course if the threshold is not reached we set $N(\gamma) = N$. In order to easily visualize the kind of informations given by this indicator, we have superposed the variation of $N(\gamma)$ as a function of x to the plot of the corresponding orbits (Fig.1). The lower value of $N(\gamma)$ correspond to the chaotic orbits, while the whole set of invariant curves, rotational tori as well as libration islands, has $N(\gamma) = N$. The values of $N(\gamma)$ have been renormalized to the interval $[0 : \pi]$ in order to be visualized in the phase space.

2.2.2. *Frequency map analysis*

The method Introduced by Laskar (1988) to understand the long time evolution of the Solar system, this analysis was then successfully used in

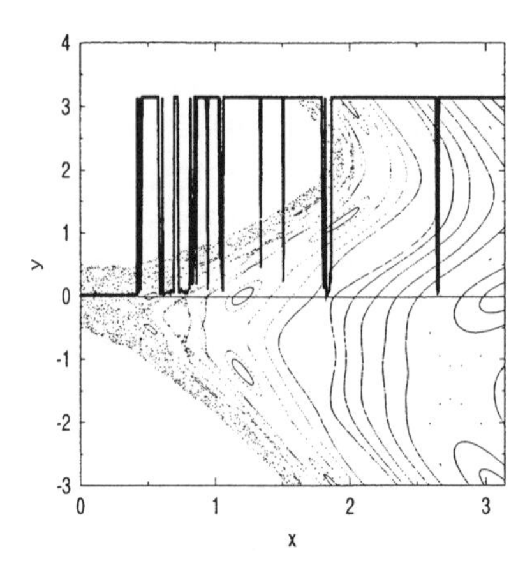

Figure 1. A set of orbits of the standard map for a value of the non linearity parameter $\epsilon = 0.8$. Variation of the time $N(\gamma)$, necessary for the Fast Lyapunov Indicator to reach a threshold value of $\gamma = 10^{10}$, for a set of 1 000 orbits on a cross-section of the x-axis for $y = 0$ (continuous line). The values of $N(\gamma)$ have been renormalized to the interval $[0 : \pi]$ in order to be visualized in the phase space.

two-dimensional mappings to determine the critical value for which the last invariant torus disappear (Laskar et al. 1992), as well as for the study of global dynamics and diffusion in multi-dimensional systems (Laskar 1993) and for the long time diffusion in particle accelerator dynamics (Dumas & Laskar 1993). More recently it has been used to explore the structure around an invariant KAM torus (Celletti & Froeschlé 1995, Lega & Froeschlé 1996).

The basic idea of the frequency map analysis is to obtain directly, in a numerical manner, the quasi-periodic approximation of the solution of an n-dimensional quasi-integrable Hamiltonian system (and therefore the set of associated frequencies ν_i, $i = 1, ..n$) without searching for an explicit change of coordinates in action angle variables.

Different methods have been developed in order to compute the frequencies associated to the quasi-periodic motion (Laskar 1988, Laskar et al. 1992, Hénon, private communication). We present them separately below in order to focus now the attention on the definition and use of the frequency map in the case of the standard map.

Let us suppose to have computed the quasi-periodic approximation for a given non-degenerated Hamiltonian system with n degrees of freedom. Following Laskar (1988, Laskar et al. 1992) we define the frequency map F as the application which associates to a vector of initial action-like variables $I_j(0)$, $j = 1, .., n-1$ the frequency vector ν_j, $j = 1, .., n$ as follows:

$$F: \quad (I(0)_1, I(0)_2, ...I(0)_{n-1}) \rightarrow (\frac{\nu_1}{\nu_n}, \frac{\nu_2}{\nu_n}, ..., \frac{\nu_{n-1}}{\nu_n}) \tag{16}$$

The angles are fixed to arbitrary values $\theta_j(0) = \theta_{j0}$ and the last action I_n is determined by the condition: $H(I, \theta) = h$.

For example, for a Hamiltonian system with two degrees of freedom the motion take place on a 3 dimension space, the surface of section restrict the study to a 2 dimension space and the frequency map further restrict the dynamics of the system to a 1 dimension space: $F: \ I_1 \to \nu = \nu_1/\nu_2$. This reduction can be very useful for the study of higher dimension problems.

One important remark is that, although frequencies are defined only on invariant tori, the frequency analysis algorithm compute numerically the frequency vector for any initial condition. On the KAM tori, this frequency vector will be a very accurate approximation of the actual frequencies, while in the weakly chaotic regions, it will provide a natural interpolation between these fixed frequencies.

As example, we consider the same set of orbits of the standard map studied for the FLI, i.e. a cross-section of the x-axis at $y = 0$. The FMA associates to each orbit its corresponding frequency $\nu = \nu_1$ ($\nu_2 = 1$). We can clearly see (Fig.2) that noisy variations of the frequency correspond to chaotic regions, while the set of rotational tori are revealed by a monotonic variation of ν and the crossing of islands is identified by a constant value of the frequency. Again a renormalization is applied in order to rescale the

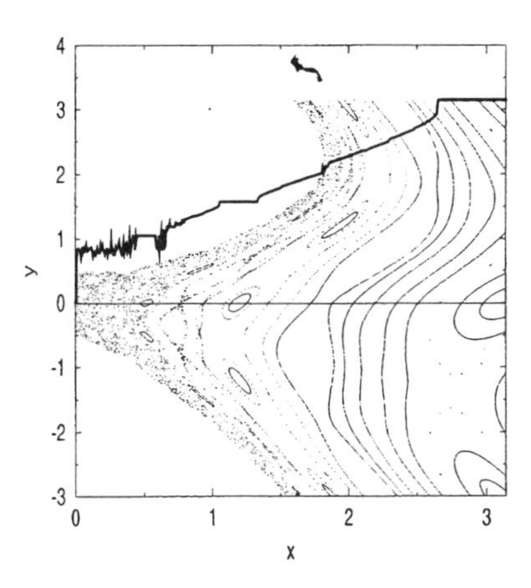

Figure 2. Same as Fig.1 for the frequency map analysis.

frequency in the domain of $[0 : \pi]$ of the phase space.

Computation of frequencies

Laskar's method The numerical analysis of the fundamental frequency (hereafter NAFF) developed by Laskar (1988,1990, Laskar et al. 1992) relies in a sophisticated use of the Fourier transform.

Since all the examples given in this paper are based on the Hénon's method we don't enter here in the detail of the Laskar's method (the reader can find a detailed explanation in particular in Laskar et al. 1992).

Hénon's method and Continued Fraction method As far as two-dimensional mappings are concerned a very easy and quick method of computation was developed by Hénon (private communication) and then used and extended by Celletti and Froeschlé (1995), Lega and Froeschlé (1996) and Celletti et al. (1996).

Given a mapping M we consider the $n-th$ iterate point $P_n = M^n(P_0) \equiv (x_n, y_n)$. Assuming the existence of invariant curves we then consider the change of variables:

$$\left\{ \begin{array}{lcl} x & = & f(X,Y) \\ y & = & h(X,Y) \end{array} \right. \tag{17}$$

such that the transformed mapping is a simple rotation around the origin by an angle $\alpha \equiv \alpha(R)$, depending on the radius R. In the new variables (X, Y) we obtain the sequence of points $Q_n \equiv (R\cos n\alpha, R\sin n\alpha)$ and the corresponding coordinates of P_n:

$$\left\{ \begin{array}{lclcl} x_n & = & f(R\cos n\alpha, R\sin n\alpha) & \equiv & F(n\alpha) \\ y_n & = & h(R\cos n\alpha, R\sin n\alpha) & \equiv & H(n\alpha) \end{array} \right. \tag{18}$$

We iterate now N times the mapping M. Following Hénon (private communication), we start to compute the fundamental frequency (also called rotation number): $\nu = \alpha/2\pi$, by selecting, among the set $(P_1, ...P_N)$, the two points which are the successively closer to P_0 and we denote by n_1 and n_2 their indices. We define the integers p_1,p_2 by the relations:

$$\left\{ \begin{array}{lcl} n_1\nu & = & p_1 + \epsilon_1 \\ n_2\nu & = & p_2 + \epsilon_2 \end{array} \right. \tag{19}$$

where the ϵ_i are small quantities. In other words the p_i's count the number of revolutions around the invariant curve. With this setting we have $x_{n_i} = F(\epsilon_i)$, $y_{n_i} = H(\epsilon_i)$ and we can expand y_{n_i} such as:

$$y_{n_i} = H(0) + H'(0)\epsilon_i + \tfrac{1}{2}H''(0)\epsilon_i^2 \quad i = 1, 2 \tag{20}$$

Keeping only the first order terms we have:

$$y_{n_i} \quad = \quad H(0) + H'(0)\epsilon_i \quad i = 1, 2 \tag{21}$$

Combining (19) and (21) by elimination of ϵ_i we get:

$$\nu \equiv \nu(y_0) = \frac{p_1(y_{n_2} - y_0) - p_2(y_{n_1} - y_0)}{n_1(y_{n_2} - y_0) - n_2(y_{n_1} - y_0)} \tag{22}$$

Examining the sequence of the results it turns out that the strength of the method lies in the fact that the sequence of points is exactly that of the points closest to P_0. We have in fact the following sequence of inequalities: $\epsilon_1 > \epsilon_2 .... > \epsilon_k$, i.e.:

$$|n_1\nu - p_1| > |n_2\nu - p_2| > ... > |n_k\nu - p_k|$$

where the n_i's are the smallest integers giving such a sequence of inequalities. Taking $n_1 = 1$, all the conditions which insure that the sequence p_k/n_k is nothing but the development in continued fractions of ν, are satisfied. We know from a theorem of Lagrange that such a development gives the best approximation of the number ν.

We also know that: $\epsilon_k < 1/n_k n_{k+1}$ and $n_{k+1} > a_k n_k + n_{k-1}$ where a_k is the $k-th$ term of the development in continued fractions. These two relations allow us to get an estimate of the precision, or in other words, of the price we have to pay in order to obtain the next closest point to P_0.

We have therefore evaluated the rotation number by the simple formula: $\nu = p_k/n_k$ (hereafter CFM for continued fraction method). We have shown (Lega and Froeschlé 1996) that the Hénon's and the CFM methods converge to the same value, and while the Hénon's gives a better approximation as far as the first terms are concerned, the CFM one allows a correct determination of the precision thanks to the properties of the development in continued fractions.

For an extension of the Hénon's method of computation to a 4 dimensional mapping the reader can refer to Celletti et al. (1996).

Twist angle method Another indicator of stochasticity, which turned out (Froeschlé and Lega 1997) to be strictly related to the fundamental frequency ν, is the twist angle (Contopoulos and Voglis 1997). Following Contopoulos and Voglis (1997) the twist angle $\Delta\phi_n$, for the particular case of a two dimensional mapping, is equal to the difference of two consecutive angles:

$$\Delta\phi_n = \phi_{n+1} - \phi_n \tag{23}$$

where ϕ_n, called by the authors "helicity angle", is the angle formed by a vector $\vec{V}_n$ with a fixed direction, say the x-axis. In the case of invariant curve, after a transient regime the image of any initial vector $\vec{V}_0$ becomes tangent to the orbit, therefore ϕ measures the deviation of the orbit from a fixed direction.

The distribution of the twist angles is shown to be invariant in the chaotic domain and depending on initial conditions for invariant orbits (Contopoulos and Voglis 1997). Therefore, the average value $\langle \Delta\phi \rangle_N = \sum_{i=1}^{N} \phi_i / N$, after a given number N of iterations, is quite the same for the chaotic domain while changes smoothly for invariant curves. In particular the authors apply their method to distinguish between invariant and chaotic orbits.

We have shown (Froeschlé and Lega 1997) that this indicator not only is able to distinguish between invariant and chaotic orbits, but even between islands and tori, which are both invariant curves. In fact the average twist angle is almost zero for a rotational torus while it has a non zero value in the case of islands. In particular, in this case the value of $\langle \Delta\phi \rangle_N$ is very similar to the frequency computed with respect to an observer situated in the center of the island. In fact, the frequency of rotation ν around the center of an island is the mean value of the angles formed by the vector $\vec{X}_{n+1}$ with $\vec{X}_n$ $\forall n = 1, ...N$ and $\langle \Delta\phi \rangle_N$ is the mean value of the angles formed by the tangent vector $\vec{V}_{n+1}$ with $\vec{V}_n$, i.e. $\nu \simeq \langle \Delta\phi \rangle_N$ (we would have $\nu \equiv \langle \Delta\phi \rangle_N$ for a pure rotation). Let us remark that, in order to get the fundamental frequency of a rotational torus with this method it is necessary to consider the system in polar coordinates.

In the case of islands we have obtained exactly the values of the frequency coupling the computation of the twist angle with the continued fraction method of computation of ν (Froeschlé and Lega 1997).

We stress the fact that, for a chain of islands, in order to obtain the same result with the previous methods of computation of the frequency, we would have to search for the center of one island and then to compute the frequency with respect to the new frame of reference. Moreover we have shown that $\langle \Delta\phi \rangle$, for secondary chain of islands, gives the composition of the rotation around the first order island with the rotation around the center of the secondary order islands.

We can see in Fig.3 that the crossing of chaotic regions is revealed by a noisy variation of the twist angle, while zero values correspond to tori and the "hat-shaped" structures give the frequency of rotation around the center of islands.

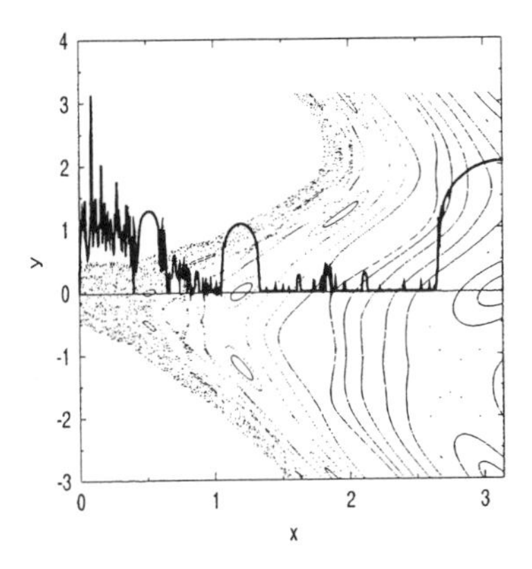

Figure 3. Same as Fig.1 for the average of the twist angles.

2.2.3. *Sup-map analysis*

The idea of the sup-map analysis (Laskar 1990, Froeschlé & Lega 1996) came from the observation that there is a strong correlation between the fundamental frequency and the variation of the action, i.e. the same monotonic variation appears when considering, instead of the frequency as a function of x, the sup of the action as a function of x. Actually, for the particular case of the standard map (Eq.7) we use the fact that:

1. The horizontal line $y = 0$ is intersecting the invariant KAM curves only once.
2. If two initial conditions are such that $x_1 < x_2$ and x_2 lead to a KAM curve then $\sup\{x_1\} < \sup\{x_2\}$ where $\{x_i\}$ stands for the set of iterates of the initial condition x_i.

Therefore any inversion of monotony, which remains unchanged when the number of iterations increases drastically (a good approximation of ∞), indicates the non existence of KAM invariant curves.

In a way similar to the frequency map, we can describe all the characteristic structures of the system through the sup-map. In fact (Fig.4) a monotonic variation of the sup corresponds to a set of invariant KAM tori, a noisy variation indicates the crossing of a chaotic zone and the crossing of nested islands gives rise to v-shaped structures.

2.3. COMPARISON BETWEEN THE DIFFERENT METHODS

2.3.1. *Application to 2 dimensional map*

As we mentioned in the introduction, the reason for the development of new tools of detection of chaos is mainly that of having computational methods faster than the computation of the LCEs, in order to explore large sets of

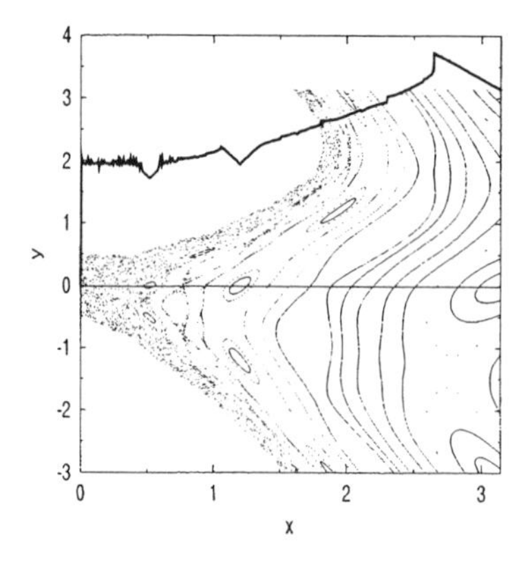

Figure 4. Same as Fig.1 for the sup-map analysis.

orbits and in particular to detect slow chaotic motion. In fact, as far as strong chaos is concerned, even the LCEs are a fast method of detection.

The problem of detection of slow chaotic layers was already raised by Laskar et al. (1992) who detected chaotic orbits looking at the variation of their frequency with time.

In order to compare the different methods of analysis we have explored the same case than Laskar et al. (1992), i.e., for the standard map with $\epsilon = -1.3$, the vicinity of an hyperbolic point connected with the 1:6 resonance (Fig. 5, the arrow indicates the zone explored). We have computed each

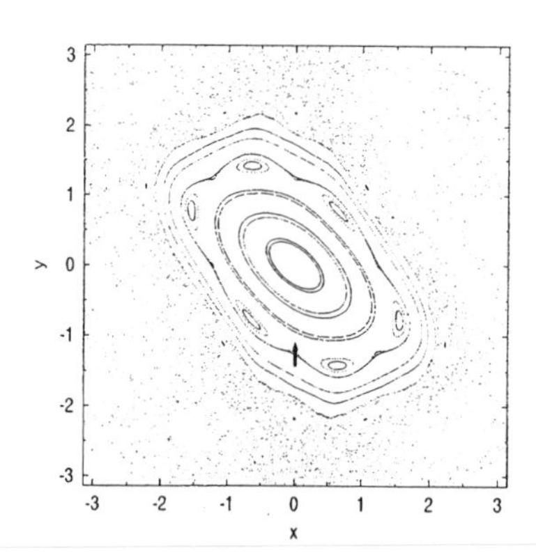

Figure 5. A set of orbits of the Standard map with perturbation parameter $\epsilon = -1.3$.

indicator for a set of 1 000 orbits at $x = 0$ on a cross-section of the y-axis in the interval $-1.223 < y < -1.220$. For one orbit in the vicinity of the hyperbolic point Laskar et al. claim that using the largest LCE: "5×10^6

iterations were necessary to clearly detect the chaotic motion, while it was already visible with the frequency analysis with 20 000 iterations".

Fig. 6a shows, for $N = 20\,000$ iterations, the variation of $\langle \Delta\phi \rangle$ and Fig. 6b the variation of the frequency. All the feature appearing in the frequency analysis also appear when looking at the variation of $\langle \Delta\phi \rangle$.

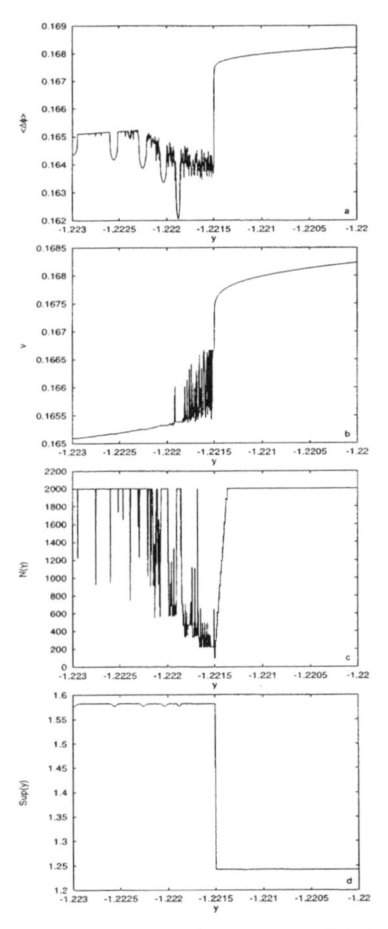

Figure 6. Cross section for a set of 1 000 orbits in the vicinity of the hyperbolic point (indicated by the arrow in Fig.5) associated to the 1/6 resonance of the standard map with perturbation parameter $\epsilon = -1.3$. a) Average values of the twist angles computed using 20 000 iterations of the mapping. b) Frequencies of the same set of orbits computed also using 20 000 iterations of the mapping. c) Number of iterations $(N(\gamma))$ necessary for the Fast Lyapunov Indicator to reach a threshold value of $\gamma = 10^{10}$. The maximum number of iterations is $N = 2\,000$. d) Supremum of y computed using $N = 20\,000$ iterations.

We remark that the variation of $\langle \Delta\phi \rangle$ in the interval $-1.2224 \leq y \leq -1.2215$ reveals that, at this resolution, all invariant curves belonging to the big island have disappeared. In fact the values of $\langle \Delta\phi \rangle$ in this interval

(Fig. 6,a) are all smaller than those obtained in the interval $-1.223 < x < -1.2224$, and, as it is for the frequency, a noisy or non monotonic variation of $\langle \Delta\phi \rangle$ corresponds to the existence of islands or chaotic orbits. It is difficult to obtain the same information with the FMA (Fig. 6,b), at least in the interval $-1.2224 < y < -1.2220$, since small noisy variations of the frequency are difficult to separate from a monotonic increasing of the frequency and even from small *plateau*. Moreover, looking at the interval $-1.223 < y < -1.222$ (left side of Fig.6a) we observe that it is easier to reveal small islands in the case of the helicity angles than in the case of the frequency analysis (the "hat-shaped" structures are easier to see than the *plateau*).

Fig.6,c shows the time $N(\gamma)$ necessary for the FLI to reach a threshold γ with $\gamma = 10^{10}$. We made the computation over a total number of $N =$ 2 000 iterations. Of course, if the threshold is not reached in less than 2 000 iterations we take $N(\gamma) = N$. All the features appearing in the frequency analysis also appear when looking at the variation of $N(\gamma)$. We stress the fact that the computation has been made using only 2 000 iterations. As we said in the introduction with this method we cannot distinguish between islands and tori.

Finally, we show the variation of the $\sup -y$ computed for the same cross-section taking again 20 000 iterations. As shown by Froeschlé and Lega (1996) the sup-map method is as sensitive as the FMA and it has the advantage of being trivial to compute and therefore to extend to higher dimensional problems. Of course, the sup-map does not carry all the informations contained in the frequency map: for example it clearly detects islands but it does not allows to say to which resonance they belong to.

The v-shaped variations of the $\sup -y$ (Fig. 6,d) correspond to islands, the jump stands for the crossing of the hyperbolic point while a monotonic variation reveal the presence of invariant curves. In Fig. 6,d the monotonic variations of the $\sup y$ are masked by the amplitude of the jump at the hyperbolic point.

Fig.7,d shows the variation of the sup-map for 1 000 orbits in the interval $-1.2214 < y < -1.22$, i.e. for orbits of the central island.

The $\sup -y$ decreases monotonically apart for some v-shaped structures which can reveal small islands or orbits with frequencies very near to a rational number (see Contopoulos et al. 1997 for a discussion of this property of the sup-map analysis). In the same interval and at the same resolution, using the other three indicators, $\langle \Delta\phi \rangle$ (Fig.7,a), ν (Fig.7,b) and $N(\gamma)$ (Fig.7,c), we do not find any small higher order resonance. This indicates that all the v – *shaped* structures of Fig.7,d are due to a cinematic effect of the sup-y related to the presence of frequencies very near to a rational number.

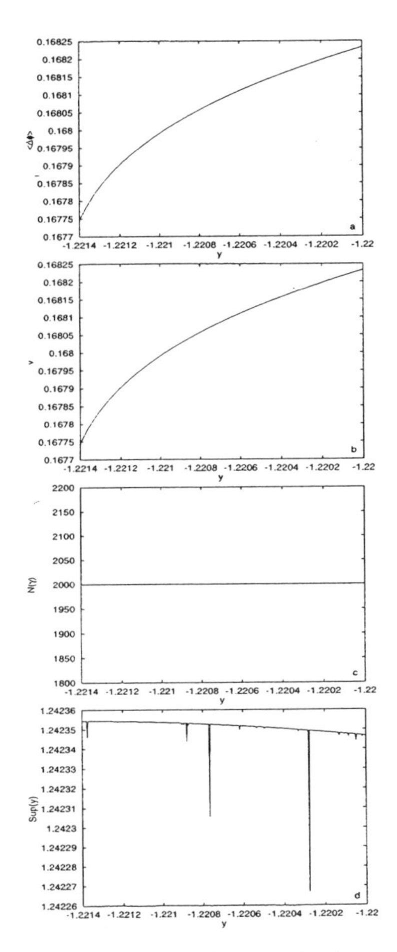

Figure 7. Same as Fig. 6 for the set of orbits in the interval $-1.2214 \leq y \leq -1.22$, i.e. for orbits in the central island. The monotonic variation of the twist angle a) and of the frequency b) as well as the constancy of $N(\gamma)$ c) show that, at this resolution, the whole interval is filled with invariant curves. At this scale it is possible to see the monotonic variation of the supremum of y d) corresponding to invariant orbits as well as some v-shaped structures which reveal the presence of invariant curves with frequencies very near to rational numbers.

In our opinion all these methods have to be used as complementary tools.

2.3.2. *Application to 4 dimensional map*

In this part we will use all the methods already introduced to study the global dynamics of the 4 dimensional symplectic map already used by several authors (Froeschlé 1971,1972; Laskar 1993). We recall the set of equations:

$$\left\{\begin{array}{lcll} x_1' & = & x_1 + \epsilon_1 \sin(x_1 + y_1) + b \sin(0.5(x_1 + y_1 + x_2 + y_2)) & \bmod (4\pi) \\ y_2' & = & x_1 + y_1 & \bmod (4\pi) \\ x_2' & = & x_2 + \epsilon_2 \sin(x_2 + y_2) + b \sin(0.5(x_1 + y_1 + x_2 + y_2)) & \bmod (4\pi) \\ y_2' & = & x_2 + y_2 & \bmod (4\pi) \end{array}\right. \quad (2$$

In the following we take for the perturbing parameters the values: $\epsilon_1 = \epsilon_2 = -1.3$. We have tested the sup-map and the FLI methods for the same set of experiments than those performed by Laskar (1993) using the Frequency Map Analysis method.

To visualize the complete phase space we map the plane defined by $x_1 = x_2 = 0$ onto the sup-plane: $\sup y_1$,$\sup y_2$ (Fig.8) for three values of the coupling parameter b: $b = 0, 0.001, 0.1$ The map are obtained for 120 orbits along y_1 and 1200 along y_2 using 516 iterations for each orbit. The initial conditions are taken for regularly spaced values of y^2 instead of y. We have also considered, using a representation on a grey-scale, the estimated values of the largest LCI for 516 iterations (Fig.9 left column) and for 10 000 iterations (Fig.9 right column) as well as the time $N(\gamma)$ for $\gamma = 10^{12}$ (Fig.10 left column) and the FLI (Fig.10 right column), both obtained using 516 iterations. The images, for the three values of the coupling parameter, are obtained on a grid of 256×256 pixels regularly spaced on y^2.

All these figures show clearly the interaction between the resonances through the Arnold web structure. Of course, as expected, when looking to the chart of the largest LCI (Fig.9 left column) the structure reveals itself only for the 10 000 iterations. We emphasize that for such a qualitative approach the sup-map method and the FLI-chart are really easy to handle.

3. Applications

3.1. EXPLORATION OF THE PRACTICAL EXTENSION OF EXISTENCE THEOREMS.

Usually in the theory of dynamical systems, the results are rigorously proved assuming that the perturbation is small enough. Therefore, it is interesting to check numerically if the properties outlined by a given theoretical description can be extended over their mathematical limit of validity and to see if results are just asymptotic, or if they concern macroscopic regions of physical interest.

3.1.1. *The KAM theorem and the persistence of invariant tori*

In this spirit Hénon (1969) showed numerically that the domain of application of KAM theorem is much larger than the theoretical prove. In fact, using the quadratic map, he showed that invariant curves exist up to a distance to the origin of the order of unit, for which the perturbation reaches

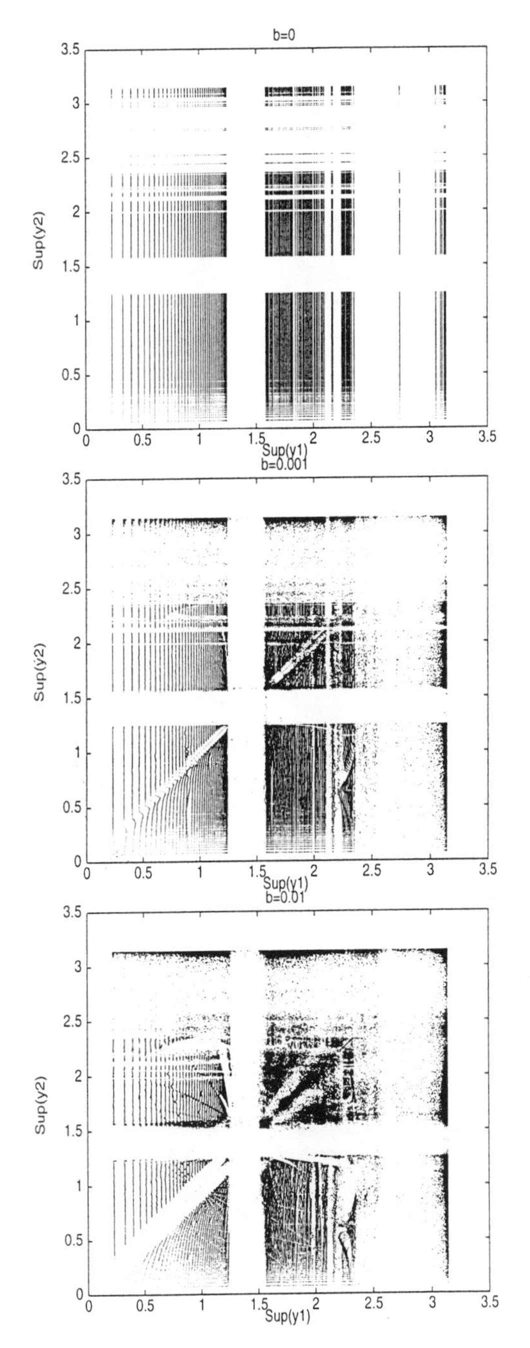

Figure 8. Visualization in the sup-plane $\sup y_1, \sup y_2$ of the sup-map for the 4D standard map with $\epsilon_1 = \epsilon_2 = -1.3$ and various values of the coupling parameter b.

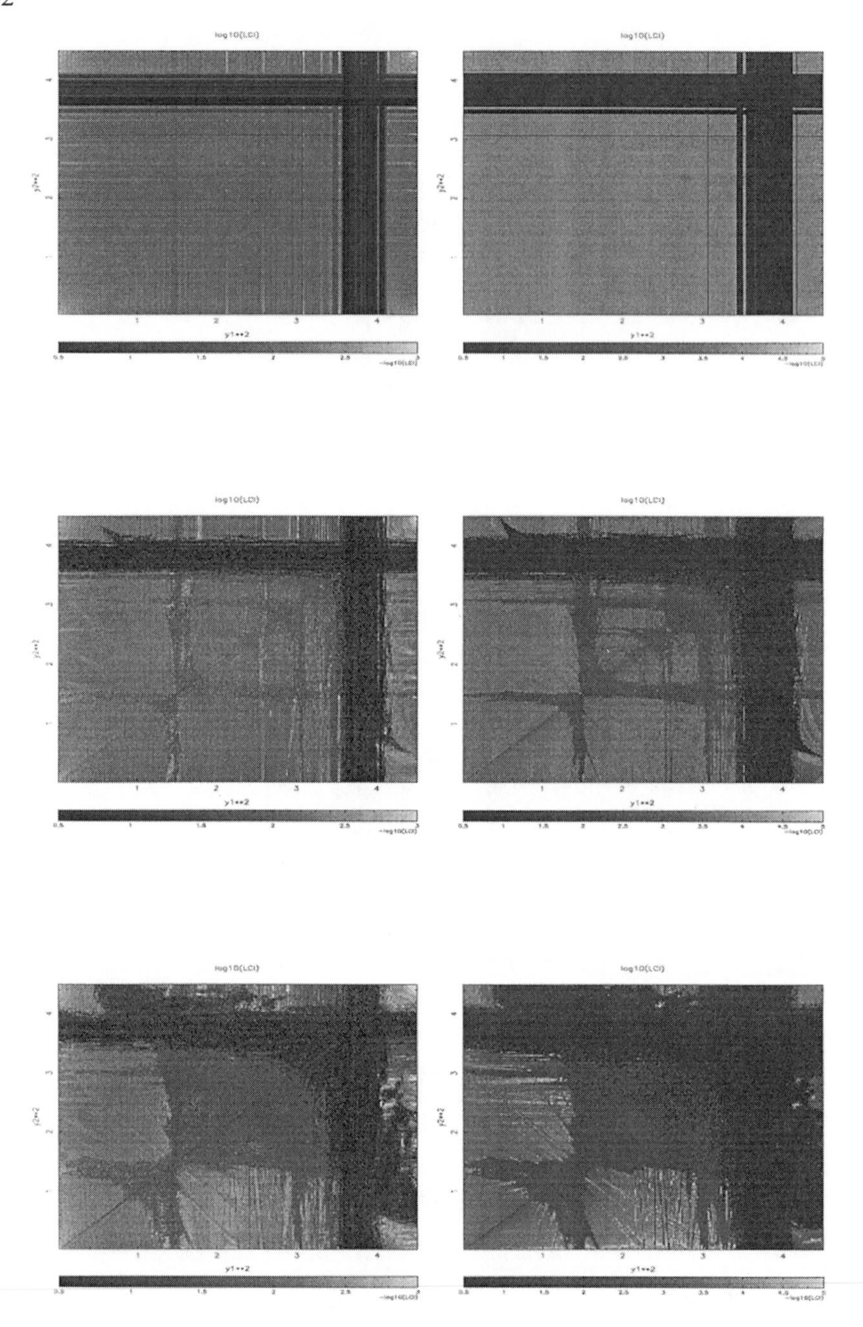

Figure 9. The logarithm of the LCI for the same zone of the phase space explored for the sup plane of Fig.8 and for the three different values of the coupling parameter b: $b = 0, 0.001, 0.1$ (from top to bottom). The darker points correspond to the chaotic zones. Left column: the computation is done using 516 iterations. Right column: the computation is done using 10 000 iterations.

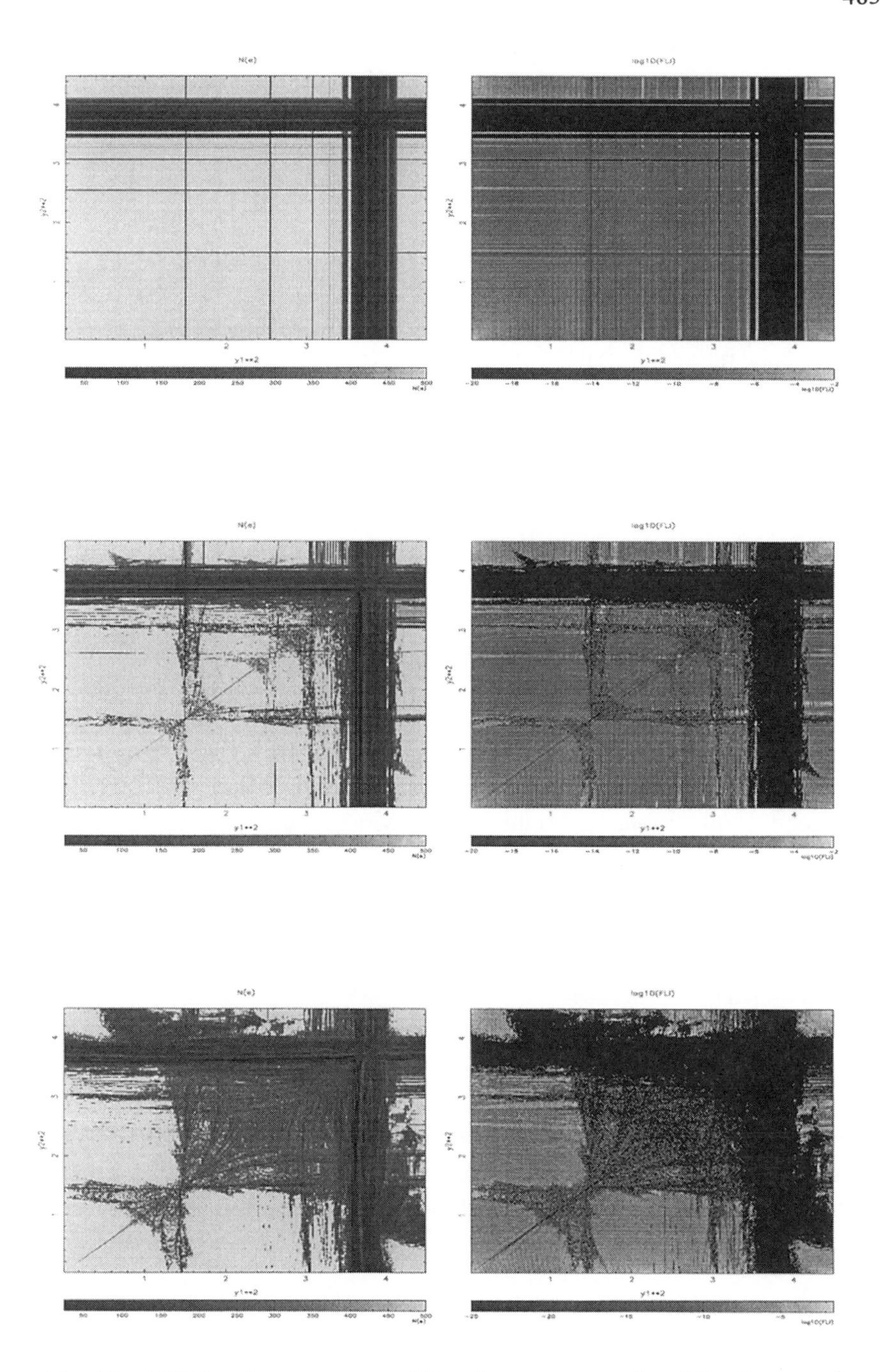

Figure 10. The FLI for the same zone of the phase space explored for the sup plane and for the three different values of the coupling parameter b: $b = 0, 0.001, 0.1$ (from top to bottom). The darker points correspond to the chaotic zones. Left column: the time $N(\gamma)$ necessary for ψ_3 to reach a threshold value of $\gamma = 10^{12}$ on a maximum number of 516 iterations. Right column: the values of the FLI after 516 iterations.

a size comparable to the main terms of the mapping. The KAM theorem applies only to a very small region around the origin: the magnitude of the perturbation, i.e. the distance to the origin, must be less than 10^{-48} ! In the same philosophy we have made a numerical exploration of the neighborhood of a KAM torus.

3.1.2. *The Morbidelli-Giorgilli theorem and the structure of a neighborhood of a KAM torus*

Starting from the now classical result of Nekhoroshev (1977), a recent theorem made by Morbidelli and Giorgilli (1995a, hereafter MG) proving the super-exponential stability of invariant KAM tori had provided a real theoretical breakthrough for the old and crucial question of the stability of a dynamical system.

This important result motivated our numerical study of the structure of the phase space in the vicinity of an invariant KAM torus. We tried to "visualize" and quantify the properties outlined by the Morbidelli and Giorgilli theorem using the Standard map as a sort of experimental laboratory.

This kind of work requires, first of all, tools for a very sensitive analysis of a lot of orbits. Using the methods of analysis introduced in Section 2 we have shown in a striking way that the description of the dynamics given, for small perturbations, by Morbidelli and Giorgilli's result is true in reality as long as the invariant torus persists (Lega & Froeschlé 1996, Froeschlé and Lega 1996, Froeschlé et al. 1996). Moreover, the size of the structure described by Morbidelli and Giorgilli around the invariant torus shrinks to 0 when the size of the perturbation ϵ tends to the critical value ϵ_c corresponding to the torus break-up. This implies that, when the perturbation magnitude is a little bit smaller than the break–up threshold, the size of such structure is macroscopic.

Actually, the dynamics in the neighborhood of a particular invariant KAM torus, the golden one, has been deeply investigated in the near critical regime by Greene and Percival (1981) and by MacKay (1982,1983,1992). Our work gives a quite different dynamical description of similar phenomena thanks to the new results of the MG theorem. We agree with their results (Lega and Froeschlé 1996) although it is beyond our purpose to discriminate between the two possible interpretations.

Theoretical background: super-exponential stability of invariant tori In their investigation of the dynamics in the vicinity of an invariant KAM torus, Morbidelli and Giorgilli started from the so called Kolmogorov normal form (Kolmogorov,1954). According to Kolmogorov's construction, one can introduce suitable action angle variables P, Q, such that, in the neighborhood of the invariant torus $P = 0$, the Hamiltonian writes:

$$H(P, Q) = \omega \cdot P + O(P^2) f(Q) .$$

The Kolmogorov normal form shows in an non-equivocally way that in the vicinity of the invariant torus the significant perturbation parameter is the distance $|P|$ from the torus itself.

Therefore, in the ball $|P| < \rho$ one can introduce new action angle variables J_ρ, ψ_ρ such as to reduce the local perturbation to its optimal size, which, assuming analytic Hamiltonians, is exponentially small with $1/\rho$, i.e.

$$H(J_\rho, \psi_\rho) = \omega \cdot J_\rho + H_0(J_\rho) + \epsilon_\rho H_1(J_\rho, \psi_\rho)$$

with H_0 quadratic in J_ρ and $\epsilon_\rho \sim \exp(-1/\rho)$.

At this point, it is enough to remark that, provided ρ is small enough, ϵ_ρ is smaller than the threshold for the applicability of Arnold's version of KAM theorem (Arnold, 1963) in the ball $|J_\rho| < \rho$. This allows to prove that in the vicinity of the central torus at $P = 0$ there exist an infinity of invariant tori, the volume of the complement decreasing to zero exponentially with $1/\rho$.

On the other hand, provided $H_0(J_\rho)$ is convex in $J_\rho = 0$, if ρ is small enough, the local perturbation parameter ϵ_ρ is also smaller than the threshold for the applicability of Nekhoroshev's theorem. This allows to prove that the diffusion of the actions J_ρ must be bounded by ϵ_ρ^b for all times up to $\exp(1/\epsilon_\rho)$, which, by substitution gives the super-exponential estimate $\exp[\exp(1/\rho)]$. The hypothesis of local convexity is a very natural one. It means indeed that on a given energy surface, the torus with given frequency ratios is locally unique.

The picture provided by Morbidelli and Giorgilli's result is therefore the following. The tori given by Kolmogorov's theory are *master* tori, surrounded by a structure of *slave* tori, which accumulate in an exponential way to the central master torus. These slave tori are all n–dimensional Diophantine ones, but they are characterized by a very small Diophantine constant γ (we recall that a frequency ω is said to be Diophantine if it satisfies the relation $|k \cdot \omega| > \gamma / |k|^{n+1}$ for all integer vectors k and some positive γ); for this reason, they could not be found directly by Kolmogorov's construction. Moreover, diffusion among this structure of slave tori is super-exponentially slow, so that chaotic orbits can enter in, or escape from, only in a time proportional to $\exp[\exp(1/\rho)]$.

The interest of this result is double. On the one hand, this makes open the set of invariant tori from all practical point of view; this is important for what concerns the compatibility of KAM theorem with the errors in initial conditions of numerical experiments. On the other hand, a direct consequence of the local super-exponential stability is that invariant tori can form, even in three or more degrees of freedom, a kind of impenetrable structure which orbits cannot penetrate for an exceedingly long time, very large even with respect to the usual Nekhoroshev's estimates.

A qualitative study of the structure of the neighborhood of a KAM torus. Although a direct numerical check of the super-exponential slow diffusion seems to be beyond the possibility of numerical experiments, the related description of structures made by *slave* tori which accumulate exponentially to a central *master* KAM torus can be checked. In fact, as we have shown in Section 2, tools like the frequency map analysis, the sup-map analysis and the fast Lyapunov indicator are perfectly adapted to a very fine exploration of the structure of the phase space.

In order to test the structure of invariant tori around a chief torus we have taken as a model problem the Chirikov formulation of standard map (Eq.11).

Actually for low values of the perturbation parameter ϵ we know that the phase space is generally full of invariant tori, in the non resonant domain: an orbit with an initial condition picked at random within this domain will show Lyapunov's exponents equal to zero. Therefore it is very difficult in such a domain to separate a chief KAM torus from the other ones. In fact, as far as the numerical approach is concerned, the neighborhood of such a chief torus probably contains other chief tori. In order to bypass this difficulty, we have taken for our numerical experiments, a value of ϵ close to the critical value for which only the so called golden torus remains. Since the golden torus corresponds to a value of the rotation number equal to the golden number $\nu_o = \frac{1}{2}(3-\sqrt{5})$ (the irrational farthest from the rationales) we know that such a torus is a chief one.

For the numerical experiments we have taken $\epsilon = 0.9715$. This value of the perturbation parameter is, in fact, closed to the critical value $\epsilon_c = 0.971635$, found by Greene and improved by MacKay (1993), for which the golden curve does not survive.

For $\epsilon = 0.9715$ we have computed the frequency map (with the Hénon's method) and the sup map for successive enlargement around the golden torus. We have also measured for each enlargement the time $N(\gamma)$ necessary for the FLI to reach a threshold $\gamma = 10^6$ on a maximum number of iterations equal to that used for the computation of the frequency and of the sup. Figure 11a,b,c show the frequency-map and Figs.12a',b',c' and 13a",b",c" show the variation of respectively the $\sup\{y_i\}$ and of $N(\gamma)$ for the same initial conditions taken in Figs.11a,b,c.

On each plot the origin corresponds to the golden torus, i.e. the torus whose rotation number is equal to the golden number $\nu_o = \frac{1}{2}(3-\sqrt{5})$.

We have indicated on each plot the values of frequencies corresponding to the Fibonacci sequence, i.e. the set of the successive terms obtained when developing the golden number through the continued fraction process.

We observe that the majority of orbits of Figs.11a, 12a', 13a" corresponds to chaotic regions and islands. The situation changes drastically in Figs.11b, 12b', 13b": the noisy variation of ν and of $N(\gamma)$ corresponding to strong chaotic regions, as well as the fuzzy plateau on the sup, have disappeared; islands and crossing of hyperbolic points are still there, but their relative measure in the action variables is now definitively smaller than the relative measure of tori. This phenomenon is strongly enhanced in the last magnification: up to a step size of $\Delta y = 1.6\,10^{-11}$ (Fig. 11c, 12c', 13c") we only see one hyperbolic point and a large continuous region of tori.

It is clear that the magnifications represented in Figs. 11b,c show a completely different regime and in Fig. 11c only the chief torus and its "slaves" appear with a density which seems to be in full agreement with the prevision of the Morbidelli Giorgilli theorem.

The same conclusions hold when looking at the variation of the sup, in particular the change of regime from Fig.12a' to Fig.12c'. Qualitatively, even with the FLI we clearly detect a change of regime from Fig.13a" to Fig.13c". However, as we explained in Section 3.4, with this indicator it is impossible to separate islands from tori and we need to separate such structures in order to obtain a quantitative measure of the exponential decay of both chaotic zones and islands.

A quantitative study of the structure of the neighborhood of a KAM torus. In order to test the exponential decrease of the volume occupied by the complement of the set of tori (V_c), as a function of the distance (ρ) to the chief torus, we have measured, using the frequency maps shown in Fig.11a,b,c, the size of the Fibonacci islands. Such islands are the largest ones and therefore they fill the major part of V_c.

Figure 14 shows the variation of the size of the Fibonacci islands as a function of the distance ρ in a log-log diagram.

The distance ρ is the absolute value: $|y_c - y_o|$ where y_c is the action of the center of the island and y_o is the action of the golden torus. In the Morbidelli-Giorgilli regime we have also taken into account the hyperbolic points corresponding to the Fibonacci chain of islands. We have made an estimate of the dimension of the corresponding islands through the jump in frequency which occurs when crossing the hyperbolic point.

On Fig.14 the change of regime is drastic: at the 19th term of the Fibonacci sequence we enter in the Morbidelli-Giorgilli regime where the size of the perturbation decreases exponentially with the distance from the golden torus. Let us emphasize that the measure has been done for a value

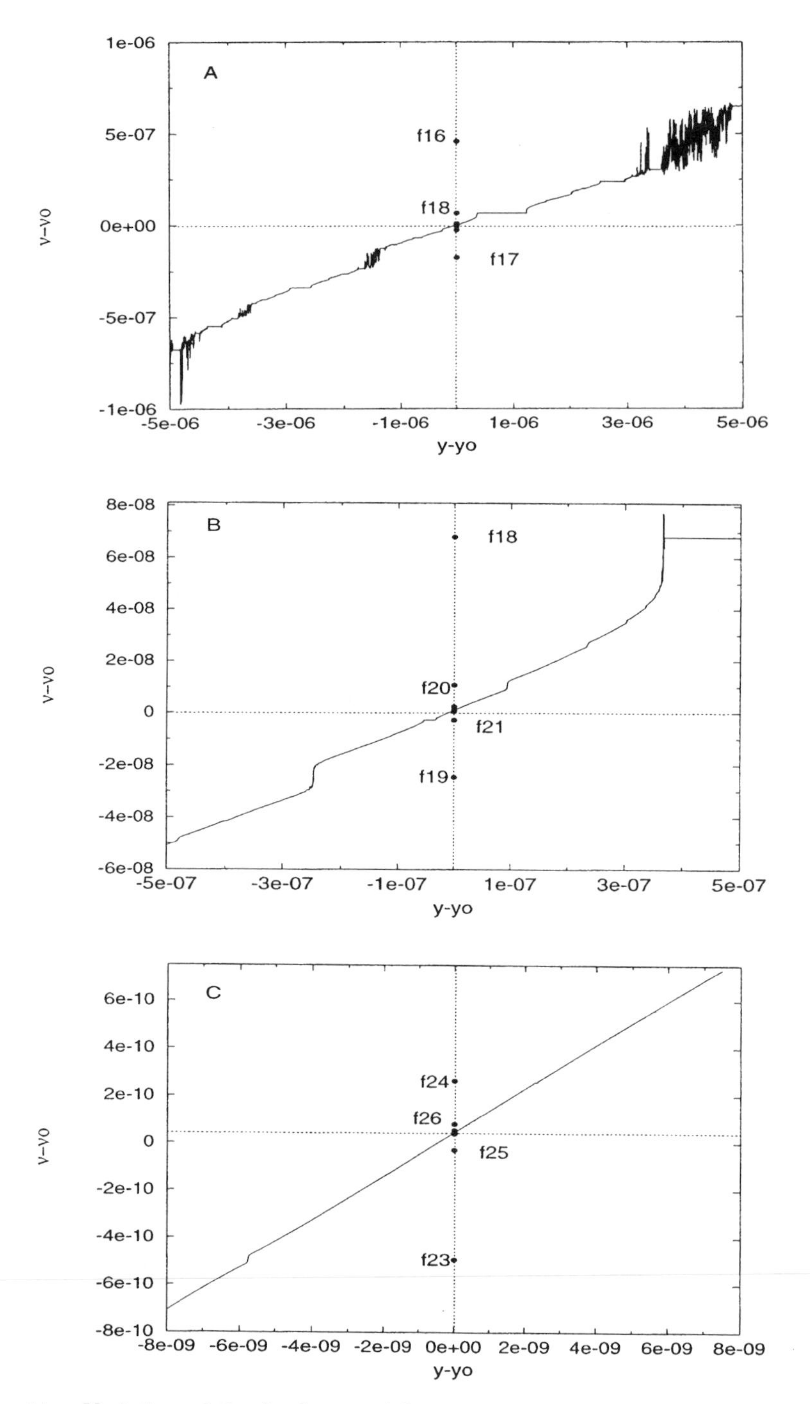

Figure 11. Variation of the fundamental frequency ν for the standard mapping with $\epsilon = 0.9715$, for successive enlargements (a,b,c) in the vicinity of the golden torus whose coordinates (y_o, ν_o) correspond to the origin of the axes. The Fibonacci terms are indicated on each figure by the set of points f_i.

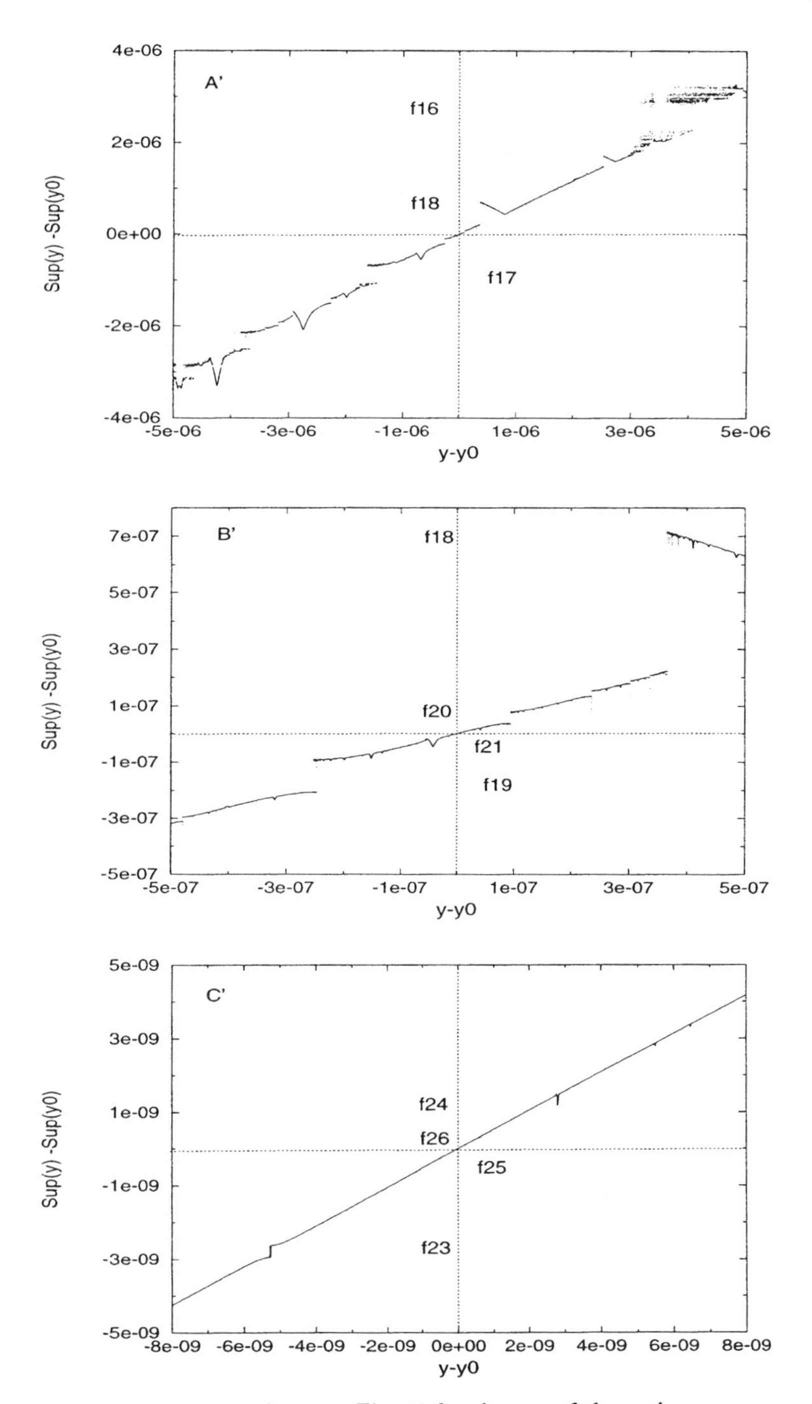

Figure 12. Same as Fig. 11 for the sup of the action.

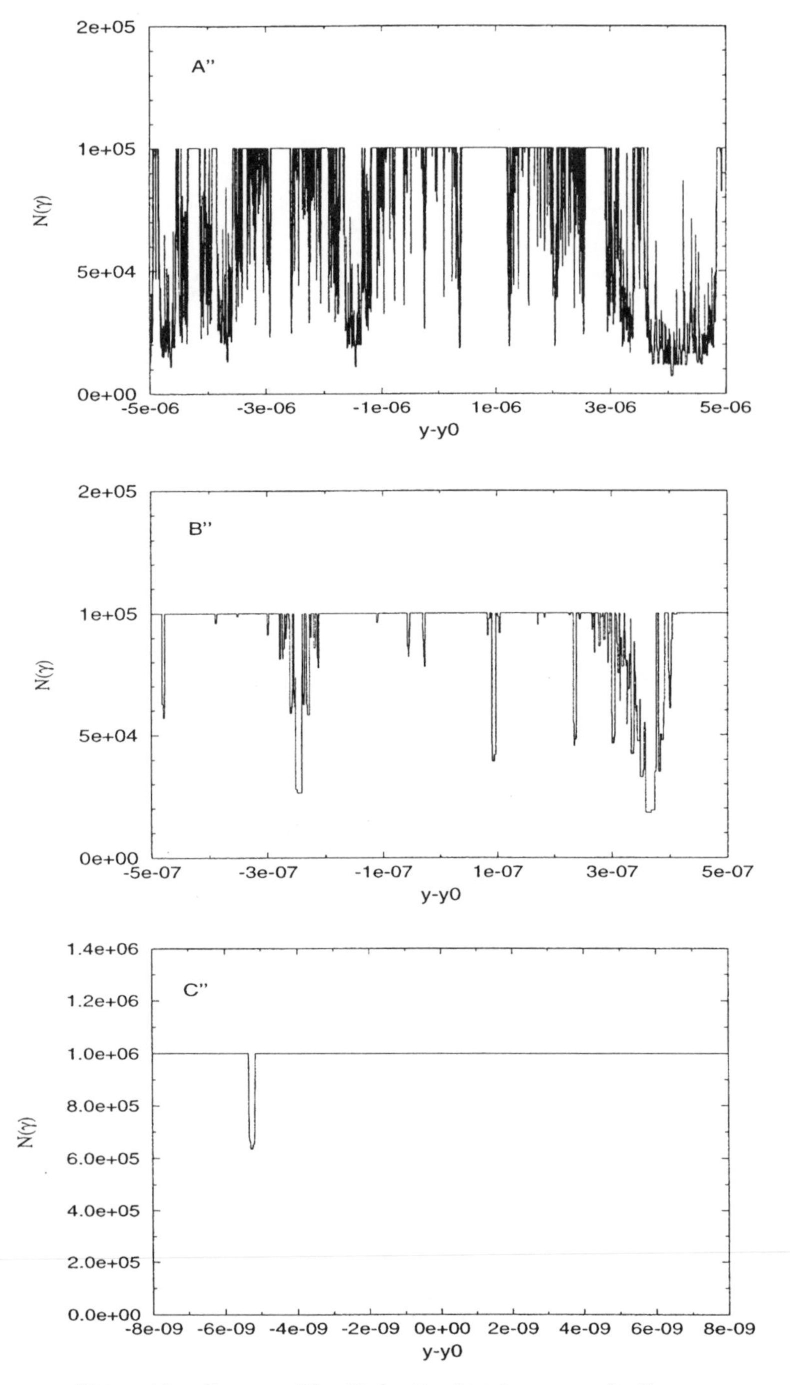

Figure 13. **Same as Fig. 11 for the fast Lyapunov Indicator.**

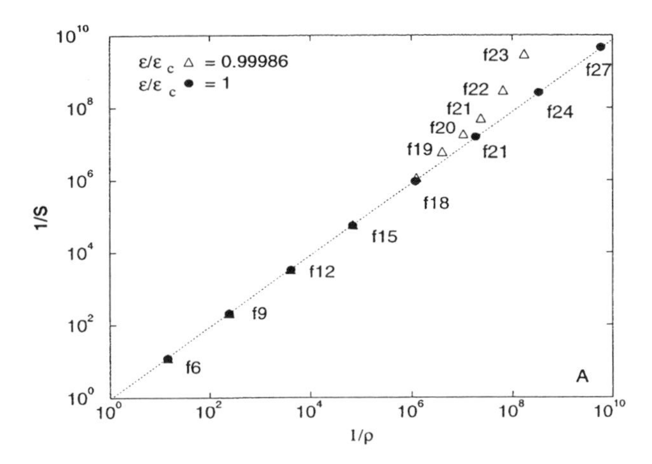

Figure 14. Variation of the size of the Fibonacci islands as a function of the distance to the golden torus for two values of the perturbation parameter: the critical threshold for the break down of the golden torus and a value just below the critical one $\epsilon = 0.9715$.

of the perturbation parameter: $\epsilon = 0.9715$ very close to the critical one: $\epsilon_c = 0.971635$. When approaching ϵ_c the exponential part bends to the linear regime. Actually when $\epsilon = \epsilon_c$ we detect, up to the precision of our computation, only the linear regime as shown on Fig.14 for which the points are equally spaced in a log-scale. This seems to corresponds to the already known (Greene 1981, Contopoulos 1993) scale invariance at ϵ_c. Moreover, the exponential decay of some quantities (the Greene residues) which result to be related to the volume of the complement of tori can be deduced by the work of Greene and Percival (1981) and was proved lately by MacKay (1992). Using the exponential part of Fig.14 we have checked (Lega and Froeschlé 1996) the law given by MacKay finding a good agreement with our results.

Using the same technique we have estimated the distance $\bar{\rho}^*$, at which we enter in the Morbidelli-Giorgilli regime, for a set of different values of the perturbation parameter. Fig.15 shows our result on the variation of $\bar{\rho}^*$ as a function of ϵ/ϵ_c.

After a linear decrease of $\bar{\rho}^*$, up to $\epsilon = 0.95$, we observe a sharp drop of $\bar{\rho}^*$ up to $\bar{\rho}^* = 8\,10^{-7}$ for $\epsilon = 0.9715$. It seems therefore that all the slave tori disappear at once, when approaching the critical value ϵ_c. Again we have found (Lega & Froeschlé 1996) a good agreement between our results and the law given by MacKay (1982).

Let us remark again the completely different description of the dynamic that we have at light of the MG theorem. We also claim that a great help is given to the comprehension by the use of tools like the frequency map,

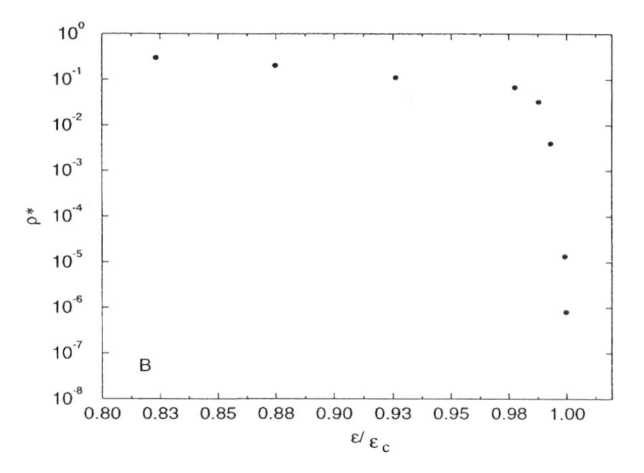

Figure 15. Variation of the threshold distance ρ^* as a function of the perturbation parameter ϵ/ϵ_c.

the sup-map and the FLI. In fact they allowed, in this case, to visualize concepts which could remain only of mathematical interest and somehow far from the reality of dynamical systems.

3.2. THE RELATIONSHIP BETWEEN LYAPUNOV TIMES AND MACROSCOPIC INSTABILITY TIMES

In this section we refer to the work of Morbidelli and Froeschlé (1996) on the use of mapping for studying the relationship between Lyapunov times and macroscopic diffusion times.

The origin of this work was a debate about this relationship for some unstable regions of the Solar System. On one hand a number of papers (Soper, Franklin and Lecar, 1990; Lecar, Franklin and Murison, 1992; Murison, Lecar and Franklin, 1994; Franklin, 1994) claimed to have found numerical evidence of a universal law connecting Lyapunov times and macroscopic instability times. More precisely, integrating several orbits of (mostly) fictitious bodies in the unstable regions of the Solar System the authors cited above found that the relationship

$$\frac{T_I}{T_0} = \alpha \left(\frac{T_L}{T_0}\right)^\beta \tag{25}$$

is generally satisfied, although with a dispersion of data of at least two orders of magnitude with respect to the exact relationship. In formula (25) T_L is the well known Lyapunov time, i.e. the inverse of the largest Lyapunov exponent; T_I is the macroscopic instability time, defined as the epoch at which occurs some strong "event", such as a close encounter with a perturbing planet, which changes completely the dynamical nature of the test particle's orbit; T_0 is an "appropriate" renormalizing period; α and β are positive constants. In particular, according to Murison et al. (1994), β should be equal to 1.74 ± 0.03. Levison and Duncan (1993) found a similar result for what concerns the unstable regions of the outer Solar System, but with a constant $\beta = 1.9$.

On the other hand, at the same time, Milani and Nobili (1992) considered the case of the asteroid 522 Helga, which has a Lyapunov time of only 6 900 yr and still is a permanent member of the asteroid belt. This is an evident violation of "law" (25). Milani and Nobili concluded that there must exist a sort of "stable chaos" characterized by hyperbolic behaviour but no macroscopic instability. Although the case of 522 Helga is the only one discussed in Milani and Nobili's work, it is now known that a significative fraction of numbered asteroids have a short Lyapunov time, in contrast with their long-time stability (Milani; Levison, private communications).

In their work, Morbidelli and Froeschlé (1996), showed that, the contradiction is not real because diffusion in Hamiltonian dynamical systems can show at least two regimes which are completely different each other. A first regime is characterized by resonance non–overlapping and by the existence of many invariant tori, and is described by the Nekhoroshev's theorem; in this case the relationship between T_I and T_L must be exponential, $T_I \sim \exp(T_L)$, and the words "stable chaos" can be appropriated. A second regime is characterized by resonance overlapping and by the absence of invariant tori; in this case the relationship between T_I and T_L can be of polynomial type as in (25), although we do not find any justification for the existence of a "universal law" concerning the value of the exponent β.

The next two subsections will be devoted to analyzing separately these two regimes.

3.2.1. *The Nekhoroshev regime*

The phase space of a Hamiltonian system can be partitioned into domains with sharply different dynamical behaviors. There are the so called "no–resonance domains", characterized by the absence of resonances of low–order; the "single–resonance domains" characterized by the presence of only one resonance of low order; and the "multi–resonance domains" characterized by at least two resonances of low order. The latter correspond to the neighborhoods of the nodes of the so–called Arnold web (i.e. the web made by the lines denoting the location of resonances in the action space).

The great contribution by Nekhoroshev has been that to give a global description of the dynamics, connecting together all the various behaviors which characterize the different domains. More precisely, the Nekhoroshev theorem considers analytic Hamiltonian systems close to integrable ones, i.e. of the form

$$H_0(p) + \epsilon H_1(p,q) \tag{26}$$

where $H_0(p)$ is an integrable Hamiltonian which depends only on the actions p. The theorem states that, if the perturbation size ϵ is sufficiently small, then, whatever is the initial condition $p(0)$, the change of the action variables is small over an exponentially long time, i.e.

$$|p(t) - p(0)| < O(\epsilon^a) \quad \text{for all } |t| < T \sim O(\exp(1/\epsilon)^b) \; ; \tag{27}$$

where a and b are suitable positive coefficients, which depend on the number n of degrees of freedom (optimal values being $a = b = 1/2n$ according to Lochak, 1993). For sake of simplicity, in the following discussion we will neglect all dependencies on the number of degrees of freedom.

At the light of Nekhoroshev theorem, therefore, one can say that, provided the perturbation is sufficiently small, the macro–instability time T_I must be at least exponentially long with respect to the inverse of the perturbation size ϵ, i.e. $T_I \sim \exp(1/\epsilon)$.

For what concerns the Lyapunov time, this varies strongly from one domain to another. In the "non–resonance domains", the Lyapunov exponent must be very close to zero. If one assumes that the threshold between low–order resonances and high order resonances is about $1/\epsilon$ as in the construction by Nekhoroshev, then the Lyapunov exponent in the non resonant domain should be smaller than $\exp(-1/\epsilon)$. Indeed, one can construct a non–resonant normal form, eliminating all Fourier terms up to order $1/\epsilon$. Due to the property of analyticity of the Hamiltonian, the remaining terms, which are the only ones which can generate a positive Lyapunov exponent, are as small as $\exp(-1/\epsilon)$. For what concerns the "single–resonance domains", the computation is more complicated and there is no rigorous mathematical result on the expected values of Lyapunov exponents. The width of a low–order resonance is of magnitude $\sqrt{\epsilon}$, while all non–resonant terms can be reduced to be exponentially small (i.e. of order $\exp(-1/\epsilon)$), as in the previous case. Thanks to this fact, Neishtadt (1984) proved that the angle of transversal intersection of the stable and the unstable manifolds of the low order resonance must be exponentially small. The Lyapunov exponent of the orbits over such manifolds is of order of $\sqrt{\epsilon}$, but these fill a region the volume of which is zero. With heuristic considerations inspired by the work of Neishtadt (1984), we can expect that the Lyapunov exponent is positive only on a region which is exponentially small, and, apart from the orbits on the stable and unstable manifolds, should be also exponentially small with respect to $1/\epsilon$.

In the "multi–resonance domains" the perturbation acting on one resonance cannot be reduced smaller than $\sqrt{\epsilon}$ because of the presence of additional resonances of approximately the same size, i.e. $\sqrt{\epsilon}$. Then, we can expect that the Lyapunov exponent is proportional to $\sqrt{\epsilon}$ over a region the size of which is of the same order of magnitude. In this most favorable case, then, one should have the shortest Lyapunov time $T_L \sim 1/\sqrt{\epsilon}$ over a region with significative size.

Since these statements on the value of Lyapunov exponents are just based on theoretical "reasonable" considerations but not on rigorous proofs, we have proceeded to the following numerical experiment.

Taking the Froeschlé map (1972), with equations:

$$\left\{ \begin{array}{lcl} y' & = & y + a\epsilon \sin(x+y) + b\epsilon \sin(\frac{x+y+p+q}{2}) \\ x' & = & x+y \\ p' & = & p + c\epsilon \sin(q+p) + b\epsilon \sin(\frac{x+y+p+q}{2}) \\ q' & = & q+p \end{array} \right. \tag{28}$$

with $a = 1, b = 0.7$ and $c = 1.3$, we have considered the multi–resonance domain in correspondence of the node between the two main resonances, the location of which is determined by the values of the actions $y = 0$ and $p = 0$, respectively. For a fixed value of the perturbation parameter ϵ, we have chosen the initial conditions of our test orbits on the action line $y = 73/50p$. Moreover, we have fixed the initial phases $x = q = 0$ in order to be on the double–hyperbolic periodic orbit at the exact double resonance $y = p = 0$, i.e. on a periodic orbit which is unstable with respect to both (x, y) and (q, p), even in the limit $\epsilon \to 0$. The result that we obtain for each value of ϵ between 10^{-1} and 10^{-6} show that the Lyapunov exponent is close to zero (at the level of the computation error), then sharply increases as soon as we enter the chaotic region around $y = 0$. In the chaotic region the value of the Lyapunov exponent is approximately constant. In Fig. 16a we plot the average value of the Lyapunov exponent in the chaotic region as a function of ϵ in Log–Log scale.

The error bars are computed as the standard deviation of the Lyapunov exponents over the chaotic region. The line which best fits the data has a slope equal to 0.47. This shows that the value of the Lyapunov exponent is almost proportional to $\sqrt{\epsilon}$ as we have deduced above. Figure 16b shows, as a function of ϵ, the size of the chaotic zone. The latter is computed as the length of the interval $(-y_a, y_b)$ where $-y_a$ and y_b are the values of y corresponding to the sharp increase of the Lyapunov exponent. In Fig. 16b the best fit of the data is provided by a straight line with slope very close to 1/2 (actually 0.52) in a Log–Log scale, in good agreement with our guess that the size of the chaotic zone is of order $\sqrt{\epsilon}$ which shows that even for small ϵ the size of the chaotic zone is not of measure zero, i.e. a sizeable portion of the phase space corresponds to this region.

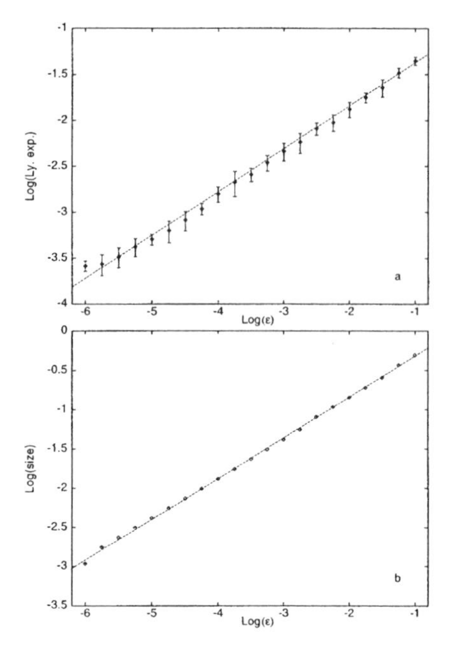

Figure 16. a) The logarithm of the Lyapunov exponent as a function of the logarithm of ϵ in the chaotic region associated to the double resonance at $y = p = 0$ in Froeschlé map. The best fit line's slope is equal to 0.47. b) The logarithm of the size of the chaotic region as a function of the logarithm of ϵ. The best fit line's slope is equal to 0.52.

These numerical results, therefore, confirm our idea that the Lyapunov exponent is proportional to $\sqrt{\epsilon}$ in the multi–resonance domains, the size of which is also of order of $\sqrt{\epsilon}$. Then, in the case where ϵ is sufficiently small for the application of Nekhoroshev theorem, one has

$$T_I \sim \exp(\frac{1}{\epsilon}) \ ; \quad T_L \sim \frac{1}{\sqrt{\epsilon}} \quad \text{i.e. } T_I \sim \exp(T_L) \tag{29}$$

This shows definitely that, at least in the limit of small perturbations, Murison et al. 'law" is wrong.

How can one know if a given system is in the Nekhoroshev regime? The usual mathematical proofs give thresholds on the size of the perturbation which are irrealistically small. But in a rigorous mathematical proof all constants are overestimated, so that the description provided by Nekhoroshev theorem usually applies also to much larger perturbations than those for which the theorem can be formally proved.

A numerical tool for checking if a given dynamical system can be in the *Nekhoroshev regime* is offered by a recent theorem on the connection between Nekhoroshev and KAM theories (Morbidelli and Giorgilli, 1995b). They show that, if Nekhoroshev theorem can be applied to an autonomous Hamiltonian system over an open part of the phase space, then there must exist many invariant tori in the "no–resonance domains". The existence of invariant tori can be easily checked numerically, since invariant tori must have all Lyapunov exponents equal to 0. So, if one finds that there are no orbits with 0–Lyapunov exponent, then the system cannot be in the *Nekhoroshev regime*; conversely, if many orbits of the system have 0–Lyapunov exponents, then it is reasonable to expect that the system is in the *Nekhoroshev regime* and that all orbits with positive Lyapunov exponents are examples of "stable chaos". Moreover, it has been proved (Morbidelli and Giorgilli, 1995a) that in open neighborhoods of invariant tori the Nekhoroshev theorem can always be applied.

3.2.2. *The resonance overlapping regime*

When low–order resonances overlap together, so that "no–resonance domains" cannot be defined, the Nekhoroshev theorem cannot be applied. In these cases, the Chirikov criterion (Chirikov, 1960) correctly determines that all invariant tori of the KAM theory have been destroyed. The door is open to fast chaotic "diffusion". We have checked the law connecting macroscopic instability time T_I and Lyapunov time T_L in this case using two different mappings. The first one is:

$$\left\{ \begin{array}{lcl} p_n & = & p_{n-1} + \epsilon \sin(K(p_{n-1} + q_{n-1})) \\ q_n & = & p_{n-1} + q_{n-1} \end{array} \right. \tag{30}$$

with $K = 200$. Then, for ϵ ranging from 0.03 to 1 we have measured

$$< p^2 > \equiv \frac{\sum_{n=1}^{N} p_n^2}{N}$$

and the Lyapunov exponent over a number of iterations $N = 10^5$. Looking in Log–Log scale $< p^2 >$ and the Lyapunov exponent as a function of ϵ we have found that the data are well fitted by straight lines and the power law between T_I and T_L is closed to $T_I \propto T_{L^3}$ which can be deduced from an idealized random-walk process.

Performing the same numerical experiments on a different mapping:

$$\left\{ \begin{array}{lcl} p_n & = & p_{n-1} + \epsilon \sum_{j=-K}^{K} \sin(p_{n-1} + q_{n-1} - 2n\frac{j}{k}\pi) \\ q_n & = & p_{n-1} + q_{n-1} \end{array} \right. \tag{31}$$

where $K = 80$. We find again a power law relationship between T_I and T_L with different values of the coefficients (not at all connected with a pure random walk) and a stronger dispersion of the data. It is clear that in the

Chirikov overlapping regime we have a power law which is strongly model dependent.

Therefore, in this work it has been shown by theoretical considerations and numerical experiments based on mappings, the existence of two very different regimes for a Hamiltonian dynamical systems. In what we call the *Nekhoroshev regime*, the relationship between macro–instability time T_I and Lyapunov time T_L must be exponential (i.e. $T_I \sim \exp(T_L)$). Conversely in the *resonance overlapping regime* the relationship is of polynomial character (i.e. $T_I \sim T_L^\beta$, with some positive β). What the author (Morbidelli and Froeschlé 1996) claim is that the power law (25) should not be used for obtaining previsions on the depletion times of specific regions of the Solar System, at least as long as the dynamical mechanisms at the origin of the chaotic phenomena are not completely understood. Indeed, even in the case of the *resonance overlapping regime*, the "constants" α and β in equation (25) can change significantly from case to case; moreover, even if the experiments are conducted in a quite uniform region as in the case by Murison et al. (1994), the dispersion of the data with respect to the fitting "law" is of a few orders of magnitude.

3.2.3. *Application to asteroidal dynamics: the Kirkwood gaps*

The asteroidal population (whose distribution is plotted in Fig.17) of the first order Jovian resonances and in particular the so called Kirkwood gaps was for a long time a puzzling problem (see Moons 1997 for an up to date review on the subject). Only in this last decade it has been understood that for the major gaps except for the 2/1 the overlapping of mean motion resonances with secular resonances leads to almost overall chaos where asteroids undergo large and wild variations in their orbital elements.

Such asteroids are either thrown directly to the Sun or, within a few millions of years, submitted to strong close encounters with the largest inner planets. This mechanism is not able to explain the 2/1 gap and the computation of the LCEs inside the gap, which turn out to be positive, does not allow to reach definitive conclusions (see Section 3.2.1). However, Nesvorný and Ferraz-Mello (1997), using the frequency map analysis to detect chaos, have recently shown that almost all the phase space of the 2/1 resonance is everywhere weakly chaotic and therefore it isn't in a Nekhoroshev regime, conversely to the 3/2 Hilda group (Fig.18). The Hilda group is mostly located at the plane where the frequency variation is under the level of 10^{-4}. There is not such stable region in the 2/1 resonance and the global difference may be considered as solving the Kirkwood gaps puzzle.

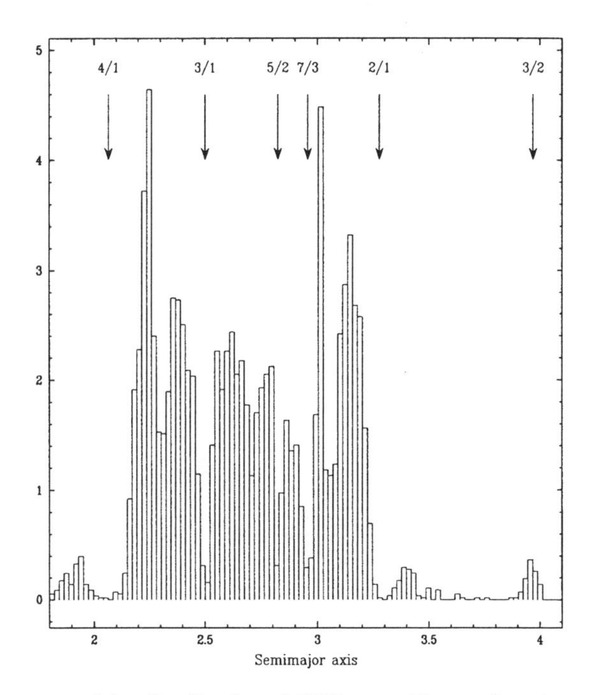

Figure 17. Histogram of the distribution of 5551 asteroids as a function of the semi-major axis. The locations of some mean motion resonances are shown.

4. Conclusion

A fine numerical study of connected regions of the phase space resulted to be necessary in order to characterize the dynamical behaviour of a dynamical system. This fact motivated the development of different tools of analysis which are sensitive to the character of the orbits and as cheap as possible in computational time. The application of such tools allowed to give physical interest to some recent theoretical results concerning the structure of the vicinity of an invariant KAM torus. At the same time a fine exploration of some particular regions of the asteroidal belt allowed to get a possible explanation for the 2/1 Kirkwood gap as well as for the existence of the 3/2 Hilda group.

References

1. V.I. Arnold. Proof of a theorem by A.N. Kolmogorov on the invariance of quasi-periodic motions under small perturbations of the Hamiltonian. *Russ. Math. Surv.*, **18**:9, (1963).
2. G. Benettin, L. Galgani, A. Giorgilli, and J.M. Strelcyn. Lyapunov characteristic exponents for smooth dynamical systems; a method for computing all of them. *Meccanica*, **15**:Part I : theory, 9–20 – Part 2 : Numerical applications, –21–30, (1980).

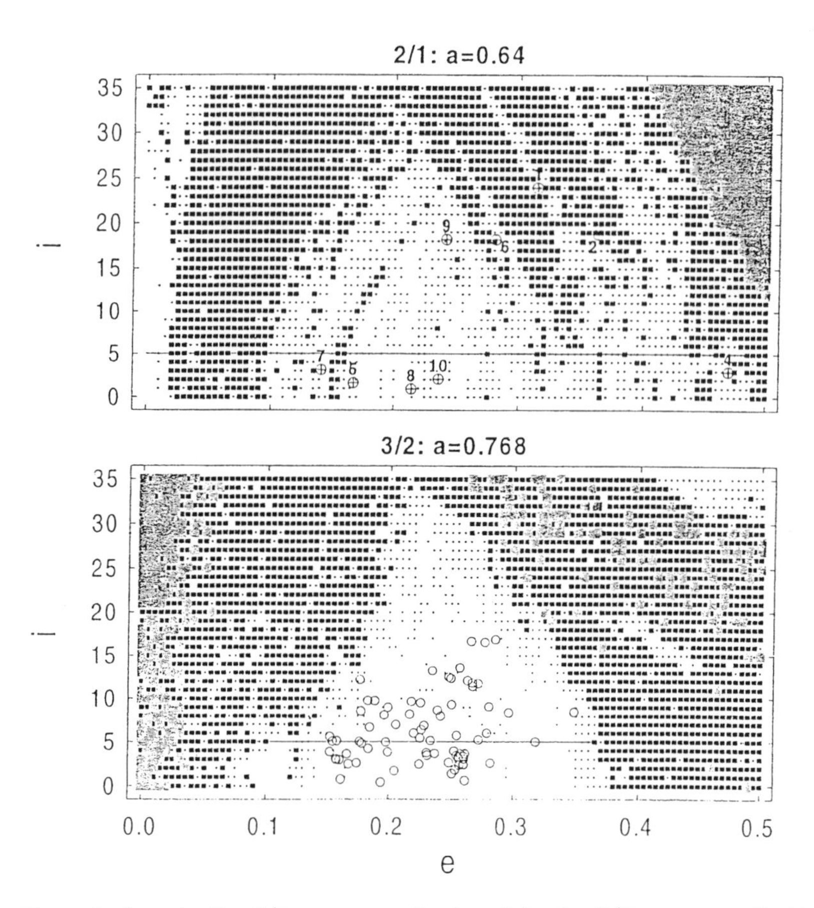

Figure 18. e,i plane in the 2/1 resonance (top) and in the 3/2 resonance (bottom). Black squares correspond to a variation of the frequency $\delta f > 10^{-2}$ in 10^5 years, small crosses to $10^{-2} > \delta f > 10^{-4}$ and voids to $\delta f < 10^{-4}$. The gray area correspond to highly chaotic escape orbits (from Nesvorny and Ferraz-Mello).

3. A. Celletti and C. Froeschlé. On the determination of the stochasticity threshold of invariant curves. *Int. J. of Bif. and Chaos*, **5**, n.6:1713–1719, (1995).
4. A. Celletti, C. Froeschlé, and E. Lega. Determination of the frequency vector in the four dimensional standard mapping. *Int. J. of Bifurcation and Chaos*, **6**, n.8:1579–1585, (1996).
5. B. V. Chirikov. *Plasma Phys.*, **1**:253, (1960).
6. B.V. Chirikov. An universal instability of many dimensional oscillator system. *Phys. Rep.*, **52**:263–379, (1979).
7. G. Contopoulos. *Galactic Dynamics and N-Body Simulations.* Springer-Verlag, (1993).
8. G. Contopoulos, E. Grousousakaou, and N. Voglis. Invariant spectra in Hamiltonian systems. *A.& A.*, **304**:374–380, (1995).
9. G. Contopoulos and N. Voglis. A fast method for distinguishing between order and chaotic orbits. *A.& A.*, to be published, (1997).
10. H.S. Dumas and J. Laskar. Global dynamics and long-time in Hamiltonian systems

via numerical frequency analysis. *Phys. Rev. Lett.*, **70**:2975–2979, (1993).

11. F. Franklin. An examination of the relation between chaotic orbits and the Kirkwood gap at the 2/1 resonance. *AJ*, **107**:1890–1899, (1994).
12. C. Froeschlé. A numerical Study of the Stocasticity of Dynamical Systems with two degrees of freedom. *A& A*, **9**:15-23,(1970).
13. C. Froeschlé. On the number of isolating integrals in systems with three degrees of freedom. *Astrophys. and Space Sciences*, **14**:110, (1971).
14. C. Froeschlé. Numerical study of a four-dimensional mapping. *A.& A.*, **16**:172, (1972).
15. C. Froeschlé. The Lyapunov characteristic exponents and applications. *Journal de Méc. théor. et apll.*, Numero spécial:101–132, (1984).
16. C. Froeschlé. The Lyapunov characteristic exponents. Applications to celestial mechanics. *Celest. Mech.*, **34**,95-115,(1984).
17. C. Froeschlé, Ch. Froeschlé, and E. Lohinger. Generalized Lyapunov characteristic indicators and corresponding Kolmogorov like entropy of the standard mapping. *Celest. Mech. and Dynamical Astron.*, **56**:307–315, (1993).
18. C. Froeschlé, A. Giorgilli, E. Lega, and A. Morbidelli. *On the measure of the structure around an invariant KAM torus.* Proceedings of the IAU symposium 172, Dynamics, Ephemerides and Astrometry of the Solar System, S. Ferraz Melloz et al. (eds), 293-298, (1996).
19. C. Froeschlé and E. Lega. On the measure of the structure around the last KAM torus before and after its break-up. *Celest. Mech. and Dynamical Astron.*, **64**:21–31, (1996).
20. C. Froeschlé and E. Lega. From discrete to continuous dynamical systems and vice-versa. *Revue Economique*, **46**:1511–1526, (1995).
21. C. Froeschlé and E. Lega. Twist angles: a fast method for distinguishing islands, tori and weak chaotic orbits. Comparison with other methods of analysis. *A & A*, submitted, (1997).
22. C. Froeschlé, E. Lega, and R. Gonczi. Fast Lyapunov indicators. Application to asteroidal motion. *Celest. Mech. and Dynam. Astron.*, forthcoming, (1997)a.
23. C. Froeschlé, R. Gonczi. and E. Lega. The fast Lyapunov indicator: a simple tool to detect weak chaos. Application to the structure of the main asteroidal belt. *Planetary and space science*, **45**:881-886, (1997)b.
24. J.M Greene. A method for determining a stochastic transition. *J. Math. Phys.*, **20**:1183, (1979).
25. J.M. Greene, J.M. MacKay, R.S. Vivaldi, and M.J. Feigenbaum. Universal behaviour in families of area-preserving maps. *Physica D*, **3**:468–486, (1981).
26. J.M Greene and C. Percival. Hamiltonian maps in the complex plane. *Physica D*, **3**:530–548, (1981).
27. M. Hénon. Numerical study of quadratic area-preserving mappings. *Quarterly of Applied Mathematics*, **27**:291–312, (1969).
28. M. Hénon. A two-dimensional mapping with a strange attractor. *Commun. math. Phys.*, **50**:69-77, (1976).
29. M. Hénon and C. Heiles. The applicability of the third integral of motion: some numerical experiments. *A. J.*, **1**:73–79, (1964).
30. M. Hénon. Numerical exploration of Hamiltonian systems. *Cours des Houches XXXVI North Holland, Amsterdam 5.*,57-168, (1981).
31. A.N. Kolmogorov. On the conservation of conditionally periodic motions under small perturbation of the Hamiltonian. *Dokl. Akad. Nauk. SSSR*, **98**,524, (1954).
32. J. Laskar. Secular evolution of the Solar system over 10 million years. *A.& A.*, **198**:341–362, (1988).
33. J. Laskar. The chaotic motion of the Solar system. A numerical estimate of the size of the chaotic zones. *Icarus*, **88**:266–291, (1990).
34. J. Laskar. Frequency analysis for multi-dimensional systems. Global dynamics and diffusion. *Physica D*, **67**:257–281, (1993).

35. J. Laskar, C. Froeschlé, and A. Celletti. The measure of chaos by the numerical analysis of the fundamental frequencies. Application to the standard mapping. *Physica D*, **56**:253, (1992).
36. M. Lecar, F. Franklin, and M. Murison. On predicting long-term orbital instability: a relation between Lyapunov time and sudden orbital transitions. *AJ*, **104**:1230–1236, (1992).
37. E. Lega and C. Froeschlé. Numerical investigations of the structure around an invariant KAM torus using the frequency map analysis. *Physica D*, **95**:97–106, (1996).
38. H. F. Levison and M. J. Duncan. The gravitational sculpting of the Kuiper belt. *Astroph.J.Lett.*, **406**:L35–L38, (1993).
39. A.J. Lichtenberg and M.A. Lieberman. *Regular and Stochastic motion.* Springer, Berlin, Heidelberg, New York, (1983).
40. P. Lochak. Hamiltonian perturbation theory: periodic orbits, resonances and intermittency. *Nonlinearity*, **6**:885–904, (1993).
41. R.S. MacKay. *Renormalisation in Area Preserving Maps.* Thesis Princeton, (1982).
42. R.S. MacKay. A renormalisation approach to invariant circles in area-preserving maps. *Physica D*, **7**:283–300, (1983).
43. R.S. MacKay. Greene's residue criterion. *Nonlinearity*, **5**:161, (1992).
44. R.S. MacKay. *Renormalisation in Area Preserving Maps.* World Scientific, (1993).
45. A. Milani and A. Nobili. An example of stable chaos in the solar system. *Nature*, **357**:569–571, (1992).
46. A. Milani, Nobili A., and Knezevic Z. Stable chaos in the asteroid belt. *Icarus*, (1995).
47. M. Moons. Review of the dynamics in the Kirkwood gaps. *Cel. Mech. and Dynam. Astr.*, **65**:175–204, (1997).
48. A. Morbidelli, Zappala E., Moons M., Cellino A., and Gonczi R. Asteroid families close to mean motion resonances: dynamical effects and physical implications. *Icarus*, **118**:132–154, (1995).
49. A. Morbidelli and C. Froeschlé. On the relationship between Lyapunov times and macroscopic instability times. *Celest. Mech. and Dynam. Astr.*, **63** n.2:227–239, (1996).
50. A. Morbidelli and A. Giorgilli. Superexponential stability of KAM tori. *J. Stat. Phys.*, **78**:1607, (1995)a.
51. A. Morbidelli and A. Giorgilli. On a connection between KAM and Nekhoroshev's theorems. *Physica D*, **86**:514–516, (1995)b.
52. M. Murison, M. Lecar, and F. Franklin. Chaotic motion in the outer asteroid belt and its relation to the age of the Solar system. *AJ*, **108**:2323–2329, (1994).
53. A. I. Neishtadt. The separation of motions in systems with rapidly rotating phase. *J. Appl. Math. Mech.*, **48**:133–139, (1984).
54. N.N. Nekhoroshev. Exponential estimates of the stability time of near-integrable Hamiltonian systems. *Russ. Math. Surveys*, **32**:1–65, (1977).
55. D. Nesvorny and S. Ferraz-Mello. On the asteroidal population of the first-order Jovian resonances. *Icarus*, in press, (1997).
56. M. Soper, F. Franklin, and M. Lecar. On the original distribution of the asteroids. *Icarus*, **87**:265–284, (1990).

ON THE RELATIONSHIP BETWEEN LOCAL LYAPUNOV CHARACTERISTIC NUMBERS, LARGEST EIGENVALUES AND MAXIMUM STRETCHING PARAMETERS

CLAUDE FROESCHLÉ

Observatoire de la Cote d'Azur CNRS/ UMR 6529 Bd. de l'Observatoire, BP 4229 06304 Nice, France

ELKE LOHINGER

Observatoire de la Cote d'Azur CNRS/ UMR 6529 Bd. de l'Observatoire, BP 4229 06304 Nice, France
Institut für Astronomie, Universität Wien, Türkenschanzstraße 17, A-1180 Vienna, Austria.

AND

ELENA LEGA

Observatoire de la Cote d'Azur CNRS/ UMR 6529 Bd. de l'Observatoire, BP 4229 06304 Nice, France
CNRS-LATAPSES, 250 Rue A. Einstein, 06560 Valbonne, France

1. Introduction

The computation of the Lyapunov Characteristic Indicators (LCI hereafter) remains the only tool which allows to quantify the chaoticity of an orbit for a given dynamical system and given initial conditions. The introduction of the Local Lyapunov Characteristic Numbers (Froeschlé et al. 1993, LLCNs hereafter), called also Stretching parameters (Contopoulos 1995), has allowed to reveal the complex structure of both the regular and the chaotic zone. Already in 1970 (Froeschlé 1970) some connection was made between the characterization of chaos and the variation of the largest eigenvalue of the tangential mapping $D(T^n)$ of the mapping T^n, where T^n maps the point P_1 to P_n through n iterations of the mapping T. Through the work of Galgani, Benettin et al. (1980) the attention has been focused on the variation of the evolution of a given initial tangential vector whose length $||\vec{v}_n||$ after n-iterations seems to be closely connected to the absolute value of the above defined eigenvalue. In fact the largest LCI is well approximated by $\ln|\lambda_n|/n$ and this corresponds also to $\ln||\vec{v}_n||/n$. In both cases a renormalization procedure was applied either on the coefficients of the tangential mapping or on the length of the vector $\vec{v}_n$. The LLCNs have been defined using the second approach, in fact if we renormalize the tangential vector at each iteration we can consider the set of its norms (the LLCNs) as a distribution whose mean is the largest Lyapunov exponent. The aim of the paper is to explore if there exists some connection between the LLCNs and the distribution of the largest eigenvalues of the mapping $D(T)$ which maps P_{k-1} to P_k. In other words we want to compare the distribution of the LLCNs, i.e. of the set $\ln(||v_{k+1}||/||v_k||)$ for $k = 1, ...N$, and the distribution of the $\ln\lambda_k$, where λ_k is the largest eigenvalue of the mapping $D(T)$. In some cases the largest eigenvalue of the mapping $D(T)$ corresponds to the largest possible increase of a tangential vector evolving under the mapping T: $\ln(\sup_\theta ||v_k(\theta)||)$ with θ varying in the interval $[0, 2\pi]$. Therefore, we will consider also the distribution of the $\ln(\sup_\theta ||v_k(\theta)||)$ which we will call Maximum Stretching Parameters (MSPs hereafter).

B.A. Steves and A.E. Roy (eds.), The Dynamics of Small Bodies in the Solar System, 503–508.

2. The standard map as a model problem

We consider the Standard map as a model problem (Froeschlé 1970, Lichtenberg & Lieberman 1983):

$$T = \begin{cases} x_{k+1} &= x_k + \epsilon \sin(x_k + y_k) \pmod{2\pi} \\ y_{k+1} &= y_k + x_k \qquad\qquad\quad \pmod{2\pi} \end{cases} \tag{1}$$

Such a mapping exhibits all the well-known typical features of problems with two degrees of freedom, such as invariant curves, "islands", and stochastic zones where the points wander in a chaotic way. We consider the tangential mapping DT:

$$DT = \begin{cases} u_{k+1} &= (1 + \epsilon \cos(x_k + y_k))u_k + \epsilon \cos(x_k + y_k)v_k \\ v_{k+1} &= u_k + v_k \end{cases} \tag{2}$$

u and v being the coordinates of a vector $\vec{V}$. Given an initial condition P_0 and an initial vector $\vec{V_0}$ we consider the distribution of the quantities:

$$\gamma_k = \ln\left(\frac{||v_k||}{||v_{k-1}||}\right) \tag{3}$$

which are the so called LLCNs (Froeschlé et al. 1993) or the stretching parameters (Contopoulos 1995). On the other hand we consider the distribution of the largest eigenvalue of the mapping DT computed on each point of the orbit:

$$\tilde{\gamma}_k = \ln|\lambda_k| \tag{4}$$

The third distribution is given by the set of largest tangential vectors:

$$\hat{\gamma}_k = \ln(\sup_\theta ||v_k(\theta)||) \quad 0 \leq \theta \leq 2\pi \tag{5}$$

The set of Figures 2,4,6,8 shows the different distributions of these quantities and associated orbits (Fig.1,3,5,7). The different cases correspond to two regular orbits for the circulation (Fig.1) and libration (Fig.3) cases and to two chaotic orbits (Fig. 5,7) for different values of the LCI. On all cases it appears clearly that there is no real connection between the distribution of the LLCNs and that of the MSPs. However, for the regular cases the distributions look very similar in shape (Fig. 2,top and Fig. 4,top) and seem to be deducible one from the other by an affine transformation. In both regular cases the LLCNs do not have all the available values between the maximum and the minimum of the stretching parameters. In the chaotic cases (Fig. 6,top and Fig. 8,top) all the available range is used and the shape do not show the same similarities. However, for $\epsilon = -10$ (Figs. 7,8), i.e. for a strong chaotic case for which not only the original invariant tori have disappeared but also the size of the islands has drastically shrunk, the distributions of the LLCNs mainly concerns the range of the maximum stretching parameters. When we look at the distribution of the eigenvalues $\tilde{\gamma}(k)$ the information that we get concerns the proximity of the point to either an elliptic or an hyperbolic fixed point. In the case of the circulation torus (Fig. 2,bottom) almost the half of the eigenvalues are complex, i.e. the orbit is locally elliptic for almost half of its life. For the libration case the eigenvalues are complex and have modulo 1. In this case (Fig. 4,bottom) we have considered the distribution of the real part of the eigenvalues which is related to the stretching of a vector, while the imaginary part is related to the rotation. In this way we obtain a distribution similar to that of the LLCNs. For the chaotic cases (Fig. 6,bottom and Fig.8,bottom), the distribution of the LLCNs and of the largest eigenvalues have the same range but they do not show any similarity in shape.

If we compute the largest LCI using the two definitions recalled in the introduction, i.e. by computing $\ln|\lambda_n|/n$ and $ln||\vec{v}_n||/n$, we get (Fig.9) exactly the same result for the chaotic case (Fig.9, top) and the same behaviour for the two regular orbits corresponding to a circulation torus (Fig.9, middle) and to a libration island (Fig.9, bottom). In this case both the eigenvalue λ_n and the vector $\vec{v}_n$ take into account all the previous history of the orbit, and this is the case also for the distribution of the LLCNs while both the maximum

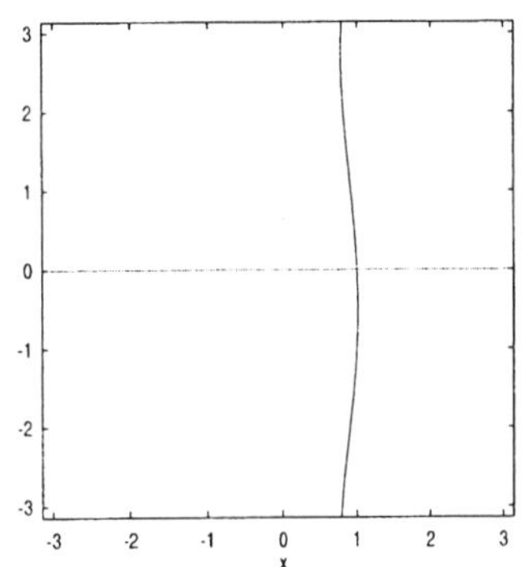

gure 1: The standard map for $\epsilon = -0.1$ for a circulation rus with initial conditions $x_0 = 1$, $y_0 = 0$.

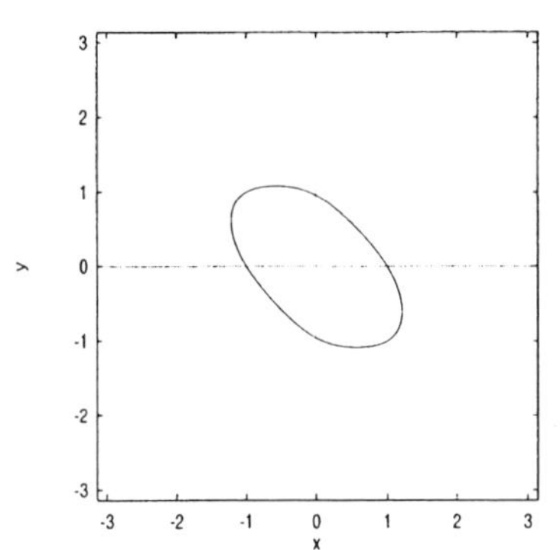

Figure 3: The standard map for $\epsilon = -1.3$ for a libration island with initial conditions $x_0 = 1$, $y_0 = 0$.

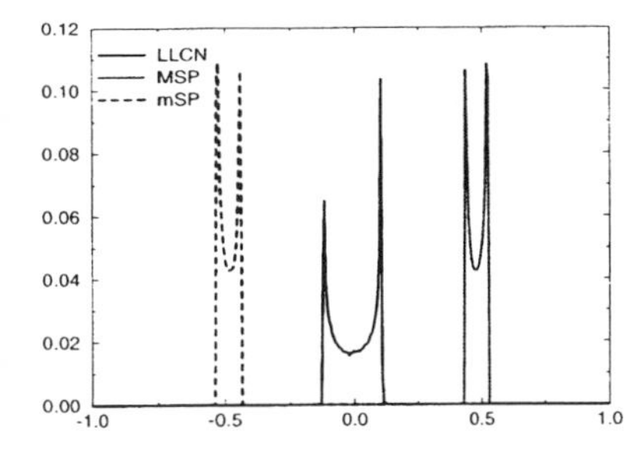

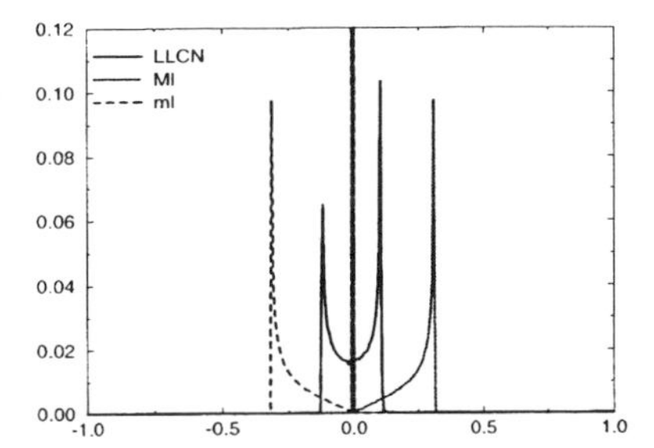

igure 2: (top) Distribution of the LLCNs and of the max- num and minimum stretching parameters (labeled respec- vely MSP and mSP) for the orbit of Fig.1. (bottom) Dis- ibution of the LLCNs and of the largest and smaller eigen- alues for the orbit of Fig.1. The distributions are plotted the same interval than Fig.2 (top) in order to compare e results. So doing we have cut the peak corresponding the set of complex eigenvalues which are almost the 45% the total.

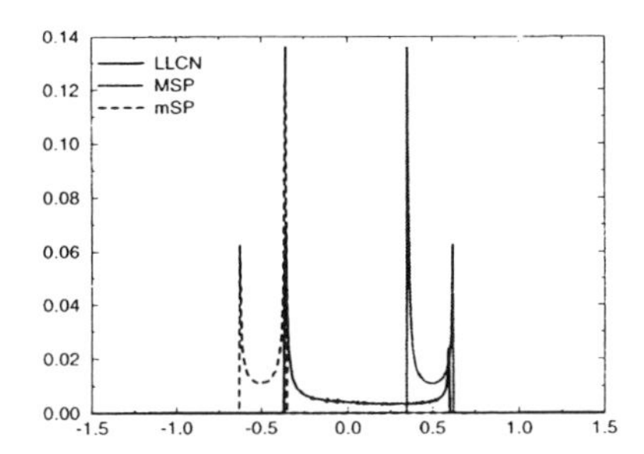

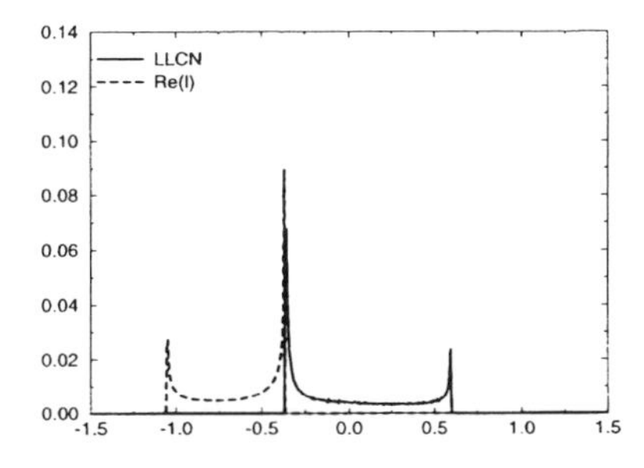

Figure 4: Distribution of the LLCNs and of the maximum and minimum stretching parameters (labeled respectively MSP and mSP) for the orbit of Fig.2. (bottom) Distribution of the LLCNs and of the eigenvalues for the orbit of Fig.2. The distributions are plotted in the same interval than Fig.4 (top). In this case all the eigenvalues are complex so we have plotted the distribution of the real part of the eigenvalues which is relate to the stretching of a vector, while the imaginary part is related to the rotation.

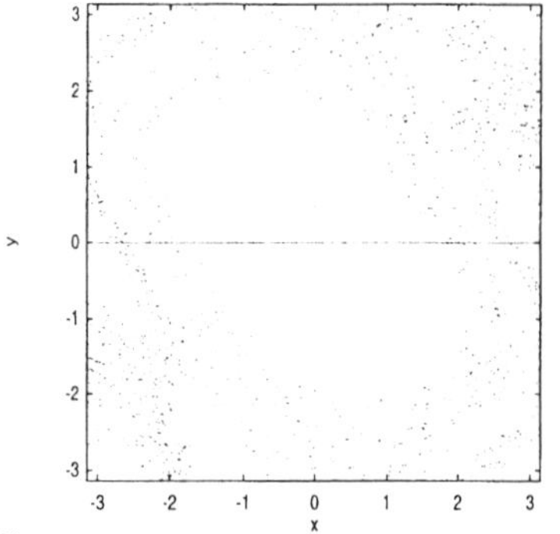

Figure 5: The standard map for $\epsilon = -1.3$ for a chaotic orbit with initial conditions $x_0 = 2$, $y_0 = 0$.

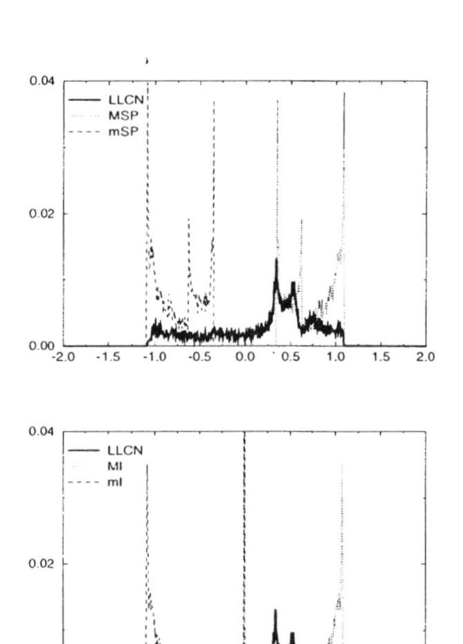

Figure 6: (top) Distribution of the LLCNs and of the maximum and minimum stretching parameters (labeled respectively MSP and mSP) for the orbit of Fig.5. (bottom) Distribution of the LLCNs and of the eigenvalues for the orbit of Fig.5. The distributions are plotted in the same interval than Fig.6 (top). So doing we have cut the peak corresponding to the set of complex eigenvalues which are almost the 35% of the total.

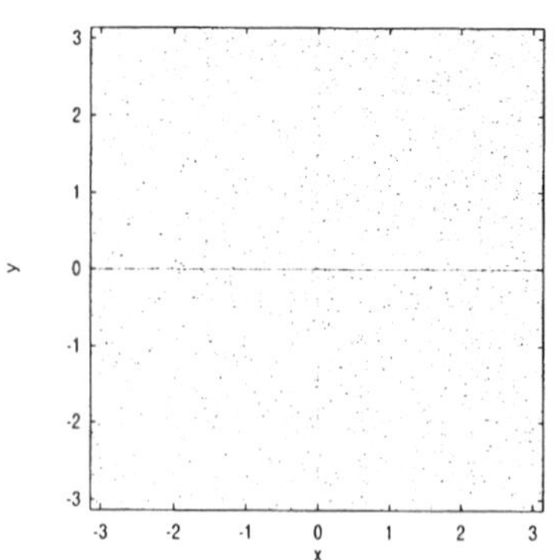

Figure 7: The standard map for $\epsilon = -10.$ for a ch orbit with initial conditions $x_0 = 1$, $y_0 = 0$.

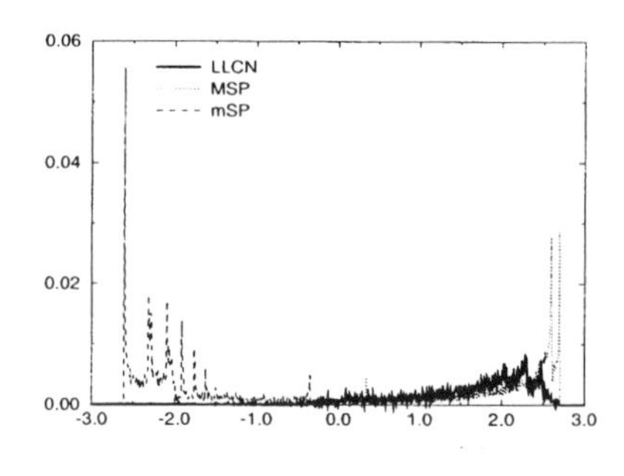

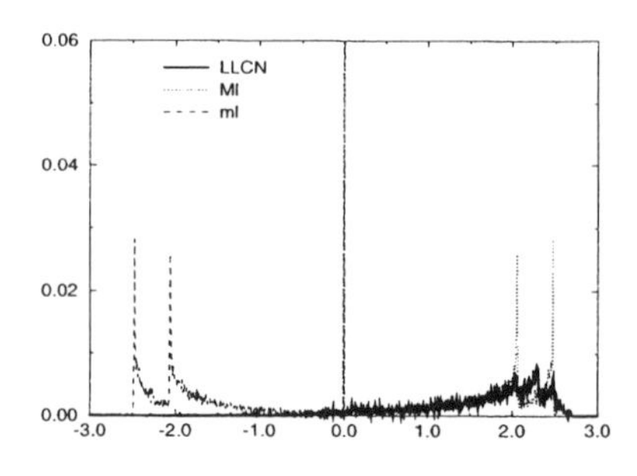

Figure 8: (top) Distribution of the LLCNs and of the r imum and minimum stretching parameters (labeled res tively MSP and mSP) for the orbit of Fig.7. (bottom) tribution of the LLCNs and of the eigenvalues for the c of Fig.7. The distributions are plotted in the same inte than Fig.8 (top). So doing we have cut the peak corresp ing to the set of complex eigenvalues which are still the of the total.

stretching parameter and the largest eigenvalue depend only of the point of the phase space where they are computed.

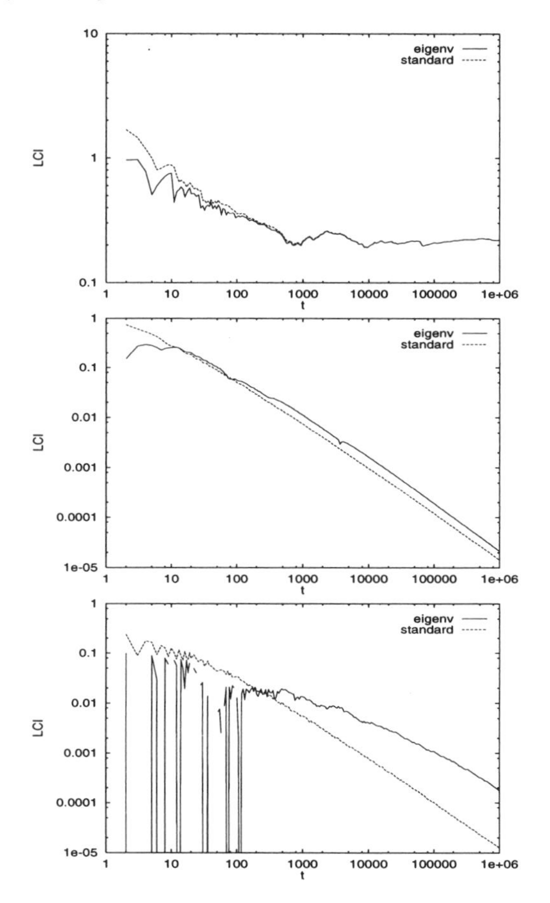

Figure 9. Variation of the LCI as a function of time for the chaotic orbit of Fig.5 (top), for the circulation torus of Fig.1 (middle) and for the libration island of Fig.3 (bottom). The LCI are obtained computing both the largest eigenvalue of the mapping T^n (label 'eigenv' in the Figures) and the norm of a vector v_n (label 'standard' in the Figures).

3. Conclusion

We have clearly shown that there is only a loose connection for very chaotic orbits between the distribution of the LLCNs and the distribution of the maximum stretching parameters and that of the largest eigenvalues. Actually, this is a consequence of the fact the LLCNs are dependent of the previous history of the orbit and not only of the point of the phase space where they are computed as it is the case for the MSPs and the largest eigenvalues.

References

G. Benettin, L. Galgani, A. Giorgilli, and J.M. Strelcyn. Lyapunov characteristic exponents for smooth dynamical systems; a method for computing all of them. *Meccanica*, **15**:Part I : theory, 9–20 – Part 2 : Numerical applications, -21–30, (1980).

G. Contopoulos, E. Grousousakaou, and N. Voglis. Invariant spectra in Hamiltonian systems. *Astron. Astrophys.*, **304**:374–380, (1995).

C. Froeschlé. A numerical study of the stocasticity of dynamical systems with two degrees of freedom. *Astron. Astrophys.*, **9**:15–23, (1970).

C. Froeschlé, Ch. Froeschlé, and E. Lohinger. Generalized Lyapunov characteristic indicators and corresponding Kolmogorov like entropy of the standard mapping. *Celest. Mech. and Dynamical Astron.*, **56**:307–315, (1993).

A.J. Lichtenberg and M.A. Lieberman. *Regular and Stochastic motion.* Springer, Berlin, Heidelberg, New York, (1983), p.218.

"STICKINESS" IN DYNAMICAL SYSTEMS

RUDOLF DVORAK
Institute of Astronomy, University of Vienna
Türkenschanzstrasse 17, A-1180 Vienna, AUSTRIA

Abstract. In the first part we develop the basic ideas of motions in discrete dynamical systems described by difference equations (mappings) and in continuous dynamical systems of 2 degrees of freedom described by differential equations (flows). Then the method of the "Poincaré Surface of Sections" is explained as a main instrument to distinguish between regular and chaotic motion. We expose the restricted three body problem to clarify how the method of surface of section works and show the complicated structure of the phase space of dynamical systems. Additionally, we mention different methods of characterizing chaotic motion briefly. Then we recall the rôle of the cantori in connection with motions close to invariant curves (quasiperiodic motion on KAM tori): they are destroyed KAM tori with an infinite number of gaps (rapidly growing with larger non-linearity parameters). We also explain how we can approximate cantori with the aid of the continuous fraction procedure using the Farey-tree.

In the second part, we concentrate on the numerical investigation of regions closed to invariant curves (islands) in the standard map with a stochasticity parameter $K = 5$, where only two islands of regular motion survive. In a numerical approach we integrated some 10^6 orbits for 10^8 iterations and checked how long they stay in a well defined region close to the last KAM curve around the major island. It was confirmed that there exist orbits – which are called "sticky" orbits – which stay for very long times near the last KAM torus before they escape into the surrounding chaotic region. Our results show that there is a clear distinction between two different types of such stickiness regions: an inner one confined by cantori and an outer one which shows a characteristic fractal chararacter.

In the third part we show results of an extensive study of motions close to the last KAM torus in the Sitnikov Problem. This dynamical system of $2\frac{1}{2}$ degrees of freedom consists of two equally massive primary bodies which move on elliptic orbits around their barycenter whereas a third massless body is oscillating along the line perpendicular to the plane of motion of

B.A. Steves and A.E. Roy (eds.), The Dynamics of Small Bodies in the Solar System, 509–534.

the primaries. It turns out that we also can distinguish between two different sticky regions with a clear fractal structure of the outer one.

1. Basic Ideas

Although **time** is a variable which is "known" to be continuous since the ancient Greek philosophers, it was primarely Newton who used **time** for the mathematical description of dynamical processes in nature in form of differential equations. But it may also be possible to describe a dynamical system with the aid of difference equations where time is treated as being discrete. Such systems can either stem from the successive intersection of a dynamical trajectory with a certain surface of section in phase space, or just be approximations to solutions which are then valid only for a limited time span or limited regions in phase space. But they may also come directly from the dynamical system under consideration. The relation between Hamiltonian differential equations – as description of dynamical systems – and area preserving mappings (in form of difference equations) can be found in the literature: e.g. in an introductory and beautifully clear written way in Percival and Richards (1989), more detailed in Lichtenberg and Liebermann (1983) and in Guckenheimer and Holmes (1990).

1.1. THE EULER MAP

Let us define the first order differential equation

$$\dot{x} = f(x,t); x(0) = x_0 \tag{1}$$

where, for every point in the plane (x,t), the slope of the curve is given by $f(x,t)$. Suppose we know the solution for an instant $t = t_n$, $x(t_n) = x_n$. After an infinitesimally small time Δt, e.g. $t_{n+1} = t_n + \Delta t$, we can compute the solution

$$x(t_n + \Delta t) = x(t_{n+1}) = x_{n+1} = x_n + \Delta t \dot{x}_n. \tag{2}$$

From this we can derive the mapping

$$x_{n+1} = x_n + \Delta t f(x_n, n.\Delta t) \tag{3}$$

with $t_n = n \cdot \Delta t$; this is known as the Euler map and corresponds to an approximate solution of eqn. (1). Solving the Euler Map and performing the limiting process $\Delta t \rightarrow 0$ we can – in principle – solve any system of ordinary differential equation of the nonautonomous form

$$\dot{x} = F(x,t) \tag{4}$$

simply by solving the mapping

$$x_{n+1} = f(T, x_n); x(0) = x_0 \tag{5}$$

where T is a certain constant time step over which we say that the solution is mapped.

The precision of the solution over a certain time span depends on the time step Δt; it deviates relatively fast from the solution of eqn. (1), unless the right hand side of this equation can be found by an explicit solution x_n for the mapping after n iterations.

1.2. HAMILTONIAN SYSTEMS

The motion of a particle in a plane which obeys Newton's laws of motion can be described by the second order differential equations

$$\ddot{x} = F(x, y.t) \tag{6}$$

$$\ddot{y} = G(x, y, t) \tag{7}$$

The forces F(x,y,t) and G(x,y,t) may always be derived from a potential V(x,y,t)

$$F(x, y) = -\frac{\partial V}{\partial x}(x, y, t) \tag{8}$$

$$G(x, y) = -\frac{\partial V}{\partial y}(x, y, t) \tag{9}$$

Defining $\vec{q} = (x, y)$ (generalized coordinates) and $\vec{p} = \dot{\vec{q}} \cdot m$ as (conjugate momenta) we call the pair $(\vec{q}, \vec{p})$ conjugate variables. The Hamiltonian function H of the system reads

$$H(\vec{q}, \vec{p}) = \frac{\vec{p}^2}{2m} + V(\vec{q}, t). \tag{10}$$

We can write the so called Hamilton's equations of motion

$$\dot{\vec{q}} = \frac{\partial H}{\partial \vec{p}}(\vec{p}, \vec{q}, t) \tag{11}$$

$$\dot{\vec{p}} = -\frac{\partial H}{\partial \vec{q}}(\vec{p}, \vec{q}, t). \tag{12}$$

When H, respectively V, does not depend explicitely on the time we speak of a conservative system and then the value of H is always conserved; in Newtonian systems it is also equal to the energy E of the system.

It can be shown (e.g. Persival and Richards loc.cit. p.57) that a conservative Hamiltonian system is equivalent to an area preserving mapping. This fact is widely used to approximate a dynamical system by a mapping which is much easier to explore than the respective differential equation. It is immediately clear, that a mapping – if we are able to find one which describes the dynamical system in question – can be explored numerically extremely fast compared to numerical integration methods of ordinary differential equations.

1.3. THE CIRCULAR RESTRICTED THREE BODY PROBLEM AND THE SURFACE OF SECTION

Now we want to treat an important problem of Celestial Mechanics and describe the motion of a massless body under the influence of two massive bodies (primaries) moving on circular orbits (in a plane). Making use of uniformly rotating coordinates ξ and η, where the primaries have fixed positions on the ξ axis, we can write the following equations for the **circular restricted three body problem** (e.g. Szebehely, 1970, Roy, 1988, Stumpff, 1970):

$$\ddot{\xi} - 2\dot{\eta} = \frac{\partial \Omega}{\partial \xi} \tag{13}$$

$$\ddot{\eta} + 2\dot{\xi} = \frac{\partial \Omega}{\partial \eta} \tag{14}$$

with $\Omega = \frac{1}{2}(\xi^2 + \eta^2) + \frac{1-\mu}{r_1} + \frac{\mu}{r_2}$ being the potential function in the rotating system; r_1 and r_2 are the actual distances between the primaries – with masses μ and $1-\mu$ – and the massless body. Due to the existence of an additional integral the system of 4th order can be reduced to third order. This means that, e.g. with the aid of the **Jacobian integral** we may replace one of the coordinates of the phase space $(\xi, \dot{\xi}, \eta, \dot{\eta})$. A further reduction of the now three dimensional phase space (e.g. $\xi, \dot{\xi}, \eta$) is possible when we introduce a **Poincaré surface of section** (SOS henceforth) setting $\eta = 0$, which is illustrated in Fig. 1.

1.4. THE STRUCTURE OF PHASE SPACE

According to Fig.1 we can distinguish the following distribution of points on the SOS:

- the consecutive points coincide always with already existing ones in the SOS, thus having a fixed period; this fact corresponds to a motion in phase space along a closed curve; we then speak of a **periodic motion** (Fig. 1b),

Figure 1. Orbit in phase space with surface of section (upper left graph), periodic (upper right), quasiperiodic (lower left) and chaotic motion (lower right graph)

- the points fill more and more a closed curve on the SOS, but never one point coincides with an already existing one; in phase space the **quasiperiodic motion** stays forever on the surface of a torus (Fig. 1c),
- the points on the SOS are more or less scattered; this is characteristic for **chaotic motion** (Fig. 1c).

A complete picture of motions on tori is developed by the KAM theory named after Kolmogoroff (1954), Arnold (1963) and Moser (1962). Several textbooks explain the complicated theory more or less in detail (e.g. Lichtenberg and Liebermann, loc.cit. p 159ff). Briefly speaking, there are tori in an integrable system which are just deformed under a sufficiently small perturbation; as a consequence there are quasiperiodic motions which remain quasiperiodic.

1.5. METHODS OF DETECTION OF CHAOTIC MOTION

We briefly list some of the methods commonly used to distinguish between the different types of motion in a dynamical system.

- The simplest method – although not a mathematically strict one – is to check the respective SOS under what category (see the former subsection) the consecutive points fall. This is sometimes difficult when the scale is not fine enough (see e.g. Lega and Froeschlé, 1997) and a "line" can turn out to be in fact a scattered chain of points along a line.
- Commonly used are the maximum Lyapunov Characteristic Numbers (LCN, sometimes also called Lyapunov Characteristic Indicators) – where d(0) is the initial deviation of two nearby orbits and d(t) characterizes the deviation at time t. These deviations are assumed to be infinitesimally small and can be computed by solving the variational equations together with the equations of motions (or the mapping). The LCNs are defined as follows (e.g. Froeschlé, 1984):

$$\chi = \chi(t) = \frac{1}{t} ln \frac{d(t)}{d(0)} \tag{15}$$

$$LCN = \lim_{t \to \infty} \chi \tag{16}$$

  A motion is chaotic when the maximum LCN is positive; otherwise the motion is regular.
- Using the "frequency map analysis" (Laskar 1993) one is able to detect chaos over shorter time scales it is also possible to estimate the size of chaotic zones in phase space.
- The rotation number (RN) is defined as the averaged angle of two consecutive points (with respect to a center, which is in general a stable periodic orbit; Lichtenberg and Liebermann loc.cit.). The RN is well defined only if the orbit is on an KAM curve. Thus, in an integrable system the RN is always defined; when the dynamical system is not integrable there are regions where the RN of an orbit is not defined; this means that the orbits under consideration are chaotic. Note that a periodic orbit has a RN determined by the multiplicity of its periodicity.
- The sup-map analysis (Froeschlé and Lega, 1996) consists in detecting the maximum of the action for a specific orbit. For close by orbits "a monotonic variation of the sup corresponds to a set of invariant KAM tori, a jump followed by a plateau indicates the crossing of a chaotic zone and the crossing of nested islands give rise to v-shaped structures."(loc.cit)
- The helicity angle (Contopoulos and Voglis, 1996, 1997) is the angle ϕ_i of the deviation $\vec{\xi}$ with a fixed direction (e.g. the x-axis in the SOS or a 2-dimensional mapping). With its aid one can quite well determine

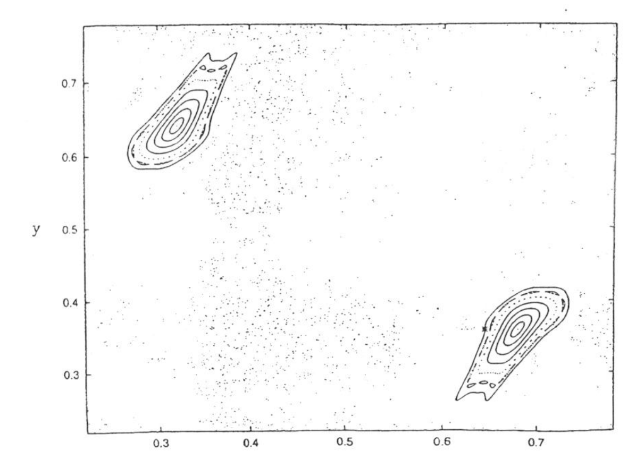

Figure 2. The Standard Map for K=5

whether orbits are chaotic or regular; this method even determines automatically the RN of orbits in secondary islands.

- The spectra of the maximum local LCN (or stretching numbers) can also be used to characterize orbits. When the distribution is symmetric it is regular orbit; when it is asymmetric (positive) the orbit is chaotic (for details see Contopoulos et al. 1997).

1.6. THE CANTORI

We already mentioned the structure of phase space of a dynamical system with – roughly speaking – three different types of orbits. To understand their complicated structure and the importance of the socalled cantori we show the standard map (SM) with a stochasticity parameter K=5, where only two islands of regular motion survive (Fig. 2).

In this mapping the perturbing parameter K (stochasticity or nonlinearity parameter) acts such that from a certain value of K on, the outer KAM tori (on the edge to the chaotic region) are destroyed to cantori. (Fig. 3 and Fig.4) Thus a cantorus is the remnant of a KAM torus, which was destroyed due to the strength of the perturbation on the edge of the island. It can be regarded a set of invariant points with the measure zero forming a cantor set. KAM tori and cantori have irrational rotation numbers and – because irrationals can be approximated by rationals with larger and larger numerators and denominators – cantori can be approximated by a series of periodic orbits with increasing "order". Consequently the cantorus itself can be regarded as a periodic orbit with an infinite period (see e.g. Aubry 1978, Percival 1979, Meiss 1992). In Fig.3, we see the island on the

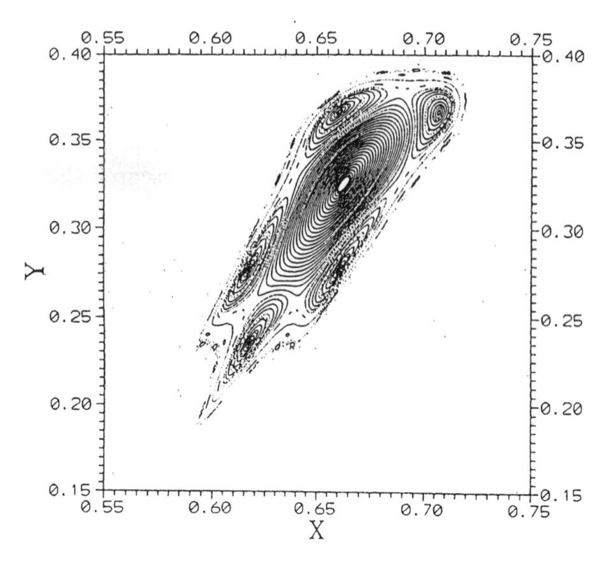

Figure 3. The Standard Map for K=4.79

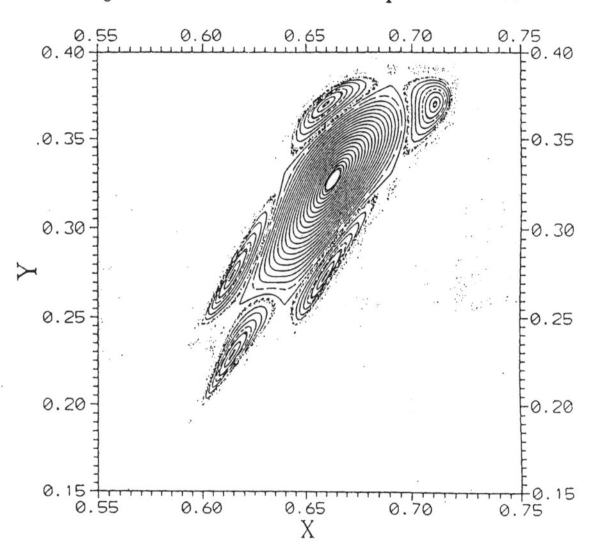

Figure 4. The Standard Map for K=4.80

right lower part of Fig. 2 for a K = 4.79. There still is a strongly deformed invariant torus around the 5 islands of the 5:3 stable periodic orbit. The region of large chaos and escapes is outside this curve. In Fig. 4 we plotted the same island for a slightly larger nonlinearity parameter (K=4.80). We recognize that the "last" KAM curve degenerated to a cantorus.

As already mentioned above KAM - tori and cantori are characterized by irrational RNs which can be approximated by rationals. The best method to approximate these irrationals is by successive truncation of their continued

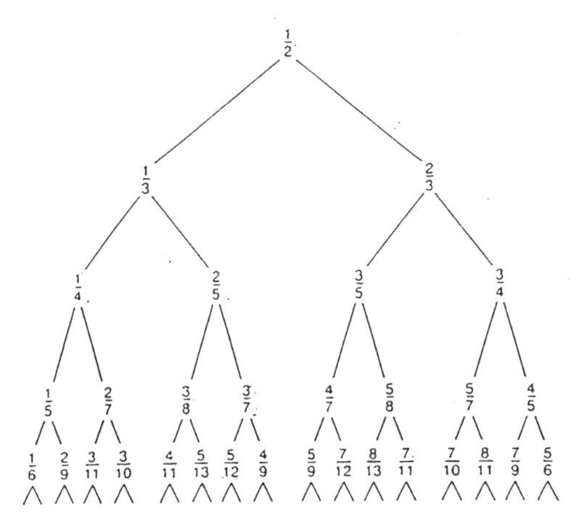

Figure 5. The Farey Tree

fraction representation. According to Greene (1979) the last KAM tori to be destroyed are those with the **noble** RNs:

$$a = [a_1, a_2, a_3, ...] \equiv \frac{1}{a_1 + \frac{1}{a_2 + \frac{1}{a_3 + ..}}} \tag{17}$$

where the a_i are integers, with $a_i = 1$ for all i above a certain number N (the order of the noble number). The $a = [2, 1, 1, 1, ...]$ is of order 1 and $[1, 1, 1, ...] = \frac{1}{2}(\sqrt{5} - 1)$, the **golden mean**, is of order 0. Some characteristics of the noble numbers are the following ones:

– for every noble number of order 1 the following relation is true

$$[1, 1, ...] > [2, 1, ...] > [3, 1, ...] > ... > [\infty, 1, ...] = 0 \tag{18}$$

– The 2nd order noble numbers are between the 1st order noble numbers

$$[1, 1, 1, ...] > \frac{1}{2} = [2, \infty, 1, ...] > ... > [2, 3, 1, ...] > \\ > [2, 2, 1, ...] > [2, 1, 1, ...] \tag{19}$$

It is possible to construct the successive truncations of the noble numbers with the aid the so called Farey tree.

As one can see from Fig.5, the following rules of construction holds: each rational of the sequence has a numerator and denominator which are the sum of the two previous rationals. Additionally the numerator multiplied with the denominator of the following rational minus the denominator of the first rational multiplied with the nominator of the following rational is always ± 1.

What is now the connection between noble numbers and the degeneration of KAM tori? As already mentioned above a KAM curve possesses an irrational RN; some of them have even noble numbers as RN and these are exactly the ones which resist the longest when the perturbations increase. It should also be noted that there is not only one KAM torus with a noble number, which then degenerates to a cantorus, but a whole sequence corresponding to one and the same order N. Every noble number can be approximated by rationals of a sequence in the Farey - tree and these rationals correspond to periodic orbits in the dynamical system. As we continue down in the Farey Tree $\swarrow\searrow\swarrow\searrow$ ... we approximate the noble number better and better. Consequently, as the order of truncation increases, the set of periodic orbits approaches closer and closer the corresponding cantori. Thus, we have found a mean to construct numerically the approximate location of a cantorus.

2. The phenomenon of Stickiness

In an early paper Contopoulos (1971) tried to find the last KAM curve around a stable periodic orbits (island) in a Hamiltonian system of 2 degrees of freedom. For orbits starting very close to the border of this island it was found that they stay for very long times close to that border before they escape into the surrounding chaotic domain. The name "stickiness" to this phenomenon was given in a paper by Shirts and Reinhardt (1982). Several interesting investigations devoted to this topic were undertaken in the following years (e.g. Greene, 1979, MacKay et al., 1984); the first systematic study were published by Contopoulos et al. (1997 = paper I) and then by Efthymiopoulos et al. (1997 = paper II). In paper I we studied the different diffusion times in different regions of the "stickiness" zone of the Standard Map, which is exponentially dependent on the distance from the last KAM curve around an island. In paper II, the problem of finding numerically the borders of the sticky regions was discussed in detail and for the first time we succeeded to find numerically an orbit crossing the holes of the bordering cantorus. Dvorak et al. (1997 = paper III) compared the results of the structure of the sticky regions in mappings with the results of a study in a dynamical system. We will discuss in the following the stickiness regions in the Standard Map (SM) and compare them also with new results concerning the structure of the regions in the Sitnikov Problem (SP), a dynamical system which will be defined later.

In the former mentioned papers we found out that the best method to study the structure of the stickiness region is a purely numerical one. The initial conditions were chosen close to the invariant curves; we then varied the initial conditions of the mapping with very small steps and examined

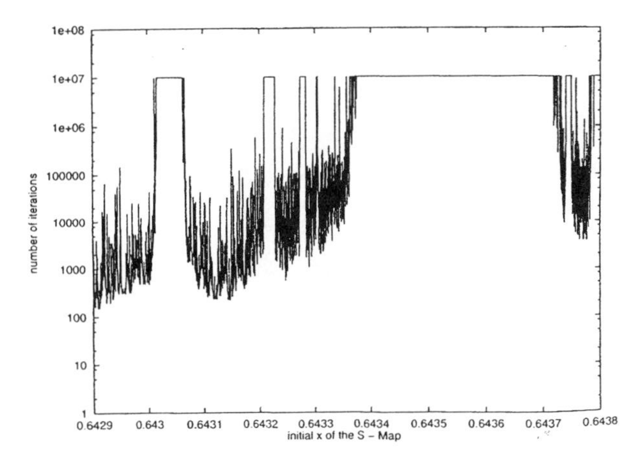

Figure 6. Stickiness Plot: Logarithm of the Number of sticked orbits versus the initial x of the Standard Map

how long (how many mapping steps or how many consecutive intersections with the SOS) an orbit spent in a clearly defined region close to the island ("main land" around the central fixed point in the dynamical system). The results are presented in the form of the following plots: initial x -value of the SM – z-value in the SP – versus the logarithm of the number of iterated points within a certain region (ellipse around the islands in the SM; distance from the central fixed point in the SOS of the SP). In Fig. 6 such a plot for a region very close to the left edge (star) of the lower right island of Fig. 2 for the SM is shown.

It should be mentioned that initial conditions of orbits inside an island are trapped there for all iterations. Thus an island manifests as a plateau on the top of the plot (see Fig.6). The other orbits – escaping after a certain number of iterations – are "visible" through many sharp peaks in that plot. After careful test calculations we fixed the following numbers: 10^7 and 10^8 iterations for the mapping and 1591 and 15915 intersecting points in the SOS of the SP. We also had to select a grid for the initial conditions to put into evidence the complicated structure of the stickiness regions. It was not a problem for the SM but it was an important point of tests for the investigation of the SP. It is a very time consuming computation to integrate orbits in phase space compared to iterations in a mapping derivable from simple arithmetic operations like eqn. (20) for the SM. In total some 10^3 hours on 3 DEC alpha stations and 2 IBM RISC 6000 workstations were necessary to derive the results which we discuss here!

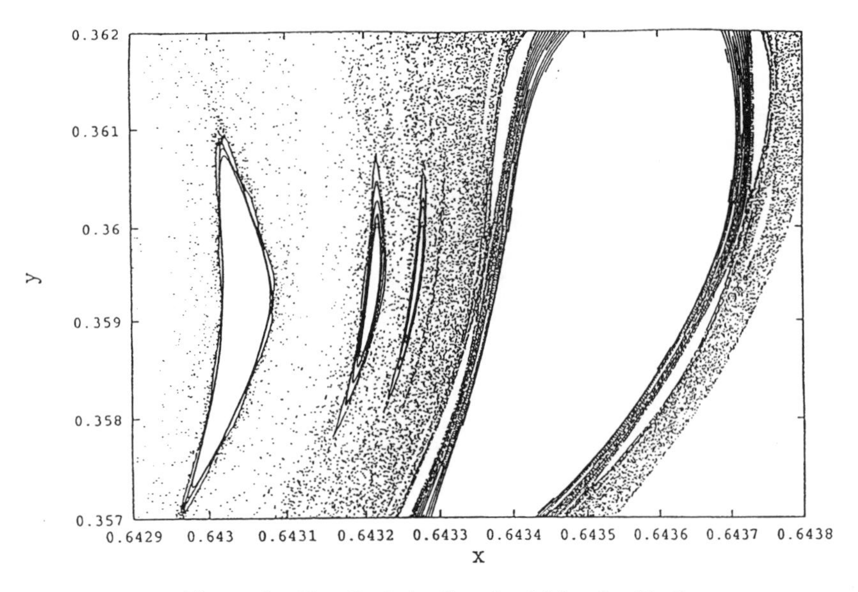

Figure 7. Detail of the Standard Map for K=5

3. The Standard Map

We investigated the standard map in the form

$$\begin{aligned} x_{i+1} &= x_i + y_{i+1} \\ y_{i+1} &= y_i + \frac{K}{2\pi} sin(2\pi x_i) \end{aligned} \qquad (20)$$

with a stochasticity parameter K=5. One can see that with such a large value of the parameter K there survive only two islands of regular motion (Fig. 2).

In Fig. 7 the mapping in the region of interest (marked by a star on the left edge of Fig. 2) is shown in detail. The line for y=0.360 corresponds to the initial conditions chosen to produce Fig. 6, where the logarithm of the number of iterated points which stay within a certain region around the island is plotted versus the initial x -value (30000 with a $\Delta x = 3.10^{-8}$) of the SM. We extended our computation to smaller x values and show the results in Fig. 8. Here we distinguish clearly between an inner sticky region I (=RI) with an exponential growth of the number of trapped orbits when approaching the island (right side of the plot) and a sticky region II (RII) on the left side. Several peaks with escape times up to 10^3 in a region of more or less immediate escape are visible for the initial conditions $0.632 < x < 0.6425$ on Fig. 8.

One can see that the RI is attached to the island itself; the larger RII is on the outside and shows an interesting structure where the peaks look

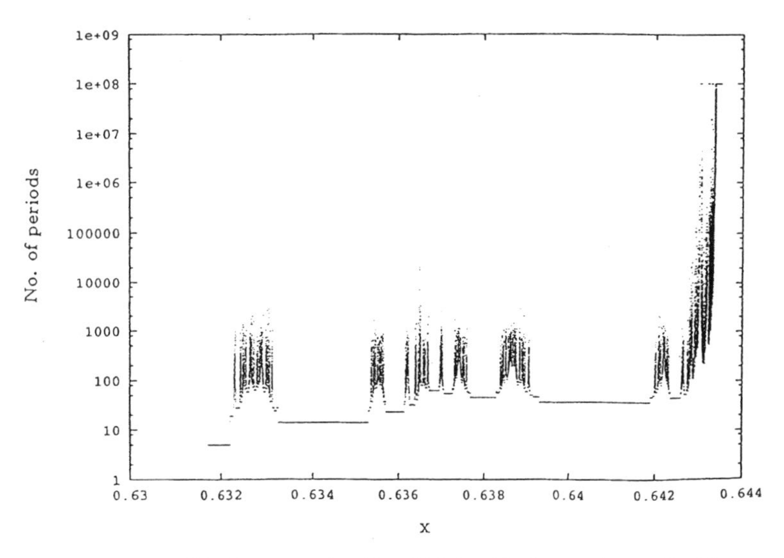

Figure 8. Stickiness region I and II for K=5; cut for y=0.360

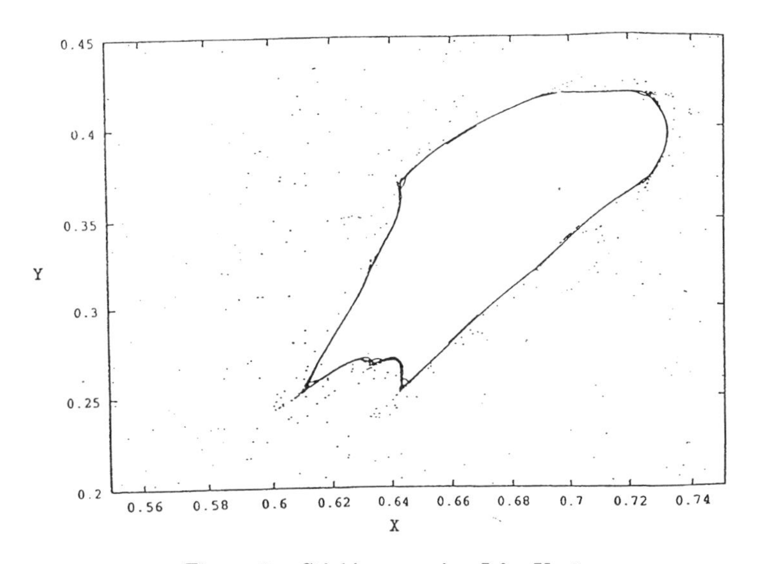

Figure 9. Stickiness region I for K=5

quite similar. In the following Figs. 9 and 10 we show the two different regions in the map (x – y plot). We always started with initial conditions inside the regions and one can see, how the RI is confined from inside by the last KAM curve and from outside by a border which still has to be determined.

Fig. 10 shows that orbits initially started in RII have a completely

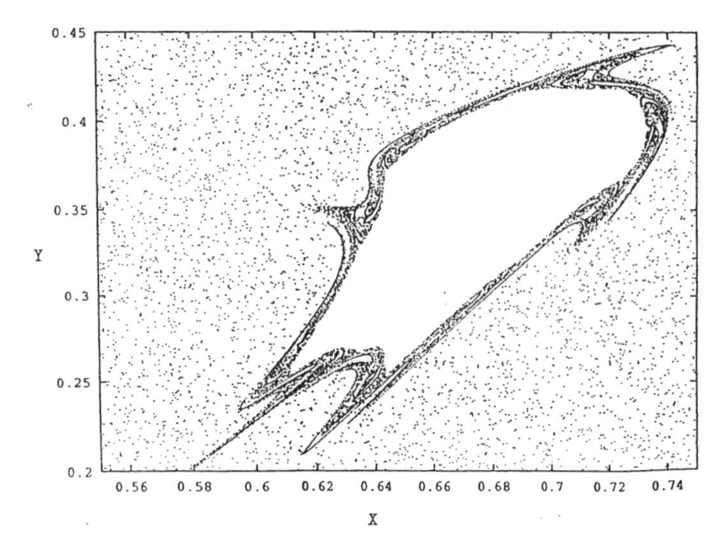

Figure 10. Stickiness region II for K=5

different behaviour than the ones started in RI. First of all the region where they stay is somewhat more extended and show structures with lobes and holes, and, as the most interesting feature, spikes imbedded in the surrounding chaotic sea. We can explain this structure when we plot the stable branch of an unstable periodic orbit on the x-y plane of the SM (Fig. 11), where we superimposed several stickiness diagrams (time of escape versus initial x for different fixed values of initial y).

The orbits with initial conditions in RII escape very fast (only few iterations) through openings between the lobes of asymptotic curves (regions A, B and C). Orbits starting in the inclined of Fig. 11 escape sometimes only after several 1000 iterations.

The self-similarity of the outer sticky region is shown in Fig. 12 using three different scalings: Fig. 12a shows an enlargement of the structure on the very left side of Fig. 8 (starting from $x = 0.632$), Fig. 12b is an enlargement of the structure on the right edge with the two peaks of Fig. 12a, Fig. 12c is again an enlargement of the structure on the right edge of Fig. 12b. From Fig. 11 it is clear that this structure is connected with the complicated form of the stable branch of the unstable periodic orbits but the self-similar character of RII is not yet understood.

In section II.6 it was explained how to construct step by step the high order periodic orbits which can be used to approximate the location of the cantori (the barriers for orbits in RI). To put into evidence what sequence of the continuous fraction represents the noble irrational RN for the most resistant cantori it is necessary to compute orbits for nonlinearity parameters

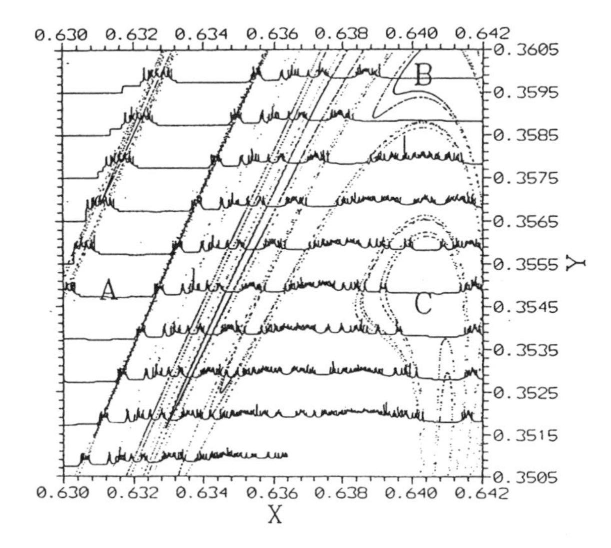

Figure 11. Detail of Stickiness region II for K=5 superimposed with stickiness diagrams: the points along inclined lines are orbits on the stable asymptotic manifold of an unstable periodic orbit; along the parallel lines y=0.3505, 0.3515... the escape times are given in a logarithmic scale

close to $K = 5$ where the corresponding KAM curves still exist. To locate – at least approximately – the cantorus in Fig. 13 the 157/348 periodic orbit is plotted, which is still outside the cantorus. An escaping orbit is visible there starting from the lower left in the graph still INSIDE the cantorus, which crosses the holes in this barrier with entangled lobes, and ends up finally OUTSIDE the cantorus. For a detailed description one should look up in paper II.

In this study we investigated escapes from RI and RII into the chaotic sea; because the SM is area preserving we checked how the inverse phenomenon could occur, namely capture into the stickiness regions. We started with arbitrarily chosen initial conditions within the chaotic sea and iterated forward. For most initial conditions orbits were in fact captured into region I (and then they again escaped), but we had to compute a great number of iterations (10^9). The results are not yet worked out in detail and will be presented in a later study.

4. Sitnikov Problem

The Sitnikov Problem (SP) is defined as follows:

- Two primaries of equal mass move on Keplerian orbits, while an infinitesimally small mass is confined to move on an axis (z-axis) perpendicular to the primaries' orbital plane.

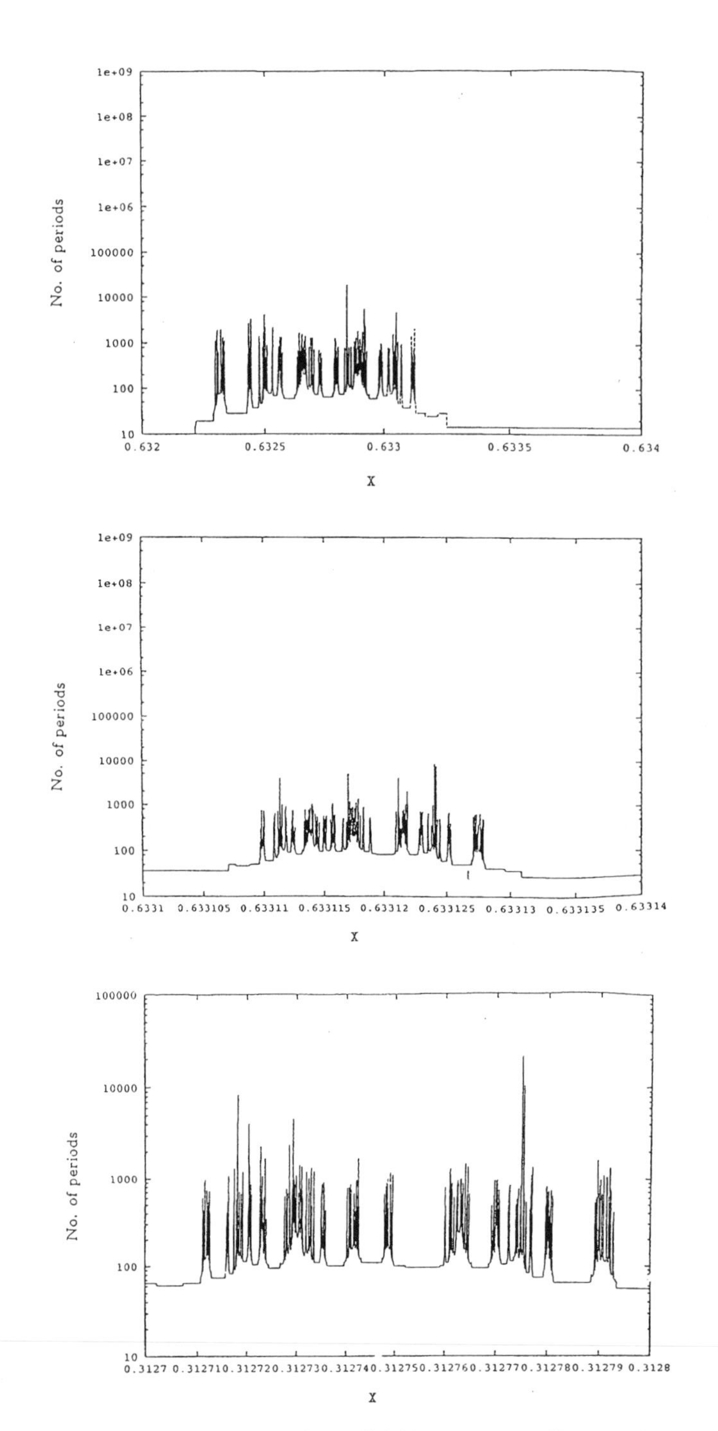

Figure 12. The fractal structure of the Stickiness region II; each lower graph is en enlargement of the graph above it.

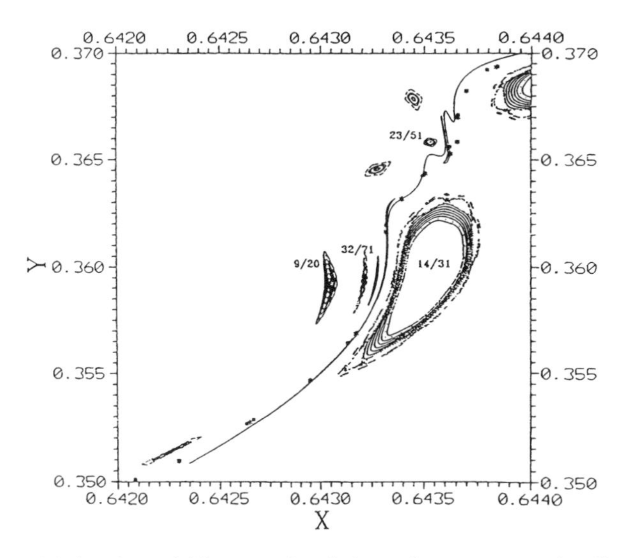

Figure 13. An orbit leaving stickiness region I through a cantorus; detail of the Standard Map for K=5 where the crosses mark the 157/348 periodic orbit.

If the primaries' orbit is circular then the motion of the massless body is integrable and is either bounded (periodic solution) or unbounded (capture-escape solution, MacMillan, 1913). For eccentric orbits, however, the situation is different: the distances of the primaries from the axis of motion of the planet vary periodically with time, that means that energy is pumped (or removed) into the planet's motion depending on the primaries' position and on the distance of the planet to them. The time interval between two successive crossings of the barycenter sensitively depends on the primaries' position at the previous crossing which is – roughly speaking – a main characteristic of chaotic motion.

The equation of motion is the following:

$$\ddot{z} + \frac{z}{\sqrt{r^2 + z^2}^3} = 0, \tag{21}$$

where z is the distance of the massless body from the barycenter of the primaries and r is the distance of the primaries from the barycenter. When r is not constant (r=r(t)) – corresponding to an eccentric orbit of the primaries – the problem was first defined and studied by Sitnikov (1960); his original article was analysed in all details by Wodnar (1992, 1993). Other results on the SP were derived by: e.g. Moser (1973), Liu and Sun (1990), Wodnar (1991), Hagel (1992), Dvorak (1993), Kallrath et al. (1997).

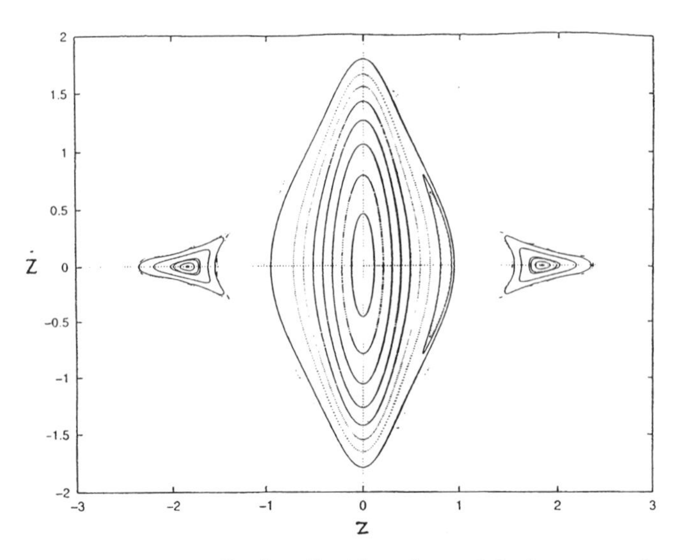

Figure 14. The SOS of the Sitnikov Problem for e=0.2; $\dot{z}$ versus z for the primaries being in their pericenter.

4.1. THE STICKINESS CLOSE TO THE CENTRAL FIXED POINT

To visualize the stickiness region we need now a 2-dimensional plane, where we can do a similar investigation as for the SM. We choose a Poincaré surface of section z versus $\dot{z}$ and plot the points whenever the primaries are in their periastron position. To integrate the equation of motion (21) the high speed - high precision Lie - series integrator (Hanslmeier and Dvorak, 1984, Kasper 1996) was used. As a typical example of the phase space structure of the SP, we show in Fig. 14 the SOS for an eccentricity $e = 0.2$: one can see the invariant curves in the vicinity of the fixed point for $z = 0$ and $\dot{z} = 0$. The fixed point ($z \sim 0.8$, $\dot{z} = 0$) in the middle of the island still inside the last KAM curve corresponds to the 1:1 resonance. The 1:2 periodic orbit (1 oscillation of the massless body for 2 complete revolutions of the primaries) is the fixed point ($z \sim 1.8$, $\dot{z} = 0$) in the center of the large island inside the chaotic sea.

We investigated a small region outside the last KAM curve ($0.98 \leq z \leq 1.1$ with $\dot{z} = 0$). Figure 15 shows the detailed results: the logarithm of the time in periods of the primaries – which corresponds to the number of intersections with the chosen SOS – is plotted versus the initial z values. As escape criterion, we used the distance $d \leq 2$ ($d = \sqrt{z^2 + \dot{z}^2}$); it means that on the y axis we plotted the logarithms of the "escape time" of an orbit. The mainland of invariant curves is visible on the left side, then there is a small zone of escape orbits which is already part of the stickiness region with

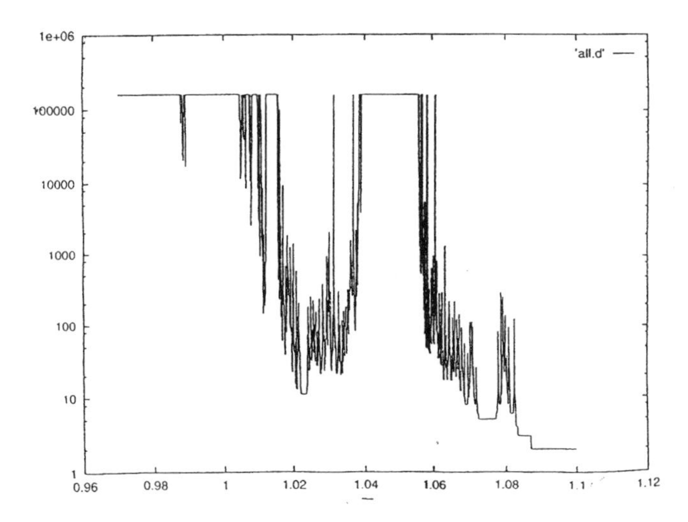

Figure 15. Stickiness diagram of the Sitnikov Problem for e=0.2; escape time versus initial z for $0.98 \leq z \leq 1.1$ with $\dot{z} = 0$.

orbits trapped for very long times. Several smaller islands of non escaping orbits (around $z_o = 1.01$) and a large island at $1.04 \leq z \leq 1.06$ with $\dot{z} = 0$ corresponding to the 3:2 periodic orbit close to the right are seen. It is very difficult to determine the limit of the inner sticky region I, because of the existence of the large island well inside it, which has itself a stickiness region. It seems that there is a limiting cantorus close to $z = 1.02$, but it is – at least for the moment – impossible for us to approximate numerically this barrier with high order periodic orbits. We can see the steep slope with the superexponential decrease outside the last KAM curve which was already found for the SM in region I.

We also plotted the SOS for initial conditions inside this region and found a well defined small zone outside the last KAM curve (Fig. 16) very similar to the one of the SM (Fig. 9).

The region outside the large island of the 3:4 periodic orbit for $1.07 \leq z \leq 1.09$ with $\dot{z} = 0$ can be identified with the region II. On both sides of a valley with more or less immediate escape we find 2 "chains of peaks" (Fig. 17) corresponding to escapes after some 10^2 intersections with the SOS; these two peaks have a very similar shape.

In fact the fractality of these structures is quite well visible in Fig. 18a, where we show an enlargement of the peaks on the right edge at initial conditions $z \sim 1.082$ with $\dot{z} = 0$ and in Fig. 18b where we show the enlarged region $1.08255 \leq z \leq 1.08265$ with $\dot{z} = 0$. In fact we observe the same phenomenon like in the SM where we succeeded to identify these structures with the asymptotic curves of an unstable periodic orbit of high

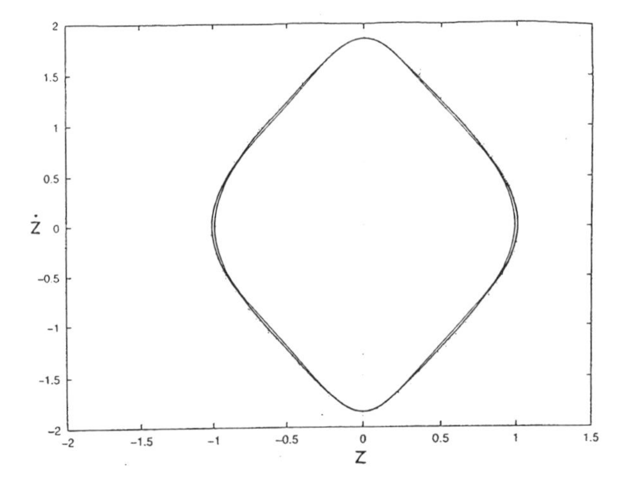

Figure 16. SOS of the Stickiness Region II of the Sitnikov Problem for e=0.2

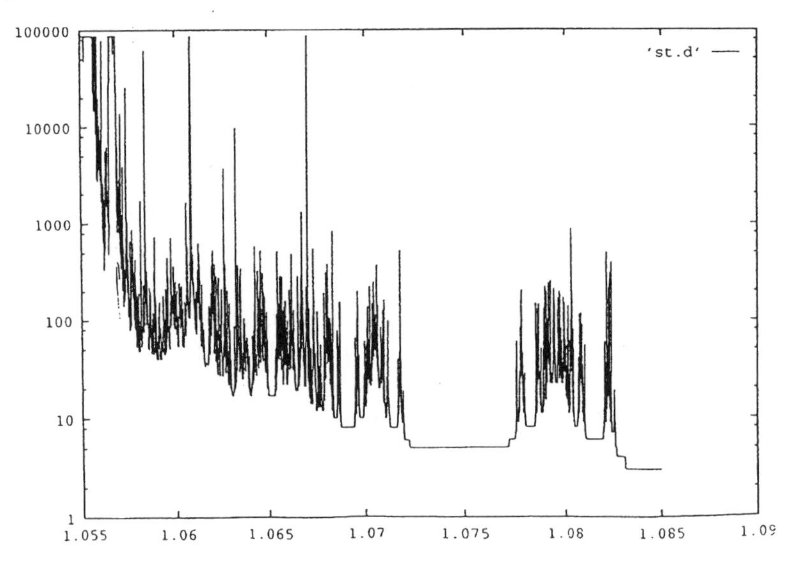

Figure 17. Stickiness diagram of the Sitnikov Problem for e=0.2; escape time versus initial z for $1.055 \leq z \leq 1.085$ with $\dot{z} = 0$

order. To locate the escape orbits in region II we computed the SOS of orbits starting inside the fractal structures. We show in Fig. 19 the two escape channels which look very similar to the cusps visible in region II of the SM (Fig. 10). A detail of Fig. 19 is shown in Fig. 20 where the lobes of the asymptotic curves are visible; unfortunately it was impossible to associate these structures with a specific high order unstable periodic orbit.

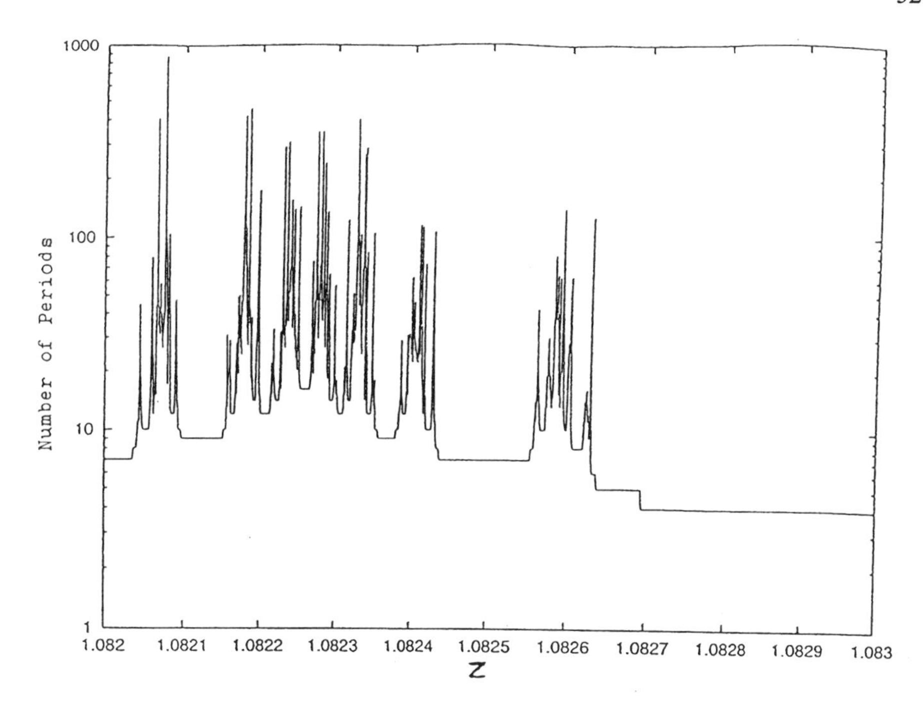

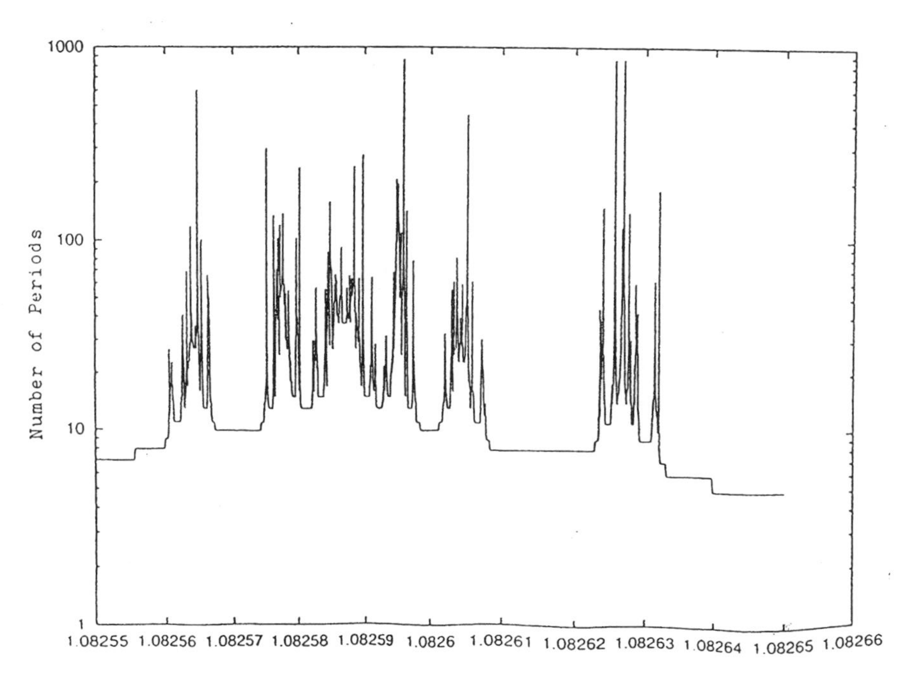

Figure 18. Stickiness diagram of the Sitnikov Problem for e=0.2; escape time versus initial z; the lower graph is en enlargement of the upper

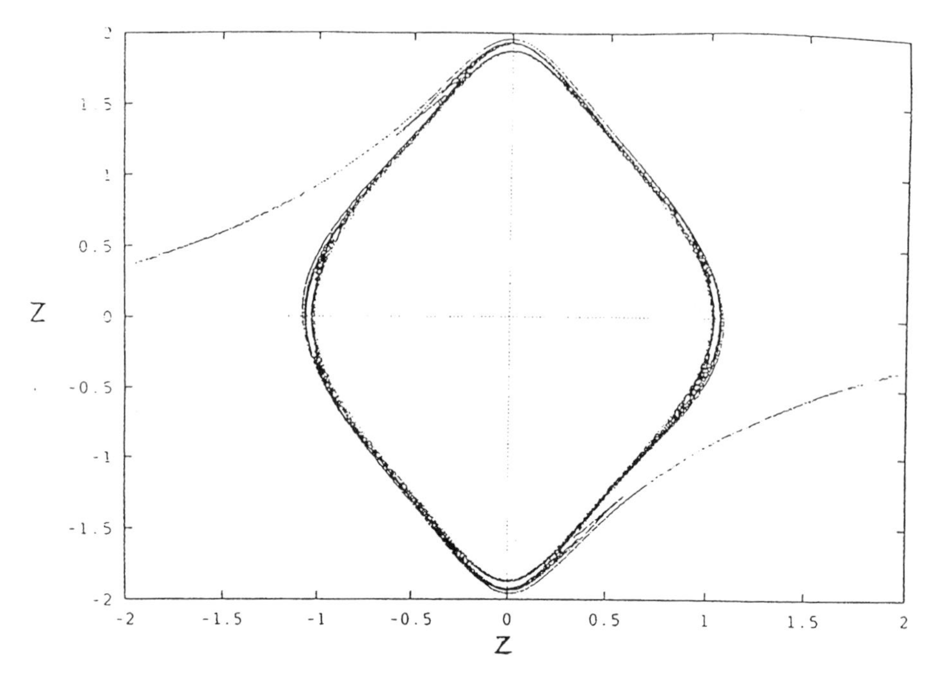

Figure 19. SOS of Stickiness Region II for e=0.2

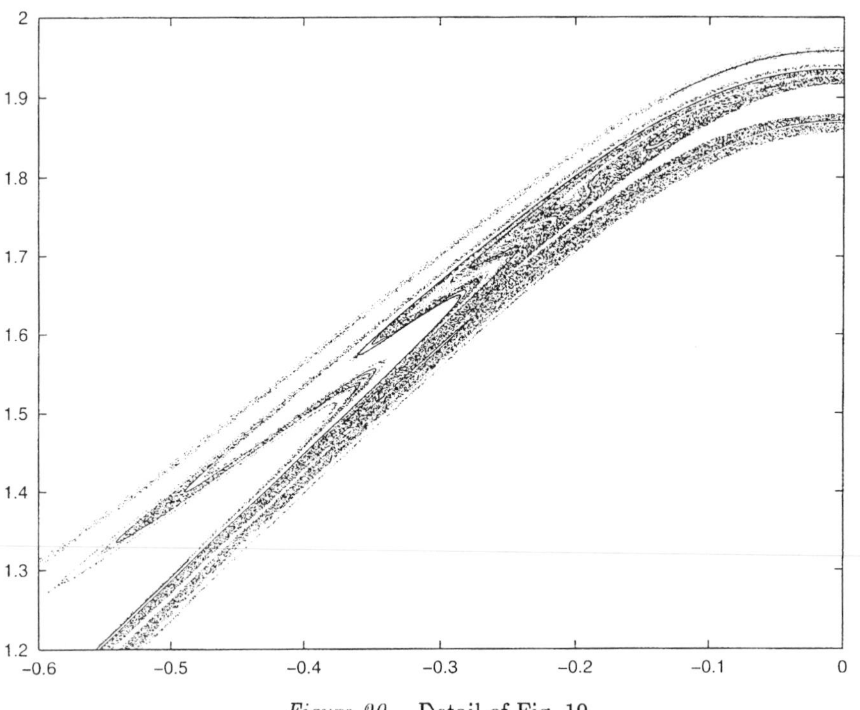

Figure 20. Detail of Fig. 19

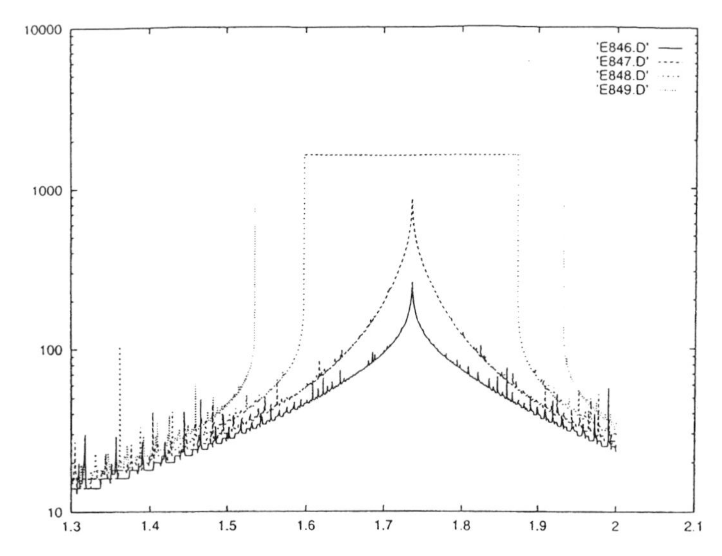

Figure 21. The formation of the 1:2 island between for $0.846 < e < 0.848$

4.2. THE 1:2 PERIODIC ORBIT

The next step of investigation of the SP was to explore systematically orbits close to the 1:2 periodic orbit located at $z_o = 1.848460, 1.836787, 1.816155, 1.786429, 1.747073$ for the eccentricities $e = 0.0, 0.2, 0.4, 0.6, 0.8$ (Kallrath et al. 1997). The respective results will be discussed in detail in a separated paper.

The most interesting feature is the disappearance – appearence - disappearence of the 2:1 island which is due to the change of the stability of the 2:1 periodic orbit for different values of the eccentricity of the primaries. Fig. 21 shows the growing of the sharp peak for values of the eccentricity e=0.846 and e=0.847 and then, for e=0.848, we see that the island is already present, and is even larger for e=0.849. So we can visualize how out from a "sticky island" the stable periodic orbit builds up. The change of a stable periodic orbit into an unstable one in the SP has already been found by Alfaro and Chiralt (1993) where they analysed the stability of the central fixed point $z = 0$, $\dot{z} = 0$. Up to e=0.54 no instability could be found, the first intervall is for $0.85586179645597 < e < 0.855863313794944$ and then many more small unstable intervalls were computed for larger eccentricities. Their results were numerically confirmed by my own calculations (unpublished) and in the present investigation we found a similar behaviour for the 2:1 periodic orbit.

5. Conclusions

After developing the basics ideas of dynamical systems described by difference equations (mappings) and by differential equations (flows) we reviewed the different methods to classify the motions in such systems. We then explained the importance of the cantori in connection with motion close to invariant curves. A cantorus is a destroyed KAM tori with an infinite number of gaps which form barriers for the orbits close to KAM tori. We also showed a procedure to approximate cantori with the aid of the continous fraction procedure using the Farey – tree.

As main topic of this lecture we presented the results of numerical investigations of orbits in stickiness regions in a mapping (SM) and in a dynamical system (SP). Sticked orbits stay for long times in the vicinity of stable KAM tori and then escape to the surrounding chaotic domain of the system under consideration. For orbits in the standard map it was found that

- there is one stickiness region close to the last KAM torus
- this region has as outer barrier a series of cantori
- outside this region there is a more extended 2nd stickiness region
- the structure of the outer region is determined by the stable branch of unstable periodic orbits
- the outer region shows a self-similar character

As model for a dynamical system the Sitnikov Problem was studied and the results for orbits close to the stable domain were the following:

- we also distinguish 2 different stickiness regions with quite the same characteristics as the map
- the outer stickiness region is also self-similar
- the change of an unstable to a stable periodic orbit is visible through an island of stickiness, where the escape times increase when increasing the nonlinearity parameter (eccentricity)

It is evident from this study, that the characteristics of mappings and dynamical systems are essentially the same in what the stickiness regions are concerned. This is not surprising as we pointed out in the introductionary chapters the correspondance between symplectic mappings and Hamiltonian systems. This work will be continued for both systems, because there are still many open questions: how is the capture into the regions of stickiness related to the stochasticity parameter, how can we explain the appearance of stable periodic orbits in connection with stickiness and how important is this stickiness phenomenon for real astrodynamical systems like the motions of asteroids and comets in the Solar System.

Acknowledgements

For the most valuable advice in preparing the manuscript for this lecture I thank Professor Ferraz-Mello, Dr. Lohinger, Prof. Kallrath and DI Machacek.

References

1. Arnold, V.I.: 1963, Proof of A.N. Kolmogorov's theorem on the preservation of quasiperiodic motions under small perturbations of the Hamiltonian, *Usp.Mat.Nauk* SSSR 18, no. 5, p.13
2. Aubry, S.: 1978, in Bishop, A.R. and Schneider, T (eds.) *Solitons and Condensed Matter Physics*, Springer, p. 264
3. Contopoulos, G.: 1971, Orbits in Highly Perturbed Systems III, *Astron. J. 76*, 147
4. Contopoulos, G., Voglis, N., Efthymiopoulos, C., Froeschlé, C., Gonczi, R., Lega, E., Dvorak, R., Lohinger, E.: 1997, Transition Spectra of Dynamical Systems, *Cel. Mech. Dyn. Astr. 67*, 393
5. Contopoulos, G., Voglis, N.: 1996, Spectra of Stretching Numbers and Helicity Angles in Dynamical Systems, *Cel. Mech. Dyn. Astr. 64*, 1
6. Contopoulos, G., Voglis, N.: 1997, A fast method for distinguishing between ordered and chaotic orbits, *Astron.Astrophys. 317*, 73
7. Dvorak, R.: 1993, Numerical Results to the Sitnikov Problem, *Cel. Mech. Dyn. Astr. 56*, 71
8. Dvorak, R., Contopoulos, G., Efthymiopoulos, C.: 1997, The Stickiness in Mappings and Dynamical systems, in Benest and Froeschlé (eds.), Chaos dans les Systèmes gravitationelles, OCA, p.55
9. Efthymiopoulos, C., Contopoulos, G.,Voglis, N., Dvorak, R.: 1997, Stickiness and Cantori, *J. Phys. A: Math. Gen 30* , 8167
10. Froeschlé, Cl.: 1984, The Lyapunov Characteristic Exponents – Applications to Celestial mechanics, *Celestial Mechanics 34*, 95
11. Froeschlé, Cl., Lega, E.: 1996, On the measure of the structure around the last KAM Torus before and after its break-up, *Cel. Mech. Dyn. Astr. 64*, 21
12. Greene, J.M.: 1979, A method for determining a stochastic transition, *J. Math. Phys. 20*, 1183
13. Guckenheimer, J, Holmes, P.: 1983, Nonlinear Oscillations, Dynmamical Systems, and Bifurcations of Vector Fields, Springer-Verlag, New York, Berlin, heidelberg, Tokyo, p. 453
14. Hanslmeier, A., Dvorak, R.: 1984, Numerical Integration with Lie–Series, *Astron.Astrophys. 132*, 203
15. Hagel, J.: 1992, A New Analytic Approach to the Sitnikov Problem, *Celestial Mechanics 53*, 267
16. Kallrath, J., Dvorak, R., Schlöder, J.: 1997, Periodic Orbits in the Sitnikov Problem, in Dvorak, R. and Henrard, J. (eds.) *The Dynamical Behaviour of our Planetary System*, Kluwer, p. 415
17. Kasper, N., 1996, private communication
18. Kolmogoroff, A.N.: 1954, Preservation of Conditionally Periodic Movements with Small Change in the Hamiltonian Function,*Dokl.Akad.Nauk* SSSR 98,527
19. Laskar, J.: 1993, Frequency analysis for multi-dimensional systems. Global dynamics and Diffusion, *Physica D 67*, p. 257
20. Lega, E. and Froeschlé, C.: 1997, Fast Lyapunov Indicators, Comparison with other Chaos Indicators, Application to two and four dimensional maps, in Dvorak, R. and Henrard, J. (eds.) *The Dynamical Behaviour of our Planetary System* Kluwer, p. 257
21. Lichtenberg, A.J., Lieberman, M.A.: 1983, *Regular and Stochastic Motion*, Springer-

Verlag, New York Heidelberg Wien
22. Jie Liu, and Yi-Sui Sun: 1990, On the Sitnikov Problem, *Celestial Mechanics 49*, 285
23. MacKay, R.S., Meiss, J.D., Percival, I.C.: 1984, Transport in Hamiltonian Systems, *Physica 13 D*, 55
24. MacMillan, W.D.: 1913, An Integrable Case in the Restricted Problem of Three Bodies, *Astronomical Journal 27*, 11
25. Meiss, J.D.: 1992, *Rev.Mod.Phys. 64*, 795
26. Moser, J.: 1962, On Invariant Curves of Area-Preserving Mappings on an Annulus, *Nachr.Akad.Wiss.Göttingen*, Math.Phys.Kl. p.1
27. Moser, J.: 1973, *Random and Stable Motion in Dynamical Systems*, Princeton University Press
28. Percival, I.C.: 1979, in Month, M. and Herrera, J.C. (eds.) *Nonlinear Dynamics and the Beam – Beam Interaction*, Amer. Inst. Phys. p. 302
29. Percival, I.C., Richards, D.: 1989, *Introduction to Dynamics*, Cambridge University Press, 228
30. Roy, Archie, E.: 1988, *Orbital Motion*, 3rd edition, A.Hilger, Bristol
31. Shirts, R.B. and Reinhardt, W.P.: 1982, *J. Chem. Phys. 77*, 5204
32. Sitnikov, K.: 1960, Existence of Oscillatory Motions for the Three-Body Problem, *Dokl. Akad. Nauk, USSR, 133/2*, 303
33. Stumpff, K.: 1970, *Himmelsmechanik II*, VEB, Deutscher Verlag der Wissenschaften, Berlin,
34. Szebehely, V.: 1970, *Theory of Orbits*, Academic Press, New York
35. Wodnar, K.: 1991, New Formulations of the Sitnikov Problem, in A.E. Roy (ed.) *Predictability, Stability and Chaos in N-Body Dynamical Systems*, Plenum Press, New York
36. Wodnar, K.: 1992, The Original Sitnikov Article – New Insights, *master's thesis*, University of Vienna
37. Wodnar, K.,1993: The Original Sitnikov Article – New Insights, *Cel. Mech. Dyn. Astr. 56*, 99

DISTRIBUTION OF PERIODIC ORBITS IN 2-D HAMILTONIAN SYSTEMS

E. GROUSOUZAKOU
Academy of Athens, Center for Theoretical Mathematics
and
Department of Astronomy, University of Athens
GR-15784, Athens, Greece

AND

G. CONTOPOULOS
Academy of Athens, Center for Astronomy
and
Department of Astronomy, University of Athens
GR-15784, Athens, Greece

Abstract

We study the distribution of periodic orbits in two Hamiltonian systems of two degrees of freedom, using Poincaré surfaces of section. We distinguish between regular periodic orbits (bifurcations of the periodic families of the unperturbed system) and irregular periodic orbits (not connected to the above). Regular orbits form characteristic lines joining periodic orbits of multiplicities following a Farey tree, while irregular periodic orbits form lines very close to the asymptotic curves of the main unstable families. We find that all irregular orbits are inside the lobes of the homoclinic tangles. These tangles have gaps between various lobes containing regular periodic orbits. There are also gaps inside some lobes containing stable irregular periodic orbits.

Description of the systems

The first dynamical system is symmetric with respect to the x-axis and has the Hamiltonian

$$H \equiv \frac{1}{2}\left(Ax^2 + By^2\right) + \frac{1}{2}\left(\dot{x}^2 + \dot{y}^2\right) - \varepsilon xy^2 \qquad (1)$$

B.A. Steves and A.E. Roy (eds.), The Dynamics of Small Bodies in the Solar System, 535–544.

Because of the symmetry of (1) all periodic orbits are either symmetric with respect to the x-axis or they appear in pairs symmetric with respect to this axis. In order to study their distribution we have chosen a $x\dot{x}$ Poincaré surface of section.

If $\varepsilon = 0$ and the ratio $\sqrt{A}/\sqrt{B}$ is rational all orbits are periodic [1]. In our experiments we have chosen $A = 1.6, B = 0.9, h = 0.00765, \sqrt{A}/\sqrt{B} = 4/3$. The values of the above parameters remain constant while we vary the value of perturbation ε. For low values of ε, there are four main families, the central family O near the origin, the boundary $\dot{x}^2 + Ax^2 = 2h$ $(y = 0)$ and also two resonant families 4/3 one stable $(\overline{O}_1\overline{O}_2\overline{O}_3)$ and one unstable $(O_1O_2O_3)$, (Fig. 1a). As the perturbation increases many more resonant families appear (Fig. 1b, 1c) and the region of invariant curves surrounding the central periodic orbit decreases (Fig. 1d) and finally disappears (Fig. 1e). It should be noted that in Fig. 1d the central periodic orbit is unstable (it becomes unstable for $\varepsilon = 4.306$) but it is still surrounded by well defined invariant curves. For $\varepsilon > \varepsilon_{esc} = 4.603$ the curves of zero velocity open and all orbits go to infinity [2].

The main families generate, by equal period or multiple period bifurcation, higher order families, which also give higher order bifurcations. All these families are connected to the main families and are called regular.

The families not connected to the above are called irregular and they are generated by tangent bifurcation. They appear in pairs, one stable and one unstable, for a certain value of ε. For some greater ε, the stable family generates a sequence of families that eventually becomes unstable.

Thus each stable family, regular or irregular, produces by successive bifurcations, an infinity of unstable families and for large ε we have mainly unstable orbits.

The second system has the Hamiltonian

$$H \equiv \frac{1}{2}\left(\dot{x}^2 + \dot{y}^2 + x^2 - y^2\right) - y^4 + x^2y^2 = h \tag{2}$$

We find the distribution of regular and irregular periodic orbits on a Poincaré surface of section $y\dot{y}$, $(x = 0)$ for various values of the energy h. The Hamiltonian (2) shows a different behaviour from the Hamiltonian (1). Its central periodic orbit is always located at $y = 0,\ \dot{y} = 0$ and it is unstable for small values of h. But as the energy increases it has infinite transitions from stability to instability (Fig. 2). This is due to the fact that the variational equations reduce to a Mathieu equation [8] of the form

$$\ddot{\xi} + (a + 2q\cos 2t) = 0 \tag{3}$$

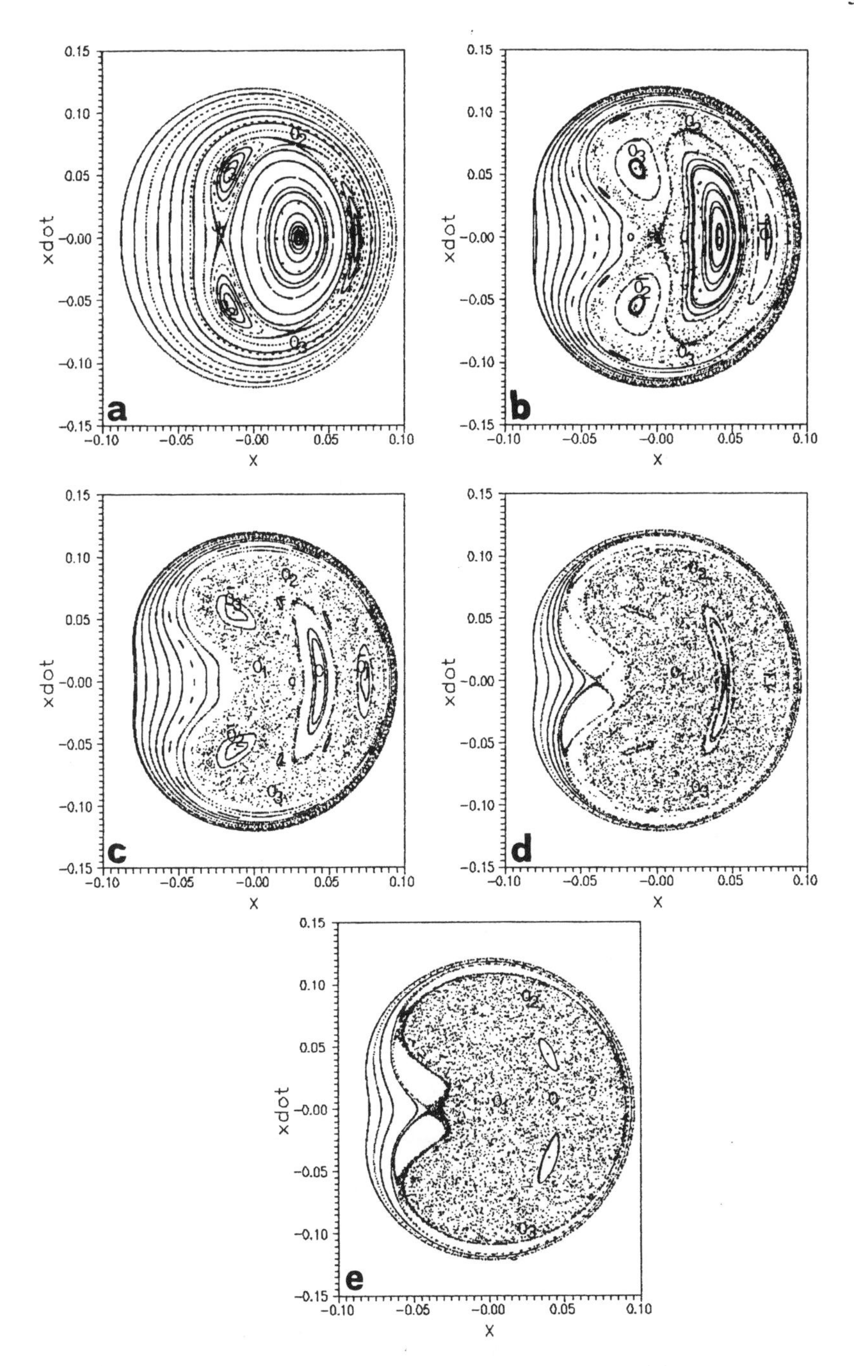

Figure 1. Poincaré surfaces of section $x\dot{x}$, for the Hamiltonian (1), with increasing ε. a) $\varepsilon = 3.0$ b) $\varepsilon = 4.0$ c) $\varepsilon = 4.3$ d) $\varepsilon = 4.4$ e) $\varepsilon = 4.5$. The outermost line represents the orbit $y = 0$. O is the central periodic orbit, $O_1O_2O_3$ is the triple unstable family, $\overline{O}_1\overline{O}_2\overline{O}_3$ is the triple stable family.

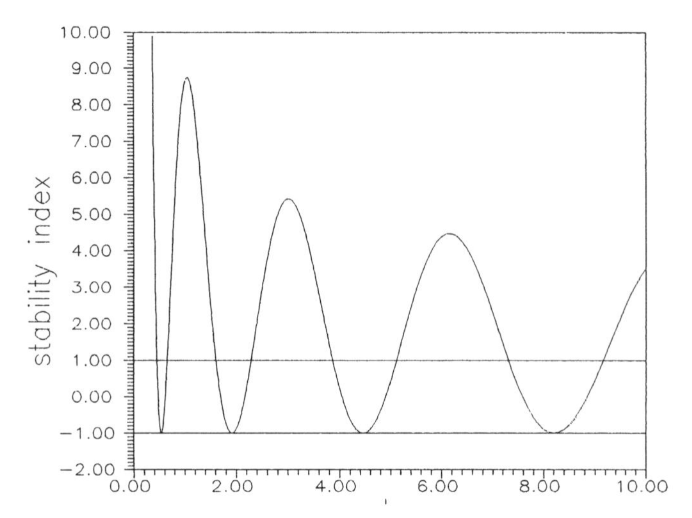

Figure 2. Hénon index of stability a versus the energy h, for the central periodic orbit of the Hamiltonian (2). We see the successive transitions from stability to instability and vice-versa as the curve crosses the line $a = 1$.

Thus the transitions from stability to instability are found [5] along the line $a = 2q - 1$ in the diagram of the stable and unstable solutions of the Mathieu equation [8].

Distribution of periodic orbits

In Fig. 3 we see the distribution of periodic orbits for the Hamiltonian (1) with increasing ε on a Poincaré surface of section $x\dot{x}$, $(y = 0)$. We have located [4] periodic orbits which intersect the x-axis perpendicularly, up to a certain multiplicity. We call multiplicity the number of times that a periodic orbit intersects the surface of section upwards $(\dot{x} > 0)$.

In Fig. 4 we see the distribution of periodic orbits, for the Hamiltonian (2) with increasing h. The selected surface of section is $y\dot{y}$, $(x = 0)$ and the periodic orbits intersect perpendicularly the y-axis, and the surface of section upwards $(\dot{y} > 0)$.

In both cases we note that:

a) The number of periodic orbits increases, as ε increases (case 1), or h increases (case 2). As an example the regular families up to a certain multiplicity m, bifurcating from the central family are of order $O(m^2)$ [1], while the number of the irregular periodic orbits is of order $O(2^m)$ [7].

b) For low values of ε,or of h, there are some well defined lines joining periodic orbits. These lines are best seen in Figs. 3a,b and 4b,c,d. They become more and more distorted with increasing ε (case 1) or h (case 2).

We call the orbits of low order "basic orbits" (e.g. the basic orbits

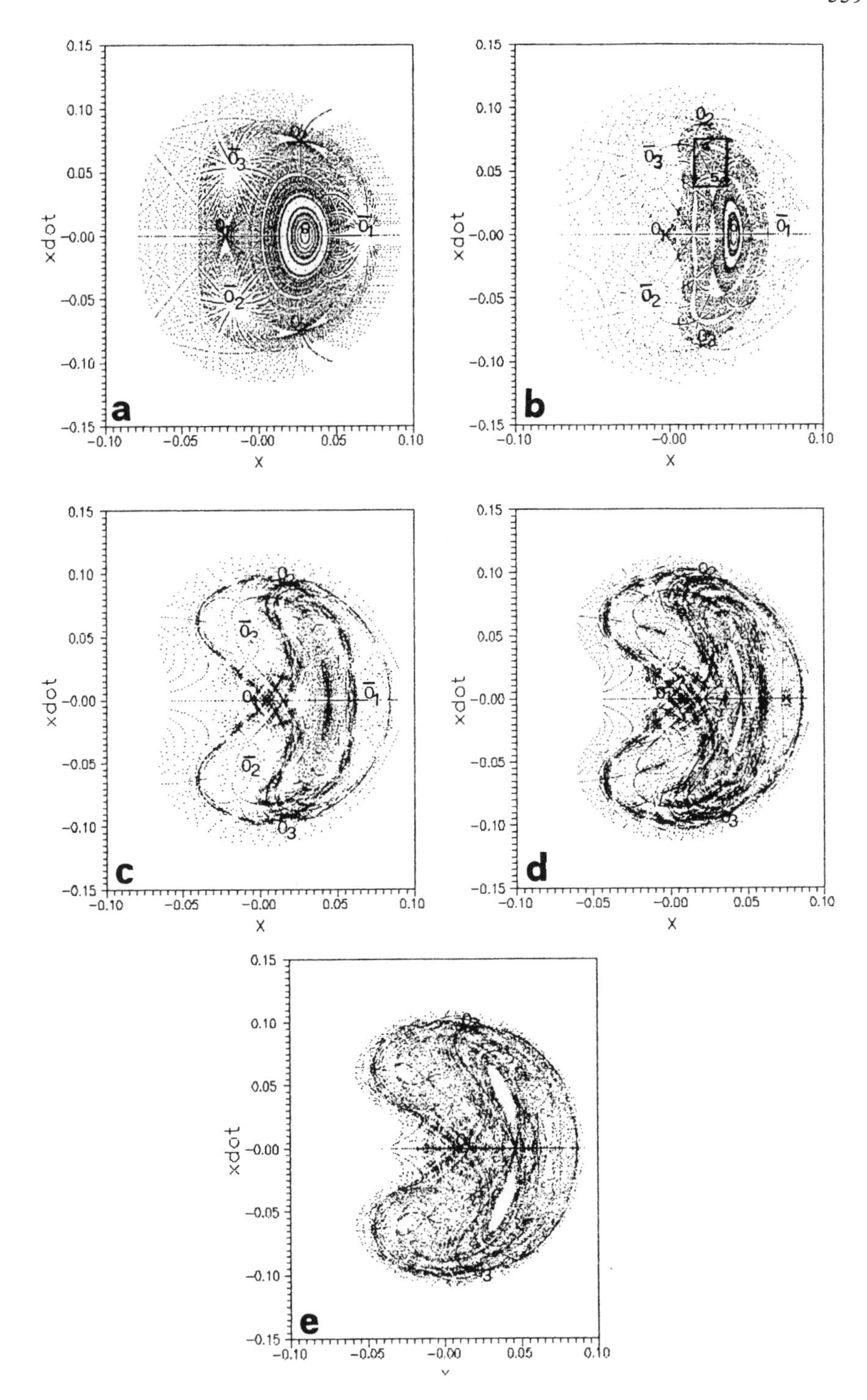

Figure 3. Distribution of periodic orbits on a Poincaré surface of section $x\dot{x}$ for the Hamiltonian (1) with increasing ε. a) $\varepsilon = 3.0$ b) $\varepsilon = 4.0$ c) $\varepsilon = 4.3$ d) $\varepsilon = 4.4$ e) $\varepsilon = 4.5$.

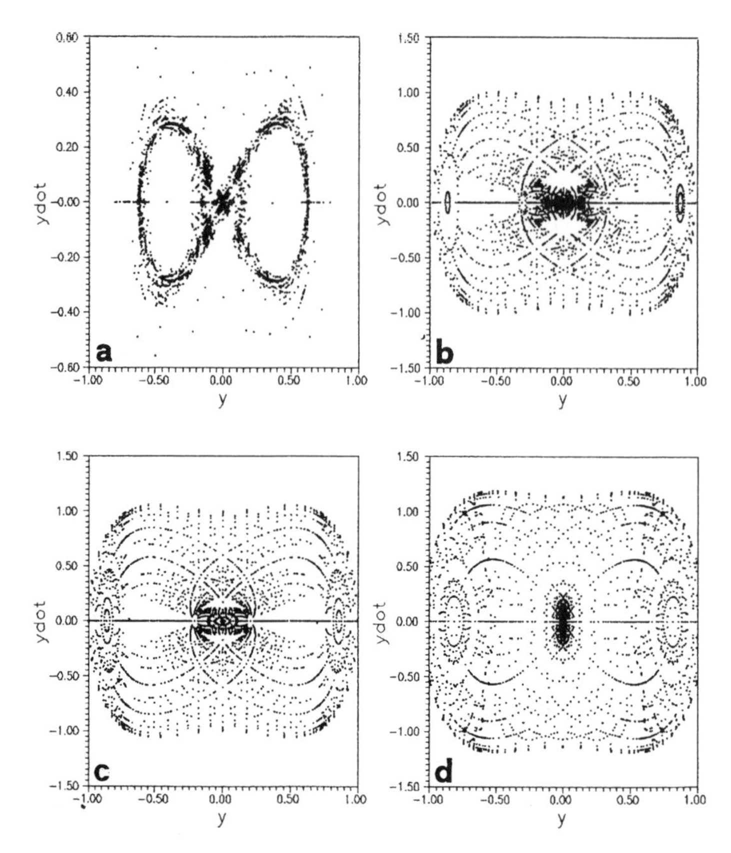

Figure 4. Distribution of periodic orbits on a Poincaré surface of section $y\dot{y}$ for the Hamiltonian (2) with increasing h. a) $h = 0.1$ (the central periodic orbit (c.p.o.) is unstable). b) $h = 0.45$ (c.p.o. on the limit stable). c) $h = 0.5$ (c.p.o. stable). d) $h = 0.651$ (c.p.o. on the limit unstable).

of the case 1 are the ones with multiplicities 1,3,4,5,6). We find that between two successive basic orbits of multiplicities m_1 and m_2 there are two basic sequences of orbits forming two arithmetic progressions, namely $m_l + m_2, m_1 + 2m_2, m_1 + 3m_2$ and $m_l + m_2, 2m_1 + m_2, 3m_1 + m_2$ with increments m_2 and m_1 approaching asymptotically the orbits m_2 and m_1 respectively.

As an example, we study the part of the line in Fig. 3b that lies inside the square where $m_1 = 4, m_2 = 5$. This section contains the following orbits, along the line joining the points, 4 and 5.

4...	37, 33, 29, 25, 21, 17, 13	9	14, 19, 24, 29, 34, 39, ...	**5**
	1st sequence		2nd sequence	

The existence of these sequences results from the fact that the rotation numbers of the periodic orbits form Farey trees [9].

These lines contain only regular orbits, unless they are images of the x-axis, in which case these lines are expected to contain also some irregular periodic orbits (those intersecting perpendicularly the x-axis). However irregular periodic orbits form also other lines, (dark areas in Fig. 3c,d,e, Fig. 4a,b and Fig. 5a) quite different from the lines containing Farey tree orbits. These lines appear in the chaotic domain, near the asymptotic curves of the main unstable families.

Periodic Orbits and the Homoclinic Tangle

In Fig. 5a,b we compare the asymptotic curves of the central unstable family ($y = 0, \dot{y} = 0$) of the second Hamiltonian and the distribution of periodic orbits. We see that many lines of periodic orbits are very close to the asymptotic curves.

This closeness is due to a general property of the unstable periodic orbits, namely that the images of any neighbourhood of an unstable point are stretched along the unstable eigenvectors and contracted along the stable eigenvectors. These lines contain many irregular periodic orbits, because all the periodic orbits inside lobes are irregular [3]. In fact such orbits cannot be continously connected with the regular orbits that are all outside the lobes. We conjecture that the opposite is also true, i.e. that all irregular orbits are inside lobes. Fig. 6 verifies our conjecture, because the pair of irregular orbits 19, which is far away from the triple unstable family (first Hamiltonian), is reached by lobes of the homoclinic tangle of the triple unstable family, if we calculate the asymptotic curves for long periods.

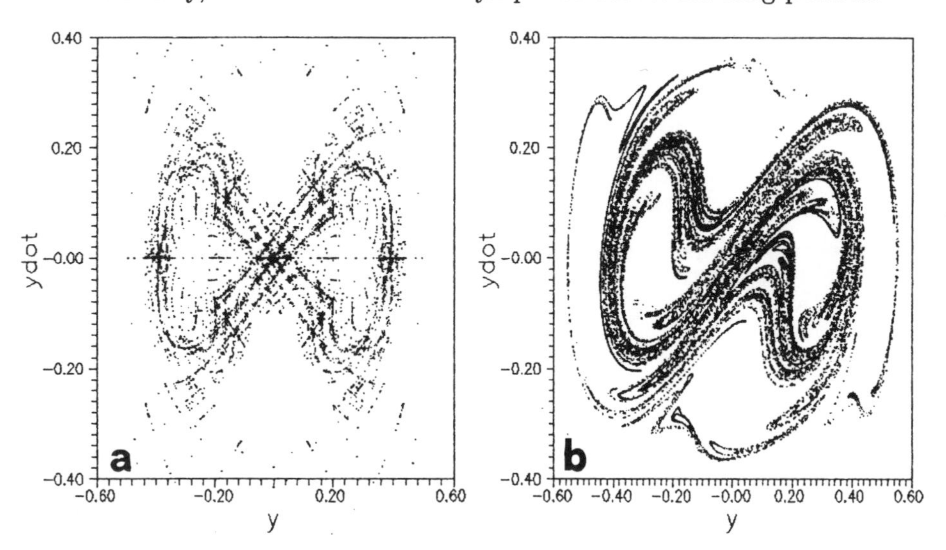

Figure 5. a) Distrbution of periodic orbits for the Hamiltonian (2) in the case $h = 0.3$ and b) the unstable manifold of the central periodic orbit.

The homoclinic tangle of each unstable family is characterized by its irregular periodic orbits. In Fig. 7 the central periodic orbit (first Hamiltonian) is unstable but there is no complete dissolution of invariant curves which surround it thus we can easily distinguish between irregular orbits that belong to the homoclinic tangle of the central unstable family and those that belong to the triple unstable family. This separation is not possible when the last KAM curve that surrounds the central periodic orbit has been destroyed and heteroclinic intersections appear (Fig. 8).

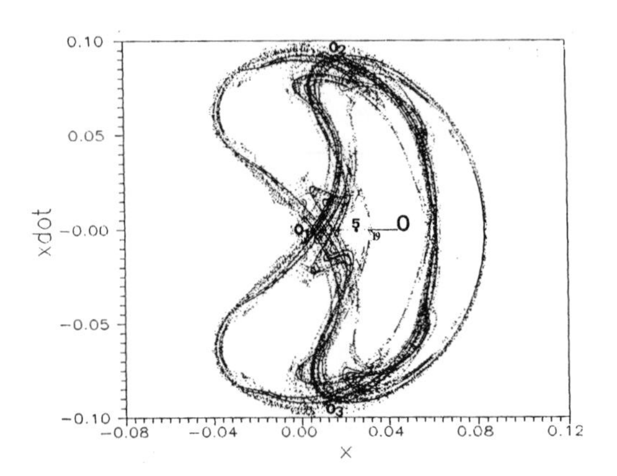

Figure 6. The pair of irregular orbits 19 is reached by the asymptotic curves of the triple unstable orbit $O_1O_2O_3$, in the case $\varepsilon = 4.3$, for the Hamiltonian (1). The regular stable periodic orbit (5) is trapped in the homoclinic tangle of the triple unstable orbit.

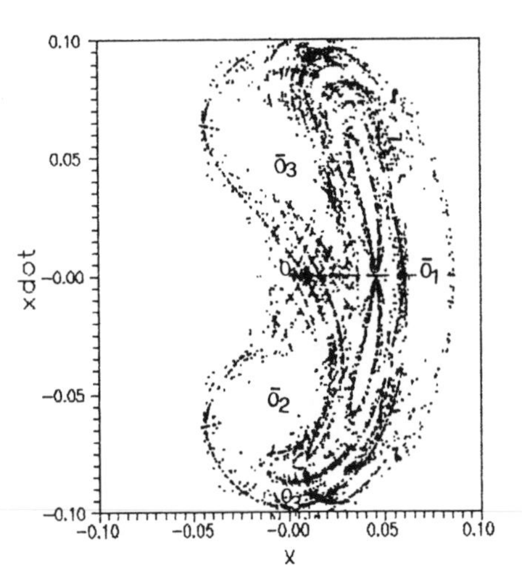

Figure 7. Distribution of irregular periodic orbits of the Hamiltonian (1) for the case $\varepsilon = 4.4$. The irregular orbits which belong to the homoclinic tangle of the central unstable family are separated from the irregular orbits that belong to the the homoclinic tangle of the triple unstable family.

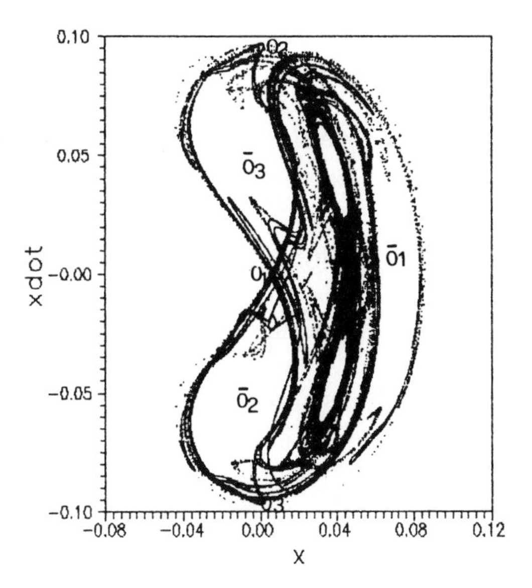

Figure 8. Distribution of irregular periodic orbits of the Hamiltonian (1) for the case $\varepsilon = 4.5$. There is resonance overlapping between the tangles of the central periodic orbit and of the triple unstable periodic orbit. Thus we cannot distinguish in which homoclinic tangle belongs each irregular periodic orbit.

Apart from the irregular orbits that we find inside the lobes of each homoclinic tangle, in the area between the lobes we find regular periodic orbits, that were generated by bifurcation from the central orbit but trapped inside the tangle as the perturbation increases (orbit 5 in Fig. 6). Some of these orbits can be stable, surrounded by small, but finite, islands of stability. Stability islands can also be found inside the lobes, caused by stable irregular periodic orbits [6]. Therefore the homoclinic tangle does not form a simply connected set. Although it contains a Λ-set, which is isomorphic to a Bernoulli shift [7], it has also gaps, i.e. it is not completely chaotic.

Conclusion

The distinction between regular and irregular periodic orbits, the Farey tree lines which join regular periodic orbits, the distribution of the irregular periodic orbits close to the asymptotic curves, and the existence of gaps inside the homoclinic tangles of the main unstable families, are quite general phenomena that appear in generic dynamical systems.

Acknowledgement

This research was supported in part by the Greek National Secretariat for Research and Technology (PENED 293/1995).

References

1. Contopoulos G. (1970) Orbits in highly perturbed systems I, *Astron. J.*, **75**, 96.
2. Contopoulos G. (1971) Orbits in highly perturbed systems III, *Astron. J.*, **76**, 147.
3. Contopoulos, G. and Grousouzakou, E. (1997) Regular and irregular periodic orbits, *Cel. Mech. Dyn. Astron.*, **65**, 33.
4. Contopoulos, G., Grousouzakou, E., and Polymilis C. (1996) Distribution of periodic orbits and the homoclinic tangle, *Cel. Mech. Dyn. Astron*, **64**, 363.
5. Grousouzakou, E. and Contopoulos, G. (1997a) Open Systems and Information Dynamics (in press).
6. Grousouzakou ,E. and Contopoulos, G. (1997b) Periodic orbits in 2-d dynamical systems, in Benest, D. and Froeschlé, C. (eds), *Analysis and Modeling of Discrete Dynamical Systems*, Gordon and Breach, p.107.
7. Guckenheimer, J. and Holmes, P. (1983) *Oscillations, Dynamical Systems and Bifurcations of Vector Fields*, Springer-Verlag.
8. MacLachlan, N.W. (1947) *Theory and applications of Mathieu functions*, Clarendon Press, Oxford.
9. Niven, I. and Zuckerman, H. (1960) *An introduction to the Theory of Numbers*, Wiley, N.York.

THE MANY FACES OF DYNAMICAL INSTABILITY

I: *The definition of "Chaos" and the (in)equivalence of different indicators*

PIERO CIPRIANI[†] AND MARIA DI BARI
Dipartimento di Fisica "E. Amaldi", Università "Roma Tre"
[†] *also* I.N.F.M. - *Istituto Nazionale di Fisica della Materia*

1. Non-universality of instability criteria.

Recently, [6, 7, 3, 8], we proposed a generalization to non Riemannian manifolds of the so-called *Geometro-Dynamical Approach* (GDA) to Chaos, [11, 2], able to widen the applicability of the method to a considerably larger class of dynamical systems (DS's). Here, we carry on our efforts on a pathway directed towards a synthetic and *a priori* characterization of the qualitative properties of generic DS's. Although being aware that this goal is very ambitious, and that, up to now, many of the trials have been discouraging, we shouldn't forget the theoretical as well practical relevance held by a possible successful attempt. Indeed, if only it would be concevaible to single out a *synthetic* indicator of (in)stability, we will be able to avoid all the consuming computations needed to *empirically discover* the nature of a particular orbit, perhaps noticeably different with respect to another one very *nearby*. Besides this, going beyond the semi-phenomenological *mere recognition* of the occurrence of dynamical instability, this approach could give deeper hints on its sources, even in those situations where the boundary between Order and Chaos tends to become more and more nuanced, and different tools seem to give conflicting answers. Lately, a renewed interest towards a concise description of dynamical instability[1] has grown, together with the feeling of the need to look deeply at the intermingled structures underlying the transition from *quasi-integrable* to stochastic motions. Within this conceptual perspective, the geometrical approach stands in a privileged position in that it offers many useful, by construction invariant and intrinsic, tools able to disentangle the possibly existing connections between the behaviour of dynamical systems and the topology, or the geometry, of the underlying *ambient manifold* (configuration, state or phase spaces), where evolution takes place. Indeed, as discussed in [3, 8], we claim that

[1] A taste of the need for a re-consideration of some paradigms of Chaos as been neatly felt at the Institute, and the talks presented by, e.g., Contopoulos, Froeschlé, Dvorak and coworkers have been witness of this.

B.A. Steves and A.E. Roy (eds.), The Dynamics of Small Bodies in the Solar System, 545–550.

the present limits to our general understanding of Chaos emerges mainly when we look to those DS's whose phase space is divided in *coexisting regions* in each of which the trajectories behave in a qualitatively very different fashion and we try to trace a definite *border* separating them. When doing this, using different tools or criteria, we are often forced to realize that such a smooth line, or surface, actually do not exists. Different *indicators* give distinct answers to the question about stability of motions. Within this scheme, the GDA can be viewed, at least, as another tool to explore those lands. We share the opinion that those discrepancies come from a too demanding requirement, which reflects, in turn, in a very strict definition of *Chaos*, regarded as *universal*, but which cannot be believed as such in all situations[2]. A different way to look at this point can be phrased, looking at the very definition of the most used tool to ascertain the occurrence of Chaos in a DS, *i.e.*, the *Lyapounov Characteristic Number's* (LCN's), observing how we have, in principle, to observe the behaviour of a system for an infinite interval of time, in order to grasp some information on its qualitative behaviour on finite lapses. An obvious question emerges at this point: for a DS with *mixed* behaviour, when an initial condition has been chosen *around* but very close to the *boundary* of an integrable region (*the last KAM torus*), we expect the phenomenon of so-called *stickness* to occur (see Contopoulos and Dvorak contributions). If there is only one connected chaotic region, we know that the true, asymptotic, (maximal) LCN will converge to the *unique* value. But, in the general case, how could we estimate the *sticky time*, *i.e.*, the interval of time required for the orbit to flow into the stochastic sea? When, monitoring the time behaviour of $LCN(t) \approx \ln \|\boldsymbol{\xi}(t)\|/t$, we see a t^{-1} trend, how could we discriminate between a *truly regular* and a *trapped* one? Incidentally, the preceding discussion rules out the general validity of the often reported (and seemingly reasonable) claim, that the convergence time, τ_r, of the LCN to its asymptotic value is of the order of its inverse, $\tau_r \sim LCN^{-1}$.

The results we obtained for what concerns both few dimensional, [8, 4, 5, 9] and many degrees of freedom systems, [2, 3], of which a partial account is given below and in the accompanying contribution, show clearly how the GDA gives the same answers obtained within the usual framework in all those situations where the regular or stochastic nature of an orbit is definitely established, whereas the outcomes differ to some extent when the behaviour

[2]The General Relativistic context is perhaps the one where some flimsiness in such an universality emerge at a first sight, due to the requested gauge invariance of the theory. We claim nevertheless that from an epistemological viewpoint the study of General Relativistic DS's represented a stimulus towards a deep thought on the *standard* definitions. A partial account on this topic is exposed in [3, 8], more details are presented in [4, 5], and a thorough exposition of the still *under construction* theory of G.R. DS's can be found in [10].

is not so clean. Moreover, the GDA allows to grasp very interesting hints on the sources of instability, and the major insights come out just from the analysis of the peculiar behaviour emerging at the boundary between order and Chaos.
Going back to the ultimate goal of the line of research this work belongs to, we could say that the pathway to locate the sources of Chaos, starting from the mere recognition of its occurrence, can be accomplished, step by step, through a hierarchy which can be schematically summarized as follows:

- A set of initial conditions for both trajectories and *disturbances* is chosen and then the equations of motion are numerically integrated simultaneously with the *second order* equations describing the evolution of perturbations along the given orbit. The instability is recognized *a posteriori*, and no attempt is made to relate its occurrence to any local or global property of the ambient space.
- Initial conditions are chosen as above, and suitable quantities, *depending on the given orbit and perturbation*, are time-averaged along the orbit itself. The occurrence of a chaotic nature of the orbit is then detected at the end of the numerical integration and the possible relationships with the computed averages are investigated, looking for a tool giving some informations on the behaviour of the particular set of i.c. chosen.
- An *a posteriori* link is searched between either microcanonical or time averages[3] of dynamic (or geometric) quantities and the global degree of stochasticity of the DS, for the corresponding values of the *macroscopic* parameters (*Toda-like* criteria). In this case, however, only a global characterization of the degree of stochasticity is possible and no information can be obtained about the behaviour of single orbits.
- A step forward is obtained if an indicator is found able to characterize the degree of instability whether the global parameters are varied or the initial state changes, irrespective of the perturbation. In this case, the indicator is evaluated along the orbit, with no assumption on the choice of perturbation; and not only it is not more necessary to integrate the equations for perturbations, but also we are forced to believe that a deep connection must exist between the indicator and the occurrence of unpredictability. Once found, this indicator allows to predict the occurrence of a possibly chaotic behaviour, [4, 5].

[3]In the first case, when use is made of $\mu-$canonical averages a strong assumption on the ergodicity of the system is believed to be understood, though usually not proved; instead, its *literal* occurrence can very often be discarded. On the same grounds, in the second case, time averages are intended to be performed over a suitable representative set of initial conditions. Nevertheless, the outcomes of all procedures strongly indicate that such a *literal* ergodicity is actuallynot needed, [2].

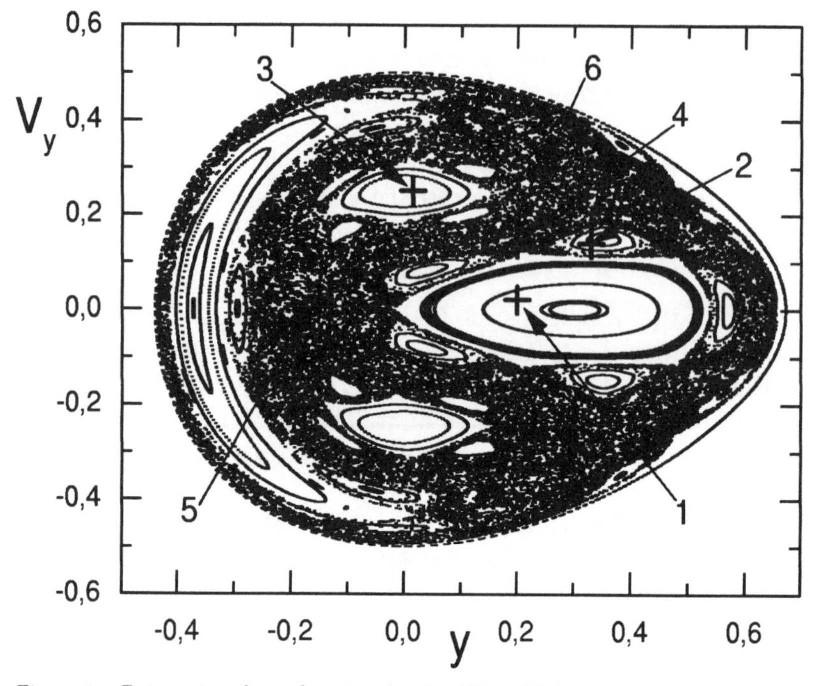

Figure 1. Poincaré surface of section for the Hénon-Heiles system at E=0.125 for a set of 169 initial conditions, each numerically evolved for T=5000 units. The intersections belonging to the orbits for which the instability exponents have been computed are marked by plus signs.

- A theoretical link is found, able to explain the implications over the dynamical features of trajectories of suitable averages of dynamical indicators of chaoticity. A threshold is predicted for the values of this indicator discriminating between regular and chaotic behaviour.
- The latest goal, whose attainment is perhaps only illusory, resides in a synthetic, *local*, indicator, able to single out the nature of any orbit simply on the basis of *puntual values* assumed in a given phase point (or, more plausibly, on a small piece of the trajectory).

The small bibliography below is obviously highly incomplete, but is intended only to give a personal account of the general setting of the GDA, and therefore do not aknowledge most important papers on the subject. The reason is that, in this contribution, we limit ourselves to display some of the results recently obtained, [7, 3, 8] and [4, 5, 9]. There, we discuss also some outcomes reported in the literature, [1], and prove the ability of the method to single out the transition between different regimes of *stochasticity*. As it is well known, the stability properties of DS's is usually investigated through the analysis of the behaviour of nearby trajectories. Within the *standard* DS's theory this analysis is performed using the *Tangent Dynamics* equations, which determine the evolution of small perturbations. The same (linear) *local* analysis is accomplished, in the GDA, through the

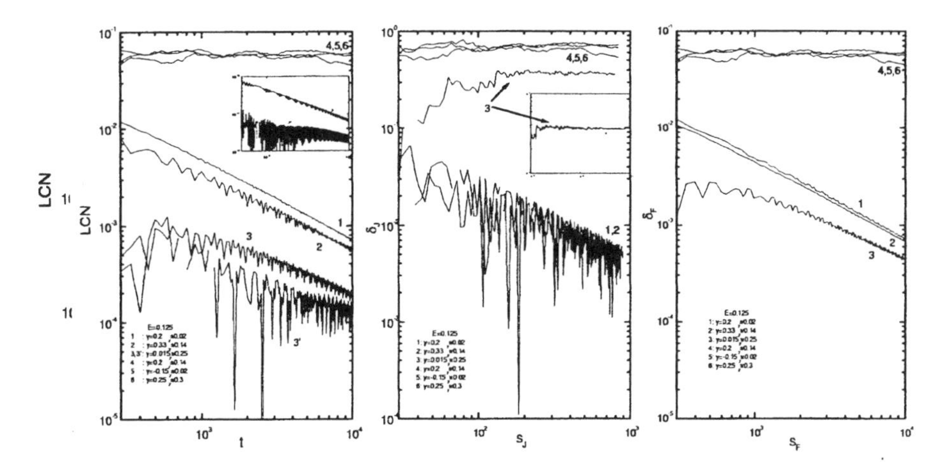

Figure 2. From left to right, LCN's, instability exponents in the Jacobi and Finsler metrics, for the orbits indicated in the previous figure. In the leftmost figure the two lines labeled as 3 and 3' refer to the same orbit but for different orientations of the disturbance vector. The small panels display the long-time behaviours of exponents relative to orbits 3 (and 3').

Jacobi–Levi-Civita (JLC) equations for geodesic deviation. Recently, [1], the tools of Riemannian geometry have been applied to investigate the behaviour of the Hénon-Heiles hamiltonian, a DS paradigmatic of Chaos and of mixed behaviour, with relative populations of regular and chaotic orbits varying with the increase of a global parameter, in this case the energy. We refer to [4, 5] for a detailed discussion of the results (see also [8]), and present here the evidence that a deeper investigation of the behaviour of orbits highlights how the results presented in [1] seem do not keep their validity in the full phase space. Instead, in a well defined region of phase space (*i.e.*, of the surface of section, see figure 1), the Riemannian GDA , within the Maupertuis-Jacobi geometry, gives answers partially clashing with those emerging from the behaviour of the LCN, see figure 2. The last box in the same figure show how a suitable extension of the GDA allows to recover a complete agreement between the approaches, and in [8] and [4, 5] we present a hopefully convincing explanation of the sources of such disagreements which also leads to single out the clear links existing between curvature and stability. Finally, in figure 3, we give a taste of the relationship between global degree of instability and geometric properties of configuration manifold, without reference to any specific choice of perturbation[4]. These results however improve only a little with respect to Toda-like criteria; we hope to have nevertheless accomplished an important step along the path outlined above, finding, [4], a *synthetic indicator of chaoticity*. To

[4]A similar result, though based on a specific choice of the orientation of the disturbance vector, has been obtained in [1].

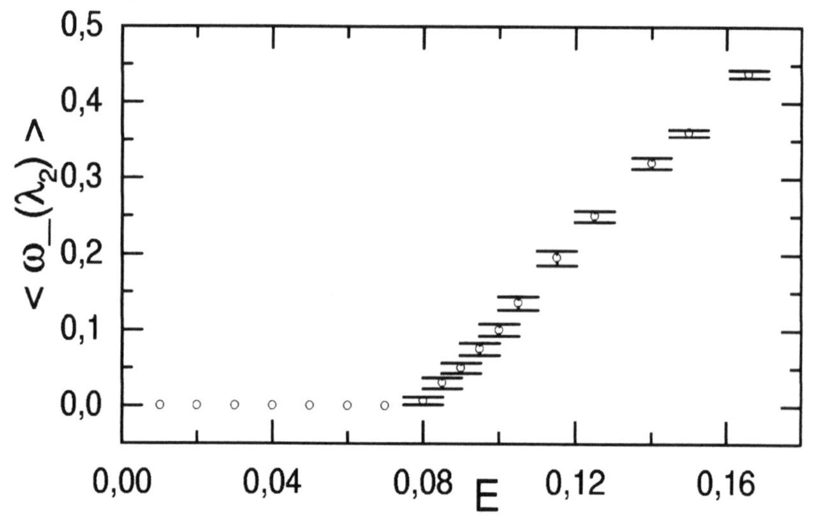

Figure 3. Transition in the geometric properties of the configuration manifold in the Finsler metric for the Hénon-Heiles DS. The frequency of negative values of one of the two non trivial principal sectional curvatures is plotted against the energy. A knee is neatly apparent at the same value of energy where stochasticity set in.

conclude, a comment on the disagreement between Jacobi geometry on one side and Finsler and Tangent Dynamics results. Figure 2 seems indeed to indicate that the JLC equation within the Jacobi metric tends to overestimate the dynamical instability, depicting as chaotic the orbits emanating from some regions of phase space, which are instead recognized as regular by the *usual* and Finsler's variational equations as well, and also from a visual side, looking at the consecutive intersections with the PSS. As we prove in [4, 5] the orbits emanating from the two islands on the $\dot{y}$-axis are characterised by a dynamical behaviour which is in some sense in between those displayed by *truly regular* and *definitely chaotic* orbits. If these orbits are *sticky*, even the qualitative information obtained by the surface of section could be misleading, as the instability can take a very long time to develop.

References

1. Cerruti-Sola, M. and Pettini, M. (1996) *Phys Rev. E*, **53**, 179.
2. Cipriani, P. (1993) *Ph.D.Thesis*, Univ. of Rome "La Sapienza", (in italian).
3. Cipriani, P. and Di Bari, M. (1997) Planetary and Space Science, *to appear*, xxx.
4. Cipriani, P. and Di Bari, M. (1997) submitted.
5. Cipriani, P. and Di Bari, M. (1997) to be submitted.
6. Di Bari, M. (1996) *Ph.D.Thesis*, Univ. of Rome "La Sapienza" (in italian).
7. Di Bari, M., Boccaletti, D., Cipriani, P. and Pucacco, G. (1997) *Phys. Rev. E*, **55**, 6448.
8. Di Bari, M. and Cipriani, P. (1997) *Planetary and Space Science to appear*, xxx.
9. Di Bari, M. and Cipriani, P. (1997) *to appear*.
10. Hobill, D., Burd, A. and Coley, A. (1994) *Deterministic Chaos in General Relativity*, Plenum.
11. Pettini, M. (1993) *Phys. Rev. E*, **47**, 828.

THE MANY FACES OF DYNAMICAL INSTABILITY

II: *The sources "Chaos" in Dynamical Systems*

MARIA DI BARI AND PIERO CIPRIANI[†]
Dipartimento di Fisica "E. Amaldi", Università "Roma Tre"
[†] *also* I.N.F.M. - *Istituto Nazionale di Fisica della Materia*

1. Introduction

Referring for an introductive discussion to the accompanying contribution, we focus here on different general aspects of the Geometrodynamical Approach, (GDA), enlightening its reliability as an alternative and in some instances preferred tool to investigate instability in dynamical systems (DS's). The approach can ultimately dated back to the beginning of this century, to the work of Ricci and Levi-Civita, and then was further developed by Synge, Eisenhart, Cartan and others (see, e.g., [19] and bibliography therein), who exploited most of the tools of the (Riemannian) differential geometry to intrinsically describe the qualitative character of the dynamics. These studies were revived within the framework of Statistical Mechanics by the work of Krylov, [15], subsequently resumed by the Russian School on Ergodic theory, [18], though these latter approaches, unlike the Krylov's studies, became more and more focused on the formal and mathematical aspects rather than on the physical implications. In the last years, however, [17, 3, 4] the *physical reliability* of the geometrical description of DS's has received considerable support from the results obtained for many degrees of freedom Hamiltonians, where a neat correlation has been found between geometrical properties of configuration manifold, qualitative features of dynamics and Statistical behaviour. With respect to few dimensional DS's, a lot of efforts has been spent, looking for similar relationships between geometry and dynamics. Very recently the main tool for the investigation of instability in the GDA, that is the Jacobi–Levi-Civita (JLC) equation for geodesic spread has been successfully applied to the investigation of the transition to Chaos in a paradigmatic two degrees of freedom DS, [2]. There, a *generic qualitative* agreement has been found between the results obtained integrating the JLC and the tangent dynamics equations. As *qualitative* we mean that both the tools predict instability almost in the same regions of phase space, while the word *generic* indicates that this agreement is not everywhere fulfilled, as shown in the previous contribution.

B.A. Steves and A.E. Roy (eds.), The Dynamics of Small Bodies in the Solar System, 551–556.

The ultimate aim of the line of research this work belongs to is very ambitious and resides in the search for an intrinsic and *a priori* characterization of Chaos, going beyond a somewhat phenomenological approach. The path towards this goal can be splitted into *easier* intermediate steps, whose hierarchy has been sketched in the previous paper. Leaving aside the perhaps only well known situation of two dimensional discrete DS's (maps), the usual procedure to detect the occurrence of chaotic behaviour in typical flows describing systems with two or more degrees freedom, consists in the exploration of the behaviour of orbits corresponding to *neighbouring initial conditions*, in order to single out the possible loss of memory of the initial state. The detailed way in which this exploration is accomplished constitutes only a matter of tools[1], whether different initial conditions are actually chosen and the corresponding trajectories numerically integrated, or the local stability properties are investigated by means of a set of equations describing the behaviour of *small perturbations* to a reference orbit and derived from equations governing the trajectories themselves[2]. The presence of Chaos is then detected looking to a possible exponential growth of *separation* between initially closeby trajectories, and, within this setting, the issue of the sources of instability is not addressed since the beginning.
A step beyond this ultimately phenomenological approach has been recently pursued, [2], looking for a qualitative correlation between the occurrence of an unstable behaviour *of a given perturbation* and the cumulative average of a dynamical quantity, though defined in terms of *the perturbation* itself. In a series of recent works, [17, 3, 5, 12], mainly within the GDA, a further improvement along the path towards the understanding of the *sources* of Chaos has been attained. There, very neat qualitative correlations between *global* instability properties of the dynamics and suitable averages (temporal or microcanonical) of geometrical quantities have been singled out, and, for DS's with many degrees of freedom, also an effective equation governing the behaviour of *small disturbances* has been derived, allowing for a semi-analytical evaluation of the maximal Lyapunov exponent. On this light, a connection between some changes in the average values of global quantities, evaluated over the constant energy surface, and the global degree of stochasticity, as measured, loosely speaking, by the probability that a *ran-*

[1]This is not exactly true, as the variational equations can in principle only give information on the *local* stability properties of the dynamics, whereas the actual evolution of two initially closeby trajectories could give hints on its global features. So the two procedure are able to give complemetary informations, and the practical and computational advantages offered by the former tool should not induce to forget the meaningful insights which can be obtained using the other method. This issue is however out of place here and will be investigated in a deeper detail elsewhere.

[2]In [12] we presented an analysis on the (in)equivalence of variational equations written within different frameworks.

domly chosen trajectory turns out to be chaotic, has been definitely found. However, when the number of d.o.f. is small, and no averaging procedure is guaranteed to faithfully describe the real behaviour, a quantitative (behind those of qualitative nature as that discussed also in the companion paper) relationship between average geometric (or geometrodynamical) quantities and the possibly chaotic nature of DS's remained, until now, rather hidden. Moreover, up to now, no general criteria have been presented able to go beyond those of *global nature* described above: it has been made possible to find out a correlation between system parameters (e.g., the energy) and the fraction of the phase space occupied by chaotic orbits, but no methods able to relate some averages along a given trajectory and the nature of the orbit itself have been found. The criticisms, [1], against the Toda criterion, [20], motivated just showing that his chaoticity indicator was unable to distinguish between regular and chaotic orbits existing over the same constant energy surface, would equally apply to these criteria.
In [6, 7] we present the first, as far as we know, synthetic indicator of chaotic behaviour, able to distinguish between regular and chaotic orbits over the same surface, which moreover do not depends at all on the perturbation chosen, but is instead a *proper* characteristic of the trajectory itself.
In order to present however a *panorama* of the major merits of the GDA, we will address here to discuss some of the results obtained applying the method to a General Relativistic (GR) model, whose peculiarities made it perhaps more interesting as a DS in itself than as a cosmological model, as it was originally. Moreover, the analysis of the properties of this model will give us the occasion to discuss some general aspects whose relevance extends beyond the specific problem.

2. An invariant definition of Chaos.

In the last years it has been widely recognized that the most used quantitative indicator to detect and measure Chaos cannot be an acceptable tool within all those frameworks where a requirement of *invariance* under arbitrary change of coordinates and/or time is present. General Relativity is perhaps the first theoretical setting coming to mind in this respect, but we claim that such a question should be asked also within classical theory of DS's. It is true that within Classical Analytical and Statistical Mechanics there are some plausible arguments to assign to Newtonian time a privileged meaning, but, from an epistemological viewpoint, the question is however legitimate.
Without going into *dangerous topics of semi-philosophical nature*, we will concentrate on the discussion of the GR cosmological model, illustrating some of the results obtained. For more details, see [10, 11, 13].
Among the homogeneous spaces classified by Bianchi, [16, §115], that numbered as IX has received a lot of attention for the interesting features it

displays when studied as a closed Universe model, [16, 14]. It represents indeed the *most general* solution of Einstein equations within the class of homogeneous spaces. From a DS's viewpoint, it constitutes therefore the most general finite-dimensional cosmological model. From an astrophysical point of view, it has for some time believed as a good alternative to the *theory of Inflation*, though now this possibility seems to be ruled out. But there are a couple of further motivations which make it interesting from a DS's side: firstly, being a prototype of GR dynamical system, it offers the chance to investigate the issue of a meaningful *gauge-invariant* definition of Chaos. Moreover, with respect to the GDA, it gives the possibility to prove that most of the *naïve* beliefs on the sources of instability of realistic geodesic flows, borrowed from the abstract Ergodic theory, are generally false. It also forced us to enlarge the class of manifolds in which the dynamics is described to an ensemble wider than Riemannian, [8, 9]. Finally, it allows to show how the mechanisms supervising to the onset of Chaos are qualitatively the same in both few dimensional and many degrees of freedom DS's, but nevertheless they are easily singled out only in the extreme cases, *i.e.*, either for two degrees of freedom systems or when the dimensionality of the space is so high that statistical arguments can be applied. The Bianchi IX model can be rephrased as a three degrees of freedom DS, whose hamiltonian is given by, [9, 13]:

$$H = \frac{1}{2}\left(\dot{\beta}_+^2 + \dot{\beta}_-^2 - \dot{\alpha}^2\right) + \mathcal{U}(\alpha, \beta_+, \beta_-) \, , \tag{1}$$

where α and $\beta_\pm$ are functions of the scale factors a,b,c of the Universe, and measure, respectively, the overall volume and the anisotropy of the 3-dimensional hypersurface $\tau = const.$; the dot indicates differentiation with respect to the time parameter τ, defined in terms of the *cosmological proper time* t through the transformation $d\tau = dt/abc$. The Hamiltonian (1) in nevertheless peculiar, in that the General Relativity imposes a constraint not encountered usually in classical DS's: Einstein equations request that H be a *null hamiltonian*, *i.e.*, $H \equiv 0$. Moreover we see that the kinetic part is not positive definite; this peculiarity is in turn responsible of the lack of reliability of the already attemped GDA's, [21], because the Jacobi metric cannot be applied in this case, [12, 13]. This DS has been extensively studied in the last two decades, and the claims about its integrability (or non-chaoticity) equal almost in number the works asserting the occurrence of stochastic behaviour. It can be easily recognized that most of the discrepancies come out from different choices of the time variable used, or from a neglecting of the amount of time elapsed between two successive *mappings*.

Without going further into details, we show in figure 1 the short and long time behaviours of the instability exponents defined within the GDA, using the Finsler metric. These have been computed numerically integrating the

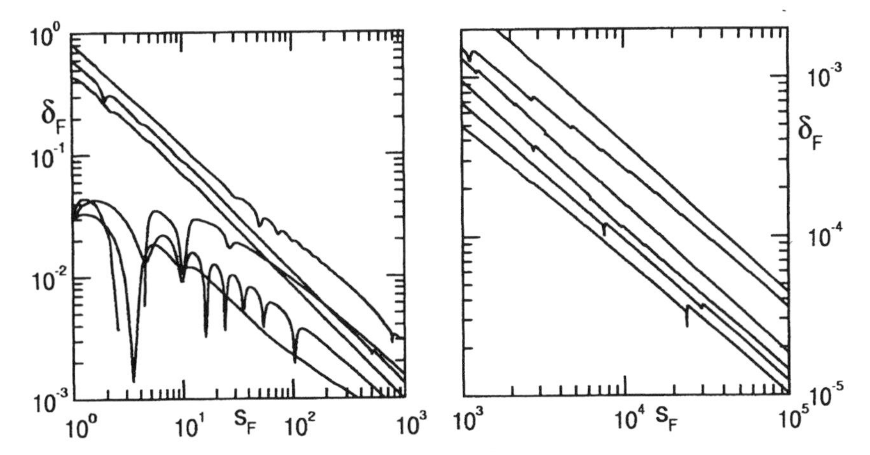

Figure 1. Short (left) and long (right) time behaviour of Instability exponents computed in the Finsler metric for a set of representative initial conditions for the Bianchi IX model.

geodesic and JLC equations of geodesic spread, [9, 5, 12]. The non chaotic nature of the model, as far as chaoticity is recognized by the presence of an exponential growth of perturbations, is manifestly excluded from these plots. It is also apparent that in the initial phases of the evolution some instability is present. This occur when the orbit *bounces* against the potential wells, but, as it is also evident from the plots (which are in a log-log scale), the interval between two successive bounces increases very quickly after the first transient stage. For this reason every transcription of the Bianchi IX dynamics in terms of discrete maps (and the analysis within the Jacobi geometry is ultimately equivalent to this, [13]) turns out to find Chaos, as it neglects the ever increasing interval of time between these unstable (if not chaotic) *scattering processes.*

What we want to stress here, however, is not a pretentious and umpteenth claim about the *regularity* of Bianchi IX dynamics, instead we want to point out, on the light of the in any case substantially unpredictable evolution of this DS, the need for a deeper reconsideration of the very definition of Chaos, in GR in particular, but also in a more general context.

As a further evidence of the links between geometry and dynamics, beyond the results of the integration of JLC equations, figure 2 shows again short and long time behaviours of *principal sectional curvatures* (PSC's) of the Finsler manifolds associated to Bianchi IX Lagrangian, for a typical initial condition, along with the Ricci curvature, (suitably normalized) given by the trace of the *stability tensor*, [3, 5]. From the figure we see that, during the first *unstable* evolutionary phases, the manifold is strongly anisotropic (an ingredient which is very important for the origin of Chaos, [6]), then it becomes more and more isotropic (two of three PSC's become virtually equal each other), evolving quickly towards flatness. So, we see that, also in this DS, instability and geometry again evolve together.

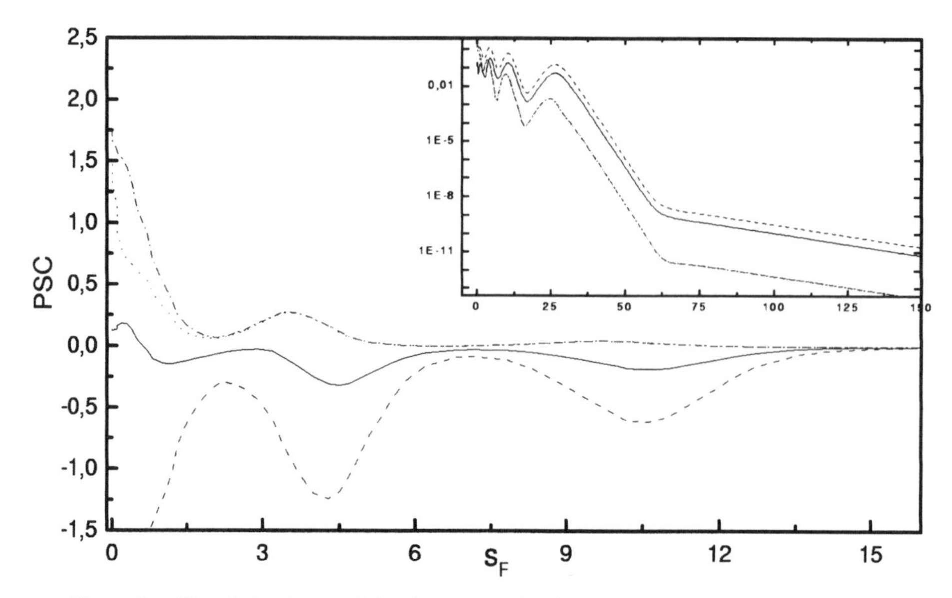

Figure 2. *Time* behaviours of the three non trivial principal sectional curvatures in the Finsler manifold and of the Ricci curvature (full curve). In the inset, the *absolute values* on longer times are plotted in a linear-log scale.

References

1. Benettin, G., Brambilla, R. and Galgani, L. (1977) *Physica*, **87A**, 381.
2. Cerruti-Sola, M. and Pettini, M. (1996) *Phys Rev. E*, **53**, 179.
3. Cipriani, P. (1993) *Ph.D.Thesis*, Univ. of Rome "La Sapienza", (in italian).
4. Cipriani, P. and Pucacco, G. (1994) *N. Cimento B*, **109**, 325.
5. Cipriani, P. and Di Bari, M. (1997), *Planetary and Space Science to appear*, xxx.
6. Cipriani, P. and Di Bari, M. (1997), submitted.
7. Cipriani, P. and Di Bari, M. (1997), to be submitted.
8. Di Bari, M. (1996) *Ph.D.Thesis*, Univ. of Rome "La Sapienza" (in italian).
9. Di Bari, M., Boccaletti, D., Cipriani, P. and Pucacco, G. (1997) *Phys. Rev. E*, **55**, 6448.
10. Di Bari, M. and Cipriani, P. (1997) in *Proc. 12th Italian Conference on General Relativity and Gravitational Physics*, World Scientific, xxx.
11. Di Bari, M. and Cipriani, P. (1997) in *Birth II - Fundamental Phyisics at the Birth of the Universe.*, F.Occhionero (ed.), World Scientific, xxx.
12. Di Bari, M. and Cipriani, P. (1997) *Planetary and Space Science to appear*, xxx.
13. Di Bari, M. and Cipriani, P. (1997) submitted.
14. Hobill, D., Burd, A. and Coley, A. (1994) *Deterministic Chaos in General Relativity*, Plenum.
15. Krylov, N.S. (1979) *Works on Foundations on Statistical Physics*, Princeton Univ.
16. Landau, L.D. and Lifsits, E.M. (1976) *Mécanique* (3rd edition), MIR.
17. Pettini, M. (1993) *Phys. Rev. E*, **47**, 828.
18. Sinai, Ya.G. (1991) (Ed.) *Dynamical Systems*, World Scientific.
19. Synge, J.L. (1926) *Phil. Trans. A* **226** 31.
20. Toda, M. (1974) *Phys. Lett*, **48A**, 335.
21. Szydlowski, M. and Krawiec, A. (1993) *Phys. Rev. D*, **47**, 5323.

SCALE RELATIVITY IN THE SOLAR SYSTEM

A review and prospects

R. P. HERMANN
Institut d'Astrophysique de Cointe – Université de Liège
B - 4000 LIEGE
Belgium
Email : R.Hermann@ulg.ac.be

Abstract. We shall consider the scale-relativistic model for describing the behaviour of chaotic systems beyond the predictability horizon. Non-differentiable mechanics will give us our main structure equation. We shall then apply it to get a model of the positions of planets and satellites in the solar system. Finally, we shall see how this method could be applied to the structure of the asteroid belt.

1. Introduction

The scale-relativity theory has been developed during the last ten years by L. Nottale. Its fundamental assumption is that resolutions are essential variables that characterise the relative state of scale of the reference system, in the same way as velocities characterise its state of motion.

Our scale laws combine resolution dependent behaviour at small and large scales, and a transition to scale-independence at intermediate scales. This means space-time is Riemannian at intermediate scales and fractal for some resolution Δx or Δt smaller and larger than some relative transition scale.

2. Giving up the differentiability hypothesis.

At intermediate scales, experiments show the physical space-time is differentiable. There is no evidence that this holds at small or large scales. Here we shall deal with large time scales.

We thus postulate non-differentiability (nearly everywhere) of space with respect to time being more general as differentiability. This implies

B.A. Steves and A.E. Roy (eds.), The Dynamics of Small Bodies in the Solar System, 557–563.

immediately scale-dependence of distances between two points, that is, a fractal space-time.

In such a space-time there are infinitely many geodesics between any two points. Figure 1 demonstrates this statement for an orthogonal fractal space (since the construction of the fractal shown must be continued for smaller resolution). So we get an infinite set of geodesics between any two points [1].

3. Consideration on chaotic trajectories.

Chaotic systems are generally characterised by exponential divergence with time t of the distance $\delta x(t)$ between initially close trajectories (where τ is the Lyapounov time) :

$$\delta x(t) = \delta x(0) \exp(t/\tau). \tag{1}$$

At large time scales the trajectories of a chaotic system form a set of non-differentiable curves in space (Figure 2 - from left to right : decreasing time resolution).

Moreover, the description can no longer be deterministic beyond the predictability horizon. We have a loss of information and the motion becomes Brownian (we shall assume fractal dimension 2). We must jump to a statistical description.

4. Structure equation.

Taking the considerations above into account leads to a new physical model. The complete development can be found in [1] (pp. 135-153), [2] and [3]. Some more mathematical results are presented in [4]. Let's just take one

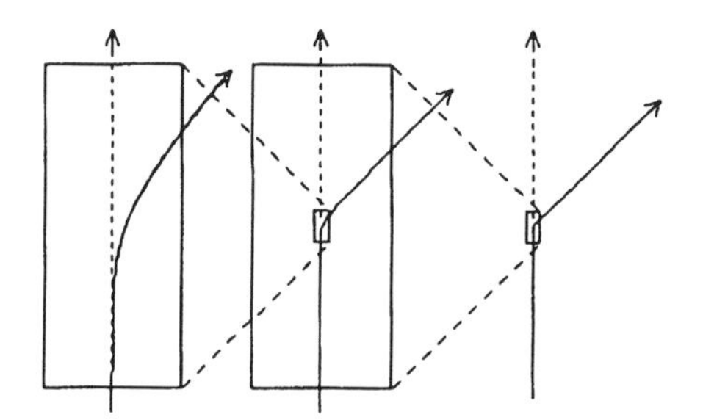

Figure 2. Divergence of chaotic trajectories : decreasing zoom.

main result : non-differentiability of space with respect to time leads to a new complex time derivative :

$$\frac{\mathrm{d}}{dt} = \left(\frac{\partial}{\partial t} - i\mathcal{D}\Delta + \mathcal{V}\cdot\nabla\right), \tag{2}$$

where $\mathcal{D}$ is dimensionally a real diffusion coefficient and $\mathcal{V}$ the complex velocity.

We have that $\mathcal{V} = \nabla\mathcal{S}/m$ where $\mathcal{S}$ is the complex action and we write $\Psi \equiv \exp(i\mathcal{S}/2m\mathcal{D})$ (thus Ψ is just another way to write the action).

Then again Newton's equation becomes (Φ is the real potential) :

$$-\nabla\frac{\Phi}{m} = \frac{\mathrm{d}}{dt}\mathcal{V} = -2\mathcal{D}\nabla\left\{i\frac{\partial}{\partial t}\ln\Psi + \mathcal{D}\frac{\Delta\Psi}{\Psi}\right\}. \tag{3}$$

Finally we get, by integration :

$$\mathcal{D}^2\Delta\Psi + i\mathcal{D}\frac{\partial}{\partial t}\Psi - \frac{\Phi}{2m}\Psi = 0. \tag{4}$$

This means Schrödinger equation is the non-differentiable Newton equation. It can be shown that $\rho = |\Psi|^2$ (where ρ is the probability density) also holds [3][5].

We can now apply this equation to describe some gravitational systems that are stationary at large time scales. Let's add three remarks. First, although some tools are in common with quantum mechanics, we do not have the whole quantum mechanical interpretation, mainly because the differentiability is recovered at small time scales. Secondly, our results do not compete with those obtained from standard celestial mechanics, which are valid below the prediction horizon while our results are only valid beyond this horizon. Thirdly, we shall only present the simplest cases here, that is problems with central symmetry.

5. The planets of the solar system and the extra-solar planets.

While its formation, motion in the solar system was probably chaotic so our model should apply. At time-scales $\Delta t \gg \tau$, the motion of the dust then becomes non-differentiable, and we apply our time-independent Schrödinger-like equation (M is the central mass, m the mass of the test particle) [1] (p. 313), [3]:

$$2\mathcal{D}^2 \Delta\Psi + \frac{E}{m}\Psi + \frac{GM}{r}\Psi = 0. \tag{5}$$

This is a hydrogenoid-like equation, so the solution for the semimajor axis and for the eccentricity are quantized. Here, we shall only consider low eccentricity solutions, so we get (a_0 is the length unit of the problem : $a_0 = 4\mathcal{D}^2/GM$, its value is obtained by the fit ; $n \in N_0$) :

$$a_n = (n^2 + n/2).a_0 \quad \text{for the density mean semimajor axis,} \tag{6}$$

$$a_n = n.a_0 \quad \text{for the density peak semimajor axis.} \tag{7}$$

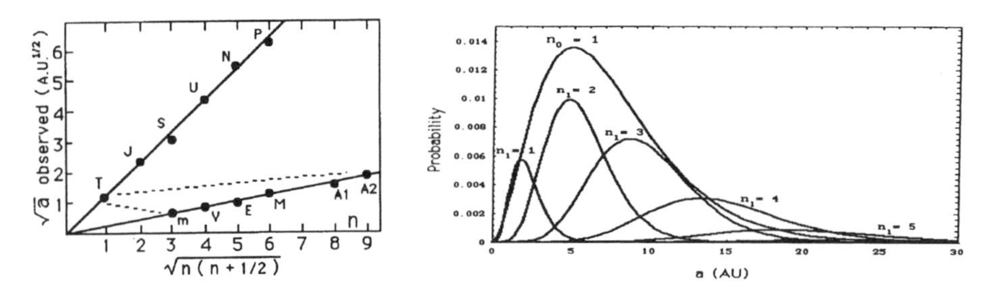

Figure 3. Solar system (left) and orbital hierarchy (right)

A distribution of planetesimals is then expected to fill the orbitals. On the other hand, for a planet there are two possibilities : first, if there is evolution from the planetesimals stage with negligible external perturbations, the energy conservation implies the planet will lay at the mean distance. Secondly, if there are perturbation the planet can fall into the peak of probability.

In the case of the solar system one sees that there are two sub-systems (Figure 3 (left)) : one with the telluric planets, the other with the outer planets. This could be related to the formation history of the system. First, planetesimals fill in the orbital $n_0 = 1$ of Figure 3 (right), then there is fragmentation of this orbital (due to the accretion of a massive body at $n_1 = 2$). The remaining dust of orbital $n_1 = 1$ then again fragments to give raise to the telluric planets. The mass distribution predicted by this model [6] is in good agreement with the mass distribution observed for both systems (Figure 4) .

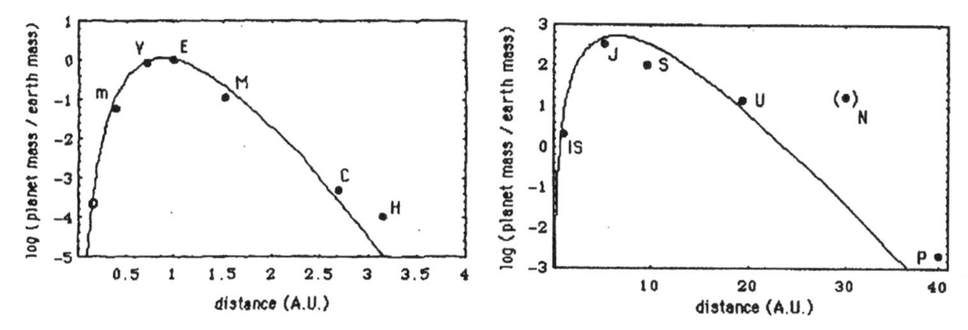

Figure 4. Mass distribution for telluric (left) and outer planets (right).

Another domain of validity of our theory is the field of the recently discovered extra-solar planets. Indeed, in this case, it is striking that the peak law :

$$a_n = GM/w_0^2.n^2, \tag{8}$$

(where $w_0 = GM/2\mathcal{D}$) is in good agreement with the position of all extra-solar planets for the same value of $w_0 = 144$ km/s [7]. This can be seen in Figure 5 (left) . The shaded zones are low probability zones, the white zones high probability zones of the theory.

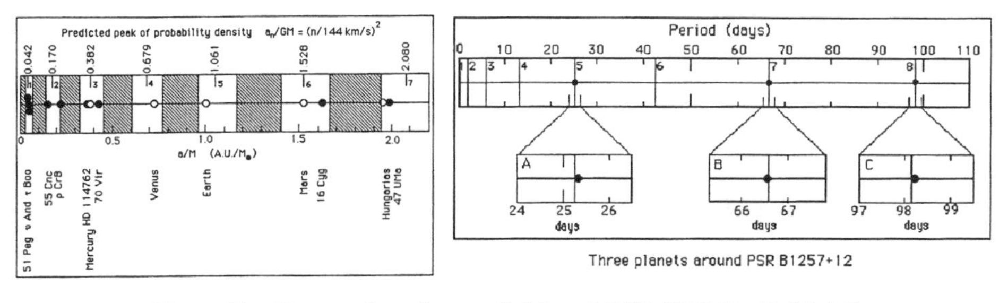

Figure 5. Extra solar planets (left) and PSR B1257+12 (right)

This agreement is even far better for the PSR B1257+12 system where there are 3 planets orbiting around the pulsar [7] (Figure 5 (right)). In this case, our theory indicates the positions for further possible planets.

Some prospects can be formulated : are there planets in the n=1 and n=2 position of the inner solar system (cf. Figure 3 (left)) [6] ? How to understand the better agreement of the 'mean' law than the 'peak' law ? One can also remark that using this model allows us to know the position of possible extra-solar planets from the mass of the star, or vice-versa.

6. The satellites of Jupiter, Saturn and Uranus.

Another extension of our model are the satellite systems of the outer planets. We take our mean law $a_n = (n^2 + n/2).a_0$ and try the best fit. Figure 6 shows the results (Y axes : root of the semimajor axis ; X axes : root $(n^2 + n/2)$). All bodies did not fit our law, like Titan and Hyperion for Saturn, but since this is only a first order model the agreement seems satisfying. Here also there is a need for a second order theory taking perturbations into account.

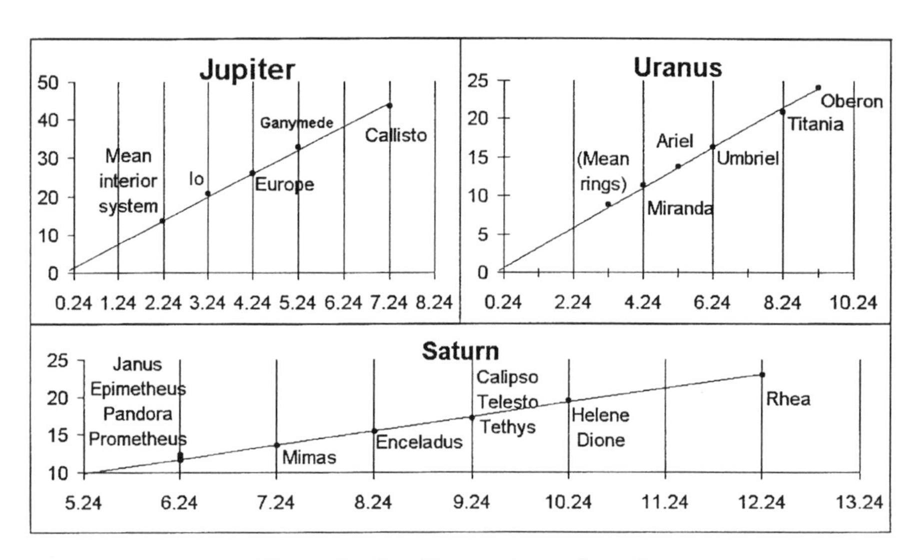

Figure 6. Satellites systems - best fits.

7. Statistical validity of the models

We now tried to fit a random distribution of bodies with our law (same number of bodies and degrees of freedom). This gave us some χ^2 distribution for random systems which we compared with the χ^2 of the real systems. The percentage of Table 1 indicate what proportion of random systems achieve a lower χ^2 than the real system.

This results take into account that some positions of our laws are left empty and that some positions are filled with more than one body. These percentage may not look impressive at first sight, but we should remember

TABLE 1.

	χ^2 percentages
External solar system	7,3 %
Internal solar system	7,9 %
Uranian satellites	8,9 %
Saturnian satellites	0,6 %
Jovian satellites	4,0 %
PSR 1257	0,3 %

that this single one-parameter law is in rather good agreement for all these 6 systems [8].

8. The asteroid belt

Asteroids correspond to n=7,8,9,10 orbitals of the inner solar system. Since there are strong interactions with Jupiter, we shall have to write a Schrödinger equation with time dependent perturbations. Our conjecture for further work is that at the time-scale of the perturbations, the orbits of the asteroids can be seen as a continuum of 'wave-functions', so that a time-dependent perturbation would induce some absorption rays.

As a conclusion, we shall remark that scale-relativity seems to be an efficient tool for describing chaotic gravitational systems at long time scales.

References

1. Nottale, L. (1993) *Fractal Space-Time and Microphysics, Towards a Theory of Scale Relativity*, World Scientific, Singapore.
2. Nottale, L. (1996) Scale Relativity and Fractal Space-Time, *Chaos, Solitons & Fractals*,**Vol. no. 7**, pp. 877–938.
3. Nottale, L. (1997) Scale relativity and quantization of the universe, Theoretical framework, *Astron. Astrophys.*, **Vol. no. 327**, p. 867
4. Pissondes, J.-C. (1997) Scale Covariant Representation of Quantum Mechanics, *to appear in Chaos, Soliton & Fractals.*
5. Hermann, R.(1997) Numerical simulation of a quantum particle in a box, *J. Phys. A*, **Vol. no. 30**, pp. 3967–3975.
6. Nottale, L., Schumacher, G., and Gay, J. (1997) Scale relativity and quantization of the Solar System, *Astron. Astrophys.*, **Vol no. 322**, p. 1018.
7. Nottale, L. (1996) Scale relativity and quantization of extrasolar planetary systems, *Astron. Astrophys. Lett.*, **Vol. no. 315**, L9–L12.
8. Hermann, R., Schumacher, G. and Guyard R. (1998) Scale relativity and quantization of the solar system, Orbit quantization of the planet's satellites, *Astron. Astrophys.*, **Vol. no. 335**, pp. 281–286.

POSSIBLE COSMOLOGICAL ORIGIN OF COUNTERROTATING GALAXIES

M. HARSOULA AND N. VOGLIS
University of Athens, Department of Physics
Section of Astronomy, Astrophysics and Mechanics,
Panepistimiopolis, Athens, Zografos 157 84

Abstract. We investigate numerically the possibility that single dissipationless cosmological collapses can lead to the formation of counterrotating objects, e.g. counterrotating galaxies. We consider that such systems start forming from initially small density excesses embeded in the environment of other density perturbations in an otherwise homogeneous and isotropic expanding early Universe. Our study is based on a version of the model proposed by Voglis and Hiotelis (1989) and Voglis et al. (1991), in which the growth of angular momentum in an aspherical (initially small) density perturbation is studied by the N-body method.

Counterrotating objects are formed rather naturally as a result of the partial mixing between the material in which positive angular momentum dominates with the material in which negative angular momentum dominates.

1. Introduction

A structure, born from small initial density peturbations in an otherwise homogeneous and isotropic expanding Universe, acquires angular momentum from the tidal field due to other small perturbations that break the isotropy of its environment (Peebles 1969, Doroshkevich 1970).

A reasonable question concerns the distribution of angular momentum along the radius of the relaxed object, particularly, whether such a distribution can be consistent with the phenomenon of counterrotation observed in galaxies. Counterrotating cores have often been observed in ellipticals, (e.g. Franx and Illingworth 1988, Bender 1988, Franx et al 1989, Bender

B.A. Steves and A.E. Roy (eds.), The Dynamics of Small Bodies in the Solar System, 565–574.

and Surma 1992). Generally, in recent years, it became evident that in probably up to 1/3 of luminous ellipticals the core regions are kinematically decoupled from the main bodies of the galaxies. A merger event can be responsible for counterrotation. N-body simulations have shown the formation of a counterrotating central disk in a merger of two gas-rich disk galaxies (Hernquist and Barnes 1991).

An alternative scenario, in which counterrotating galaxies may have been formed in a single collapse, is also possible as we will see below. In this case, the two counterrotating parts reflect to some extend the surviving memory of cosmological initial conditions.

In the field of small fluctuations in an otherwise homogeneous and isotropic expanding early Universe (at decoupling) consider a density excess that makes bound a mass M, destine to detach from the general expansion and collapse to form a galaxy. Let S be the surface surrounding this mass (Fig. 1). Inside this surface there may be several local density peaks of various mass scales smaller than M. These peaks lead to subclumps which merge during the collapse to form a single object of mass M. The mass M cannot be isolated. Other density perturbations of various mass scales outside S create an anisotropic environment and therefore an external tidal field exerting a torque on M. Thus the material inside S acquires angular momentum. The external torque vanishes when the mass M collapses provided that the environment still expands. Thus the total angular momentum that has been transferred to M up to this time remains constant.

This picture is applicable to any mass scale in the hierarchical clustering scenario. The most important, however, is the galactic mass scale. For this reason our terminology here on refers to galaxies.

The distribution of angular momentum along the radius of the relaxed object cannot be unique. It depends on the position in space of the various subclumps relative to the center of mass M and relative to the direction of the external tidal field. Let the inertia tensor of the mass M has a principal axis A. Let a coordinate system OXYZ has its origin O at the center of mass of M, the axis Y oriented along the direction of stronger forces of the external tidal field and the axis Z of the plane of Y and A as in figure 1. Then, particles on the first and the third quadrant acquire negative angular momentum, while particles on the second and the fourth quadrant acquire positive angular momentum because of the corresponding signs of the torque. As a consequence, the formed galaxy has a mixture of two components of angular momentum (positive and negative) oriented along the X axis. The net rotation is due to the fact that the two components have not exactly equal absolute values and therefore the system has a net angular momentum. In general, the two components of angular momentum have also different distributions along the radius of a galaxy and it may happen that one component is in excess in the inner parts, while the other component is in excess in the outer parts of a galaxy. In this case, the galaxy presents counterrotation.

We can initiate such a scenario, by cosmological initial conditions such as in figure 1. Consider the density perturbation from which the galaxy is formed having a central bar inside S oriented along the direction second-fourth quadrant, while the rest of the bound mass has a quadrupole moment with major axis along the direction fist-third quadrant. The material in the bar will acquire positive net angular momentum while the bound material outside the bar will acquire negative net angular momentum. If the mixing of the material along the radius after the collapse is not complete a counterrotating galaxy can be formed.

In the next section we describe in brief such a model of initial conditions for N-body simulations based on the model of initial conditions proposed by Voglis and Hiotelis 1989, Voglis et al 1991. Our results are given in section 3 and our conclusions are summarized in section 4.

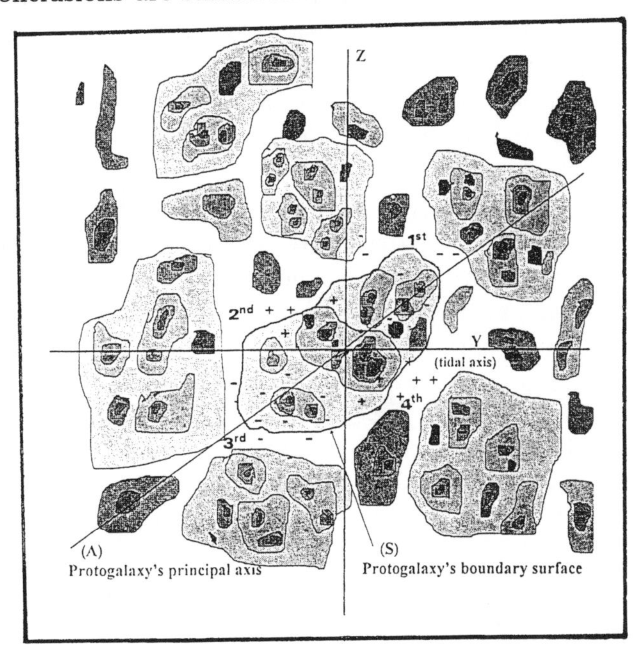

Figure 1. In a hierarchical clustering scenario A schematic representation of density excesses in a hierarchical clustering scenario. A density excess inside the surface S at decoupling exposed to the tidal field of its environment is destined to detach from the general expansion and collapse to form a rotatinggalaxy. Plus and minus signs indicate the sign of the tidal torque acting on individual particles

2. The model

The basic idea in the model of initial conditions is to consider in the early Universe (e.g. at a redshift $z \approx 1000$), a sphere with total mass of a few galactic masses, called G-scale, at the center of a spherical extensive mass environment (hundreds of galactic mass), called E-scale. The unpertubed system expands as an Einstein-de Sitter Universe.

A bar-like (prolate spheroid) density perturbation is imposed at the center of the G-scale with its major axis on the Y-Z plane along a direction making an angle $\Theta = -45^o$ with the Y axis (second-fourth quadrant). This density perturbation makes bound not only the mass inside the bar but also part of the mass outside the bar having a quadrupole moment with major axis on the Y-Z plane perpendicular to the bar. As we will show below such a model of initial conditions in the G-scale exposed to the tidal field due to other density perturbations of the environment can create a counterrotating galaxy.

In this model a cosmological tidal field is realised by imposing a density perturbation on the E-scale causing tidal forces along the Y-axis.

The G-scale is resolved into 2176 particles of equal masses in a cubic grid. The total mass of the G-scale is $2M_u$, where M_u is the unit of mass considered to be equal to the typical mass of a galaxy e.g. $M_u = 10^{12}$ solar masses.

The E-scale is resolved into 664 paricles in a cubic grid. Each particle has a mass equal to $1M_u$. Special care has been taken so that the torque on the protogalaxy from the environment is in agreement with the torque predicted by the linear theory of evolution of cosmological density perturbations (cosmological torque).

These initial conditions are evolved according to the Zeldovich approximation (Zeldovich 1970) to a redshift of about $z \approx 40$, when the perturbations start being non-linear. A projection on the Y-Z plane of the whole system at this redshift is shown in figure 2. The sphere at the center of the figure corresponds to the G-scale with the central bar-like perturbation in it.

The time evolution of both scales (G and E) is followed simultaneously by Aarseth's N-body2 code, for a time comparable to the age of the Universe.

The scaling units of length and time are defined in Voglis and Hiotelis (1989). Namely, if the unit of mass is $M_u = 10^{12}$ solar masses, the unit of length is $r_u = 1.5\beta Kpc$ and the unit of time is $t_u = 0.86\beta^{\frac{3}{2}} Myear$. The velocity unit is $v_u = 1677\beta^{-\frac{1}{2}} Km/sec$. In these expressions β is a rescaling parameter depending on the precise values of the amplitude of perturbations (that determine the epoch of galaxy formation and hence the

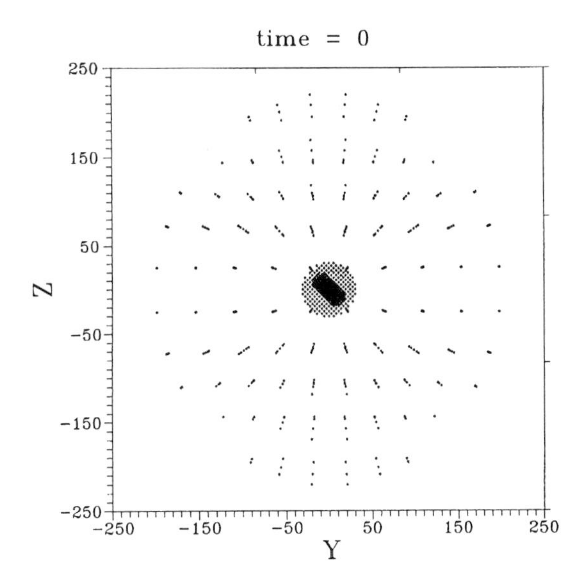

Figure 2. The projection of the initial configuration on the Y-Z plane at a time=0.

virial velocity in dissipationless collapse and relaxation). If in an experiment a mass $M = 0.5M_u$ is virialized in a radius $R = x$ the virial velocity (estimated by $v_{virial}^2 = \frac{M}{2R}v_u^2$) is $v_{virial} = \frac{1}{2\sqrt{x}}v_u = 1677(4x\beta)^{-\frac{1}{2}}Km/sec$. Therefore if x is known β could be evaluated from the velocity dispersion observed in galaxies today.

3. Results

In the experiments presented below the major axis of the central bar collapses at about a time $t = 200$, while at a time about $t = 2000$ the main part of the bound material has been relaxed.

The bound mass of the final object is estimated by examining the time evolution of the radii $R(M)$ containing various amounts M of mass. Then we adopt as bound mass M_b the mass in the maximum radius that remains constant between $t = 2000$ and $t = 6000$ while larger radii still expand. In the experiments presented below the bound mass varies by a fraction less than 20% around the value of $1M_u$ i.e. $N_u = 1088$ particles.

The rotation curve of the resulting galaxy is found as follows. We define 10 successive coaxial cylinders with their axis along the X direction (the direction of rotation) passing through the center of the bound mass. The radii of these cylinders are so defined that 10% of the bound mass is contained between any two successive cylinders.

Let f =1,2,...10 be the number of a cylinder. If J_{fx} is the total angular momentum of particles between the cylinders $f-1$, f and $I_{fyz} = \Sigma m_i (y_i^2 + z_i^2)^{1/2}$ is their first moment the rotational velocity of this material, corresponding to the mean radius between the two cylinders, is estimated by the formula:

$$V_f = \frac{J_{fx}}{I_{fyz}} \tag{1}$$

In figure 3a,b,c the rotation curves are shown in three cases at a time $t = 6000$. The only difference in the initial conditions in these three cases is the axial ratio of the central bar in the G-scale i.e. $k = 0.4, 0.5, 0.8$ as indicated in the figure. It is remarkable that the rotational velocity curves are so sensitive in k. A small initial axial ratio of the bar $k = 0.4$ leads to a negative rotation curve (Fig. 3a) at $t = 6000$, while a large $k = 0.8$ to a positive one (Fig. 3c). An intermediate $k = 0.5$ leads to a counterrotating galaxy (Fig. 3b).

This figure shows that the idea of a central bar model of initial condition in principle works. An important question is whether such a distribution of angular momentum can survive for a Hubble time. For this reason we have extended our simulation up to a time $t = 16000$. This time is comparable to the age of the Universe if the parameter β in the scaling unit of time is close to 1.

In figure 4a,b,c the time evolution of the rotational velocity is given up to this time for the fractions $f = 20\%, 40\%, 80\%, 100\%$ of cummulative mass for the cases of figure 3. The total angular momentum ($f = 100\%$) in the three cases remains almost constant throught the whole evolution, but it is negative in (a) and (b), while it is positive in (c).

A remarkable feature in Fig.4a is that the inner parts (20%,40%) have initially positive rotation until $t = 2000-3000$ but their rotation is inverted and remains negative until the end of the simulation. This effect is due to the secondary infall of loosely bound particles passing from the central region. Most of these particles carry relatively large amount of negative angular momentum which is transferred to the central region of the galaxy. Thus, the positive rotation of this region is inverted. However, a small core at the center could preserve its positive rotation, because the pericenters of the infalling particles with considerable amount of angular momentum cannot be arbitrarily small. A small counterrotating core is even more probable in a dissipative scenario. Therefore, if counterrotating galaxies observed today are formed according to this mechanism a very small counterrotating core should be considered as a natural event.

The effect of inversion of rotation by the secondary infall in the inner parts is also seen in the case of $k = 0.8$ (Fig. 4c). This happens rather suddenly shortly after $t = 6000$. In comparison with Fig. 4a this event happens at a later time because it is due to a different reason which is related to the two-body relaxation mechanism as we show in Harsoula and Voglis (1998).

The inversion of the rotation in the central part of a galaxy by the secondary infall depends also on the total amount of angular momentum that already exists in this part. Thus in the case of $k = 0.5$, where the central parts carry relatively more angular, counterrotation survives throughout the whole simulation (Fig. 4b).

In figure 5a,b,c we give the rotational velocities of the three cases versus the cumulative fraction of bound mass from the center. The rotational velocity V at every mass fraction is evaluated as in figure 3 but now the values of V are the time averages in the intervals $\Delta t_1 = (4000, 8000)$, $\Delta t_2 = (8000, 12000)$, $\Delta t_3 = (12000, 16000)$. These values of V are indicated in figures 5a,b,c by stars, squares and cycles respectively.

In these figures we see three different types of rotational velocity curves i.e. a simple negative rotation and two counterrotations, formed by different time evolutions. In the first case a galaxy that counterrotates after the collapse of the main part becomes later on simply rotating, except perhaps a very small counterrotatingg core. In the second case counterrotation is preserved. In the third case a galaxy that is simply rotating after the collapse of the main part becomes counterrotating later on. These differences show that the axial ratio of the central bar, which is the only difference in the initial conditions, is an important parameter.

From Figs 5a,b,c it is also clear that the rotational velocity curves (and hence the corresponding distribution of angular momentum), established after the secondary infall is completed, do not change considerably in time. This means that, in the dissipationless scenario, other mechanisms able to redistribute angular momentum are not so important.

4. Conclusions

Our conclusions are summarized as follows. We have shown by N-body simulations that:

1) Cosmological initial conditions can create counterrtating galaxies.

2) This effect can be initiated simply by a central bar-like density perturbation in the early Universe.

3) The axial ratio of the central bar perturbation is important for counterrotation.

4) The secondary infall can considerably alter the distribution of angular momentum along the radius established after the collapse of the main part of the galaxy.

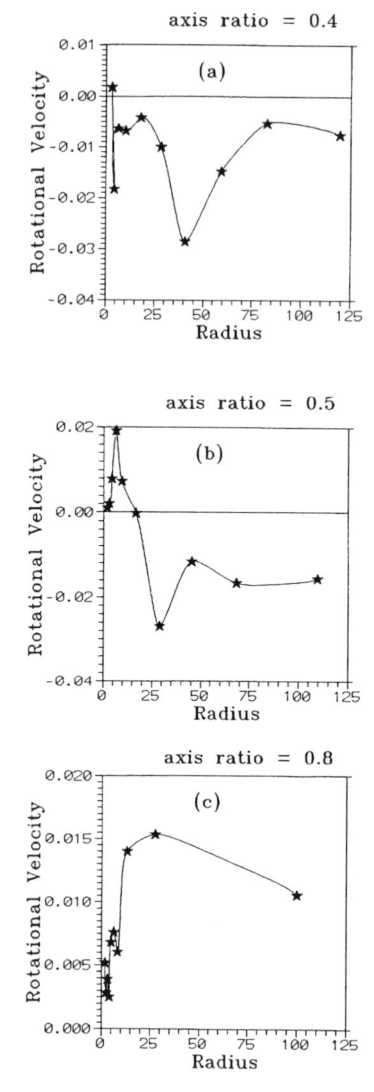

Figure 3. The rotation curves for time=6000 of the experiments with axial ratio $k = 0.4$ (a), $k = 0.5$ (b) and $k = 0.8$ (c).

Acknowledgments

We would like to thank Dr. Sverre Aarseth for his kind offer of the "Ahmad Cohen version of his N-body code" and the useful accompanying instructions.

This research was supported in part by the General Secreteriat for Research and Technology under PENED grant 293/1995.

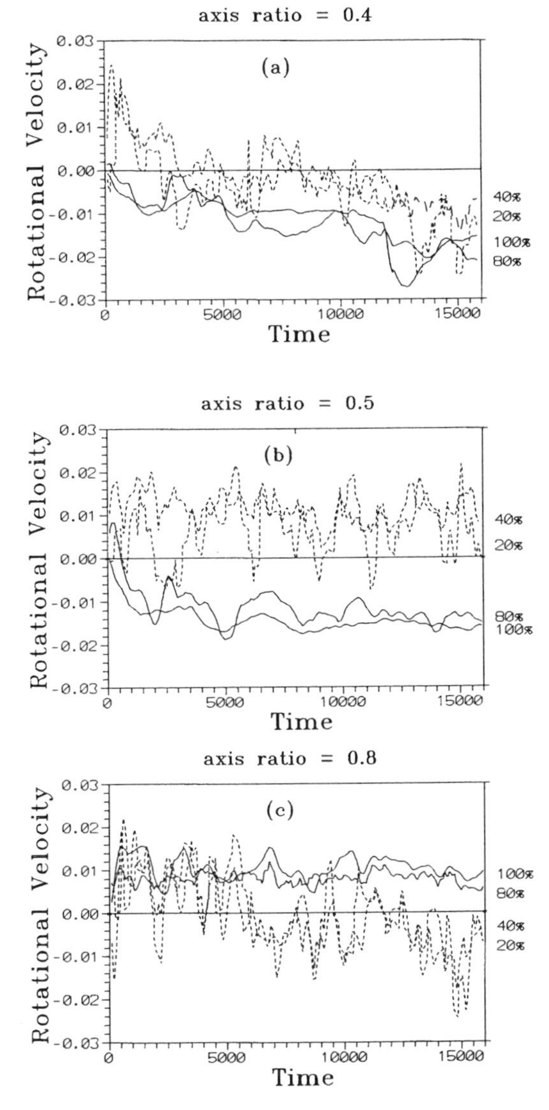

Figure 4. The time evolution of the rotational velocity for the fraction $f = 20\%, 40\%, 80\%, 100\%$ of cumulative mass for the cases of figure 3.

References

1. Bender R., 1988, Astron. Astrophys. , **202**, L5
2. Bender R. Surma P., 1992, Astron. Astrophys., **258**,250
3. Bender R., 1988, Astron. Astrophys., **202**, L5
4. Bender R., Surma,P., 1992, Astron. Astrophys., **258**, 250
5. Doroshkevich A.G., 1970, Astrofisika, **6**, 581 Astrophys.J., **292**,371
6. Franx M., Illingworth G., 1988, Astroph. J., **327**,L55

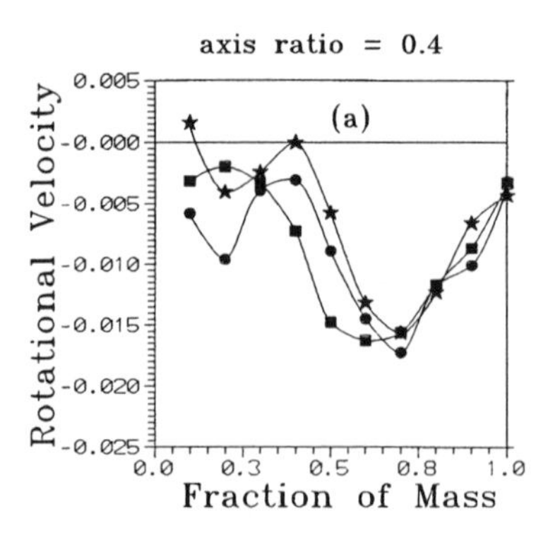

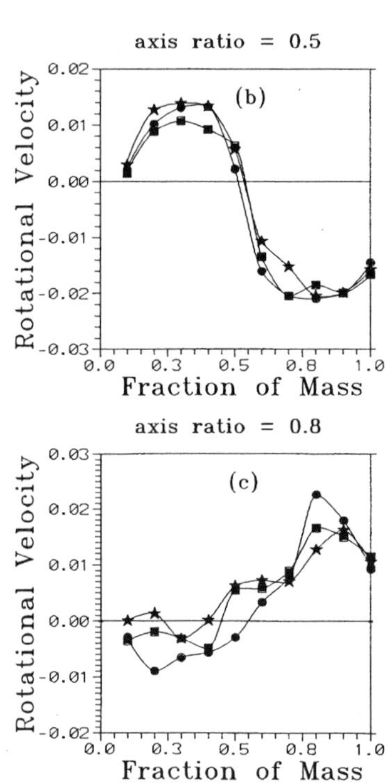

Figure 5. The rotational velocities as a function of the cumulative fraction of the bound mass. The stars, squares and circles indicate the time averages in the intervals $\Delta t_1 = (4000, 8000)$, $\Delta t_2 = (8000, 12000)$ and $\Delta t_3 = (12000, 16000)$ respectively.

7. Harsoula M., Voglis N., 1998, Astron. Astrophys., **335**
8. Hernquist L., Barnes J., 1991, Nature, **354**,210
9. Peebles P.J.E., 1969, Astroph. J., **155**,393
10. Voglis N., Hiotelis N., 1989, Astron. Astrophys., **218**,1
11. Voglis N., Hiotelis N. and Hoflich P., 1991 Astron. Astrophys., **249**,5
12. Zeldovich Ya.B., 1970, Astro. Astrophys., **5**,84

GYLDEN–LIKE SYSTEMS AND DEF–MAPPINGS

IGNACIO APARICIO AND LUIS FLORÍA
Grupo de Mecánica Celeste I. Universidad de Valladolid.
Dept. de Matemática Aplicada a la Ingeniería, E. T. S. I. I.
E – 47 011 Valladolid, Spain.

Abstract. We apply the linearizing Deprit–Elipe–Ferrer (DEF) mapping (including time t into the transformation equations) to a two–body problem with time–varying Keplerian parameter (a Gylden–type system), whose Hamiltonian is presented in homogeneous canonical formulation. After changing the independent variable by a true–like anomaly, we derive regularized second–order equations for the coordinates, resembling harmonic equations. Reduction to harmonic oscillators may not be achieved.

Key words: linearizing transformations, regularization, variable Keplerian parameter, perturbed Gylden Hamiltonian system, DEF focal variables.

1. Introduction and Scope of the Paper

On a *Gylden system* (Deprit 1983, Deprit *et al.* 1989), we superimpose time–dependent perturbing effects. Operating on the extended phase space, we study an enlarged *DEF linearizing transformation to redundant focal–type variables* (Deprit *et al.* 1994, §§4.1), when applied to that perturbed, non-stationary two–body problems [with a sufficiently regular *time–dependent Keplerian parameter* $\mu(t)$], with the aim of *regularizing* the equations of motion issued from the Gylden–like Hamiltonian formulated in *homogeneous canonical formalism.* We resort to the *focal method canonical approach* due to Deprit, Elipe and Ferrer (1994). Our version of the DEF–mapping includes time t, and we follow their analytical treatment. The presumed time–dependence and the presence of a perturbing part lead us to use adequate notations to exploit analogies with the stationary case. From the canonical system derived from the Hamiltonian, we obtain regularized *second–order*

B.A. Steves and A.E. Roy (eds.), The Dynamics of Small Bodies in the Solar System, 575–580.

equations *resembling* those of harmonic oscillators, although linearization may not be achieved. Our developments, which are *independent of the energy of the system*, are uniformly valid for *any* type of orbit.

The Gylden model applies to the motion of a particle in a central gravitational field with a isotropically time–varying central mass. As far as we know (Deprit 1983; Deprit *et al.* 1989; Polyakhova 1994), the motivation of two–body problems with changing reduced gravitational parameter comes from the study of the effect of meteoric accretion on the evolution of the Earth–Moon orbit and the explanation of secular inequalities in the Moon's motion (Deprit, 1983). The nonstationary two–body problem has also been used as a model for describing double– or multiple–star evolution at an isotropic secular mass variation. Another model is provided by the heliocentric motion of a comet that continuously loses its mass throughout its revolution, or comets that burn out while passing their perihelion points.

A Levi–Cività regularization of nonstationary planar two–body systems in homogeneous Hamiltonian formalism is due to Guillaume (1974). Aparicio & Floría (1997a) perform a focal–type (see below) BF–treatment and regularization of certain perturbed Gylden systems. In the stationary case, Aparicio & Floría (1997b) point out conservative non–planar perturbed Keplerian systems admitting *exact linearization* in DEF–formulation.

2. Fundamentals of the Focal Method Approach

Ideas concerning this method are found in Burdet (1969) and Deprit *et al.* (1994). In the Kepler problem, the Newtonian equations of motion are singular, nonlinear and Ljapunov–unstable, and unperturbed solutions to the two–body problem are *not* closed–form, explicit functions of time. On the contrary, the harmonic oscillator is stable and easily solvable. But, after application of certain transformations introducing redundant variables (see below), *the spatial Kepler problem is equivalent to* a linear and stable differential system corresponding to *four harmonic oscillators.*

The two–body relative position vector $\mathbf{x}$ can be replaced by four coordinates: the distance $r = ||\mathbf{x}||$ and the direction $\mathbf{u} = \mathbf{x}/r$. These parameters, introduced by a dimension–raising point–transformation, are homogeneous Cartesian coordinates in a projective space. This option allows one to define a 4–dimensional harmonic oscillator for the variables $\mathbf{u}$ and $1/r$, the oscillation centre being at $\mathbf{x} = \mathbf{0}$ (the centre of gravitational attraction, which coincides with a focus of the orbits of the two–body problem), and the same set of formulae yields a unified treatment of *any* type of Keplerian orbit, and reduce it to the *oscillator* form. This reduction also requires the use of a new *focal fictitious time*, a true–like anomaly, given by a Sundman–type reparametrizing transformation. From the standpoint of analytical treatments, this pseudo–time regularizes the pole–type singularity at the origin of the gravitational field. From the viewpoint of the accuracy of numerical integration, the time transformation introduces an automatic analytical step–size regulation as a function of the distance (here, the step–size would essentially vary as r^2.) Moreover, regularized equations usually admit a larger step–size during the numerical integration, and computing requirements are reduced. Redundant variables increase the number of equations; but the *order* of the system is less important than the *type* (here, linear; or, at least, quasi–linear, in perturbed cases) and *behaviour* of the equations, specially regarding qualitative properties and accuracy of numerical integrations. On that score, redundant equations are often well behaved.

Burdet (1969) applied the focal method to the Kepler problem within a Newtonian, non–Hamiltonian framework. In defining the conjugate momenta, *Ferrándiz* (1986, 1988) extended the focal coordinates to a (weakly) canonical transformation increasing the number of variables, and proposed eight redundant canonical variables of the focal type, the *BF–variables*, which *reduce the Kepler problem to four uncoupled and unperturbed harmonic oscillators.* As for the exact linearization of conservative *perturbed* Keplerian systems in BF–variables, Ferrándiz & Fernández–Ferreirós (1991) and Aparicio & Floría (1996) have identified linearizable perturbations. Deprit, Elipe & Ferrer (1994) revisited the subject of linearization of Keplerian systems, clarified the *weakly canonical* nature of the BF–mapping, and other questions concerning linearizing transformations (e.g. the classical projective decomposition and its canonical or "weakly" canonical extensions into the momentum space.) That paper contains *further extensions of the focal method*: other transformations (the DEF– and the D–mapping) to focal variables are obtained. They also show how the use of these new sets of variables allows one to exactly linearize the standard Kepler problem.

3. Sets of Variables. The DEF–Transformation

We present the variables and some notations that we shall use later. All these sets are universally applicable to *any* type of orbit. As in Deprit *et al.* (1994), we will work out our developments in compact vector notation.

In the extended phase space we consider the *polar nodal* variables, $(r, \theta, \nu, R, \Theta, N)$, completed by the conjugate pair (t, T), namely: time t and the negative of the total energy of the system. In *homogeneous canonical formalism*, time t is introduced as a coordinate. Extended phase–space formulation facilitates the introduction of new independent variables other that t (Stiefel & Scheifele 1971, Ch. VIII, §30 and §34, Ch. X, §37.)

We shall also employ the *extended Cartesian variables*, with coordinates $\mathbf{x} = (x_1, x_2, x_3)$ and momenta $\mathbf{X} = (X_1, X_2, X_3)$, along with its trivial extension to the enlarged phase space by means of the above couple (t,T).

The redundant set of *DEF–variables* $(u_0, \mathbf{u}, U_0, \mathbf{U})$ is defined from Cartesian variables by the *DEF–mapping* (Deprit, Elipe and Ferrer 1994, §§4.1):

$$\mathbf{x} = u_0\mathbf{u}, \quad u_0 \in \mathbb{R}_+, \quad r = \|\mathbf{x}\| = u_0\|\mathbf{u}\|, \tag{1}$$

$$\mathbf{X} = U_0\mathbf{u} + (1/u_0)(\mathbf{u}\times\mathbf{U})\times\mathbf{u}, \quad U_0 \in \mathbb{R}, \tag{2}$$

and the extension $t = u_4$, $T = U_4$, to the enlarged phase space. We also give notations and expressions of the momenta Θ and N in these sets:

$$\Theta^2 = \|\mathbf{x}\times\mathbf{X}\|^2 = \|\mathbf{u}\|^4\|\mathbf{Q}\|^2 = \beta^4\|\mathbf{Q}\|^2 = c^2, \tag{3}$$

$$N = x_1X_2 - x_2X_1 = \|\mathbf{u}\|^2(u_1U_2 - u_2U_1) = \|\mathbf{u}\|^2 n, \tag{4}$$

$$\mathbf{Q} = \mathbf{u}\times\mathbf{U}, \quad \beta = \|\mathbf{u}\|, \quad n = u_1U_2 - u_2U_1. \tag{5}$$

This transformation increases the number of variables by two, linearizes and regularizes three–dimensional Keplerian systems. Other properties (e.g., *weak* canonicity) require the mapping to be restricted to the manifold $\beta = \|\mathbf{u}\| = 1$. Additional features are analysed in Deprit *et al.* (1994), §4.

4. The Hamiltonian and Its Expressions. Canonical Equations

We consider the motion of a particle in a time–varying gravitational field under the effect of an additional time–dependent perturbation, that is, the Newtonian interaction under a potential that can be split into a pure predominant *Gylden term* (Deprit 1983; Deprit *et al.* 1989) plus a time–dependent *perturbing part* proportional to r^{-2}, the coefficient being some sufficiently regular function of time and canonical momenta that can also be expanded in powers of a small parameter ε. In *polar nodal variables*,

$$\mathcal{H} = (1/2)\left[R^2 + \left(\Theta^2/r^2\right)\right] - (\mu(t)/r) + V\left(t;\Theta^2,N,T;\varepsilon\right)/r^2 + T \tag{6}$$

where $\mu(t)$ is the time–varying Keplerian parameter of the system, a sufficiently regular function of t. In the nonstationary case, the problem still possesses the *first–integral* of the angular momentum (per unit of mass.)

The form of (6) is intended as an approximation to analytically account for the *asphericity of the gravitational field* due to an oblate primary (Deprit 1981). Thanks to some knowledge on *exact* linearization of this perturbation in the stationary case (Burdet 1969; Ferrándiz & Fernández–Ferreirós 1991; Aparicio & Floría 1996, 1997b), we only consider r^{-2}–perturbations.

In *extended Cartesian variables*, this Hamiltonian takes the expression

$$\mathcal{H} = (1/2)\|\mathbf{X}\|^2 - (\mu(t)/r) + \left(1/r^2\right)V\left(t;\Theta^2,N,T;\varepsilon\right) + T. \tag{7}$$

Finally, in *DEF-formulation* we have the Hamiltonian function

$$\mathcal{K} = \frac{\beta^2}{2}\left(U_0^2 + \frac{\|\mathbf{Q}\|^2}{u_0^2}\right) - \frac{\mu(u_4)}{\beta u_0} + \frac{1}{\beta^2 u_0^2} V\left(u_4; c^2, N, U_4; \varepsilon\right) + U_4 \quad (8)$$

Formulae (3)–(5) provide expressions of $\Theta^2 = c^2$ and N for (7) and (8).

Our next considerations are inspired in the treatment of Keplerian systems (Deprit *et al.* 1994, §§4.1.) We introduce quantities and vectors

$$a = 1 + 2\left(\partial V/\partial c^2\right), \quad k = \left(1/\beta^2\|\mathbf{Q}\|\right)(\partial V/\partial N), \quad (9)$$

$$A = \left(u_0^2/\beta^4\|\mathbf{Q}\|\right)\left[2\mathcal{K} + (3\mu(u_4)/\beta u_0) - \left(4V/\beta^2 u_0^2\right) - 2U_4\right], \quad (10)$$

$$B = \left(1/\beta^4\|\mathbf{Q}\|\right)\left[4\beta^2\|\mathbf{Q}\|^2\left(\partial V/\partial c^2\right) + 2n(\partial V/\partial N)\right], \quad (11)$$

$$\tilde{\mathbf{u}} = (-u_2, u_1, 0), \quad \tilde{\mathbf{U}} = (U_2, -U_1, 0). \quad (12)$$

From the canonical equations issued from $\mathcal{K}$ (we skip this step), we create a system of second–order equations for the position–like variables. To simplify the equations of motion, we replace time t by a regularizing pseudo–time (a kind of true anomaly) by means of a *generalized Sundman transformation*

$$t \rightarrow f: \ \beta^2\|\mathbf{Q}\|\,dt = u_0^2\,df. \quad (13)$$

The canonical equations of motion with respect to this fictitous time are

$$u_0' = u_0^2 U_0/\|\mathbf{Q}\|, \quad \mathbf{u}' = (a/\|\mathbf{Q}\|)(\mathbf{Q}\times\mathbf{u}) + k\tilde{\mathbf{u}}, \quad (14)$$

$$u_4' = \left(1/\beta^4\|\mathbf{Q}\|\right)(\partial V/\partial U_4) + \left(u_0^2/\beta^2\|\mathbf{Q}\|\right), \quad (15)$$

$$U_0' = (1/\|\mathbf{Q}\|)\left[\left(\|\mathbf{Q}\|^2/u_0\right) - \left(\mu(u_4)/\beta^3\right) + \left(2V/\beta^4 u_0\right)\right], \quad (16)$$

$$\mathbf{U}' = (a/\|\mathbf{Q}\|)(\mathbf{Q}\times\mathbf{U}) - (A+B)\mathbf{u} - k\tilde{\mathbf{U}}, \quad (17)$$

$$U_4' = \left(u_0/\beta^2\|\mathbf{Q}\|\right)(d\mu/du_4) - \left(1/\beta^4\|\mathbf{Q}\|\right)(\partial V/\partial u_4). \quad (18)$$

5. Second–Order Equations for the New Coordinates

We form the f–derivatives of Eqs. (14)–(15), bearing in mind (14)–(18) and (9)–(12), and introduce some notations: a vector $\tilde{\mathbf{P}}$, two 3×3 matrices E and I_3, and a *new dependent variable* σ to replace the scalar variable u_0:

$$\tilde{\mathbf{P}} = (a'/\|\mathbf{Q}\|)(\mathbf{Q}\times\mathbf{u}) + k'\tilde{\mathbf{u}} = (a'/a)\,\mathbf{u}' - \left[(a'/a)\,k - k'\right]\tilde{\mathbf{u}}, \quad (19)$$

$$E = \mathrm{diag}(1,1,0), \quad I_3 = \mathrm{diag}(1,1,1), \quad \sigma = \|\mathbf{Q}\|^2/u_0. \quad (20)$$

After manipulating the equations according to Deprit *et al.* (1994) §§4.1, we arrive at the following second–order differential equations for the new coordinates $(\sigma, \mathbf{u})$, with the pseudo–time f as the independent variable:

$$\mathbf{u}'' = -\left(a^2 I_3 - k^2 E\right)\mathbf{u} + 2k\widetilde{\mathbf{u}}' + \widetilde{\mathbf{P}}\,, \tag{21}$$

$$\sigma'' = -\left[1 + \left(2V/\beta^4\|\mathbf{Q}\|^2\right)\right]\sigma + \left(\mu\left(u_4\right)/\beta^3\right)\,, \tag{22}$$

which are *regular*, their form resembling that of harmonic equations. Our results are *independent of the type of the orbit*. In particular, for a pure Gylden system (say, $V \equiv 0$) there results $a \equiv 1$, $k \equiv 0$ and $\widetilde{\mathbf{P}} \equiv \mathbf{0}$, and the equations for $\mathbf{u}$ are those of three uncoupled and unperturbed oscillators.

Work on other analytical and qualitative aspects is in progress.

Acknowledgements

Partial financial support came from the DGES of Spain, Grant PB 95–0807.

References

Aparicio, I. and Floría, L. (1996) On Perturbed Two–Body Problems and Harmonic Oscillators, *C. R. Acad. Sci. Paris* série **II b**, **323**, pp. 71–76.

Aparicio, I. and Floría, L. (1997a) On Reduction of the Nonstationary Two–Body Problem to Oscillator Form. In: J. Docobo, A. Elipe and H. McCalister (Eds.), *Visual Double Stars: Formation, Dynamics and Evolutionary Tracks*. Kluwer A.P. (In press.)

Aparicio, I. and Floría, L. (1997b) An Application of the Linearizing DEF–Transformation. *Proc. Fourth Intern. Workshop on Positional Astronomy and Celestial Mechanics*, Peñíscola (Spain). (In press: Obs. Astron. Univ. Valencia.)

Burdet, C. A. (1969) Le mouvement Képlérien et les oscillateurs harmoniques, *Journal für die reine und angewandte Mathematik*, **238**, pp. 71–84.

Deprit, A. (1981) The Elimination of the Parallax in Satellite Theory, *Celest. Mech.*, **24**, pp. 111–153.

Deprit, A. (1983) The Secular Acceleration in Gylden's Problem, *Celest. Mech.*, **31**, pp. 1–22.

Deprit, A., Elipe, A. and Ferrer, S. (1994) Linearization: Laplace vs. Stiefel, *Celest. Mech.* **58**, pp. 151–201.

Deprit, A., Miller, B. and Williams, C. A. (1989) Gylden Systems: Rotation of Pericenters, *Astrophysics and Space Science*, **159**, pp. 239–270.

Ferrándiz, J.M. (1986) A New Set of Conical Variables for Orbit Calculation. *Proc. Second Intern. Symp. on Spacecraft Flight Dynamics*, Darmstadt. ESA SP–255, pp. 361-364.

Ferrándiz, J.M. (1988) A General Canonical Transformation Increasing the Number of Variables with Application to the Two–Body Problem, *Celest. Mech.*, **41**, pp. 343–357.

Ferrándiz, J.M. and Fernández–Ferreirós, A. (1991) Exact Linearization of Non–Planar Intermediary Orbits in the Satellite Theory, *Celest. Mech.*, **52**, pp. 1–12.

Guillaume, P. (1974) Regularization of the Two–Body Problem with Variable Mass, *Celest. Mech.*, **10**, pp. 141–149.

Polyakhova, E. N. (1994) A Two–Body Variable–Mass Problem in Celestial Mechanics: The Current State, *Astronomy Reports*, **38**, pp. 283–291.

Stiefel, E. L. and Scheifele, G. (1971) *Linear and Regular Celestial Mechanics*. Springer.

UNIFORM DEVELOPMENT OF A TR–TRANSFORMATION TO GENERALIZED UNIVERSAL VARIABLES

LUIS FLORÍA
Grupo de Mecánica Celeste I. Universidad de Valladolid.
Dept. de Matemática Aplicada a la Ingeniería, E. T. S. I. I.
E - 47 011 Valladolid, Spain.

Abstract. We investigate a general approach to the derivation of the Delaunay–Similar (DS) canonical TR–variables (with the true anomaly as the independent variable), originally introduced by Scheifele, in order to render the construction applicable to the study of any kind of two–body orbit. Our analysis also allows for certain classes of perturbations compatible with a basic Keplerian–like dynamical structure. Accordingly, universal TR–elements are obtained for a wide class of perturbed Keplerian systems.

Key words: perturbed Keplerian systems, Delaunay–Scheifele mapping, canonical transformations, TR–elements, universal variables, universal functions, uniform treatment of two–body motion.

1. Introduction and Statement of the Problem

We study a canonical transformation in extended phase space that achieves the transition from polar nodal variables to new variables, of the Delaunay–Similar (DS) type, apt to the *uniform description and treatment of any kind of perturbed two–body motion* if certain perturbations (Deprit 1981b, §7; Floría 1993) are considered. Extended phase–space formulation facilitates the canonical incorporation of new independent variables different from time t. For additional details, see Stiefel & Scheifele (1971), Chapter VIII (§30 and §34), and Chapter X, §37, or Scheifele & Stiefel (1972) pp. 28–33.

Scheifele (1970) proposed a set of eight canonical orbital elements for *elliptic Keplerian motion*, based on the *true* anomaly as the independent variable. The true anomaly was essentially (say, proportional to) a canonical

B.A. Steves and A.E. Roy (eds.), The Dynamics of Small Bodies in the Solar System, 581–586.

coordinate. Thus, in terms of these Delaunay–Similar TR–elements, perturbations proportional to negative powers of the distance are described by *trigonometric–like polynomials* (say, finite Fourier–like expansions) in the DS angles, and so one avoids infinite and slowly convergent Fourier series.

Scheifele derived his DS set in the extended phase space by formulating the Keplerian Hamiltonian in polar spherical variables and integrating the Hamilton–Jacobi equation by separation of variables. In his turn, operating on the extended phase space of the enlarged polar nodal variables, *Deprit* (1981a) gave an alternative interpretation of Scheifele's reducing TR–transformation and the subsequent derivation of the DS elements.

We apply Deprit's technique toward enlarging Scheifele's construction of elliptic Delaunay–Similar TR–elements. The first step consists in defining a time–dependent generating function [(5) below] which involves a function γ of the new momenta. Another function of momenta, (29), enters at the second step where we transform the independent variable. For elliptic motion, in Floría (1994) appropriate selection of these functions yields back the special cases considered by Scheifele and his co–workers (1970, 1972), Deprit (1981a) and other authors. These authors worked out their transformations in order to find integrable approximations of non integrable Hamiltonian systems (namely: the *Main Problem* in Artificial Satellite Theory.)

Obtaining the final transformation equations will require the calculation of certain quadratures over the radial variable r, whose evaluation is achieved after introducing two *auxiliary variables*: one of them is a true–like anomaly, while the other one will adopt (in certain cases) the meaning of an eccentric–like anomaly. This last variable is involved in the argument of the universal *Stumpff* c_n functions and the *Battin* U_n functions, (Battin 1987, §4.5, §4.6; Stiefel & Scheifele 1971, §11; Stumpff 1959, §37, §40, §41; 1965, §183, §184), in terms of which one will establish the *equation for the radial distance* and the *generalized Kepler equation* [see (12) below.]

Irrespective of the specific type of the orbit, the concept of *pericentre distance* will play an essential role in the derivation of the new variables.

Unlike other sets of variables and orbital elements, the polar nodal set is applicable to *any* two–body orbit. Starting from extended Hill–Whittaker polar nodal variables, $(r, \theta, \nu, t; p_r, p_\theta, p_\nu, p_0)$, p_0 being the negative of the energy, we aim at a *canonical reduction* of perturbed Keplerian systems governed by a generic *homogeneous Hamiltonian* (see also Floría 1993, §2)

$$\mathcal{H} = \mathcal{H}_0\,(r\,; p_r\,, p_\theta\,) + V\,(r\,; p_\theta\,, p_\nu\,, p_0\,) + p_0\,, \tag{1}$$

$$\mathcal{H}_0 = \frac{1}{2}\left[p_r^2 + \left(p_\theta^2/r^2\right)\right] - \frac{\mu}{r} \quad \text{(standard Keplerian Hamiltonian)}, \tag{2}$$

$$V = \sum_{j=0}^{2} r^{-j}\, V_j\,(p_\theta\,, p_\nu\,, p_0\,)\,. \tag{3}$$

To this end, we construct *canonical elements of a DS type and universal applicability* to the system given by $\mathcal{H}$, which will incorporate perturbation terms due to V, and obtain a *universal Keplerian-like representation* of the solution to $\mathcal{H}$. Intermediate quadratures are carried out with the help of *subsidiary quantities* and *auxiliary variables* whose introduction and treatment require a judicious adaptation of the procedure applied in Deprit (1981a), or Floría (1994), to develop the classical Delaunay-Similar elements.

2. The Transformation: A Delaunay–Scheifele Mapping

To construct our DS set, and solve the dynamical problem posed by $\mathcal{H}$, we perform a canonical transformation of the extended phase space,

$$(r\,,\,\theta\,,\,\nu\,,\,t\,;\,p_r\,,\,p_\theta\,,\,p_\nu\,,\,p_0) \ \xrightarrow{S}\ (q_\Phi\,,\,q_L\,,\,q_G\,,\,q_N\,;\,\Phi\,,\,L\,,\,G\,,\,N)\,, \tag{4}$$

from polar nodal variables, defined by a generating function

$$S \ \equiv\ S\,(r,\theta,\nu,t;\Phi,L,G,N) = \theta\,G + \nu\,N + t\,L + \int_{r_0}^{r} \sqrt{Q}\,d\,r\,, \tag{5}$$

$$Q \ \equiv\ Q\,(r;\Phi,L,G,N) \ =\ \frac{2\mu}{r} - 2L - \frac{\gamma^2}{r^2} - 2V_0 - \frac{2}{r}\,V_1 - \frac{2}{r^2}\,V_2. \tag{6}$$

Here $V_j \equiv V_j\,(\,G\,,\,N\,,\,L\,)$. According to Scheifele (1970, 1972) and Deprit (1981a), we take $\gamma = G - \Phi$. And r_0 is the lowest positive root of the equation $Q(r;\Phi,L,G,N) = 0$, such that $Q(r;\Phi,L,G,N) > 0$ for $r > r_0$.

The implicit equations of the transformation generated by S read

$$p_r \ =\ \frac{\partial S}{\partial r} \ =\ \sqrt{Q}\,, \tag{7}$$

$$p_\theta \ =\ \frac{\partial S}{\partial \theta} \ =\ G\,, \qquad p_\nu \ =\ \frac{\partial S}{\partial \nu} \ =\ N\,, \qquad p_0 \ =\ \frac{\partial S}{\partial t} \ =\ L\,, \tag{8}$$

$$q_\Phi \ =\ \frac{\partial S}{\partial \Phi} \ =\ \gamma\,\hat{I}_2\,, \tag{9}$$

$$q_G \ =\ \frac{\partial S}{\partial G} \ =\ \theta \ -\ \frac{\partial V_0}{\partial G}\,\hat{I}_0 \ -\ \frac{\partial V_1}{\partial G}\,\hat{I}_1 \ -\ \left[\gamma + \frac{\partial V_2}{\partial G}\right]\,\hat{I}_2\,, \tag{10}$$

$$q_N \ =\ \frac{\partial S}{\partial N} \ =\ \nu \ -\ \frac{\partial V_0}{\partial N}\,\hat{I}_0 \ -\ \frac{\partial V_1}{\partial N}\,\hat{I}_1 \ -\ \frac{\partial V_2}{\partial N}\,\hat{I}_2\,, \tag{11}$$

$$q_L \ =\ \frac{\partial S}{\partial L} \ =\ t \ -\ \left[1 + \frac{\partial V_0}{\partial L}\right]\,\hat{I}_0 \ -\ \frac{\partial V_1}{\partial L}\,\hat{I}_1 \ -\ \frac{\partial V_2}{\partial L}\,\hat{I}_2\,, \tag{12}$$

with the following notation for the quadratures:

$$\hat{I}_m \ =\ \int_{r_0}^{r} \frac{d\,r}{r^m\,\sqrt{Q}}\,, \qquad m \ =\ 0\,,1\,,2\,. \tag{13}$$

3. Quadratures: Subsidiary Quantities and Auxiliary Variables

To make future calculations and their interpretation easier, we introduce some *subsidiary quantities* μ^*, q, e, and κ by means of the formulae

$$\mu^* = \mu - V_1, \quad q = \frac{\mu^*(1-e)}{2(L+V_0)}, \quad \gamma^2 + 2V_2 = \mu^* q(1+e) \equiv \kappa^2, \tag{14}$$

which allows us to represent Q in the form

$$Q = \mu^*(r-q)[r(e-1) + q(e+1)]/(qr^2). \tag{15}$$

Accordingly, by setting $r = q$, we see that $Q(q; \Phi, L, G, N) = 0$, and we can postulate *a fictitious Keplerian motion* with a *hypothetic* Hamiltonian

$$\mathcal{H}^* = (1/2)\left[p_r^2 + \left(\kappa^2/r^2\right)\right] - (\mu^*/r) + (p_0 + V_0). \tag{16}$$

To calculate $\hat{I}_2$ with $r_0 = q$, we define (as in Stiefel & Scheifele 1971, p. 48, Formula [57]) an auxiliary variable $f = f(r; \Phi, L, G, N)$ such that

$$r = q(1+e)/(1+e\cos f), \tag{17}$$

which looks like the *polar equation* of a conic–section with the true–like anomaly f as the polar angle. With the change of integration variable $r \to f$ we have $r = r_0 = q \Longrightarrow f_0 = f(r_0) = 0$. From (17) we also obtain

$$d\left(\frac{1}{r}\right) = -\frac{dr}{r^2} = -\frac{e\sin f\, df}{q(1+e)}, \quad Q = \frac{\mu^* e^2 \sin^2 f}{q(1+e)} \Rightarrow \hat{I}_2 = \frac{f}{\kappa}. \tag{18}$$

Next we take our cue from Stiefel & Scheifele (1971) p. 51, Formulae [67] and [68]. After using (17) and (18), for the determination of the quadrature

$$\hat{I}_0 = \int_{r_0}^{r} \frac{dr}{\sqrt{Q}} = \frac{1}{\sqrt{\mu^* q(1+e)}} \int_0^f r^2\, df = \frac{1}{\kappa}\int_0^f r^2\, df \tag{19}$$

we introduce the *notation* $\Lambda = L + V_0$ for a new quantity that, now, will play the same role as the negative of the Keplerian energy in the classical, unperturbed case, and consider a *new auxiliary variable* s such that

$$\sqrt{r}\cos\frac{f}{2} = \sqrt{q}\, c_0\left(\frac{\Lambda}{2}s^2\right) = \sqrt{q}\, U_0\left(\frac{s}{2}, 2\Lambda\right), \tag{20}$$

$$\sqrt{r}\sin\frac{f}{2} = \sqrt{\mu^*(1+e)}\,\frac{s}{2}\,c_1\left(\frac{\Lambda}{2}s^2\right) = \sqrt{\mu^*(1+e)}\,U_1\left(\frac{s}{2}, 2\Lambda\right) \tag{21}$$

The c_n are the *Stumpff universal functions* and the U_n are those of *Battin* (Stiefel & Scheifele 1971, §11, p. 43; Stumpff 1959, §41, pp. 205–207; Battin

1987, §4.5 and §4.6). By adding up the squares of (20) and (21), applying the relation between Λ and e in (14), and $s^2(\Lambda/2) = 2\Lambda(s/2)^2$, we obtain

$$r = q + \mu^* e s^2 c_2 \left(2\Lambda s^2\right) = q + \mu^* e U_2(s, 2\Lambda), \quad (22)$$

$$dr = \mu^* e s c_1 \left(2\Lambda s^2\right) = \mu^* e U_1(s, 2\Lambda)\, ds, \quad (23)$$

after using properties of the Stumpff functions (Battin 1987, p. 187, Problem 4–33; Stiefel & Scheifele 1971, p. 45, Formulae [45].) Finally, since (Battin 1987, §4.5, p. 181, Problem 4–23; Stiefel & Scheifele 1971, §11, p. 43, Formula [36]) $U_n(s, 2\Lambda) = s^n c_n(2\Lambda s^2)$, we arrive at our Formula (22).

In addition to this, from the above Formulae (20) and (21), by virtue of Battin (1987), p. 176, Formula (4.77), we deduce (23).

On the other hand, by multiplying Formulae (20) and (21) and using the properties (see Battin 1987, §4.6, p. 183, Formulae [4.89] and p. 187, Problem 4–33; Stiefel & Scheifele 1971, p. 45, Formulae [45]) we obtain

$$r \sin f = 2\sqrt{\mu^* q(1+e)}\; U_0(s/2, 2\Lambda)\, U_1(s/2, 2\Lambda) \quad (24)$$

$$= \sqrt{\mu^* q(1+e)}\; U_1(s, 2\Lambda) \implies df = (\kappa/r)\, ds, \quad (25)$$

$$\hat{I}_0 = \frac{1}{\kappa}\int_0^f r^2\, df = \int_0^s r\, ds = q s + \mu^* e U_3(s, 2\Lambda). \quad (26)$$

Finally, to calculate $\hat{I}_1$, we can use (17) and (18) and apply the relation between df and ds given in (25) to arrive at

$$\hat{I}_1 = \sqrt{\frac{p}{\mu^*}}\int_0^f \frac{df}{1 + e\cos f} = \sqrt{\frac{p}{\mu^*}}\int_0^s \frac{r}{p}\kappa\frac{ds}{r} = \int_0^s ds = s. \quad (27)$$

4. Final Transformation Equations and Parametrical Solution

Under the effect of our transformation, Hamiltonian (1) becomes

$$\mathcal{H}(r, -, -, -; p_r, p_\theta, p_\nu, p_0) \longrightarrow \tilde{\mathcal{H}} = (G + \gamma)\Phi/(2r^2). \quad (28)$$

We *reparametrize the motion* by a new independent variable, proportional to the true anomaly of the *fictitious* Keplerian orbit, and obtain a simplified Hamiltonian. The *pseudo-time* τ is introduced by the differential relation

$$dt = \tilde{f}\, d\tau = \left(r^2/\mathcal{G}\right) d\tau, \quad (29)$$

$\mathcal{G}$ being a function of the new momenta. The corresponding Hamiltonian is

$$\mathcal{K} = \tilde{\mathcal{H}}\tilde{f} = (G + \gamma)\Phi/(2\mathcal{G}) = (2G - \Phi)\Phi/(2\mathcal{G}). \quad (30)$$

An obvious choice of function $\mathcal{G}$ to considerably simplify $\mathcal{K}$ is

$$\mathcal{G} = (G + \gamma)/2 = (2G - \Phi)/2, \tag{31}$$

and $\mathcal{K}(\Phi) = \Phi$. We obtain a *simple canonical solution*, parametrized by τ: $q_\Phi = \tau + \text{const.}$, and the remaining elements are constants of the motion.

To conclude, we propose the following *solution* in terms of two anomaly-like parameters s and f *along the auxiliary fictitious motion*:

$$r = p/(1 + e\cos f) = q + \mu^* e\, U_2(s, 2\Lambda), \qquad p = q(1+e),$$

$$p_r = \sqrt{\mu^*/p}\; e \sin f = \mu^* e\, U_1(s, 2\Lambda)/[q + \mu^* e\, U_2(s, 2\Lambda)],$$

$$p_\theta = G = \text{const.}, \qquad p_\nu = N = \text{const.}, \qquad p_0 = L = \text{const.},$$

$$\theta = q_G + \frac{\partial V_0}{\partial G}\,[q s + \mu^* e\, U_3(s, 2\Lambda)] + \frac{\partial V_1}{\partial G}\, s + \left[\gamma + \frac{\partial V_2}{\partial G}\right]\frac{f}{\kappa},$$

$$\nu = q_N + \frac{\partial V_0}{\partial N}\,[q s + \mu^* e\, U_3(s, 2\Lambda)] + \frac{\partial V_1}{\partial N}\, s + \frac{\partial V_2}{\partial N}\frac{f}{\kappa},$$

$$t = q_L + \left[1 + \frac{\partial V_0}{\partial L}\right][q s + \mu^* e\, U_3(s, 2\Lambda)] + \frac{\partial V_1}{\partial L}\, s + \frac{\partial V_2}{\partial L}\frac{f}{\kappa},$$

along with $q_\Phi = \gamma f/\kappa = \tau + \text{const.}$, $q_G = \text{const.}$, $q_N = \text{const.}$, $q_L = \text{const.}$

Acknowledgements

Research supported in part by DGES (MEC) of Spain, Grant PB 95–0807.

References

Battin, R. H. (1987) *An Introduction to the Mathematics and Methods of Astrodynamics*. AIAA Education Series, American Institute of Aeronautics and Astronautics, N.Y.

Deprit, A. (1981a) A Note Concerning the TR–Transformation, *Celest. Mech.*, **23**, 299–305.

Deprit, A. (1981b) The Elimination of the Parallax in Satellite Theory, *Celest. Mech.*, **24**, 111–153.

Floría, L. (1993) Canonical Elements and Keplerian–like Solutions for Intermediary Orbits of Satellites of an Oblate Planet, *Celest. Mech. and Dyn. Astron.*, **57**, 203–223.

Floría, L. (1994) On the Definition of the Delaunay–Similar Canonical Variables of Scheifele, *Mechanics Research Communications*, **21**, 409–414.

Scheifele, G. (1970) Généralisation des éléments de Delaunay en Mécanique Céleste. Application au mouvement d'un satellite artificiel, *Comptes rendus de l'Académie des Sciencies de Paris*, Série **A**, **271**, 729–732.

Scheifele, G. and Stiefel, E. (1972) *Canonical Satellite Theory Based on Independent Variables Different from Time*. Report to ESRO under ESOC–Contract No. 219/70/AR, ETH–Zürich.

Stiefel, E. L. and Scheifele, G. (1971) *Linear and Regular Celestial Mechanics*. Springer–Verlag, Berlin–Heidelberg–New York.

Stumpff, K. (1959, 1965) *Himmelsmechanik* (vols. I and II). VEB Deutscher Verlag der Wissenschaften, Berlin.

EPILOGUE:
THE SHAPE OF THINGS TO COME

FUTURE PLANETARY MISSIONS

A. A. CHRISTOU
Astronomy Unit, Queen Mary and Westfield College,
London E1 4NS, United Kingdom

1. Introduction

The 1997 NATO ASI meeting at Maratea coincided with the first science return from NASA's "faster–cheaper–better" Discovery program. The string of recent planetary successes, namely the flyby of asteroid (253) Mathilde by the NEAR spacecraft, the successful Mars Pathfinder landing and the data-gathering Lunar Prospector and Mars Global Surveyor missions have boosted the concept of small, simple and frequent planetary missions with focused scientific objectives at least as far as the inner solar system is concerned. The upcoming investigation of the asteroid (433) Eros by NEAR and the saturnian system by Cassini can only serve to increase scientists' appetites for more ambitious undertakings. The following years should see a profusion of such missions and an end to the long term hiatus in planetary exploration that was the legacy of the large, billion-dollar projects and modern time economic realities. Here we present a list of the missions that have been approved or proposed for launch over the next 15 years. Individual entries contain the funding status (A for Approved or P for Proposed), agencies involved (see references for associated web sites), scheduled launch date and a short description of the mission. The potential scientific value of each target body or class of bodies is summarised.

2. The Moon

Despite the large number of manned and unmanned lunar missions during the 1960s and 1970s there are still numerous aspects of lunar science that have remained largely unaddressed. Furthermore, the development of

B.A. Steves and A.E. Roy (eds.), The Dynamics of Small Bodies in the Solar System, 587–593.

advanced instrumentation and techniques in the past 20 years holds great promise for new discoveries concerning the history and present state of the Moon.

Lunar-A (A) ISAS Jul 1999
Postponed until the summer of 1999 due to penetrator battery problems, this project involves placing 2 or 3 instrumented penetrators into the lunar subsurface to conduct seismic and heat flow studies. The relay orbiter is also equipped with an imaging camera.

SELENE (A) ISAS/NASDA 2003
Multi-instrumented orbiter/lander to conduct global multi-spectral mapping of the lunar surface, investigate the lunar environment and test new technologies for future missions. After a year of remote sensing activities the propulsion module is to detach from the rest of the spacecraft and attempt a radar-guided soft landing, the first since the Luna 24 sample return mission which took place in 1976.

Russian Lunar Mission (P) RKA 2000?
A small-scale lunar mission very similar in objectives and design to the Japanese Lunar-A mission. A number of penetrators, similar in design to those carried by the ill-fated Mars 96 mission, will constitute a lunar seismic and heat flow measurement network. The current budgetary status of the russian planetary exploration program makes this mission highly uncertain, however.

Chinese Lunar Mission (P) CAST 2000?
Several recent news reports have indicated that the Chinese are considering a lunar mission as part of their ambitious space plans. Currently there is no information concerning the objectives of this mission or its status other than a projected launch date, assuming government approval of the project, sometime in the beginning of the next decade

Earthrise (P) SSTL 2001?
A project under consideration by Surrey Satellite Technologies Limited, this is a technology demonstration of a low cost ($\sim$ $10m) science platform for deep space missions. The current plan is to insert the spacecraft in orbit around the Moon and dispatch a small instrumented probe at the lunar south pole. Depends on funding from the UK and/or other European countries.

3. Mars

Since the success of the Viking missions 20 years ago there has been a 10-year lull in Mars exploration followed by a string of unsuccessful spacecraft attempts to study the Red Planet and its environment (Phobos 1 & 2, Mars Observer, Mars 96). In recent years the United States has initiated an aggressive program of orbiters and landers to be launched every 2 years culminating in a projected 2005 sample return mission. Other countries and consortia are also including Mars missions in their space exploration agendas signalling the beginning of a global cooperative effort to understand and, perhaps one day set foot upon, our intriguing planetary neighbour.

Planet-B (A) CSA/ISAS/NASA Jul 1998

The first Japanese Mars mission is aimed at investigating the planet's magnetic environment, the interaction of the upper atmosphere with the solar wind and the distribution of dust on and in orbit around Mars. The spacecraft will carry 15 instruments from Japan, the USA, Canada, Sweden and Germany.

Mars Surveyor 98 Orbiter (A) NASA Dec 1998

Taking advantage of the 1998–99 Mars launch window, this is the orbiter part of the 1998 Mars Surveyor mission to the Red Planet. The main payload is a copy of the Pressure Modulator Infrared Radiometer instrument that was lost with Mars Observer in 1993. A colour camera is also included.

Mars Surveyor 98 Lander (A) NASA Jan 1999

The lander element of Mars Surveyor 98 is to attempt a landing at the planet's south polar region. Once on the surface, it will investigate the physics and chemistry of ices on and below the surface by means of the Mars Volatiles and Climate Surveyor instrument package. A Russian-provided LIDAR instrument will also conduct measurements of atmospheric haze and dust. A late addition to the payload has been a simple microphone funded by the California-based Planetary Society.

Deep Space 2 (A) NASA Jan 1999

The second New Millenium technology validation mission is composed of two micro-penetrators to be launched piggyback on Mars Surveyor Lander 98. Following a hard landing and emplacement into the subsurface, each probe will conduct an analysis of any existing water ice caches. Expected operating lifetime of the probes is about two days but the two probes could go on operating for up to 2 weeks.

Mars Surveyor 01 Orbiter (A) NASA Mar 2001

Due to a Mars Surveyor 2001 mission rescoping currently in progress the information given below for the Orbiter and Lander portions should not be considered as definitive. The 2001 Mars Surveyor Orbiter craft is intended to carry the last component of the lost Mars Observer payload to be reflown, the Gamma Ray Spectrometer, along with the THEMIS infrared mapper. The spacecraft will enter orbit around Mars by means of a novel aerocapture manoeuver during which a grazing entry into the Martian atmosphere will reduce the probe's hyperbolic approach velocity.

Mars Surveyor 01 Lander (A) NASA Apr 2001

The 2001 Mars Surveyor landing attempt was to support the 2005 sample return effort. However, the extended range Athena rover which was to play a key part in that plan has now been delayed to 2003 and other changes in instrumentation are also expected. The lander itself will probably carry instruments to determine the effect of the Martian environment on future human activities as well as an *in situ* propellant manufacture technology demostrator.

Mars 2001 (P) NASA/RKA 2001

A project involving a Russian long range rover on the surface of Mars and a US orbiter, probably one of the Mars Surveyor craft, for telecommunications support. As in the case of the proposed russian lunar mission, its future is very uncertain due to funding difficulties.

Mars Express (P) ESA/RKA? May/June 2003

In an attempt to salvage the science lost with the demise of Mars 96, ESA is currently pushing for an orbiter mission carrying copies of instruments that were on board the russian spacecraft. A small lander is also a possibility but its inclusion hinges on funding by agencies of individual ESA member states rather than ESA itself which is providing the orbiter. An intriguing lander proposal by a team of researchers led by Open University's Colin Pillinger involves the biological analysis of a subsurface soil sample.

4. Small Bodies

Once considered by astronomers to be "vermin of the skies", these bodies are now known to hold key information concerning the formation and evolution of the solar system. This realisation is reflected in the first asteroid flybys (951 Gaspra and 243 Ida) in 40 years of space exploration and the recent funding approval for numerous missions dedicated to studying these objects.

Deep Space 1 (A) NASA Oct 1998

The first mission developed under the New Millenium initiative is to fly by asteroid 1992 KD on Jul 28, 1999. Although the main purpose of this mission is to demonstrate new technologies to bring down the cost of space science missions such as ion propulsion, autonomous navigation, etc the science return is expected to be quite respectable. Depending on funding, an extended mission could involve a flyby of comet 19P/Borelly in 2001.

Stardust (A) NASA Feb 1999

Stardust is the fourth discovery mission to be selected. Following launch in February 1999, its first task will be to collect samples of interstellar dust particles passing through our solar system by using aerogel plates. The same technique will be applied during a 2003 flyby of comet 81P/Wild 2 in an attempt to collect cometary coma material. The samples will be returned to Earth in 2006.

SMART 1 (A) ESA late 2001

ESA has recently decided to develop and launch a small mission as a technological precursor to the Mercury Orbiter Cornerstone. Designed to test advanced technology equipment such as ion engines, miniaturized cameras/spectrometers etc. it was decided that a scientific target would be a worthwhile addition to the mission plan. Although the identification of the specific target body is not expected for several months yet, priority has been given to certain unusual types of near-Earth objects.

Muses-C (A) ISAS/NASA Jan 2002

A mission to rendezvous with near Earth asteroid (4660) Nereus in 2003, collect samples and return them to Earth in 2006. A recent cooperation agreement between the United States and Japan has resulted in the inclusion of a US-built nano-rover to perform multiple site analyses. Due to spacecraft weight constraints the mission may have to be diverted to the secondary target, 1989 ML.

COmet Nucleus TOUR (A) NASA Aug 2002

The latest Discovery mission to be selected along with the Genesis solar wind sample return aims at flying by three near-Earth comets in the first decade of the 21st century. Following launch in mid-2002, repeated Earth/Moon flybys are used to direct CONTOUR to comets 2P/Encke (11/03), 73P/Schwassmann-Wachmann 3 (6/06) and 6P/d'Arrest (8/08). At each of these opportunities, high resolution imaging and *in situ* gas/dust compositional analysis will be carried out.

Rosetta (A) ESA/NASA Jan 2003

This ambitious ESA Cornerstone mission will be launched atop an Ariane 5 in 2003 and, after a circuitous cruise through the inner solar system, rendezvous with comet 46P/Wirtanen in 2011. Once there it will follow the comet through perihelion, measuring its basic properties (mass, rotation rate, density, etc) and monitoring its response to varying solar heat flux. A surface science package will be placed on the nucleus and conduct various scientific investigations over a period of a few days. Also envisioned are flybys of one or two main belt asteroids, a current favourite being C-type (160) Siwa, during the long cruise phase.

Champollion/Deep Space 4 (P) NASA May 2003

Originally a part of the Rosetta mission, the Champollion lander might become the primary element of a US project aiming to rendezvous with comet 9P/Tempel 1 in 2005. Champollion will land on the nucleus and, after 84 hours of surface operations, take off with a 0.1 kg cometary sample. Following a successful docking with the carrier spacecraft, the sample will be brought to Earth in 2010.

5. Outer Solar System

The special problems related to missions to the outer solar system such as long transit times, alternative power sources, etc. raise the cost of such ventures to levels higher than what the current "faster–cheaper–better" trend dictates. New technologies are now being developed and tested (ion propulsion, solar concentrator arrays etc) that will hopefully make these missions affordable under today's budgetary constraints.

Europa Orbiter (P) NASA 2003?

A project devised as an affordable continuation to the investigations currently conducted by Galileo in orbit around Jupiter. The main objective is to investigate the existence of a subsurface ocean of water on Jupiter's second moon by means of radar sounding and laser altimetry. As part of the FY98-approved Outer Planets/Solar Probe initiative, the Europa Orbiter mission has now received a further boost by being included as a "new start" in President Clinton's proposed FY99 budet. If launched in 2003 as currently scheduled, the following two missions are expected to leave Earth a few years later.

Pluto–Kuiper Express (P) NASA 2004?

Proposed as a key part of the Ice and Fire initiative, this is a twin spacecraft attempt to reach the Pluto/Charon system sometime in the 2010–2020 timeframe. A possible extended mission to one or more Edgeworth–Kuiper Belt objects is also under consideration.

Solar Probe (P) NASA 2007?

A mission utilising a Jupiter gravity assist in order to perform an extremely close pass (~4 solar radii) of the Sun and conduct *in situ* investigations of the near-solar environment.

6. Other Planetary Systems

In view of the recent observational results concerning planetary systems other than our own it is no surprise that several exoplanet search proposals have been submitted to research funding agencies. Some of these concern space-based observation platforms which take advantage of the diffraction-limited seeing and broad wavelength access that the space environment offers.

COROT (A) CNES 2001–02

This was originally an asteroseismology mission, now given the additional task of monitoring photometrically a number of stars for occultations by orbiting planets.

Deep Space 3 (A) NASA late 2001

A technology demonstration flight of a space-based optical interferometer with free-flying components moving in formation.

Space Interferometry Mission (P) NASA 2004
An Earth-orbiting interferometric array that will be used for micro-arcsec astrometry, milli-arcsec synthesis imaging and interferometric nulling experiments.
DARWIN (P) ESA 2010s
A candidate cornerstone mission for ESA's Horizons 2000 program. It involves a multimirror interferometer that will be placed in the vicinity of Jupiter's orbit and search for life-bearing extrasolar planets through direct ultra high resolution imaging and spectroscopy.

Acknowledgements

The author wishes to thank Carl Murray for numerous helpful suggestions concerning the presentation style of this paper.

References

1. ASI, Italian Space Agency, *http://hp835.mt.asi.it/welcome.html*
2. Aviation Week and Space Technology, *http://www.awgnet.com/*
3. CAST, Chinese Academy of Space Technology, *http://www.cast.ac.cn/*
4. CNES, Centre National d'Etudes Spatiales, France, *http://www.cnes.fr/*
5. CSA, Canadian Space Agency, *http://www.space.gc.ca/*
6. ESA, European Space Agency, *http://www.esrin.esa.it/*
7. IKI, Space Research Institute, Russia, *http://www.iki.rssi.ru/Welcome.html*
8. ISAS, Institute of Space and Astronautical Science, Japan, *http://www.isas.ac.jp/*
9. JPL, Jet Propulsion Laboratory, USA, *http://www.jpl.nasa.gov/*
10. NASDA, National Space Development Agency, Japan, *http://www.nasda.go.jp/welcome_e.html*
11. NASA, National Aeronautics and Space Administration, USA, *http://www.gsfc.nasa.gov/NASA_homepage.html*
12. RKA, Russian Space Agency, *http://liftoff.msfc.nasa.gov/rsa/rsa.html*
13. SSTL, Surrey Satellite Technologies Limited, United Kingdom, *http://www.ee.surrey.ac.uk/Research/CSER/UOSAT/index.html*

PARTICIPANTS AND SPEAKERS

NATO ADVANCED STUDY INSTITUTE ON
THE DYNAMICS OF SMALL BODIES IN THE SOLAR SYSTEM
A Major Key to Solar System Studies
Held June 29 to July 12, 1997 in Maratea, Italy

Aarseth, S.	UK	Institute of Astronomy Cambridge, CB3 0HA, UK sverre@ast.cam.ac.uk
Andreu, M.	SPAIN	Matematica Aplicada i Analisi, ETSII Universitat de Barcelona Barcelona, 08071, SPAIN mangel@maia.ub.es
Aparicio, I.	SPAIN	Matematica Aplicada a la Ingenieria Universidad de Valladolid Valladolid, E-47011, SPAIN ignacio@dali.eis.uva.es
Baille, P.	CANADA	Mathematics Dept, Royal Military College Kingston, Ontario, K7K 5L0, CANADA baille_p@rmc.ca
Barnett, A.	UK	Dept of Mathematics Glasgow Caledonian University Glasgow, G4 0BA, UK adba@gcal.ac.uk
Belbruno, E.	USA.	Innovative Orbital Design Ardsley, New York, 10502, USA. belbruno@iorbit.com
Boyce, W.	USA	Dept. of Aerospace Engineering University of Texas at Austin Austin,TX, USA bboyce@mail.utexas.edu
Broucke, R.	USA	Dept of Aerospace Engineering & Engineering Mechanics The University of Texas at Austin Austin, TX, 78712-1085, USA broucke@uts.cc.utexas.edu
Casotto, S.	ITALY	Department of Astronomy University of Padua Padova, 35122, ITALY casotto@astrpd.pd.astro.it
Celletti, A.	ITALY	Dipartimento di Matematica Pura ed Applicata, Universita degli Studi di L'Aquila L'Aquila, 67100, ITALY CELLETTI@aquila.infn.it

Chesley, S.	USA	Department of Aerospace Engineering and Engineering Mechanics University of Texas at Austin Austin, Texas, 78712, USA chesley@chaos.ae.utexas.edu
Christou, A.	UK	Astronomy Unit, School of Mathematical Sciences, Queen Mary and Westfield College London, E1 4NS, UK A.Christou@qmw.ac.uk
Cipriani, P.	ITALY	Physics Dept. "E.Amaldi", Universita "Roma 3" Roma, 00146, ITALY cipriani@amaldi.fis.uniroma3.it
Contopoulos, G.	GREECE	Dept of Astronomy, University of Athens Zografos Athens, GR - 157 63, GREECE gcontop@atlas.uoa.gr
Conway, B.	USA	Dept of Aeronautica & Astronautical Engineering University of Illinois at Urbana-Champaign Urbana, IL,USA, 61801-2935 bconway@ uiuc.edu
Della Penna, M.	FRANCE	Dpt. Cassini, Observatoire de la Cote d' Azur Nice Cedex4, 06304, FRANCE gabry@purcell.obs-nice.fr
Di Bari, M.	ITALY	Department of Physics Universita "La Sapienza" (Roma) Roma, 00185, ITALY dibari@amaldi.fis.uniroma3.it
Dvorak, R.	AUSTRIA	Institut fur Astronomie, Universitat Wien Wien, A-1180, AUSTRIA DVORAK@astro.ast.univie.ac.at
Efthymiopoulos, C.	GREECE	Department of Physics Section of Astrophysics,Astronomy and Mechanics University of Athens Athens, GR-15784, GREECE cefthim@atlas.uoa.gr
Eggleton, P.	UK	Institute of Astronomy University of Cambridge Cambridge, CB3 0HA, UK ppe@ast.cam.ac.uk

Érdi, B.	HUNGARY	Department of Astronomy Eotvos University, Budapest, 1083, HUNGARY erdi@innin.elte.hu
Ferraz-Mello, S.	BRAZIL	Instituto Astronomico e Geofisico Universidade de Sao Paulo Sao Paulo, 01065-970, BRAZIL sylvio@orion.iagusp.usp.br
Floría, L.	SPAIN	Departamento de Matematica Aplicada a la Ingenieria, ETSII, Universidad de Valladolid Valladolid,SPAIN, E-47011
Froeschlé, Cl.	FRANCE	Observatoire de la Cote d'Azur Nice Cedex 4, 06304, FRANCE claude@rameau.obs-nice.fr
Froeschlé, Ch.	FRANCE	Observatoire de Nice Nice Cedex 4, 06304, FRANCE froesch@obs-nice.fr
Fulconis, M.	FRANCE	Observatoire de la Cote d'Azur Nice Cedex 4, 06304, FRANCE fulconis@obs-nice.fr
Gomatam, J.	UK	Dept of Mathematics Glasgow Caledonian University Glasgow, G4 0BA, UK j.gomatam@gcal.ac.uk
Gozdziewski, K.	POLAND	Torun Centre for Astronomy (TCfA) Torun, PL-87-100, POLAND chris@astri.uni.torun.pl
Grousouzakou, E.	GREECE	Department of Physics,University of Athens Athens, GR-15784, GREECE hskokos@atlas.uoa.gr
Hadjifotinou, K.	GREECE	Dept. of Mathematics, Univ. of Thessaloniki Thessaloniki, 540 06, GREECE khad@math.auth.gr
Harsoula, M.	GREECE	Department of Physics, University of Athens Athens, GR-15784, GREECE mharsoul@atlas.uoa.gr

Hermann, R.	BELGIUM	Institut d'Astrophysique de Cointe - Universite de Liege Liege, B-4000, BELGIUM R.Hermann@ulg.ac.be
Irigoyen, M.	FRANCE	Universite Paris 2, Paris, 75006, FRANCE irigoyen@math.jussieu.fr
Janssens, F.	NETHER-LANDS	WMM, ESTEC, Noordwijk, 2200 AG NETHERLANDS FJANSSEN@estec.esa.nl
Kiseleva, L.	UK	Institute of Astronomy Cambridge, CB3 0HA, UK lgk@ast.cam.ac.uk
Leftaki, M.	GREECE	Dept of Mathematics, University of Patras Patras, 26110, GREECE leftaki@math.upatras.gr
Maciejewski, A.	POLAND	Torun Centre for Astronomy Nicolaus Copernicus University Torun, 87-100, POLAND maciejka@astri.uni.torun.pl
Manara, A.	ITALY	Brera Observatory, Milan, 20121, ITALY manara@brera.mi.astro.it
Manley, S.	UK	Armagh Observatory Armagh, BT61 9DG, UK spm@star.arm.ac.uk
Marchal, C.	FRANCE	D.E.S. ONERA Chatillon Cedex, 92322, FRANCE
Mardling, R.	AUSTRALIA	Mathematics, Monash University Clayton, Victoria, 3168, AUSTRALIA r.mardling@maths.monash.edu.au
Message, J.	UK	Division of Applied Mathematics, Dept. of Mathematical Sciences, University of Liverpool Liverpool, L69 3BX, UK SX20@LIVERPOOL.AC.UK
Michel, P.	FRANCE	UMR 6529 Cassini, Observatoire de Nice Nice Cedex 4, 06304, FRANCE michel@obs-nice.fr

Michtchenko, T.	BRAZIL	Instituto Astronomico e Geofisico Sao Paulo, 01065-970, BRAZIL tatiana@orion.iagusp.usp.br
Migliorini, F.	UK	Armagh Observatory, Armagh, BT61 9DG, UK pat@star.arm.ac.uk
Milani, A.	ITALY	Dipartimento di Matematica Universita di Pisa Pisa, 56127, ITALY milani@dm.unipi.it
Morais, H.	UK	Astronomy Unit, School of Mathematical Sciences, Queen Mary and Westfield College London, E1 4NS, UK M.H.Morais@qmw.ac.uk
Muinonen, K.	FINLAND	Observatory, University of Helsinki Helsinki, FIN-00014, FINLAND karri@gstar.astro.Helsinki.FI
Murray, C.	UK	Astronomy Unit, School of Mathematical Sciences, Queen Mary and Westfield College London, E1 4NS, UK C.D.Murray@qmw.ac.uk
Nesvorný, D.	FRANCE	Observatoire de Nice Nice Cedex 4, 06304, FRANCE david@obs-nice.fr
Perozzi, E.	ITALY	Servizi e Sistemi Spaziali, Nuova Telespazio Roma, 00141, ITALY eperozzi@sni.telespazio.it
Pretka, H.	POLAND	Astronomical Observatory Adam Mickiewicz University Poznan, 60-286, POLAND pretka@phys.amu.edu.pl
Probin, J.	UK	Dept. of maths, University of Liverpool Liverpool, L69 3BX, UK J.Probin@liverpool.ac.uk
Puel, F.	FRANCE	Observatoire de Besancon Universite de Franche-Comte Besancon, 25010, FRANCE puel@fennec.obs-besancon.fr

Roig, F.	BRAZIL	Instituto Astronomico e Geofisico da Universidade de Sao Paulo Sao Paulo, CEP 04301-904, BRAZIL froig@orion.iagusp.usp.br
Rossi, A.	ITALY	CNUCE - CNR, Pisa, 56126, ITALY a.rossi@cnuce.cnr.it
Roy, A.	UK	Dept. of Physics and Astronomy University of Glasgow Glasgow, G12 8QQ, UK daphne@astro.gla.ac.uk
Sansaturio, M.	SPAIN	E.T.S. de Ingenieros Industriales University of Valladolid Valladolid, E-47011, SPAIN marsan@wmatem.eis.uva.es
Shefer, V.	RUSSIA	Research Institute of Applied Mathematics and Mechanics, Tomsk, 634050, RUSSIA shefer@urania.tomsk.su
Šidlichovský, M.	CZECH REPUBLIC	Astronomical Institute, Academy of Sciences Prague, 141 31, CZECH REPUBLIC sidli@ig.cas.cz
Sidorenko, V.	RUSSIA	Keldysh Institute of Applied Mathematics Moscow , 125047, RUSSIA sidorenk@spp.keldysh.ru
Simó, C.	SPAIN	Departament de Matematica Aplicada i Analisi Universitat de Barcelona Barcelona, 08071, SPAIN carles@maia.ub.es
Simula, A.	BRAZIL	Instituto Astronomico e Geofisico da Universidade de Sao Paulo Sao Paulo, CEP 04301-904, BRAZIL simula@orion.iagusp.usp.br
Stavliotis, Ch.	GREECE	Dept of Physics, University of Patras Patras, 26110, GREECE
Steves, B.	UK	Dept of Mathematics Glasgow Caledonian University Glasgow, G4 0BA, UK bst@gcal.ac.uk

Szebehely, V.	USA	Dept of Aerospace Engineering WRW415D The University of Texas at Austin Austin, TX, 78712-1085, USA szebehely@mail.utexas.edu
Taidakova, T.	UKRAINE	Simeiz Observatory Crimean Astrophysical Observatory Simeiz, Yalta, 334242, Crimea, UKRAINE tat@astro.crao.crimea.ua
Valsecchi, G.	ITALY	I. A. S. - Planetologia Roma, 00185, ITALY giovanni@saturn.ias.fra.cnr.it
Vigo-Aguiar, I.	SPAIN	Matematica Aplicada E P S, Universidad de Alicante Alicante, 03080, SPAIN vigo@aitana.cpd.ua.es
Waldvogel, J.	SWITZER- LAND	Applied Mathematics, ETH - Zentrum Zürich, CH-8092, SWITZERLAND waldvoge@sam.math.ethz.ch
Williams, I.	UK	Astronomy Unit, School of Mathematical Sciences, Queen Mary & Westfield College London, E1 4NS, UK I.P.Williams@qmw.ac.uk
Wnuk, E.	POLAND	Astronomical Observatory A. Mickiewicz University Poznan, 60-286, POLAND wnuk@phys.amu.edu.pl
Yokoyama, T.	BRAZIL	Estatistica Matematica Aplicada e Computacional, Instituto de Geociencias e ciencias Exatas de Rio Claro, Rio Claro, Sao Paulo, CEP 13.500-970, BRAZIL tadashi@demac.igce.unesp.br

SUBJECT INDEX

D

E

F

G

AUTHOR INDEX